GERARD J. TORTORA
ROBERT J. AMITRANO

Bergen Community College, Paramus, New Jersey

ANATOMY AND PHYSIOLOGY
Laboratory Manual

SIXTH EDITION

Prentice Hall

Upper Saddle River, New Jersey 07458

Library of Congress Cataloging-in-Publication Data

Tortora, Gerard J.
    Anatomy and physiology laboratory manual / Gerard J. Tortora.—6th ed.
        p. cm.
    Includes index.
    ISBN 0-13-089670-5
        1. Human physiology—Laboratory manuals.   2. Human anatomy—Laboratory manuals.
I. Title.

QP44.T66 2001
612'.0078—dc21

00-058445

Editor in Chief for Biology: Sheri Snavely
Senior Acquisitions Editor: Halee Dinsey
Editorial Project Manager: Don O'Neal
Assistant Vice President and Director of ESM Production: David W. Riccardi
Special Projects Manager: Barbara A. Murray
Editorial Production: Clarinda Publication Services
Text Illustrations: Mary Dersh; Publication Services, Academy Artworks
Cover Design: Joseph Sengotta

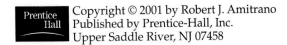

Earlier editions copyright © 1998 by Biological Science Textbooks, Inc. Copyright
© 1994 by Biological Science Textbooks, Inc. and A & P Textbooks, Inc.
Copyright © 1990 by Biological Science Textbooks, Inc.,
A & P Textbooks, Inc., and Elia-Sparta, Inc. Copyright © 1986 by Biological Science
Textbooks, Inc., Elia-Sparta Corporation, Inc. Copyright © 1980 by Gerard J. Tortora
and Nicholas P. Anagnostakos. The First Edition appeared under the title Laboratory
Exercises in Anatomy and Physiology: Brief Edition.

Printed in the United States of America

ISBN 0-13-089670-5 NBZI

10 9 8 7 6 5 4 3 2

Prentice-Hall International (UK) Limited, *London*
Prentice-Hall of Australia Pty. Limited, *Sydney*
Prentice-Hall Canada Inc., *Toronto*
Prentice-Hall Hispanoamericana, S.A., *Mexico*
Prentice-Hall of India Private Limited, *New Delhi*
Prentice-Hall of Japan, Inc., *Tokyo*
Prentice-Hall of Asia, Private Limited, *Singapore*
Editora Prentice Hall do Brazil, Ltda., *Rio de Janeiro*

Preface

Anatomy and Physiology Laboratory Manual, Sixth Edition, has been written to guide students in the laboratory study of introductory anatomy and physiology. The manual was written to accompany most of the leading anatomy and physiology textbooks.

COMPREHENSIVENESS

This manual examines virtually every structure and function of the human body that is typically studied in an introductory anatomy and physiology course. Because of its detail, the need for supplemental handouts is minimized; the manual is a strong teaching device in itself.

USE OF THE SCIENTIFIC METHOD

Anatomy (the science of structure) and physiology (the science of function) cannot be understood without the practical experience of laboratory work. The exercises in this manual challenge students to understand the way scientists work by asking them to make microscopic examinations and evaluations of cells and tissues, to observe and interpret chemical reactions, to record data, to make gross examinations of organs and systems, to dissect, and to conduct physiological laboratory work and interpret and apply the results of this work.

ILLUSTRATIONS

The manual contains a large number and variety of illustrations. The illustrations of the body systems of the human have been carefully drawn to depict structures that are essential to students' understanding of anatomy and physiology. Numerous photographs, photomicrographs, and scanning electron micrographs are presented to show students how the structures of the body actually look. We feel that this laboratory manual has better and more complete illustrations than any other anatomy and physiology manual.

IMPORTANT FEATURES

Among the key features of this manual are (1) dissection of the white rat, and selected mammalian organs; (2) numerous physiological experiments; (3) emphasis on the study of anatomy through histology; (4) lists of appropriate terms accompanying drawings and photographs to be labeled; (5) inclusion of numerous scanning electron micrographs and specimen photos; (6) phonetic pronunciations and derivations for the vast majority of anatomical and physiological terms; (7) diagrams of commonly used laboratory equipment; (8) laboratory report questions and reports at the end of each exercise that can be filled in, removed, and turned in for grading if the instructor so desires; (9) three appendixes dealing with units of measurement, a periodic table of elements, and eponyms used in the laboratory manual; and (10) emphasis on laboratory safety throughout the manual.

NEW TO SIXTH EDITION

Numerous changes have been made in the sixth edition of this manual in response to suggestions from instructors and students. We have added some new physiology experiments, line drawings, cadaver photographs, photomicrographs, and phonetic pronunciations and derivations. The number of orientation diagrams has also been greatly expanded. Virtually all black-and-white photomicrographs have been replaced with color ones. The principal additions to various exercises are as follows:

In Exercise 5, "Integumentary System," there is a new illustration of the skin.

In Exercise 7, "Bones," there is a new section on types of ribs, and joint names have been added for the various articulations.

In Exercise 8, "Articulations," joints have been re-organized by structure. The section on synovial joints has been expanded. The exercise includes a new sesction on axes of movements at synovial joints, and a new illustration on movements at synovial joints.

In Exercise 10, "Skeletal Muscles," there is a new section on synergists, an expanded table on naming skeletal muscles, greatly expanded overviews in all muscle tables, and a new table on intrinsic muscles of the foot.

Exercise 11, "Surface Anatomy," has been completely rewritten and expanded, and the black-and-white photos have been replaced with color photos. In addition, several new photos have been added.

Exercise 13, "Nervous System," now contains new color photos of the spinal cord and brain, and new illustrations on cerebral white matter and the cerebellum.

In Exercise 14, "General Senses and Sensory Motor Pathways," the discussions of generator potentials and cutaneous receptors have been expanded.

In Exercise 17, "Blood," there is a new section with an accompanying illustration dealing with hemopoiesis.

Several new illustrations have been added to Exercise 18, "The Heart." Also, there are new discussions on the borders and surface projection of the heart. The discussion of the pericardium, chambers of the heart, valves of the heart, and blood vessels of the heart have also been revised.

In Exercise 19, "Blood Vessels," descriptions of arteries and veins in the tables have been greatly expanded, several new flow diagrams have been added, and several new illustrations have been added.

In Exercise 22, "Respiratory System," new art has also been added for the trachea, along with a new photo of the lungs.

In Exercise 23, "Digestive System," the introduction is new, and sections dealing with the pharynx, stomach, small intestines, and large intestines have been expanded. Several new illustrations dealing with histology have also been added.

In Exercise 24, "Urinary System," there is a new illustration on the histology of a nephron.

CHANGES IN TERMINOLOGY

In recent years, the use of eponyms for anatomical terms has been minimized or eliminated. Anatomical eponyms are terms named after various individuals. Examples include Fallopian tube (after Gabriello Fallopio) and Eustachian tube (after Bartolommeo Eustachio).

Anatomical eponyms are often vague and nondescriptive and do not necessarily mean that the person whose name is applied contributed anything very original. For these reasons, we have also decided to minimize their use. However, because some still prevail, we have provided eponyms, in parentheses, after the first reference in each chapter to the more acceptable synonym. Thus, you will expect to see terms such as *uterine (Fallopian) tube* or *auditory (Eustachian) tube*. See Appendix C.

INSTRUCTOR'S GUIDE

A complementary instructor's guide by the authors to accompany the manual is available from the publisher. This comprehensive guide contains: (1) a listing of materials needed to complete each exercise, (2) suggested audiovisual materials, (3) answers to illustrations and questions within the exercises, and (4) answers to laboratory report questions.

Gerard J. Tortora
Robert J. Amitrano
Science and Health, S229
Bergen Community College
400 Paramus Road
Paramus, NJ 07652

About the Authors

Gerard J. Tortora is Professor of Biology at Bergen Community College in Paramus, New Jersey, where he teaches human anatomy and physiology, as well as microbiology. The author of several best selling science textbooks and laboratory manuals, he was named Distinguish Faculty Scholar at Bergen Community College and received a National Institute for Staff and Organizational Development (NISOD) excellence award from the University of Texas in 1996.

Jerry received his bachelor's degree in biology from Fairleigh Dickinson University and his master's degree in science education from Montclair State College. He is a member of many professional organizations such as the Human Anatomy and Physiology Society (HAPS), the American Society of Microbiology (ASM), the American Association for the Advancement of Science (AAAS), and the Metropolitan Association of College and University Biologists (MACUB).

Robert J. Amitrano is an Associate Professor of Biology at Bergen Community College in Paramus, New Jersey. His teaching assignments for the past twelve years include human anatomy and physiology, human biology, comparative anatomy, and embryology. He has published several learning guides and laboratory manuals, and twice received awards from Who's Who Among America's Teachers. Bob is currently serving as the Biological Sciences coordinator at Bergen Community College.

Bob received his bachelor's degree in biology from Seton Hall University and a Doctorate of Chiropractic from the New York Chiropractic College. He is a member of the Human Anatomy and Physiology Society (HAPS), and the Metropolitan Association of College and University Biologists (MACUB).

Contents

Laboratory Safety*

In 1989, The Centers for Disease Control and Prevention (CDC) published "Guidelines for Prevention of Transmission of Human Immunodeficiency Virus and Hepatitis B Virus to Health-Care and Public-Safety Workers" (*MMWR,* vol. 36, No. 6S). The CDC guidelines recommend precautions to protect health care and public safety workers from exposure to human immunodeficiency virus (HIV), the causative agent of acquired immunodeficiency syndrome (AIDS), and hepatitis B virus (HBV), the causative agent of hepatitis B. These guidelines are presented to reaffirm the basic principles involved in the transmission of not only the AIDS and hepatitis B viruses, but also any disease-producing organism.

Based on the CDC guidelines for health care workers, as well as on other standard additional laboratory precautions and procedures, the following list has been developed for your safety in the laboratory. Although specific cautions and warnings concerning laboratory safety are indicated throughout the manual, read the following *before* performing any experiments.

A. GENERAL SAFETY PRECAUTIONS AND PROCEDURES

1. Arrive on time. Laboratory directions and procedures are given at the beginning of the laboratory period.
2. Read all experiments before you come to class to be sure that you understand all the procedures and safety precautions. Ask the instructor about any procedure you do not understand exactly. Do not improvise any procedure.

3. Protective eyewear and laboratory coats or aprons must be worn by all students performing or observing experiments.
4. Do not perform any unauthorized experiments.
5. Do not bring any unnecessary items to the laboratory and do not place any personal items (pocketbooks, bookbags, coats, umbrellas, etc.) on the laboratory table or at your feet.
6. Make sure each apparatus is supported and squarely on the table.
7. Tie back long hair to prevent it from becoming a laboratory fire hazard.
8. Never remove equipment, chemicals, biological materials, or any other materials from the laboratory.
9. Do not operate any equipment until you are instructed in its proper use. If you are unsure of the procedures, ask the instructor.
10. Dispose of chemicals, biological materials, used apparatus, and waste materials according to your instructor's directions. Not all liquids are to be disposed of in the sink.
11. Some exercises in the laboratory manual are designed to induce some degree of cardiovascular stress. Students should not participate in these exercises if they are pregnant or have hypertension or any other known or suspected condition that might compromise health. Before you perform any of these exercises, check with your physician.
12. Do not put anything in your mouth while in the laboratory. Never eat, drink, taste chemicals, lick labels, smoke, or store food in the laboratory.
13. Your instructor will show you the location of emergency equipment such as fire extinguishers, fire blankets, and first-aid kits as well as eyewash stations. Memorize their locations and know how to use them.
14. Wash your hands before leaving the laboratory. Because bar soaps can become contaminated, liquid or powdered soaps should be used. Before leaving the laboratory, remove any protective clothing, such as laboratory coats or aprons, gloves, and eyewear.

*The authors and publisher urge consultation with each instructor's institutional policies concerning laboratory safety and first-aid procedures.

B. PRECAUTIONS RELATED TO WORKING WITH BLOOD, BLOOD PRODUCTS, OR OTHER BODY FLUIDS

1. Work only with *your own* body fluids, such as blood, saliva, urine, tears, and other secretions and excretions; blood from a clinical laboratory that has been tested and certified as noninfectious; or blood from a mammal (other than a human).

2. Wear gloves when touching another person's blood or other body fluids.

3. Wear safety goggles when working with another person's blood.

4. Wear a mask and protective eyewear or a face shield during procedures that are likely to generate droplets of blood or other body fluids.

5. Wear a gown or an apron during procedures that are likely to generate splashes of blood or other body fluids.

6. Wash your hands immediately and thoroughly if contaminated with blood or other body fluids. Hands can be rapidly disinfected by using (1) a phenol disinfectant-detergent for 20 to 30 seconds (sec) and then rinsing with water, or (2) alcohol (50 to 70%) for 20 to 30 sec, followed by a soap scrub of 10 to 15 sec and rinsing with water.

7. Spills of blood, urine, or other body fluids onto bench tops can be disinfected by flooding them with a disinfectant-detergent. The spill should be covered with disinfectant for 20 minutes (min) before being cleaned up.

8. Potentially infectious wastes, including human body secretions and fluids, and objects such as slides, syringes, bandages, gloves, and cotton balls contaminated with those substances, should be placed in an autoclave container. Sharp objects (including broken glass) should be placed in a puncture-proof sharps container. Contaminated glassware should be placed in a container of disinfectant and autoclaved before it is washed.

9. Use only single-use, disposable lancets, and needles. Never recap, bend, or break the lancet once it has been used. Place used lancets, needles, and other sharp instruments in a *fresh* 1:10 dilution of household bleach (sodium hypochlorite) or other disinfectant such as phenols (Amphyl), aldehydes (glutaraldehyde, 1%), and 70% ethyl alcohol and then dispose of the instruments in a puncture-proof container. These disinfectants disrupt the envelope of HIV and HBV. The fresh household bleach solution or other disinfectant should be prepared for *each* laboratory session.

10. All reusable instruments, such as hemocytometers, well slides, and reusable pipettes, should be disinfected with a *fresh* 1:10 solution of household bleach or other disinfectant and thoroughly washed with soap and hot water. The fresh household bleach solution or other disinfectant should be prepared for *each* laboratory session.

11. A laboratory disinfectant should be used to clean laboratory surfaces *before* and after procedures, and should be available for quick cleanup of any blood spills.

12. Mouth pipetting should never be done. Use mechanical pipetting devices for manipulating all liquids in the laboratory.

13. All procedures and manipulations that have a high potential for creating aerosols or infectious droplets (such as centrifuging, sonicating, and blending) should be performed carefully. In such instances, a biological safety cabinet or other primary containment device is required.

C. PRECAUTIONS RELATED TO WORKING WITH REAGENTS

1. Use extreme care when working with reagents. Should any reagents make contact with your eyes, flush with water for 15 min; or, if they make contact with your skin, flush with water for 5 min. Notify your instructor immediately should a reagent make contact with your eyes or skin, and seek immediate medical attention.

2. Report all accidents to your instructor, no matter how minor they may appear.

3. When you are working with chemicals or preserved specimens, the room should be well ventilated. Avoid breathing fumes for any extended period of time.

4. Never point the opening of a test tube containing a reacting mixture (especially when heating it) toward yourself or another person.

5. Exercise care in noting the odor of fumes. Use "wafting" if you are directed to note an odor. Your instructor will demonstrate this procedure.

6. Do not force glass tubing or a thermometer into rubber stoppers. Lubricate the tubing and introduce it gradually and gently into the stopper. Protect your hands with toweling when inserting the tubing or thermometer into the stopper.

7. Never heat a flammable liquid over or near an open flame.

8. Use only glassware marked Pyrex or Kimax. Other glassware may shatter when heated. Handle hot glassware with test-tube holders.

9. If you have to dilute an acid, always add acid (AAA) to water.

10. When shaking a test tube or bottle to mix its contents, do not use your fingers as a stopper.

11. Read the label on a chemical twice before using it.

12. Replace caps or stoppers on bottles immediately after using them. Return spatulas to their correct place immediately after using them and do not mix them up.

13. Mouth pipetting should never be done. Use mechanical pipetting devices for manipulating all liquids in the laboratory.

D. PRECAUTIONS RELATED TO DISSECTION

1. When you are working with chemicals or preserved specimens, the room should be well ventilated. Avoid breathing fumes for any extended period of time.

2. Wear rubber gloves when dissecting.

3. To reduce the irritating effects of chemical preservatives to your skin, eyes, and nose, soak or wrap your specimen in a substance such as "Biostat." If this is not available, hold your specimen under running water for several minutes to wash away excess preservative and dilute what remains.

4. When dissecting, there is always the possibility of skin cuts or punctures from dissecting equipment or the specimens themselves, such as the teeth or claws of an animal. Should you sustain a cut or puncture in this manner, wash your hands with disinfectant soap, notify your instructor, and seek immediate medical attention to decrease the possibility of infection. A first-aid kit should be readily available for your use.

5. When cleaning dissecting instruments, always hold the sharp edges away from you.

6. Dispose of any damaged or worn-out dissecting equipment in an appropriate container supplied by your instructor.

SELECTED LABORATORY SAFTEY SIGNS/LABELS

⚠ CAUTION

Protect eyes.
Wear goggles at
all times.

⚠ CAUTION

Hot surface.
Do not touch.

⚠ CAUTION

Cancer suspect
agent. Trained
personnel only.

⚠ CAUTION

Radiation area.
Authorized
personnel only.

⚠ CAUTION

Biological hazard.
Authorized
personnel only.

⚠ DANGER

Highly toxic.
Handle with
care.

⚠ DANGER

Do not smoke
in this area.

⚠ DANGER

Do not smoke,
eat or drink
in this area.

⚠ DANGER

Do not pipet
liquids by
mouth.

⚠ DANGER

Corrosive. Avoid
contact with eyes
and skin.

⚠ DANGER

Flammable
material. Keep
fire away.

EMERGENCY

Eye Wash
Station.
Keep area clear.

EMERGENCY

Safety Shower.
Keep area clear.

EMERGENCY

First Aid Station.

Fire extinguisher.
Remove pin and
squeeze trigger.

COMMONLY USED LABORATORY EQUIPMENT

Beaker

Erlenmeyer flask

Florence flask

funnel

Graduated cylinder

Pipet

Mortar and pestle

Watch glass

Stirring rod

Test tube

Test tube brush

Test tube holder

Test tube rack

Ring stand and ring

Pinch clamp

Utility clamp

Tripod

Clay triangle

Wire gauze

Crucible tongs

Beaker tongs

Forceps

Medicine dropper

Nichrome wire

Spatula

Pronunciation Key

A unique feature of this revised manual is the phonetic pronunciations given for many anatomical and physiological terms. The pronunciations are given in parentheses immediately after the particular term is introduced. The following key explains the essential features of the pronunciations.

1. The syllable with the strongest accent appears in capital letters; for example, bilateral (bī-LAT-er-al) and diagnosis (dī-ag-NŌ-sis).
2. A secondary accent is denoted by a single quote mark ('); for example, constitution (kon'-sti-TOO-shun) and physiology (fiz'-ē-OL-ō-jē). Additional secondary accents are also noted by a single quotation mark; for example, decarboxylation (dē'-kar-bok'-si-LĀ-shun).

3. Vowels marked with a line above the letter are pronounced with the long sound, as in the following common words:

   ā as in *māke*   ī as in *īvy*
   ē as in *bē*   ō as in *pōle*

4. Unmarked vowels are pronounced with the short sound, as in the following words:

   e as in *bet*   o as in *not*
   i as in *sip*   u as in *bud*

5. Other phonetic symbols are used to indicate the following sounds:

   a as in *above*   yoo as in *cute*
   oo as in *soon*   oy as in *oil*

Microscopy

Note: Before you begin any laboratory exercises in this manual, please read the section on LABORATORY SAFETY on page xi.

One of the most important instruments that you will use in your anatomy and physiology course is a compound light microscope. In this instrument, the lenses are arranged so that images of objects too small to be seen with the naked eye can become highly magnified; that is, apparent size can be increased, and their minute details can be revealed. Before you actually learn the parts of a compound light microscope and how to use it properly, discussion of some of the principles employed in light microscopy (mī-KROS-kō-pē) will be helpful.

A. COMPOUND LIGHT MICROSCOPE

A *compound light microscope* uses two sets of lenses, ocular and objective, and employs light as its source of illumination. Magnification is achieved as follows. Light rays from an illuminator are passed through a condenser, which directs the light rays through the specimen under observation; from here, light rays pass into the objective lens, the magnifying lens that is closest to the specimen; the image of the specimen then forms on a prism and is magnified again by the ocular lens.

A general principle of microscopy is that the shorter the wavelength of light used in the instrument, the greater the resolution. *Resolution*, or *resolving power*, is the ability of the lenses to distinguish fine detail and structure, that is, to distinguish between two points as separate objects. As an example, a microscope with a resolving power of 0.3 micrometers (mī-KROM-e-ters), symbolized μm, is capable of distinguishing two points as separate objects if they are at least 0.3 μm apart. 1 μm = 0.000001 or 10^{-6} m. (See Appendix A.) The light used in a compound light microscope has a relatively long wavelength and cannot resolve structures smaller than 0.3 μm. This fact, as well as practical considerations, means that even the best compound light microscopes can magnify images only about 2000 times.

A *photomicrograph* (fō-tō-MĪ-krō'-graf), a photograph of a specimen taken through a compound light microscope, is shown in Figure 4.1. In later exercises you will be asked to examine photomicrographs of various specimens of the body before you actually view them yourself through the microscope.

1. Parts of the Microscope

Carefully carry the microscope from the cabinet to your desk by placing one hand around the arm and the other hand firmly under the base. Gently place it on your desk, directly in front of you, with the arm facing you. Locate the following parts of the microscope and, as you read about each part, label Figure 1.1 by placing the correct numbers in the spaces next to the list of terms that accompanies the figure.

1. *Base* The bottom portion on which the microscope rests.
2. *Body tube* The portion that receives the ocular.
3. *Arm* The angular or curved part of the frame.
4. *Inclination joint* A movable hinge in some microscopes that allows the instrument to be tilted to a comfortable viewing position.
5. *Stage* A platform on which microscope slides or other objects to be studied are placed. The opening in the center, called the *stage opening*, allows light to pass from below through the specimen being examined. Some microscopes have a *mechanical stage*. An adjustor knob below the stage moves the stage forward and backward and from side to side. With a mechanical stage, the slide and the stage move simultaneously. A mechanical stage permits a smooth, precise movement of a slide. Sometimes a mechanical stage is fitted with calibrations that permit the numerical "mapping" of a specimen on a slide.
6. *Stage (spring) clips* Two clips mounted on the stage that hold the microscope slide securely in place.

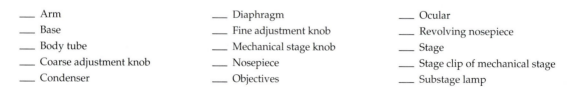

___ Arm   ___ Diaphragm   ___ Ocular

___ Base   ___ Fine adjustment knob   ___ Revolving nosepiece

___ Body tube   ___ Mechanical stage knob   ___ Stage

___ Coarse adjustment knob   ___ Nosepiece   ___ Stage clip of mechanical stage

___ Condenser   ___ Objectives   ___ Substage lamp

FIGURE 1.1 Olympus CH-2 microscope.

7. *Substage lamp* The source of illumination for some light microscopes with a built-in lamp.
8. *Mirror* A feature found in some microscopes below the stage. The mirror directs light from its source through the stage opening and through the lenses. If the light source is built-in, a mirror is not necessary.
9. *Condenser* A lens located beneath the stage opening that concentrates the light beam on the specimen.
10. *Condenser adjustment knob* A knob that functions to raise and lower the condenser. In its highest position, it allows full illumination and thus can be used to adjust illumination.
11. *Diaphragm* (DĪ-a-fram) A device located below the condenser that regulates light intensity passing through the condenser and lenses to the observer's eyes. Such regulation is needed because transparent or very thin specimens cannot be seen in bright light. One of two types of diaphragms is usually used. An *iris diaphragm*, as found in cameras, is a series of sliding leaves that vary the size of the opening and thus the amount of light entering the lenses. The leaves are moved by a *diaphragm lever* to regulate the diameter of a central opening. A *disc diaphragm* consists of a plate with a graded series of holes, any of which can be rotated into position.
12. *Coarse adjustment knob* A usually larger knob that raises and lowers the body tube (or stage) to bring a specimen into general view.
13. *Fine adjustment knob* A usually smaller knob found below or external to the coarse adjustment knob and used for fine or final focusing. Some microscopes have both coarse and fine adjustment knobs combined into one.
14. *Nosepiece* A plate, usually circular, at the bottom of the body tube.
15. *Revolving nosepiece* The lower, movable part of the nosepiece that contains the various objective lenses.
16. *Scanning objective* A lens, marked 5× on most microscopes (× means the same as "times"); it is the shortest objective and is not present on all microscopes.
17. *Low-power objective* A lens, marked 10× on most microscopes; it is the next longer objective.
18. *High-power objective* A lens, marked 40×, 43× or 45× on most microscopes; also called a *high-dry objective*; it is an even longer objective.
19. *Oil-immersion objective* A lens, marked 100× on most microscopes and distinguished by an etched colored circle (special instructions for this objective are discussed later); it is the longest objective.
20. *Ocular (eyepiece)* A removable lens at the top of the body tube, marked 10× on most microscopes. An ocular is sometimes fitted with a pointer or measuring scale.

2. Rules of Microscopy

You must observe certain basic rules at all times to obtain maximum efficiency and provide proper care for your microscope.

1. Keep all parts of the microscope clean, especially the lenses of the ocular, objectives, condenser, and also the mirror. *You should use the special lens paper that is provided and never use paper towels or cloths, because these tend to scratch the delicate glass surfaces.* When using lens paper, use the same area on the paper only once. As you wipe the lens, change the position of the paper as you go.
2. Do not permit the objectives to get wet, especially when observing a *wet mount*. You must use a *cover slip* when you examine a wet mount or the image becomes distorted.
3. Consult your instructor if any mechanical or optical difficulties rise. *Do not try to solve these problems yourself.*
4. Keep *both* eyes open at all times while observing objects through the microscope. This is difficult at first, but with practice becomes natural. This important technique will help you to draw and observe microscopic specimens without moving your head. Only your eyes will move.
5. Always use either the scanning or low-power objective first to locate an object; then, if necessary, switch to a higher power.
6. If you are using the high-power or oil-immersion objectives, *never focus using the coarse adjustment knob*. The distance between these objectives and the slide, called *working distance*, is very small and you may break the cover slip and the slide and scratch the lens.
7. Some microscopes have a stage that moves while focusing, others have a body tube that moves while focusing. Be sure you are familiar with which type you are using. Look at your microscope from the side and using the scanning or low power objective, gently turn the coarse adjustment knob. Which moves? The stage or the body tube? *Never focus downward if the microscope's body tube moves when focusing. Never*

focus upward if the microscope's stage moves when focusing. By observing from one side you can see that the objectives do not make contact with the cover slip or slide.

8. Make sure that you raise the body tube before placing a slide on the stage or before removing a slide.

3. Setting up the Microscope

PROCEDURE

1. Place the microscope on the table with the ocular toward you and with the back of the base at least 1 inch (in.) from the edge of the table.
2. Position yourself and the microscope so that you can look into the ocular comfortably.
3. Wipe the objectives, the top lens of the ocular, the condenser, and the mirror with lens paper. Clean the most delicate and the least dirty lens first. Apply xylol or alcohol to the lens paper only to remove grease and oil from the lenses and microscope slides.
4. Position the low-power objective in line with the body tube. When it is in its proper position, it will click. Lower the body tube using the coarse adjustment knob until the bottom of the lens is approximately 1/4 in. from the stage.
5. Admit the maximum amount of light by opening the diaphragm, if it is an iris diaphragm, or turning the disc to its largest opening, if it is a disc diaphragm.
6. Place your eye to the ocular, and adjust the light. When a uniform circle (the *microscopic field*) appears without any shadows, the microscope is ready for use.

4. Using the Microscope

PROCEDURE

1. Using the coarse adjustment knob, raise the body tube to its highest fixed position.
2. Make a temporary mount using a single letter of newsprint, or use a slide that has been specially prepared with a letter, usually the letter "e." If you prepare such a slide, cut a single letter—"a," "b," or "e"—from the smallest print available and place this letter in the correct position to be read with the naked eye. Your instructor will provide directions for preparing the slide.
3. Place the slide on the stage, making sure that the letter is centered over the stage opening, directly over the condenser. Secure the slide in place with the stage clips.
4. Align the low-power objective with the body tube.

5. Lower the body tube or raise the stage as far as it will go *while you watch it from the side*, taking care not to touch the slide. The tube should reach an automatic stop that prevents the low-power objective from hitting the slide.
6. While looking through the ocular, turn the coarse adjustment knob counterclockwise, raising the body tube. Or, turn the coarse adjustment knob clockwise, lowering the stage. When focusing, always *raise* the body tube or *lower* the stage. Watch for the object to suddenly appear in the microscopic field. If it is in proper focus, the low-power objective is about 1/2 in. above the slide. When focusing, always *raise* the body tube.
7. Use the fine adjustment knob to complete the focusing; you will usually use a counterclockwise motion once again.
8. Compare the position of the letter as originally seen with the naked eye to its appearance under the microscope.

   Has the position of the letter been changed?

   ———————————————

9. While looking at the slide through the ocular, move the slide by using your thumbs, or, if the microscope is equipped with them, the mechanical stage knobs. This exercise teaches you to move your specimen in various directions quickly and efficiently.

   In which direction does the letter move when you move the slide to the left?

   ———————————————

This procedure, called "scanning" a slide, will be useful for examining living objects and for centering specimens so you can observe them easily.

   Make a drawing of the letter as it appears under low power in the microscopic field in the space below on the left side.

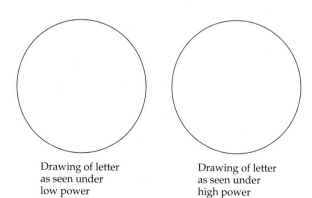

Drawing of letter as seen under low power

Drawing of letter as seen under high power

10. Change your magnification from low to high power by carrying out the following steps:

   a. Place the letter in the center of the field under low power. Centering is important because you are now focusing on a smaller area of the microscopic field. As you will see, *microscopic field size decreases with higher magnifications.*
   b. Make sure the illumination is at its maximum. Illumination must be increased at higher magnifications because the amount of light entering the lens decreases as the size of the objective lens increases.
   c. The letter should be in focus, and if the microscope is parfocal, the high-power objective can be switched into line with the body tube without changing focus. *Parfocal* means that when clear focus has been attained using any objective at random, revolving the nosepiece results in a change in magnification but leaves the specimen still in focus. If it is not completely in focus after switching the lens, a slight turn of the fine adjustment knob will focus it.
   d. If your microscope is not parfocal, observe the stage from one side and carefully switch the high-power objective in line with the body tube.
   e. While still observing from the side and using the coarse adjustment knob, *carefully* lower the objective or raise the stage until the objective almost touches the slide.
   f. Look through the ocular and focus up slowly. Finish focusing by turning the fine adjustment knob.
   g. If your microscope has an oil-immersion objective, you must follow special procedures. Place a drop of special *immersion oil* directly over the letter on the microscope slide, and lower the oil-immersion objective until it just contacts the oil. If your microscope is parfocal, you do not have to raise or lower the objectives. For example, if you are using the high-power objective and the specimen is in focus, just switch the high-power objective out of line with the body tube. Then add the oil and switch the oil-immersion objective into position; the specimen should be in focus. The same holds true when you switch from low power to high power. The special light-transmitting properties of the oil are such that light is refracted (bent) toward the specimen, per-

mitting the use of powerful objectives in a relatively narrow field of vision. This objective is extremely close to the slide being examined, so when it is in position, take precautions *never to focus downward* while you are looking through the ocular. Whenever you finish using immersion oil, be sure to saturate a piece of lens paper with xylol or alcohol and clean the oil-immersion objective and the slide if it is to be used again.

Is as much of the letter visible under high power as under low power? Explain. ＿＿＿＿＿

＿＿＿＿＿＿＿＿＿＿＿＿＿

Make a drawing of the letter as it appears under high power in the microscopic field in the space next to the drawing you made for low power following step 9 on page 4.

11. Now select a prepared slide of three different-colored threads. Examination will show that a specimen mounted on a slide has depth as well as length and width. At lower magnification the amount of depth of the specimen that is clearly in focus, the depth of field, is greater than that at higher magnification. You must focus at different depths to determine the position (depth) of each thread.

   After you make your observation under low power and high power, answer the following questions about the location of the different threads:
   What color is at the bottom, closest to the slide?

＿＿＿＿＿＿＿＿＿＿＿＿＿

On top, closest to the cover slip? ＿＿＿＿＿

＿＿＿＿＿＿＿＿＿＿＿＿＿

In the middle? ＿＿＿＿＿＿＿＿＿＿＿

＿＿＿＿＿＿＿＿＿＿＿＿＿

12. Your instructor might want you to prepare a wet mount as part of your introduction to microscopy. If so, the directions are given in Exercise 4, A.2.f, on page 56.
13. When you are finished using the microscope
   a. Remove the slide from the stage.
   b. Clean all lenses with lens paper.
   c. Align the mechanical stage so that it does not protrude.
   d. Leave the scanning or low-power objective in place.
   e. Lower the body tube or raise the stage as far as it will go.

f. Wrap the cord according to your instructor's directions.

g. Replace the dust cover or place the microscope in a cabinet.

5. Magnification

The total magnification of your microscope is calculated by multiplying the magnification of the ocular by the magnification of the objective used. Example: An ocular of $10\times$ used with an objective of $5\times$ gives a total magnification of $50\times$. Calculate the total magnification of each of the objectives on your microscope:

1. Ocular _____ $\times$ _____ Objective = _____

2. Ocular _____ $\times$ _____ Objective = _____

3. Ocular _____ $\times$ _____ Objective = _____

4. Ocular _____ $\times$ _____ Objective = _____

Note: A section of *LABORATORY REPORT QUESTIONS* is located at the end of each exercise. These questions can be answered by the student and handed in for grading at the discretion of the instructor. Even if the instructor does not require you to answer these questions, we recommend that you do so anyway to check your understanding.

Some exercises also have a section of *LABORATORY REPORT RESULTS*, in which students can record results of laboratory exercises, in addition to laboratory report questions. As with the laboratory report questions, the laboratory report results are located at the end of selected exercises and can be handed in as the instructor directs. Instructions in the manual tell students when and where to record laboratory results.

ANSWER THE LABORATORY REPORT QUESTIONS AT THE END OF THE EXERCISE.

Microscopy 1

Student _____   Date _____

Laboratory Section _____   Score/Grade _____

PART 1. Multiple Choice

_____ 1. The amount of light entering a microscope may be adjusted by regulating the (a) ocular (b) diaphragm (c) fine adjustment knob (d) nosepiece

_____ 2. If the ocular on a microscope is marked $10\times$ and the low-power objective is marked $15\times$, the total magnification is (a) $50\times$ (b) $25\times$ (c) $150\times$ (d) $1500\times$

_____ 3. The size of the light beam that passes through a microscope is regulated by the (a) revolving nosepiece (b) coarse adjustment knob (c) ocular (d) condenser

_____ 4. Parfocal means that (a) the microscope employs only one lens (b) final focusing can be done only with the fine adjustment knob (c) changing objectives by revolving the nosepiece will still keep the specimen in focus (d) the highest magnification attainable is $1000\times$

_____ 5. Which of these is *not* true when changing magnification from low power to high power? (a) the specimen should be centered (b) illumination should be decreased (c) the specimen should be in clear focus (d) the high-power objective should be in line with the body tube

_____ 6. The ability of a microscope to distinguish between two points as separate objects is called (a) parfocal focusing (b) working distance (c) diffraction (d) resolution

PART 2. Completion

7. The advantage of using immersion oil is that it has special _____ properties that permit the use of a powerful objective in a narrow field of vision.

8. The uniform circle of light that appears when one looks into the ocular is called the

_____ .

9. In determining the position (depth) of the colored threads, the _____ (red, blue, yellow) colored thread was in the middle.

10. If you move your slide to the right, the specimen moves to the _____ as you are viewing it microscopically.

11. After switching from low power to high power, _____ (more or less) of the specimen will be visible.

12. Microscopic field size _____ (increases or decreases) with higher magnifications.

13. The distance between the objectives and the slide is called the _____ .

14. A photograph of a specimen taken through a compound light microscope is called a(n)

_____ .

15. An ocular of 10× used with an objective of 40× gives a total magnification of

_____ ×.

PART **3. Matching**

_____ **16.** Ocular

_____ **17.** Stage

_____ **18.** Arm

_____ **19.** Condenser

_____ **20.** Revolving nosepiece

_____ **21.** Low-power objective

_____ **22.** Fine adjustment knob

_____ **23.** Diaphragm

_____ **24.** Coarse adjustment knob

_____ **25.** High-power objective

A. Platform on which slide is placed

B. Mounting for objectives

C. Lens below stage opening

D. Brings specimen into sharp focus

E. Eyepiece

F. An objective usually marked 40×, 43× or 45×

G. An objective usually marked 10×

H. Angular or curved part of frame

I. Brings specimen into general focus

J. Regulates light intensity

Introduction to the Human Body

2

In this exercise, you will be introduced to the organization of the human body through a study of the principal subdivisions of anatomy and physiology, levels of structural organization, principal body systems, the anatomical position, regional names, directional terms, planes of the body, body cavities, abdominopelvic regions, and abdominopelvic quadrants.

A. ANATOMY AND PHYSIOLOGY

Whereas *anatomy* (a-NAT-o-mē; *ana* = up; *tome* = cutting) refers to the study of *structure* and the relationships among structures, *physiology* (fiz-ē-OL-ō-jē; *phys* = nature; *ology* = study of) deals with the *functions* of body parts, that is, how they work. Each structure of the body is designed to carry out a particular function.

Following are selected subdivisions of anatomy and physiology. In the spaces provided, define each term.

SUBDIVISIONS OF ANATOMY

Surface anatomy _____

_____

_____

_____

Gross (macroscopic) anatomy _____

_____

_____

_____

Systemic anatomy _____

_____

_____

_____

Regional anatomy _____

_____

_____

_____

Radiographic (rā'-dē-ō-GRAF-ik) *anatomy* _____

_____

_____

_____

Developmental anatomy _____

_____

_____

_____

Embryology (em'-brē-OL-ō-jē) _____

_____

_____

_____

Histology (hiss'-TOL-ō-jē) _____

_____

_____

_____

Cytology (sū-TOL-ō-jē) _____

_____

_____

_____

Pathological (path'-ō-LOJ-i-kal) *anatomy* _____

_____

_____

_____

SUBDIVISIONS OF PHYSIOLOGY

Cell physiology _____

_____

_____

_____

Pathophysiology (PATH-ō-fiz-ē-ol'-ō-jē) _____

_____

_____

_____

Exercise physiology _____

_____

_____

_____

Neurophysiology (NOO-rō-fiz-ē-ol'-ō-jē) _____

_____

_____

_____

Endocrinology (en'-dō-kri-NOL-ō-jē) _____

_____

_____

_____

Cardiovascular (kar-dē-ō-VAS-kyoo-lar) *physiol-*

ogy _____

_____

_____

_____

Immunology (im'-yoo-NOL-ō-jē) _____

_____

_____

_____

Respiratory (re-SPĪ-ra-tō-rē) *physiology* _____

_____

_____

_____

Renal (RĒ-nal) *physiology* _____

_____

_____

_____

B. LEVELS OF STRUCTURAL ORGANIZATION

The human body is composed of several levels of structural organization associated with one another in various ways:

1. *Chemical level* Composed of all atoms and molecules necessary to maintain life.
2. *Cellular level* Consists of cells, the basic structural and functional units of the body.
3. *Tissue level* Formed by tissues, groups of cells (and their intercellular material) that usually arise from common ancestor cells and work together to perform a particular function.
4. *Organ level* Consists of organs, structures composed of two or more different tissues, having specific functions and usually having recognizable shapes.
5. *System level* Formed by systems, associations of organs that have a common function. The systems together constitute an *organism,* a total living individual.

C. SYSTEMS OF THE BODY

Using an anatomy and physiology textbook, torso, wall chart, and any other materials that might be available to you, identify the principal organs that compose the following body systems.[1] In the spaces that follow, indicate the organs and functions of the systems.

Integumentary

Organs _____

_____

_____

_____

---

[1]You will probably need other sources, plus any aids the instructor might provide, to label many of the figures and answer some questions in this manual. You are encouraged to use other sources as you find necessary.

Functions _____

_____

_____

_____

Skeletal

Organs _____

_____

_____

_____

Functions _____

_____

_____

_____

Muscular

Organs _____

_____

_____

_____

Functions _____

_____

_____

_____

Nervous

Organs _____

_____

_____

_____

Functions _____

_____

_____

_____

Endocrine

Organs _____

_____

_____

_____

Functions _____

_____

_____

_____

Cardiovascular

Organs _____

_____

_____

_____

Functions _____

_____

_____

_____

Lymphatic and immune

Organs _____

_____

_____

_____

Functions _____

_____

_____

_____

Respiratory

Organs _____

_____

_____

_____

Functions _____

_____

_____

_____

Digestive

Organs _____

_____

_____

_____

Functions _____

_____

_____

_____

Urinary

Organs _____

_____

_____

_____

Functions _____

_____

_____

_____

Reproductive

Organs _____

_____

_____

_____

Functions _____

_____

_____

_____

D. LIFE PROCESSES

All living forms carry on certain processes that distinguish them from nonliving things. Using an anatomy and physiology textbook, define the following life processes of humans.

Metabolism _____

_____

_____

_____

Responsiveness _____

_____

_____

Movement _____

_____

_____

_____

Growth _____

_____

_____

_____

Differentiation _____

_____

_____

_____

Reproduction _____

_____

_____

_____

E. HOMEOSTASIS

Homeostasis (hō-mē-ō-STĀ-sis) is a condition in which the body's internal environment (interstitial fluid) remains within certain physiological limits. Homeostasis is achieved through the operation of feedback systems. A *feedback system* involves a cycle of events in which information about the status of a condition is continually monitored and fed back to a central control region.

Using an anatomy and physiology textbook, define the following components of a feedback system.

Stimulus _____

_____

_____

_____

Controlled condition _____

_____

_____

_____

Receptor _____

_____

_____

_____

Input _____

_____

_____

_____

Control center _____

_____

_____

_____

Output _____

_____

_____

_____

Effector _____

_____

_____

_____

Response _____

_____

_____

_____

If the response of the body reverses the original stimulus, the system is a *negative feedback system*. If the response enhances or intensifies the original stimulus, the system is a *positive feedback system*.

Negative feedback systems tend to maintain conditions that require frequent monitoring and adjustments within physiological limits, such as body temperature or blood sugar level. (See Figure 2.1.) Most feedback systems in the body are negative. Positive feedback systems, on the other hand, are important for conditions that do not require continual fine-tuning. Since positive feedback systems tend to intensify or amplify a controlled condition, they usually are shut off by some mechanism outside the system if they are part of a normal physiological response. Positive feedback systems can be destructive and result in various disorders, yet some are normal and beneficial. For example, during blood clotting, which helps stop loss of blood from a cut, the initial signal is amplified until the blood clot forms and bleeding is under control.

Then, other substances help turn off the clotting response. Positive feedback mechanisms also contribute during birth of a baby to strengthen labor contractions and during immune responses to provide defense against pathogens.

Label Figure 2.1, the control of blood pressure by a negative feedback system.

F. ANATOMICAL POSITION AND REGIONAL NAMES

Figure 2.2 shows anterior and posterior views of a subject in the *anatomical position*. The subject is standing erect (upright position) and facing the observer with the head, eyes, and toes facing forward

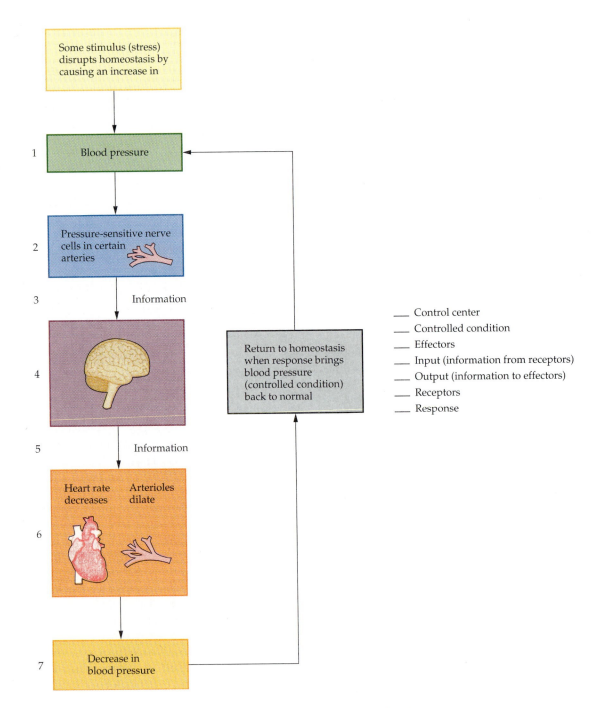

____ Control center
____ Controlled condition
____ Effectors
____ Input (information from receptors)
____ Output (information to effectors)
____ Receptors
____ Response

FIGURE 2.1   The control of blood pressure by a negative feedback system.

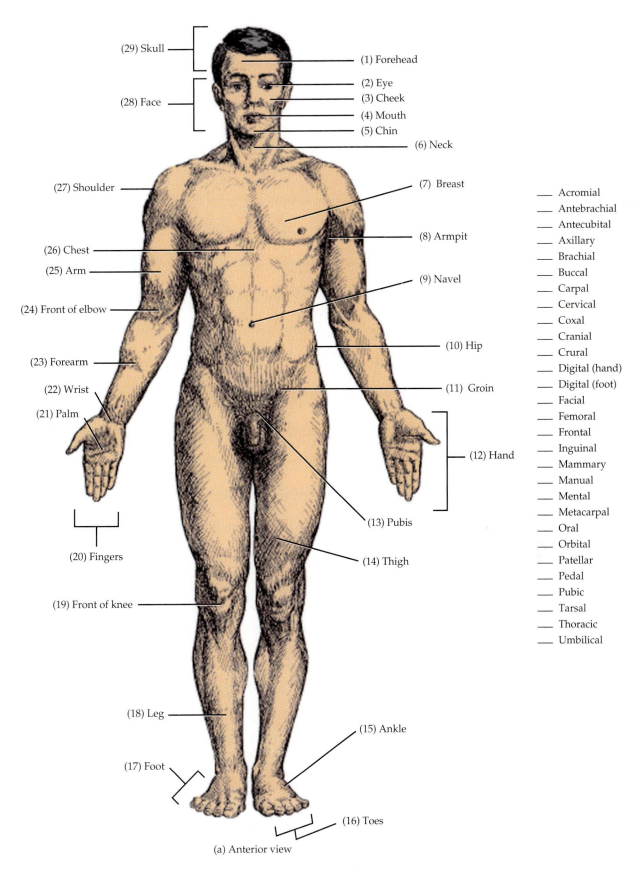

(29) Skull
(28) Face
(27) Shoulder
(26) Chest
(25) Arm
(24) Front of elbow
(23) Forearm
(22) Wrist
(21) Palm
(20) Fingers
(19) Front of knee
(18) Leg
(17) Foot

(1) Forehead
(2) Eye
(3) Cheek
(4) Mouth
(5) Chin
(6) Neck
(7) Breast
(8) Armpit
(9) Navel
(10) Hip
(11) Groin
(12) Hand
(13) Pubis
(14) Thigh
(15) Ankle
(16) Toes

____ Acromial
____ Antebrachial
____ Antecubital
____ Axillary
____ Brachial
____ Buccal
____ Carpal
____ Cervical
____ Coxal
____ Cranial
____ Crural
____ Digital (hand)
____ Digital (foot)
____ Facial
____ Femoral
____ Frontal
____ Inguinal
____ Mammary
____ Manual
____ Mental
____ Metacarpal
____ Oral
____ Orbital
____ Patellar
____ Pedal
____ Pubic
____ Tarsal
____ Thoracic
____ Umbilical

(a) Anterior view

FIGURE 2.2 The anatomical position.

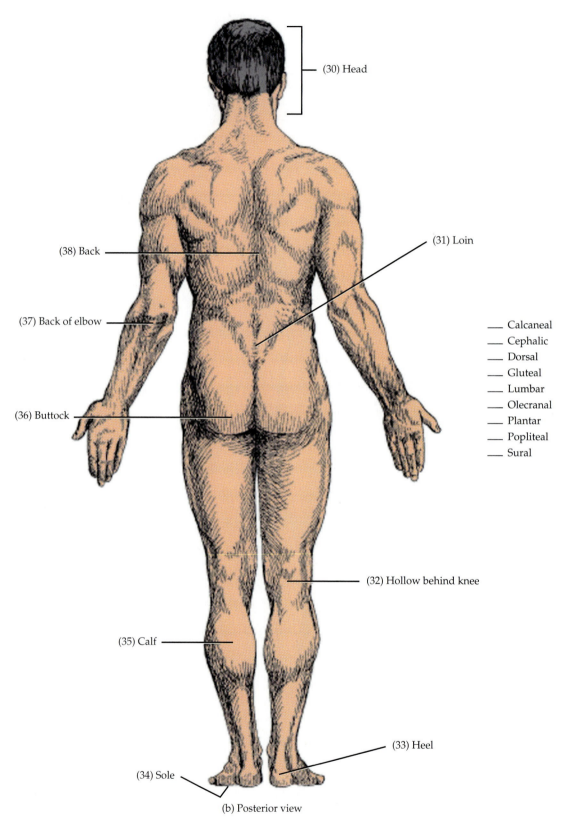

(30) Head

(31) Loin

(38) Back

Calcaneal
Cephalic
Dorsal
Gluteal
Lumbar
Olecranal
Plantar
Popliteal
Sural

(37) Back of elbow

(36) Buttock

(32) Hollow behind knee

(35) Calf

(33) Heel

(34) Sole

(b) Posterior view

FIGURE 2.2 (*Continued*) The anatomical position.

when in anterior view, the upper limbs (extremities) are at the sides with the palms facing forward, and the feet are flat on the floor and directed forward. The figure also shows the common names for various regions of the body. When you as the observer make reference to the left and right sides of the subject you are studying, this refers to the *subject's* left and right sides. In the spaces next to the list of terms in Figure 2.2 write the number of each common term next to each corresponding anatomical term. For example, the skull (29) is cranial, so write the number *29* next to the term *Cranial*.

G. EXTERNAL FEATURES OF THE BODY

Referring to your textbook and human models, identify the following external features of the body:

1. *Head (cephalic region* or *caput)* This is divided into the *cranium* (brain case) and *face*. The cranium (skull) encloses and protects the brain; the face is the anterior portion of the head that includes the eyes, nose, mouth, forehead, cheeks and chin.
2. *Neck (collum)* This region supports the head and is called the cervical region.
3. *Trunk* This region is also called the torso and is divided into the *back* (dorsum), *chest* (thorax), *abdomen* (venter), and *pelvis*.
4. *Upper limb (extremity)* This consists of the *armpit* (axilla), *shoulder* (acromial region or omos), *arm* (brachium), *elbow* (cubitus), *forearm* (antebrachium), and *hand* (manus). The hand, in turn, consists of the *wrist* (carpus), *palm* (metacarpus), and *fingers* (digits). Individual bones of a digit (finger or toe) are called *phalanges. Phalanx* is singular.
5. *Lower limb (extremity)* This consists of the *buttocks* (gluteal region), *thigh* (femoral region), *knee* (genu), *leg* (crus), and *foot* (pes). The foot includes the *ankle* (tarsus), *sole* (metatarsus), and *toes* (digits). The *groin* is the part of attachment between the lower limb and trunk.

H. DIRECTIONAL TERMS

To explain exactly where a structure of the body is located, it is a standard procedure to use *directional terms*. Such terms are very precise and avoid the use of unnecessary words. Commonly used directional terms for humans are as follows:

1. *Superior* (soo'-PEER-ē-or) *(cephalic* or *cranial)* Toward the head or the upper part of a structure; generally refers to structures in the trunk.
2. *Inferior* (in'-FEER-ē-or) *(caudal)* Away from the head or toward the lower part of a structure; generally refers to structures in the trunk.
3. *Anterior* (an-TEER-ē-or) *(ventral)* Nearer to or at the front surface of the body. In the *prone position*, the body lies anterior side down; in the *supine position*, the body lies anterior side up.
4. *Posterior* (pos-TEER-ē-or) *(dorsal)* Nearer to or at the back or backbone surface of the body.
5. *Medial* (MĒ-dē-al) Nearer the midsagittal plane. The *midline* is an imaginary vertical line that divides the body into equal left and right sides.
6. *Lateral* (LAT-er-al) Farther from the midsagittal plane.
7. *Intermediate* (in'-ter-MĒ-dē-at) Between two structures.
8. *Ipsilateral* (ip-si-LAT-er-al) On the same side of the midline of the body.
9. *Contralateral* (CON-tra-lat-er-al) On the opposite side of the midline of the body.
10. *Proximal* (PROK-si-mal) Nearer the attachment of a limb to the trunk; nearer to the point of origin.
11. *Distal* (DIS-tal) Farther from the attachment of a limb to the trunk; farther from the point of origin.
12. *Superficial* (soo'-per-FISH-al) *(external)* Toward or on the surface of the body.
13. *Deep* (DĒP) *(internal)* Away from the surface of the body.

Using a torso and an articulated skeleton, and consulting with your instructor as necessary, describe the location of the following by inserting the proper directional term. Please use each term only once.

1. The ulna is on the _____ side of the forearm.

2. The lungs are _____ to the heart.

3. The heart is _____ to the liver.

4. The muscles of the arm are _____ to the skin of the arm.

5. The sternum is _____ to the heart.

6. The humerus is _____ to the radius.

7. The stomach is _____ to the lungs.

8. The muscles of the thoracic wall are _____ to the viscera in the thoracic cavity.

9. The esophagus is _____ to the trachea.

10. The phalanges are _____ to the carpals.

11. The ring finger is _____ between the little (medial) and middle (lateral) fingers.

12. The ascending colon of the large intestine and the gallbladder are _____.

13. The ascending and descending colons of the large intestine are _____.

I. PLANES OF THE BODY

The structural plan of the human body may be described with respect to *planes* (imaginary flat surfaces) passing through it. Planes are frequently used to show the anatomical relationship of several structures in a region to one another.

Commonly used planes are as follows:

1. *Midsagittal* (mid-SAJ-it-tal; *sagittalis* = arrow) or *median* A vertical plane that passes through the midline of the body and divides the body or an organ into *equal* right and left sides.
2. *Parasagittal* (par-a-SAJ-it-tal; *para* = near) A vertical plane that does not pass through the midline of the body and divides the body or an organ into *unequal* right and left sides.
3. *Frontal* (*coronal;* kō-RŌ-nal; *corona* = crown) A vertical plane that divides the body or an organ into anterior (front) and posterior (back) portions.
4. *Transverse* (*cross-sectional* or *horizontal*) A plane that divides the body or an organ into superior (top) and inferior (bottom) portions.
5. *Oblique* (ō-BLĒK) A plane that passes through the body or an organ at an angle between the transverse plane and either the midsagittal, parasagittal, or frontal plane.

Refer to Figure 2.3 and label the planes shown.

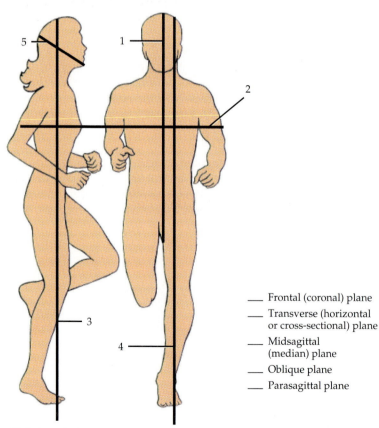

_____ Frontal (coronal) plane
_____ Transverse (horizontal or cross-sectional) plane
_____ Midsagittal (median) plane
_____ Oblique plane
_____ Parasagittal plane

Right lateral view   Anterior view

FIGURE 2.3   Planes of the body.

J. BODY CAVITIES

Spaces within the body that help protect, separate, and support internal organs are called *body cavities*. One way of organizing the principal body cavities follows:

> *Dorsal body cavity*
> > *Cranial (KRĀ-nē-al) cavity*
> > *Vertebral (VER-te-bral)* or *spinal cavity*
>
> *Ventral body cavity*
> > *Thoracic (thor-AS-ik) cavity*
> > > Right pleural (PLOOR-al)
> > > Left pleural
> > > Pericardial (per'-i-KAR-dē-al)
> >
> > *Abdominopelvic cavity*
> > > Abdominal
> > > Pelvic

The mass of tissue in the thoracic cavity between the coverings (pleurae) of the lungs and extending from the sternum (breast bone) to the backbone and from the neck to diaphragm is called the *mediastinum* (mē'-dē-as-TĪ-num *media* = middle; *stare* = stand in). It contains all structures in the thoracic cavity, except the lungs themselves. Included are the heart, thymus gland, esophagus, trachea, and many large blood and lymphatic vessels.

Label the body cavities shown in Figure 2.4. Then examine a torso or wall chart, or both, and determine which organs lie within each cavity.

Using *T* (for thoracic), *A* (for abdominal), and *P* (for pelvic), indicate which organs are found in their respective cavities.

1. _____ Urinary bladder
2. _____ Stomach
3. _____ Spleen
4. _____ Lungs
5. _____ Liver
6. _____ Internal reproductive organs
7. _____ Small intestine
8. _____ Heart
9. _____ Gallbladder
10. _____ Small portion of large intestine

K. ABDOMINOPELVIC REGIONS

To describe the location of viscera more easily, the abdominopelvic cavity may be divided into *nine regions* by using four imaginary lines: (1) an upper horizontal *subcostal* (sub-KOS-tal) *line* that passes just below the bottom of the rib cage through the lower portion of the stomach, (2) a lower horizontal

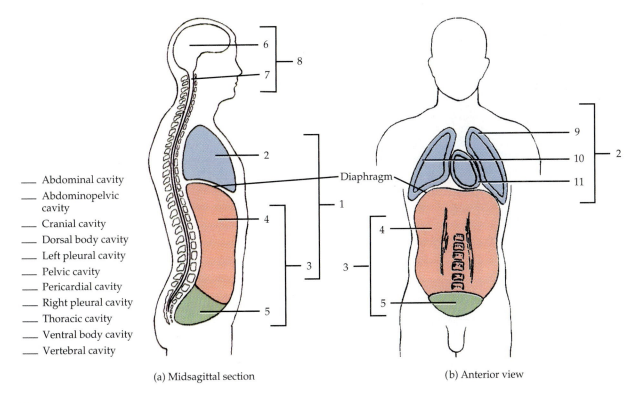

_____ Abdominal cavity
_____ Abdominopelvic cavity
_____ Cranial cavity
_____ Dorsal body cavity
_____ Left pleural cavity
_____ Pelvic cavity
_____ Pericardial cavity
_____ Right pleural cavity
_____ Thoracic cavity
_____ Ventral body cavity
_____ Vertebral cavity

Diaphragm

(a) Midsagittal section

(b) Anterior view

FIGURE 2.4   Body cavities.

line, the ***transtubercular*** (trans-too-BER-kyoo'-lar) ***line***, just below the top surfaces of the hipbones, (3) a ***right midclavicular*** (mid-kla-VIK-yoo'-lar) ***line*** drawn through the midpoint of the right clavicle slightly medial to the right nipple, and (4) a ***left midclavicular line*** drawn through the midpoint of the left clavicle slightly medial to the left nipple.

The four imaginary lines divide the abdominopelvic cavity into the following nine regions: (1) ***umbilical*** (um-BIL-i-kul) ***region,*** which is centrally located; (2) ***left lumbar*** (*lumbus* = loin) ***region,*** to the left of the umbilical region; (3) ***right lumbar,*** to the right of the umbilical region; (4) ***epigastric*** (ep-i-GAS-trik; *epi* = above; *gaster* = stomach) ***region,*** directly above the umbilical region; (5) ***left hypochondriac*** (hī'-pō-KON-drē-ak; *hypo* = under;

chondro = cartilage) ***region,*** to the left of the epigastric region; (6) ***right hypochondriac region,*** to the right of the epigastric region; (7) ***hypogastric (pubic) region,*** directly below the umbilical region; (8) ***left iliac*** (IL-ē-ak; *iliacus* = superior part of hip bone) or ***inguinal region,*** to the left of the hypogastric (pubic) region; and (9) ***right iliac (inguinal) region,*** to the right of the hypogastric (pubic) region.

Label Figure 2.5 by indicating the names of the four imaginary lines and the nine abdominopelvic regions.

Examine a torso and determine which organs or parts of organs lie within each of the nine abdominopelvic regions.

In the space provided on page 21, list several organs or parts of organs found in the following abdominopelvic regions:

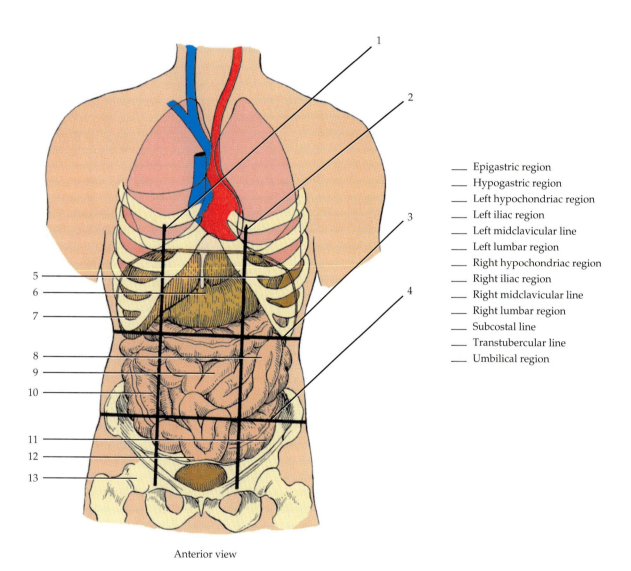

_____ Epigastric region
_____ Hypogastric region
_____ Left hypochondriac region
_____ Left iliac region
_____ Left midclavicular line
_____ Left lumbar region
_____ Right hypochondriac region
_____ Right iliac region
_____ Right midclavicular line
_____ Right lumbar region
_____ Subcostal line
_____ Transtubercular line
_____ Umbilical region

Anterior view

FIGURE 2.5   Abdominopelvic regions.

Right hypochondriac _____

_____

_____

Epigastric _____

_____

_____

Left hypochondriac _____

_____

_____

Right lumbar _____

_____

_____

Umbilical _____

_____

_____

Left lumbar _____

_____

_____

Right iliac _____

_____

_____

Hypogastric _____

_____

_____

Left iliac _____

_____

_____

L. ABDOMINOPELVIC QUADRANTS

Another way to divide the abdominopelvic cavity is into *quadrants* by passing one horizontal line and one vertical line through the umbilicus (navel). The two lines thus divide the abdominopelvic

cavity into a *right upper quadrant (RUQ), left upper quadrant (LUQ), right lower quadrant (RLQ)*, and *left lower quadrant (LLQ)*. Quadrant names are frequently used by health care professionals for locating the site of an abdominopelvic pain, tumor, or other abnormality.

Examine a torso or wall chart, or both, and determine which organs or parts of organs lie within each of the abdominopelvic quadrants.

M. DISSECTION OF WHITE RAT

Now that you have some idea of the names of the various body systems and the principal organs that comprise each, you can actually observe some of these organs by dissecting a white rat. **Dissect** means "to separate." This dissection gives you an excellent opportunity to see the different sizes, shapes, locations, and relationships of organs and to compare the different textures and external features of organs. In addition, this exercise will introduce you to the general procedure for dissection before you dissect in later exercises.

CAUTION! *Please reread Section D, "Precautions Related to Dissection" at the beginning of the laboratory manual on page xiii before you begin your dissection.*

PROCEDURE

1. Place the rat on its backbone on a wax dissecting pan (tray). Using dissecting pins, anchor each of the four limbs to the wax (Figure 2.6a).
2. To expose the contents of the thoracic, abdominal, and pelvic cavities, you will have to first make a midline incision. This is done by lifting the abdominal skin with a forceps to separate the skin from the underlying connective tissue and muscles. While lifting the abdominal skin, cut through it with scissors and make an incision that extends from the lower jaw to the anus (Figure 2.6a).
3. Now make four lateral incisions that extend from the midline incision into the four limbs (Figure 2.6a).
4. Peel the skin back and pin the flaps to the wax to expose the superficial muscles (Figure 2.6b).
5. Next, lift the abdominal muscles with a forceps and cut through the muscle layer, being careful not to damage any underlying organs. Keep the scissors parallel to the rat's backbone. Extend this incision from the anus to a point just below the bottom of the rib cage (Figure 2.6b). Make

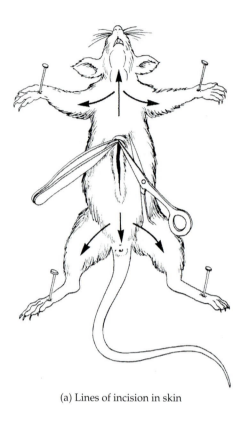

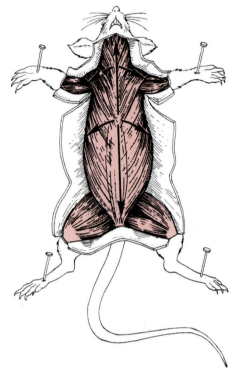

(a) Lines of incision in skin

(b) Peeling back skin and lines of incision in muscles

FIGURE 2.6   Dissection procedure for exposing thoracic and abdominopelvic viscera of the white rat for examination.

two lateral incisions just below the rib cage and fold back the muscle flaps to expose the abdominal and pelvic viscera (Figure 2.6b).

6. To expose the thoracic viscera, cut through the ribs on either side of the sternum. This incision should extend from the diaphragm to the neck (Figure 2.6b). The *diaphragm* is the thin muscular partition that separates the thoracic from the abdominal cavity. Again make lateral incisions in the chest wall so that you can lift the ribs to view the thoracic contents.

1. Examination of Thoracic Viscera

You will first examine the thoracic viscera (large internal organs). As you dissect and observe the various structures, palpate (feel with the hand) them so that you can compare their texture. Use Figure 2.7 as a guide.

a. ***Thymus gland***   An irregular mass of glandular tissue superior to the heart and superficial to the trachea. Push the thymus gland aside or remove it.

b. ***Heart***   A structure located in the midline, deep to the thymus gland and between the lungs. The sac covering the heart is the ***pericardium***, which

may be removed. The large vein that returns blood from the lower regions of the body is the ***inferior vena cava***; the large vein that returns blood to the heart from the upper regions of the body is the ***superior vena cava***. The large artery that carries blood from the heart to most parts of the body is the ***aorta***.

c. ***Lungs***   Reddish, spongy structures on either side of the heart. Note that the lungs are divided into regions called ***lobes***.

d. ***Trachea***   A tubelike passageway superior to the heart and deep to the thymus gland. Note that the wall of the trachea consists of rings of cartilage. Identify the ***larynx*** (voice box) at the superior end of the trachea and the ***thyroid gland***, a bilobed structure on either side of the larynx. The lobes of the thyroid gland are connected by a band of thyroid tissue, the isthmus.

e. ***Bronchial tubes***   Trace the trachea inferiorly and note that it divides into bronchial tubes that enter the lungs and continue to divide within them.

f. ***Esophagus***   A muscular tube posterior to the trachea that transports food from the throat into the stomach. Trace the esophagus inferiorly to see where it passes through the diaphragm to join the stomach.

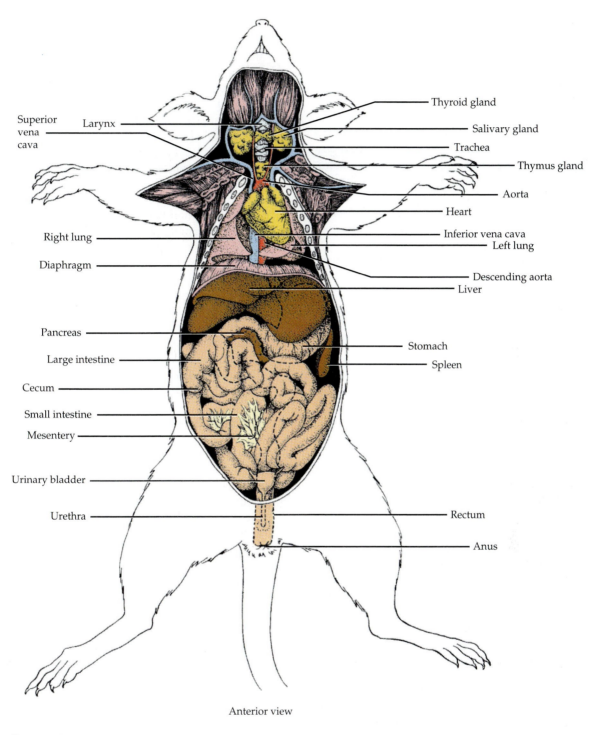

Thyroid gland

Superior vena cava

Larynx

Salivary gland

Trachea

Thymus gland

Aorta

Heart

Right lung

Inferior vena cava

Left lung

Diaphragm

Descending aorta

Liver

Pancreas

Stomach

Large intestine

Spleen

Cecum

Small intestine

Mesentery

Urinary bladder

Urethra

Rectum

Anus

Anterior view

FIGURE 2.7 Superficial structures of the thoracic and abdominopelvic cavities of the white rat.

2. Examination of Abdominopelvic Viscera

You will now examine the principal viscera of the abdomen and pelvis. As you do so, again refer to Figure 2.7.

a. *Stomach* An organ located on the left side of the abdomen and in contact with the liver. The digestive organs are attached to the posterior abdominal wall by a membrane called the *mesentery*. Note the blood vessels in the mesentery.

b. *Small intestine* An extensively coiled tube that extends from the stomach to the first portion of the large intestine called the *cecum*.

c. *Large intestine* A wider tube than the small intestine that begins at the cecum and ends at the rectum. The cecum is a large, saclike structure. In humans, the appendix arises from the cecum.

d. *Rectum* A muscular passageway, located on the midline in the pelvic cavity, that terminates in the anus.

e. *Anus* Terminal opening of the digestive system to the exterior.

f. *Pancreas* A pale gray, glandular organ posterior and inferior to the stomach.

g. *Spleen* A small, dark red organ lateral to the stomach.

h. *Liver* A large, brownish-red organ directly inferior to the diaphragm. The rat does not have a gallbladder, a structure associated with the liver. To locate the remaining viscera, either move the superficial viscera aside or remove them. Use Figure 2.8 as a guide.

i. *Kidneys* Bean-shaped organs embedded in fat and attached to the posterior abdominal wall on either side of the backbone. As will be explained later, the kidneys and a few other structures are behind the membrane that lines the abdomen (*peritoneum*). Such structures are referred to as *retroperitoneal* and are not actually within the abdominal cavity. See if you can find the *abdominal aorta*, the large artery located along the midline behind the inferior vena cava. Also, locate the *renal arteries* branching off the abdominal aorta to enter the kidneys.

j. *Adrenal (suprarenal) glands* Glandular structures. One is located on top of each kidney.

k. *Ureters* Tubes that extend from the medial surface of the kidneys inferiorly to the urinary bladder.

l. *Urinary bladder* A saclike structure in the pelvic cavity that stores urine.

m. *Urethra* A tube that extends from the urinary bladder to the exterior. Its opening to the exterior is called the *urethral orifice*. In male rats, the urethra extends through the penis; in female rats, the tube is separate from the reproductive tract.

If your specimen is female (no visible scrotum anterior to the anus), identify the following:

n. *Ovaries* Small, dark structures inferior to the kidneys.

o. *Uterus* An organ located near the urinary bladder consisting of two sides (horns) that join separately into the vagina.

p. *Vagina* A tube that leads from the uterus to the external vaginal opening, the *vaginal orifice*. This orifice is in front of the anus and behind the urethral orifice.

If your specimen is male, identify the following:

q. *Scrotum* Large sac anterior to the anus that contains the testes.

r. *Testes* Egg-shaped glands in the scrotum. Make a slit into the scrotum and carefully remove one testis. See if you can find a coiled duct attached to the testis (*epididymis*) and a duct that leads from the epididymis into the abdominal cavity (*vas deferens*).

s. *Penis* Organ of copulation medial to the testes.

When you have finished your dissection, store or dispose of your specimen according to your instructor's directions. Wash your dissecting pan and dissecting instruments with laboratory detergent, dry them, and return them to their storage areas.

ANSWER THE LABORATORY REPORT QUESTIONS AT THE END OF THE EXERCISE.

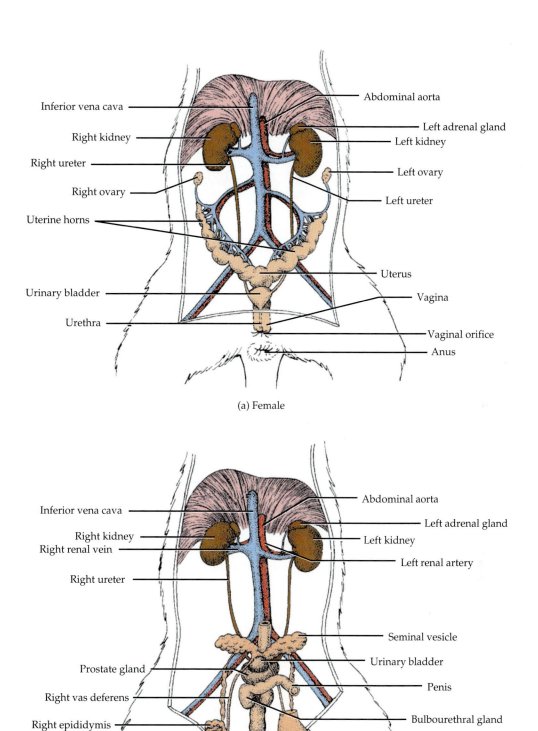

Inferior vena cava

Right kidney

Right ureter

Right ovary

Uterine horns

Urinary bladder

Urethra

Abdominal aorta

Left adrenal gland
Left kidney

Left ovary

Left ureter

Uterus

Vagina

Vaginal orifice
Anus

(a) Female

Inferior vena cava

Right kidney
Right renal vein

Right ureter

Prostate gland

Right vas deferens

Right epididymis

Right testis

Abdominal aorta

Left adrenal gland
Left kidney

Left renal artery

Seminal vesicle

Urinary bladder

Penis

Bulbourethral gland

Anus

Left scrotum

(b) Male

FIGURE 2.8   Deep structures of the abdominopelvic cavity of the white rat.

25

Introduction to the Human Body 2

Student _____   Date _____

Laboratory Section _____   Score/Grade _____

PART 1. Multiple Choice

_____ 1. The directional term that best describes the eyes in relation to the nose is (a) distal (b) superficial (c) anterior (d) lateral

_____ 2. Which does *not* belong with the others? (a) right pleural cavity (b) pericardial cavity (c) vertebral cavity (d) left pleural cavity

_____ 3. Which plane divides the brain into an anterior and a posterior portion? (a) frontal (b) median (c) sagittal (d) transverse

_____ 4. The urinary bladder lies in which region? (a) umbilical (b) hypogastric (c) epigastric (d) left iliac

_____ 5. Which is *not* a characteristic of the anatomical position? (a) the subject is erect (b) the subject faces the observer (c) the palms face backward (d) the upper limbs are at the sides

_____ 6. The abdominopelvic region that is bordered by all four imaginary lines is the (a) hypogastric (b) epigastric (c) left hypochondriac (d) umbilical

_____ 7. Which directional term best describes the position of the phalanges with respect to the carpals? (a) lateral (b) distal (c) anterior (d) proximal

_____ 8. The pancreas is found in which body cavity? (a) abdominal (b) pericardial (c) pelvic (d) vertebral

_____ 9. The anatomical term for the leg is (a) brachial (b) tarsal (c) crural (d) sural

_____ 10. In which abdominopelvic region is the spleen located? (a) left lumbar (b) right lumbar (c) epigastric (d) left hypochondriac

_____ 11. Which of the following represents the most complex level of structural organization? (a) organ (b) cellular (c) tissue (d) chemical

_____ 12. Which body system is concerned with support, protection, leverage, blood-cell production, and mineral storage? (a) cardiovascular (b) integumentary (c) skeletal (d) digestive

_____ 13. The skin and structures derived from it, such as nails, hair, sweat glands, and oil glands, are components of which system? (a) respiratory (b) integumentary (c) muscular (d) digestive

_____ 14. Hormone-producing glands belong to which body system? (a) cardiovascular (b) lymphatic and immune (c) endocrine (d) digestive

_____ 15. Which body system brings about movement, maintains posture, and produces heat? (a) skeletal (b) respiratory (c) reproductive (d) muscular

_____ 16. Which abdominopelvic quadrant contains most of the liver? (a) RUQ (b) RLQ (c) LUQ (d) LLQ

_____ **17.** The physical and chemical breakdown of food for use by body cells and the elimination of solid wastes are accomplished by which body system? (a) respiratory (b) urinary (c) cardiovascular (d) digestive

_____ **18.** The ability of an organism to detect and respond to environmental changes is called (a) metabolism (b) differentiation (c) responsiveness (d) respiration

_____ **19.** In a feedback system, the component that produces a response is the (a) effector (b) receptor (c) input (d) output

PART 2. Completion

20. The tibia is _____ to the fibula.

21. The ovaries are found in the _____ body cavity.

22. The upper horizontal line that helps divide the abdominopelvic cavity into nine regions is the

_____ line.

23. The anatomical term for the hollow behind the knee is _____.

24. A plane that divides the stomach into a superior and an inferior portion is a(n)

_____ plane.

25. The wrist is described as _____ to the elbow.

26. The heart is located in the _____ cavity within the thoracic cavity.

27. The abdominopelvic region that contains the rectum is the _____ region.

28. A plane that divides the body into unequal left and right sides is the _____ plane.

29. The spinal cord is located within the _____ cavity.

30. The body system that removes carbon dioxide from body cells, delivers oxygen to body cells, helps maintain acid-base balance, helps protect against disease, helps regulate body temperature, and prevents hemorrhage by forming clots is the _____ system.

31. The _____ abdominopelvic quadrant contains the descending colon of the large intestine.

32. Which body system returns proteins and plasma to the cardiovascular system, transports lipids from the digestive system to the cardiovascular system, filters blood, protects against disease, and produces

white blood cells? _____.

33. The sum of all chemical processes that occur in the body is called _____.

34. A structure that monitors changes in a controlled condition and sends the information to the control

center is the _____.

PART 3. Matching

_____ 35. Right hypochondriac region

_____ 36. Hypogastric region

_____ 37. Left iliac region

_____ 38. Right lumbar region

_____ 39. Epigastric region

_____ 40. Left hypochondriac region

_____ 41. Right iliac region

_____ 42. Umbilical region

_____ 43. Left lumbar region

A. Junction of descending and sigmoid colons of large intestine

B. Descending colon of large intestine

C. Spleen

D. Most of right lobe of liver

E. Appendix

F. Ascending colon of large intestine

G. Middle of transverse colon of large intestine

H. Adrenal (suprarenal) glands

I. Sigmoid colon of large intestine

PART 4. Matching

_____ 44. Anterior

_____ 45. Skull

_____ 46. Transtubercular line

_____ 47. Armpit

_____ 48. Umbilical region

_____ 49. Medial

_____ 50. Cranial cavity

_____ 51. Front of knee

_____ 52. Breast

_____ 53. Chest

_____ 54. Buttock

_____ 55. Superior

_____ 56. Groin

_____ 57. Vertebral cavity

_____ 58. Cheek

_____ 59. Front of neck

_____ 60. Distal

_____ 61. Pericardial cavity

_____ 62. Forearm

_____ 63. Plantar

_____ 64. Mouth

A. Passes through iliac crests

B. Contains spinal cord

C. Nearer the midline

D. Thoracic

E. Cervical

F. Axillary

G. Contains the heart

H. Cranial

I. Antebrachial

J. Gluteal

K. Mammary

L. Sole

M. Contains navel

N. Buccal

O. Farther from the attachment of a limb

P. Patellar

Q. Toward the head

R. Nearer to or at the front of the body

S. Oral

T. Contains brain

U. Inguinal

Cells

3

A *cell* is the basic living structural and functional unit of the body. The study of the structure of cells is called *cytology* (sī-TOL-ō-jē; *cyto* = cell; *logos* = study of). The study of the functions of cells is called *cell physiology.* The different kinds of cells—blood, nerve, bone, muscle, epithelial, and others—perform specific functions and differ from one another in shape, size, and structure. You will start your study of cells by learning the important components of a theoretical, generalized cell.

A. CELL PARTS

Refer to Figure 3.1, a generalized cell based on electron micrograph studies. With the aid of your textbook and any other items made available by your instructor, label the parts of the cell indicated. In the spaces that follow, describe the function of the cellular structures indicated:

1. *Plasma (cell) membrane* _____

_____

_____

2. *Cytoplasm* (SĪ-tō-plazm') _____

_____

_____

_____

3. *Nucleus* (NOO-klē-us) _____

_____

_____

_____

4. *Endoplasmic reticulum* (en'-dō-PLAS-mik re-TIK-yoo-lum) or *ER* _____

_____

_____

_____

5. *Ribosome* (RĪ-bō-sōm) _____

_____

_____

_____

6. *Golgi* (GOL-jē) *complex* _____

_____

_____

_____

7. *Mitochondrion* (mī'-tō-KON-drē-on) _____

_____

_____

_____

8. *Lysosome* (LĪ-sō-sōm) _____

_____

_____

9. *Peroxisome* (pe-ROKS-i-sōm) _____

_____

_____

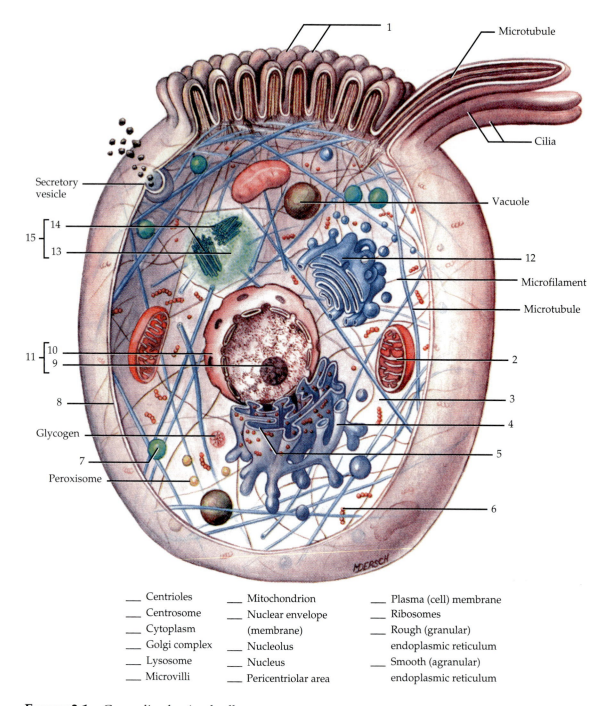

1 — Microtubule

Cilia

Secretory vesicle

Vacuole

15 — 14, 13

12

Microfilament

Microtubule

11 — 10, 9

2

8

3

4

Glycogen

5

7

6

Peroxisome

MDERSCH

___ Centrioles   ___ Mitochondrion   ___ Plasma (cell) membrane
___ Centrosome   ___ Nuclear envelope   ___ Ribosomes
___ Cytoplasm   (membrane)   ___ Rough (granular)
___ Golgi complex   ___ Nucleolus   endoplasmic reticulum
___ Lysosome   ___ Nucleus   ___ Smooth (agranular)
___ Microvilli   ___ Pericentriolar area   endoplasmic reticulum

FIGURE 3.1   Generalized animal cell.

10. *Cytoskeleton* _____

_____

_____

_____

11. *Centrosome* (SEN-trō-sōm') _____

_____

_____

_____

12. *Cilium* (SIL-ē-um) _____

_____

_____

_____

13. *Flagellum* (fla-JEL-um) _____

_____

_____

_____

Nerve cell

Muscle cell

B. DIVERSITY OF CELLS

Now obtain prepared slides of the following types of cells and examine them under the magnifications suggested:

1. Ciliated columnar epithelial cells (high power)
2. Sperm cells (oil immersion)
3. Nerve cells (high power)
4. Muscle cells (high power)

After you have made your examination, draw an example of each of these kinds of cells in the spaces provided and under each cell indicate how each is adapted to its particular function.

Ciliated columnar epithelial cell

Sperm cell

C. MOVEMENT OF SUBSTANCES ACROSS AND THROUGH PLASMA MEMBRANES

Biological membranes serve as permeability barriers. They maintain both the internal integrity of the cell and the solute concentrations within the cell that are considerably different from those found in the extracellular environment. In general, most substances in the body have a high solubility in polar liquids, such as water, and low solubility in nonpolar liquids, such as alcohol. Because cellular membranes are composed of a combination of polar and nonpolar components, they present a formidable barrier to the movement of water-soluble substances into and out of a cell. In general, substances move across plasma membranes by two principal kinds of transport processes—*passive transport processes*, which do not require the expenditure of cellular energy, and *active transport processes*, which do require the expenditure of cellular energy, usually via the chemical breakdown of adenosine triphosphate (ATP) into adenosine diphosphate (ADP), inorganic phosphate, and energy. In passive transport processes, which are physical processes, substances move because of differences in concentration (or pressure) from areas of higher concentration (or pressure) to areas of lower concentration (or pressure). The movement continues until the concentration of substances (or pressure) reaches equilibrium. Passive transport processes are the result of the kinetic energy (energy of motion) of the substances themselves and the cell does not expend energy to move the substances. Examples of passive transport processes are simple diffusion, facilitated diffusion, osmosis, filtration, and dialysis.

By contrast, in active transport processes, which are physiological processes, substances move against their concentration gradient, from areas of lower concentration to areas of higher concentration. Moreover, cells must expend energy (ATP) to carry on active transport processes. Examples of active transport processes are active transport and endocytosis (phagocytosis and pinocytosis).

CAUTION! *Please reread Section A, "General Safety Precautions and Procedures," on page xi, and Section C, "Precautions Related to Working with Reagents," on page xii, at the beginning of the laboratory manual before you begin any of the following experiments. Read the experiments before you perform them to be sure that you understand all the procedures and safety cautions.*

1. Passive Transport Processes

Before looking at examples of passive transport processes, we will first examine how molecules move.

a. BROWNIAN MOVEMENT

At temperatures above absolute zero ($-273°C$ or $-460°F$), all molecules are in constant random motion because of their inherent kinetic energy. This phenomenon is called *Brownian movement.* Less energy is required to move small molecules or particles than large ones. Because all molecules are constantly bombarded by the molecules surrounding them, the smaller the particle, the greater its random motion in terms of speed and distance.

PROCEDURE

1. With a medicine dropper, place a drop of dilute detergent solution in a depression (concave) slide.
2. Using another medicine dropper, add a drop of dilute India ink to the first drop in the depression slide.
3. Stir the two solutions with a toothpick and cover the depression with a cover slip.

4. Let the slide stand for 10 min and then place it on your microscope stage and observe it under high power.
5. Using forceps, place the slide on a hot plate and warm it for 15 sec. Then remove the slide with forceps and observe it again under high power.
6. Record your observations in Section C.1.a of the LABORATORY REPORT RESULTS at the end of the exercise.

b. SIMPLE DIFFUSION

Simple diffusion is the net (greater) movement of molecules or ions from a region of higher concentration to a region of lower concentration until they are evenly distributed (in equilibrium). An example in the human body is the movement of oxygen and carbon dioxide between body cells and blood.

The following two experiments illustrate simple diffusion. Either or both may be performed.

PROCEDURE

1. To demonstrate simple diffusion of a solid in a liquid, *using forceps, carefully* place a large crystal of potassium permanganate ($KMnO_4$) into a test tube filled with water.

CAUTION! *Avoid contact of $KMnO_4$ with your skin by using acid- or caustic-resistant gloves.*

2. Place the tube in a rack against a white background where it will not be disturbed.
3. Note the diffusion of the crystal material through the water at 15-min intervals for 2 hr.
4. Record the diffusion of the crystal in millimeters (mm) per minute at 15-min intervals in Section C.1.b of the LABORATORY REPORT RESULTS at the end of the exercise. Simply measure the distance of diffusion using a millimeter ruler.

PROCEDURE

1. To demonstrate simple diffusion of a solid in a solid, *using forceps, carefully* place a large crystal of methylene blue on the surface of agar in the center of a petri plate.
2. Note the diffusion of the crystal through the agar at 15-min intervals for 2 hr.
3. Record the diffusion of the crystal in millimeters (mm) per minute at 15-min intervals, using a millimeter ruler, in Section C.1.b of the LABORATORY REPORT RESULTS at the end of the exercise.

c. FACILITATED DIFFUSION

Facilitated diffusion is the movement of a substance across a selectively permeable membrane from a region of higher concentration to a region of lower concentration with the assistance of integral proteins that serve as water-filled channels or transporters (carriers) in the plasma membrane for each type of substance. This process does not require the expenditure of cellular energy. Different sugars, especially glucose, cross plasma membranes by facilitated diffusion.

In this experiment, the substance undergoing facilitated diffusion is neutral red, which is red at a pH just below 7 but yellow at a pH between 7 and 8.

PROCEDURE

1. In a 125-mL flask, combine one gram (1 g) of bakers' yeast with 25 mL 0.75% Na_2CO_3 (sodium carbonate) solution. Swirl until the yeast is evenly suspended in the solution.
2. Divide the suspension into two large test tubes marked "U1" (unboiled) and "B1" (boiled) with a wax pencil.
3. Place tube "B1" in a boiling water bath for 2 to 3 min. Be sure that all the suspension is below water level so that all the yeast cells will be killed.

CAUTION! *Make sure that the mouth of the test tube is pointed away from you and all other persons in the area.*

4. Using a test tube holder, place both test tubes into a rack and add 7.5 mL of 0.02% neutral red to each.
5. Record the color of the solution in both test tubes in Section C.1.c of the LABORATORY REPORT RESULTS at the end of the exercise.
6. After 15 min, place 1 mL from each test tube into test tubes marked "U2" and "B2."
7. Add enough drops of 0.75% acetic acid to tube "B2" to make its color identical to that of tube "U2."
8. Mark two microscope slides "U" and "B"; examine a sample from both tubes under high power (using a cover slip) and note which cells are stained. _____

_____

9. Filter half of the solution remaining in tube "B1" and examine the filtrate. Repeat for tube "U1."

10. Observe and record the color of the filtrate and that of the yeast cells on the filter paper.

_____

_____

11. To the suspension remaining in tubes "U1" and "B1," add about 9 mL 0.75% acetic acid. Filter again according to the instructions in step 9.
12. Record the color of these "U1" and "B1" filtrates and cells in Section C.1.c of the LABORATORY REPORT RESULTS at the end of the exercise.
13. How does boiling affect the facilitated diffusion of neutral red by the cells? _____

_____

Describe the membrane's permeability to neutral red. _____

_____

Did acetic acid enter the living cells? _____

_____

Did sodium carbonate enter the living cells?

_____

d. OSMOSIS

Osmosis (oz-MŌ-sis) is the net movement of *water* through a selectively permeable membrane from a region of higher water (lower solute) concentration to a region of lower water (higher solute) concentration. In the body, fluids move between cells as a result of osmosis.

PROCEDURE

1. Refer to the osmosis apparatus in Figure 3.2.
2. Tie a knot very tightly at one end of a 4-in. piece of cellophane dialysis tubing that has been soaking in water for a few minutes. Fill the dialysis tubing with a 10% sugar (sucrose) solution that has been colored with Congo red (red food coloring can also be used).
3. Close the open end of the dialysis tubing with a one-hole rubber stopper into which a glass tube has already been inserted by a laboratory assistant or your instructor.

CAUTION! *If you are unfamiliar with the procedure, do not attempt to insert the glass tube into the stopper yourself because it may break and result in serious injury.*

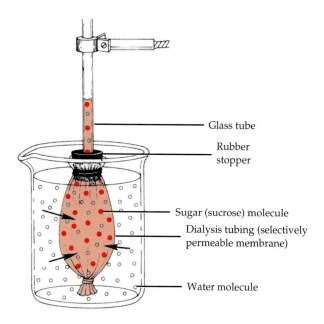

Glass tube

Rubber stopper

Sugar (sucrose) molecule

Dialysis tubing (selectively permeable membrane)

Water molecule

FIGURE 3.2   Osmosis apparatus.

4. Tie a piece of string tightly around the dialysis tubing to secure it to the stopper.
5. Secure and suspend the glass tube and dialysis tubing by means of a clamp attached to a ring stand.
6. Insert the dialysis tubing into a beaker or flask of water until the water comes up to the bottom of the rubber stopper.
7. As soon as the sugar solution becomes visible in the glass tube, mark its height with a wax pencil and note the time.
8. Mark the height of liquid in the glass tube after 10-, 20-, and 30-min intervals by using your millimeter (mm) ruler and record your results in Section C.1.d of the LABORATORY REPORT RESULTS at the end of the exercise.

e. HEMOLYSIS AND CRENATION

Osmosis can be demonstrated by noting the effects of different water concentrations on red blood cells. Red blood cells maintain their normal shape when placed in an *isotonic* (*iso* = same) *solution.* An isotonic solution has the same salt concentration (0.90% NaCl) as that found in red blood cells. If, however, red blood cells are placed in a *hypotonic* (*hypo* = lower) *solution* (a salt solution with less than 0.90% NaCl), a net movement of water into the cells occurs, causing the cells to swell and possibly burst. The red blood cells swell because the difference in osmotic pressure inside the cell compared to outside the cell results in a net movement of water into the cell. The rupture of blood cells in

this manner and the resulting loss of hemoglobin into the surrounding liquid is termed *hemolysis* (hē-MOL-i-sis). If, instead, red blood cells are placed in a *hypertonic* (*hyper* = higher) *solution* (a salt solution with more than 0.90% NaCl), a net movement of water out of the cells occurs, causing the cells to shrink. The red blood cells undergo this shrinkage, known as *crenation* (kri-NĀ-shun) because the difference in osmotic pressure inside the cell compared to outside the cell results in the net movement of water out of the cell.

PROCEDURE

CAUTION! *Please reread Section B, "Precautions Related to Working with Blood, Blood Products, or Other Body Fluids," on page xii, at the beginning of the laboratory manual, before you begin any of the following experiments. Read the experiments before you perform them to be sure that you understand all the procedures and safety precautions. When working with whole blood, take care to avoid any kind of contact with an open sore, cut, or wound. Wear tight-fitting surgical gloves and safety goggles.*

When you finish this part of the exercise, place the reusable items in a fresh bleach solution and the discarded items in a biohazard container.

1. With a wax marking pencil, mark three microscope slides as follows: 0.90%, DW (distilled water), and 3%.
2. Using a medicine dropper, place three drops of fresh (uncoagulated) ox blood on a microscope slide that contains 2 mL of a 0.90% NaCl solution (isotonic solution). Mix gently and thoroughly with a clean toothpick.
3. Using a medicine dropper, place three drops of fresh ox blood on a microscope slide that contains 2 mL of distilled water (hypotonic solution). Mix gently and thoroughly with a clean toothpick.
4. Now, using a medicine dropper, add three drops of fresh ox blood to a microscope slide that contains 2 mL of a 3% NaCl solution (hypertonic solution). Mix gently and thoroughly with a clean toothpick.
5. Using a medicine dropper, place two drops of the red blood cells in the isotonic solution on another microscope slide, cover with a cover slip, and examine the red blood cells under high power. Reduce your illumination.

What is the shape of the cells?_____

_____

Explain their shape._____

_____

6. Using a medicine dropper, place two drops of the red blood cells in the hypotonic solution on another microscope slide, cover with a cover slip, and examine the red blood cells under high power. Reduce your illumination.

   What is the shape of the cells?_____

   _____

   Explain their shape._____

   _____

7. Using a medicine dropper, place two drops of the red blood cells in the hypertonic solution on another microscope slide, cover with a cover slip, and examine the red blood cells under high power. Reduce your illumination.

   What is the shape of the cells?_____

   _____

   Explain their shape._____

   _____

ALTERNATE PROCEDURE

1. Obtain three pieces of raw potato that have an *identical* weight.
2. Immerse one piece in a beaker that contains an isotonic solution; immerse a second piece in a hypotonic solution; immerse the third piece in a hypertonic solution.
3. Continue the experiment for 1 hr. Record the time.
4. At the end of 1 hr, remove the pieces of potato and weigh them separately.

   Weight of potato in isotonic solution_____

   _____

   Weight of potato in hypotonic solution_____

   _____

   Weight of potato in hypertonic solution_____

   _____

   Explain the differences in the weights of the

   three pieces of potato._____

   _____

   _____

_____

_____

_____

_____

f. FILTRATION

Filtration is the movement of solvents and dissolved substances across a selectively permeable membrane from regions of higher pressure to regions of lower pressure. Movement of the solvents and dissolved substances occurs under the influence of gravity and the pressure exerted by the solvent, which is termed **hydrostatic pressure.** The selectively permeable membrane prevents molecules with higher molecular weights from passing through the membrane, while the solvent and substances with lower molecular weights easily pass through the selectively permeable membrane. In general, any substance having a molecular weight of less than 100 is filtered, because the pores in the filter paper are larger than the molecules of the substance. Filtration is one mechanism by which the kidneys regulate the chemical composition of the blood.

PROCEDURE

1. Refer to Figure 3.3, which shows the filtration apparatus. In this apparatus, the filter paper represents the selectively permeable membrane of a cell.

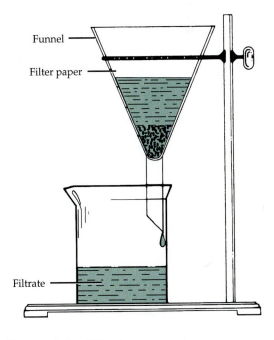

FIGURE 3.3   Filtration apparatus.

2. Fold a piece of filter paper in half and then in half again.

3. Open it into a cone, place it in a funnel, and place the funnel over the beaker.

4. Shake a mixture of a few particles of powdered wood charcoal (black), 1% copper sulfate (blue), boiled starch (white), and water, and slowly pour it into the funnel until the mixture almost reaches the top of the filter paper. Gravity will pull the particles through the pores of the filter paper.

5. Count the number of drops passing through the funnel for the following time intervals: 10, 30, 60, 90, and 120 sec. Record your observations in Section C.1.f of the LABORATORY REPORT RESULTS at the end of the exercise.

6. Observe which substances passed through the filter paper by noting their color in the filtered fluid in the beaker.

7. Examine the filter paper to determine whether any colored particles were not filtered.

8. To determine if any starch is in the liquid (termed the *filtrate*) in the beaker, add several drops of 0.01 M IKI solution. A blue-black color reaction indicates the presence of starch.

g. DIALYSIS

Dialysis (dī-AL-i-sis) is the separation of smaller molecules from larger ones by a selectively permeable membrane. Such a membrane permits diffusion of the small molecules but not the large ones. Although dialysis does not occur in the human body, it is employed in artificial kidneys.

PROCEDURE

1. Refer to the dialysis apparatus in Figure 3.4.

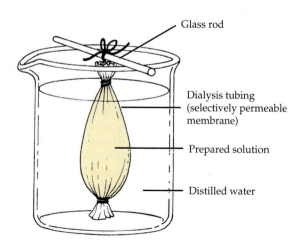

FIGURE 3.4 Dialysis apparatus.

2. Tie off one end of a piece of dialysis tubing that has been soaking in water. Place a prepared solution containing starch, sodium chloride, 5% glucose, and albumin into the dialysis tubing.

3. Tie off the other end of the dialysis tubing and immerse it in a beaker of distilled water.

4. After 1 hr, test the solution in the beaker for the presence of each of the substances in the tubing, as follows, and record your observations in Section C.1.g of the LABORATORY REPORT RESULTS at the end of the exercise.

CAUTION! *Be extremely careful using nitric acid. It can severely damage your eyes and skin. Use acid- or caustic-resistant gloves.*

  a. *Albumin*—Carefully add several drops of concentrated nitric acid to a test tube containing 2 mL of the solution in the beaker. Positive reaction = white coagulate.

  b. *Sugar*—Test 5 mL of the solution in the beaker in a test tube with 5 mL of Benedict's solution. Place the test tube in a boiling water bath for 3 min.

CAUTION! *Make sure that the mouth of the test tube is pointed away from you and everyone else in the area.*

Using a test tube holder, remove the test tube from the water bath. Note the color. Positive reaction = green, yellow, orange, or red precipitate.

  c. *Starch*—Add several drops of IKI solution to 2 mL of the solution in the beaker in a test tube. Note the color. Positive reaction = blue-black color.

  d. *Sodium chloride*—Place 2 mL of the solution in the beaker in a test tube and add several drops of 1% silver nitrate. Note the color. Positive reaction = white precipitate.

2. Active Transport Processes

a. ACTIVE TRANSPORT

There are two types of *active transport.* In *primary active transport,* energy derived from splitting ATP *directly* moves, or "pumps," a substance across a plasma membrane. The cell uses energy from ATP to change the shape of integral membrane proteins. An example of primary active transport is the sodium pump, which maintains a low concentration of sodium ions (Na$^+$) in the cytosol, the fluid portion of the cytoplasm, by pumping sodium ions out against their concentration gradient. The pump also moves potassium ions (K$^+$) into cells against their concentration gradient. In **secondary active transport,** the energy stored in ion concentration gradients (differences) drives substances across a plasma membrane. Since ion gradients are established by primary active transport, secondary active transport *indirectly* uses energy obtained from splitting ATP. An example of secondary active transport is the movement of an amino acid and Na$^+$ in the same direction across a plasma membrane with the assistance of an integral membrane protein. Another example is the movement of calcium ions (Ca^{2+}) and Na$^+$ in opposite directions across a membrane with the assistance of an integral membrane protein.

Because reliable results are difficult to demonstrate simply, you will not be asked to demonstrate active transport.

b. ENDOCYTOSIS

Endocytosis (*endo* = into) refers to the passage of large molecules and particles across a plasma membrane, in which a segment of the membrane surrounds the substance, encloses it, and brings it into the cell. Here we will consider two types of endocytosis—phagocytosis and pinocytosis.

(1) Phagocytosis

Phagocytosis (fag'-ō-sī-TŌ-sis; *phagein* = to eat), or cell "eating," is the engulfment of solid particles or organisms by *pseudopods,* temporary fingerlike projections of the cytoplasm of the cell. Once the particle is surrounded by the membrane, the membrane folds inward, pinches off from the rest of the plasma membrane, and forms a *phagosome* around the particle. The particle within the phagosome is digested either by the secretion of enzymes into the vesicle or by the combining of the vesicle with an enzyme-containing lysosome.

Phagocytosis can be demonstrated by observing the feeding of an amoeba, a unicellular organism whose movement and ingestion are similar to those of human leukocytes (white blood cells).

PROCEDURE

1. Using a medicine dropper, place a drop of culture containing amoebas that have been starved for 48 hr into the well of a depression slide and cover the well with a cover slip. Cultures containing *Chaos chaos* or *Amoeba proteus* should be used. (Your instructor may wish to use the hanging-drop method instead. If so, she or he will give you verbal instructions.)

2. Examine the amoebas under low power, and be sure that your light is reduced considerably.
3. Observe the locomotion of an amoeba for several minutes. Pay particular attention to the pseudopods that appear to flow out of the cell.
4. To observe phagocytosis, use a medicine dropper and add a drop containing small unicellular animals called *Tetrahymena pyriformis* to the culture containing the amoebas.
5. Examine under low power, and observe the ingestion of *Tetrahymena pyriformis* by an amoeba. Note the action of the pseudopods and the formation of the phagosome around the ingested organism.

(2) Pinocytosis

Pinocytosis (pi-nō-sī-TŌ-sis; *pinein* = to drink), or cell "drinking," is the engulfment of a liquid. The liquid is attracted to the surface of the membrane; the membrane folds inward and surrounds the liquid and detaches from the rest of the intact membrane forming a *pinocytic vesicle.*

D. EXTRACELLULAR MATERIALS

Substances that lie outside the plasma membranes of body cells are referred to as *extracellular materials.* They include body fluids, such as interstitial fluid and plasma, which provide a medium for dissolving, mixing, and transporting substances. Extracellular materials also include special substances in which some cells are embedded.

Some extracellular materials are produced by certain cells and deposited outside their plasma membranes where they support cells, bind them together, and provide strength and elasticity. They have no definite shape and are referred to as *amorphous.* These include hyaluronic (hī-a-loo-RON-ik) acid and chondroitin (kon-DROY-tin) sulfate. Others are *fibrous* (threadlike). Examples include collagen, reticular, and elastic fibers.

Using your textbook as a reference, indicate the location and function for each of the following extracellular materials:

Hyaluronic acid

Location _____

_____

_____

_____

Function_____

_____

_____

_____

Chondroitin sulfate

Location_____

_____

_____

_____

Function_____

_____

_____

_____

Collagen fibers

Location_____

_____

_____

_____

Function_____

_____

_____

_____

Reticular fibers

Location_____

_____

_____

_____

Function_____

_____

_____

_____

Elastic fibers

Location_____

_____

_____

_____

Function_____

_____

_____

_____

E. CELL DIVISION

Cell division is the basic mechanism by which cells reproduce themselves. It consists of a nuclear division and a cytoplasmic division. Because nuclear division can be of two types, two kinds of cell division are recognized. In the first type, called *somatic cell division,* a single starting cell called a *parent cell* duplicates itself and the result is two identical cells called *daughter cells.* Somatic cell division consists of a nuclear division called *mitosis* (mī-TŌ-sis) and a cytoplasmic division called *cytokinesis* (sī-tō-ki-NĒ-sis; *cyto* = cell; *kinesis* = motion). It provides the body with a means of growth and of replacement of diseased or damaged cells (Figure 3.5). The second type of cell division is called *reproductive cell division* and is the mechanism by which sperm and ova are produced (Exercise 26). Reproductive cell division consists of a nuclear division called *meiosis* and two cytoplasmic divisions (cytokinesis), and it results in the development of four nonidentical daughter cells.

In order to study somatic cell division, obtain a prepared slide of a whitefish blastula and examine it under high power.

A cell between divisions is said to be in *interphase* of the cell cycle. Interphase is the longest part of the cell cycle and is the period of time during which a cell carries on its physiological activities. One of the most important activities of interphase is the replication of DNA so that the two daughter cells that eventually form will each have the same kind and amount of DNA as the parent cell. In addition, the proteins needed to produce structures required for doubling all cellular components are manufactured. Scan your slide and find a cell in interphase. Such a

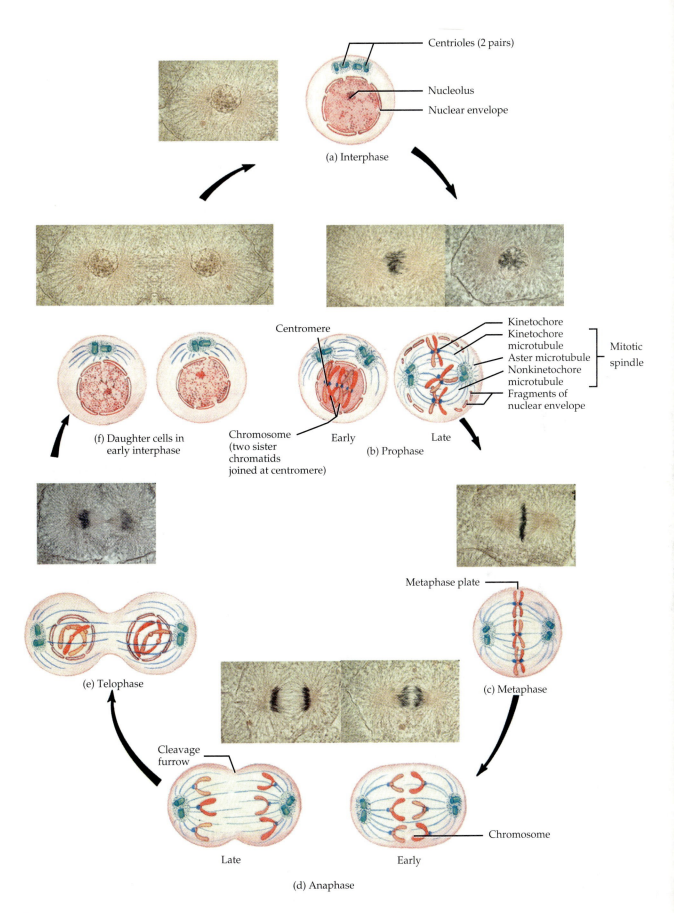

FIGURE 3.5 Cell division: mitosis and cytokinesis. Diagrams and photomicrographs (450×) of the various stages of cell division in whitefish eggs.

parent cell is characterized by a clearly defined nuclear envelope. Within the nucleus, look for the nucleolus (or nucleoli) and *chromatin,* DNA that is associated with protein in the form of a granular substance. Also locate the centrosomes.

Draw a labeled diagram of an interphase cell in the space provided.

Once a cell completes its interphase activities, mitosis begins. Mitosis is the distribution of two sets of chromosomes into two separate and equal nuclei after replication of the chromosomes of the parent cell, an event that takes place in the interphase preceding mitosis. Although a continuous process, mitosis is divided into four stages for purposes of study: prophase, metaphase, anaphase, and telophase.

1. *Prophase*—The first stage of mitosis is called *prophase* (*pro* = before). During early prophase, the chromatin condenses and shortens into visible chromosomes. Because DNA replication took place during interphase, each prophase chromosome contains a pair of identical double-stranded DNA molecules called *chromatids.* Each chromatid pair is held together by a constricted region body called a *centromere* that is required for the proper segregation of chromosomes. Attached to the outside of each centromere is a protein complex known as the *kinetochore* (ki-NET-ō-kor), whose function will be described shortly. Later in prophase, the nucleolus (or nucleoli) disappears, and the nuclear envelope breaks up. In addition, each centrosome with its pair of centrioles, moves to an opposite pole (end) of the cell. As they do so, the pericentriolar areas of the centrosomes start to form the *mitotic spindle,* a football-shaped assembly of microtubules that are responsible for the movement of chromosomes.

    The lengthening of microtubules between centrosomes pushes the centrosomes to the poles of the cell so that the spindle extends from pole to pole. As the mitotic spindle continues to develop, three types of microtubules form: (a) *nonkinetochore microtubules,* which grow from centrosomes and extend inward, but do not bind to kinetochores; (b) *kinetochore microtubules,* which grow from centrosomes, extend inward, and attach to kinetochores; and (c) *aster microtubules,* which grow out of centrosomes but radiate outward from the mitotic spindle. Overall, the spindle is an attachment site for chromosomes. It also distributes chromosomes to opposite poles of the cell. Draw and label a cell in prophase in the space provided.

2. *Metaphase*—During *metaphase* (*meta* = after), the second stage of mitosis, the kinetochore microtubules line up the centromeres of the chromatid pairs at the exact center of the mitotic spindle. This midpoint region is called the *metaphase plate,* or *equatorial plane region.* Draw and label a cell in metaphase in the space provided.

3. *Anaphase*—The third stage of mitosis, *anaphase* (*ana* = upward), is characterized by the splitting and separation of the centromeres (and kinetochores) and the movement of the two sister chromatids of each pair toward opposite poles of the cell. Once separated, the sister chromatids are referred to as *chromosomes.* The movement of chromosomes is the result of the shortening of kinetochore microtubules and elongation of the nonkinetochore microtubules, processes that increase the distance between separated chromo-

somes. As the chromosomes move during anaphase, they appear V-shaped. Draw and label a cell in anaphase in the space provided.

4. *Telophase*—The final stage of mitosis, *telophase* (*telo* = far or end), begins as soon as chromosomal movement stops. Telophase is essentially the opposite of prophase. During telophase, the identical sets of chromosomes at opposite poles of the cell uncoil and revert to their threadlike chromatin form; kinetochore microtubules disappear; nonkinetochore microtubules elongate even more; a new nuclear envelope re-forms around each chromatin mass; new nucleoli reappear in the daughter nuclei; and eventually the mitotic spindle breaks up. Draw and label a cell in telophase in the space provided.

Cytokinesis begins during late anaphase or early telophase with the formation of a *cleavage furrow,* a slight indentation of the plasma membrane that extends around the center of the cell. The furrow gradually deepens until opposite surfaces of the cell make contact and the cell is split in two. The result is two separated daughter cells, each with separate portions of cytoplasm and organelles and its own set of identical chromosomes.

Following cytokinesis, each daughter cell re-turns to interphase. Each cell in most tissues of the body eventually grows and undergoes mitosis and cytokinesis, and a new divisional cycle begins. Examine your telophase cell again and be sure that it contains a cleavage furrow.

Using high power and starting at 12 o'clock, move around the blastula and count the number of cells in interphase and in each mitotic phase. It will be easier to do this if you imagine lines dividing the blastula into quadrants. Count the interphase cells in each quadrant, then assign each dividing cell to a specific mitotic stage. It will be hard to assign some cells to a phase—e.g., to distinguish late anaphase from early telophase. If you cannot make a decision, assign one cell to the earlier phase in question and the next cell to the later phase.

Divide the number of cells in each stage by the total number of cells counted and multiply by 100 to determine the percent of the cells in each mitotic stage at a given point in time. Record your results in Section F of the LABORATORY REPORT RESULTS at the end of the exercise.

ANSWER THE LABORATORY REPORT QUESTIONS AT THE END OF THE EXERCISE.

Cells 3

Student _____   Date _____

Laboratory Section _____   Score/Grade _____

SECTION C. MOVEMENT OF SUBSTANCES ACROSS PLASMA MEMBRANES

1. Passive Transport Processes

a. BROWNIAN MOVEMENT

Describe the movement of the India ink particles on the unheated slide. _____

_____

How does this movement differ from that on the heated slide? _____

_____

b. SIMPLE DIFFUSION

Solid in Liquid		Solid in Solid	
Time, min	Distance, mm	Time, min	Distance, mm
15	_____	15	_____
30	_____	30	_____
45	_____	45	_____
60	_____	60	_____
75	_____	75	_____
90	_____	90	_____
105	_____	105	_____
120	_____	120	_____

c. FACILITATED DIFFUSION

Neutral Red		Acetic Acid	
Tube "U1"	_____	Tube "U1"	_____
Tube "B1"	_____	Tube "B1"	_____

d. OSMOSIS

Time, min   Height of liquid, mm

10   _____

20   _____

30   _____

Explain what happened. _____

_____

f. FILTRATION

10 sec _____

30 sec _____

60 sec _____

90 sec _____

120 sec _____

g. DIALYSIS
Place a check in the appropriate place to indicate if the following tests are positive (+) or negative (−).

	(+)	(−)
Albumin	_____	_____
Sugar	_____	_____
Starch	_____	_____
Sodium chloride	_____	_____

SECTION E. Cell Division

Percent of cells in interphase   _____

Percent of cells in prophase   _____

Percent of cells in metaphase   _____

Percent of cells in anaphase   _____

Percent of cells in telophase   _____

Cells 3

Student _____   Date _____

Laboratory Section _____   Score/Grade _____

PART 1. Multiple Choice

_____ 1. The portion of the cell that forms part of the mitotic spindle during division is the (a) endoplasmic reticulum (b) Golgi complex (c) cytoplasm (d) pericentriolar area of the centrosome

_____ 2. Movement of molecules or ions from a region of higher concentration to a region of lower concentration via a process that does not require cellular energy is called (a) phagocytosis (b) simple diffusion (c) active transport (d) pinocytosis

_____ 3. If red blood cells are placed in a hypertonic solution of sodium chloride, they will (a) swell (b) burst (c) shrink (d) remain the same

_____ 4. The reagent used to test for the presence of sugar is (a) silver nitrate (b) nitric acid (c) IKI (d) Benedict's solution

_____ 5. A cell that carries on a great deal of digestion also contains a large number of (a) lysosomes (b) centrosome (c) mitochondria (d) nuclei

_____ 6. Which process does *not* belong with the others? (a) active transport (b) dialysis (c) phagocytosis (d) pinocytosis

_____ 7. Movement of oxygen and carbon dioxide between blood and body cells is an example of (a) osmosis (b) active transport (c) simple diffusion (d) facilitated diffusion

_____ 8. Which type of solution will cause hemolysis? (a) isotonic (b) hypotonic (c) isometric (d) hypertonic

_____ 9. In addition to active transport and simple diffusion, the kidneys regulate the chemical composition of blood by utilizing the process of (a) phagocytosis (b) osmosis (c) filtration (d) pinocytosis

_____ 10. Engulfment of solid particles or organisms by pseudopods is called (a) active transport (b) dialysis (c) phagocytosis (d) filtration

_____ 11. Rupture of red blood cells with subsequent loss of hemoglobin into the surrounding medium is called (a) hemolysis (b) plasmolysis (c) plasmoptysis (d) hemoglobinuria

_____ 12. The area of the cell between the plasma membrane and nucleus where chemical reactions occur is the (a) centrosome (b) vacuole (c) peroxisome (d) cytoplasm

_____ 13. The "powerhouses" of the cell where ATP is produced are the (a) ribosomes (b) mitochondria (c) centrosomes (d) lysosomes

_____ 14. The sites of protein synthesis in the cell are (a) peroxisomes (b) flagella (c) ribosomes (d) centrosomes

_____ 15. Which process does *not* belong with the others? (a) simple diffusion (b) phagocytosis (c) active transport (d) pinocytosis

——————   **16.** The study of the structure of cells is called (a) surface anatomy (b) cytology (c) cell physiology (d) embryology

——————   **17.** Which extracellular material is found in ligaments and tendons? (a) elastic fibers (b) chondroitin sulfate (c) collagen fibers (d) mucus

——————   **18.** The organelles that contain enzymes for the metabolism of hydrogen peroxide are (a) lysosomes (b) mitochondria (c) Golgi complexes (d) peroxisomes

——————   **19.** The framework of cilia, flagella, centrioles, and the mitotic spindle is formed by (a) endoplasmic reticulum (b) collagen fibers (c) chondroitin sulfate (d) microtubules

——————   **20.** A viscous fluidlike substance that binds cells together, lubricates joints, and maintains the shape of the eyeballs is (a) elastin (b) hyaluronic acid (c) mucus (d) plasmin

PART **2. Completion**

21. The external boundary of the cell through which substances enter and exit is called the

————————————.

22. The cytoskeleton is formed by microtubules, intermediate filaments, and ————————————.

23. The portion of the cell that contains hereditary information is the ————————————.

24. The tail of a sperm cell is a long whiplike structure called a(n) ————————————.

25. Cells placed in a ———————————— solution will undergo hemolysis.

26. Division of the cytoplasm is referred to as ————————————.

27. Lipid and protein secretion, formation of lysosomes and secretory vesicles, and assembly of glycoproteins are functions of the ————————————.

28. Storage of digestive enzymes is accomplished by the ———————————— of a cell.

29. Projections of cells that move substances along their surfaces are called ————————————.

30. The ———————————— is the site of fatty acid, phospholipid, and steroid synthesis, and a temporary storage area for newly synthesized molecules.

31. The main function of a ———————————— cell is contraction.

32. A jellylike substance that supports cartilage, bone, the skin, and blood vessels is

————————————.

33. The framework of many soft organs is formed by ———————————— fibers.

34. The net movement of water through a selectively permeable membrane from a region of higher concentration of water to a region of lower concentration of water is known as ————————————.

35. The principle of ———————————— is employed in the operation of an artificial kidney.

36. In an interphase cell, DNA is in the form of a granular substance called ————————————.

37. Distribution of chromosomes into separate and equal nuclei is referred to as ————————————.

38. The constant random motion of molecules caused by their inherent kinetic energy is called

————————————.

PART 3. Matching

_____	**39.** Anaphase
_____	**40.** Metaphase
_____	**41.** Interphase
_____	**42.** Telophase
_____	**43.** Prophase

A. Mitotic spindle appears

B. Movement of chromosome sets to opposite poles of cell

C. Centromeres line up on metaphase plate

D. Formation of two identical nuclei

E. Phase between divisions

Tissues

4

A **tissue** (*texere* = to weave) is a group of similar cells that usually have the same embryological origin and function together to perform a specific function. The study of tissues is called **histology** (hiss-TOL-ō-jē; *histio* = tissue; *logos* = study of). The various body tissues can be categorized into four principal kinds: (1) epithelial, (2) connective, (3) muscular, and (4) nervous. In this exercise you will examine the structure and functions of epithelial and connective tissues, except for bone or blood. Other tissues will be studied later as parts of the systems to which they belong.

A. EPITHELIAL TISSUE

Epithelial (ep'-i-THĒ-lē-al) **tissue,** or **epithelium,** may be divided into two types: (1) covering and lining and (2) glandular. Covering and lining epithelium forms the outer layer of the skin and some internal organs; forms the inner lining of blood vessels, ducts, body cavities, and many internal organs; and helps make up special sense organs for smell, hearing, vision, and touch. Glandular epithelium constitutes the secreting portion of glands.

1. Characteristics

Following are the general characteristics of epithelial tissue:

a. Epithelium consists largely or entirely of closely packed cells with little extracellular material between cells.
b. Epithelial cells are arranged in continuous sheets, in either single or multiple layers.
c. Epithelial cells have an **apical** (free) **surface** that is exposed to a body cavity, lining of an internal organ, or the exterior of the body and a **basal surface** that is attached to the basement membrane (described shortly).
d. Cell junctions (points of contact between cells) are plentiful, providing secure attachments among the cells.

e. Epithelia are **avascular** (*a* = without; *vascular* = blood vessels). The vessels that supply nutrients and remove wastes are located in the adjacent connective tissue. The exchange of materials between epithelium and connective tissue is by diffusion.
f. Epithelia adhere firmly to nearby connective tissue, which holds the epithelium in position and prevents it from being torn. The attachment between the epithelium and the connective tissue is a thin extracellular layer called the **basement membrane**. It consists of two layers. The **basal lamina** contains collagen, laminin, and proteoglycans secreted by the epithelium. Cells in the connective tissue secrete the second layer, the **reticular lamina**, which contains reticular fibers, fibronectin, and glycoproteins. The basement membrane provides physical support for epithelium, provides for cell attachment, serves as a filter in the kidneys, and guides cell migration during development and tissue repair.
g. Epithelial tissue has a nerve supply.
h. Epithelial tissue is the only tissue that makes direct contact with the external environment.
i. Since epithelium is subject to a certain amount of wear and tear and injury, it has a high capacity for renewal (high mitotic rate).
j. Epithelia are diverse in origin. They are derived from all three primary germ layers (ectoderm, mesoderm, and endoderm).
k. Functions of epithelia include protection, filtration, lubrication, secretion, digestion, absorption, transportation, excretion, sensory reception, and reproduction.

2. Covering and Lining Epithelium

Before you start your microscopic examination of epithelial tissues, refer to Figure 4.1. Study the tissues carefully to familiarize yourself with their general structural characteristics. For each of the types of epithelium listed, obtain a prepared slide and, unless otherwise specified by your instructor, examine each under high power. In conjunction with

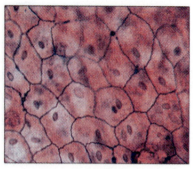

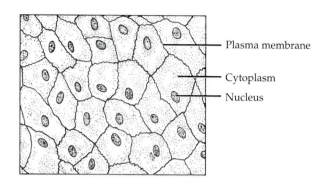

Plasma membrane

Cytoplasm

Nucleus

Surface view of mesothelial lining of
peritoneal cavity (240×)

(a) Simple squamous epithelium

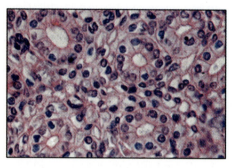

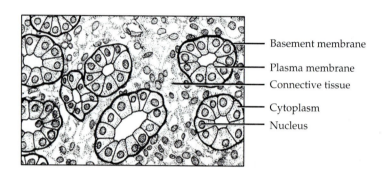

Basement membrane

Plasma membrane

Connective tissue

Cytoplasm

Nucleus

Sectional view of kidney tubules (400×)

(b) Simple cuboidal epithelium

Mucus-producing   Connective   Basement   Nucleus of
goblet cell   tissue   membrane   absorptive cell

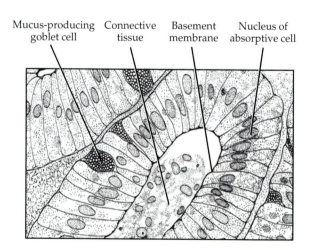

Sectional view of colonic glands (140×)

(c) Simple columnar (nonciliated) epithelium

FIGURE 4.1 Epithelial tissues. Photomicrographs are unlabeled; the line drawings of the same tissues are labeled.

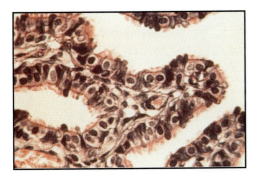 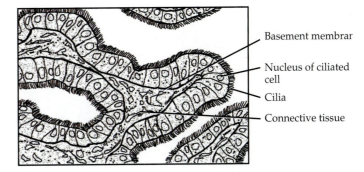

Basement membrar

Nucleus of ciliated cell

Cilia

Connective tissue

Sectional view of uterine (Fallopian) tube (175×)

(d) Simple columnar (ciliated) epithelium

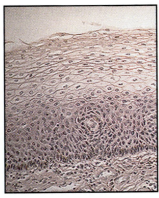 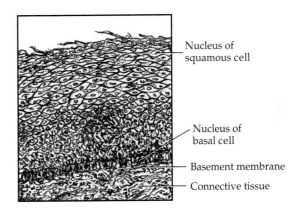

Nucleus of squamous cell

Nucleus of basal cell

Basement membrane

Connective tissue

Sectional view of pharynx (67×)

(e) Stratified squamous epithelium

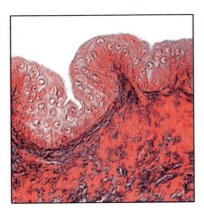 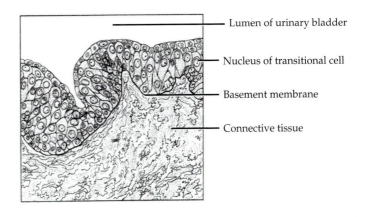

Lumen of urinary bladder

Nucleus of transitional cell

Basement membrane

Connective tissue

Sectional view of urinary bladder in relaxed state (100×)

(f) Transitional epithelium

FIGURE 4.1 *(Continued)*   Epithelial tissues.

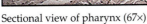

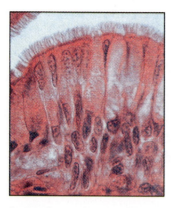

Sectional view of trachea (250×)

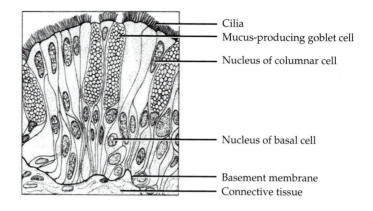

(g) Pseudostratified columnar epithelium

FIGURE 4.1 *(Continued)*   Epithelial tissues.

your examination, consult a textbook of anatomy and physiology.

a. **Simple squamous** (SKWĀ-mus; *squama* = flat) **epithelium** This tissue consists of a single layer of flat cells and is highly adapted for diffusion and filtration because of its thinness. Simple squamous epithelium lines the air sacs (alveoli) of the lungs, glomerular (Bowman's) capsules (filtering units) of the kidneys, and inner surface of the tympanic membrane (eardrum) of the ear. Simple squamous epithelium that lines the heart, blood vessels, and lymphatic vessels is called **endothelium** (*endo* = within; *thelium* = covering). Simple squamous epithelium that forms the epithelial layer of a serous membrane is called **mesothelium** (*meso* = middle). Serous membranes line the thoracic and abdominopelvic cavities and cover viscera within the cavities. After you make your microscopic examination, draw several cells in the space that follows and label plasma membrane, cytoplasm, and nucleus.

Simple squamous epithelium

b. **Simple cuboidal epithelium** This tissue consists of a single layer of cube-shaped cells. When the tissue is viewed from the side, its cuboidal nature is obvious. Highly adapted for secretion and absorption, cuboidal tissue covers the surface of the ovaries; lines the anterior surface of the lens capsule of the eye; and forms the pigmented epithelium of the retina of the eye, part of the tubules of the kidneys and smaller ducts of many glands and the secreting units of other glands, such as the thyroid gland.

After you make your microscopic examination, draw several cells in the space that follows and label plasma membrane, cytoplasm, nucleus, basement membrane, and connective tissue layer.

Simple cuboidal epithelium

c. **Simple columnar (nonciliated) epithelium** This tissue consists of a single layer of columnar cells, and, when viewed from the side these cells appear as rectangles. Adapted for secretion and absorption, this tissue lines the gastrointestinal tract from the stomach to the anus, gallbladder, and ducts of many glands. Some columnar cells are modified in that the plasma membranes are folded into microscopic fingerlike cytoplasmic projections called *microvilli* (*micro* = small; *villus* = tuft of hair) that increase the surface area for absorption. Other cells are *goblet cells*, modified columnar cells that secrete and store mucus to protect the lining of the gastrointestinal tract. After you make your microscopic examination, draw several cells in the space that follows and label plasma membrane, cytoplasm, nucleus, goblet cell, absorptive cell, basement membrane, and connective tissue layer.

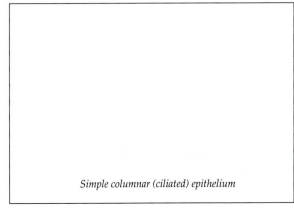

Simple columnar (ciliated) epithelium

Simple columnar (nonciliated) epithelium

d. Simple columnar (ciliated) epithelium This type of epithelium consists of a single layer of columnar absorptive, goblet, and ciliated cells. *Cilia* (*cilia* = eyelashes) are hairlike processes that move substances over the surfaces of cells. Simple columnar (ciliated) epithelium lines portions of the upper respiratory tract, uterine (Fallopian) tubes, uterus, some paranasal sinuses, and the central canal of the spinal cord. Mucus produced by goblet cells forms a thin film over the surface of the tissue, and movements of the cilia propel the mucus and the trapped substances over the surface of the tissue. After you make your microscopic examination, draw several cells in the space that follows and label plasma membrane, cytoplasm, nucleus, cilia, goblet cell, basement membrane, and connective tissue layer.

e. Stratified squamous epithelium This tissue consists of several layers of cells and affords considerable protection against friction. The superficial cells are flat whereas cells of the deep layers vary in shape from cuboidal to columnar. The basal (bottom) cells continually multiply by

cell division. As surface cells are sloughed off, new cells replace them from the basal layer. The surface cells of **keratinized stratified squamous epithelium** contain a protective protein called **keratin** (*kerato* = horny) that also resists friction and bacterial invasion. The keratinized variety forms the outer layer of the skin. Surface cells of **nonkeratinized stratified squamous epithelium** do not contain keratin and remain moist. The nonkeratinized variety lines wet surfaces such as the mouth, esophagus, vagina, and part of the epiglottis, and it covers the tongue. After you make your microscopic examination, draw several cells in the space that follows and label plasma membrane, cytoplasm, nucleus, squamous surface cells, basal cells, basement membrane, and connective tissue layer.

Stratified squamous epithelium

f. Stratified squamous epithelium (student prepared) Before examining the next slide, prepare a smear of cheek cells from the epithelial lining of the mouth. As noted previously, epithelium that lines the mouth is nonkeratinized stratified squamous epithelium. However, you will be examining surface cells only, and these will appear similar to simple squamous epithelium.

PROCEDURE

CAUTION! *Please reread Section B, "Precautions Related to Working with Blood, Blood Products, or Other Body Fluids" on page xii at the beginning of the laboratory manual before you begin any of the following experiments. You should also read the experiments before you perform them to be sure that you understand all the procedures and safety precautions. When you finish this part of the exercise, place the reusable items in a fresh bleach solution and the discarded items in a biohazard container.*

   a. Using the blunt end of a toothpick, *gently* scrape the lining of your cheek several times to collect some surface cells of the stratified squamous epithelium.

   b. Now move the toothpick across a clean glass microscope slide until a thin layer of scrapings is left on the slide.

   c. Allow the preparation to air dry.

   d. Next, cover the smear with several drops of 1% methylene blue stain. After about 1 min, gently rinse the slide in cold tap water or distilled water to remove excess stain.

   e. *Gently* blot the slide dry using a paper towel.

   f. Examine the slide under low and high power. See if you can identify the plasma membrane, cytoplasm, nuclear membrane, and nucleoli. Some bacteria are commonly found on the slide and usually appear as very small rods or spheres.

   g. *Transitional epithelium* This tissue resembles nonkeratinized stratified squamous epithelium, except that the superficial cells are larger and more rounded. When stretched, the surface cells are drawn out into squamouslike cells. This drawing out permits the tissue to stretch without the outer cells breaking apart from one another. The tissue lines parts of the urinary system that are subject to expansion from within, such as the urinary bladder, parts of the ureters, and urethra. After you have made your microscopic examination, draw several cells in the space that follows and label plasma membrane, cytoplasm, nucleus, surface cells, basement membrane, and connective tissue layer.

   h. *Pseudostratified columnar epithelium* Nuclei of cells in this tissue are at varying depths, and, although all the cells are attached to the basement membrane in a single layer, some do not reach the surface. This arrangement gives the impression of a multilayered tissue when sectioned, thus the name *pseudostratified* (*pseudo* = false). In *pseudostratified ciliated columnar epithelium,* the cells that reach the surface either

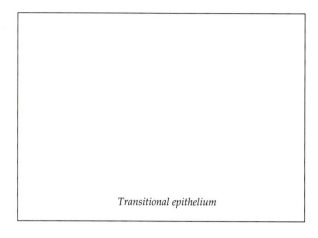

Transitional epithelium

secrete mucus (goblet cells) or bear cilia that sweep away mucus and trap particles for elimination from the body. This tissue lines most of the upper respiratory tract. In *pseudostratified non-ciliated columnar epithelium,* there are no goblet cells or cilia. This tissue lines the larger ducts of many glands, epididymis, and part of the male urethra. After you have made your microscopic examination, draw and label in the space that follows basement membrane, a cell that reaches the surface, a cell that does not reach the surface, and nuclei of each cell.

Pseudostratified columnar epithelium

3. Glandular Epithelium

A *gland* may consist of a single epithelial cell or a group of highly specialized epithelial cells that secrete various substances. Glands that have no ducts (ductless), secrete hormones, and release their secretions into the blood are called *endocrine* (*endo* = within) *glands*. Examples include the pituitary gland and thyroid gland (Exercise 16). Glands that secrete their products into ducts are called *exocrine* (*exo* = outside) *glands*. Examples include sweat glands and salivary glands.

B. CONNECTIVE TISSUE

Connective tissue, the most abundant tissue in the body, functions by protecting, supporting, and separating structures (e.g., skeletal muscles), and binding structures together.

1. Characteristics

Following are the general characteristics of connective tissue.

a. Connective tissue consists of three basic elements: cells, ground substance, and fibers. Together, the ground substance (part of a connective tissue that occupies the space between the cells and fibers) and fibers, both of which are outside the cells, form the *matrix*. Unlike epithelial cells, connective tissue cells rarely touch one another; they are separated by a considerable amount of matrix.

b. In contrast to epithelia, connective tissues do not usually occur on free surfaces, such as the surfaces of a body cavity or the external surface of the body.

c. Except for cartilage, connective tissue, like epithelium, has a nerve supply.

d. Unlike epithelium, connective tissue usually is highly vascular (has a rich blood supply). Exceptions include cartilage, which is avascular, and tendons, which have a scanty blood supply.

e. The matrix of a connective tissue, which may be fluid, semifluid, gelatinous, fibrous, or calcified, is usually secreted by the connective tissue cells and adjacent cells and determines the tissue's qualities. In blood the matrix, which is not secreted by blood cells, is fluid. In cartilage it is firm but pliable. In bone it is considerably harder and not pliable.

2. Connective Tissue Cells

Following are some of the cells contained in various types of connective tissue. The specific tissues to which they belong will be described shortly.

a. *Fibroblasts* (FĪ-brō-blasts; *fibro* = fiber) are large, flat, spindle-shaped cells with branching processes; they secrete the fibers and ground substance of the matrix.

b. **Macrophages** (MAK-rō-fā-jez; *macro* = large; *phagein* = to eat), or *histiocytes*, develop from *monocytes*, a type of white blood cell. Macrophages have an irregular shape with short branching projections and are capable of engulfing bacteria and cellular debris by phagocytosis. Thus, they provide a vital defense for the body.

c. *Plasma cells* are small and either round or irregular in shape. They develop from a type of white blood cell called a *B lymphocyte (B cell)*. Plasma cells secrete specific antibodies and, accordingly, provide a defense mechanism through immunity.

d. *Mast cells* are abundant alongside blood vessels. They produce histamine, a chemical that dilates small blood vessels and increases their permeability during inflammation.

e. Other cells in connective tissue include *adipocytes (fat cells)* and *white blood cells (leukocytes)*.

3. Connective Tissue Ground Substance

The *ground substance* is the component of a connective tissue between the cells and fibers. It is amorphous, meaning that it has no specific shape and may be a fluid, gel, or solid. Fibroblasts produce the ground substance and deposit it in the space between the cells.

Several examples of ground substance are as follows. *Hyaluronic* (hī-a-loo-RON-ik) acid is a viscous, slippery substance that binds cells together, lubricates joints, and helps maintain the shape of the eyeballs. It also appears to play a role in helping phagocytes migrate through connective tissue during development and wound repair. *Chondroitin* (kon-DROY-tin) *sulfate* is a jellylike substance that provides support and adhesiveness in cartilage, bone, the skin, and blood vessels. The skin, tendons, blood vessels, and heart valves contain *dermatan sulfate,* while bone, cartilage, and the cornea of the eye contain *keratan sulfate. Adhesion proteins* (fibronectin, laminin, collagen, and fibrinogen) interact with plasma membrane receptors to anchor cells in position.

The ground substance supports cells and binds them together and provides a medium through which substances are exchanged between the blood and cells. Until recently, the ground substance was thought to function mainly as an inert scaffolding to support tissues. Now it is clear that the ground substance is quite active in tissue development, migration, proliferation, shape, and even metabolic functions.

4. Connective Tissue Fibers

Fibers in the matrix are secreted by fibroblasts and provide strength and support for tissues. Three types of fibers are embedded in the matrix between the cells of connective tissue: collagen, elastic, and reticular fibers.

a. *Collagen* (*kolla* = glue) *fibers*, of which there are at least five different types, are very tough and resistant to a pulling force, yet allow some flexibility in the tissue because they are not taut. These fibers often occur in bundles made up of many minute fibrils lying parallel to one another. The bundle arrangement affords great strength. Chemically, collagen fibers consist of the protein *collagen*. This is the most abundant protein in your body, representing about 25% of the total protein. Collagen fibers are found in most types of connective tissues, especially bone, cartilage, tendons, and ligaments.

b. *Elastic fibers* are smaller than collagen fibers and freely branch and rejoin one another. They consist of a protein called *elastin*. Like collagen fibers, elastic fibers provide strength. In addition, they can be stretched 150% of their relaxed length without breaking. Elastic fibers are plentiful in the skin, blood vessels, and lungs.

c. *Reticular* (*rete* = net) *fibers* consisting of the protein collagen and a coating of glycoprotein, provide support in the walls of blood vessels and form a network around fat cells, nerve fibers, and skeletal and smooth muscle cells. They are much thinner than collagen fibers and form branching networks. Like collagen fibers, reticular fibers provide support and strength and also form the *stroma* (framework) of many soft-tissue organs, such as the spleen and lymph nodes. These fibers also help form the basement membrane.

5. Types

Before you start your microscopic examination of connective tissues, refer to Figure 4.2. Study the tissues carefully to familiarize yourself with their general structural characteristics. For each type of connective tissue listed, obtain a prepared slide and, unless otherwise specified by your instructor, examine each under high power.

Here, we will concentrate on various types of *mature connective tissues,* meaning connective tissues that are present in the newborn and that do not change afterward. The types of mature connective tissue are *loose connective tissue, dense connective tissue, cartilage, bone,* and *blood.*

a. LOOSE CONNECTIVE TISSUE

In this general type of connective tissue, the fibers are *loosely* woven and there are many cells.

1. *Areolar* (a-RĒ-ō-lar; *areola* = small space) *connective tissue* This is one of the most widely distributed connective tissues in the body. It contains at one time or another all cells normally found in connective tissue, including fibroblasts, macrophages, plasma cells, mast cells, adipocytes, and a few white blood cells. All three types of fibers—collagen, elastic, and reticular—are present and randomly arranged. The fluid, semifluid, or gelatinous ground substance contains hyaluronic acid, chondroitin sulfate, dermatan sulfate, and keratan sulfate. Areolar connective tissue is present in many mucous membranes, the superficial region of the dermis of the skin, around blood vessels, nerves, and organs, and, together with adipose tissue, forms the *subcutaneous* (sub'-kyoo-TĀ-nē-us) *layer* or *superficial fascia* (FASH-ē-a). This layer is located between the skin and underlying tissues. After you make your microscopic examination, draw a small area of the tissue in the space that follows and label the elastic fibers, collagen fibers, fibroblasts, and mast cells.

Areolar connective tissue

2. *Adipose tissue* This is fat tissue in which cells derived from fibroblasts, called *adipocytes* (*adeps* = fat), are modified for triglyceride (fat) storage. The cytoplasm and nuclei of the cells are pushed to the edge. The tissue is found wherever areolar connective tissue is located and around the kidneys and heart, in the yellow bone marrow of long bones, around joints, and behind the eyeball. It provides insulation, energy reserve, support, and protection. After your microscopic examination, draw several

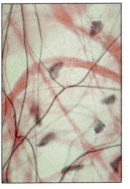

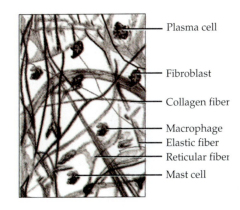

Plasma cell

Fibroblast

Collagen fiber

Macrophage
Elastic fiber
Reticular fiber

Mast cell

Surface view of subcutaneous
tissue (160×)

(a) Areolar connective tissue

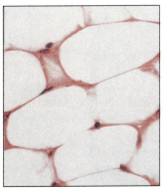

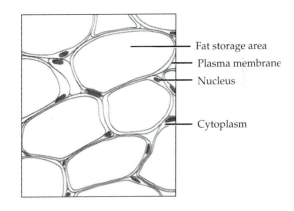

Fat storage area
Plasma membrane
Nucleus

Cytoplasm

Sectional view of white fat of
pancreas (1,600×)

(b) Adipose tissue

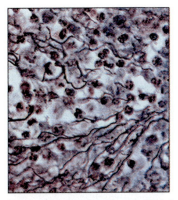

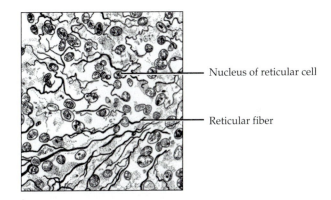

Nucleus of reticular cell

Reticular fiber

Sectional view of lymph node (250×)

(c) Reticular connective tissue

FIGURE 4.2   Connective tissues. Photomicrographs are unlabeled; the line drawings of the same tissues are labeled.

Sectional view of capsule of adrenal gland (250×)

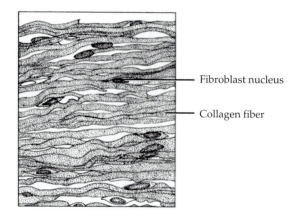

Fibroblast nucleus

Collagen fiber

(d) Dense regular connective tissue

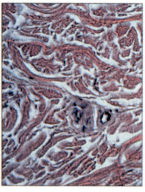

Sectional view of dermis of skin (275×)

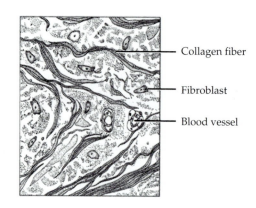

Collagen fiber

Fibroblast

Blood vessel

(e) Dense irregular connective tissue

Sectional view of ligamentum nuchae (400×)

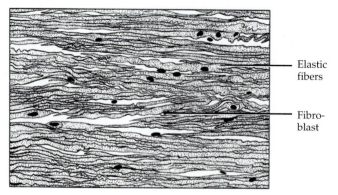

Elastic fibers

Fibro-blast

(f) Elastic connective tissue

FIGURE 4.2 *(Continued)*   Connective tissues.

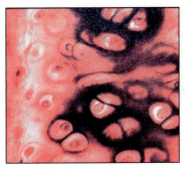

Sectional view of hyaline cartilage from trachea (160×)

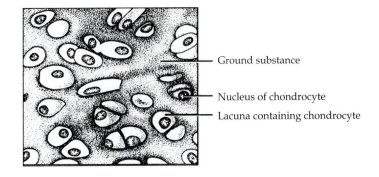

Ground substance

Nucleus of chondrocyte

Lacuna containing chondrocyte

(g) Hyaline cartilage

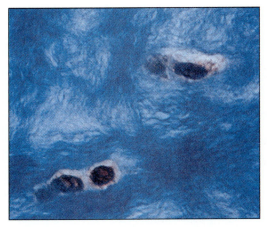

Sectional view of fibrocartilage from medial meniscus of knee (315×)

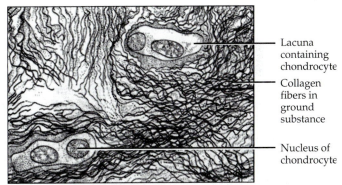

Lacuna containing chondrocyte

Collagen fibers in ground substance

Nucleus of chondrocyte

(h) Fibrocartilage

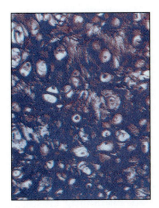

Sectional view of elastic cartilage from auricle (pinna) of external ear (175×)

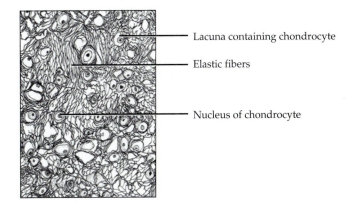

Lacuna containing chondrocyte

Elastic fibers

Nucleus of chondrocyte

(i) Elastic cartilage

FIGURE 4.2 *(Continued)* Connective tissues.

cells in the space that follows and label the fat storage area, cytoplasm, nucleus, and plasma membrane.

Adipose tissue

3. ***Reticular connective tissue*** This tissue consists of fine interlacing reticular fibers in which reticular cells are interspersed between the fibers. It provides the stroma (framework) of certain organs and helps bind certain cells together. It is found in the liver, spleen, and lymph nodes; in a portion of the basement membrane; and in red bone marrow, and around blood vessels and muscles. After you make your microscopic examination, draw a sample of the tissue in the space that follows and label the reticular fibers and cells of the organ.

Reticular connective tissue

b. DENSE CONNECTIVE TISSUE

In this general type of connective tissue, the fibers are more numerous, *thicker,* and *densely* packed and there are fewer cells than in loose connective tissue.

1. ***Dense regular connective tissue*** In this tissue, bundles of collagen fibers have a regular paral-

lel arrangement that confers great strength. The tissue structure withstands pulling in one direction. Fibroblasts, which produce the fibers and ground substance, appear in rows between the fibers. The tissue is silvery white, tough, yet somewhat pliable. Because of its great strength, it is the principal component of ***tendons***, which attach muscles to bones; ***aponeuroses*** (ap'-ō-noo-RO-sēz), which are sheetlike tendons connecting one muscle with another or with bone; and most ***ligaments*** (collagen ligaments), which hold bones together at joints. After you make your microscopic examination, draw a sample of the tissue in the space that follows and label the collagen fibers and fibroblasts.

Dense regular connective tissue

2. ***Dense irregular connective tissue*** This tissue contains collagen fibers that are *irregularly* arranged (without regular orientation) and is found in parts of the body where tensions are exerted in various directions. The tissue usually occurs in sheets. It forms some fasciae, the reticular (deeper) region of the dermis of the skin, the pericardium of the heart, the periosteum of bone, the perichondrium of cartilage, joint capsules, heart valves, and the membrane (fibrous) capsules around organs, such as the kidneys, liver, testes, and lymph nodes. After you make your microscopic examination, draw a sample of the tissue in the space that follows and label the collagen fibers.

3. ***Elastic connective tissue*** This tissue has a predominance of freely branching elastic fibers. These fibers give the unstained tissue a yellowish color. Fibroblasts are present in the spaces between fibers. Elastic connective tissue can be stretched and will snap back into shape (elasticity). It is a component of the walls of elastic

component of the ground substance. Whereas the strength of cartilage is due to its collagen fibers, its resilience (ability to assume its original shape after deformation) is due to chondroitin sulfate.

The cells of mature cartilage, called *chondrocytes* (KON-drō-sīts; *chondros* = cartilage), occur singly or in groups within spaces called *lacunae* (la-KOO-nē; *lacuna* = little lake) in the matrix. The surface of cartilage, except for fibrocartilage and some hyaline cartilage, is surrounded by dense irregular connective tissue called the *perichondrium* (per'-i-KON-drē-um; *peri* = around). There are three kinds of cartilage: hyaline cartilage, fibrocartilage, and elastic cartilage.

1. *Hyaline cartilage* This cartilage, also called *gristle,* contains a resilient gel as its ground substance and appears in the body as a bluish-white, shiny substance. The fine collagen fibers, although present, are not visible with ordinary staining techniques, and the prominent chondrocytes are found in lacunae. Most hyaline cartilage is surrounded by a perichondrium. Hyaline cartilage is the most abundant cartilage in the body. It is found at joints over the ends of the long bones (articular cartilage) and at the anterior ends of the ribs (costal cartilage). Hyaline cartilage also helps to support the nose, larynx, trachea, bronchi, and bronchial tubes leading to the lungs. Most of the embryonic skeleton consists of hyaline cartilage, which gradually becomes calcified and develops into bone. Hyaline cartilage affords flexibility and support and, at joints, reduces friction and absorbs shock. After you make your microscopic examination, draw a sample of the tissue in the space that follows and label the perichondrium, chondrocytes, lacunae, and ground substance.

Dense irregular connective tissue

arteries, the trachea, bronchial tubes of the lungs, and the lungs themselves. Elastic connective tissue provides stretch and strength, allowing structures to perform their functions efficiently. Yellow elastic ligaments, as contrasted with collagen ligaments, are composed mostly of elastic fibers; they form the ligamenta flava of the vertebrae (ligaments between successive vertebrae), the suspensory ligament of the penis, and the true vocal cords. After you make your microscopic examination, draw a sample of the tissue in the space that follows and label the elastic fibers and fibroblasts.

Elastic connective tissue

C. CARTILAGE

Cartilage is capable of enduring considerably more stress than the tissues just discussed. Unlike other connective tissues, cartilage has no blood vessels or nerves, except for those in the perichondrium (membranous covering). Cartilage consists of a dense network of collagen fibers and elastic fibers firmly embedded in chondroitin sulfate, a rubbery

Hyaline cartilage

2. *Fibrocartilage* Chondrocytes are scattered among clearly visible bundles of collagen fibers within the matrix of this type of cartilage. Fibrocartilage does not have a perichondrium. Fibrocartilage forms the pubic symphysis, the point where the hipbones fuse anteriorly at the midline. It is also found in the intervertebral discs between vertebrae, and the menisci (cartilage pads) of the knee. This tissue combines strength and rigidity. After you make your microscopic examination, draw a sample of the tissue in the space that follows and label the chondrocytes, lacunae, ground substance, and collagen fibers.

Fibrocartilage

3. *Elastic cartilage* In this tissue, chondrocytes are located in a threadlike network of elastic fibers within the matrix. Elastic cartilage has a perichondrium and provides strength and elasticity and maintains the shape of organs—the epiglottis of the larynx, the external part of the ear (auricle), and the auditory (Eustachian) tubes. After you make your microscopic examination, draw a sample of the tissue in the space that follows and label the perichondrium, chondrocytes, lacunae, ground substance, and elastic fibers.

Elastic cartilage

C. MEMBRANES

Membranes are flat sheets of pliable tissue that cover or line a part of the body. The combination of an epithelial layer and an underlying layer of connective tissue constitutes an *epithelial membrane*. Examples are mucous, serous, and cutaneous membranes (skin). Another kind of membrane, a synovial membrane, has no epithelium. It contains only connective tissue. *Mucous membranes*, also called the *mucosa*, line body cavities that open directly to the exterior, such as the gastrointestinal, respiratory, urinary, and reproductive tracts. The surface tissue of a mucous membrane consists of epithelium and has a variety of functions, depending on location. Accordingly, the epithelial layer secretes mucus but may also secrete enzymes, filter dust, and have a protective and absorbent action. The underlying connective tissue layer of a mucous membrane, called the *lamina propria* (LAM-i-na PRŌ-prē-a), binds the epithelial layer in place, protects underlying tissues, provides the epithelium with nutrients and oxygen and removes wastes, and holds blood vessels in place.

Serous (*serous* = watery) *membranes*, also called the *serosa*, line body cavities that do not open to the exterior and cover organs that lie within the cavities. Serous membranes consist of a surface layer of mesothelium (simple squamous epithelium) and an underlying layer of areolar connective tissue. The mesothelium secretes a lubricating fluid. Serous membranes consist of two layers. The layer attached to the cavity wall is called the *parietal* (pa-RĪ-e-tal; *paries* = wall) *layer*; the layer that covers the organs in the cavity is called the *visceral* (*viscus* = body organ) *layer*. Examples of serous membranes are the pleurae, pericardium, and peritoneum.

The *cutaneous membrane*, or skin, is the principal component of the integumentary system, which will be considered in the next exercise.

Synovial (sin-Ō-vē-al) *membranes* line joint cavities. They do not contain epithelium but rather consist of areolar connective tissue, adipose tissue, and elastic fibers. Synovial membranes produce synovial fluid, which lubricates the ends of bones as they move at joints and nourishes the articular cartilage around the ends of bones.

ANSWER THE LABORATORY REPORT QUESTIONS AT THE END OF THE EXERCISE.

Tissues 4

Student _____ Date _____

Laboratory Section _____ Score/Grade _____

PART 1. Multiple Choice

_____ 1. In parts of the body such as the urinary bladder, where considerable distention (stretching) occurs, you can expect to find which epithelial tissue? (a) pseudostratified columnar (b) cuboidal (c) columnar (d) transitional

_____ 2. Stratified epithelium is usually found in areas of the body where the principal activity is (a) filtration (b) absorption (c) protection (d) diffusion

_____ 3. Ciliated epithelium destroyed by disease would cause malfunction in which system? (a) digestive (b) respiratory (c) skeletal (d) cardiovascular

_____ 4. The tissue that provides the skin with resistance to wear and tear and serves to waterproof it is (a) keratinized stratified squamous (b) pseudostratified columnar (c) transitional (d) simple columnar

_____ 5. The connective tissue cell that would most likely increase its activity during an infection is the (a) melanocyte (b) macrophage (c) adipocyte (d) fibroblast

_____ 6. Torn ligaments would involve damage to which tissue? (a) dense regular (b) reticular (c) elastic (d) areolar

_____ 7. Simple squamous epithelial tissue that lines the heart, blood vessels, and lymphatic vessels is called (a) transitional (b) adipose (c) endothelium (d) mesothelium

_____ 8. Microvilli and goblet cells are associated with which tissue? (a) hyaline cartilage (b) simple columnar nonciliated (c) transitional (d) stratified squamous

_____ 9. Superficial fascia contains which tissue? (a) elastic (b) reticular (c) fibrocartilage (d) areolar connective tissue

_____ 10. Which tissue forms articular cartilage and costal cartilage? (a) fibrocartilage (b) elastic cartilage (c) adipose (d) hyaline cartilage

_____ 11. Membranes that line cavities that open directly to the exterior are called (a) synovial (b) serous (c) mucous (d) cutaneous

_____ 12. Which statement about connective tissue is false? (a) Cells are always very closely packeed together. (b) Connective tissue always has an abundant blood supply. (c) Matrix is always present in large amounts. (d) It is the most abundant tissue in the body.

_____ 13. A group of similar cells that has a similar embryological origin and operates together to perform a specialized activity is called a(n) (a) organ (b) tissue (c) system (d) organ system

_____ 14. Which statement best describes covering and lining epithelium? (a) It is always arranged in a single layer of cells. (b) It contains large amounts of intercellular substance. (c) It has an abundant blood supply. (d) Its free surface is exposed to the exterior of the body or to the interior of a hollow structure.

_____ **15.** Which statement best describes connective tissue? (a) It usually contains a large amount of matrix. (b) It's always arranged in a single layer of cells. (c) It's primarily concerned with secretion. (d) It usually lines a body cavity.

_____ **16.** A gland (a) is either exocrine or endocrine (b) may be single celled or multicellular (c) consists of epithelial tissue (d) is described by all of the preceding statements.

_____ **17.** Which of the following statements is not correct? (a) Simple squamous epithelium lines blood vessels. (b) Endothelium is composed of cuboidal cells. (c) Ciliated epithelium is found in the respiratory system. (d) Transitional epithelium is found in the urinary bladder.

PART 2. Completion

18. Cells found in epithelium that secrete mucus are called_____cells.

19. A type of epithelium that appears to consist of several layers but actually contains only one layer of

cells is_____.

20. The cell in connective tissue that forms new fibers is the_____.

21. Histamine, a substance that dilates small blood vessels during inflammation, is secreted

by_____cells.

22. Cartilage cells called_____are found in lacunae.

23. The simple squamous epithelium of a serous membrane that covers viscera is

called_____.

24. The tissue that provides insulation, support, protection, and serves as a food reserve

is_____.

25. _____tissue forms the stroma of organs such as the liver and spleen.

26. The cartilage that provides support for the larynx and external ear is_____.

27. The ground substance that helps lubricate joints and binds cells together is_____.

28. Ductless glands that secrete hormones are called_____glands.

29. _____membranes consist of parietal and visceral layers and line cavities that do not open to the exterior.

30. Membranes that line joint cavities are called_____membranes.

31. The structure that attaches epithelium to underlying connective tissue is called

the_____.

PART 3. Matching

_____ **32.** Lines inner surface of the stomach and intestine

_____ **33.** Lines urinary tract, as in urinary bladder, permitting distention

_____ **34.** Lines mouth; present on outer surface of skin

_____ **35.** Single layer of cube-shaped cells; found in kidney tubules and ducts of some glands

_____ **36.** Lines air sacs of lungs where thin cells are required for diffusion of gases into blood

_____ **37.** Not a true stratified tissue; all cells on basement membrane, but some do not reach surface

_____ **38.** Derived from lymphocyte, gives rise to antibodies and so is helpful in defense

_____ **39.** Phagocytic cell; engulfs bacteria and cleans up debris; important during infection

_____ **40.** Believed to form collagen and elastic fibers in injured tissue

_____ **41.** Abundant along walls of blood vessels; produces histamine, which dilates blood vessels

_____ **42.** Contains lacunae and chondrocytes

_____ **43.** Forms fasciae and dermis of skin

_____ **44.** Stores fat and provides insulation

A. Transitional epithelium

B. Fibroblast

C. Pseudostratified columnar epithelium

D. Dense irregular connective tissue

E. Simple columnar epithelium

F. Macrophage

G. Stratified squamous epithelium

H. Adipose

I. Simple cuboidal epithelium

J. Plasma cell

K. Simple squamous epithelium

L. Mast cell

M. Cartilage

Integumentary System

5

The skin and its accessory structures derived from it (hair, nails, and glands), and several specialized receptors constitute the ***integumentary*** (in-teg-yoo-MEN-tar-ē; *integumentum* = covering) ***system,*** which you will study in this exercise. An ***organ*** is an aggregation of tissues of definite form and usually recognizable shape that performs a specific function; a ***system*** is a group of organs that operate together to perform specialized functions.

A. SKIN

The ***skin*** is one of the largest organs of the body in terms of surface area and weight, occupying a surface area of about 2 square meters (2 m²) (22 ft²) and weighing about 4.5 to 5 kg (10 to 11 lbs) about 16% of total body weight. Among the functions performed by the skin are regulation of body temperature; protection of underlying tissues from physical abrasion, microorganisms, dehydration, and ultraviolet (UV) radiation; excretion of water and salts and several organic compounds; synthesis of vitamin D in the presence of sunlight; reception of stimuli for touch, pressure, pain, and temperature change sensations; serving as a blood reservoir; and immunity.

The skin consists of an outer, thinner ***epidermis*** (*epi* = above), which is avascular, and an inner, thicker ***dermis*** (*derm* = skin), which is vascular. Below the dermis and not part of the skin is the ***subcutaneous layer (superficial fascia or hypodermis)*** that attaches the skin to underlying tissues and organs.

1. Epidermis

The epidermis consists of four principal kinds of cells. ***Keratinocytes*** (ker-a-TIN-ō-sīts; *kerato* = horny) are the most numerous cells. They produce a protein called keratin, which helps protect the skin and underlying tissue from light, heat, microbes, and many chemicals. Keratocytes undergo keratinization, that is, newly formed cells produced in the basal layers are pushed up to the surface and in the process synthesize keratin. ***Melanocytes*** (MEL-a-nō-sīts; *melan* = black) are pigment cells that impart color to the skin. The third type of cell in the epidermis is called a ***Langerhans*** (LANG-er-hans) ***cell.*** These cells arise from red bone marrow and migrate to the epidermis and other stratified squamous epithelial tissue in the body. They are sensitive to UV radiation and lie above the basal layer of keratinocytes. Langerhans cells interact with white blood cells called ***helper T cells*** to assist in the immune response. ***Merkel cells*** are found in the bottom layer of the epidermis. Their bases are in contact with flattened portions of the terminations of sensory nerves ***(Merkel discs)*** and function as receptors for touch. At this point, we will concentrate only on keratinocytes.

Obtain a prepared slide of human thick skin and carefully examine the epidermis. Identify the following layers from the inside outward:

a. ***Stratum basale*** (*basale* = base) Single layer of cuboidal to columnar cells that constantly undergo division. This layer contains tactile (Merkel) discs, receptors sensitive to touch, melanocytes and Langerhans cells. Also called the ***stratum germinativum.***

b. ***Stratum spinosum*** (*spinosum* = thornlike) 8 to 10 rows of polyhedral (many-sided) cells.

c. ***Stratum granulosum*** (*granulum* = little grain) 3 to 5 rows of flat cells that contain ***lamellar granules*** which produce a lipid waterproof sealant.

d. ***Stratum lucidum*** (*lucidus* = clear) Several rows of clear, flat, dead cells; only apparent in the thick skin of the palms and soles.

e. ***Stratum corneum*** (*corneum* = horny) 25 to 30 rows of flat, dead cells that are filled with intermediate filaments and lipids from lamellar granules; these cells are continuously shed and replaced by cells from deeper strata. This layer is an effective water repellent barrier and protects underlying layers.

Label the epidermal layers in Figure 5.1.

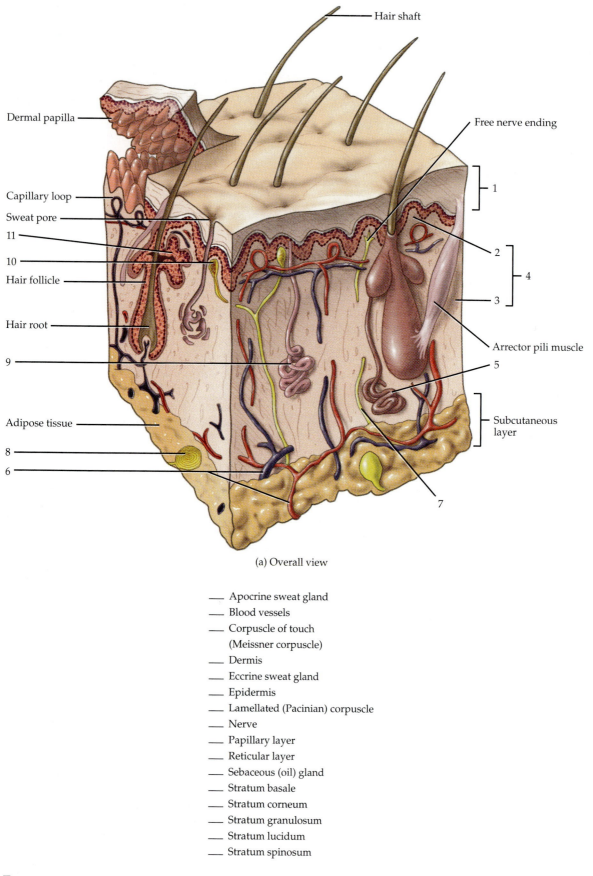

Hair shaft

Dermal papilla

Free nerve ending

Capillary loop

1

Sweat pore

11

2

10

4

Hair follicle

3

Hair root

Arrector pili muscle

9

5

Adipose tissue

Subcutaneous layer

8

6

7

(a) Overall view

—— Apocrine sweat gland
—— Blood vessels
—— Corpuscle of touch
    (Meissner corpuscle)
—— Dermis
—— Eccrine sweat gland
—— Epidermis
—— Lamellated (Pacinian) corpuscle
—— Nerve
—— Papillary layer
—— Reticular layer
—— Sebaceous (oil) gland
—— Stratum basale
—— Stratum corneum
—— Stratum granulosum
—— Stratum lucidum
—— Stratum spinosum

FIGURE 5.1 Structure of the skin.

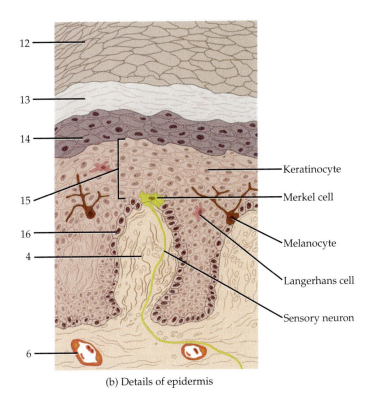

12 ⎯⎯

13 ⎯⎯

14 ⎯⎯

15 ⎯⎯

16 ⎯⎯

4 ⎯⎯

6 ⎯⎯

Keratinocyte

Merkel cell

Melanocyte

Langerhans cell

Sensory neuron

(b) Details of epidermis

FIGURE 5.1 *(Continued)* Structure of the skin.

2. Dermis

The dermis is divided into two regions and is composed of connective tissue containing collagen and elastic fibers and a number of other structures. The superficial region of the dermis *(papillary layer)* is composed of areolar connective tissue containing fine elastic fibers. This layer contains fingerlike projections, the *dermal papillae* (pa-PIL-ē; *papilla* = nipple). Some papillae enclose blood capillaries; others contain *corpuscles of touch (Meissner corpuscles)*, nerve endings sensitive to touch; others contain free nerve endings that convey sensations of warmth, coolness, pain, tickling and itching. The deeper region of the dermis *(reticular layer)* consists of dense, irregular connective tissue with interlacing bundles of larger collagen and some coarse elastic fibers. Spaces between the fibers may be occupied by *hair follicles, sebaceous (oil) glands, bundles of smooth muscle (arrector pili muscle), sudoriferous (sweat) glands, blood vessels*, and *nerves.*

The reticular layer of the dermis is attached to the underlying structures (bones and muscles) by the subcutaneous layer. This layer also contains nerve endings sensitive to pressure called *lamellated* or *Pacinian* (pa-SIN-ē-an) *corpuscles.*

Carefully examine the dermis and subcutaneous layer on your microscope slide. Label the following structures in Figure 5.1: papillary layer, reticular layer, corpuscle of touch, blood vessels, nerves, sebaceous gland, sudoriferous glands (apocrine and eccrine), and lamellated corpuscle. Also label the epidermis, dermis, and blood vessels in Figure 5.2.

If a model of the skin and subcutaneous layer is available, examine it to see the three-dimensional relationship of the structures to one another.

3. Skin Color

The color of skin results from (1) *hemoglobin in red blood cells in capillaries* of the dermis (beneath the epidermis); (2) *carotene* (KAR-o-tēn; *keraton* = carrot), a yellow-orange pigment in the stratum corneum of the epidermis and fatty areas of the dermis and subcutaneous layer; and (3) *melanin* (MEL-a-nin), a pale yellow to black pigment found primarily in the melanocytes in the stratum basale and spinosum of the epidermis. Whereas hemoglobin in red blood cells in capillaries imparts a pink color to Caucasian skin, carotene imparts a yellowish color to skin. Because the number of *melanocytes* (MEL-a-nō-sīts), or melanin-producing cells, is about the same in all races,

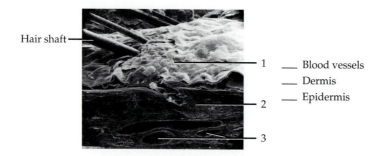

Hair shaft

1 ___ Blood vessels
___ Dermis
___ Epidermis

2

3

FIGURE 5.2   Scanning electron micrograph of the skin and several hairs at a magnification of 260×. (Reproduced by permission from R. G. Kessel and R. H. Kardon, *Tissues and Organs: A Text Atlas of Scanning Electron Microscopy,* W. H. Freeman, 1979.)

most differences in skin color are due to the amount of melanin that the melanocytes synthesize and disperse. Exposure to UV radiation increases melanin synthesis, resulting in darkening (tanning) of the skin to protect the body against further UV radiation.

An inherited inability of a person of any race to produce melanin results in *albinism* (AL-bi-nizm). The pigment is absent from the hair and eyes as well as from the skin, and the individual is referred to as an *albino.* In some people, melanin tends to accumulate in patches called *freckles.* Others inherit patches of skin that lack pigment, a condition called *vitiligo* (vit-i-LĪ-gō).

B. HAIR

Hairs (pili) develop from the epidermis and are variously distributed over the body. Each hair is composed of columns of dead, keratinized cells and consists of a *shaft,* most of which is visible above the surface of the skin, and a *root,* the portion mostly below the surface that penetrates deep into the dermis and even into the subcutaneous layer.

The shaft of a coarse hair consists of the following parts:

1. *Medulla* Inner region composed of several rows of polyhedral cells containing pigment and air spaces. Poorly developed or not present in fine hairs.
2. *Cortex* Middle layer; contains several rows of dark cells surrounding the medulla; contains pigment in dark hair and mostly air in white hair.
3. *Cuticle of the hair* Outermost layer; consists of a single layer of flat, keratinized cells arranged like shingles on a house.

The root of a hair also contains a medulla, cortex, and cuticle of the hair along with the following associated parts:

1. *Hair follicle* Structure surrounding the root that consists of an external root sheath and an internal root sheath. These epidermally derived layers are surrounded by a dermal layer of connective tissue.
2. *External root sheath* Downward continuation of the epidermis.
3. *Internal root sheath* Cellular tubular sheath that separates the hair from the external root sheath; consists of (a) the *cuticle of the internal root sheath,* an inner single layer of flattened cells with atrophied nuclei, (b) *granular (Huxley's) layer,* a middle layer of one to three rows of cells with flattened nuclei, and (c) *pallid (Henle's) layer,* an outer single layer of cuboidal cells with flattened nuclei.
4. *Bulb* Enlarged, onion-shaped structure at the base of the hair follicle.
5. *Papilla of the hair* Dermal indentation into the bulb; contains areolar connective tissue and blood vessels to nourish the hair.
6. *Matrix* Region of cells at the base of the bulb derived from the stratum basale and that divides to produce new hair.
7. *Arrector (arrector = to raise) pili muscle* Bundle of smooth muscle extending from the superficial dermis of the skin to the side of the hair follicle; its contraction, under the influence of fright or cold, causes the hair to move into a vertical position, producing "goose bumps."
8. *Hair root plexus* Nerve endings around each hair follicle that are sensitive to touch and respond when the hair shaft is moved.

___ Bulb
___ Cortex
___ Cuticle of hair
___ Cuticle of internal root sheath
___ External root sheath
___ Granular (Huxley's) layer
___ Internal root sheath
___ Matrix
___ Medulla
___ Pallid (Henle's) layer
___ Papilla of hair

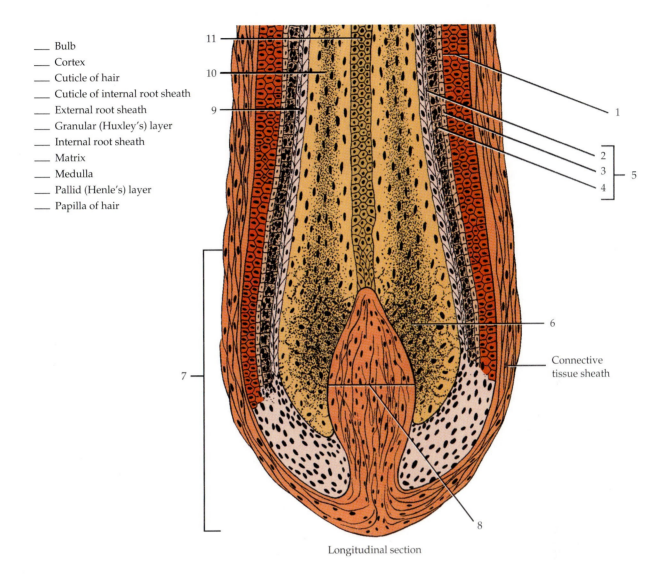

Connective
tissue sheath

Longitudinal section

FIGURE 5.3   Hair root.

Obtain a prepared slide of a transverse (cross) section and a longitudinal section of a hair root and identify as many parts as you can. Using your textbook as a reference, label Figure 5.3. Also label the parts of a hair shown in Figure 5.4.

C. GLANDS

Sebaceous (se-BĀ-shus; *sebo* = grease) or *oil glands,* with few exceptions, are connected to hair follicles (see Figures 5.1 and 5.4). They are simple, branched acinar glands and secrete an oily substance called *sebum* (SĒ-bum), a mixture of fats, cholesterol, proteins, salts, and pheromones. Sebaceous glands are absent in the skin of the palms and soles but are numerous in the skin of the face, neck, upper chest, and breasts. Sebum

helps prevent hair from drying and forms a protective film over the skin that prevents excessive evaporation and keeps the skin soft and pliable. Sebum also inhibits the growth of certain bacteria.

Sudoriferous (soo'-dor-IF-er-us; *sudor* = sweat; *ferre* = to bear) or *sweat glands* are separated into two principal types on the basis of structure, location, and secretion. *Apocrine sweat glands* are simple, branched tubular glands found primarily in the skin of the axilla (armpit), pubic region, areolae (pigmented areas) of the breasts and bearded regions of the face in adult males. Their secretory portion is located in the dermis or subcutaneous layer; the excretory duct opens into hair follicles. Apocrine sweat glands begin to function at puberty and produce a more viscous secretion than the other type of sweat gland. Apocrine sweat glands are stimulated

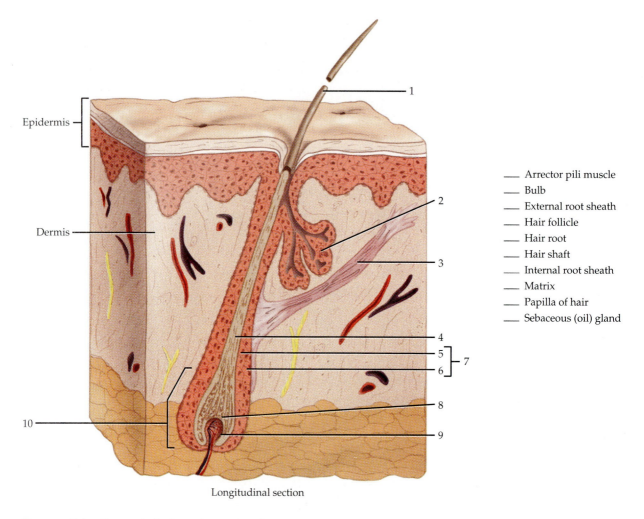

Epidermis

Dermis

10

—— Arrector pili muscle
—— Bulb
—— External root sheath
—— Hair follicle
—— Hair root
—— Hair shaft
—— Internal root sheath
—— Matrix
—— Papilla of hair
—— Sebaceous (oil) gland

Longitudinal section

FIGURE 5.4   Parts of a hair and associated structures.

during emotional stress and sexual excitement; their secretion is commonly called "cold sweat". *Eccrine sweat glands* are simple, coiled tubular glands found throughout the skin, except for the margins of the lips, nail beds of the fingers and toes, glans penis, glans clitoris, and eardrums. They are most numerous in the skin of the palms and soles. The secretory portion of these glands is in the subcutaneous layer; the excretory duct projects upward and terminates at a pore at the surface of the epidermis (see Figure 5.1). Eccrine sweat glands function throughout life and produce a more watery secretion than the apocrine glands. Sudoriferous glands produce *perspiration,* a mixture of water, salt, urea, uric acid, amino acids, ammonia, glucose, lactic acid, and ascorbic acid. The evaporation of perspiration helps to maintain normal body temperature.

Ceruminous (se-ROO-mi-nus; *cera* = wax) *glands* are modified sudoriferous glands in the external auditory canal (ear canal). They are simple, coiled

tubular glands. The combined secretion of ceruminous and sudoriferous glands is called *cerumen* (earwax). Cerumen, together with hairs in the external auditory canal, provides a sticky barrier that prevents foreign bodies from reaching the eardrum.

D. NAILS

Nails are plates of tightly packed, hard, keratinized epidermal cells that form a clear, solid covering over the dorsal surfaces of the terminal portions of the fingers and toes. Each nail consists of the following parts:

1. *Nail body* Portion that is visible.
2. *Free edge* Part that may project beyond the distal end of the digit.
3. *Nail root* Portion hidden in nail groove (see item 7).

4. ***Lunula*** (LOO-nyoo-la; *lunula* = little moon) Whitish semilunar area at proximal end of body.
5. ***Nail fold*** Fold of skin that extends around the proximal end and lateral borders of the nail.
6. ***Nail bed*** Strata basale and spinosum of the epidermis beneath the nail.
7. ***Nail groove*** Furrow between the nail fold and nail bed.
8. ***Eponychium*** (ep'-ō-NIK-ē-um) Cuticle; a narrow band of epidermis.
9. ***Hyponychium*** Thickened area of stratum corneum below the free edge of the nail.
10. ***Nail matrix*** Epithelium of the proximal part of the nail bed; division of the cells brings about growth of nails.

Using your textbook as a reference, label the parts of a nail shown in Figure 5.5. Also, identify the parts that are visible on your own nails.

E. HOMEOSTASIS OF BODY TEMPERATURE

One of the best examples of homeostasis in humans is the regulation of body temperature by the skin. As warm-blooded animals, we are able to maintain a remarkably constant body temperature of 37°C (98.6°F) even though the environmental temperature varies greatly.

Suppose you are in an environment where the temperature is 38°C (101°F). Heat (the stimulus) continually flows from the environment to your body, raising body temperature. To counteract these changes in a controlled condition, a sequence of events is set into operation. Temperature-sensitive receptors (nerve endings) in the skin called ***thermoreceptors*** detect the stimulus and send nerve impulses (input) to your brain (control center). A temperature control region of the brain (called the

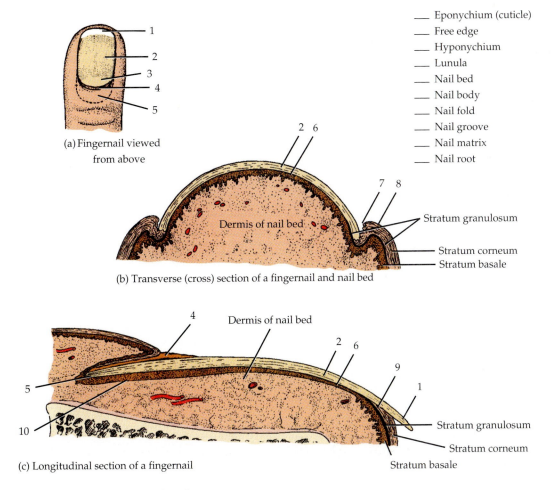

___ Eponychium (cuticle)
___ Free edge
___ Hyponychium
___ Lunula
___ Nail bed
___ Nail body
___ Nail fold
___ Nail groove
___ Nail matrix
___ Nail root

(a) Fingernail viewed from above

(b) Transverse (cross) section of a fingernail and nail bed

(c) Longitudinal section of a fingernail

FIGURE 5.5   Structure of nails.

hypothalamus) then sends nerve impulses (output) to the sudoriferous glands (effectors), which produce perspiration more rapidly. As the sweat evaporates from the surface of your skin, heat is lost and your body temperature decreases (response). This cycle continues until body temperature drops to normal (returns to homeostasis). When environmental temperature is low, sweat glands produce less perspiration.

Your brain also sends output to blood vessels (a second set of effectors), dilating (widening) those in the dermis so that skin blood flow increases. As more warm blood flows through capillaries close to the body surface, more heat can be lost to the environment, which lowers body temperature. Thus heat is lost from the body, and body temperature falls to the normal value to restore homeostasis. In response to low environmental temperature, blood vessels in the dermis constrict, blood flow decreases, and heat is lost by radiation.

This temperature regulation involves a *negative feedback system* because the response (cooling) is opposite to the stimulus (heating) that started the cycle. Also, the thermoreceptors continually monitor body temperature and feed this information back to the brain. The brain, in turn, continues to send impulses to the sweat glands and blood vessels until the temperature returns to 37°C (98.6°F).

Regulating the rate of sweating and changing dermal blood flow are only two mechanisms by which body temperature can be adjusted. Other mechanisms include regulating metabolic rate (a slower metabolic rate reduces heat production) and regulating skeletal muscle contractions (decreased muscle tone results in less heat production).

Label Figure 5.6, the role of the skin in regulating the homeostasis of body temperature.

ANSWER THE LABORATORY REPORT QUESTIONS AT THE END OF THE EXERCISE.

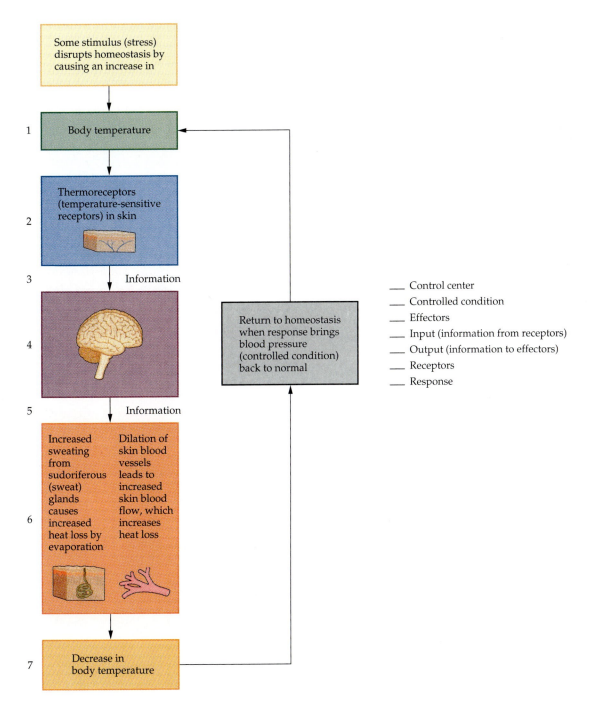

1. Body temperature

2. Thermoreceptors (temperature-sensitive receptors) in skin

3. Information

4.

Return to homeostasis when response brings blood pressure (controlled condition) back to normal

5. Information

6. Increased sweating from sudoriferous (sweat) glands causes increased heat loss by evaporation   Dilation of skin blood vessels leads to increased skin blood flow, which increases heat loss

7. Decrease in body temperature

Some stimulus (stress) disrupts homeostasis by causing an increase in

___ Control center
___ Controlled condition
___ Effectors
___ Input (information from receptors)
___ Output (information to effectors)
___ Receptors
___ Response

FIGURE 5.6 Role of the skin in regulating the homeostasis of body temperature.

Integumentary System 5

Student _____ Date _____

Laboratory Section _____ Score/Grade _____

PART 1. Multiple Choice

_____ 1. The waterproofing quality of skin is due to the presence of (a) melanin (b) carotene (c) lamellar granules (d) receptors

_____ 2. Sebaceous glands (a) produce a watery solution called sweat (b) produce an oily substance that prevents excessive water evaporation from the skin (c) are associated with mucous membranes (d) are part of the subcutaneous layer

_____ 3. Which of the following is the proper sequence of layering of the epidermis, going from the free surface toward the underlying tissues? (a) basale, spinosum, granulosum, corneum (b) spinosum, basale, granulosum, corneum (c) corneum, lucidum, granulosum, spinosum, basale (d) corneum, granulosum, lucidum, spinosum

_____ 4. Skin color is *not* determined by the presence or absence of (a) melanin (b) carotene (c) keratin (d) hemoglobin in red blood cells in capillaries in the dermis

_____ 5. Destruction of what part of a single hair would result in its inability to grow? (a) sebaceous gland (b) arrector pili muscle (c) matrix (d) bulb

_____ 6. One would expect to find relatively few, if any, sebaceous glands in the skin of the (a) palms (b) face (c) neck (d) upper chest

_____ 7. Which of the following sequences, from outside to inside, is correct? (a) epidermis, reticular layer, papillary layer, subcutaneous layer (b) epidermis, subcutaneous layer, reticular layer, papillary layer (c) epidermis, reticular layer, subcutaneous layer, papillary layer (d) epidermis, papillary layer, reticular layer, subcutaneous layer

_____ 8. The attached visible portion of a nail is called the (a) nail bed (b) nail root (c) nail fold (d) nail body

_____ 9. Nerve endings sensitive to touch are called (a) corpuscles of touch (Meissner's corpuscles) (b) papillae (c) lamellated (Pacinian) corpuscles (d) follicles

_____ 10. The cuticle of a nail is referred to as the (a) matrix (b) eponychium (c) hyponychium (d) fold

_____ 11. One would *not* expect to find sudoriferous glands associated with the (a) forehead (b) axilla (c) palms (d) nail beds

_____ 12. Fingerlike projections of the dermis that contain loops of capillaries and receptors are called (a) dermal papillae (b) nodules (c) polyps (d) pili

_____ 13. Which of the following statements about the function of skin is *not* true? (a) it helps control body temperature (b) it prevents excessive water loss (c) it synthesizes several compounds (d) it absorbs water and salts

_____   **14.** Which is *not* part of the internal root sheath? (a) granular (Huxley's) layer (b) cortex (c) pallid (Henle's) layer (d) cuticle of the internal root sheath

_____   **15.** Growth in the length of nails is the result of the activity of the (a) eponychium (b) nail matrix (c) hyponychium (d) nail fold

PART 2. Completion

16. A group of tissues that performs a definite function is called a(n) _____.

17. The outer, thinner layer of the skin is known as the _____.

18. The skin is attached to underlying structures by the _____.

19. A group of organs that operate together to perform a specialized function is called a(n)

_____.

20. The epidermal layer that is more apparent in the palms and soles is the stratum

_____.

21. The epidermal layer that produces new cells is the stratum _____.

22. The smooth muscle attached to a hair follicle is called the _____ muscle.

23. An inherited inability to produce melanin is called _____.

24. Nerve endings sensitive to deep pressure are referred to as _____ corpuscles.

25. The inner region of a hair shaft and root is the _____.

26. The portion of a hair containing areolar connective tissue and blood vessels is the

_____.

27. Modified sweat glands that line the external auditory canal are called _____ glands.

28. The whitish semilunar area at the proximal end of the nail body is referred to as the

_____.

29. The secretory product of sudoriferous glands is called _____.

30. Melanin is synthesized in cells called _____.

31. In the control of body temperature, the effectors are blood vessels in the dermis and

_____ glands.

Bone Tissue

Structurally, the *skeletal system* consists of two types of connective tissue: cartilage and bone. The microscopic structure of cartilage has been discussed in Exercise 4. In this exercise the gross structure of a typical bone and the histology of *bone (osseous) tissue* will be studied. *Osteology* (os-tē-OL-ō-jē; *osteo* = bone; *logos* = study of) is the study of bone structure and the treatment of bone disorders.

A. FUNCTIONS OF BONE

The skeletal system has the following basic functions:

1. *Support* It provides a supporting framework for the soft tissues, maintaining the body's shape and posture, and provides points of attachment for many skeletal muscles.
2. *Protection* It protects delicate structures such as the brain, spinal cord, heart, lungs, major blood vessels in the chest, and pelvic viscera.
3. *Assistance in movement* When skeletal muscles contract, they pull on bones to help produce body movements.
4. *Mineral homeostasis* Bone tissue stores several minerals, especially calcium and phosphorus, which are important in muscle contraction and nerve activity, among other functions. On demand, bone releases minerals into the blood to maintain critical mineral balances and for distribution to other parts of the body.
5. *Blood cell production* or *hemopoiesis* (hē'-mō-poy-Ē-sis) *Red bone marrow* in certain parts of bones consists of primitive blood cells in immature stages, adipose cells, and macrophages. It is responsible for producing red blood cells, white blood cells, and platelets.
6. *Storage of triglycerides Yellow bone marrow* consists of mostly adipose cells and a few scattered blood cells. Triglycerides stored in cells of yellow bone marrow are an important source of a chemical energy reserve.

B. GROSS STRUCTURE OF A LONG BONE

Examine the external features of a fresh long bone and locate the following structures:

1. *Diaphysis* (dī-AF-i-sis; *dia* = through; *physis* = growth) Elongated shaft or body of a bone; the long cylindrical main portion of a bone.
2. *Epiphysis* (e-PIF-i-sis; *epi* = above) End or extremity of a bone; the epiphyses are referred to as proximal and distal.
3. *Metaphysis* (me-TAF-i-sis; *meta* = after or beyond) In mature bone, the region where the diaphysis joins the epiphysis; in growing bone, the region that includes the epiphyseal plate where calcified cartilage is replaced by bone as the bone lengthens.
4. *Articular cartilage* Thin layer of hyaline cartilage covering the ends of the bone where joints are formed. It reduces friction and absorbs shock at freely movable joints.
5. *Periosteum* (per'-ē-OS-tē-um; *peri* = around; *osteo* = bone) Connective tissue membrane covering the surface of the bone, except for areas covered by articular cartilage. The outer *fibrous layer* is composed of dense, irregular connective tissue and contains blood vessels, lymphatic vessels, and nerves that pass into the bone; the inner *osteogenic* (os'-te-ō-JEN-ik) *layer* contains elastic fibers, blood vessels, *osteogenic* (os'-te-ō-JEN-ik, *pro* = precursor, *gen* = to produce) *cells* (stem cells derived from mesenchyme that differentiate into osteoblasts), and *osteoblasts* (OS-tē-ō-blasts; *blast* = germ or bud), cells that secrete the organic components and mineral salts involved in bone formation. The periosteum functions in bone growth, nutrition, and repair, and as an attachment site for tendons and ligaments.
6. *Medullary* (MED-yoo-lar-ē; *medulla* = central part) or *marrow cavity* Cavity within the diaphysis that contains yellow bone marrow in adults.
7. *Endosteum* (end-OS-tē-um; *endo* = within) Membrane that lines the medullary cavity and

contains osteoprogenitor cells and *osteoclasts* (OS-tē-ō-clasts; *clast* = to break), cells that remove bone by destroying the matrix, a process called *resorption*.

Label the parts of a long bone indicated in Figure 6.1.

C. HISTOLOGY OF BONE

Bone tissue, like other connective tissues, contains a large amount of matrix that surrounds widely separated cells. Unlike other connective tissues, the matrix of bone is very hard. This hardness results from the presence of mineral salts, mainly *tricalcium phosphate* ($Ca_3(PO_4)_2 \cdot [OH]_2$), called *hydroxyapatite*, and some calcium carbonate ($CaCO_3$). Mineral salts compose about 50% of the weight of bone. Despite its hardness, bone is also flexible, a characteristic that enables it to resist various forces. The flexibility of bone comes from organic substances in its matrix, especially collagen fibers. Organic materials compose about 25% of the weight of bone. The remaining 25% of the bone matrix is water. The cells in bone tissue include osteogenic cells, osteoblasts, osteoclasts, and *osteocytes* (OS-tē-ō-sīts; *cyte* = cell), mature bone cells that maintain daily activities of bone tissue.

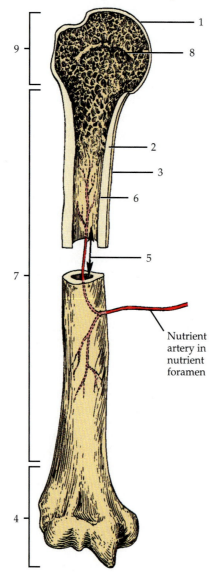

1 Articular cartilage
2 Compact bone tissue
7 Diaphysis
4 Distal epiphysis
6 Endosteum
5 Medullary (marrow) cavity
3 Periosteum
9 Proximal epiphysis
8 Spongy bone tissue

Nutrient artery in nutrient foramen

Partially sectioned long bone

FIGURE 6.1   Parts of a long bone.

Depending on the size and distribution of spaces between its hard components, bone tissue may be categorized as spongy (cancellous) or compact (dense). *Spongy (cancellous) bone tissue* composes most of the bone tissue of short, flat, and irregularly shaped bones and most of the epiphyses of long bones. The spaces within the spongy bone tissue of some bones contain red bone marrow. *Compact (dense) bone tissue,* which contains few spaces, forms the external layer of all bones of the body and the bulk of the diaphyses of long bones.

Label the spongy and compact bone tissue in Figure 6.1.

Spongy bone tissue is composed of concentric layers of hardened matrix, called *lamellae* (la-MEL-ē), that are arranged in an irregular latticework of thin columns of bone called *trabeculae* (tra-BEK-yoo-lē). See Figure 6.2. Trabeculae are the microscopic units of spongy bone tissue.

Compact bone tissue is composed of microscopic units called *osteons* (*Haversian systems*).

Obtain a prepared slide of compact bone tissue in which several osteons (Haversian systems) are shown in transverse (cross) section. Observe under high power. Look for the following structures:

1. *Central (Haversian) canal* Circular canal in the center of an osteon (Haversian system) that runs longitudinally through the bone; the canal contains blood vessels, lymphatic vessels, and nerves.
2. *Concentric lamellae* Rings of hard calcified matrix.
3. *Lacunae* (la-KOO-nē; *lacuna* = little lake) Spaces or cavities between lamellae that contain osteocytes.
4. *Canaliculi* (kan'-a-LIK-yoo-lē; *canaliculi* = small canal) Minute canals that radiate in all directions from the lacunae and interconnect with each other; contain slender processes of osteocytes; canaliculi provide routes so that nutrients can reach osteocytes and wastes can be removed from them.
5. *Osteocyte* Mature bone cell located within a lacuna.
6. *Osteon (Haversian system)* Microscopic structural unit of compact bone made up of a central (Haversian) canal plus its surrounding lamellae, lacunae, canaliculi, and osteocytes.

Label the parts of the osteon (Haversian system) shown in Figure 6.2.

Now obtain a prepared slide of a longitudinal section of compact bone tissue and examine under high power. Locate the following:

1. *Perforating (Volkmann's) canals* Canals that extend obliquely or horizontally inward from the periosteum and contain blood vessels, lymphatic vessels, and nerves; they extend into the central (Haversian) canals and medullary cavity.
2. *Endosteum*
3. *Medullary (marrow) cavity*
4. *Concentric lamellae*
5. *Lacunae*
6. *Canaliculi*
7. *Osteocytes*

Label the parts indicated in the microscopic view of bone in Figure 6.2.

D. CHEMISTRY OF BONE

PROCEDURE

1. Obtain a bone that has been baked. How does this compare to an untreated one? _____

   _____

   _____

   _____

   What substances does baking remove from the bone (inorganic or organic)? _____

   _____

   _____

2. Now obtain a bone that has already been soaked in nitric acid by your instructor. How does this bone compare to an untreated one?

   _____

   _____

   _____

   What substances does nitric acid treatment remove from the bone (inorganic or organic)?

   _____

   _____

   _____

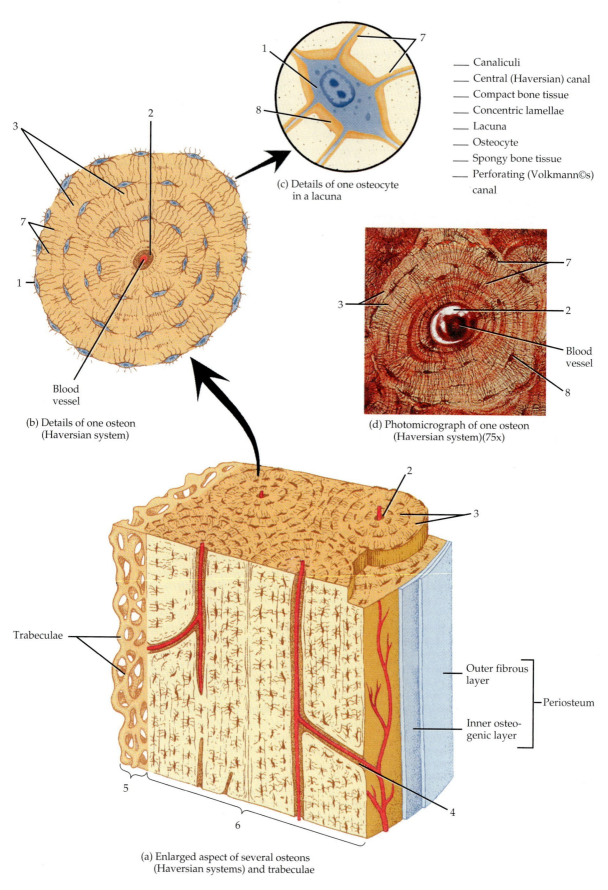

(c) Details of one osteocyte
in a lacuna

___ Canaliculi
___ Central (Haversian) canal
___ Compact bone tissue
___ Concentric lamellae
___ Lacuna
___ Osteocyte
___ Spongy bone tissue
___ Perforating (Volkmann©s)
    canal

(b) Details of one osteon
(Haversian system)

Blood
vessel

(d) Photomicrograph of one osteon
(Haversian system)(75x)

Blood
vessel

Trabeculae

Outer fibrous
layer

Inner osteo-
genic layer

Periosteum

(a) Enlarged aspect of several osteons
(Haversian systems) and trabeculae

FIGURE 6.2   Histology of bone.

E. BONE FORMATION: OSSIFICATION

The process by which bone forms is called ***ossification*** (os'-i-fi-KĀ-shun; *facere* = to make). The "skeleton" of a human embryo is composed of fibrous connective tissue membranes formed by embryonic connective tissue (mesenchyme) or hyaline cartilage that are loosely shaped like bones. They provide the supporting structures for ossification. Ossification begins around the sixth or seventh week of embryonic life and follows one of two patterns.

1. ***Intramembranous*** (in'-tra-MEM-bra-nus; *intra* = within; *membranous* = membrane) ***ossification*** refers to the formation of bone directly on or within loose fibrous connective tissue membranes. Such bones form *directly* from mesenchyme without first going through a cartilage stage. The flat bones of the skull and mandible (lower jawbone) form by this process. As you will see later, the fontanels ("soft spots") of an infant's skull, which are composed of loose fibrous connective tissue membranes, are also eventually replaced by bone through intramembranous ossification.
2. ***Endochondral*** (en'-dō-KON-dral; *endo* = within; *chondro* = cartilage) ***ossification*** refers to the formation of bone within hyaline cartilage. In this process, mesenchyme is transformed into chondroblasts which produce a hyaline cartilage matrix that is gradually replaced by bone. Most bones of the body form by this process.

These two kinds of ossification do *not* lead to differences in the gross structure of mature bones. They are simply different methods of bone formation. Both mechanisms involve the replacement of a preexisting connective tissue with bone.

The first stage in the development of bone is the migration of embryonic mesenchymal cells into the area where bone formation is about to begin. These cells increase in number and size and become osteoprogenitor cells. In some skeletal structures where capillaries are lacking, they become chondroblasts; in others where capillaries are present, they become osteoblasts. The **chondroblasts** are responsible for cartilage formation. Osteoblasts form bone tissue by intramembranous or endochondral ossification.

F. BONE GROWTH

During childhood, bones throughout the body grow in diameter by appositional growth (deposition of matrix on the surface) and long bones lengthen by interstitial growth (the addition of bone material at the epiphyseal plate). Growth in length of bones normally ceases by age 25, although bones may continue to thicken.

1. Growth in Length

To understand how a bone grows in length, you will need to know some of the details of the structure of the epiphyseal plate.

The ***epiphyseal*** (ep'-i-FIZ-ē-al; *epiphyein* = to grow upon) ***plate*** is a layer of hyaline cartilage in a growing bone that consists of four zones. The ***zone of resting cartilage*** is near the epiphysis and consists of small, scattered chondrocytes. The cells do not function in bone growth (thus the term "resting"); they anchor the epiphyseal plate to the bone of the epiphysis.

The ***zone of proliferating cartilage*** consists of slightly larger chondrocytes arranged like stacks of coins. Chondrocytes divide to replace those that die at the diaphyseal surface of the epiphyseal plate.

The zone of ***hypertrophic*** (hī-per-TRŌF-ik) or ***maturing cartilage*** consists of even larger chondrocytes that are also arranged in columns. The lengthwise expansion of the epiphyseal plate is the result of cell divisions in the zone of proliferating cartilage and maturation of the cells in the zone of hypertrophic cartilage.

The ***zone of calcified cartilage*** is only a few cells thick and consists mostly of dead cells because the matrix around them has calcified. The calcified matrix is taken up by osteoclasts, and the area is invaded by osteoblasts and capillaries from the bone in the diaphysis. These cells lay down bone on the calcified cartilage that persists. As a result, the diaphyseal border of the epiphyseal plate is firmly cemented to the bone of the diaphysis.

Label the various zones in the epiphyseal plate in Figure 6.3.

The activity of the epiphyseal plate is the only mechanism by which the diaphysis can increase in length. Unlike cartilage, which can grow by both interstitial and appositional growth, bone can grow in diameter only by appositional growth. Eventually, the epiphyseal cartilage cells stop dividing and bone replaces the cartilage. The newly formed bony structure is called the ***epiphyseal line***, a remnant of the once active epiphyseal plate. With the appearance of the epiphyseal line, bone stops growing in length. In general, lengthwise growth in bones in females is completed before that in males.

Epiphyseal side

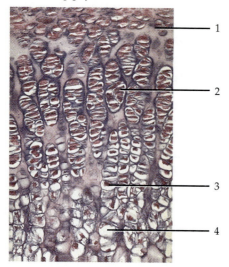

Diaphyseal side

Photomicrograph of epiphyseal plate (100x)

___ Zone of calcified cartilage
___ Zone of hypertrophic cartilage
___ Zone of proliferating cartilage
___ Zone of resting cartilage

FIGURE 6.3   Histology of the epiphyseal plate.

2. Growth in Thickness

Enlargement of bone thickness or diameter is by appositional growth and occurs as follows. First, the bone lining the medullary cavity is destroyed by osteoclasts in the endosteum so that the cavity increases in diameter. At the same time, osteoblasts from the periosteum add new bone tissue to the outer surface. Initially, diaphyseal and epiphyseal ossification produce only spongy bone. Later, the outer region of spongy bone is reorganized into compact bone.

G. FRACTURES

A *fracture* is any break in a bone. Usually, the fractured ends of a bone can be reduced (aligned to their normal positions) by manipulation without surgery. This procedure of setting a fracture is called *closed reduction.* In other cases, the fracture must be exposed by surgery before the break is rejoined. This procedure is known as *open reduction.*

Using your textbook as a reference, define the fractures listed in Table 6.1 and label the fractures indicated in Figure 6.4.

H. TYPES OF BONES

The 206 named bones of the body may be classified by shape into five principal types:

1. *Long* Have greater length than width, consist of a diaphysis and a variable number of epiphyses, and have a medullary cavity; contain more compact than spongy bone tissue, and are slightly curved for strength. Example: humerus.
2. *Short* Somewhat cube-shaped, nearly equal in length and width, and contain more spongy than compact bone tissue. Example: wrist bones.
3. *Flat* Generally thin and flat and composed of two more-or-less parallel plates of compact bone tissue enclosing a layer of spongy bone tissue. Example: sternum.
4. *Irregular* Very complex shapes; cannot be grouped into any of the three categories just described. Example: vertebrae.
5. *Sesamoid* *Sesamoid* means "resembling a sesame seed." Small bones that develop in tendons; variable in number; the only constant sesamoid bones are the paired kneecaps.

Other bones that are recognized, although not considered in the structural classification, include *Sutural* (SOO-chur-al) *bone* small bones in sutures between certain cranial bones. They are variable in number.

Examine the disarticulated skeleton, Beauchene (disarticulated) skull, and articulated skeleton and find several examples of long, short, flat, and irregular bones. List examples of each type you find.

1. *Long* _____

_____

_____

_____

2. *Short* _____

_____

_____

_____

TABLE 6.1
Summary of selected fractures.

Type of fracture	Definition
Partial	
Complete	
Closed (simple)	
Open (compound)	
Comminuted (KOM-i-nyoo'-ted)	
Greenstick	
Spiral	
Transverse	
Impacted	
Displaced	
Nondisplaced	
Stress	
Pathologic	
Pott's	
Colles' (KOL-ez)	

3. Flat _____

_____

_____

_____

4. Irregular _____

_____

_____

_____

I. BONE SURFACE MARKINGS

The surfaces of bones contain various structural features that have specific functions. These features are called **bone surface markings** and are listed in Table 6.2. Knowledge of the bone surface markings will be very useful when you learn the bones of the body in Exercise 7.

Next to each marking listed in Table 6.2, write its definition, using your textbook as a reference.

ANSWER THE LABORATORY REPORT QUESTIONS AT THE END OF THE EXERCISE.

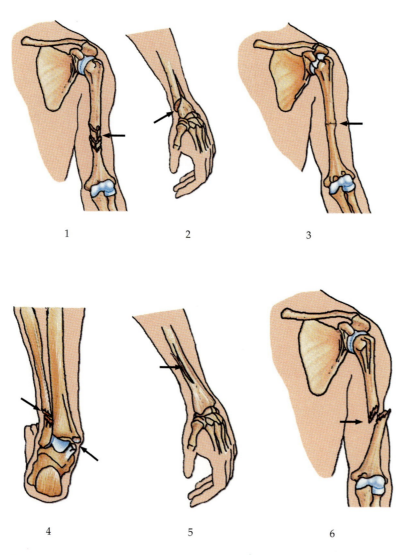

2	Colles' fracture
1	Comminuted fracture
5	Greenstick fracture
3	Impacted fracture
6	Open fracture
4	Pott's fracture

FIGURE 6.4 Types of fractures.

TABLE 6.2
Bone surface markings.

Marking	Description
DEPRESSIONS AND OPENINGS	
Fissure (FISH-ur)	
Foramen (fō-RĀ-men; *foramen* = hole)	
Meatus (mē-Ā-tus; *meatus* = canal)	
Paranasal sinus (*sin* = cavity)	
Sulcus (*sulcus* = ditchlike groove)	
Fossa (*fossa* = basinlike depression)	
PROCESSES (PROJECTIONS) THAT FORM JOINTS	
Condyle (KON-dīl; *condulus* = knucklelike process)	
Head	
Facet	
PROCESSES (PROJECTIONS) TO WHICH TENDONS, LIGAMENTS, AND OTHER CONNECTIVE TISSUES ATTACH	
Tubercle (TOO-ber-kul; *tube* = knob)	
Tuberosity	
Trochanter (trō-KAN-ter)	
Crest	
Line	
Spinous process (spine)	
Epicondyle (*epi* = above)	

Bone Tissue 6

Student _____   Date _____

Laboratory Section _____   Score/Grade _____

PART 1. Completion

1. Small clusters of bones in sutures between certain cranial bones are referred to as

   _____ bones.

2. The technical name for a mature bone cell is a(n) _osteocyte_.

3. Canals that extend obliquely inward or horizontally from the bone surface and contain blood vessels

   and lymphatic vessels are called _Volkmann's_ canals.

4. The end, or extremity, of a bone is referred to as the _epiphysis_.

5. Cube-shaped bones that contain more spongy bone tissue than compact bone tissue are known as

   short bones.

6. The cavity within the shaft of a bone that contains yellow bone marrow in the adult is the

   medullary cavity.

7. The thin layer of hyaline cartilage covering the end of a bone where joints are formed is called

   articular cartilage.

8. Minute canals that connect lacunae are called _Canaliculi_.

9. The membrane around the surface of a bone, except for the areas covered by cartilage, is the

   periosteum

10. The shaft of a bone is referred to as the _diaphysis_.

11. The _concentric lamellae_ are rings of calcified matrix.

12. The membrane that lines the medullary cavity and contains osteoblasts and a few osteoclasts is the

    endosteum.

13. The technical name for bone tissue is _osseous_ tissue.

14. In a mature bone, the region where the shaft joins the extremity is called the _joint_.

15. The microscopic structural unit of compact bone tissue is called a(n) _osteon_.

16. The hardness of bone is primarily due to the mineral salt _hydroxyapatite_.

17. The process by which bone is formed is called ___ossification___.

18. The zone of _____ is closest to the diaphysis of the bone.

19. Growth in diameter of bones occurs by ___appositional___ growth.

20. The destruction of matrix by osteoclasts is called ___resorption___.

21. Most bones of the body form by which type of ossification? ___endochondral___.

22. The lengthwise expansion of the epiphyseal plate is partly the result of cell divisions in the zone of

_____.

23. Any break in a bone is called a ___fracture___.

24. The term ___trabeculae___ refers to the irregular latticework of thin columns of spongy bone.

25. On the basis of shape, vertebrae are classified as ___irregular___ bones.

Bones

7

The 206 named bones of the adult skeleton are grouped into two divisions: axial and appendicular. The *axial skeleton* consists of bones that compose the longitudinal axis of the body. The longitudinal axis is an imaginary straight line that runs through the center of gravity of the body, through the head, and down to the space between the feet. The *appendicular skeleton* consists of the bones of the limbs or extremities (upper and lower) and the girdles (pectoral and pelvic), which connect the limbs to the axial skeleton.

In this exercise you will study the names and locations of bones and their markings by examining various regions of the adult skeleton:

Region	Number of bones
Axial skeleton	
Skull	
Cranium	8
Face	14
Hyoid (above the larynx)	1
Auditory ossicles, 3 in each ear	6
Vertebral column	26
Thorax	
Sternum	1
Ribs	24
	Subtotal = 80
Appendicular skeleton	
Pectoral (shoulder) girdles	
Clavicle	2
Scapula	2
Upper limbs (extremities)	
Humerus	2
Ulna	2
Radius	2
Carpals	16
Metacarpals	10
Phalanges	28
Pelvic (hip) girdles	
Hipbone (pelvic or coxal bone)	2

Region	Number of bones
Appendicular skeleton (Continued)	
Lower limbs (extremities)	
Femur	2
Fibula	2
Tibia	2
Patella	2
Tarsals	14
Metatarsals	10
Phalanges	28
	Subtotal = 126
	Total = 206

A. BONES OF ADULT SKULL

The *skull* is composed of two sets of bones—cranial and facial. The 8 *cranial* (*cranium* = brain case) *bones* form the cranial cavity and enclose and protect the brain. The cranial bones are 1 *frontal*, 2 *parietals* (pa-RĪ-i-tals), 2 *temporals*, 1 *occipital* (ok-SIP-i-tal), 1 *sphenoid* (SFĒ-noyd), and 1 *ethmoid*. The 14 *facial bones* form the face and include 2 *nasals*, 2 *maxillae* (mak-SIL-ē), 2 *zygomatics*, 1 *mandible*, 2 *lacrimals* (LAK-ri-mals), 2 *palatines* (PAL-a-tīns), 2 *inferior conchae* (KONG-kē), or *turbinates*, and 1 *vomer*. These bones are indicated by *arrows* in Figure 7.1. Using the illustrations in Figure 7.1 for reference, locate the cranial and facial bones on both a Beauchene (disarticulated) and an articulated skull.

Obtain a Beauchene skull and an articulated skull and observe them in anterior view. Using Figure 7.1a for reference, locate the parts indicated in the figure.

Turn the Beauchene skull and articulated skull so that you are looking at the right side. Using Figure 7.1b for reference, locate the parts indicated in the figure. Note also the **hyoid bone** below the mandible. This is not a bone of the skull; it is noted here because of its proximity to the skull.

If a skull in median section is available, use Figure 7.1c for reference and locate the parts indicated in the figure.

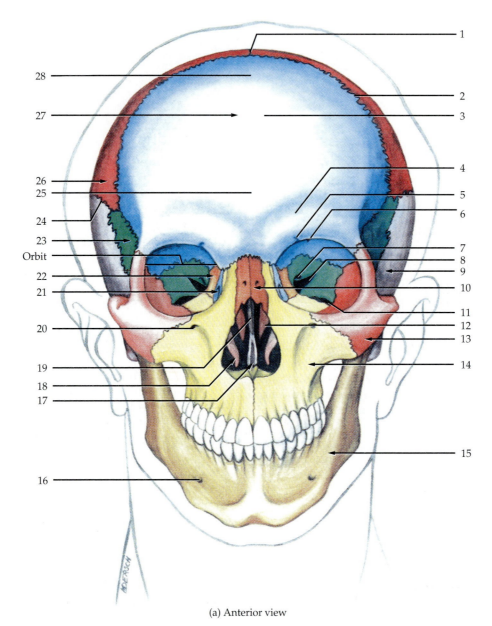

(a) Anterior view

1. Sagittal suture	11. Inferior orbital fissure	19. Perpendicular plate
2. Coronal suture	12. Middle nasal concha	20. Infraorbital foramen
3. Frontal squama	(turbinate)	21. Lacrimal bone
4. Superciliary arch	13. Zygomatic bone	22. Ethmoid bone
5. Supraorbital foramen	14. Maxilla	23. Sphenoid bone
6. Supraorbital margin	15. Mandible	24. Squamous suture
7. Optic foramen	16. Mental foramen	25. Glabella
8. Superior orbital fissure	17. Vomer	26. Parietal bone
9. Temporal bone	18. Inferior nasal concha	27. Frontal bone
10. Nasal bone	(turbinate)	28. Frontal eminence

FIGURE 7.1   Skull.

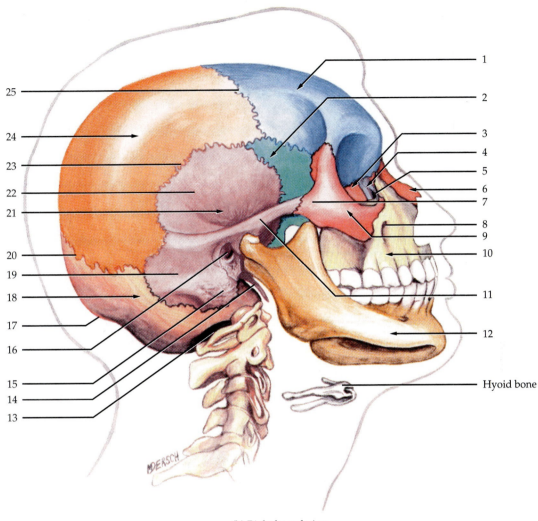

(b) Right lateral view

1. Frontal bone	11. Zygomatic process	18. Occipital bone
2. Sphenoid bone	12. Mandible	19. Mastoid portion
3. Ethmoid bone	13. Foramen magnum	20. Lambdoid suture
4. Lacrimal bone	14. Styloid process	21. Temporal bone
5. Lacrimal foramen	15. Mastoid process	22. Temporal squama
6. Nasal bone	16. External auditory	23. Squamous suture
7. Temporal process	(acoustic) meatus	24. Parietal bone
8. Infraorbital foramen	17. External occipital	25. Coronal suture
9. Zygomatic bone	protuberance	
10. Maxilla		

FIGURE 7.1 (Continued)   Skull.

Take an articulated skull and turn it upside down so that you are looking at the inferior surface. Using Figure 7.1d for reference, locate the parts indicated in the figure.

Obtain an articulated skull with a removable crown. Using Figure 7.1e for reference, locate the parts indicated in the figure.

Examine the right orbit of an articulated skull. Using Figure 7.2 for reference, locate the parts indicated in the figure.

Obtain a mandible and, using Figure 7.3 for reference, identify the parts indicated in the figure.

Before you move on, refer to Table 7.1, "Summary of Foramina of the Skull" on page 100. Complete

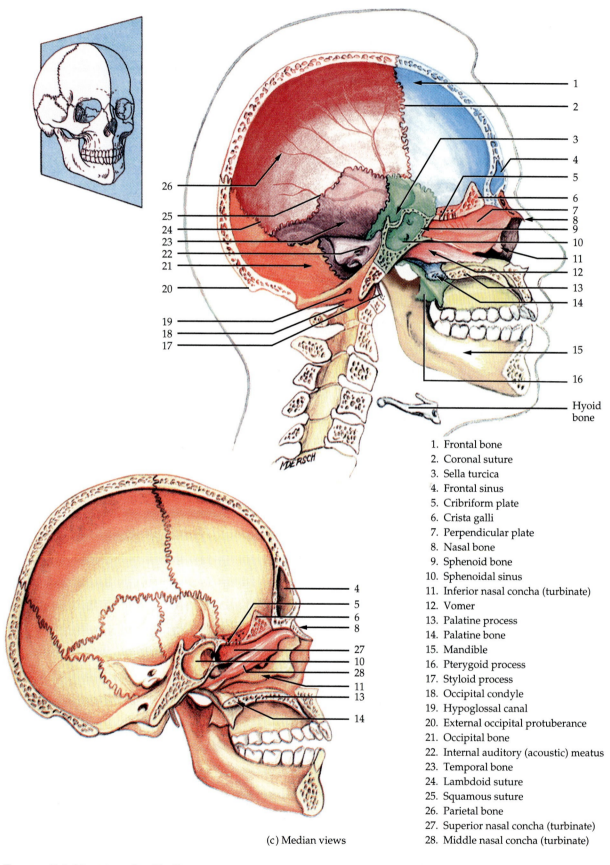

1. Frontal bone
2. Coronal suture
3. Sella turcica
4. Frontal sinus
5. Cribriform plate
6. Crista galli
7. Perpendicular plate
8. Nasal bone
9. Sphenoid bone
10. Sphenoidal sinus
11. Inferior nasal concha (turbinate)
12. Vomer
13. Palatine process
14. Palatine bone
15. Mandible
16. Pterygoid process
17. Styloid process
18. Occipital condyle
19. Hypoglossal canal
20. External occipital protuberance
21. Occipital bone
22. Internal auditory (acoustic) meatus
23. Temporal bone
24. Lambdoid suture
25. Squamous suture
26. Parietal bone
27. Superior nasal concha (turbinate)
28. Middle nasal concha (turbinate)

Hyoid bone

(c) Median views

FIGURE 7.1 *(Continued)*   Skull.

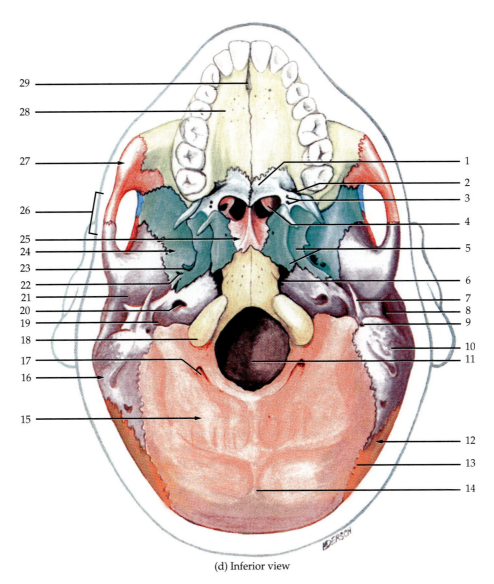

(d) Inferior view

1. Horizontal plate
2. Greater palatine foramen
3. Lesser palatine foramina
4. Middle nasal concha (turbinate)
5. Pterygoid process
6. Foramen lacerum
7. Styloid process
8. External auditory (acoustic) meatus
9. Stylomastoid foramen
10. Mastoid process
11. Foramen magnum
12. Parietal bone
13. Lambdoid suture
14. External occipital protuberance
15. Occipital bone

16. Temporal bone
17. Condylar canal
18. Occipital condyle
19. Jugular foramen
20. Carotid foramen
21. Mandibular fossa
22. Foramen spinosum
23. Foramen ovale
24. Sphenoid bone
25. Vomer
26. Zygomatic arch
27. Zygomatic bone
28. Palatine process
29. Incisive foramen

FIGURE 7.1 (*Continued*)   Skull.

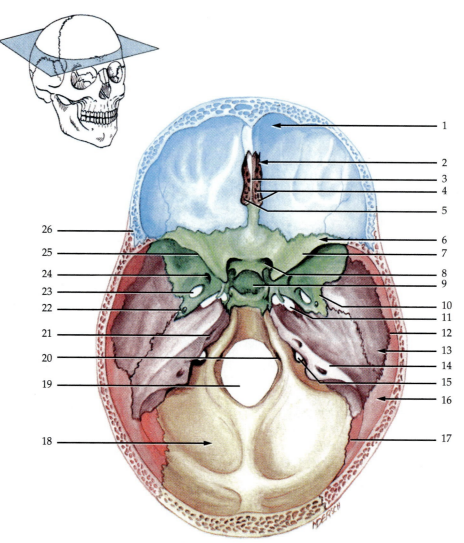

(e) Superior view of floor of cranium

1. Frontal bone
2. Ethmoid bone
3. Crista galli
4. Olfactory foramina
5. Cribriform plate
6. Sphenoid bone
7. Lesser wing
8. Optic foramen
9. Sella turcica
10. Greater wing
11. Foramen lacerum
12. Squamous suture
13. Temporal bone
14. Petrous portion
15. Jugular foramen
16. Parietal bone
17. Lambdoid suture
18. Occipital bone
19. Foramen magnum
20. Hypoglossal canal
21. Internal auditory (acoustic) meatus
22. Foramen spinosum
23. Foramen ovale
24. Foramen rotundum
25. Superior orbital fissure
26. Coronal suture

FIGURE 7.1 *(Continued)*   Skull.

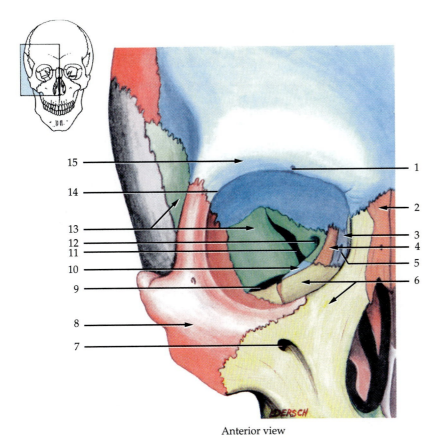

1. Supraorbital foramen
2. Nasal bone
3. Lacrimal bone
4. Ethmoid bone
5. Lacrimal foramen
6. Maxilla
7. Infraorbital foramen
8. Zygomatic bone
9. Inferior orbital fissure
10. Palatine bone
11. Superior orbital fissure
12. Optic foramen
13. Sphenoid bone
14. Supraorbital margin
15. Frontal bone

Anterior view

FIGURE 7.2 Right orbit.

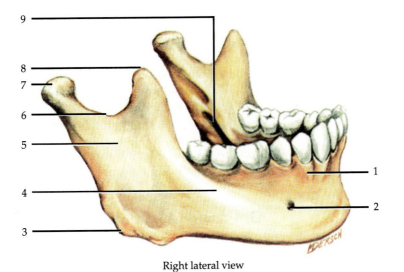

1. Alveolar process
2. Mental foramen
3. Angle
4. Body
5. Ramus
6. Mandibular notch
7. Condylar process
8. Coronoid process
9. Mandibular foramen

Right lateral view

FIGURE 7.3 Mandible.

TABLE 7.1
Summary of foramina of the skull

Foramen	Structures passing through
Carotid (relating to carotid artery in neck)	
Greater palatine (*palatum* = palate)	
Hypoglossal (*hypo* = under; *glossus* = tongue)	
Incisive (*incisive* = pertaining to incisor teeth)	
Infraorbital (*infra* = below)	
Jugular (*jugular* = pertaining to jugular vein)	
Lacerum (*lacerum* = lacerated)	
Lesser palatine (*palatum* = palate)	
Magnum (*magnum* = large)	
Mandibular (*mandere* = to chew)	
Mastoid (*mastoid* = breast-shaped)	
Mental (*mentum* = chin)	
Olfactory (*olfacere* = to smell)	
Optic (*optikas* = eye)	
Ovale (*ovale* = oval)	
Rotundum (*rotundum* = round opening)	
Spinosum (*spinosum* = resembling a spine)	
Stylomastoid (*stylo* = stake or pole)	
Supraorbital (*supra* = above)	
Zygomaticofacial (*zygoma* = cheekbone)	

the table by indicating the structures that pass through the foramina listed.

B. SUTURES OF SKULL

A *suture* (SOO-chur; *sutura* = seam) is an immovable joint found only between skull bones. Suture hold skull bones together. The four prominent sutures are the *coronal* (*corona* = crown), *sagittal* (SAJ-i-tal; *sagitta* = arrow), *lambdoid* (LAM-doyd), and *squamous* (SKWĀ-mos; *squama* = flat). Using Figures 7.1 and 7.4 for reference, locate these sutures on an articulated skull.

C. FONTANELS OF SKULL

At birth, the skull bones are separated by fibrous connective-tissue membrane-filled spaces called *fontanels* (fon'-ta-NELZ; *fontanelle* = little fountain). They (1) enable the fetal skull to compress as it passes through the birth canal, (2) permit rapid growth of the brain during infancy, (3) facilitate determination of the degree of brain development by their state of closure, (4) serve as landmarks (anterior fontanel) for withdrawal of blood from the superior sagittal sinus, and (5) aid in determining the position of the fetal head prior to birth. The principal fontanels are the *anterior (frontal), posterior (occipital), anterolateral (sphenoidal),* and *posterolateral (mastoid) fontanels.* Using Figure 7.4 for reference, locate the fontanels on the skull of a newborn infant.

D. PARANASAL SINUSES OF SKULL

A *paranasal* (*para* = beside) *sinus,* or simply *sinus,* is a cavity in a bone located near the nasal cavity. Paired paranasal sinuses are found in the *maxillae* and the *frontal, sphenoid,* and *ethmoid* bones. Locate the paranasal sinuses on the Beauchene skull or other demonstration models that may be available. Label the paranasal sinuses shown in Figure 7.5.

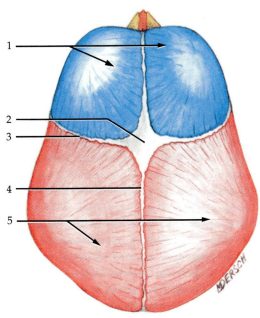

(a) Superior view

1. Frontal bones
2. Anterior (frontal) fontanel
3. Coronal suture
4. Sagittal suture
5. Parietal bones

(b) Right lateral view

1. Parietal bone
2. Anterior (frontal) fontanel
3. Coronal suture
4. Frontal bone
5. Anterolateral (sphenoid) fontanel
6. Sphenoid bone
7. Temporal bone
8. Squamous suture
9. Posterolateral (mastoid) fontanel
10. Occipital bone
11. Lambdoid suture

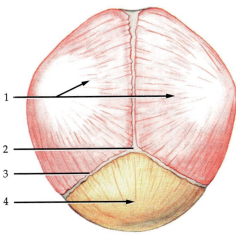

(c) Posterior view

1. Parietal bones
2. Posterior (occipital) fontanel
3. Lambdoid suture
4. Occipital bone

FIGURE 7.4   Fontanels.

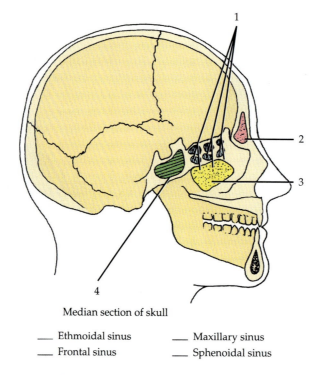

Median section of skull

___ Ethmoidal sinus   ___ Maxillary sinus
___ Frontal sinus   ___ Sphenoidal sinus

FIGURE 7.5   Paranasal sinuses.

E. VERTEBRAL COLUMN

The *vertebral column* (*backbone* or *spine*) is composed of a series of bones called *vertebrae*. The vertebrae of the adult column are distributed as follows: 7 *cervical* (SER-vi-kal) (neck), 12 *thoracic* (thō-RAS-ik) (chest), 5 *lumbar* (lower back), 5 *sacral* (fused into one bone), the *sacrum* (SĀ-krum) (between the hipbones), and usually 4 *coccygeal* (kok-SIJ-ē-al) (fused into one bone, the *coccyx* [KOK-six] forming the tail of the column). Locate each of these regions on the articulated skeleton. Label the same regions in Figure 7.6a, the anterior view.

Examine the vertebral column on the articulated skeleton and identify the cervical, thoracic, lumbar, and sacral (sacrococcygeal) curves. Label the curves in Figure 7.6b, the right lateral view.

F. VERTEBRAE

A typical *vertebra* consists of the following parts:

1. *Body* Thick, disc-shaped anterior portion.
2. *Vertebral (neural) arch* Posterior extension from the body that surrounds the spinal cord and consists of the following parts:
   a. *Pedicles* (PED-i-kuls; *pediculus* = little feet) Two short, thick processes that project poste-

riorly; each has a superior and inferior notch (*vertebral notch*) and, when successive vertebrae are fitted together, the adjoining notches form an *intervertebral foramen* through which a spinal nerve and blood vessels pass.
   b. *Laminae* (LAM-i-nē; *lamina* = thin layer) Flat portions that form the posterior wall of the vertebral arch.
   c. *Vertebral foramen* Opening through which the spinal cord passes; when all the vertebrae are fitted together, the foramina form a canal, the *vertebral canal*.
3. *Processes* Seven processes arise from the vertebral arch:
   a. Two *transverse processes* Lateral extensions where the laminae and pedicles join.
   b. One *spinous process* (*spine*) Posterior projection of the lamina.
   c. Two *superior articular processes* Articulate with the vertebra above. Their top surfaces, called *superior articular facets* (*facet* = little face), articulate with the vertebra above.
   d. Two *inferior articular processes* Articulate with the vertebra below. Their bottom surfaces, called *inferior articular facets*, articulate with the vertebra below.

Obtain a thoracic vertebra and locate each part just described. Now label the vertebra in Figure 7.7a.

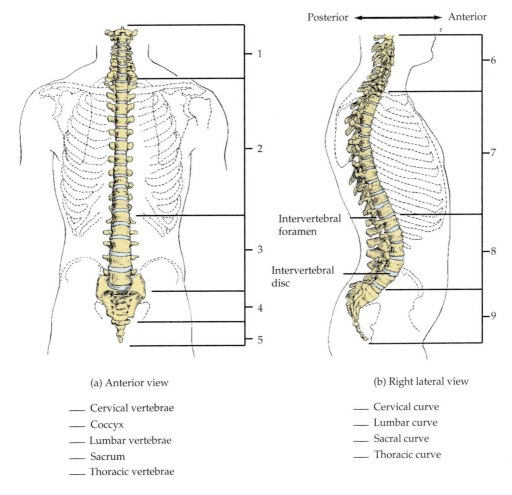

(a) Anterior view

      (b) Right lateral view

___ Cervical vertebrae
___ Coccyx
___ Lumbar vertebrae
___ Sacrum
___ Thoracic vertebrae

___ Cervical curve
___ Lumbar curve
___ Sacral curve
___ Thoracic curve

FIGURE 7.6   Vertebral column.

You should be able to distinguish the general parts on all the different vertebrae that contain them.

Although vertebrae have the same basic design, those of a given region have special distinguishing features. Obtain examples of the following vertebrae and identify their distinguishing features.

1. *Cervical vertebrae (C1–C7)*
   a. *Atlas (C1)* First cervical vertebra (Figure 7.7b)
     Transverse foramen Opening in transverse process through which an artery, a vein, and a branch of a spinal nerve pass.
     Anterior arch Anterior wall of vertebral foramen.
     Posterior arch Posterior wall of vertebral foramen.
     Lateral mass Side wall of vertebral foramen.

Label the other indicated parts.

   b. *Axis (C2)* Second cervical vertebra (Figure 7.7c)

Dens (*dens* = tooth) Superior projection of body that articulates with atlas.

Label the other indicated parts.

   c. *Cervicals 3 through 6* (Figure 7.7d)
     Bifid spinous process (C3–C6) Cleft in spinous processes of cervical vertebrae 2 through 6.

Label the other indicated parts.

   d. *Vertebra prominens (C7)* Seventh cervical vertebra; contains a nonbifid and long spinous process.
2. *Thoracic vertebrae (T1–T12)* (Figure 7.7e)
   a. *Facets and demifacets* For articulation with the tubercle and head of a rib; found on body and transverse processes.
   b. *Spinous processes* Usually long, pointed, downward projections.

Posterior

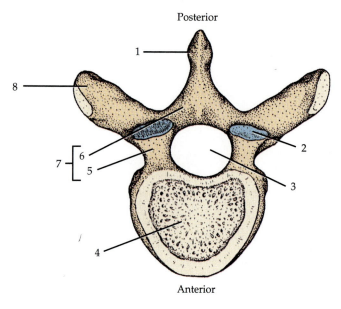

1

8

7 { 6
     5

2

3

4

Anterior

(a) Superior view of a typical vertebra

___ Body
___ Lamina
___ Pedicle
___ Spinous process
___ Superior articular facet
___ Transverse process
___ Vertebral arch
___ Vertebral foramen

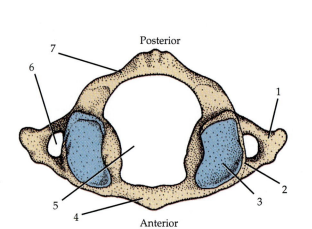

Posterior

7

6

1

5

4

2

3

Anterior

(b) Superior view of the atlas

___ Anterior arch
___ Lateral mass
___ Posterior arch
___ Superior articular facet
___ Transverse foramen
___ Transverse process
___ Vertebral foramen

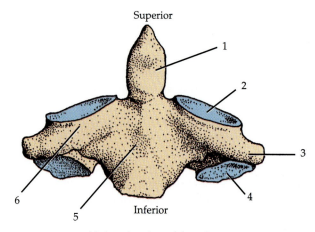

Superior

1

2

3

6

5

4

Inferior

(c) Anterior view of the axis

___ Body
___ Dens
___ Inferior articular facet
___ Lateral mass
___ Superior articular facet
___ Transverse process

FIGURE 7.7   Vertebrae.

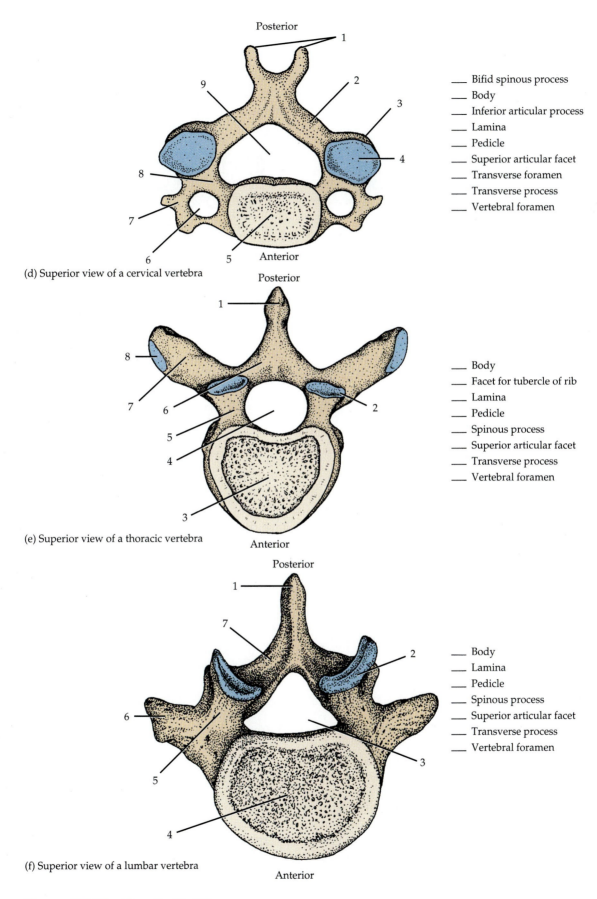

Posterior

1

9

2

3

4

8

7

6   5   Anterior

(d) Superior view of a cervical vertebra

___ Bifid spinous process
___ Body
___ Inferior articular process
___ Lamina
___ Pedicle
___ Superior articular facet
___ Transverse foramen
___ Transverse process
___ Vertebral foramen

Posterior

1

8

7

6

5

4

2

3

(e) Superior view of a thoracic vertebra

Anterior

___ Body
___ Facet for tubercle of rib
___ Lamina
___ Pedicle
___ Spinous process
___ Superior articular facet
___ Transverse process
___ Vertebral foramen

Posterior

1

7

2

6

5

3

4

(f) Superior view of a lumbar vertebra

Anterior

___ Body
___ Lamina
___ Pedicle
___ Spinous process
___ Superior articular facet
___ Transverse process
___ Vertebral foramen

FIGURE 7.7 (Continued) Vertebrae.

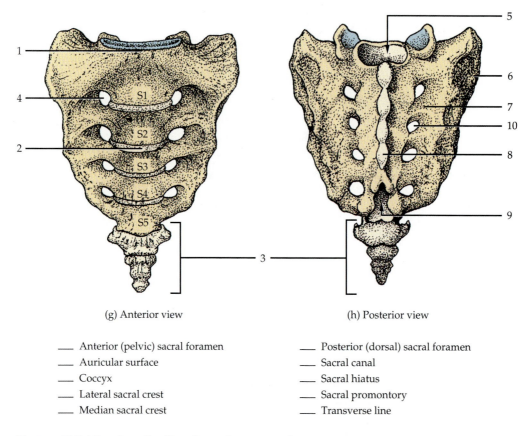

(g) Anterior view   (h) Posterior view

___ Anterior (pelvic) sacral foramen   ___ Posterior (dorsal) sacral foramen
___ Auricular surface   ___ Sacral canal
___ Coccyx   ___ Sacral hiatus
___ Lateral sacral crest   ___ Sacral promontory
___ Median sacral crest   ___ Transverse line

FIGURE 7.7 *(Continued)*   Vertebrae. Sacrum and coccyx.

Label the other indicated parts.

3. *Lumbar vertebrae (L1-L5)* (Figure 7.7f)
  a. *Spinous processes* Broad, blunt.
  b. *Superior articular processes* Directed medially, not superiorly.
  c. *Inferior articular processes* Directed laterally, not inferiorly.

Label the other indicated parts.

4. *Sacrum* (Figure 7.7g and h). Formed by the fusion of five sacral vertebrae.
  a. *Transverse lines* Areas where bodies of sacral vertebrae are joined.
  b. *Anterior (pelvic) sacral foramina* Four pairs of foramina that communicate with posterior sacral foramina; passages for blood vessels and nerves.
  c. *Median sacral crest* Fused spinous processes of upper sacral vertebrae.
  d. *Lateral sacral crest* Fused transverse processes of sacral vertebrae.

  e. *Posterior (dorsal) sacral foramina* Four pairs of foramina that communicate with anterior foramina; passages for blood vessels and nerves.
  f. *Sacral canal* Continuation of vertebral canal.
  g. *Sacral hiatus* (hi-Ā-tus) Inferior entrance to sacral canal where laminae of S5, and sometimes S4, fail to meet.
  h. *Sacral promontory* (PROM-on-tō'-rē) Superior, anterior projecting border.
  i. *Auricular surface* Articulates with ilium of hipbone to form sacroiliac joint.
5. *Coccyx* (Figure 7.7g and h) Formed by the fusion of usually four coccygeal vertebrae.

Label the coccyx and the parts of the sacrum in Figure 7.7g and h.

G. STERNUM AND RIBS

The skeleton of the *thorax* consists of the *sternum, costal cartilages, ribs,* and bodies of the *thoracic vertebrae.*

Examine the articulated skeleton and disarticulated bones and identify the following:

1. *Sternum (breastbone)* Flat bone in midline of anterior thorax.
   a. *Manubrium* (ma-NOO-brē-um; *manubrium* = handle-like) Superior portion.
   b. *Body* Middle, largest portion.
   c. *Sternal angle* Junction of the manubrium and body.
   d. *Xiphoid* (ZI-foyd; *xipho* = sword-like) *process* Inferior, smallest portion.
   e. *Suprasternal notch* Depression on the superior surface of the manubrium that may be felt.

f. *Clavicular notches* Articular surfaces lateral to suprasternal notches that articulate with the clavicles to form the sternoclavicular joints.

Label the parts of the sternum in Figure 7.8a.

2. *Ribs* The first through seventh pairs of ribs have a direct attachment to the sternum by a strip of hyaline cartilage called *costal cartilage* (*costa* = rib). Such ribs are called *true (vertebrosternal) ribs.* The remaining ribs are called *false ribs* because their costal cartilages either attach indirectly to the sternum or do not attach to the sternum at all. For example, the cartilages of the eighth, ninth, and

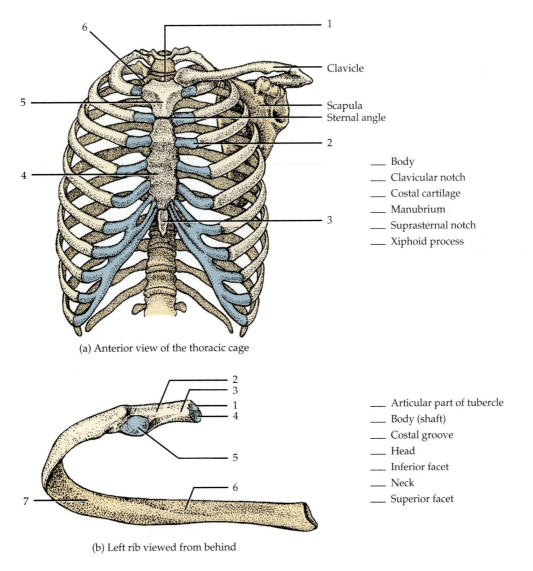

(a) Anterior view of the thoracic cage

Clavicle

Scapula
Sternal angle

___ Body
___ Clavicular notch
___ Costal cartilage
___ Manubrium
___ Suprasternal notch
___ Xiphoid process

(b) Left rib viewed from behind

___ Articular part of tubercle
___ Body (shaft)
___ Costal groove
___ Head
___ Inferior facet
___ Neck
___ Superior facet

FIGURE 7.8   Bones of the thorax.

tenth pairs of ribs attach to each other and then to the cartilages of the seventh pair of ribs. These false ribs are called *vertebrochondral ribs.* The 11th and 12th pairs of ribs are false ribs designated as *floating ribs* because their costal cartilage does not attach to the sternum at all. These ribs attach only posteriorly to the thoracic vertebrae.

Parts of a typical rib (third through ninth) include

a. *Body* Shaft, main part of rib.
b. *Head* Posterior projection.
c. *Neck* Constricted portion lateral to head.
d. *Tubercle* (TOO-ber-kul) Knoblike elevation just below neck; consists of a *nonarticular part* that affords attachment for a ligament and an *articular part* that articulates with an inferior vertebra.
e. *Costal groove* Depression on the inner surface containing blood vessels and a nerve.
f. *Superior facet* Articulates with facet on superior vertebra.
g. *Inferior facet* Articulates with facet on inferior vertebra.

Label the parts of a rib in Figure 7.8b.

H. PECTORAL (SHOULDER) GIRDLES

Each *pectoral* (PEK-tō-ral) or *shoulder girdle* consists of two bones—*clavicle* (collar bone) and *scapula* (shoulder blade). Its purpose is to attach the bones of the upper limb to the axial skeleton.

Examine the articulated skeleton and disarticulated bones and identify the following:

1. *Clavicle* (KLAV-i-kul; *clavus* = key) Slender bone with a double curvature; lies horizontally in superior and anterior part of the thorax.
   a. *Sternal extremity* Rounded, medial end that articulates with manubrium of sternum.
   b. *Acromial* (a-KRŌ-mē-al) *extremity* Broad, flat, lateral end that articulates with the acromion of the scapula to form acromioclavicular joint.
   c. *Conoid tubercle* (TOO-ber-kul; *konos* = cone) Projection on the inferior, lateral surface for attachment of a ligament.

Label the parts of the clavicle in Figure 7.9a.

2. *Scapula* (SCAP-yoo-la) Large, flat triangular bone in dorsal thorax between the levels of ribs 2 through 7.

a. *Body* Flattened, triangular portion.
b. *Spine* Ridge across posterior surface.
c. *Acromion* (a-KRŌ-mē-on; *acro* = top or summit) Flattened, expanded process of spine.
d. *Medial (vertebral) border* Edge of body near vertebral column.
e. *Lateral (axillary) border* Edge of body near arm.
f. *Inferior angle* Bottom of body where medial and lateral borders join.
g. *Glenoid cavity* Depression below acromion that articulates with head of humerus to form the shoulder joint.
h. *Coracoid* (KOR-a-koyd; *korakodes* = like a crow's beak) *process* Projection at lateral end of superior border.
i. *Supraspinous* (soo'-pra-SPĪ-nus) *fossa* Surface for muscle attachment above spine.
j. *Infraspinous fossa* Surface for muscle attachment below spine.
k. *Superior border* Superior edge of the body.
l. *Superior angle* Top of body where superior and medial borders join.

Label the parts of the scapula in Figure 7.9b, c, and d.

I. UPPER LIMBS

The skeleton of the *upper limbs* consists of a humerus in each arm, an ulna and radius in each forearm, carpals in each wrist, metacarpals in each palm, and phalanges in the fingers.

Examine the articulated skeleton and disarticulated bones and identify the following:

1. *Humerus* (HYOO-mer-us) Arm bone; longest and largest bone of the upper limb.
   a. *Head* Articulates with the glenoid cavity of scapula to form the glenohumeral (shoulder) joint.
   b. *Anatomical neck* Oblique groove below the head; the former site of the epiphyseal plate.
   c. *Greater tubercle* Lateral projection below the anatomical neck.
   d. *Lesser tubercle* Anterior projection.
   e. *Intertubercular sulcus (bicipital groove)* Between the tubercles.
   f. *Surgical neck* Constricted portion below the tubercles.
   g. *Body* Shaft.
   h. *Deltoid tuberosity* Roughened, V-shaped area about midway down the lateral surface of shaft.

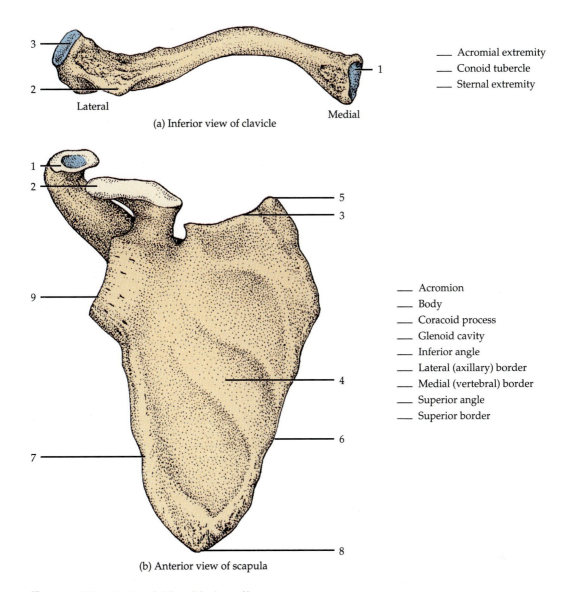

3 ——

1 ——

2 ——

Lateral

Medial

—— Acromial extremity
—— Conoid tubercle
—— Sternal extremity

(a) Inferior view of clavicle

1 ——
2 ——

5 ——
3 ——

9 ——

—— Acromion
—— Body
—— Coracoid process
—— Glenoid cavity
—— Inferior angle
—— Lateral (axillary) border
—— Medial (vertebral) border
—— Superior angle
—— Superior border

4 ——

7 ——

6 ——

8 ——

(b) Anterior view of scapula

FIGURE 7.9   Pectoral (shoulder) girdle.

i. *Capitulum* (ka-PIT-yoo-lum) Rounded knob that articulates with head of radius.

j. *Radial fossa* Anterior lateral depression that receives head of radius when forearm is flexed.

k. *Trochlea* (TRŌK-lē-a) Projection that articulates with the ulna.

l. *Coronoid* (KOR-ō-noyd; *korne* = crown-shaped) *fossa* Anterior medial depression that receives part of the ulna when the forearm is flexed.

m. *Olecranon* (ō-LEK-ra-non) *fossa* Posterior depression that receives the olecranon of the ulna when the forearm is extended.

n. *Medial epicondyle* Projection on medial side of distal end.

o. *Lateral epicondyle* Projection on lateral side of distal end.

Label the parts of the humerus in Figure 7.10a and b.

2. *Ulna* Medial bone of forearm.

a. *Olecranon (olecranon process)* Prominence of elbow at proximal end.

b. *Coronoid process* Anterior projection that, with olecranon, receives trochlea of humerus.

c. *Trochlear (semilunar) notch* Curved area between olecranon and coronoid process into which trochlea of humerus fits.

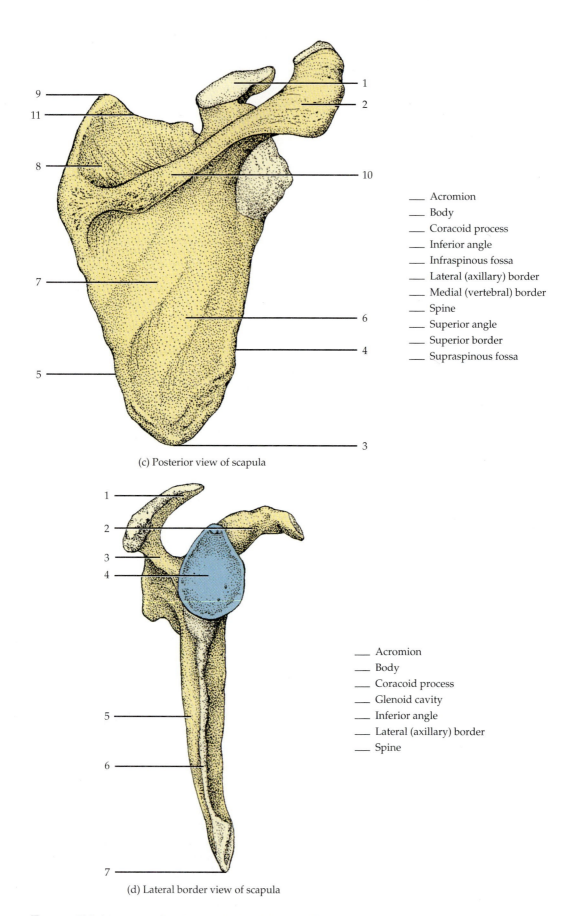

(c) Posterior view of scapula

___ Acromion
___ Body
___ Coracoid process
___ Inferior angle
___ Infraspinous fossa
___ Lateral (axillary) border
___ Medial (vertebral) border
___ Spine
___ Superior angle
___ Superior border
___ Supraspinous fossa

___ Acromion
___ Body
___ Coracoid process
___ Glenoid cavity
___ Inferior angle
___ Lateral (axillary) border
___ Spine

(d) Lateral border view of scapula

FIGURE 7.9 (Continued) Pectoral (shoulder) girdle.

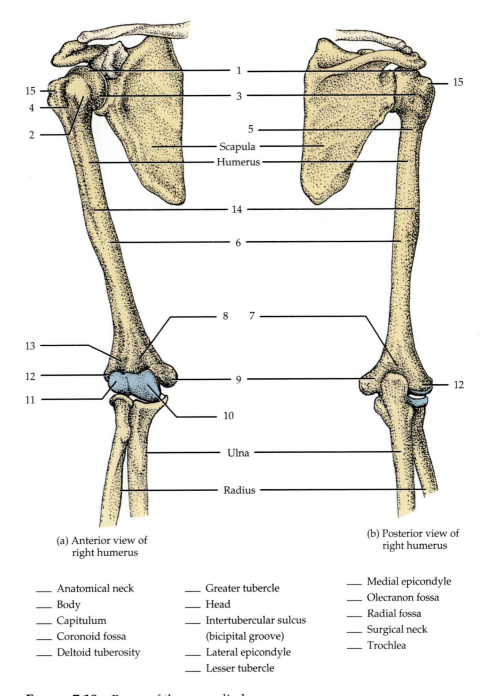

1

3

15

5

Scapula

Humerus

14

6

8   7

9

10

Ulna

Radius

15

4

2

13

12

11

15

12

(a) Anterior view of
right humerus

(b) Posterior view of
right humerus

___ Anatomical neck

___ Body

___ Capitulum

___ Coronoid fossa

___ Deltoid tuberosity

___ Greater tubercle

___ Head

___ Intertubercular sulcus
(bicipital groove)

___ Lateral epicondyle

___ Lesser tubercle

___ Medial epicondyle

___ Olecranon fossa

___ Radial fossa

___ Surgical neck

___ Trochlea

FIGURE 7.10   Bones of the upper limb.

d. **Radial notch** Depression lateral and inferior to trochlear notch that receives the head
of the radius to form the proximal radioulnar joint.

e. **Head** Rounded portion at distal end.

f. **Styloid** (*stylo* = pillar) **process** Projection
on posterior side of distal end.

Label the parts of the ulna in Figure 7.10c and d.

3. Radius Lateral bone of forearm.

a. **Head** Disc-shaped process at proximal end.

b. **Radial tuberosity** Medial projection for insertion of the biceps brachii muscle.

c. **Styloid process** Projection on lateral side of
distal end.

d. **Ulnar notch** Medial, concave depression
for articulation with head of ulna to form
distal radioulnar joint.

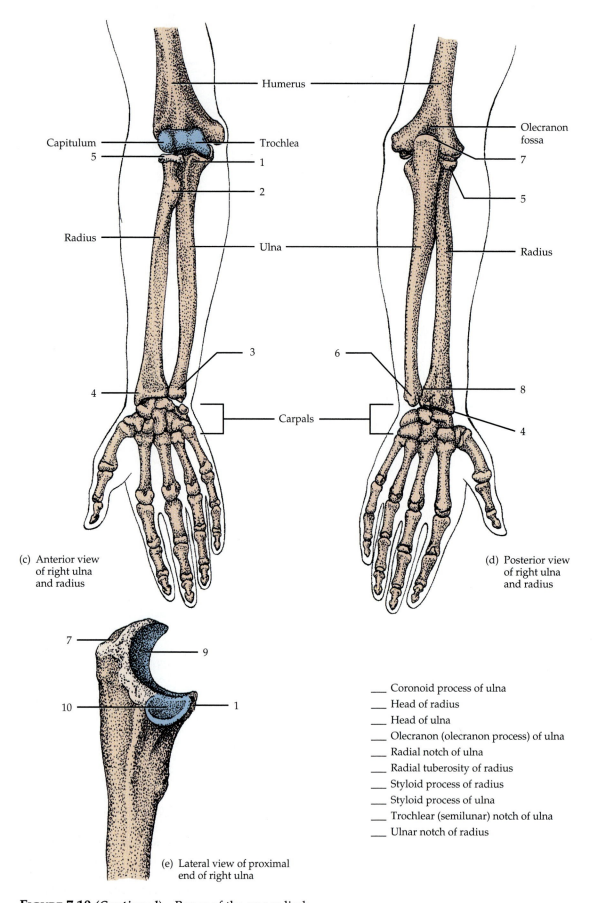

Capitulum

5

Radius

Humerus

Trochlea

1

2

Ulna

3

4

Carpals

Olecranon fossa

7

5

Radius

6

8

4

(c) Anterior view of right ulna and radius

(d) Posterior view of right ulna and radius

7

9

10

1

(e) Lateral view of proximal end of right ulna

___ Coronoid process of ulna
___ Head of radius
___ Head of ulna
___ Olecranon (olecranon process) of ulna
___ Radial notch of ulna
___ Radial tuberosity of radius
___ Styloid process of radius
___ Styloid process of ulna
___ Trochlear (semilunar) notch of ulna
___ Ulnar notch of radius

FIGURE 7.10 (Continued) Bones of the upper limb.

112

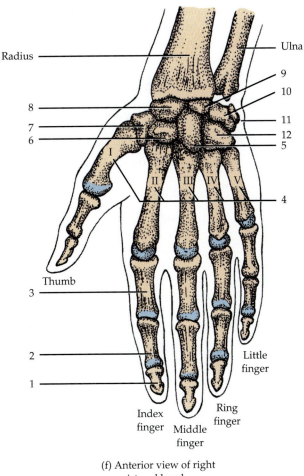

Radius

Ulna

9
10
8
7
11
6
12
5

I
II III IV V

4

Thumb
3

2

1

Little
finger

Index
finger

Middle
finger

Ring
finger

___ Capitate
___ Distal phalanx
___ Hamate
___ Lunate
___ Metacarpal
___ Middle phalanx
___ Pisiform
___ Proximal phalanx
___ Scaphoid
___ Trapezium
___ Trapezoid
___ Triquetrum

(f) Anterior view of right
wrist and hand

FIGURE 7.10 (Continued)   Bones of the upper limb.

Label the parts of the radius in Figure 7.10e.

4. **Carpus** Wrist, consists of eight small bones, called *carpals*; joints between carpal bones are called intercarpal joints.
   a. **Proximal row** From lateral to medial are called **schaphoid, lunate, triquetrum,** and **pisiform.** The distal shaft of the radius articulates with the schaphoid, lunate, and triquetrum to form the radiocarpal (wrist) joint.
   b. **Distal row** From lateral to medial are called **trapezium, trapezoid, capitate,** and **hamate.**
5. **Metacarpus** (*meta* = after or beyond) Consist of five bones of the palm called metacarpals. They are numbered as follows beginning with thumb side: I, II, III, IV, and V (or 1-5) metacarpals.
6. **Phalanges** (fa-LAN-jēz; *phalanx* = closely knit row) Bones of the fingers; two in each thumb

(proximal and distal) and three in each finger (proximal, middle, and distal). The singular of phalanges is *phalanx.* The metacarpals articulate with proximal phalanges to form the metacarpophalangeal joints. Joints between phalanges are called interphalangeal joints.

Label the parts of the carpus, metacarpus, and phalanges in Figure 7.10f.

J. PELVIC (HIP) GIRDLE

The *pelvic (hip) girdle* consists of the two hipbones or *coxal* (KOK-sal) *bones.* It provides a strong and stable support for the vertebral column and pelvic viscera and attaches the lower limbs to the axial skeleton.

Examine the articulated skeleton and disarticulated bones and identify the following parts of the hipbone:

1. *Ilium* Superior flattened portion.
   a. *Iliac crest* Superior border of ilium.
   b. *Anterior superior iliac spine* Anterior projection of iliac crest.
   c. *Anterior inferior iliac spine* Projection under anterior superior iliac spine.
   d. *Posterior superior iliac spine* Posterior projection of iliac crest.
   e. *Posterior inferior iliac spine* Projection below posterior superior iliac spine.
   f. *Greater sciatic* (sī-AT-ik) *notch* Concavity under posterior inferior iliac spine.
   g. *Iliac fossa* Medial concavity for attachment of iliacus muscle.
   h. *Iliac tuberosity* Point of attachment for sacroiliac ligament posterior to iliac fossa.
   i. *Auricular* (*auricula* = little ear) *surface* Point of articulation with sacrum (sacroiliac joint).
2. *Ischium* (IS-kē-um, meaning hip) Lower, posterior portion.
   a. *Ischial spine* Posterior projection of ischium.
   b. *Lesser sciatic notch* Concavity under ischial spine.
   c. *Ischial tuberosity* Roughened projection.
   d. *Ramus* Portion of ischium that joins the pubis and surrounds the *obturator* (OB-too-rā-ter) *foramen.*
3. *Pubis* (meaning pubic hair) Anterior, inferior portion.
   a. *Superior ramus* Upper portion of pubis.
   b. *Inferior ramus* Lower portion of pubis.
   c. *Pubic symphysis* (SIM-fi-sis) Joint between left and right hipbones.
4. *Acetabulum* (as'-e-TAB-yoo-lum) Socket that receives the head of the femur to form the hip joint; formed by the ilium, ischium, and pubis.

Label Figure 7.11.

Again, examine the articulated skeleton. This time compare the male and female pelvis. The *pelvis* consists of the two hipbones, sacrum, and coccyx. Identify the following:

1. *False (greater) pelvis* Expanded portion situated above the pelvic brim; bounded laterally by the ilia (plural of ilium) and posteriorly by the upper sacrum.
2. *True (lesser) pelvis* Below and behind the pelvic brim; constructed of parts of the ilium, pubis, sacrum, and coccyx; contains an opening above, the *pelvic inlet,* and an opening below, the *pelvic outlet.*

K. LOWER LIMBS

The bones of the *lower limbs* consist of a femur in each thigh, a patella in front of each knee joint, a tibia and fibula in each leg, tarsals in each ankle, metatarsals in each foot, and phalanges in the toes.

Examine the articulated skeleton and disarticulated bones and identify the following:

1. *Femur* Thigh bone; longest and heaviest bone in the body.
   a. *Head* Rounded projection at proximal end that articulates with acetabulum of hipbone to form the hip (coxal) joint.
   b. *Neck* Constricted portion below head.
   c. *Greater trochanter* (trō-KAN-ter) Prominence on lateral side.
   d. *Lesser trochanter* Prominence on posteromedial side.
   e. *Intertrochanteric line* Ridge on anterior surface.
   f. *Intertrochanteric crest* Ridge on posterior surface.
   g. *Linea aspera* (LIN-ē-a AS-per-a) Vertical ridge on posterior surface.
   h. *Medial condyle* Medial posterior projection on distal end that articulates with tibia to help form tibiofemoral (knee) joint.
   i. *Lateral condyle* Lateral posterior projection on distal end that articulates with tibia to help form tibiofemoral (knee) joint.
   j. *Intercondylar* (in'-ter-KON-di-lar) *fossa* Depressed area between condyles on posterior surface.
   k. *Medial epicondyle* Projection above medial condyle.
   l. *Lateral epicondyle* Projection above lateral condyle.
   m. *Patellar surface* Between condyles on the anterior surface; articulates with the posterior surface of the patella to help form the tibiofemoral (knee) joint.

Label the parts of the femur in Figure 7.12a and b.

2. *Patella* (*patera* = shallow dish) Kneecap; a sesamoid bone that develops in tendon of quadriceps femoris muscle.
   a. *Base* Broad superior portion.
   b. *Apex* Pointed inferior portion.

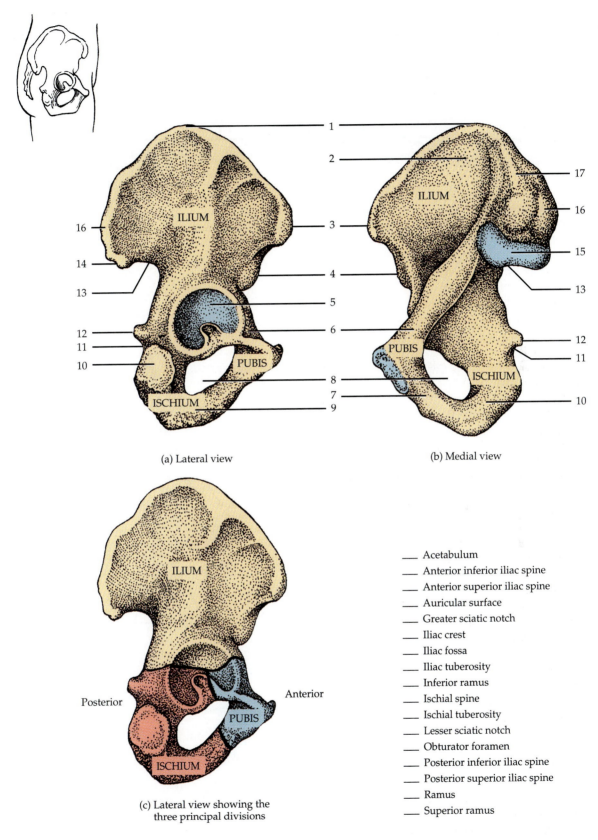

(a) Lateral view

(b) Medial view

(c) Lateral view showing the
three principal divisions

___ Acetabulum
___ Anterior inferior iliac spine
___ Anterior superior iliac spine
___ Auricular surface
___ Greater sciatic notch
___ Iliac crest
___ Iliac fossa
___ Iliac tuberosity
___ Inferior ramus
___ Ischial spine
___ Ischial tuberosity
___ Lesser sciatic notch
___ Obturator foramen
___ Posterior inferior iliac spine
___ Posterior superior iliac spine
___ Ramus
___ Superior ramus

FIGURE 7.11   Right hipbone.

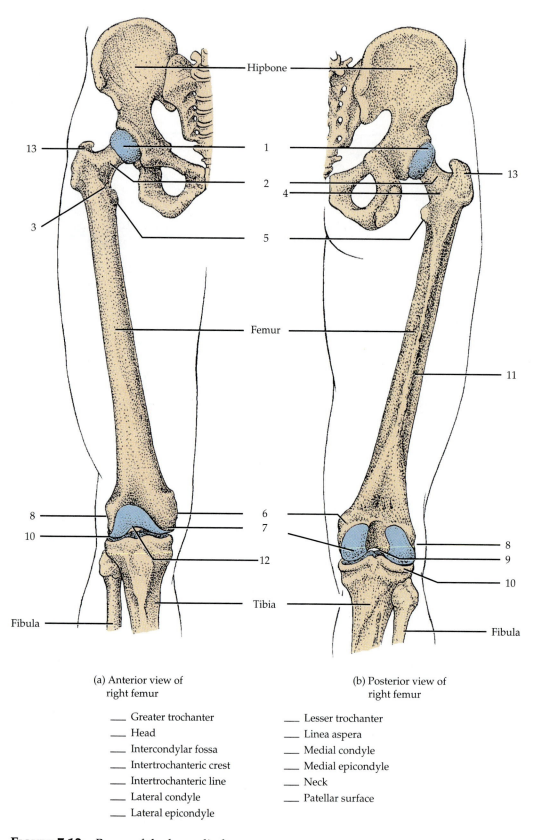

(a) Anterior view of
   right femur

(b) Posterior view of
   right femur

___ Greater trochanter   ___ Lesser trochanter
___ Head   ___ Linea aspera
___ Intercondylar fossa   ___ Medial condyle
___ Intertrochanteric crest   ___ Medial epicondyle
___ Intertrochanteric line   ___ Neck
___ Lateral condyle   ___ Patellar surface
___ Lateral epicondyle

FIGURE 7.12   Bones of the lower limb.

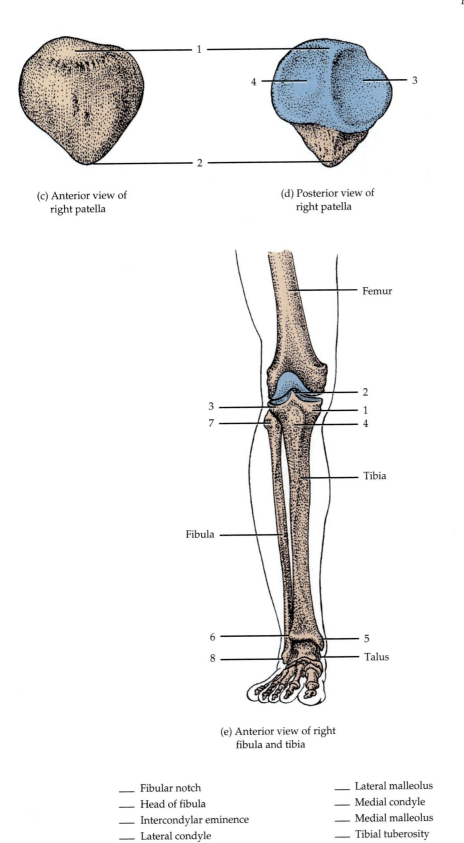

(c) Anterior view of
right patella

(d) Posterior view of
right patella

___ Apex
___ Articular facet for
lateral femoral
condyle
___ Articular facet for
medial femoral
condyle
___ Base

(e) Anterior view of right
fibula and tibia

___ Fibular notch   ___ Lateral malleolus
___ Head of fibula   ___ Medial condyle
___ Intercondylar eminence ___ Medial malleolus
___ Lateral condyle   ___ Tibial tuberosity

FIGURE 7.12 *(Continued)*   Bones of the lower limb.

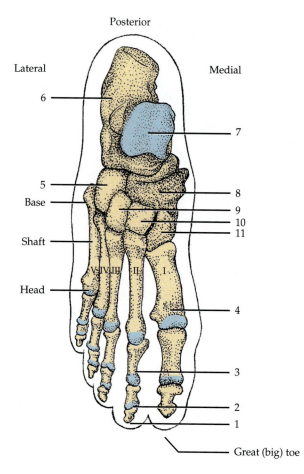

Posterior

Lateral

Medial

6

7

5

Base

8

9

10

11

Shaft

V IV III II I

Head

4

3

2

1

Great (big) toe

_____ Calcaneus
_____ Cuboid
_____ Distal phalanx
_____ First (medial) cuneiform
_____ Metatarsal
_____ Middle phalanx
_____ Navicular
_____ Proximal phalanx
_____ Second (intermediate) cuneiform
_____ Talus
_____ Third (lateral) cuneiform

(f) Superior view of right foot

FIGURE 7.12 (Continued)   Bones of the lower limb.

c. *Articular facets* Articulating surfaces on posterior surface for medial and lateral condyles of femur.

Label the parts of the patella in Figure 7.12c and d.

3. *Tibia* Shinbone; medial bone of leg.
   a. *Lateral condyle* Articulates with lateral condyle of femur to help form tibiofemoral (knee) joint.
   b. *Medial condyle* Articulates with medial condyle of femur to help form tibiofemoral (knee) joint.
   c. *Intercondylar eminence* Upward projection between condyles.
   d. *Tibial tuberosity* Anterior projection for attachment to patellar ligament.
   e. *Medial malleolus* (mal-LĒ-ō-lus; *malleus =* little hammer) Distal projection that articulates with talus bone of ankle to form part of the talocrural (ankle) joint.

   f. *Fibular notch* Distal depression that articulates with the fibula to form the distal tibiofibular joint.
4. *Fibula* Lateral bone of leg.
   a. *Head* Proximal projection that articulates with tibia to form the proximal talofibular joint.
   b. *Lateral malleolus* Projection at distal end that articulates with the talus bone of ankle.

Label the parts of the tibia and fibula in Figure 7.12e.

5. *Tarsus* Seven bones of the ankle called *tarsals.* Joints between tarsal bones are called intertarsal joints.
   a. *Posterior bones—talus* (TĀ-lus) and *calcaneus* (kal-KĀ-nē-us) (heel bone).
   b. *Anterior bones—cuboid, navicular (scaphoid),* and three *cuneiforms* called the *first (medial), second (intermediate),* and *third (lateral) cuneiforms.*

6. *Metatarsus* Consists of five bones of the foot called metatarsals, numbered as follows, beginning on the medial (great toe) side: I, II, III, IV, and V metatarsals.

7. *Phalanges* Bones of the toes, comparable to phalanges of fingers; two in each great toe (proximal and distal) and three in each small toe (proximal, middle, and distal). The metatarsals articulate with the proximal phalanges to form the metatarsophalangeal joints. Joints between phalanges are called interphalangeal joints.

Label the parts of the tarsus, metatarsus, and phalanges in Figure 7.12f.

L. ARTICULATED SKELETON

Now that you have studied all the bones of the body, label the entire articulated skeleton in Figure 7.13.

ANSWER THE LABORATORY REPORT QUESTIONS AT THE END OF THE EXERCISE.

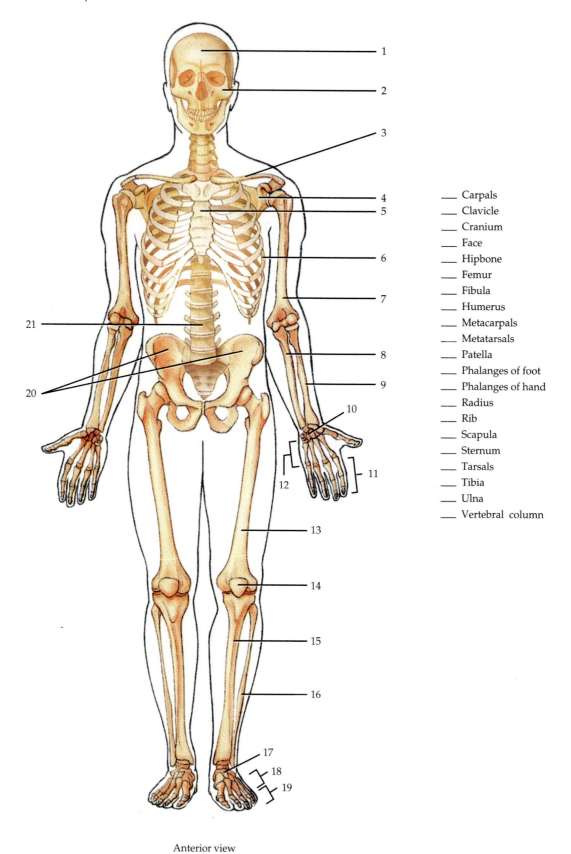

Carpals

Clavicle

Cranium

Face

Hipbone

Femur

Fibula

Humerus

Metacarpals

Metatarsals

Patella

Phalanges of foot

Phalanges of hand

Radius

Rib

Scapula

Sternum

Tarsals

Tibia

Ulna

Vertebral column

Anterior view

FIGURE 7.13   Entire skeleton.

Bones 7

Student _____   Date _____

Laboratory Section _____   Score/Grade _____

PART 1. Multiple Choice

_____ 1. The suture between the parietal and temporal bones is the (a) lambdoid (b) coronal (c) squamous (d) sagittal

_____ 2. Which bone does *not* contain a paranasal sinus? (a) ethmoid (b) maxilla (c) sphenoid (d) occipital

_____ 3. Which is the superior, concave curve in the vertebral column? (a) thoracic (b) lumbar (c) cervical (d) sacral

_____ 4. The fontanel between the parietal and occipital bones is the (a) anterolateral (b) anterior (c) posterior (d) posterolateral

_____ 5. Which is *not* a component of the upper limb? (a) radius (b) femur (c) carpus (d) humerus

_____ 6. All are components of the appendicular skeleton *except* the (a) humerus (b) occipital bone (c) calcaneus (d) triquetral

_____ 7. Which bone does *not* belong with the others? (a) occipital (b) frontal (c) parietal (d) mandible

_____ 8. Which region of the vertebral column is closer to the skull? (a) thoracic (b) lumbar (c) cervical (d) sacral

_____ 9. Of the following bones, the one that does *not* help form part of the orbit is the (a) sphenoid (b) frontal (c) occipital (d) lacrimal

_____ 10. Which bone does *not* form a border for a fontanel? (a) maxilla (b) temporal (c) occipital (d) parietal

PART 2. Identification

For each surface marking listed, identify the skull bone to which it belongs:

11. Glabella _____

12. Mastoid process _____

13. Sella turcica _____

14. Cribriform plate _____

15. Foramen magnum _____

16. Mental foramen _____

17. Infraorbital foramen _____

18. Crista galli _____

19. Foramen ovale _____

20. Horizontal plate _____

21. Optic foramen _____

22. Superior nasal concha _____

23. Zygomatic process _____

24. Styloid process _____

25. Mandibular fossa _____

PART **3.** **Matching**

_____	**26.** Iliac crest	A. Inferior portion of sternum
_____	**27.** Capitulum	B. Medial bone of distal carpals
_____	**28.** Medial malleolus	C. Distal projection of tibia
_____	**29.** Laminae	D. Lateral end of clavicle
_____	**30.** Vertebral foramen	E. Points where bodies of sacral vertebrae join
_____	**31.** Talus	F. Portion of rib that contains blood vessels
_____	**32.** Olecranon	G. Prominence of elbow
_____	**33.** Pisiform	H. Lateral projection of humerus
_____	**34.** Acromial extremity	I. Medial projection for insertion of biceps brachii muscle
_____	**35.** Pubic symphysis	J. Lower posterior portion of hipbone
_____	**36.** Hamate	K. Articulates with head of radius
_____	**37.** Costal groove	L. Prominence on lateral side of femur
_____	**38.** Xiphoid process	M. Distal projection of fibula
_____	**39.** Greater trochanter	N. Superior border of ilium
_____	**40.** Transverse lines	O. Articulates with head of humerus
_____	**41.** Radial tuberosity	P. Anterior joint between hipbones
_____	**42.** Greater tubercle	Q. Opening through which spinal cord passes
_____	**43.** Ischium	R. Form posterior wall of vertebral arch
_____	**44.** Glenoid cavity	S. Medial bone of proximal carpals
_____	**45.** Lateral malleolus	T. Component of tarsus

Articulations

8

An *articulation* (ar-tik'-yoo-LĀ-shun), or *joint*, is a point of contact between bones, cartilage and bones, or teeth and bones. When we say that one bone *articulates* with another, we mean that one bone forms a joint with another bone. The scientific study of joints is called *arthrology* (ar-THROL-ō-jē; *arthro* = joint; *logos* = study of).

Some joints permit no movement, others permit a slight degree of movement, and still others permit free movement. In this exercise you will study the structure and action of joints.

A. KINDS OF JOINTS

The joints of the body may be classified into several principal kinds on the basis of their structure and function (degree of movement they permit).

The structural classification of joints is based on the presence or absence of a space between articulating bones called a *synovial* or *joint cavity* and the type of connective tissue that binds the bones together. Structurally, a joint is classified as follows:

1. *Fibrous joint* (FĪ-brus) There is no synovial cavity and the bones are held together by fibrous (collagenous) connective tissue.
2. *Cartilaginous* (kar-ti-LAJ-i-nus) *joint* There is no synovial cavity and the bones are held together by cartilage.
3. *Synovial* (si-NŌ-vē-al) *joint* There is a synovial cavity and the bones forming the joint are united by a surrounding articular capsule and frequently by accessory ligaments (described in detail later).

The functional classification of joints is based on the degree of movement they permit and is as follows:

1. *Synarthroses* (sin'-ar-THRŌ-sēz; *syn* = together; *arthros* = joint) Immovable joints.
2. *Amphiarthroses* (am'-fē-ar-THRŌ-sēz; *amphi* = on both sides) Partially movable joints.

3. *Diarthroses* (dī-ar-THRŌ-sēz; *diarthros* = movable joint) Freely movable joints.

We will discuss the joints of the body on the basis of their structural classification, referring to their functional classification as well.

B. FIBROUS JOINTS

Allow little or no movement; do not contain synovial cavity; articulating bones held together by fibrous connective tissue.

1. *Sutures* (SOO-cherz; *sutura* = seam) A fibrous joint composed of a thin layer of dense fibrous connective tissue that unites the bones of the skull. Sutures are immovable. Example: coronal suture between the frontal and parietal bones.
2. *Syndesmoses* (sin'-dez-MŌ-sēz; *syndesmo* = band or ligament) A fibrous joint in which there is considerably more fibrous connective tissue than in a suture; the fibrous connective tissue forms an interosseous membrane or ligament that permits partial movement. Example: distal tibiofibular joint (distal articulation of tibia and fibula.
3. *Gomphoses* (gom-FŌ-sēz; *gomphosis* = to bolt together) A type of fibrous joint in which a cone-shaped peg fits into a socket. Gomphoses are immovable. Example: articulations of the roots of the teeth with the alveoli (sockets) of the maxillae and mandible in which the dense connective tissue between the tow is the periodontal ligament.

C. CARTILAGINOUS JOINTS

Allow little or no movement; do not contain synovial cavity; articulating bones held together by cartilage.

1. *Synchondroses* (sin'-kon-DRŌ-sēz; *syn* = together; *chondro* = cartilage) A cartilaginous joint in which the connecting material is hyaline cartilage. Synchondroses are immovable. Example: epiphyseal plate between the epiphysis and diaphysis of a growing bone.

2. *Symphyses* (SIM-fi-sēz; *symphysis* = growing together) A cartilaginous joint in which the connecting material is a broad, flat disc of fibrocartilage. Symphyses are partially movable. Example: intervertebral discs (between the bodies of vertebrae) and the pubic symphysis (between the anterior surfaces of the hipbones).

D. SYNOVIAL JOINTS

Synovial joints (si-NŌ-vē-al) have a variety of shapes and permit several different types of movements. First we will discuss the general structure of a synovial joint and then consider the types of synovial joints and their movements.

1. Structure of a Synovial Joint

Synovial joints have a space called a *synovial (joint) cavity* between the articulating bones (see Figure 8.1). The structure of these joints allows them to be freely movable. The bones at a synovial are covered by *articular cartilage*, which usually is hyaline cartilage, but sometimes is fibrocartilage. Although the cartilage covers each of the articulating bones with a smooth, slippery surface, it does not bind them together. Functionally, articular cartilage reduces friction at a synovial joint when the bones move and helps absorb shock.

A sleevelike *articular capsule* surrounds a synovial joint. It encloses the synovial cavity and unites the articulating bones. The outer layer of the articular capsule, the *fibrous capsule*, usually consists of dense, irregular connective tissue. It attaches to the periosteum of the articulating bones. The fibers of some fibrous capsules are arranged in parallel bundles called *ligaments* (*ligare* = to bind) and are given special names. The strength of the ligaments is one of the principal factors in holding bones together. The inner layer of the articular capsule is called the *synovial membrane* and is composed of areolar connective tissue with elastic fibers. At many synovial joints, the synovial membrane includes accumulations of adipose tissue, called *articular fat pads* (see Figure 8.4f).

The synovial membrane secretes *synovial fluid* (*syn* = together; *ovum* = egg), which forms a viscous (thick) film over the surfaces within the articular capsule. Synovial fluid consists of hyaluronic acid and interstitial fluid filtered from blood plasma. Functionally, it lubricates and reduces friction in the joint and supplies nutrients to and removes metabolic wastes given off by chondrocytes of the articular cartilage. (Recall that cartilage is an avascular tissue.) Synovial fluid also contains phagocytes that remove microbes and debris resulting from wear and tear in the joint. When there is no joint movement, the synovial fluid is quite viscous, but as joint movement increases, the fluid becomes less viscous. One of the benefits of a "warm-up" prior to exercise is that it stimulates the production and secretion of synovial fluid.

Associated with many synovial joints are *accessory ligaments*, called extracapsular ligaments and intracapsular ligaments. *Extracapsular ligaments* are located outside the articular capsule. Examples include the fibular (lateral) and tibial (medial) collateral ligaments of the knee joint (see Figure 8.4b). *Intracapsular ligaments* are located within the articular capsule but are isolated from the synovial cavity by folds of the synovial membrane. Examples are the anterior and posterior cruciate ligaments of the knee joint (see Figure 8.4b).

Within some synovial joints, like the knee, there are pads of fibrocartilage that lie between the articular surfaces of the bones and are attached to the fibrous capsule. These fibrocartilage pads are called *articular discs* or *menisci* (men-IS-ī; singular is *meniscus*). Shown in Figure 8.4b are the medial and lateral menisci in the knee joint. Articular discs alter the shape of the joint surfaces of the articulating bones, thus permitting bones of different shapes to fit more tightly. Articular discs also help to maintain the stability of the joint and direct the flow of synovial fluid to areas of greatest friction. A tearing of called *torn cartilage* and occurs often among athletes. Such damaged cartilage requires surgical removal (meniscectomy) or it will begin to wear and may precipitate arthritis.

The many and varied motions of the body create friction between moving parts. Saclike structures called *bursae* (*bursa* = pouch or purse) are strategically situated in some joints such as the shoulder and knee joints to alleviate friction (see Figure 8.4f). Bursae consist of connective tissue lined by a synovial membrane and are filled with a fluid similar to synovial fluid. Bursae are located between the skin and bone in places where skin rubs over bone, between tendons and bones, muscles and bones, ligaments and bones, and within articular capsules. Bursae cushion the movement of one part of the body over another. An acute or chronic inflammation of a bursa is called *bursitis*.

Label Figure 8.1, the principal parts of a synovial joint.

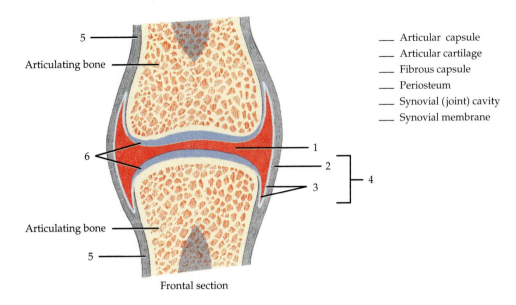

5

Articulating bone

6

Articulating bone

5

____ Articular capsule
____ Articular cartilage
____ Fibrous capsule
____ Periosteum
____ Synovial (joint) cavity
____ Synovial membrane

1
2
3
4

Frontal section

FIGURE 8.1   Parts of a synovial joint.

2. Axes of Movements at Synovial Joints

When a movement occurs at a joint, one bone remains relatively fixed while the other moves around an axis. If there is movement around a *single* axis, the joint is *monaxial* (mon-AKS-ē-al). An example is bending your lower limb at the knee and then straightening it. If there is movement around two perpendicular axes, the joint is *biaxial* (bī-AKS-ē-al). This can be demonstrated by bending the hand at the wrist and then straightening it (the first axis) and then moving the hand from side to side at the wrist (the second axis). A *multiaxial* or *triaxial* (trī-AKS-ē-al) joint, such as the shoulder joint, permits movement around *three* perpendicular axes and multiple axes in between. For example, at the shoulder joint you can swing your arm forward and backward as occurs during walking (movement around the first axis), move your arm straight out from your side and then back again (movement along the second axis), and turn your palm anteriorly and posteriorly by turning the arm (movement around the third axis).

3. Types of Synovial Joints

Based on the shape of the articulating surfaces, there are six subtypes of synovial joints: planar, hinge, pivot, condyloid, saddle, and ball-and-socket joints.

a. PLANAR JOINT

In a *planar joint* the articulating surfaces of the bones are flat or slightly curved. Examples of planar joints are the intercarpal joints (between carpal bones at the wrist) and intertarsal joints (between tarsal bones at the ankle). Planar joints permit mainly side-to-side and back-and-forth movements. Because this motion is not around an axis, planar joints are said to be *nonaxial*.

b. HINGE JOINT

In a *hinge joint* the convex surface of one bone fits into the concave surface of another bone. Examples of hinge joints are the knee (tibiofemoral), elbow, ankle (talocrural), and interphalangeal (between phalanges) joints. Hinge joints produce an angular, opening-and-closing motion like that of a hinged door and they are monoaxial because they typically allow motion around a single axis.

c. PIVOT JOINT

In a *pivot joint* the articulation is between the rounded or pointed surface of one bone and a ring formed partly by another bone and partly by a ligament. The principal movement permitted at a pivot joint is rotation around its own longitudinal axis and thus the joint is monoaxial. Examples of pivot joints are the atlanto-axial joint (between the atlas and axis) in which the atlas rotates around the axis and permits the head to turn from side to side as in signifying "no" and the radioulnar joints (between the radius and ulna) that allow us to turn the palms anteriorly and posteriorly.

d. CONDYLOID JOINT

In a *condyloid joint* (KON-di-loyd; *kondylos* = knuckle) or ellipsoidal joint, the oval-shaped projection of one bone fits into the oval-shaped joint depression of another bone. Examples are the wrist or radiocarpal joint (between radius and carpus) and metacarpophalangeal joints (between the metacarpals and phalanges of the second through fifth digits). Because the movement permitted by a condyloid joint is around two axes, it is biaxial.

e. SADDLE JOINT

In a *saddle joint,* the articular surface of one bone is saddle-shaped, and the articular surface of the other bone fits into the "saddle" as a sitting rider would. One example of a saddle joint is the carpometacarpal joint (between trapezium of the carpus and metacarpal of the thumb). Movements at a saddle joint are side to side and back and forth. Such joints are biaxial.

f. BALL-AND-SOCKET JOINT

A *ball-and-socket joint* consist of the ball-like surface of one bone that fits into a cuplike depression of another bone. Examples of ball-and-socket joints are the shoulder (glenohumeral) and hip (coxal)

joints. Such joints permit movement around three axes, and are thus triaxial.

Examine the articulated skeleton and find as many examples as you can of the joints just described. As part of your examination, be sure to note the shapes of the articular surfaces and the movements possible at each joint.

Label Figure 8.2, the types of synovial joints.

4. Movements at Synovial Joints

Specific terminology is used to indicate specific kinds of movements at synovial joints. These terms may reflect the form of motion, the direction of movement, or the relationship of one body part to another during movement. Specific movements are grouped into four main categories: (1) gliding, (2) angular movements, (3) rotation, and (4) special movements. This last category includes movements that occur only at certain joints.

a. GLIDING

Gliding is a simple movement in which relatively flat bone surfaces move back and forth and from side to side with respect to one another. There is no

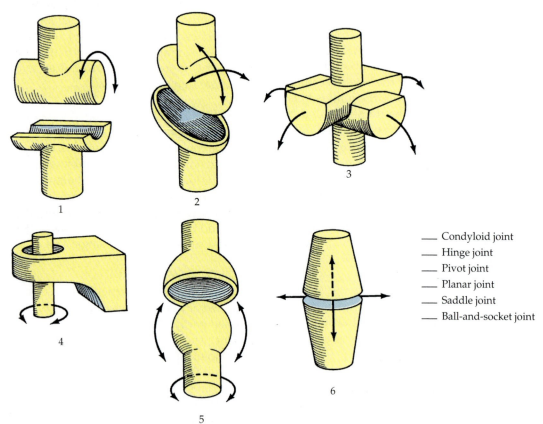

___ Condyloid joint
___ Hinge joint
___ Pivot joint
___ Planar joint
___ Saddle joint
___ Ball-and-socket joint

FIGURE 8.2 Types of synovial joints.

significant change in the angle between the bones. Gliding occurs at planar joints—the intercarpal, intertarsal, sternoclavicular, acromioclavicular, sternocostal, and vertebrocostal joints.

b. ANGULAR MOVEMENTS

In *angular movements,* there is an increase or a decrease in the angle between articulating bones. The principal angular movements are flexion, extension, lateral flexion, hyperextension, abduction, and adduction. Please remember that as these and all other movements are discussed it is assumed that the body is in the anatomical position.

Flexion (FLEK-shun; *flexus* = to bend) involves a decrease in the angle between articulating bones. *Extension* (eks-TEN-shun; *extensio* = to stretch out) involves an increase in the angle between articulating bones, frequently to restore a body part to the anatomical position after it has been flexed. Flexion and extension usually occur in the sagittal plane. Hinge, pivot, condyloid, saddle, and ball-and-socket joints all permit flexion and extension. Following are examples of flexion: (1) bending the head toward the chest at the atlanto-occipital joint (between the atlas and occipital bone) and the cervical intervertebral joints (between the cervical vertebrae) (see Figure 8.3a); (2) moving the humerus forward at the shoulder (glenohumeral) joint as in swinging the arms forward while walking; (3) moving the forearm toward the arm at the elbow joint, between the humerus, ulna, and radius; (4) moving the palm toward the forearm at the wrist (radiocarpal) joint (see Figure 8.3h); (5) bending the digits of the hand or feet at the interphalangeal joints; (6) moving the femur forward at the hip (coxal) joint, as in walking (see Figure 8.3g); and (7) moving the leg toward the thigh at the knee (tibiofemoral) joint as occurs when bending the knee.

Although flexion and extension usually occur in the sagittal plane, there are a few examples where this is not so. For instance, movement of the trunk sideways to the right or left at the waist occurs in the frontal plane. This movement involves the intervertebral joints and is called *lateral flexion* (see Figure 8.3k).

Continuation of extension beyond the anatomical position is called *hyperextension* (*hyper* = beyond). Examples of hyperextension include the following: (1) bending the head backward at the atlanto-occipital and cervical intervertebral joints (see Figure 8.3a); (2) moving the humerus backward at the shoulder (glenohumeral) joint as in swinging the arms while walking; (3) moving the palm backward

at the wrist (radiocarpal) joint (see Figure 8.3h); and (4) moving the femur backward at the hip (coxal) joint as in walking (see Figure 8.3g).

Hyperextension of other joints, such as the elbow, interphalangeal, and knee (tibiofemoral) joints, is usually prevented by the arrangement of ligaments and the anatomical alignment of the bones.

Abduction (ab-DUK-shun) refers to the movement of a bone away from the midline, whereas *adduction* (ad-DUK-shun) refers to the movement of a bone toward the midline. Both movements usually occur in the frontal plane. Condyloid, saddle, and ball-and-socket joints permit abduction and adduction. Examples of abduction include moving the humerus laterally at the shoulder (glenohumeral) joint (see Figure 8.3q) moving the palm laterally at the wrist (radiocarpal) joint (see Figure 8.3n), and moving the femur laterally at the hip (coxal) joint (see Figure 8.3c). The movement that returns each of these body parts to the anatomical position is adduction.

With regard to the fingers and toes, the midline is not the point of reference for abduction and adduction. When abducting the fingers (but not the thumb), an imaginary line drawn through the middle (longest) finger is the point of reference, and the fingers move away (spread out) from the middle finger (see Figure 8.3p). In abduction of the thumb, the thumb moves away from the palm in the sagittal plane (see Figure 8.3o). When abducting the toes, an imaginary line drawn through the second toe is the point of reference. As with abduction, adduction of the fingers and toes is relative to an imaginary line through the middle finger and second toe. Adduction of the fingers involves bringing the spread fingers together. Adduction of the thumb moves the thumb toward the palm in the sagittal plane.

Circumduction (ser-kum-DUK-shun) refers to movement of the distal end of a part of the body in a circle. It occurs as a result of a continuous sequence of flexion, abduction, extension, and adduction. Condyloid, saddle, and ball-and-socket joints allow circumduction. Because circumduction is a combined movement, it is not considered to be a separate axis of movement. Examples of circumduction are moving the humerus in a circle at the shoulder (glenohumeral) joint (see Figure 8.3g), moving the hand in a circle at the wrist joint, moving the thumb in a circle at the carpometacarpal joint, moving the fingers in a circle at the metacarpophalangeal joints, and moving the femur in a circle at the hip (coxal) joint (see Figure 8.3i). Although both the shoulder and hip joints permit circumduction, it occurs more easily in the shoulder joints because flexion, abduction, extension, and adduction are more limited in the hip joints due to the tension on certain ligaments and muscles.

c. ROTATION

In *rotation* (rō-TĀ-shun; *rotare* = to revolve) a bone revolves around its own longitudinal axis. Pivot and ball-and-socket joints permit rotation. One example is turning the head from side to side at the atlanto-axial joint, as in signifying "no" (see Figure 8.3b). Another is turning the trunk from side to side at the intervertebral joints while keeping the hips and lower limbs in the anatomical position (see Figure 8.3r). In the limbs, rotation is defined relative to the midline, and specific qualifying terms are used. If the anterior surface of a bone of the limb is turned toward the midline, the movement is called *medial rotation* (see Figure 8.3m). You can medially rotate the humerus at the shoulder joint as follows. Starting in the anatomical position, flex your elbow and then draw your palm across the chest. Medial rotation of the forearm at the radioulnar joints (between the radius and ulna) involves turning the palm medially from the anatomical position. You can medially rotate the femur at the hip joint as follows. Lie on your back, bend your knee, and then move your foot laterally from the midline. Although you are moving your foot laterally, the femur is rotating medially. Medial rotation of the leg at the knee joint can be produced by sitting on a chair, bending your knee, raising your lower limb off the floor, and turning your toes medially (see Figure 8.3t). If the anterior surface of the bone of a limb is turned away from the midline, the movement is called *lateral rotation*.

d. SPECIAL MOVEMENTS

As noted previously, *special movements* occur only at certain joints. They include elevation, depression, protraction, retraction, inversion, eversion, dorsiflexion, plantar flexion, supination, pronation, and opposition.

Elevation (el-e-VĀ-shun; *elevare* = to lift up) is an upward movement of a part of the body. Examples are closing the mouth at the temporomandibular joint (between the temporal bone and mandible) to elevate the mandible and shrugging the shoulders at the acromioclavicular joint (between the scapula and clavicle) to elevate the scapula.

Depression (dē-PRESH-un; *deprimere* = to press down) is a downward movement of a part of the body. For example, opening the mouth to depress the mandible (see Figure 8.3j), or returning shrugged shoulders to the anatomical position to depress the scapula.

Protraction (prō-TRAK-shun; *protractare* = to draw forth) is an anterior movement of a part of the body in the transverse plane. You protract your mandible at the temporomandibular joint by thrusting it outward (see Figure 8.3e) or protract your clavicles at the acromioclavicular and sternoclavicular joints by crossing your arms.

Retraction (rē-TRAK-shun; *retractare* = to draw back) is a movement of a protracted part of the body back to the anatomical position.

Inversion (in-VER-zhun; *invertere* = to turn inward) is movement of the soles medially at the intertarsal joints so that they face each other.

Eversion (ē-VER-zhun; *evertere* = to turn outward) is a movement of the soles laterally at the intertarsal joints so that they face away from each other (see Figure 8.3f).

Dorsiflexion (dor-si-FLEK-shun) refers to bending of the foot at the ankle (talocrural) joint in the direction of the dorsum (superior surface). Dorsiflexion occurs when you stand on your heels.

Plantar flexion involves bending of the foot at the ankle (talocrural) joint in the direction of the plantar surface (see Figure 8.3d), as when standing on your toes.

Supination (soo-pi-NĀ-shun) is a movement of the forearm at the proximal and distal radioulnar joints in which the palm is turned anteriorly or superiorly (see Figure 8.3m). Recall that this position of the palms is one of the defining features of the anatomical position.

Pronation (prō-NĀ-shun) is a movement of the forearm at the proximal and distal radioulnar joints in which the distal end of the radius crosses over the distal end of the ulna and the palm is turned posteriorly or inferiorly (see Figure 8.3m).

Opposition (op-ō-ZISH-un) is the movement of the thumb at the carpometacarpal joint in which the thumb moves across the palm to touch the tips of the fingers on the same hand. This is the single most distinctive digital movement that gives humans and other primates the ability to grasp and manipulate objects precisely.

Label the various movements illustrated in Figure 8.3.

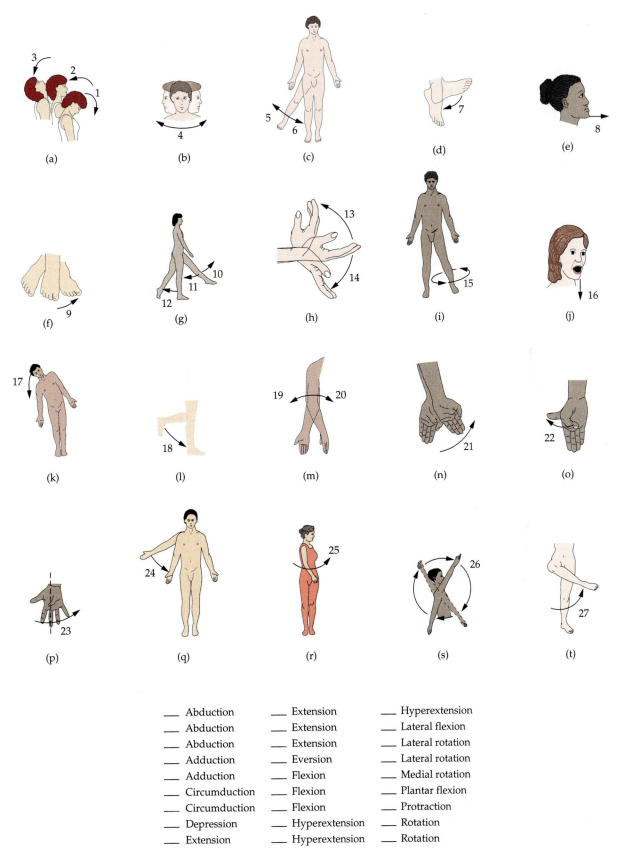

FIGURE 8.3   Movements at synovial joints.

___ Abduction	___ Extension	___ Hyperextension
___ Abduction	___ Extension	___ Lateral flexion
___ Abduction	___ Extension	___ Lateral rotation
___ Adduction	___ Eversion	___ Lateral rotation
___ Adduction	___ Flexion	___ Medial rotation
___ Circumduction	___ Flexion	___ Plantar flexion
___ Circumduction	___ Flexion	___ Protraction
___ Depression	___ Hyperextension	___ Rotation
___ Extension	___ Hyperextension	___ Rotation

E. KNEE JOINT

The knee joint is the largest and most complex joint of the body. It actually consists of three joints within a single synovial cavity: (1) an intermediate patello-femoral joint between the patella and the patellar surface of the femur, which is a planar joint; (2) a lateral tibiofemoral joint between the lateral condyle of the femur, lateral meniscus, and lateral condyle of the tibia, which is a modified hinge joint; and (3) a medial tibiofemoral joint between the medial condyle of the femur, medial meniscus, and medial condyle of the tibia, which is also a modified hinge joint.

The knee joint illustrates the basic structure of a synovial joint and the limitations on its movement. Some of the structures associated with the knee joint are as follows:

1. *Tendon of quadriceps femoris muscle* Strengthens joint anteriorly and externally.

2. *Gastrocnemius muscle* Strengthens joint posteriorly and externally.

3. *Patellar ligament* Continuation of the tendon of insertion of the quadriceps femoris muscle that extends from the patella to the tibial tuberosity. The ligament strengthens anterior portion of joint and prevents leg from being flexed too far backward.

4. *Fibular (lateral) collateral ligament* Strong, rounded ligament on the lateral aspect of the joint, between femur and fibula; strengthens the lateral side of the joint and prohibits side-to-side movement at the joint.

5. *Tibial (medial) collateral ligament* Broad, flat ligament on the medial aspect of the joint, between femur and tibia; strengthens the medial side of the joint and prohibits side-to-side movement at the joint.

6. *Oblique popliteal ligament* Broad, flat ligament that starts in a tendon that lies over the tibia and

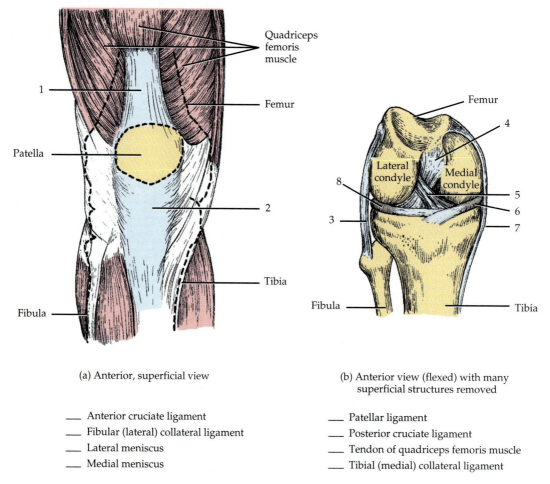

(a) Anterior, superficial view

(b) Anterior view (flexed) with many superficial structures removed

____ Anterior cruciate ligament
____ Fibular (lateral) collateral ligament
____ Lateral meniscus
____ Medial meniscus

____ Patellar ligament
____ Posterior cruciate ligament
____ Tendon of quadriceps femoris muscle
____ Tibial (medial) collateral ligament

FIGURE 8.4   Ligaments, tendons, and menisci of right knee joint.

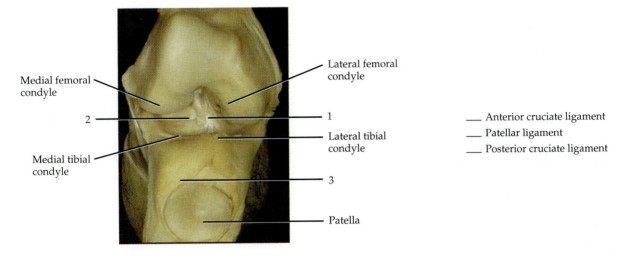

Medial femoral condyle

Lateral femoral condyle

2

1

Medial tibial condyle

Lateral tibial condyle

3

Patella

___ Anterior cruciate ligament
___ Patellar ligament
___ Posterior cruciate ligament

(c) Photograph of internal structure

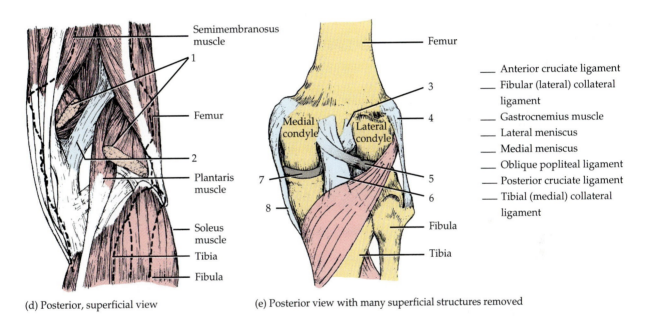

Semimembranosus muscle

Femur

1

Femur

Medial condyle

Lateral condyle

3

4

2

Plantaris muscle

7

5

8

6

Soleus muscle

Fibula

Tibia

Tibia

Fibula

___ Anterior cruciate ligament
___ Fibular (lateral) collateral ligament
___ Gastrocnemius muscle
___ Lateral meniscus
___ Medial meniscus
___ Oblique popliteal ligament
___ Posterior cruciate ligament
___ Tibial (medial) collateral ligament

(d) Posterior, superficial view

(e) Posterior view with many superficial structures removed

FIGURE 8.4 *(Continued)*   Ligaments, tendons, bursae, and menisci of right knee joint.

runs upward and laterally to the lateral side of the femur; supports the posterior surface of the knee.

7. ***Anterior cruciate ligament (ACL)*** Passes posteriorly and laterally from the tibia and attaches to the femur; strengthens the joint internally and may help stabilize the knee during its movements.

8. ***Posterior cruciate ligament (PCL)*** Passes anteriorly and medially from the tibia and attaches to the femur; strengthens the joint internally and may help stabilize the knee during its movements.

9. ***Articular discs (menisci)*** Two fibrocartilage discs between the tibial and femoral condyles

that help compensate for the irregular shapes of the bones and circulate synovial fluid.
   a. ***Medial meniscus.*** Semicircular piece of fibrocartilage (C-shaped) attached to the tibia.
   b. ***Lateral meniscus.*** Nearly circular piece of fibrocartilage (approaches an incomplete O in shape) attached to the tibia. The medial and lateral menisci are connected to each other by the *transverse ligament.*

10. The more important *bursae* of the knee include the following:
   a. ***Prepatellar bursa*** between the patella and skin.

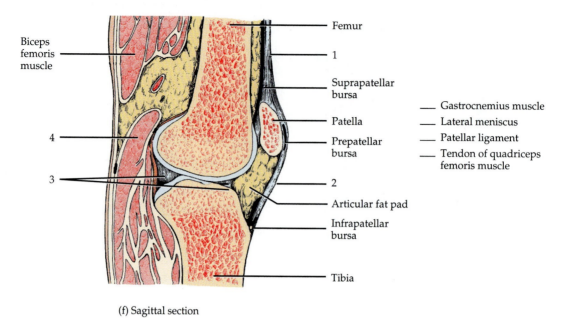

Biceps
femoris
muscle

Femur

1

Suprapatellar
bursa

Patella

Prepatellar
bursa

2

4

3

___ Gastrocnemius muscle
___ Lateral meniscus
___ Patellar ligament
___ Tendon of quadriceps
   femoris muscle

Articular fat pad

Infrapatellar
bursa

Tibia

(f) Sagittal section

FIGURE 8.4 *(Continued)*   Ligaments, tendons, bursae, and menisci of right knee joint.

b. *Infrapatellar bursa* between superior part of tibia and patellar ligament.
c. *Suprapatellar bursa* between inferior part of femur and deep surface of quadriceps femoris muscle.

Label the structures associated with the knee joint in Figure 8.4.

If a longitudinally sectioned knee joint of a cow or lamb is available, examine it and see how many structures you can identify.

ANSWER THE LABORATORY REPORT QUESTIONS AT THE END OF THE EXERCISE.

Articulations 8

Student _____   Date _____

Laboratory Section _____   Score/Grade _____

PART 1. Multiple Choice

_____ 1. A joint united by dense fibrous tissue that permits a slight degree of movement is a (a) suture (b) syndesmosis (c) symphysis (d) synchondrosis

_____ 2. A joint that contains a broad flat disc of fibrocartilage is classified as a (a) ball-and-socket joint (b) suture (c) symphysis (d) gliding joint

_____ 3. The following characteristics define what type of joint? Presence of a synovial cavity, articular cartilage, synovial membrane, and ligaments. (a) suture (b) synchondrosis (c) syndesmosis (d) hinge

_____ 4. Which joints are slightly movable? (a) diarthroses (b) amphiarthroses (c) synovial (d) synarthroses

_____ 5. Which type of joint is immovable? (a) synarthrosis (b) syndesmosis (c) symphysis (d) diarthrosis

_____ 6. What type of joint provides triaxial movement? (a) hinge (b) ball-and-socket (c) saddle (d) condyloid

_____ 7. Which ligament provides strength on the medial side of the knee joint? (a) oblique popliteal (b) posterior cruciate (c) fibular collateral (d) tibial collateral

_____ 8. On the basis of structure, which joint is fibrous? (a) symphysis (b) synchondrosis (c) pivot (d) syndesmosis

_____ 9. The elbow, knee, and interphalangeal joints are examples of which type of joint? (a) pivot (b) hinge (c) gliding (d) saddle

_____ 10. Functionally, which joint provides the greatest degree of movement? (a) diarthrosis (b) synarthrosis (c) amphiarthrosis (d) syndesmosis

PART 2. Completion

11. The thin layer of hyaline cartilage on articulating surfaces of bones is called _____ cartilage.

12. The synovial membrane and fibrous capsule together form the _____ capsule.

13. Pads of fibrocartilage between the articular surfaces of bones that maintain stability of the joint are

    called _____.

14. Fluid-filled connective tissue sacs that cushion movements of one body part over another are referred

to as _____.

15. The _____ ligament supports the back of the knee and helps to prevent hyperextension.

16. Movement of the trunk in the frontal plane is called _____.

17. In extension of the thumb, the thumb moves _____ away from the palm.

18. A joint that moves in two places is referred to as a _____ joint.

PART 3. Matching

_____ **19.** Circumduction

_____ **20.** Adduction

_____ **21.** Flexion

_____ **22.** Pronation

_____ **23.** Elevation

_____ **24.** Protraction

_____ **25.** Rotation

_____ **26.** Plantar flexion

_____ **27.** Dorsiflexion

_____ **28.** Inversion

_____ **29.** Hyperextension

A. Decrease in the angle between articulating bones usually in the sagittal plane

B. Moving a part superiorly

C. Bending the foot in the direction of the dorsum (upper surface)

D. Forward movement parallel to the ground

E. Movement of the sole inward at the ankle joint

F. Movement toward the midline in the frontal plane

G. Bending the foot in the direction of the plantar surface (sole)

H. Turning the palm posteriorly

I. Movement of a bone around its own axis

J. Distal end of a bone moves in a circle while the proximal end remains relatively stable

K. Extension beyond the anatomical position

Muscle Tissue

Although bones provide leverage and form the framework of the body, they cannot move the body by themselves. Motion results from alternating contraction and relaxation of muscles, which constitute 40 to 50% of the total body weight. The prime function of muscle is to change chemical energy (in the form of ATP) into mechanical energy to generate force, perform work, and produce movement. Muscle tissue also stabilizes body positions, regulates organ volume, and generates heat. The scientific study of muscles is known as *myology* (mī-OL-ō-jē; *myo* = muscle; *logos* = study of).

Muscle tissue has five principal characteristics that enable it to carry out its functions:

1. *Excitability (irritability),* a property of both muscle cells and nerve cells (neurons), is the ability to respond to certain stimuli by producing electrical signals called *action potentials* (impulses). For muscle, the stimuli that trigger action potentials are chemicals—neurotransmitters, released by neurons, or hormones distributed by the blood.
2. *Conductivity* is the ability of a cell, especially a muscle cell or neuron, to propagate or conduct action potentials along the plasma membrane.
3. *Contractility* is the capacity of the contractile elements in muscle tissue to shorten and develop tension. Sometimes the contraction is visible and involves shortening and thickening of the muscle while the tension remains the same. At other times, there is no visible shortening and thickening, but the tension increases greatly.
4. *Extensibility* means that muscle can be extended (stretched) without damaging the tissue. Most skeletal muscles are arranged in opposing pairs. While one is contracting, the other not only is relaxed but usually is being stretched.
5. *Elasticity* means that muscle tissue tends to return to its original shape after contraction or extension.

In this exercise you will examine the histological structure of muscle tissue and conduct exercises on the physiology of muscle.

A. TYPES OF MUSCLE TISSUE

The body has three kinds of muscle tissue—skeletal, cardiac, and smooth—that differ from one another in their location, microscopic anatomy, and control by the nervous and endocrine systems.

1. *Skeletal muscle tissue* is attached primarily to bones, and it moves parts of the skeleton. (Some skeletal muscles are also attached to skin, other muscles, or connective tissue.) Skeletal muscle tissue is said to be *striated* because alternating light and dark bands *(striations)* are visible when the tissue is examined under a microscope (see Figure 9.1). Skeletal muscle tissue works in a *voluntary* manner because it can be made to contract and relax by conscious control.
2. *Cardiac muscle tissue* forms most of the heart. It is also *striated* muscle, but its action is *involuntary;* that is, its contraction is usually not under conscious control.
3. *Smooth muscle tissue* is located in the walls of hollow internal structures, such as blood vessels, the stomach, and the intestines, as well as most other abdominal organs. It is also found in the skin attached to hair follicles. Under a microscope, this tissue looks *nonstriated* or *smooth.* The action of smooth muscle is usually *involuntary.*

B. STRUCTURE OF SKELETAL MUSCLE TISSUE

A skeletal muscle consists of hundreds or thousands of long, cylindrical cells called *muscle fibers* or *myofibers* that lie parallel to one another (see Figure 9.1). The typical length of a muscle fiber is 100 μm, but some are as long as 30 cm (12 in.); the diameter

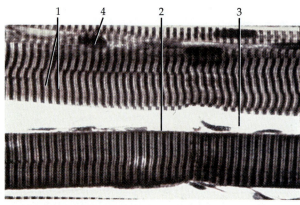

Photomicrograph of several muscle fibers
in longitudinal section (900x)

___ Endomysium
___ Nucleus
___ Sarcolemma
___ Striations

FIGURE 9.1   Histology of skeletal muscle tissue.

ranges from 10–100 μm. There are several nuclei in each muscle fiber, and they are located toward the edge of the cell. Within the cytoplasm are thread-like structures called *myofibrils* that extend length-wise within the fiber. Myofibrils are composed of even smaller structures called *filaments* that are arranged in compartments called sarcomeres, the functional units of striated muscle tissue.

Examine a prepared slide of skeletal muscle tissue in longitudinal and transverse (cross) section under high power. Look for the following:

1. *Sarcolemma* (*sarco* = flesh; *lemma* = sheath) Plasma membrane of the muscle fiber.
2. *Sarcoplasm* Cytoplasm of the muscle fiber.
3. *Nuclei* Several in each muscle fiber lying close to the sarcolemma.
4. *Striations* Alternating light and dark bands in each muscle fiber (described on page 136).
5. *Epimysium* (ep'-i-MĪZ-ē-um; *epi* = upon) Fibrous connective tissue that surrounds the entire skeletal muscle.
6. *Perimysium* (per'-i-MĪZ-ē-um; *peri* = around) Fibrous connective tissue surrounding a bundle (fascicle) of muscle fibers.
7. *Endomysium* (en'-dō-MĪZ-ē-um; *endo* = within) Fibrous connective tissue surrounding individual muscle fibers.

Refer to Figure 9.1 and label the structures indicated.

With the use of an electron microscope, additional details of skeletal muscle tissue may be noted. Among these are the following:

1. *Mitochondria* Organelles that have a smooth outer membrane and folded inner membrane in which ATP is generated.

2. *Sarcoplasmic reticulum* (sar'-kō-PLAZ-mik re-TIK-yoo-lum), or *SR* Network of fluid-filled cisterns similar to the endoplasmic reticulum of nonmuscle cells; stores calcium ions in relaxed muscle fibers. The dilated end sacs of sarco-plasmic reticulum on both sides of a tubule are called **terminal cisterns.**
3. *Transverse (T) tubules* Deep tunnel-like infold-ings of sarcolemma that run perpendicular to and connect with the sarcoplasmic reticulum; open to the outside of the muscle fiber.
4. *Triad* Transverse tubule and the terminal cis-terns on either side of it.
5. *Myofibrils* Threadlike structures that run lengthwise through a fiber and consist of *thin filaments* composed of the protein actin and *thick filaments* composed of the protein my-osin; the contractile elements of muscle tissue.
6. *Sarcomere* (*meros* = part) Contractile unit of a muscle fiber; compartment within a muscle fiber separated from other sarcomeres by dense material called *Z discs (lines).*
7. *A (anisotropic) band* Dark region in a sarcomere that consists mostly of thick filaments and por-tions of thin filaments where they overlap the thick filaments.
8. *I (isotropic) band* Light region in a sarcomere composed of the rest of the thin filaments but no thick filaments. The combination of alter-nating dark A bands and light I bands gives the muscle fiber its striated (striped) appear-ance.
9. *H zone* Region in the center of the A band of a sarcomere consisting of thick filaments only.
10. *M line* Series of fine threads in the center of the H zone formed by proteins that connect adja-cent thick filaments.

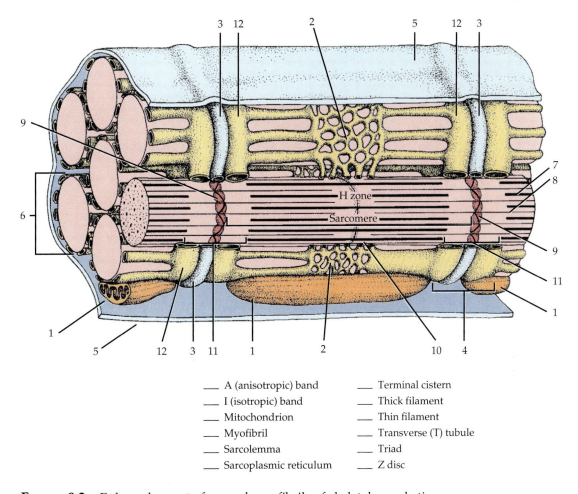

FIGURE 9.2   Enlarged aspect of several myofibrils of skeletal muscle tissue based on an electron micrograph.

___ A (anisotropic) band   ___ Terminal cistern
___ I (isotropic) band   ___ Thick filament
___ Mitochondrion   ___ Thin filament
___ Myofibril   ___ Transverse (T) tubule
___ Sarcolemma   ___ Triad
___ Sarcoplasmic reticulum   ___ Z disc

Figure 9.2 is a diagram of skeletal muscle tissue based on electron micrographic studies. Label the structures shown.

C. CONTRACTION OF SKELETAL MUSCLE TISSUE

Skeletal muscle contraction is a process of electrical, chemical, and mechanical events. *Excitation-contraction coupling* is the linkage between the electrochemical events and the actual mechanical event we call a skeletal muscle contraction. Muscle contraction occurs only if the muscle is stimulated with a stimulus of threshold or suprathreshold intensity. A *threshold stimulus* is defined as the minimum strength stimulus necessary to initiate a contraction. Under normal physiological conditions, this is accomplished by a nerve impulse's being transmitted to the skeletal muscle fiber via a nerve cell called a *motor neuron.* A motor neuron and all of the muscle fibers it innervates is termed a *motor unit.*

Each motor neuron possesses a single long process called an *axon,* extending from the nerve cell body to the muscle fiber. As the axon enters the endomysium, it branches into fine processes termed *axon terminals.* Each axon terminal enlarges to form the bulblike structures termed *synaptic end bulbs.* The structure formed by the invagination of the skeletal muscle sarcolemma and these end bulbs is termed a *neuromuscular (NMJ) or myoneural junction* (Figure 9.3). The portion of the skeletal muscle fiber immediately deep to the axon terminal demonstrates a large number of anatomical specializations and is termed the *motor end plate.*

Within each synaptic end bulb are numerous membrane-closed sacs, called *synaptic vesicles.* Each vesicle contains a store of chemicals termed *neurotransmitter substance.* The space between the axon terminal and sarcolemma is termed the *synaptic*

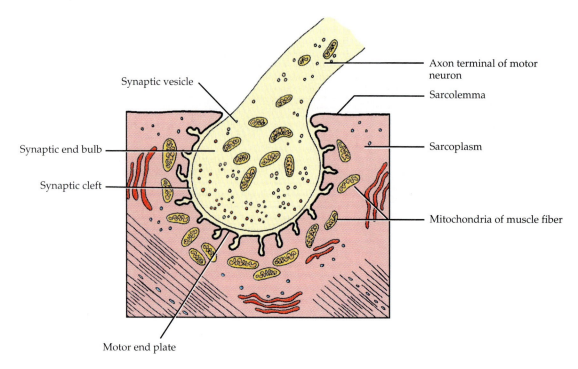

Synaptic vesicle

Synaptic end bulb

Synaptic cleft

Axon terminal of motor neuron

Sarcolemma

Sarcoplasm

Mitochondria of muscle fiber

Motor end plate

FIGURE 9.3   Neuromuscular junction.

cleft. The neurotransmitter substance diffuses across the synaptic cleft and bind to receptors on the motor end plate, initiating the excitation-contraction coupling process.

At the neuromuscular junction, a wave of depolarization (nerve impulse) spreads along the axon opening calcium specific channels. Extracellular calcium ions (Ca^{2+}) enter through these channels and bind to the synaptic vesicles. This binding effect leads to the fusion of the vesicles with the synaptic end bulb membrane, releasing the neurotransmitter substance called *acetylcholine* (as'-ē-til-KŌ-lēn), or *Ach.* Acetylcholine diffuses across the synaptic cleft and binds to receptors on the motor end plate. The binding of Ach with the receptors, open ligand (chemical) gated channels on the sarcolemma, allowing an influx of sodium into the muscle, thus initiating depolarization of the muscle cell membrane or an *action potential.* As the action potential spreads along the sarcolemma it passes into the transverse (T) tubules. The action potential then travels into the sarcoplasmic reticulum (SR) causing the release of intracellular calcium ions into the sarcoplasm. Calcium can now bind to *troponin,* on the *troponin-tropomyosin complex.* This binding effect leads to the reconfiguration of the troponin-tropomyosin complex thereby exposing the *myosin binding sites* on the

thin myofilaments, allowing the myosin cross bridges to attach to the thin myofilament. The attachment between the myosin cross bridges and the receptors on the actin causes ATP to be split by myosin ATPase, an enzyme found on the thick filament. The energy released by the splitting of ATP into ADP and inorganic phosphate causes the myosin cross bridges to move, and the thin filaments slide inward toward the H zone, causing the muscle fiber to shorten (Figure 9.4). As this is occurring, newly synthesized ATP displaces ADP from the myosin molecule. If Ca^{2+} has been taken back into the sarcoplasmic reticulum by the calcium pump, the myosin cross bridges release from the actin receptors and relaxation occurs. However, if Ca^{2+} is still present within the sarcoplasm, the entire process repeats and further contraction occurs. This theory of skeletal muscle contraction is termed the *sliding filament theory of skeletal muscle contraction.*

A contracting skeletal muscle fiber follows the *all-or-none principle.* This principle states that a skeletal muscle fiber will respond maximally or not at all to a stimulus. If the stimulus is of threshold or greater intensity the fiber will respond maximally; if the stimulus is subthreshold in intensity the fiber will not respond. This principle, however, does not imply that the entire skeletal muscle must be either

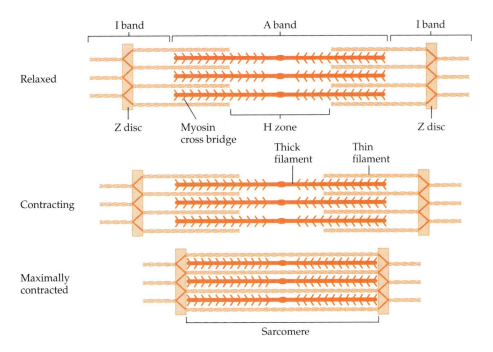

FIGURE 9.4 Sliding filament theory of a skeletal muscle contraction. The position is of the various parts of two sarcomeres in relaxed, contracting, and maximally contracted states are shown. Note the movement of thin filaments and relative size of H zone.

fully relaxed or fully contracted, because individual fibers within a muscle possess varying thresholds for stimulation. Therefore, the muscle as a whole can have graded contractions.

With these facts in mind, you will now perform a few tests that will illustrate the physiology of skeletal muscle contraction.

D. LABORATORY TESTS ON SKELETAL MUSCLE CONTRACTION

Skeletal muscles produce different kinds of contractions depending on the intensity and frequency of the stimulus applied. Twitch contractions do not occur in the body but are worth demonstrating because they show the different phases of muscle contraction quite clearly. Any record of a muscle contraction is called a *myogram.*

a. PROCEDURE USING PHYSIOFINDER
PHYSIOFINDER is an interactive computer program that permits laboratory simulations which do not involve the use of animals or advanced or expensive laboratory equipment. It permits students to perform experiments, analyze data, draw conclusions, and form hypotheses on the basis of collected data. The program is available from Addison-Wesley Longman, Inc., Customer Service (1-800-922-0579).

For activities related to skeletal muscle contraction, select the appropriate experiments from PHYSIOFINDER.

Module 1—The Action Potential
Module 2—Synaptic Integration
Module 3—The Neuromuscular Junction

b. PROCEDURE USING PHYSIOGRIP™
Physiogrip™ is a computer hardware and software system designed to demonstrate the contractile characteristics of human skeletal muscle. Through the use of a specially designed displacement transducer, Physiogrip™ provides an alternative to vivisection for studying classic muscle physiology by allowing students to experience their own flexor digitorum superficialis muscle responding to motor point stimulation. The program is available from INTELITOOL® (1-800-227-3805).

E. BIOCHEMISTRY OF SKELETAL MUSCLE CONTRACTION

You will examine the effect of the following solutions on the contraction of glycerinated skeletal muscle fibers:[1] (1) ATP solution, (2) mineral ion solution, and (3) ATP plus mineral ion solution.

PROCEDURE

1. Using 7× or 10× magnification and glass needles or clean stainless steel forceps, gently tease the muscle into very thin groups of myofibers. Single fibers or thin groups must be used because strands thicker than a silk thread curl when they contract.

2. Using a Pasteur pipette or medicine dropper, transfer one strand into a drop of glycerol on a clean glass slide and cover the preparation with a cover slip. Examine the strand under low and high power and note the striations. Also note that each fiber has several nuclei.

3. Transfer one of the thinnest strands to a drop of glycerol on a second microscope slide. Do not add a cover slip. If the amount of glycerol on the slide is more than a small drop, soak the excess into a piece of lens paper held at the edge of the glycerol farthest from the fibers. Using a dissecting microscope and a millimeter ruler held beneath the slide, measure the length of one of the fibers.

4. Now flood the fibers with the solution containing only ATP and observe their reaction. After 30 sec or more, remeasure the same fiber.

   How much did the fiber contract? _____ mm

5. Using clean slides and medicine droppers, and being especially sure to use clean teasing needles or forceps, transfer other fibers to a drop of glycerol on a slide. Again measure the length of one fiber. Next flood the fibers with a solution containing mineral ions, observe their reaction, and remeasure the fiber.

   How much did the fiber contract? _____ mm

6. Repeat the exercise, this time using a solution containing a combination of ATP and ions.

7. Observe a contracted fiber under low and high power, and look for differences in appearance between muscle in a contracted state and muscle in a relaxed state (see Figure 9.4).

[1]Glycerinated muscle preparation and solutions are supplied by the Carolina Biological Supply Company, Burlington, North Carolina 27215.

F. ELECTROMYOGRAPHY

During the contraction of a single muscle fiber, an action potential is generated that lasts between 1 and 4 msec. This electrical activity is dissipated throughout the surrounding tissues. In order to produce a smooth muscle contraction, motor units fire asynchronously, thereby causing muscle fibers to contract at different times. This asynchronous firing of motor units also prolongs the electrical activity resulting from skeletal muscle contraction. By placing two electrodes on the skin or directly within the muscle an electrical recording, termed an *electromyograph*, or *EMG*, of this muscular activity may be obtained when the muscle is stimulated (Figure 9.5). EMGs are utilized clinically to distinguish between peripheral neurological and muscular diseases, and for differentiating abnormalities resulting in reductions of either muscular strength or sensation that may have been caused by either dysfunction or peripheral nervous tissue or central nervous system (CNS) centers. Typically EMGs are recorded under three different activity levels: complete inactivity, slight muscular activity, and extreme (maximal) muscular activity. Abnormal differences in records obtained under these three conditions may also indicate an abnormality in motor unit recruitment.

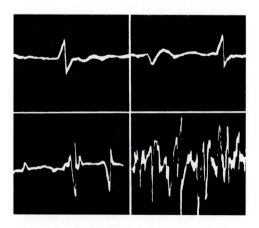

FIGURE 9.5 Diagrammatic representation of normal electromyograms. The single potential in the upper left corner has a measured amplitude of 0.8 mV and a duration of 7 msec.

1. Subject Preparation

PROCEDURE

1. Locate the *flexor digitorum superficialis* muscle (see Figure 10.12a) on the anterior surface of the forearm and the *triceps brachii* muscle (see Figure 10.11b) on the posterior surface of the subject's arm. EMG electrodes (or ECG electrodes) will be placed on the skin superficial to these muscles.
2. Prior to placing the electrodes on the skin, *gently* scrub the area with a dishwashing pad in order to remove dead epithelial cells and to facilitate recording. *Make sure that the capillaries are not damaged and no bleeding occurs.*
3. Apply a small amount of electrode gel to each electrode and apply them to the proper locations.
4. Connect the two recording electrodes (placed on the anterior surface of the forearm) and the ground electrodes (placed on the posterior surface of the arm) to the preamplifier (high gain coupler) and set the gain to × 100, sensitivity at an appropriate setting between 20 and 100, time constant to 0.03, and paper speed at high.

2. Recording of Spontaneous Muscle Activity

PROCEDURE

1. With the subject completely relaxed and his or her arm placed horizontally on a laboratory table, record any spontaneously occurring electrical activity within the flexor digitorum superficialis over a time period of 1 to 3 min.
2. Record your observations in Section F.1 of the LABORATORY REPORT RESULTS at the end of the exercise.

3. Recruitment of Motor Units

PROCEDURE

1. Have the subject gently flex her or his digits while recording, and note the electrical activity.
2. Have the subject relax his or her digits completely, followed by a more forceful contraction. Note any change in the EMG.

3. Place a tennis ball in your subject's hand and ask him or her to squeeze the ball once or twice as strongly as possible. Note the EMG recording, and then let your subject rest thoroughly for at least 5 min.
4. Record your observations in Section F.2 of the LABORATORY REPORT RESULTS at the end of the exercise.

4. Effect of Fatigue

PROCEDURE

1. *Without recording* an EMG on the polygraph, have your subject repeatedly squeeze the tennis ball as strongly as possible until no longer able to squeeze the ball.
2. When fatigue has been achieved *quickly* start recording and ask the subject to attempt to squeeze the ball as strongly as possible four or five times in quick succession.
3. Note the EMG recording and record your observations in Section F.3 of the LABORATORY REPORT RESULTS at the end of the exercise.

G. CARDIAC MUSCLE TISSUE

Examine a prepared slide of cardiac muscle tissue in longitudinal and transverse section under high power. Locate the following structures: *sarcolemma, endomysium, nuclei, striations,* and *intercalated discs.* Label Figure 9.6.

H. SMOOTH (VISCERAL) MUSCLE TISSUE

Examine a prepared slide of smooth muscle tissue in longitudinal and transverse section under high power. Locate and label the following structures in Figure 9.7: *sarcolemma, sarcoplasm, nucleus,* and *muscle fiber.*

ANSWER THE LABORATORY REPORT QUESTIONS AT THE END OF THE EXERCISE.

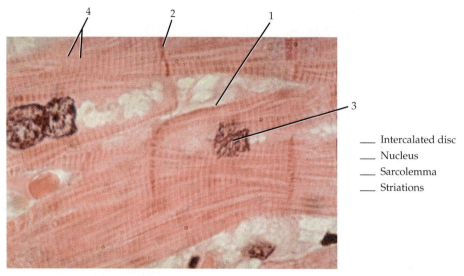

Photomicrograph of several muscle
fibers in longitudinal section (1000×)

___ Intercalated disc
___ Nucleus
___ Sarcolemma
___ Striations

FIGURE 9.6   Histology of cardiac muscle tissue.

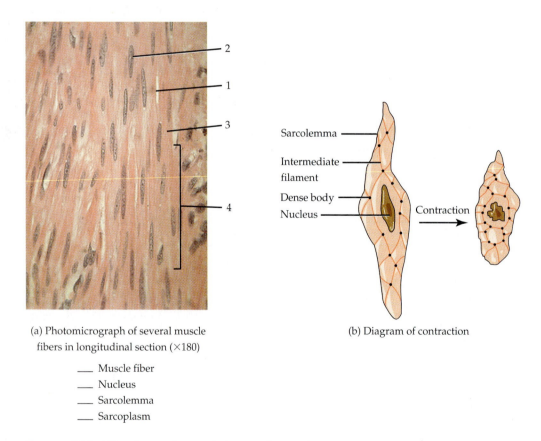

(a) Photomicrograph of several muscle
fibers in longitudinal section (×180)

(b) Diagram of contraction

___ Muscle fiber
___ Nucleus
___ Sarcolemma
___ Sarcoplasm

FIGURE 9.7   Histology of smooth (visceral) muscle tissue.

Muscle Tissue   9

Student _____   **Date** _____

Laboratory Section _____   **Score/Grade** _____

SECTION F. ELECTROMYOGRAPHY

1. Include your record of spontaneous electrical activity recorded while the subject was resting. Give a physiological explanation. _____

   _____

2. Include your record of the EMG recorded during the recruitment exercise. Give a physiological explanation and possible function of recruitment. _____

   _____

3. Include your record of the EMG recorded following fatigue, and give a physiological explanation of your observations. _____

   _____

Muscle Tissue 9

Student _____   Date _____

Laboratory Section _____   Score/Grade _____

PART 1. Multiple Choice

_____ 1. The ability of muscle tissue to return to its original shape after contraction or extension is called (a) excitability (b) elasticity (c) extension (d) tetanus

_____ 2. Which of the following is striated and voluntary? (a) skeletal muscle tissue (b) cardiac muscle tissue (c) visceral muscle tissue (d) smooth muscle tissue

_____ 3. The portion of a sarcomere composed of thin filaments only is the (a) H zone (b) A band (c) I band (d) Z disc

_____ 4. The area of contact between a motor axon terminal and a its associated skeletal muscle fiber sarcolemma (motor end plate) is called the (a) A band (b) filament (c) transverse tubule (d) neuromuscular junction

_____ 5. The connective tissue layer surrounding bundles of muscle fibers is called the (a) perimysium (b) endomysium (c) ectomysium (d) myomysium

_____ 6. Which of the following is striated and involuntary? (a) smooth muscle tissue (b) skeletal muscle tissue (c) cardiac muscle tissue (d) visceral muscle tissue

_____ 7. The portion of a sarcomere that consists mostly of thick filaments and portions of filaments where thin and thick filaments overlap is called the (a) H zone (b) triad (c) A band (d) I band

_____ 8. Intercalated discs are characteristic of which type of muscle tissue? (a) cardiac muscle tissue (b) skeletal muscle tissue (c) smooth muscle tissue (d) visceral muscle tissue

_____ 9. The space between an axon terminal and sarcolemma is called the (a) end plate (b) synaptic cleft (c) synaptic gutter (d) synaptic end bulb

PART 2. Completion

10. The ability of muscle tissue to receive and respond to stimuli is called _____.

11. Fibrous connective tissue located between muscle fibers is known as _____.

12. The regions of a muscle fiber separated by Z discs are called _____.

13. The plasma membrane surrounding a muscle fiber is called the _____.

14. The region in a sarcomere consisting of thick filaments only is known as the _____.

15. Muscle tissue that is nonstriated and involuntary is _____.

16. The ability of muscle tissue to stretch when pulled is called _____.

17. The phenomenon by which a muscle fiber contracts to its fullest extent or not at all is known as the

_____.

18. A record of muscle contraction is referred to as a(n) _____.

19. The region of a muscle fiber sarcolemma adjacent to an axon terminal is called a

_____ .

20. The binding of _____ to troponin removes troponin from _____ on thin filaments to expose myosin binding site.

21. The energy released from the splitting of _____ causes myosin cross bridges to pivot inward to pull the thin filaments inward toward the H zone.

22. In relaxed skeletal muscle fibers, calcium ions are stored in the _____.

23. Whereas thin filaments are composed of the protein actin, thick filaments are composed of the protein

_____.

24. The combination of alternating dark A bands and light _____ bands gives a muscle fiber its striated appearance.

25. A motor neuron and all the muscle fibers it innervates is called a(n) _____.

26. Neurotransmitters are stored in _____, which are located within synaptic end bulbs.

Skeletal Muscles 10

The term *muscle tissue* refers to all contractile tissues of the body: skeletal, cardiac, and smooth muscle. The *muscular system,* however, refers to the voluntary *skeletal* muscle system: the skeletal muscle tissue and connective tissues that make up individual muscle organs, such as the bicep brachii muscle. In this exercise you will learn the names, locations, and actions of the principal skeletal muscles of the body.

A. HOW SKELETAL MUSCLES PRODUCE MOVEMENT

1. Origin and Insertion

Skeletal muscles produce movements by exerting force on tendons, which in turn pull on bones or other structures, such as the skin. Most muscles cross at least one joint and are usually attached to the articulating bones that form the joint. When such a muscle contracts, it draws one of the two articulating bone toward the other. The two articulating bones usually do not move equally in response to the contraction. One bone remains near its original position because other muscles contract to stabilize that bone by pulling in the opposite direction or because its structure makes it less movable. Ordinarily, the attachment of a muscle tendon to the stationary bone is called the *origin.* The attachment of the other muscle tendon to the movable bone is the *insertion.* A good analogy is a spring on a door. The part of the spring attached to the door represents the insertion; the part attached to the frame is the origin. The fleshy portion of the muscle between the tendons of the origin and insertion is called the *belly (gaster).* The origin is usually proximal and the insertion distal, especially in the limbs. In addition, muscles that move a body part generally do not cover the moving part. For example, although contraction of the biceps brachii muscle moves the forearm, the belly of the muscle lies over the humerus.

2. Group Actions

Most movements require several skeletal muscles acting in groups rather than individually. Also, most skeletal muscles are arranged in opposing pairs at joints, that is, flexors-extensors, abductors-adductors, and so on. Consider flexing the forearm at the elbow, for example. A muscle that causes a desired action is referred to as the *prime mover* or *agonist* (*agogos* = leader). In this instance, the biceps brachii is the prime mover (see Figure 10.11a). Simultaneously with the contraction of the biceps brachii, another muscle, called the *antagonist* (*antiagonistes* = opponent), is stretching. In this movement, the triceps brachii serves as the antagonist (see Figure 10.11b). The antagonist has an action that is opposite to that of the prime mover; that is, the antagonist stretches and yields to the movement of the prime mover. You should not assume, however, that the biceps brachii is always the prime mover and the triceps brachii is always the antagonist. For example, when extending the forearm at the elbow, the triceps brachii serves as the prime mover, and the biceps brachii functions as the antagonist; their roles are reversed. Note that if the prime mover and antagonist contracted simultaneously with equal force, there would be no movement.

There are many examples in the body where a prime mover crosses several joints before it reaches the joint at which its primary action occurs. In order to prevent unwanted movements in intermediate joints, or otherwise aid the movement of the prime mover, muscles called *synergists* (SIN-er-gists; *syn* = together; *ergon* = work) contract and stabilize the intermediate joints. As an example, muscles that flex the fingers (prime movers) cross the intercarpal and radiocarpal joints. If movement at these joints was unrestrained, you would not be able to flex your fingers without also flexing the wrist at the same time. Synergistic contraction of the wrist extensors stabilizes the wrist joint and prevents it from moving (unwanted movement), while the flexor muscles of the fingers contract to bring about efficient flexion of the fingers (primary action). During flexion of the fingers, extensor muscles of the fingers serve as antagonists (see Fig. 10.12c, d).

Some muscles in a group also act as *fixators*, which stabilize the origin of the prime mover so that the prime mover can act more efficiently. Fixators steady the proximal end of a limb while movements occur at the distal end. For example, the scapula is a freely movable bone in the pectoral (shoulder) girdle that serves as an origin for several muscles that move the arm. However, for the scapula to serve as a firm origin for muscles that move the arm, it must be held steady. This is accomplished by fixator muscles that hold the scapula firmly against the back of the chest. In abduction of the arm, the deltoid muscle serves as the prime mover, whereas fixators (pectoralis minor, rhomboideus major, rhomboideus minor, trapezius, subclavius, and serratus anterior muscles) hold the scapula firmly (see Figure 10.9). These fixators stabilize the scapula that serves as the attachment site for the origin of the deltoid muscle while the insertion of the muscle pulls on the humerus to abduct the arm. Under different conditions and depending on the movement and which point is fixed, many muscles act, at various times, as prime movers, antagonists, synergists, or fixators.

B. ARRANGEMENT OF FASCICLES

Recall from Exercise 9 that skeletal muscle fibers (cells) are arranged within the muscle in bundles called *fascicles (fasciculi).* The muscle fibers are arranged in a parallel fashion within each bundle, but the arrangement of the fascicles with respect to the tendons may take several characteristic patterns.

Table 10.1 describes the major patterns of fascicles. Using your textbook as a guide, provide an example of each.

C. NAMING SKELETAL MUSCLES

Most of the almost 700 skeletal muscles of the body are named on the basis of one or more distinctive characteristics. If you understand these characteristics, you will find it much easier to learn and remember the names of individual muscles.

Table 10.2 describes the major characteristics that are used to name skeletal muscles. Using your textbook as a guide, provide an example for each.

D. CONNECTIVE TISSUE COMPONENTS

Skeletal muscles are protected, strengthened, and attached to other structures by several connective tissue components. For example, the entire muscle is usually wrapped with a dense, irregular connective tissue called the *epimysium* (ep'-i-MĪZ-ē-um; *epi* = upon). When the muscle is cut in transverse (cross) section, invaginations of the epimysium divide the muscle into fascicles. These invaginations of the epimysium are called the *perimysium* (per'-i-MĪZ-ē-um; *peri* = around). In turn, invaginations of the perimysium, called *endomysium* (en'-dō-MĪZ-ē-um; *endo* = within), penetrate into the interior of each fascicle and separate individual muscle fibers from one another. The epimysium, perimysium, and endomysium are all extensions of deep fascia and are all continuous with the connective tissue that attaches the muscle to another structure, such as bone or other muscle. All three elements may be extended beyond the muscle fibers as a *tendon* (*tendere* = to stretch out)—a cord of connective tissue that attaches a muscle to the periosteum of bone. The connective tissue may also extend as a broad, flat band of tendons called an *aponeurosis* (*apo* = from; *neuron* = a tendon). Aponeuroses also attach to the coverings of a bone or another muscle. When a muscle contracts, the tendon and its corresponding bone or muscle are pulled toward the contracting muscle. In this way skeletal muscles produce movement.

In Figure 10.1, label the epimysium, perimysium, endomysium, fascicles, and muscle fibers.

E. PRINCIPAL SKELETAL MUSCLES

In the pages that follow, a series of tables has been provided for you to learn the principal skeletal muscles by region.[1] Use each table as follows:

1. First read the *overview.* This information provides a general orientation to the muscles under consideration and emphasizes how the muscles are organized within various regions. The discussion also highlights any distinguishing or interesting features about the muscles.

---

[1]A few of the muscles listed are not illustrated in the diagrams. Please consult your textbook to locate these muscles.

TABLE 10.1
Arrangements of Fascicles.

Arrangement	Description	Example
PARALLEL	Fascicles are parallel with longitudinal axis of muscle and terminate at either end in flat tendons.	
FUSIFORM	Fascicles are nearly parallel with longitudinal axis of muscle and terminate at either end in flat tendons, but muscle tapers toward tendons where the diameter is less than that of the belly.	
PENNATE	Fascicles are short in relation to muscle length and the tendon extends nearly the entire length of the muscle.	
Unipennate	Fascicles are arranged on only one side of tendon.	
Bipennate	Fascicles are arranged on both sides of a centrally positioned tendon.	
Multipennate	Fascicles attach obliquely from many directions to several tendons.	
CIRCULAR	Fascicles are arranged in a circular pattern and enclose an orifice (opening).	
TRIANGULAR	Fascicles attached to a broad tendon converge to give the muscle a triangular appearance.	

2. Take each muscle, in sequence, and study the *phonetic pronounciations* and *derivations* that indicate how the muscles are named. They appear in parentheses after the name of the muscle, and will help you to pronounce the name of a muscle and understand the reason for giving a muscle its name.

3. As you learn the name of each muscle, determine its origin, insertion, and action and write these in the spaces provided in the table. Consult your textbook if necessary.

4. Again, using your textbook as a guide, label the diagram referred to in the table.

5. Try to visualize what happens when the muscle contracts so that you will understand its action.

6. Do steps 1 through 5 for each muscle in the table. Before moving to the next table, examine a torso or chart of the skeletal system so that you can compare and approximate the positions of the muscles.

7. When possible, try to feel each muscle on your own body.

Refer to Tables 10.3 through 10.19 and Figures 10.2 through 10.15.

TABLE 10.2
Characteristics Used for Naming Skeletal Muscles.

Characteristic	Description	Example
Direction of muscle fibers	Direction of muscle fibers relative to the midline of the body. **Rectus** means the fibers run parallel to the midline. **Transverse** means the fibers run perpendicular to the midline. **Oblique** means the fibers run diagonally to the midline.	
Location	Structure near which a muscle is found.	
Size	Relative size of the muscle. **Maximus** means largest. **Minimus** means smallest. **Longus** means longest. **Brevis** means short. **Latissimus** means widest. **Longissimus** means longest. **Magnus** means large. **Major** means larger. **Minor** means smaller. **Vastus** means great.	
Number of origins	Number of tendons of origin. **Biceps** means two origins. **Triceps** means three origins. **Quadriceps** means four origins.	
Shape	Relative shape of the muscle. **Deltoid** means triangular. **Trapezius** means trapezoidal. **Serratus** means saw-toothed. **Rhomboideus** means rhomboid- or diamond-shaped. **Orbicularis** means circle. **Pectinate** means comblike. **Piriformis** means pear shaped. **Platys** means flat. **Quadratus** means square. **Gracilis** means slender.	
Origin and insertion	Sites where muscle originates and inserts.	
Action	Principal action of the muscle. **Flexor** (FLEK-sor): decreases the angle at a joint. **Extensor** (eks-TEN-sor): increases the angle at a joint. **Abductor** (ab-DUK-tor): moves a bone away from the midline. **Adductor** (ad-DUK-tor): moves a bone closer to the midline. **Levator** (le-VĀ-tor): produces an upward movement. **Depressor** (de-PRES-or): produces a downward movement. **Supinator** (soo'-pi-NĀ-tor): turns the palm upward or anteriorly. **Pronator** (prō-NĀ-tor): turns the palm downward or posteriorly. **Sphincter** (SFINGK-ter): decreases the size of an opening. **Tensor** (TEN-sor): makes a body part more rigid. **Rotator** (RŌ-tāt-or): moves a bone around its longitudinal axis.	

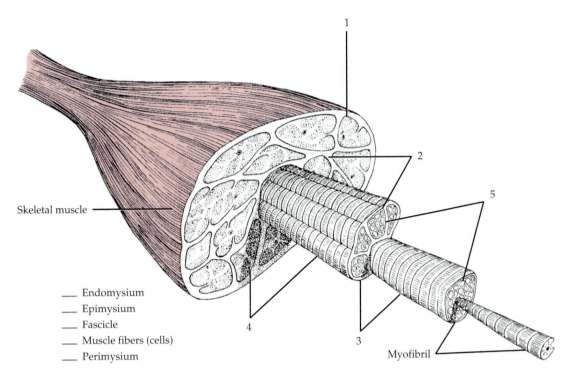

Skeletal muscle

____ Endomysium
____ Epimysium
____ Fascicle
____ Muscle fibers (cells)
____ Perimysium

Myofibril

FIGURE 10.1   Connective tissue components of a skeletal muscle.

TABLE 10.3
Muscles of Facial Expression (After completing the table, label Figure 10.2.)

OVERVIEW: The muscles in this group provide humans with the ability to express a
wide variety of emotions, including grief, surprise, fear, and happiness. The muscles
themselves lie within the layers of superficial fascia. As a rule, they arise from the fascia
or bones of the skull and insert into the skin. Because of their insertions, the muscles of
facial expression move the skin rather than a joint when they contract.

Among the noteworthy muscles in this group are those that surround the orifices (openings) of the head such as the eyes and mouth. These muscles function as *sphincters,*
which close the orifices, and *dilators,* which open the orifices. For example, the orbicularis oculi muscle closes the eye, while the levator palpebrae superioris muscle opens
the eye. As another example, the orbicularis oris muscle closes the mouth, while several
other muscles (zygomaticus major, levator labii superioris, depressor labii inferioris,
mentalis, and risorius) radiate out from the lips and open the mouth. The epicranius is
an unusual muscle in this group because it is made up of two parts: an anterior part
called the frontalis, which is superficial to the frontal bone, and a posterior part called
the occipitalis, which is superficial to the occipital bone. The two muscular portions are
held together by a strong aponeurosis (sheetlike tendon), the **galea aponeurotica**
(GĀ-lē-a ap-ō-noo'-RŌ-ti-ka; *galea* = helmet) or **epicranial aponeurosis,** which covers
the superior and lateral surfaces of the skull. The frontalis and occipitalis are treated as
separate muscles in this table. The buccinator muscle forms the major muscular portion
of the cheek. It functions in whistling, blowing, and sucking and also assists in chewing.
The duct of the parotid gland (salivary gland) pierces the buccinator muscle to reach the
oral cavity. The buccinator muscle was given its name because it compresses the cheeks
(*bucc* = cheek) during blowing, for example, when a musician plays a wind instrument
such as a trumpet. Some trumpeters have stretched their buccinator muscles so much
that their cheeks bulge out considerably when they blow forcibly.

TABLE 10.3 *(Continued)*

Muscle	Origin	Insertion	Action
Epicranius (ep-i-KRĀ-nē-us; *epi* = above; *crani* = skull)	This muscle is divisible into two portions: the frontalis, over the frontal bone, and the occipitalis, over the occipital bone. The two muscles are united by a strong aponeurosis, the galea aponeurotica (epicranial aponeurosis), which covers the superior and lateral surfaces of the skull.		
Frontalis (fron-TA-lis; *front* = forehead)			
Occipitalis (ok-si'-pi-TA-lis; *occipito* = base of skull)			
Orbicularis oris (or-bi'-kyoo-LAR-is OR-is; *orb* = circular; *or* = mouth)			
Zygomaticus (zī-gō-MA-ti-kus) **major** (*zygomatic* = cheek bone; *major* = greater)			
Levator labii superioris (le-VĀ-ter LA-bē-ī soo-per'-ē-OR-is; *levator* = raises or elevates; *labii* = lip; *superioris* = upper)			
Depressor labii inferioris (de-PRE-ser LA-bē-ī in-fer'-ē-OR-is; *depressor* = depresses or lowers; *inferioris* = lower)			
Buccinator (BUK-si-nā'-tor; *bucca* = cheek)			
Mentalis (men-TA-lis; *mentum* = chin)			
Platysma (pla-TIZ-ma; *platy* = flat, broad)			
Risorius (ri-ZOR-ē-us; *risor* = laughter)			
Orbicularis oculi (or-bi'-kyoo-LAR-is O-kyoo-lī; *oculus* = eye)			
Corrugator supercilii (KOR-a-gā'-tor soo-per-SI-lē-ī; *corrugo* = to wrinkle; *supercilium* = eyebrow)			
Levator palpebrae superioris (le-VĀ-tor PAL-pe-brē soo-per'-ē-OR-is; *palpebrae* = eyelids) (See Figure 10.4)			

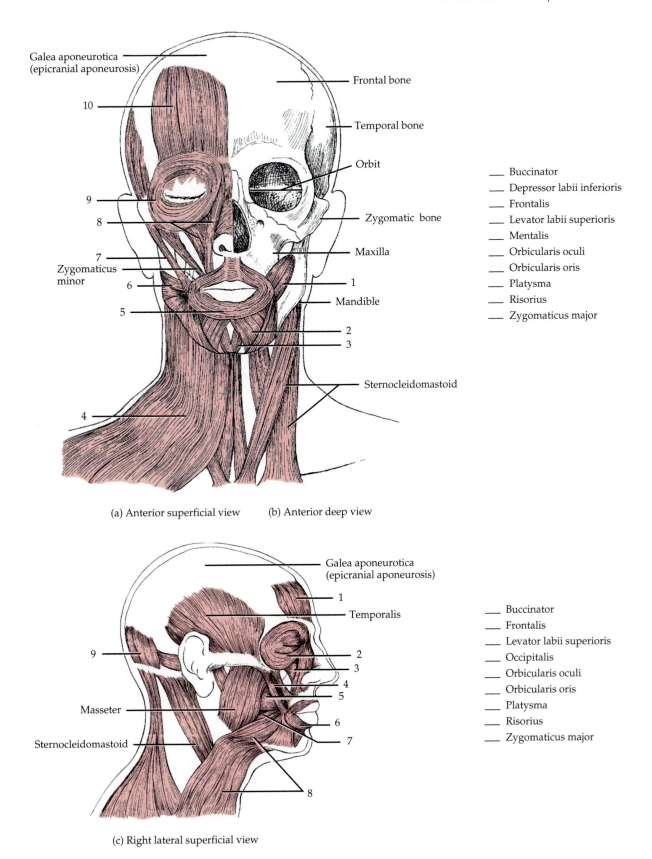

Galea aponeurotica (epicranial aponeurosis)

10

9

8

7

Zygomaticus minor

6

5

4

Frontal bone

Temporal bone

Orbit

Zygomatic bone

Maxilla

1

Mandible

2

3

Sternocleidomastoid

___ Buccinator
___ Depressor labii inferioris
___ Frontalis
___ Levator labii superioris
___ Mentalis
___ Orbicularis oculi
___ Orbicularis oris
___ Platysma
___ Risorius
___ Zygomaticus major

(a) Anterior superficial view   (b) Anterior deep view

Galea aponeurotica (epicranial aponeurosis)

1

Temporalis

9

2

3

4

5

Masseter

6

Sternocleidomastoid

7

8

___ Buccinator
___ Frontalis
___ Levator labii superioris
___ Occipitalis
___ Orbicularis oculi
___ Orbicularis oris
___ Platysma
___ Risorius
___ Zygomaticus major

(c) Right lateral superficial view

FIGURE 10.2 Muscles of facial expression.

TABLE 10.4
Muscles That Move the Mandible (Lower Jaw) (After completing the table, label Figure 10.3.)

OVERVIEW: The muscles that move the mandible (lower jaw) at the temporomandibular joint (TMJ) are known as the muscles of mastication because they are involved in chewing (mastication). Of the four pairs of muscles involved in mastication, three are powerful closers of the jaw and account for the strength of the bite: masseter, temporalis, and medial pterygoid. Of these, the masseter is the strongest muscle of mastication. The medial and lateral pterygoid muscles assist in mastication by moving the mandible from side to side to help grind food. Additionally, these muscles protrude the mandible.

In 1996, researchers at the University of Maryland reported that they had identified a new muscle in the skull, tentatively named the ***sphenomandibularis muscle.*** It extends from the sphenoid bone to the mandible. The muscle is believed to be either a fifth muscle of mastication or a previously unidentified component of an already identified muscle (temporalis or medial pterygoid). The sphenomandibularis muscle is innervated by the maxillary branch of the trigeminal (V) cranial nerve.

Muscle	Origin	Insertion	Action
Masseter (MA-se-ter; *masseter* = chewer)			
Temporalis (tem'-por-A-lis; *tempora* = temples)			
Medial pterygoid (TER-i-goid; *medial* = closer to midline; *pterygoid* = like a wing)			
Lateral pterygoid (TER-i-goid; *lateral* = farther from midline)			

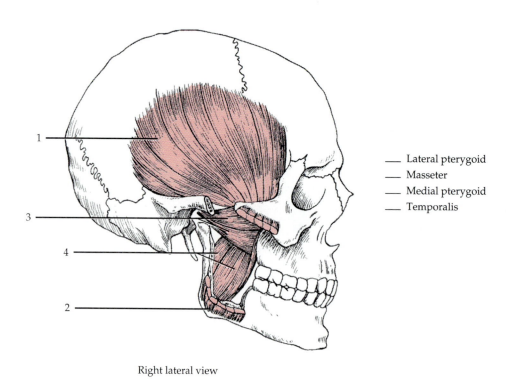

___ Lateral pterygoid
___ Masseter
___ Medial pterygoid
___ Temporalis

Right lateral view

FIGURE 10.3   Muscles that move mandible (lower jaw).

TABLE 10.5
Muscles That Move the Eyeballs—The Extrinsic Muscles* (After completing the table, label Figure 10.4.)

OVERVIEW: Muscles associated with the eyeballs are of two principal types: extrinsic and intrinsic. *Extrinsic muscles* originate outside the eyeballs and are inserted on their outer surfaces (sclera). They move the eyeballs in various directions. *Intrinsic muscles* originate and insert entirely within the eyeballs. They move structures within the eyeballs.

Movements of the eyeballs are controlled by three pairs of extrinsic muscles. (1) superior and inferior recti, (2) lateral and medial recti, and (3) superior and inferior oblique. The extrinsic muscles of the eyeballs are among the fastest contracting and most precisely controlled skeletal muscles in the body. The four recti muscles (superior, inferior, lateral, and medial) arise from a tendinous ring in the orbit and insert into the sclera of the eye. Whereas the superior and inferior recti lie in the same vertical plane, the medial and lateral recti lie in the same horizontal plane. The actions of the recti muscles can be deduced from their insertions on the sclera. The superior and inferior recti move the eyeballs superiorly and inferiorly, respectively; the lateral and medial recti move the eyeballs laterally and medially, respectively. It should be noted that neither the superior nor the inferior rectus muscle pulls directly parallel to the long axis of the eyeballs; as a result, both muscles also move the eyeballs medially.

It is not easy to deduce the actions of the oblique muscles (superior and infereior) because of their path through the orbits. For example, the superior oblique muscle originates posteriorly near the tendinous ring and then passes anteriorly and ends in a round tendon, which runs through a pulleylike loop called the *trochlea* (*trochlea* = pulley) in the anterior and medial part of the roof of the orbit. The tendon then turns and inserts on the posterolateral aspect of the eyeball. Accordingly, the superior oblique muscle moves the eyeballs inferiorly and laterally. The inferior oblique muscle originates on the maxilla at the anteromedial aspect of the floor of the orbit. It then passes posteriorly and laterally and inserts on the posterolateral aspect of the eyeballs. Because of this arrangement, the inferior oblique muscles move the eyeballs superiorly and laterally.

Muscle	Origin	Insertion	Action
Superior rectus (REK-tus; *superior* = above; *rectus* = in this case, muscle fibers running parallel to long axis of eyeball)			
Inferior rectus (REK-tus; *inferior* = below)			
Lateral rectus (REK-tus)			
Medial rectus (REK-tus)			
Superior oblique (ō-BLĒK; *oblique* = in this case, muscle fibers running diagonally to long axis of eyeball)			
Inferior oblique (ō-BLĒK)			

*Muscles situated on the outside of the eyeballs.

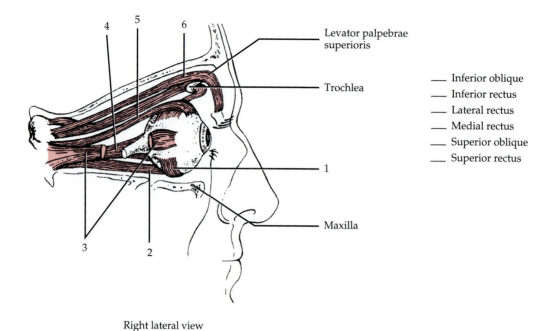

Right lateral view

FIGURE 10.4   Extrinsic muscles of eyeballs.

TABLE 10.6
Muscles That Move the Tongue—The Extrinsic Muscles (After completing the table, label Figure 10.5.)

OVERVIEW: The tongue is a highly mobile structure that is vital to digestive functions such as mastication, perception of taste, and deglutition (swallowing). It is also important in speech. The mobility of the tongue is greatly aided by its suspension from the mandible, styloid process of the temporal bone, and hyoid bone.

The tongue is divided into lateral halves by a median fibrous septum. The septum extends throughout the length of the tongue and is attached inferiorly to the hyoid bone. Like the muscles of the eyeballs, muscles of the tongue are of two principal types—extrinsic and intrinsic. *Extrinsic muscles* originate outside the tongue and insert into it. They move the entire tongue in various directions, such as anteriorly, posteriorly, and laterally. *Intrinsic muscles* originate and insert within the tongue. These muscles alter the shape of the tongue rather than move the entire tongue. The extrinsic and intrinsic muscles of the tongue are arranged in both lateral halves of the tongue.

When you study the extrinsic muscles of the tongue, you will notice that all of the names end in *glossus,* meaning tongue. You will also notice that the actions of the muscles are obvious considering the position of the mandible, styloid process, hyoid bone, and soft palate, which serve as origins for these muscles. For example, the genioglossus (originates on the mandible) pulls the tongue downward and forward, the styloglossus (originates on the sytloid process) pulls the tongue upward and backward, the hyoglossus (originates on the hyoid bone) pulls the tongue downward and flattens it, and the palatoglossus (originates on the soft palate) raises the back portion of the tongue.

When general anesthesia is administered during surgery, a total relaxation of the genioglossus muscle results. This will cause the tongue to fall posteriorly, which may obstruct the airway to the lungs. To avoid this, the mandible is either manually thrust forward and held in place, or a tube is inserted from the lips through the laryngopharynx (inferior portion of the throat) into the trachea (endotracheal intubation).

TABLE 10.6 (*Continued*)

Muscle	Origin	Insertion	Action
Genioglossus (jē'-nē-ō-GLOS-us; *geneion* = chin *glossus* = tongue)			
Styloglossus (stī'-lō-GLOS-us; *stylo* = stake or pole)			
Palatoglossus (pal'-a-tō-GLOS-us; *palato* = palate)			
Hyoglossus (hī-ō-GLOS-us)			

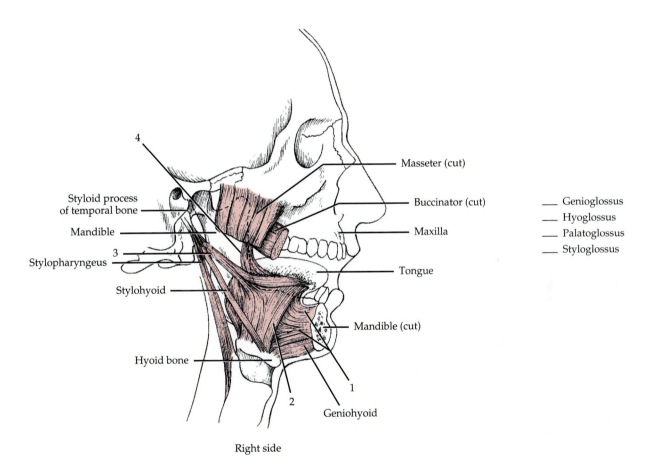

Styloid process of temporal bone

Mandible

Stylopharyngeus

Stylohyoid

Hyoid bone

Geniohyoid

Masseter (cut)

Buccinator (cut)

Maxilla

Tongue

Mandible (cut)

___ Genioglossus
___ Hyoglossus
___ Palatoglossus
___ Styloglossus

Right side

FIGURE 10.5   Muscles that move the tongue.

TABLE 10.7
Muscles That Move the Head

OVERVIEW: The head is attached to the vertebral column at the atlanto-occipital joint formed by the atlas and occipital bone. Balance and movement of the head on the vertebral column involve the action of several neck muscles. For example, contraction of the two sternocleidomastoid muscles together (bilaterally) flexes the cervical portion of the vertebral column and head. Acting singly (unilaterally), the muscle laterally flexes and rotates the head. Bilateral contraction of the semispinalis capitis, splenius capitis, and longissimus capitis muscles extends the head. However, when these same muscles contract unilaterally, their actions are quite different, involving primarily rotation of the head.

The cervical (neck) region is divided by the sternocleidomastoid muscle into two principal triangles: anterior and posterior. Within each of the principal triangles are subsidiary triangles. The *anterior triangle* is bordered superiorly by the mandible, inferiorly by the sternum, medially by the cervical midline, and laterally by the anterior border of the sternocleidomastoid muscle. The anterior triangle is subdivided into an unpaired submental triangle and three paired triangles: submandibular, carotid, and muscular. The *posterior triangle* is bordered inferiorly by the clavicle, anteriorly by the posterior border of the sternocleidomastoid muscle, and posteriorly by the anterior border of the trapezius muscle. The posterior triangle is subdivided into two triangles, occipital and supraclavicular, by the inferior belly of the omohyoid muscle. The triangles are discussed in detail in Chapter 11.

Muscle	Origin	Insertion	Action

Sternocleidomastoid
(ster'-nō-klī'-dō-MAS-toid;
sternum = breastbone;
cleido = clavicle;
mastoid = mastoid process
of temporal bone)
(label this muscle in
Figure 10.9)			

Semispinalis capitis
(se'-mē-spi-NA-lis KAP-i-tis;
semi = half;
spine = spinous process;
caput = head)			

Splenius capitis (SPLĒ-nē-us KAP-i-tis; *splenion* = bandage)			
Longissimus capitis (lon-JIS-i-mus KAP-i-tis; *longissimus* = longest)			

TABLE 10.8
Muscles That Act on the Anterior Abdominal Wall (After completing the table, label Figure 10.6.)

OVERVIEW: The anterolateral abdominal wall is composed of skin, fascia, and four
pairs of muscles: the external oblique, the internal oblique, transversus abdominis, and
rectus abdominis. The first three muscles are flat muscles; the last is a straplike vertical
muscle. The external oblique is the external flat muscle with its fibers directed inferiorly
and medially. The internal oblique is the intermediate flat muscle with its fibers directed
at right angles to those of the external oblique. The transversus abdominis is the deepest
of the flat muscles, with most of its fibers directed horizontally around the abdominal
wall. Together, the external oblique, internal oblique, and transversus abdominus form
three layers of muscle around the abdomen. The muscle fibers of each layer run cross-
directionally to one another, a structural arrangement that affords considerable protection to the abdominal viscera, especially when the muscles have good tone.

The rectus abdominis muscle is a long, flat muscle that extends the entire length of the
anterior abdominal wall, from the pubic crest and pubic symphysis to the cartilages of
ribs 5–7 and the xiphoid process of the sternum. The anterior surface of the muscle is
interrupted by three transverse fibrous bands of tissue called ***tendinous intersections,***
believed to be remnants of septa that separated myotomes during embryological development.

As a group the muscles of the anterolateral abdominal wall help contain and protect the
abdominal viscera; flex, laterally flex, and rotate the vertebral column at the intervertebral joints; compress the abdomen during forced expiration; and produce the force
required for defecation, urination, and childbirth.

The aponeuroses of the external oblique, internal oblique, and transversus abdominis
muscles form the ***rectus sheath,*** which encloses the rectus abdominis muscles and meet
at the midline to form the ***linea alba*** (= white line), a tough, fibrous band that extends
from the xiphoid process of the sternum to the pubic symphysis. In the latter stages of
pregnancy, the linea alba stretches to increase the distance between the rectus abdominis
muscles. The inferior free border of the external oblique aponeurosis, plus some collagen fibers, forms the ***inguinal ligament,*** which runs from the anterior superior iliac
spine to the pubic tubercle (see Figure 10.13). Just superior to the medial end of the
inguinal ligament is a triangular slit in the aponeurosis referred to as the ***superficial
inguinal ring,*** the outer opening of the ***inguinal canal.*** The canal contains the spermatic
cord and ilioinguinal nerve in males and round ligament of the uterus and ilioinguinal
nerve in females.

The posterior abdominal wall is formed by the lumbar vertebrae, parts of the ilia of the
hipbones, psoas major and iliacus muscles (described in Table 10.17), and quadratus
lumborum muscle. Whereas the anterolateral abdominal wall can contract and distend,
the posterior abdominal wall is bulky and stable by comparison.

Muscle	Origin	Insertion	Action
Rectus abdominis (REK-tus ab-do-MIN-is; *rectus* = fibers parallel to midline; *abdomino* = abdomen)			

External oblique (ō-BLĒK;
external = closer to
   the surface;
oblique = fibers diagonal
   to midline)			

Internal oblique (ō-BLĒK; *internal* = farther from the surface)			

TABLE 10.8 *(Continued)*

Muscle	Origin	Insertion	Action

Transversus abdominis
(tranz-VER-sus ab-do-MIN-is;
transverse = fibers
  perpendicular
  to midline)			

Quadratus lumborum (kwod-RĀ-tus lum-BOR-um; *quad* = four; *lumbo* = lumbar region)			

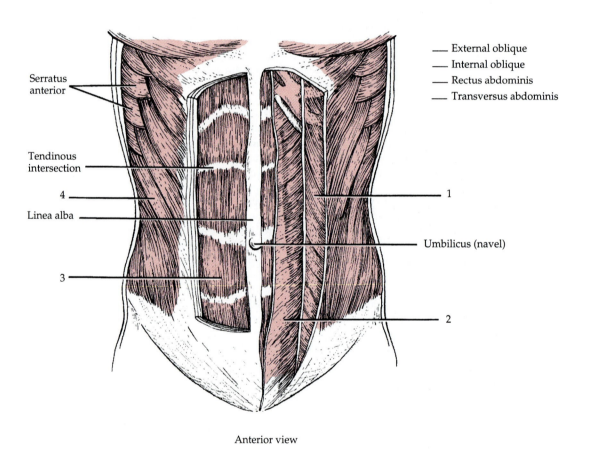

Serratus anterior

Tendinous intersection

4

Linea alba

3

External oblique
Internal oblique
Rectus abdominis
Transversus abdominis

1

Umbilicus (navel)

2

Anterior view

FIGURE 10.6   Muscles of the anterior abdominal wall.

TABLE 10.9
Muscles Used in Ventilation (breathing) (After completing the table, label Figure 10.7.)

OVERVIEW: The muscles described here alter the size of the thoracic cavity so that ventilation (breathing) can occur. Ventilation consists of two phases called inspiration (inhalation) and expiration (exhalation). Essentially, inspiration occurs when the thoracic cavity increases in size, and expiration occurs when the thoracic cavity decreases in size.

The diaphragm is the most important muscle in ventilation. It is a dome-shaped musculotendinous partitition that separates the thoracic and abdominal cavities. Although the diaphragm descends during inspiration, only its dome moves downward because the peripheral portions are attached to the sternum, costal cartilages, and lumbar vertebrae. The diaphragm is composed of two parts: a peripheral muscular portion and a central portion called the **central tendon.** The central tendon is a strong aponeurosis that serves as the tendon of insertion for all the peripheral muscular fibers of the diaphragm. It fuses with the inferior surface of the fibrous pericardium, the external covering of the heart, and the parietal pleurae, the external coverings of the lungs.

In addition to its function in inspiration, movements of the diaphragm help to return venous blood to the heart as it passes through the abdomen. Together with the anterolateral abdominal muscles, the diaphragm helps to increase intra-abdominal pressure to evacuate the pelvic contents during defecation, urination, and childbirth. This mechanism is further assisted when you take a deep breath and close the rima glottidis (the space between vocal folds). The trapped air in the respiratory system prevents the diaphragm from elevating. The increase in the intra-abdominal pressure as just described will also help support the vertebral column and prevent flexion during weight lifting. This greatly assists the back muscles in lifting a heavy weight.

The diaphragm has three major openings through which various structures pass between the thorax and abdomen. These structures include the aorta along with the thoracic duct and azygos vein, which pass through the **aortic hiatus;** the esophagus with accompanying vagus (X) cranial nerves, which pass through the **esophageal hiatus;** and the inferior vena cava, which passes through the **foramen for the vena cava.** In a condition called a hiatus hernia, the stomach protrudes superiorly through the esophageal hiatus.

The other muscles involved in ventilation are called intercostal muscles and occupy the intercostal spaces, the spaces between ribs. There are 11 external intercostal muscles, and their fibers run obliquely inferiorly and anteriorly from the rib above to the rib below. Their role in respiration is to elevate the ribs during inspiration to help increase the size of the thoracic cavity. There are also 11 internal intercostal muscles, and their fibers run obliquely inferiorly and posteriorly from the rib above to the rib below. Their function is to draw adjacent ribs together during forced expiration to help decrease the size of the thoracic cavity.

Muscle	Origin	Insertion	Action
Diaphragm (DĪ-a-fram; *dia* = across, between; *phragma* = wall)			
External intercostals (in'-ter-KOS-tals; *inter* = between; *costa* = rib)			
Internal intercostals (in'-ter-KOS-tals; *internal* = farther from surface)			

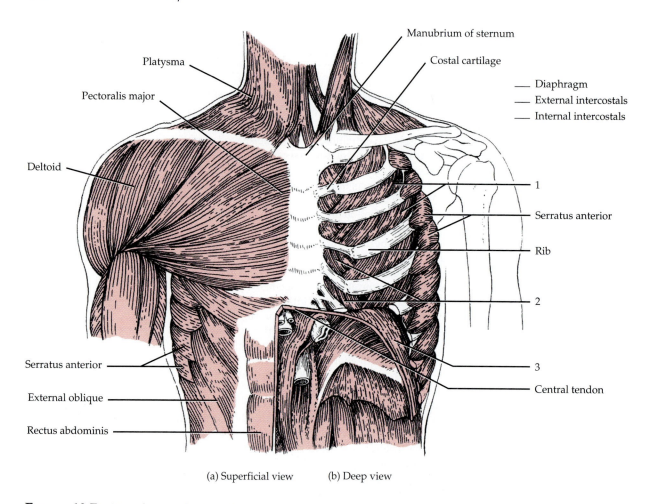

Platysma

Pectoralis major

Deltoid

Serratus anterior

External oblique

Rectus abdominis

Manubrium of sternum

Costal cartilage

___ Diaphragm
___ External intercostals
___ Internal intercostals

1

Serratus anterior

Rib

2

3

Central tendon

(a) Superficial view   (b) Deep view

FIGURE 10.7   Muscles used in breathing.

TABLE 10.10
Muscles of the Pelvic Floor (After completing the table, label Figure 10.8.)

OVERVIEW: The muscles of the pelvic floor are the levator ani and coccygeus. Together
   with the fascia covering their internal and external surfaces, these muscles are referred
   to as the *pelvic diaphragm.* It stretches from the pubis anteriorly to the coccyx posteriorly, and from one lateral wall of the pelvis to the other. This arrangment gives the pelvic
   diaphragm the appearance of a funnel suspended from its attachements. The pelvic
   diaphragm is pierced by the anal canal and urethra in both sexes and also by the vagina
   in the female.

The levator ani muscle is divisible into two muscles called the pubococcygeus and iliococcygeus. The levator ani is the largest and most important muscle of the pelvic floor. It
   supports the pelvic viscera and resists the inferior thrust that accompanies increases in
   intra-abdominal pressure during functions such as forced expiration, coughing, vomiting, urination, and defecation. The muscle also functions as a sphincter at the anorectal
   junction, urethra, and vagina. During childbirth, the levator ani muscles support the
   head of the fetus, and the muscle may be injured during a difficult childbirth or trauma-
   tized during an *episiotomy* (a cut made with surgical scissors to prevent tearing of the
   perineum during birth of a baby). This may cause urinary stress incontinence, in which
   there is a leakage of urine whenever intra-abdominal pressure is increased, for example
   during coughing. In addition to assisting the levator ani in supporting the pelvic viscera
   and resisting increases in intra-abdominal pressure, the coccygeus muscle pulls the
   coccyx anteriorly after it has been pushed posteriorly following defecation or childbirth.

TABLE 10.10 *(Continued)*

Muscle	Origin	Insertion	Action
Levator ani (le-VĀ-tor Ā-nē; *levator* = raises; *ani* = anus)	This muscle is divisible into two parts: the pubococcygeus muscle and the iliococcygeus muscle.		
Pubococcygeus (pu'-bō-kok-SIJ-ē-us; *pubo* = pubis; *coccygeus* = coccyx)			
Iliococcygeus (il'-ē-ō-kok-SIJ-ē-us; *ilio* = ilium)			
Coccygeus* (kok-SIJ-ē-us)			

*Not illustrated

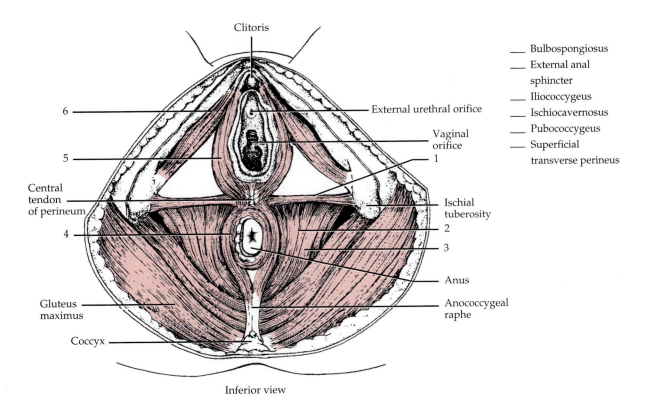

Inferior view

FIGURE 10.8   Muscles of female pelvic floor and perineum.

TABLE 10.11
Muscles of the Perineum (After completing the table, label Figure 10.8.)

OVERVIEW: The *perineum* is the region of the trunk inferior to the pelvic diaphragm. It
is a diamond-shaped area that extends from the pubic symphysis anteriorly, to the coccyx posteriorly, and to the ischial tuberosities laterally. A transverse line drawn between
the ischial tuberosities divides the perineum into an anterior *urogenital triangle* that
contains the external genitals and a posterior *anal triangle* that contains the anus. In the
center of the perineum is a wedge-shaped mass of fibrous tissue called the *central tendon (perineal body).* It is a strong tendon into which several perineal muscles insert.
The muscles of the perineum are arranged in two layers: *superficial* and *deep.* The muscles
of the superficial layer are the superficial transverse perineus, bulbospongiosus, and
ischiocavernosus. The superficial transverse perineus muscle is a narrow muscle that
passes more or less transversely across the perineum anterior to the anus. It joins with
the external sphincter posteriorly. The bulbospongiosus muscle assists in urination and
erection of the penis in males and in erection of the clitoris in females. The ischiocavernosus muscle assists in maintaining erection of the penis and clitoris. The deep muscles
of the perineum are the deep transverse perineus muscle and external urethral sphincter. The deep transverse perineus, external urethral sphincter, and their fascia are
known as the *urogenital diaphragm.* The muscles of this diaphragm assist in urination
and ejaculation in males and urination in females. The external anal sphincter closely
adheres to the skin around the margin of the anus and keeps the anal canal and anus
closed except during defecation.

Muscle	Origin	Insertion	Action
Superficial transverse perineus (per-i-NĒ-us; *superficial* = near surface; *transverse* = across; *perineus* = perineum)			
Bulbospongiosus (bul'-bō-spon'-jē-Ō-sus; *bulbus* = bulb; *spongio* = sponge)			
Ischiocavernosus is'-kē-ō-ka'-ver-NŌ-sus; *ischion* = hip)			
Deep transverse perineus* (per-i-NĒ-us; *deep* = farther from surface)			
External urethral sphincter (yoo-RĒ-thral SFINGK-ter; *urethral* = pertaining to urethra; *sphincter* = circular muscle that decreases the size of an opening)			
External anal (Ā-nal) **sphincter**			

*Not illustrated

TABLE 10.12

Muscles That Move the Pectoral (Shoulder) Girdle (After completing the table, label Figure 10.9.)

OVERVIEW: The principal action of the muscles that move the pectoral girdle is to stabilize the scapula so that it can function as a stable point of origin for most of the muscles that move the humerus. Since scapular movements usually accompany humeral
movements in the same direction, the muscles also move the scapula to increase the
range of movements of the humerus. For example, it would not be possible to abduct
the humerus past the horizontal position if the scapula did not move with the humerus.
During abduction, the scapula follows the humerus by rotating upward.

Muscles that move the pectoral girdle can be classified into two groups based on their
location in the thorax: *anterior* and *posterior* thoracic muscles. The anterior thoracic
muscles are the subclavius, pectoralis minor, and serratus anterior. The subclavius is a
small, cylindrical muscle under the clavicle that extends from the clavicle to the first rib.
It steadies the clavicle during movements of the pectoral girdle. The pectoralis minor is
a thin, flat, triangular muscle that is deep to the pectoralis major. In addition to its role
in movements of the scapula, the pectoralis minor muscle also assists in forced inspiration. The serratus anterior is a large, flat, fan-shaped muscle between the ribs and
scapula. It is named because of the saw-toothed appearance of its origins on the ribs.

The posterior thoracic muscles are the trapezius, levator scapulae, rhomboideus major,
and rhomboideus minor. The trapezius is a large, flat, triangular sheet of muscle extending from the skull and vertebral column medially to the pectoral girdle laterally. It is the
most superficial back muscle and covers the posterior neck region and superior portion
of the trunk. The two trapezius muscles form a trapezium (diamond-shaped quadran-
gle), thus its name. The levator scapulae is a narrow, elongated muscle in the posterior
portion of the neck. It is deep to the sternocleidomastoid and trapezius muscles. As its
name suggests, one of its actions is to elevate the scapula. The rhomboideus major and
rhomboideus minor muscles lie deep to the trapezius and are not always distinct from
each other. They appear as parallel bands that pass inferolaterally from the vertebrae to
the scapula. They are named on the basis of their shape, that is, a rhomboid (an oblique
parallelogram). The rhomboideus major is about two times wider than the rhomboideus
minor. Both muscles are used when forcibly lowering the raised upper limbs, as in driving a stake with a sledgehammer.

In order to understand the action of muscles that move the scapula, it will first be helpful
to describe the various movements of the scapula.

Elevation   Superior movement of the scapula, such as shrugging the shoulders or lifting
a weight over the head.

Depression   Inferior movement of the scapula, as in doing a "pull-up."

Abduction (protraction)   Movement of the scapula laterally and anteriorly, as in doing a
"push-up" or punching.

Adduction (retraction)   Movement of the scapula medially and posterially, as in pulling
the oars in a rowboat.

Upward rotation   Movement of the inferior angle of the scapula laterally so that the
glenoid cavity is moved upward. This movement is required to abduct the humerus
past the horizontal position.

Downward rotation   Movement of the inferior angle of the scapula medially so that the
glenoid cavity is moved downward. This movement is seen when the weight of the
body is supported on the hands by a gymnast on parallel bars.

Muscle	Origin	Insertion	Action
ANTERIOR MUSCLES **Subclavius** (sub-KLĀ-vē-us; *sub* = under; *clavius* = clavicle)			

TABLE 10.12 *(Continued)*

Muscle	Origin	Insertion	Action
Pectoralis (pek'-tor-A-lis) **minor** (*pectus* = breast, chest, thorax; *minor* = lesser)			
Serratus (ser-Ā-tus) **anterior** (*serratus* = sawtoothed; *anterior* = front)			
POSTERIOR MUSCLES **Trapezius** (tra-PĒ-zē-us; *trapezoides* = trapezoid-shaped)			
Levator scapulae (le-VĀ-tor SKA-pyoo-lē; *levator* = raises; *scapulae* = scapula)			
Rhomboideus (rom-BOID-ē-us) **major** (*rhomboides* = rhomboid- or diamond-shaped)			
Rhomboideus (rom-BOID-ē-us) **minor**			

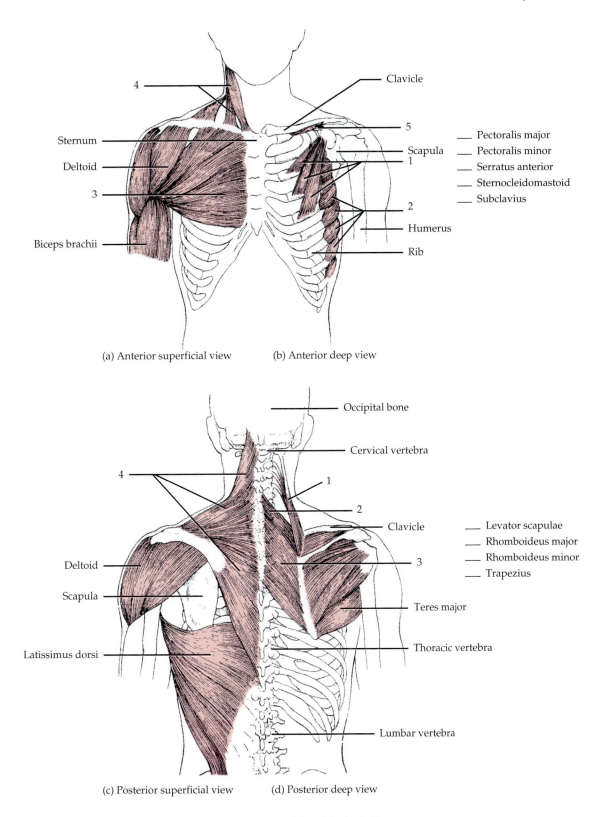

FIGURE 10.9   Muscles that move the pectoral (shoulder) girdle.

TABLE 10.13
Muscles That Move the Humerus (Arm) (After completing the table, label Figure 10.10.)

OVERVIEW: The muscles that move the humerus (arm) cross the shoulder joint. Of the nine muscles that cross the shoulder joint, only two of them (pectoralis major and latissimus dorsi) do not originate on the scapula. These two muscles are thus designated as *axial muscles*, since they originate on the axial skeleton. The remaining seven muscles, the *scapular muscles*, arise from the scapula.

Of the two axial muscles that move the humerus, the pectoralis major is a large, thick, and fan-shaped muscle that covers the superior part of the thorax. It has two origins: a smaller clavicular head and a larger sternocostal head. The latissimus dorsi is a broad, triangular muscle located on the inferior part of the back. It is commonly called the "swimmer's muscle" because its many actions are used while swimming.

Among the scapular muscles, the deltoid is a thick, powerful shoulder muscle that covers the shoulder joint and forms the rounded contour of the shoulder. This muscle is a frequent site of intramuscular injections. As you study the deltoid, note that its fibers originate from three different points and that each group of fibers moves the humerus differently. The subscapularis is a large, triangular muscle that fills the subscapular fossa of the scapula and forms part of the posterior wall of the axilla. The supraspinatus is a rounded muscle, named for its location in the supraspinous fossa of the scapula. It lies deep to the trapezius. The infraspinatus is a triangular muscle, also named for its location in the infraspinous fossa of the scapula. The teres minor is a cylindrical, elongated muscle, often inseparable from the infraspinatus, which lies along its inferior border. The teres major is a thick, flattened muscle inferior to the teres minor and also helps to form part of the posterior wall of the axilla. The coracobrachialis is an elongated, narrow muscle in the arm that is pierced by the musculocutaneous nerve.

The strength and stability of the shoulder joint are not provided by the shape of the articulating bones or its ligaments. Instead, four deep muscles of the shoulder—subscapularis, supraspinatus, infraspinatus, and teres minor—strengthen and stabilize the shoulder joint. These muscles join the scapula to the humerus. Their flat tendons fuse together to form a nearly complete circle around the shoulder joint like a cuff on a shirt sleeve. This arrangement is referred to as the *rotator (musculotendinous) cuff*. The supraspinatus muscle is especially predisposed to wear and tear because of its location between the head of the humerus and acromion of the scapula, which compresses its tendon during shoulder movements.

After you have studied the muscles in this table, arrange them according to the following actions: flexion, extension, abduction, adduction, medial rotation, and lateral rotation. (The same muscle can be used more than once.)

Muscle	Origin	Insertion	Action
AXIAL MUSCLES **Pectoralis** (pek'-tor-A-lis) **major** (label this muscle in Figure 10.9)			
Latissimus dorsi (la-TIS-i-mus DOR-sī; *dorsum* = back)			
SCAPULAR MUSCLES **Deltoid** (DEL-toyd; *delta* = triangular)			
Subscapularis (sub-scap'-yoo-LA-ris; *sub* = below; *scapularis* = scapula)			
Supraspinatus (soo'-pra-spi-NĀ-tus; *supra* = above; *spinatus* = spine of scapula)			
Infraspinatus (in'-fra-spi-NĀ-tus; *infra* = below)			
Teres (TE-rēz) **major** (*teres* = long and round)			
Teres (TE-rēz) **minor**			
Coracobrachialis (kor'-a-kō-BRĀ-kē-a'-lis; *coraco* = coracoid process)			

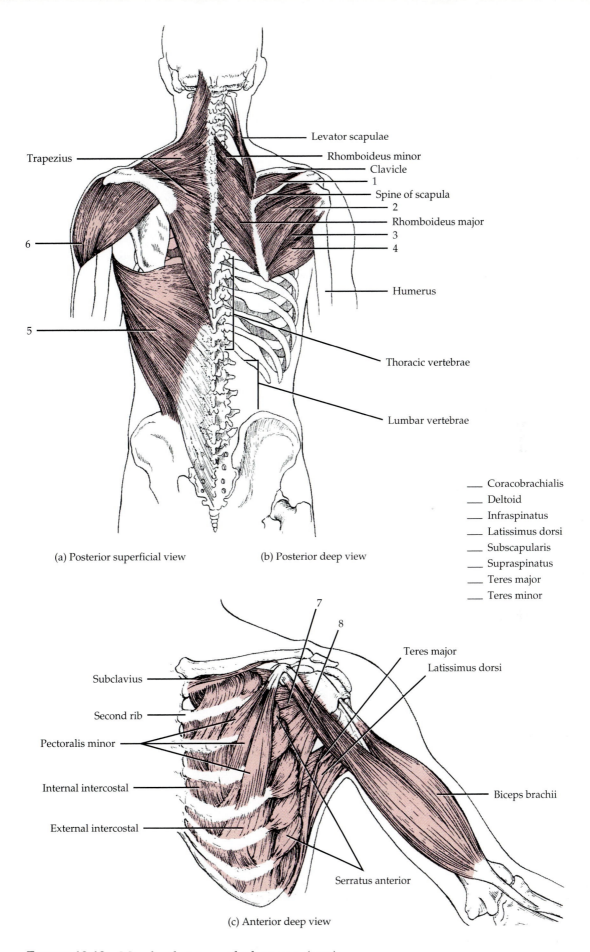

Levator scapulae

Trapezius

Rhomboideus minor

Clavicle

1

Spine of scapula

2

Rhomboideus major

3

4

6

Humerus

5

Thoracic vertebrae

Lumbar vertebrae

(a) Posterior superficial view   (b) Posterior deep view

___ Coracobrachialis
___ Deltoid
___ Infraspinatus
___ Latissimus dorsi
___ Subscapularis
___ Supraspinatus
___ Teres major
___ Teres minor

7

8

Teres major
Latissimus dorsi

Subclavius

Second rib

Pectoralis minor

Biceps brachii

Internal intercostal

External intercostal

Serratus anterior

(c) Anterior deep view

FIGURE 10.10   Muscles that move the humerus (arm).

TABLE **10.14**
Muscles That Move the Radius and Ulna (Forearm) (After completing the table, label Figure 10.11.)

OVERVIEW: Most of the muscles that move the radius and ulna (forearm) are involved
   in flexion and extension at the elbow, which is a hinge joint. The biceps brachii,
   brachialis, and brachioradialis muscles are the flexor muscles. The extensor muscles are
   the triceps brachii and the anconeus.
The biceps brachii is a large muscle located on the anterior surface of the arm. As indicated by its name, it has two heads of origin (long and short), both from the scapula.
   The muscle spans both the shoulder and elbow joints. In addition to its role in flexing
   the forearm at the elbow joint, it also supinates the forearm at the radioulnar joints and
   flexes the arm at the shoulder joint. The brachialis is deep to the biceps brachii muscle.
   It is the most powerful flexor of the forearm at the elbow joint. For this reason, it is
   called the "workhorse" of the elbow flexors. The brachioradialis flexes the forearm at
   the elbow joint, especially when a quick movement is required or when a weight is
   lifted slowly during flexion of the forearm.
The triceps brachii is the large muscle located on the posterior surface of the arm. It is the
   more powerful of the extensors of the forearm at the elbow joint. As its name implies, it
   has three heads of origin, one from the scapula (long head) and two from the humerus
   (lateral and medial heads). The long head crosses the shoulder joint; the other heads do
   not. The anconeus is a small muscle located on the lateral part of the posterior aspect of
   the elbow that assists the triceps brachii in extending the forearm at the elbow joint.
Some muscles that move the radius and ulna are involved in pronation and supination at
   the radioulnar joints. The pronators, as suggested by their names, are the pronator teres
   and pronator quadratus muscles. The supinator of the forearm is aptly named the supinator muscle. You use the powerful action of the supinator when you twist a corkscrew
   or turn a screw with a screwdriver.
In the limbs, functionally related skeletal muscles and their associated blood vessels and
   nerves are grouped together by fascia into regions called **compartments.** In the arm, the
   biceps brachii, brachialis, and coracobrachialis muscles constitute the *flexor compartment;*
   the triceps brachii muscle forms the *extensor compartment.*

Muscle	Origin	Insertion	Action
FLEXORS **Biceps brachii** (BĪ-ceps BRĀ-kē-ī; *biceps* = two heads of origin; *brachion* = arm)			
Brachialis (brā'-kē-A-lis)			
Brachioradialis (bra'-kē-ō-rā'-dē-A-lis; *radialis* = radius) (See also Figure 10.12a)			
EXTENSORS **Triceps brachii** (TRĪ-ceps BRĀ-kē-ī; *triceps* = three heads of origin)			

Anconeus (an-KŌ-nē-us;
anconeal = pertaining to the elbow)
   (label this muscle
   in Figure 10.12)			

TABLE 10.14 *(Continued)*

Muscle	Origin	Insertion	Action
PRONATORS **Pronator teres** (PRŌ-na'-tor TE-rēz; (*pronation* = turning palm downward or posteriorly)			
Pronator quadratus (PRŌ-na'-tor kwod-RĀ-tus; *quadratus* = squared, four-sided) (see Figure 10.12)			
SUPINATOR **Supinator** (SUP-pi-nā-tor; *supination* = turning palm upward or anteriorly)			

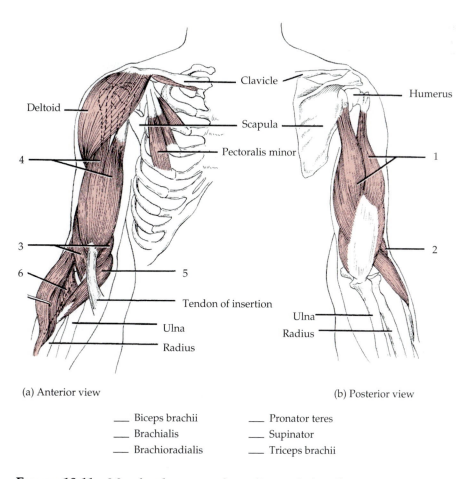

(a) Anterior view   (b) Posterior view

___ Biceps brachii   ___ Pronator teres
___ Brachialis   ___ Supinator
___ Brachioradialis   ___ Triceps brachii

FIGURE 10.11   Muscles that move the radius and ulna (forearm).

TABLE 10.15
Muscles That Move the Wrist, Hand, and Digits (Figure 10.12)

OVERVIEW: Muscles that move the wrist, hand, and digits are many and varied. However, as you will see, their names usually give some indication of their origin, insertion, or action. On the basis of location and function, the muscles are divided into two groups—anterior and posterior compartments. The *anterior compartment muscles* function as flexors. They originate on the humerus and typically insert on the carpals, metacarpals, and phalanges. The bellies of these muscles form the bulk of the proximal forearm. The *posterior compartment muscles* function as extensors. These muscles arise on the humerus and insert on the metacarpals and phalanges. Each of the two principal groups is also divided into superficial and deep muscles.

The superficial anterior compartment muscles are arranged in the following order from lateral to medial: flexor carpi radialis, palmaris longus (absent in about 10% of the population), and flexor carpi ulnaris (the ulnar nerve and artery are just lateral to the tendon of this muscle at the wrist). The flexor digitorum superficialis muscle is actually deep to the other three muscles and is the largest superficial muscle in the forearm.

The deep anterior compartment muscles are arranged in the following order from lateral to medial: flexor pollicis longus (the only flexor of the distal phalanx of the thumb) and flexor digitorum profundus (ends in four tendons that insert into the distal phalanges of the fingers).

The superficial posterior compartment muscles are arranged in the following order from lateral to medial: extensor carpi radialis longus, extensor carpi radialis brevis, extensor digitorum (occupies most of the posterior surface of the forearm and divides into four tendons that insert into the middle and distal phalanges of the fingers), extensor digiti minimi (a slender muscle generally connected to the extensor digitorum), and extensor carpi ulnaris.

The deep posterior compartment muscles are arranged in the following order from lateral to medial: abductor pollicis longus, extensor pollicis brevis, extensor pollicis longus, and extensor indicis.

The tendons of the muscles of the forearm that attach to the wrist or continue into the hand, along with blood vessels and nerves, are held close to bones by strong fascial structures. The tendons are also surrounded by tendon sheaths. At the wrist, the deep fascia is thickened into fibrous bands called *retinacula* (*retinere* = retain). The *flexor retinaculum (transverse carpal ligament)* is located over the palmar surface of the carpal bones. Through it pass the long flexor tendons of the digits and wrist and median nerve. The *extensor retinaculum (dorsal carpal ligament)* is located over the posterior surface of the carpal bones. Through it pass the extensor tendons of the wrist and digits.

After you have studied the muscles in the table, arrange them according to the following actions: flexion, extension, abduction, adduction, supination, and pronation. (The same muscles can be used more than once.)

Muscle	Origin	Insertion	Action
ANTERIOR GROUP (flexors)			
Superficial			
Flexor carpi radialis (FLEK-sor KAR-pē rā′-dē-A-lis; *flexor* = decreases angle at a joint; *carpus* = wrist; *radialis* = radius)			
Palmaris longus (pal-MA-ris LON-gus; *palma* = palm *longus* = long)			
Flexor carpi ulnaris (FLEK-sor KAR-pē ul-NAR-is; *ulnaris* = ulna)			
Flexor digitorum superficialis (FLEK-sor di′-ji-TOR-um soo′-per-fish′-ē-A-lis; *digit* = finger or toe; *superficialis* = closer to surface)			

TABLE 10.15 (*Continued*)

Muscle	Origin	Insertion	Action
Deep			
Flexor digitorum profundus (FLEK-sor di'-ji-TOR-um pro-FUN-dus; *profundus* = deep)			
Flexor pollicis longus (FLEK-sor POL-li-kis LON-gus; *pollex* = thumb)			
POSTERIOR GROUP (extensors) Superficial			
Extensor carpi radialis longus (eks-TEN-sor KAR-pē rā'-dē-A-lis LON-gus; *extensor* = increases angle at a joint)			
Extensor carpi radialis brevis (eks-TEN-sor KAR-pē rā'-dē-A-lis BREV-is; *brevis* = short)			
Extensor digitorum (eks-TEN-sor di'-ji-TOR-um)			
Extensor digiti minimi (eks-TEN-sor DIJ-i-tē MIN-i-mē; *digiti* = digit; *minimi* = finger)			
Extensor carpi ulnaris (eks-TEN-sor KAR-pē ul-NAR-is)			
Deep			
Abductor pollicis longus (ab-DUK-tor POL-li-kis LON-gus; *abductor* = moves a part away from midline)			
Extensor pollicis brevis (eks-TEN-sor POL-li-kis BREV-is)			

TABLE 10.15 (*Continued*)

Muscle	Origin	Insertion	Action
Extensor pollicis longus (eks-TEN-sor POL-li-kis LON-gus)			
Extensor indicis (eks-TEN-sor IN-di-kis; *indicis* = index)			

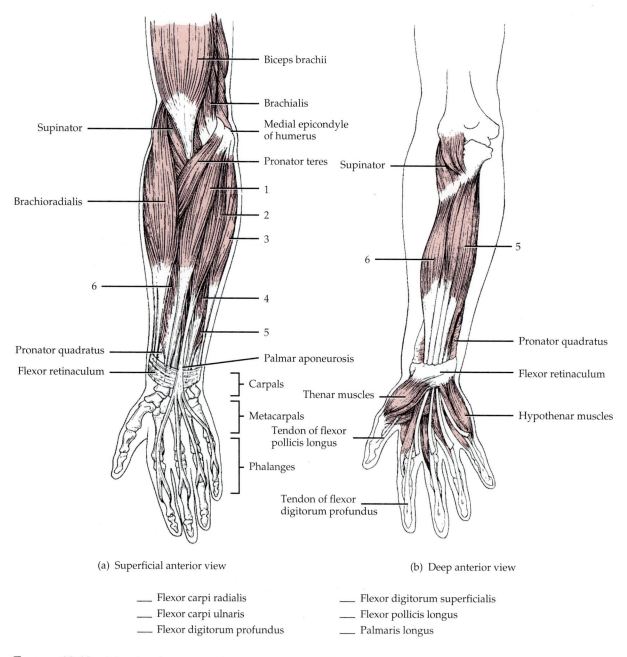

(a) Superficial anterior view   (b) Deep anterior view

___ Flexor carpi radialis   ___ Flexor digitorum superficialis
___ Flexor carpi ulnaris   ___ Flexor pollicis longus
___ Flexor digitorum profundus   ___ Palmaris longus

FIGURE 10.12   Muscles that move the wrist, hand, and digits.

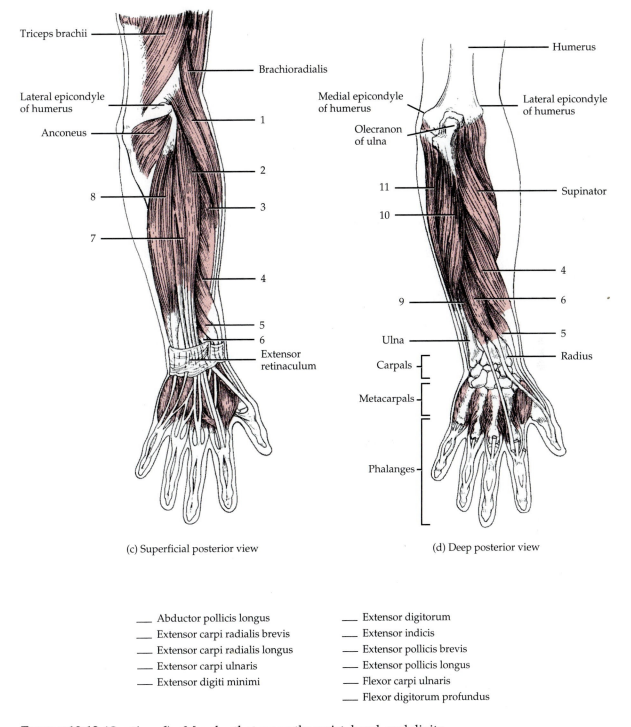

Triceps brachii

Brachioradialis

Lateral epicondyle of humerus

Anconeus

1

2

3

8

7

4

5

6

Extensor retinaculum

(c) Superficial posterior view

Humerus

Medial epicondyle of humerus

Olecranon of ulna

Lateral epicondyle of humerus

Supinator

11

10

4

9

6

5

Ulna

Radius

Carpals

Metacarpals

Phalanges

(d) Deep posterior view

___ Abductor pollicis longus

___ Extensor carpi radialis brevis

___ Extensor carpi radialis longus

___ Extensor carpi ulnaris

___ Extensor digiti minimi

___ Extensor digitorum

___ Extensor indicis

___ Extensor pollicis brevis

___ Extensor pollicis longus

___ Flexor carpi ulnaris

___ Flexor digitorum profundus

FIGURE 10.12 (Continued) Muscles that move the wrist, hand, and digits.

TABLE 10.16
Intrinsic Muscles of the Hand.

OVERVIEW: Several of the muscles discussed in Table 10.15 move the digits in various ways and are known as *extrinsic muscles.* They produce the powerful but crude movements of the digits. The *intrinsic muscles* in the palm produce weak but intricate and precise movements of the digits that characterize the human hand. The muscles in this group are so named because their origins and insertions are *within* the hands.

The intrinsic muscles of the hand are divided into three groups: (1) *thenar* (THĒ-nar), (2) *hypothenar* (HĪ-pō-thē-nar), and (3) *intermediate.* The four thenar muscles shown in Figure 10.12, act on the thumb and form the *thenar eminence* (see Figure 11.8), the lateral rounded contour on the palm that is also called the ball of the thumb. The thenar muscles include the abductor pollicis brevis, opponens pollicis, flexor pollicis brevis, and adductor pollicis. The abductor pollicis brevis muscle is a thin, short, relatively broad superficial muscle on the lateral side of the thenar eminence. The opponens pollicus muscle is a small, triangular muscle that is deep to the abductor pollicis brevis muscle. The flexor pollicis brevis muscle is a short, wide muscle that is medial to the abductor pollicis brevis muscle. The adductor pollicis muscle is fan-shaped and has two heads (oblique and transverse) separated by a gap through which the radial artery passes.

The three hypothenar muscles, shown in Figure 10.12, act on the little finger and form the *hypothenar eminence* (see Figure 11.8), the medial rounded contour on the palm that is also called the ball of the little finger. The hypothenar muscles are the abductor digiti minimi, flexor digiti minimi brevis, and opponens digiti minimi. The abductor digiti minimi muscle is a short, wide muscle and is the most superficial of the hypothenar muscles. It is a powerful muscle that plays an important role in grasping a large object with outspread fingers. The flexor digiti minimi brevis muscle is also short and wide and is lateral to the abductor digiti minimi muscle. The opponens digiti minimi muscle is triangular and deep to the other two hypothenar muscles.

The 12 intermediate (midpalmar) muscles act on all the digits except the thumb. The intermediate muscles include the lumbricals, palmar interossei, and dorsal interossei. The lumbricals, as their name indicates, are worm-shaped. They originate from and insert into the tendons of other muscles. The palmar interossei muscles are the superficial and smaller of the interossei muscles. The dorsal interossei muscles are the deep interossei muscles. Both sets of interossei muscles are located between the metacarpals and are important in abduction, adduction, flexion, and extension of the fingers, important movements in skilled activities such as writing, typing, and playing a piano.

The functional importance of the hand is readily apparent when one considers that certain hand injuries can result in permanent disability. Most of the dexterity of the hand depends on the movements of the thumb. The general activities of the hand are free motion, power grip (forcible movement of the fingers and thumb against the palm, as in squeezing), precision handling (a change in position of a handled object that requires exact control of finger and thumb positions, as in winding a watch or threading a needle), and pinch (compression between the thumb and index finger or between the thumb and first two fingers).

Movements of the thumb are very important in the precise activities of the hand, and they are defined in different planes from comparable movements of other digits because the thumb is positioned at a right angle to the other digits. The five principal movements of the thumb, illustrated below, are flexion (movement of the thumb medially across the palm), extension (movement of the thumb laterally away from the palm), abduction (movement of the thumb in an anteroposterior plane away from the palm), adduction (movement of the thumb in an anteroposterior plane toward the palm), and opposition (movement of the thumb across the palm so that the tip of the thumb meets the tips of a finger). Opposition is the single most distinctive digital movement that gives humans and other primates the ability to precisely grasp and manipulate objects.

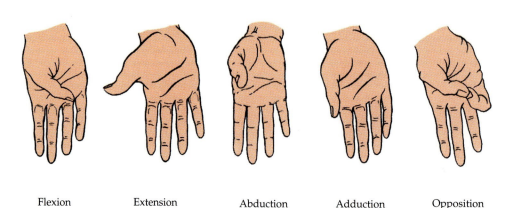

Flexion   Extension   Abduction   Adduction   Opposition

TABLE 10.17
Muscles That Move the Femur (Thigh) (After completing the table, label Figure 10.13.)

OVERVIEW: As you will see, muscles of the lower limbs are larger and more powerful
than those of the upper limbs since lower limb muscles function in stability, locomotion,
and maintenance of posture. Upper limb muscles are characterized by versatility of
movement. In addition, muscles of the lower limbs frequently cross two joints and act
equally on both.

The majority of muscles that move the femur originate from the pelvic girdle and insert on
the femur. The psoas major and iliacus muscles are together referred to as the iliopsoas
(il'-ē -ō-SŌ- as) muscle because they share a common insertion. There are three gluteal
muscles: gluteus maximus, gluteus medius, and gluteus minimus. The gluteus maximus
is the largest and heaviest of the three muscles and is one of the largest muscles in the
body. It is the chief extensor of the femur. The gluteus medius is mostly deep to the
gluteus maximus and is a powerful abductor of the femur at the hip joint. This muscle is
a common site for an intramuscular injection. The gluteus minimus muscle is the smallest of the gluteal muscles and lies deep to the gluteus medius.

The tensor fasciae latae muscle is located on the lateral surface of the thigh. There is a
layer of deep fascia composed of dense connective tissue that encircles the entire thigh
and is referred to as the *fascia lata*. It is well developed laterally where, together with
the tendons of the tensor fasciae and gluteus maximus muscles, it forms a structure
called the *iliotibial tract*. The tract inserts into the lateral condyle of the tibia.

The piriformis, obturator internus, obturator externus, superior gemellus, inferior gemellus, and quadratus femoris muscles are all deep to the gluteus maximus muscle and
function as lateral rotators of the femur at the hip joint.

Three muscles on the medial aspect of the thigh are the adductor longus, adductor brevis,
and adductor magnus. They originate on the pubic bone and insert on the femur. All
three muscles adduct, flex, and medially rotate the femur at the hip joint. The pectineus
muscle also adducts and flexes the femur at the hip joint.

Technically, the adductor muscles and pectineus muscles are components of the medial
compartment of the thigh and could also be included in Table 10.18. However, they are
included here because they act on the femur.

After you have studied the muscles in this table, arrange them according to the following
actions: flexion, extension, abduction, adduction, medial rotation, and lateral rotation.
(The same muscles can be used more than once.)

Muscle	Origin	Insertion	Action
Psoas (SŌ-as) major (*psoa* = muscle of loin)			
Iliacus (il'-ē-AK-us; *iliac* = ilium)			
Gluteus maximus (GLOO-tē-us MAK-si-mus; *glutos* = buttock; *maximus* = largest)			
Gluteus medius (GLOO-tē-us MĒ-dē-us; *media* = middle)			
Gluteus minimus (GLOO-tē-us MIN-i-mus; *minimus* = smallest)			

TABLE 10.17 (*Continued*)

Muscle	Origin	Insertion	Action
Tensor fasciae latae (TEN-sor FA-shē-ē LĀ-tē; *tensor* = makes tense; *fascia* = band; *latus* = wide)			
Piriformis (pir-i-FOR-mis; *pirum* = pear; *forma* = shape)			
Obturator internus* (OB-too-rā'-tor in-TER-nus; *obturator* = obturator foramen; *internus* = inside)			
Obturator externus (OB-too-rā'-tor ex-TER-nus; *externus* = outside)			
Superior gemellus (jem-EL-lus; *superior* = above; *gemellus* = twins)			
Inferior gemellus (jem-EL-lus; *inferior* = below)			
Quadratus femoris (kwod-RĀ-tus FEM-or-is; *quad* = four; *femoris* = femur)			
Adductor longus (LONG-us; *adductor* = moves part closer to midline; *longus* = long)			
Adductor brevis (BREV-is; *brevis* = short)			
Adductor magnus (MAG-nus; *magnus* = large)			
Pectineus (pek-TIN-ē-us; *pecten* = comb-shaped)			

*Not illustrated

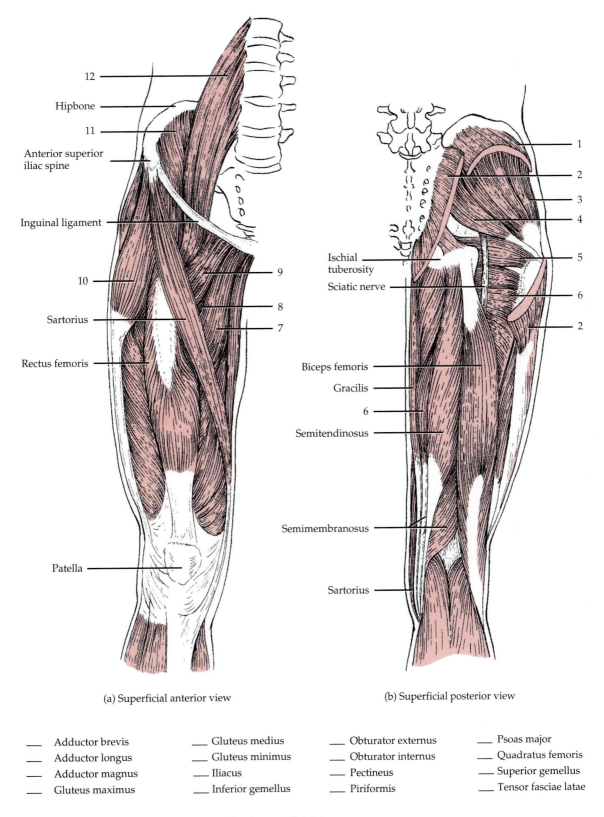

12

Hipbone

11

Anterior superior
iliac spine

Inguinal ligament

10

Sartorius

Rectus femoris

Patella

9

8

7

1

2

3

4

5

6

2

Ischial
tuberosity

Sciatic nerve

Biceps femoris

Gracilis

6

Semitendinosus

Semimembranosus

Sartorius

(a) Superficial anterior view

(b) Superficial posterior view

___ Adductor brevis

___ Adductor longus

___ Adductor magnus

___ Gluteus maximus

___ Gluteus medius

___ Gluteus minimus

___ Iliacus

___ Inferior gemellus

___ Obturator externus

___ Obturator internus

___ Pectineus

___ Piriformis

___ Psoas major

___ Quadratus femoris

___ Superior gemellus

___ Tensor fasciae latae

FIGURE 10.13   Muscles that move the femur (thigh).

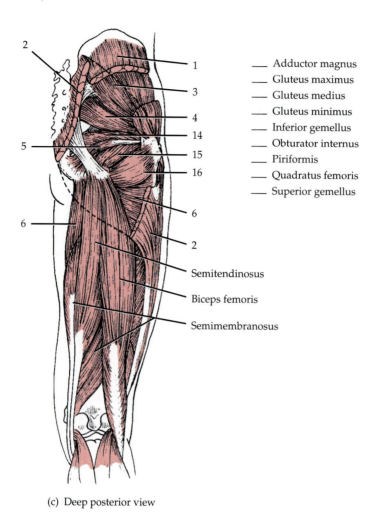

____ Adductor magnus
____ Gluteus maximus
____ Gluteus medius
____ Gluteus minimus
____ Inferior gemellus
____ Obturator internus
____ Piriformis
____ Quadratus femoris
____ Superior gemellus

Semitendinosus

Biceps femoris

Semimembranosus

(c) Deep posterior view

FIGURE 10.13 (*Continued*)   Muscles that move the femur (thigh).

TABLE 10.18
Muscles That Act on the Tibia and Fibula (Leg) (After completing the table, label Figure 10.14.)

OVERVIEW: The muscles that act on the femur (thigh) and tibia and fibula (leg) are
separated into three compartments by deep fascia: medial, anterior, and posterior. The
medial (adductor) compartment is so named because its muscles adduct the femur at
the hip joint. As noted earlier, the adductor magnus, adductor longus, adductor brevis,
and pectineus, components of the medial compartment, are included in Table 10.20
because they act on the femur. The gracilis, the other muscle in the medial compartment, not only adducts the thigh but also flexes the leg at the knee joint. For this reason,
it is included in this table. The gracilis is a long, straplike muscle that lies on the medial
aspect of the thigh and knee.

The *anterior (extensor) compartment* is so designated because its muscles extend the leg
at the knee joint (and also flex the thigh at the hip joint). This compartment is composed
of the quadriceps femoris and sartorius muscles. The quadriceps femoris muscle is the
biggest muscle in the body, covering almost all of the anterior surface and sides of the
thigh. The muscle is actually a composite muscle that includes four distinct parts, usually described as four separate muscles: (1) rectus femoris, on the anterior aspect of the
thigh; (2) vastus lateralis, on the lateral aspect of the thigh; (3) vastus medialis, on the
medial aspect of the thigh; and (4) vastus intermedius, located deep to the rectus
femoris between the vastus lateralis and medialis. The common tendon for the four
muscles is known as the *quadriceps tendon,* which inserts into the patella. The tendon
continues inferior to the patella as the *patellar ligament,* which attaches to the tibial
tuberosity. The quadriceps femoris muscle is the great extensor muscle of the leg at the
knee joint. The sartorius muscle is a long, narrow muscle that forms a band across the
thigh from the ilium of the hipbone to the medial side of the tibia. The various movements it produces help effect the cross-legged sitting position of tailors in which the heel
of one limb is placed on the knee of the opposite limb. It is known as the tailor's muscle.

The *posterior (flexor) compartment* is so named because its muscles flex the leg at the
knee joint (and also extend the thigh at the hip joint). This compartment is composed of
three muscles collectively called the hamstrings: (1) biceps femoris, (2) semitendinosus,
and (3) semimembranosus. The hamstrings are so named because their tendons are long
and stringlike in the popliteal area and from an old practice of butchers in which they
hung hams for smoking by these long tendons. Since the hamstrings span two joints
(hip and knee), they are both extensors of the thigh at the hip joint and flexors of the leg
at the knee joint. The *popliteal fossa* is a diamond-shaped space on the posterior aspect
of the knee bordered laterally by the tendons of the biceps femoris muscle and medially
by the semitendinosus and semimembranosus muscles (see also Figure 11.11).

Muscle	Origin	Insertion	Action
MEDIAL (ADDUCTOR) COMPARTMENT			
Adductor magnus (MAG-nus)			
Adductor longus (LONG-us)			
Adductor brevis (BREV-is)			
Pectineus (pek-TIN-ē-us)			
Gracilis (gra-SIL-is; *gracilis* = slender)			

TABLE 10.18 (*Continued*)

Muscle	Origin	Insertion	Action
ANTERIOR (EXTENSOR) COMPARTMENT **Quadriceps femoris** (KWOD-ri-ceps FEM-or-is; *quadriceps* = four heads of origin; *femoris* = femur)			
Rectus femoris (REK-tus FEM-or-is; *rectus* = fibers parallel to midline)			
Vastus lateralis (VAS-tus lat'-er-A-lis; *vastus* = large; *lateralis* = lateral)			
Vastus medialis (VAS-tus mē-dē-A-lis; *medialis* = medial)			
Vastus intermedius (VAS-tus in'-ter-MĒ-dē-us; *intermedius* = middle)			
Sartorius (sar-TOR-ē-us; *sartor* = tailor; refers to cross-legged position of tailors)			
POSTERIOR (FLEXOR) COMPARTMENT **Hamstrings** **Biceps femoris** (BĪ-ceps FEM-or-is; *biceps* = two heads of origin)			
Semitendinosus (sem'-ē-TEN-di-nō-sus; *semi* = half; *tendo* = tendon)			
Semimembranosus (sem'-ē-MEM-bra-nō-sus; *membran* = membrane)			

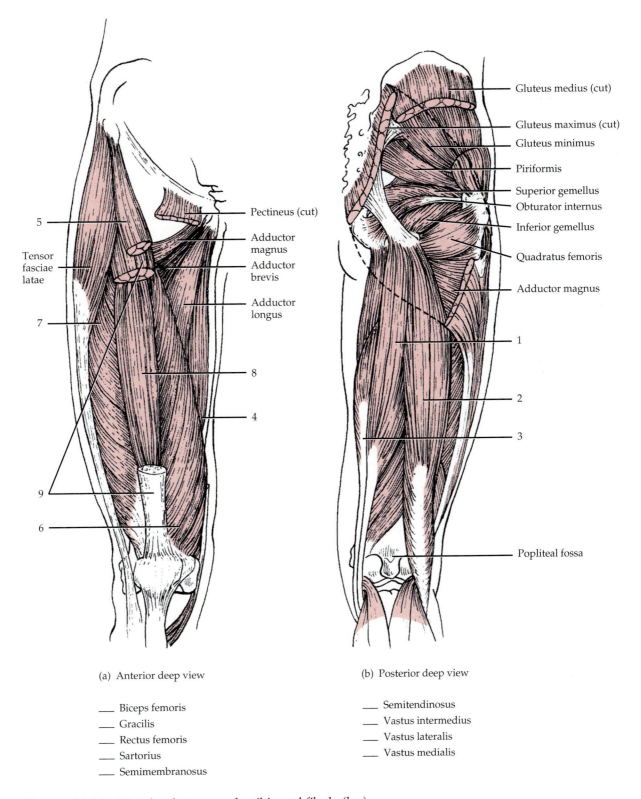

Gluteus medius (cut)

Gluteus maximus (cut)

Gluteus minimus

Piriformis

Superior gemellus

Obturator internus

Inferior gemellus

Quadratus femoris

Adductor magnus

Pectineus (cut)

Adductor magnus

Adductor brevis

Adductor longus

Tensor fasciae latae

5

7

8

4

9

6

1

2

3

Popliteal fossa

(a) Anterior deep view

(b) Posterior deep view

___ Biceps femoris

___ Gracilis

___ Rectus femoris

___ Sartorius

___ Semimembranosus

___ Semitendinosus

___ Vastus intermedius

___ Vastus lateralis

___ Vastus medialis

FIGURE 10.14   Muscles that act on the tibia and fibula (leg).

TABLE 10.19

Muscles That Move the Foot and Toes (After completing the table, label Figure 10.15.)

OVERVIEW: The musculature of the leg, like that of the thigh, is divided into three compartments by deep fascia: anterior, lateral, and posterior. The *anterior compartment*
consists of muscles that dorsiflex the foot. In a situation analogous to the wrist, the
tendons of the muscles of the anterior compartment are held firmly to the ankle by
thickenings of deep fascia called the *superior extensor retinaculum (transverse ligament
of the ankle)* and *inferior extensor retinaculum (cruciate ligament of the ankle).*
Within the anterior compartment, the tibialis anterior muscle is a long, thick muscle
against the anterolateral surface of the tibia, where it is easy to palpate. The extensor
hallucis muscle is a thin muscle that lies deep to the tibialis anterior and extensor digitorum longus muscles. This latter muscle is featherlike and lies lateral to the tibialis anterior muscle, where it can easily be palpated. The peroneus tertius muscle is actually part
of the extensor digitorum longus muscle, with which it shares a common origin.
The *lateral (peroneal) compartment* contains two muscles that plantar flex and evert the
foot: peroneus longus and peroneus brevis.
The *posterior compartment* consists of muscles that are divisible into superficial and deep
groups. The superficial muscles share a common tendon of insertion, the *calcaneal
(Achilles) tendon,* the strongest tendon of the body, that inserts into the calcaneus bone
of the ankle. The superficial muscles and most deep muscles plantar flex the foot at the
ankle joint.
The superficial muscles of the posterior compartment are the gastrocnemius, soleus, and
plantaris, the so-called calf muscles. The large size of these muscles is directly related to
our upright stance, a characteristic of humans. The gastrocnemius muscle is the most
superficial muscle and forms the prominence of the calf. The soleus is a broad, flat muscle, named because of its resemblance to a flat fish (sole) and lies deep to the gastrocnemius. The plantaris is a small muscle that may be absent or even sometimes present as a
double. It runs obliquely between the gastrocnemius and soleus muscles.
The deep muscles of the posterior compartment are the popliteus, tibialis posterior, flexor
digitorum longus, and flexor hallucis longus. The popliteus muscle is triangular and
forms the floor of the popliteal fossa. The tibialis posterior muscle is the deepest muscle
in the posterior compartment. It lies between the flexor digitorum longus and flexor
hallucis longus muscles. The flexor digitorum muscle is smaller than the flexor hallucis
longus muscle, even though the former flexes four toes, while the latter flexes only the
great toe at the interphalangeal joints.

Muscle	Origin	Insertion	Action
ANTERIOR COMPARTMENT **Tibialis** (tib'-ē-A-lis) **anterior** (*tibialis* = tibia; *anterior* = front)			
Extensor hallucis longus (HAL-u-kis LON-gus; *extensor* = increases angle at joint; *hallucis* = hallux or great toe; *longus* = long)			
Extensor digitorum longus (di'-ji-TOR-um LON-gus)			
Peroneus tertius (per'-ō-NĒ-us TER-shus; *perone* = fibula; *tertius* = third)			

TABLE 10.19 (*Continued*)

Muscle	Origin	Insertion	Action
LATERAL (PERONEAL) COMPARTMENT			
Peroneus longus (per'-ō-NĒ-us LON-gus)			
Peroneus brevis (per'-ō-NĒ-us BREV-is; *brevis* = short)			
POSTERIOR COMPARTMENT			
Superficial			
Gastrocnemius (gas'-trok-NĒ-mē-us; *gaster* = belly; *kneme* = leg)			
Soleus (SŌ-lē-us; *soleus* = sole)			
Plantaris (plan-TA-ris; *plantar* = sole)			
Deep			
Popliteus (pop-LIT-ē-us; *poples* = posterior surface of knee)			
Tibialis (tib'-ē-A-lis) **posterior** (*posterior* = back)			
Flexor digitorum longus (di'-ji-TOR-um LON-gus; *digitorum* = finger or toe)			
Flexor hallucis longus (HAL-u-kis LON-gus; *flexor* = decreases angle at joint)			

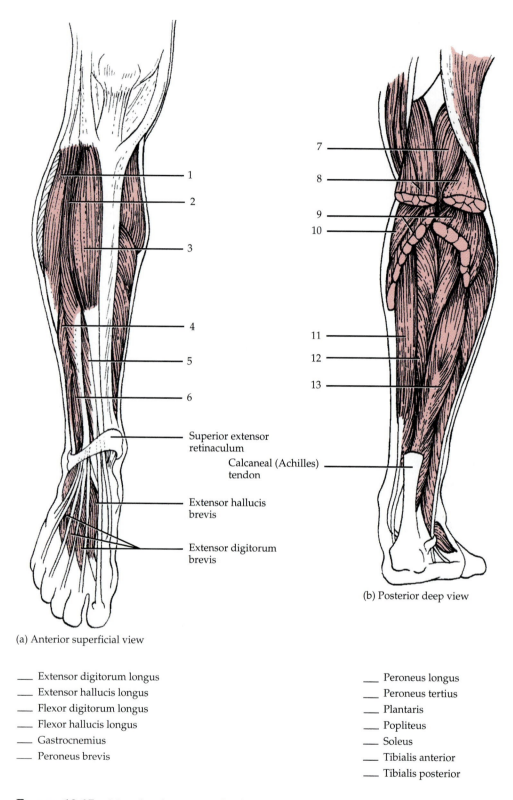

(a) Anterior superficial view

Superior extensor retinaculum

Calcaneal (Achilles) tendon

Extensor hallucis brevis

Extensor digitorum brevis

(b) Posterior deep view

___ Extensor digitorum longus
___ Extensor hallucis longus
___ Flexor digitorum longus
___ Flexor hallucis longus
___ Gastrocnemius
___ Peroneus brevis

___ Peroneus longus
___ Peroneus tertius
___ Plantaris
___ Popliteus
___ Soleus
___ Tibialis anterior
___ Tibialis posterior

FIGURE 10.15   Muscles that move the foot and toes.

F. COMPOSITE MUSCULAR SYSTEM

Now that you have studied the muscles of the body by region, label the composite diagram shown in Figure 10.16.

ANSWER THE LABORATORY REPORT QUESTIONS AT THE END OF THE EXERCISE.

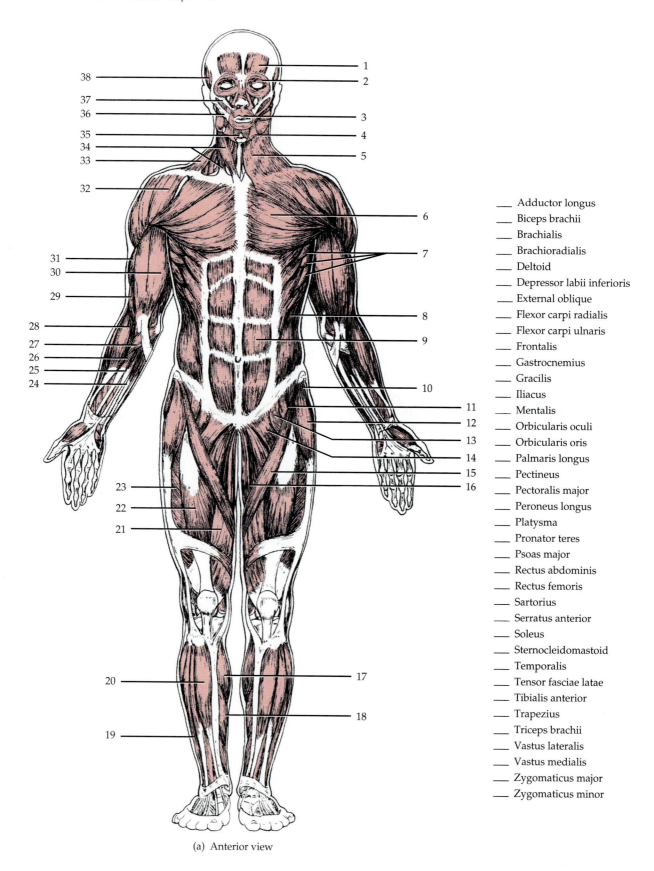

___ Adductor longus
___ Biceps brachii
___ Brachialis
___ Brachioradialis
___ Deltoid
___ Depressor labii inferioris
___ External oblique
___ Flexor carpi radialis
___ Flexor carpi ulnaris
___ Frontalis
___ Gastrocnemius
___ Gracilis
___ Iliacus
___ Mentalis
___ Orbicularis oculi
___ Orbicularis oris
___ Palmaris longus
___ Pectineus
___ Pectoralis major
___ Peroneus longus
___ Platysma
___ Pronator teres
___ Psoas major
___ Rectus abdominis
___ Rectus femoris
___ Sartorius
___ Serratus anterior
___ Soleus
___ Sternocleidomastoid
___ Temporalis
___ Tensor fasciae latae
___ Tibialis anterior
___ Trapezius
___ Triceps brachii
___ Vastus lateralis
___ Vastus medialis
___ Zygomaticus major
___ Zygomaticus minor

(a) Anterior view

FIGURE 10.16   Principal superficial muscles.

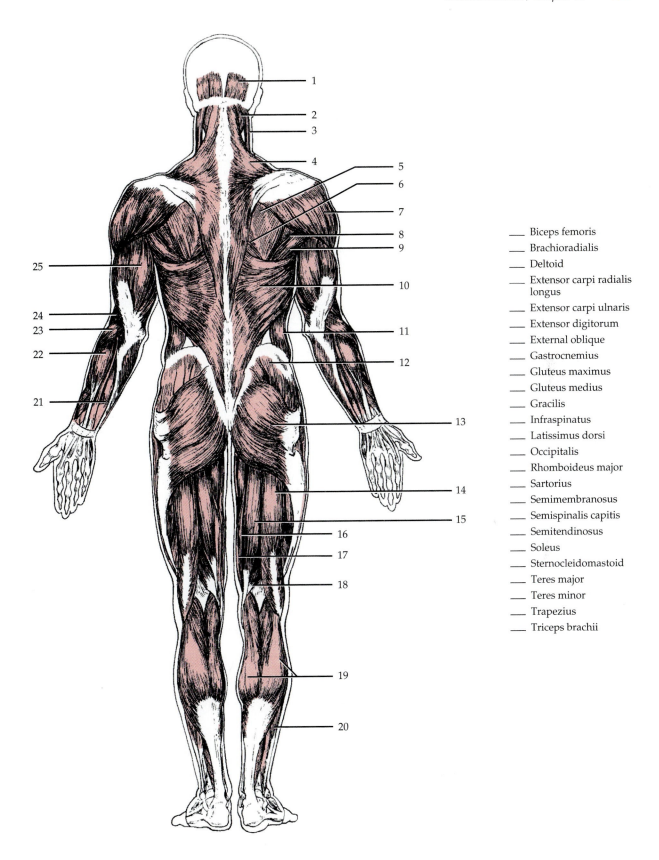

Biceps femoris
Brachioradialis
Deltoid
Extensor carpi radialis longus
Extensor carpi ulnaris
Extensor digitorum
External oblique
Gastrocnemius
Gluteus maximus
Gluteus medius
Gracilis
Infraspinatus
Latissimus dorsi
Occipitalis
Rhomboideus major
Sartorius
Semimembranosus
Semispinalis capitis
Semitendinosus
Soleus
Sternocleidomastoid
Teres major
Teres minor
Trapezius
Triceps brachii

(b) Posterior view

FIGURE 10.16 (*Continued*) Principal superficial muscles.

Skeletal Muscles 10

Student _____   Date _____

Laboratory Section _____   Score/Grade _____

PART 1. Multiple Choice

_____ 1. The connective tissue covering that encloses the entire skeletal muscle is the (a) perimysium (b) endomysium (c) epimysium (d) mesomysium

_____ 2. A cord of connective tissue that attaches a skeletal muscle to the periosteum of bone is called a(n) (a) ligament (b) aponeurosis (c) perichondrium (d) tendon

_____ 3. A skeletal muscle that decreases the angle at a joint is referred to as a(n) (a) flexor (b) abductor (c) pronator (d) evertor

_____ 4. The name *abductor* means that a muscle (a) produces a downward movement (b) moves a part away from the midline (c) elevates a body part (d) increases the angle at a joint

_____ 5. Which connective tissue layer directly encircles the fascicles of skeletal muscles? (a) epimysium (b) endomysium (c) perimysium (d) mesomysium

_____ 6. Which muscle is *not* associated with a movement of the eyeball? (a) superior rectus (b) superior oblique (c) medial rectus (d) external oblique

_____ 7. Of the following, which muscle is involved in compression of the abdomen? (a) external oblique (b) superior oblique (c) medial rectus (d) genioglossus

_____ 8. A muscle directly concerned with breathing is the (a) sternocleidomastoid (b) mentalis (c) brachialis (d) external intercostal

_____ 9. Which muscle is *not* related to movement of the wrist? (a) extensor carpi ulnaris (b) flexor carpi radialis (c) supinator (d) flexor carpi ulnaris

_____ 10. A muscle that helps move the thigh is the (a) piriformis (b) triceps brachii (c) hypoglossus (d) peroneus tertius

_____ 11. Which muscle is *not* related to mastication? (a) temporalis (b) masseter (c) lateral rectus (d) medial pterygoid

_____ 12. Which muscle elevates the tongue? (a) genioglossus (b) styloglossus (c) hyoglossus (d) omohyoid

_____ 13. Which muscle is *not* a component of the anterolateral abdominal wall? (a) external oblique (b) psoas major (c) rectus abdominis (d) internal oblique

_____ 14. All are components of the pelvic diaphragm *except* the (a) anconeus (b) coccygeus (c) iliococcygeus (d) pubococcygeus

_____ 15. Which is *not* a flexor of the forearm? (a) biceps brachii (b) brachialis (c) brachioradialis (d) triceps brachii

_____ 16. Which muscle flexes the wrist? (a) palmaris longus (b) extensor carpi radialis longus (c) supinator (d) extensor indicis

_____ 17. Which muscle is *not* involved in flexion of the thigh? (a) rectus femoris (b) sartorius (c) biceps femoris (d) vastus intermedius

_____ 18. Of the muscles that move the foot and toes, the anterior compartment muscles are involved in (a) plantar flexion (b) dorsiflexion (c) abduction (d) adduction

PART 2. Matching

Identify the characteristic(s) used to name the following muscles:

_____ 19. Supinator	A. Location
_____ 20. Deltoid	B. Shape
_____ 21. Stylohyoid	C. Size
_____ 22. Flexor carpi radialis	D. Direction of fibers
_____ 23. Gluteus maximus	E. Action
_____ 24. External oblique	F. Number of origins
_____ 25. Triceps brachii	G. Insertion and origin
_____ 26. Adductor longus	
_____ 27. Temporalis	
_____ 28. Trapezius	

PART 3. Completion

29. The principal muscle used in compression of the cheek is the _____.

30. The muscle that protracts the tongue is the _____.

31. The eye muscle that rolls the eyeball downward is the _____.

32. The _____ muscle flexes the neck on the chest.

33. The abdominal muscle that flexes the vertebral column is the _____.

34. The muscle of the pectoral (shoulder) girdle that depresses the clavicle is the

_____.

35. Flexion, adduction, and medial rotation of the arm are accomplished by the _____ muscle.

36. The _____ muscle is the most important extensor of the forearm.

37. The muscle that flexes and abducts the wrist is the _____.

38. The four muscles that extend the legs are the vastus lateralis, vastus medialis, vastus intermedius, and

_____.

39. The neck region is divided into two principal triangles by the _____ muscle.

40. Muscles that move the pectoral (shoulder) girdle originate on the axial skeleton and insert on the

clavicle or _____.

41. The muscles that move the humerus (arm) and do not originate on the scapula are called

    _____ muscles.

42. Together, the subscapularis, supraspinatus, infraspinatus, and teres major muscles form the

    _____.

43. The posterior muscles involved in moving the wrist, hand, and fingers function in extension and

    _____.

44. The posterior muscles of the thigh are involved in _____ of the leg.

45. Together, the biceps femoris, semitendinosus, and semimembranosus are referred to as the

    _____ muscles.

46. _____ fascicles attach obliquely from many directions to several tendons.

47. Movement of the thumb medially across the palm is called _____.

Surface Anatomy

11

Now that you have studied the skeletal and muscular systems, you will be introduced to the study of *surface anatomy*, the study of the anatomical landmarks on the surface (exterior) of the body.[1] A knowledge of surface anatomy will help you identify certain superficial structures by visual inspection and palpation through the skin. *Palpation* (pal-PĀ-shun) means using the sense of touch to determine the location of an internal part of the body through the skin. Knowledge of surface anatomy is important in health-related activities such as taking a pulse and blood pressure, listening to internal organs, drawing blood, and inserting needles and tubes.

A convenient way to study surface anatomy is first to divide the body into its principal regions: head, neck, trunk, and upper and lower limbs. These may be reviewed in Figure 2.2.

A. HEAD

The *head* (cephalic region or caput) is divisible into the cranium and face. The *cranium* surrounds and protects the brain. The *face* is the exterior portion of the head. The head also contains the sense organs—eyes, ears, nose, and tongue. Several surface features of the various regions of the head are

1. *Cranium (skull, or brain case)*

   a. *Frontal region* Front of skull that includes frontal bone.
   b. *Parietal region* Crown of skull that includes the parietal bones.
   c. *Temporal region* Side of skull that includes the temporal bones.
   d. *Occipital region* Base of skull that includes the occipital bones.

2. *Face*

   a. *Orbital* or *ocular region* Includes eyeballs, eyebrows, and eyelids.
   b. *Nasal region* Nose.
   c. *Infraorbital region* Inferior to orbit.
   d. *Oral region* Mouth.
   e. *Mental region* Anterior part of mandible.
   f. *Buccal region* Cheek.
   g. *Zygomatic region* Inferolateral to orbit.
   h. *Auricular region* Ear.

Using your textbook as an aid, label Figure 11.1.

Using a mirror, examine the various features of the head just described. Working with a partner, be sure that you can identify the regions by both common *and* anatomical names.

The surface anatomy features of the eyeball and accessory structures are presented in Figure 15.5, of the ear in Figure 15.9, and of the nose in Figure 22.2.

B. NECK

The *neck* (collum) can be divided into an *anterior cervical region*, two *lateral cervical regions*, and a *posterior (nuchal) region*. Among the surface features of the neck are

1. *Thyroid cartilage (Adam's apple)* Triangular laryngeal cartilage in the midline of the anterior cervical region.
2. *Hyoid bone* First resistant structure palpated in the midline inferior to the chin, lying just superior to the thyroid cartilage opposite the superior border of C4.
3. *Cricoid cartilage* Inferior laryngeal cartilage that attaches larynx to trachea. This structure can be palpated by running your fingertip down from your chin over the thyroid cartilage. (After you pass the cricoid cartilage, your fingertip sinks in.) This cartilage is used as a landmark in locating the rings of cartilage in the trachea

---

[1]At the discretion of your instructor, surface anatomy may be studied either before or in conjunction with your study of various body systems. This exercise can be used as an excellent review of many topics already studied.

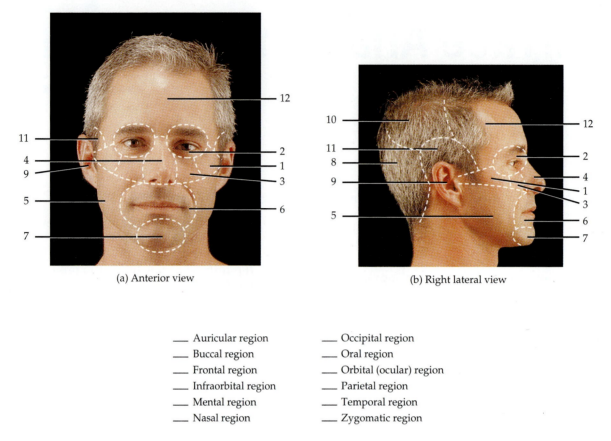

(a) Anterior view   (b) Right lateral view

___ Auricular region   ___ Occipital region
___ Buccal region   ___ Oral region
___ Frontal region   ___ Orbital (ocular) region
___ Infraorbital region   ___ Parietal region
___ Mental region   ___ Temporal region
___ Nasal region   ___ Zygomatic region

FIGURE 11.1   Regions of the cranium and face.

(windpipe) when performing a tracheostomy. The incision is made through the second, third, or fourth tracheal rings and a tube is inserted to assist breathing.

4. *Thyroid gland* Two lobed gland, one on either side of the trachea.

5. *Sternocleidomastoid muscles* Form major portion of lateral cervical regions, extending from mastoid process of temporal bone (felt as bump behind ear) to sternum and clavicle.

6. *Carotid arteries* The common carotid artery is deep to the sternocleidomastoid muscle. At the level of the superior margin of the thyroid cartilage, it divides into internal and external carotid arteries. At this point, the carotid pulse can be detected.

7. *External jugular veins* Prominent veins along lateral cervical regions superficial to the sternocleidomastoid muscles, readily seen when a person is angry or a collar fits too tightly.

8. *Trapezius muscles* Form portion of lateral cervical region, extending inferiorly and laterally

from base of skull. "Stiff neck" is frequently associated with inflammation of these muscles.

9. *Vertebral spines* The spinous processes of the cervical vertebrae which may be felt along the midline of the posterior aspect of the neck. Especially pominant at the base of the neck is the spinous process of C7.

The sternocleidomestoid muscle divides the neck into two major triangles. The anterior triangle is bordered superiorly by the mandible, inferiorly by the sternum, medially by the cervical midline, and laterally by the anterior border of the sternocleidomastoid muscle. The posterior triangle is bordered inferiorly by the clavicle anteriorly by the posterior border of the sternocleidomastoid muscle and posteriorly by the anterior border of the trapezius muscle.

Using a mirror and working with a partner, use your textbook as an aid in identifying the surface features of the neck just described. Then label Figure 11.2.

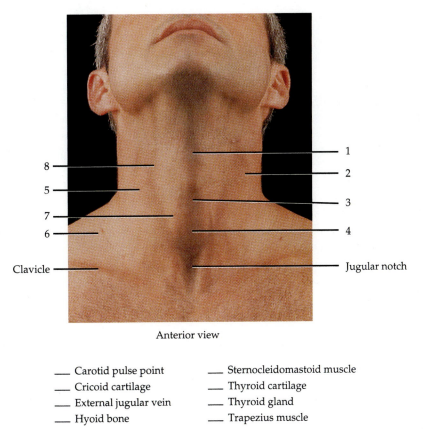

Anterior view

_____ Carotid pulse point   _____ Sternocleidomastoid muscle
_____ Cricoid cartilage   _____ Thyroid cartilage
_____ External jugular vein   _____ Thyroid gland
_____ Hyoid bone   _____ Trapezius muscle

FIGURE 11.2   Surface anatomy of the neck.

C. TRUNK

The *trunk* is divided into the back, chest, abdomen, and pelvis. Its surface features include
Read the descriptions of the surface features of the *back* that follow and then label Figure 11.3.

1. *Spinous processes (spines)* Posteriorly pointed projections of vertebrae that are more prominent when the vertebral column is flexed. The spinous process of C7 *(vertebra prominens)* is the superior of the two prominences found at the base of the neck; the spinous process of T1 is the lower prominence at the base of the neck; the spinous process of T3 is at about the same level as the spinous process of the scapula; the spinous process of T7 is about opposite the inferior angle of the scapula; a line passing through the highest points of the iliac crests, called the *supracristal line,* passes through the spinous process of L4.
2. *Scapula* Shoulder blade. They lie between ribs 2 and 7. Depending on how lean a person is, it might be possible to palpate various parts of the scapulae such as the vertebral border, axillary border, inferior angle, spine, and acromion.
3. *Muscles* Among the visible superficial back muscles are the *latissimus dorsi* (covers lower half of back), *erector spinae* (on either side of vertebral column), *infraspinatus* (inferior to spine of scapula), *trapezius*, and *teres major* (inferior to infraspinatus).
4. *Posterior axillary fold* Formed by the latissimus dorsi and teres major muscles; it can be palpated between the finger and thumb.
5. *Triangle of auscultation* A region of the back just medial to the inferior part of the scapula where the rib cage is not covered by superficial muscles. It is triangular and is formed by the latissimus dorsi, trapezius, and vertebral border of scapula. The space between the muscles in the region permits respiratory sounds to be heard clearly with a stethoscope.

Using your textbook as an aid, label Figure 11.3.
Read the descriptions of the surface features of the chest (thorax) that follows and then label Figure 11.4.

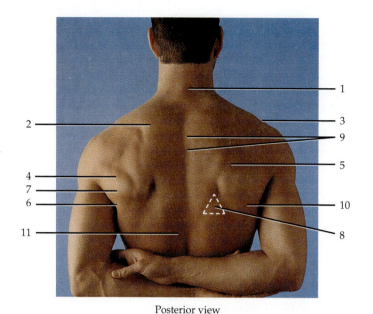

Posterior view

FIGURE 11.3   Surface anatomy of the back.

____ Acromion of scapula
____ Erector spinae muscle
____ Infraspinatus muscle
____ Latissimus dorsi muscle
____ Posterior axillary fold
____ Scapula
____ Spinous processes (spines)
____ Teres major muscle
____ Trapezius muscle
____ Triangle of auscultation
____ Vertebra prominens

1. **Clavicles** Collarbones. These lie in superior region of thorax and can be palpated along their entire length.
2. **Sternum** Breastbone. Lies in midline of chest. The following parts of the sternum are important surface features:
   Suprasternal (jugular) notch Depression on superior surface of manubrium of sternum between medial ends of clavicles. The trachea can be palpated in the notch.
   Manubrium of sternum Superior portion of sternum at the same levels as the bodies of the third and fourth thoracic vertebrae and anterior to the arch of the aorta.

Body of sternum Midportion of sternum anterior to heart and the vertebral bodies of T5–T8.
Sternal angle Formed by junction of manubrium and body of sternum, about 4 cm (1½ in.) inferior to suprasternal notch. This is palpable under the skin, locates the costal cartilage of the second rib, and is the starting point from which the ribs are counted.
Xiphoid process of sternum Inferior portion of sternum medial to the seventh costal cartilages. The heart lies on the diaphragm deep to the **xiphisternal joint** (joint between the xiphoid process and body of sternum).

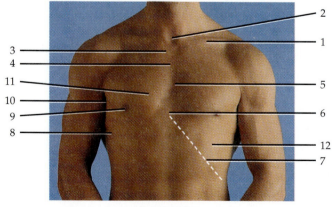

Anterior view

____ Anterior axillary fold
____ Body of sternum
____ Clavicle
____ Costal margin
____ Manubrium of sternum
____ Nipple
____ Pectoralis major muscle
____ Rib
____ Sternal angle of sternum
____ Suprasternal (jugular) notch of sternum
____ Xiphoid process of sternum
____ Serratus anterior muscle

FIGURE 11.4   Surface anatomy of the chest. (The serratus anterior muscle is shown in Figure 11.5).

3. *Ribs* Form bony cage of thoracic cavity. The apex beat of the heart in adults is heard in the left fifth intercostal space, just medial to the left midclavicular line.
4. *Costal margins* Inferior edges of costal cartilages of ribs 7 through 10. The first costal cartilage lies inferior to the medial end of the clavicle; the seventh costal cartilage is the most inferior to articulate directly with the sternum; the tenth costal cartilage forms the most inferior part of the costal margin, when viewed anteriorly.
5. *Muscles* Among the superficial chest muscles that can be seen are the *pectoralis major* (principal upper chest muscle) and *serratus anterior* (inferior and lateral to pectoralis major).
6. *Mammary glands* Accessory organs of the female reproductive system located inside the breasts. They overlie the pectoralis major muscle (two-thirds) and serratus anterior muscle (one-third). After puberty, they enlarge to their hemispherical shape, and in young adult females, they extend from the second through sixth ribs and from the lateral margin of the sternum to the *midaxillary line* (an imaginery line that extends downward from the center of the maxilla along the lateral thoracic wall).
7. *Nipples* Superficial to fourth intercostal space or fifth rib about 10 cm (4 in.) from the midline in males and most females. The position of the nipples in females is variable, depending on the size and pendulousness of the breasts. The right dome of the diaphragm is just inferior to the right nipple, the left dome is about 2-3 cm (1 in.) inferior to the left nipple, and the central tendon is at the level of the junction of the body and xiphoid process of the sternum.

8. *Anterior axillary fold* Formed by the lateral border of the pectoralis major muscle; can be palpated between the fingers and thumb.

Using your textbook as an aid, label Figure 11.4. Read the descriptions of the surface features of the abdomen and pelvis that follow and then label Figure 11.5.

1. *Umbilicus* Also called *navel*; previous site of attachment of umbilical cord to fetus. It is level with the intervertebral disc between the bodies of L3 and L4 and is the most obvious surface marking on the abdomen of most individuals. The *abdominal aorta* bifurcates (branches) into the right and left common iliac arteries anterior to the body of vertebra L4. It can be palpated through the upper part of the anterior abdominal wall just to the left of the midline. The *inferior vena cava* lies to the right of the abdominal aorta and is wider; it rises anterior to the body of vertebra L5.
2. *Muscles* Among the superficial abdominal muscles are the *external oblique* (inferior to serratus anterior) and *rectus abdominis* (just lateral to midline of abdomen).
3. *Linea alba* Flat, tendonous raphe forming a furrow along midline between rectus abdominis muscles. The furrow extends from the xiphoid process to the pubic symphysis. It is particularly obvious in thin, muscular individuals. It is broad superior to the umbilicus and narrow inferior to it. The linea alba is a frequently selected site for abdominal surgery since an incision through it severs no muscles and only a few blood vessels and nerves. The *linea semilunaris* is the lateral edge of the rectus

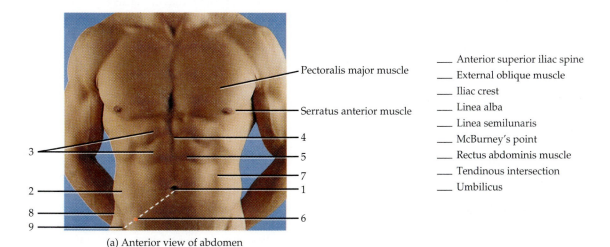

Pectoralis major muscle

Serratus anterior muscle

___ Anterior superior iliac spine
___ External oblique muscle
___ Iliac crest
___ Linea alba
___ Linea semilunaris
___ McBurney's point
___ Rectus abdominis muscle
___ Tendinous intersection
___ Umbilicus

(a) Anterior view of abdomen

FIGURE 11.5   Surface anatomy of abdomen.

abdominis muscle that crosses the costal margins at the tip of the ninth costal cartilage.

4. *Tendinous intersections* Fibrous bands that run transversely across the rectus abdominis muscle. Three or more are visible in muscular individuals. One intersection is at the level of the umbilicus, one at the level of the xiphoid process, and one midway between.

5. *Pubic symphysis* Anterior joint of hipbones. This structure is palpated as a firm resistance in the midline at the inferior portion of the anterior abdominal wall.

6. *McBurney's point* Located two-thirds of the way down an imaginary line drawn between the umbilicus and anterior superior iliac spine. An oblique incision through this point is made for an appendectomy (removal of the appendix). Pressure of the finger on this point produces tenderness in acute appendicitis, inflammation of the appendix.

7. *Iliac crest* Superior margin of the ilium of the hipbone that forms the outline of the superior portion of the buttock. When you rest your hands on your hips, they rest on the iliac crests. A horizontal line drawn across the highest point of each iliac crest is called the *supracristal line,* which intersects the spinous process of the fourth lumbar vertebra. This vertebra is a landmark for performing a lumbar puncture.

8. *Anterior superior iliac spine* The anterior end of the iliac crest that lies at the upper lateral end of the fold of the groin.

9. *Posterior superior iliac spine* The posterior end of the iliac crest that is indicated by a dimple in the skin that coincides with the middle of the sacroiliac joint where the hipbone attaches to the sacrum.

10. *Pubic tubercle* Projection on the superior border of the pubis of the hipbone. Attached to it is the medial end of the inguinal ligament. The lateral end is attached to the anterior superior iliac spine. The *inguinal ligament* is the inferior free edge of the aponeurosis of the external oblique muscle that forms the *inguinal canal.* Through the canal pass the spermatic cord in males and the round ligament of the uterus in females.

11. *Mons pubis* An elevation of adipose tissue covered by skin and pubic hair that is anterior to the pubic symphysis.

12. *Sacrum* The median sacral crest, the fused spinous processes of the sacrum, can be palpated beneath the skin in the superior portion of the gluteal cleft, a depression along the midline that separates the buttocks (described shortly).

13. *Coccyx* The inferior surface of the tip of the coccyx can be palpated in the gluteal cleft, about 2.5 cm (1 in.) posterior to the anus.

Using your textbook as an aid, label Figure 11.5.

D. UPPER LIMB (EXTREMITY)

The *upper limb (extremity)* consists of the armpit, shoulder, arm, elbow, forearm, wrist, and hand (palm and fingers).

Read the descriptions of the surface features of the *shoulder (acromial)* region in the list that follows and then label Figure 11.6.

1. *Acromioclavicular joint* Slight elevation at lateral end of clavicle. It is the joint between the acromion of the scaupula and the clavicle.

2. *Acromion* Expanded lateral end of spine of scapula. This is clearly visible in some individuals and can be palpated about 2.5 cm (1 in.) distal to acromioclavicular joint.

3. *Humerus* The *greater tubercle* of the humerus may be palpated on the superior aspect of the shoulder. It is the most lateral palpable bony structure (see Figure 11.6).

4. *Deltoid muscle* Triangular muscle that forms rounded prominence of shoulder. This is a frequent site for intramuscular injections (see Figure 11.3).

Read the descriptions of the surface features of the *arm (brachium)* and *elbow (cubitus)* in the list that follows and then label Figure 11.7.

1. *Humerus* This may be palpated along its entire length, especially at the elbow.

2. *Biceps brachii muscle* Forms bulk of anterior surface of arm. On the medial side of the muscle is a groove that contains the brachial artery; the artery is frequently used to take blood pressure. Pressure may be applied to it in cases of severe hemorrhage in the forearm and hand.

3. *Triceps brachii muscle* Forms bulk of posterior surface of arm.

4. *Medial epicondyle* Medial projection at distal end of humerus.

5. *Ulnar nerve* Can be palpated as a rounded cord in a groove posterior to the medial epicondyle. The "funny bone" is the region where the ulnar nerve rests against the medial epicondyle.

6. *Lateral epicondyle* Lateral projection at distal end of humerus.

7. *Olecranon* Projection of proximal end of ulna that lies between and slightly superior to epicondyles when forearm is extended; it forms elbow.

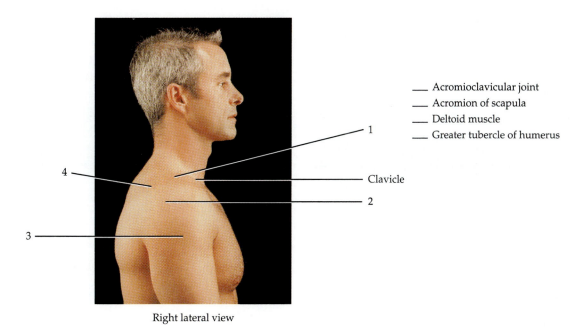

Right lateral view

___ Acromioclavicular joint
___ Acromion of scapula
___ Deltoid muscle
___ Greater tubercle of humerus

FIGURE 11.6   Surface anatomy of the shoulder.

8. *Cubital fossa* Triangular space in anterior region of elbow bounded proximally by an imaginary line between humeral epicondyles, laterally by the medial border of the brachioradialis muscle, and medially by the lateral border of the pronator teres muscle; contains tendon of biceps brachii muscle, brachial artery and its terminal branches (radial and ulnar arteries), medial cubital vein, and parts of median and radial nerves. Pulse can be detected in the brachial artery in the cubital fossa.

9. *Median cubital vein* Crosses cubital fossa obliquely. This vein is frequently selected for removal of blood, or introduction of substances such as medications, contrast media for radiographic procedures, nutrients, and blood cells and/or plasma for transfusions.

10. *Bicipital aponeurosis* An aponeurotic band that inserts the biceps brachii muscle into the deep fascia in the medial aspect of the forearm. It can be felt when the muscle contracts.

Read the descriptions of the surface features of the *forearm (antebrachium)* and *wrist (carpus)* that follow and then label Figure 11.8.

1. *Ulna* Medial bone of the forearm. It can be palpated along its entire length from the olecranon to the *styloid process,* a projection on the distal end of the bone at the medial side of the wrist. The *head of the ulna* is a conspicuous enlargement just proximal to the styloid process.

2. *Radius* When the forearm is rotated, the distal half of the radius can be palpated; the proximal half is covered by muscles. The *styloid process* of the radius is a projection on the distal end of the bone at the lateral side of the wrist.

3. *Muscles* Because of their close proximity, it is difficult to identify muscles of the forearm. However, it is easy to identify the tendons of some of the muscles as they approach the wrist and then trace them proximally to the muscles.

Brachioradialis muscle Located at superior and lateral aspect of forearm.

Flexor carpi radialis muscle The tendon of this muscle is about 1 cm medial to the styloid process of the radius on the lateral side of the forearm.

Palmaris longus muscle The tendon of this muscle is medial to the flexor carpi radialis tendon and can be seen if the wrist is slightly flexed and the base of the thumb and little finger are drawn together.

Flexor digitorum superficialis muscle The tendon of this muscle is medial to the palmaris longus tendon and can be palpated by flexing the fingers at the metacarpophalangeal and proximal interphalangeal joints.

Flexor carpi ulnaris muscle The tendon of this muscle is on the medial aspect of the forearm.

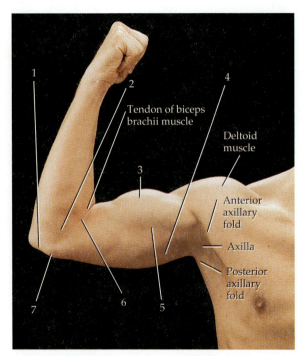

(a) Medial view of arm

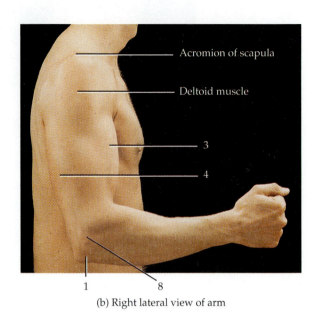

(b) Right lateral view of arm

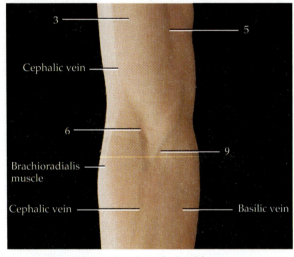

(c) Anterior view of cubital fossa

_____ Biceps brachii muscle
_____ Bicipital aponeurosis
_____ Cubital fossa
_____ Groove for brachial artery
_____ Lateral epicondyle of humerus
_____ Medial epicondyle of humerus
_____ Median cubital vein
_____ Olecranon of ulna
_____ Triceps brachii muscle

FIGURE 11.7   Surface anatomy of the arm and elbow.

4. *Radial artery* Located on the lateral aspect of the wrist between the flexor carpi radialis muscle tendon and styloid process of the radius. It is frequently used to take a pulse.

5. *Pisiform bone* Medial bone of proximal carpals. The bone is easily palpated as a projection distal and anterior to styloid process of ulna.

6. *"Anatomical snuffbox"* Triangular depression between tendons of extensor pollicis brevis and extensor pollicis longus muscles. Styloid process

of the radius, the base of the first metacarpal, trapezium, scaphoid, and radial artery can all be palpated in the depression.

7. *Wrist creases* Three more or less constant lines on anterior aspect of wrist (named proximal, middle, and distal) where skin is firmly attached to underlying deep fascia.

Read the description of the surface features of the *hand (manus)* that follows and then label Figure 11.9.

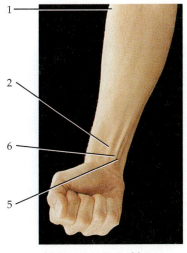

(a) Anterior view of forearm
and wrist

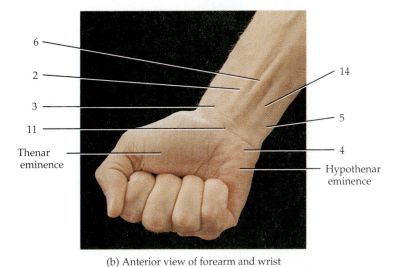

Thenar
eminence

Hypothenar
eminence

(b) Anterior view of forearm and wrist

Cephalic vein

(c) Posterolateral view of wrist

___ "Anatomical snuffbox"

___ Brachioradialis muscle

___ Head of ulna

___ Pisiform bone

___ Site for palpation of radial
artery

___ Styloid process of radius

___ Styloid process of ulna

___ Tendon of extensor pollicis
brevis muscle

___ Tendon of extensor pollicis
longus muscle

___ Tendon of flexor carpi
radialis muscle

___ Tendon of flexor carpi ulnaris
muscle

___ Tendon of flexor digitorum
superficialis muscle

___ Tendon of palmaris longus
muscle

___ Wrist creases

FIGURE 11.8   Surface anatomy of forearm and wrist.

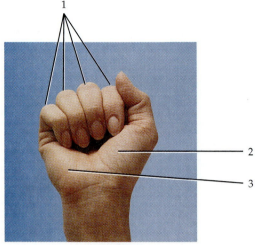

1

2

3

(a) Anterior and posterior view

1

(b) Posterior view

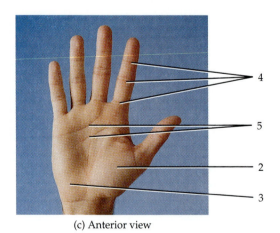

(c) Anterior view

4

5

2

3

(d) Posterior view

8

6

7

— Digital flexion creases   — Palmar flexion creases
— Dorsal venous arch   — Tendon of extensor digiti minimi muscle
— Hypothenar eminence   — Tendon of extensor digitorum muscle
— "Knuckles"   — Thenar eminence

FIGURE 11.9   Surface anatomy of hand.

1. *"Knuckles"* Commonly refers to dorsal aspects of distal ends of metacarpals II–V (or 2–5), but also includes dorsal aspects of metacarpophalangeal and interphalangeal joints.
2. *Dorsal venous arch* Superficial veins on dorsum surface of hand that drain blood into the cephalic vein. It can be displayed by compressing the blood vessels at the wrist for a few minutes as the hand is opened and closed.
3. *Extensor tendons* Besides the tendons of the extensor pollicis brevis and extensor pollicis longus muscles associated with the thumb, the following extensor tendons are also visible on the posterior aspect of the hand: *extensor digiti minimi tendon* in line with phalanx of the little finger and *extensor digitorum* in line with phalanges of the ring, middle, and index finger.
4. *Thenar eminence* Larger rounded contour on the lateral aspect of the palm formed by muscles that move the thumb.
5. *Hypothenar eminence* Smaller rounded contour on medial aspect of the palm formed by muscles that move the little finger.
6. *Skin creases* Several more or less constant lines on the anterior aspect of the palm *(palmar flexion creases)* and digits *(digital flexion creases)* where skin is firmly attached to underlying deep fascia.

E. LOWER LIMB (EXTREMITY)

The *lower limb (extremity)* consists of the buttocks, thigh, knee, leg, ankle, and foot.

Read the descriptions of the surface features of the *buttocks (gluteal region)* that follow and label Figure 11.10.

1. *Gluteus maximus muscle* Forms major portion of prominence of buttock. The sciatic nerve is deep to this muscle.
2. *Gluteus medius muscle* Superolateral to gluteus maximus. This is a frequent site for intramuscular injections.
3. *Gluteal (natal) cleft* Depression along midline that separates the buttocks; it extends as high as the fourth or third sacral vertebra.
4. *Gluteal fold* Inferior limit of buttock formed by inferior margin of gluteus maximus muscle.
5. *Ischial tuberosity* Bony prominence of ischium of hipbone just superior to the medial side of the gluteal fold that bears weight of body when seated.
6. *Greater trochanter* Projection of proximal end of femur on lateral surface of thigh felt and seen in front of hollow on side of hip. This can be palpated about 20 cm (8 in.) inferior to iliac crest.

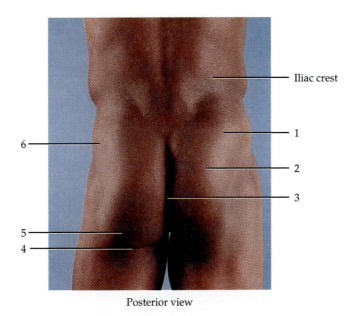

Iliac crest

1
2
3
6
5
4

Posterior view

___ Gluteal (natal) cleft
___ Gluteal fold
___ Gluteus maximus muscle
___ Gluteus medius muscle
___ Greater trochanter of femur
___ Site of ischial tuberosity

FIGURE 11.10   Surface anatomy of the buttocks.

Read the descriptions of the surface anatomy features of the ***thigh (femoral region)*** and ***knee (genu)*** that follow and label Figure 11.11.

1. ***Sartorius muscle*** Superficial anterior muscle that can be traced from the lateral aspect of the thigh to the medial aspect of the knee.
2. ***Quadriceps femoris muscle*** Three of the four components of this muscle can be seen.

   Rectus femoris Located at the midportion of the anterior aspect of the thigh.
   Vastus medialis Located at the anteromedial aspect of the thigh.
   Vastus lateralis Located at the anterolateral aspect of the thigh. This muscle is a frequent intramuscular injection site for diabetics.

3. ***Adductor longus muscle*** Located at the superior aspect of the thigh, medial to the sartorius.
4. ***Hamstring muscles*** Superficial, posterior thigh muscles located inferior to the gluteal folds. They are the

   Biceps femoris Lies more laterally as it passes inferiorly to the knee.
   Semitendinosus and ***semimembranosus*** Lie medially as they pass inferiorly to the knee.

5. ***Femoral triangle*** A large space formed by the inguinal ligament superiorly, the satorius muscle laterally, and the adductor longus muscle medially. The triangle contains the femoral artery, vein, and nerve; inguinal lymph nodes; and the terminal portion of the great saphenous vein. The triangle is an important arterial pressure point in cases of severe hemorrhage of the lower limb. Hernias occur frequently in this area.
6. ***Patella*** Kneecap. A large sesamoid bone located within the tendon of the quadriceps femoris muscle on the anterior surface of the knee along the midline.
7. ***Patellar ligament*** Continuation of quadriceps femoris tendon inferior to patella.
8. ***Medial condyles of femur and tibia*** Medial projections just inferior to patella. Superior part of projection belongs to distal end of femur; inferior part of projection belongs to proximal end of tibia.
9. ***Lateral condyles of femur and tibia*** Lateral projections just inferior to patella. Superior part of projections belongs to distal end of femur. Inferior part of projections belongs to proximal end of tibia.

10. ***Popliteal fossa*** Diamond-shaped space on posterior aspect of knee visible when knee is flexed. Fossa is bordered superolaterally by the biceps femoris muscle, superomedially by the semimembranosus and semitendinosus muscles, and inferolaterally and inferomedially by the lateral and medial heads of the gastrocnemius muscle, respectively. The ***head of the fibula*** can easily be palpated on the lateral side of the popliteal fossa. The fossa also contains the popliteal artery and vein. It is sometimes possible to detect a pulse in the popliteal artery.

Read the descriptions of the surface anatomy features of the ***leg (crus), ankle (tarsus),*** and ***foot*** that follows and then label Figure 11.12.

1. ***Tibial tuberosity*** Bony prominence of tibia inferior to patella into which patellar ligament inserts.
2. ***Tibialis anterior muscle*** Lies against the lateral surface of the tibia where it is easy to palpate.
3. ***Tibia*** The medial surface and anterior border (shin) of the tibia are subcutaneous and can be palpated throughout the length of the bone.
4. ***Peroneus longus muscle*** A superficial lateral muscle that overlies the fibula.
5. ***Gastrocnemius muscle*** Forms the bulk of the midportion and superior portion of the posterior aspect of the leg. The medial and lateral heads can be clearly seen when standing on the toes.
6. ***Soleus muscle*** Located deep to the gastrocnemius muscle and together with the gastrocnemius are referred to as the calf muscles.
7. ***Calcaneal (Achilles) tendon*** Prominent tendon of the gastrocnemius and soleus muscles on the posterior aspect of the ankle that inserts into the ***calcaneus*** (heel) bone of the foot.
8. ***Lateral malleolus of fibula*** Projection of the distal end of the fibula that forms the lateral prominence of the ankle. The head of the fibula, at the proximal end of the bone, lies at the same level as the tibial tuberosity.
9. ***Medial malleolus of tibia*** Projection of the distal end of the tibia that forms the medial prominence of the ankle.
10. ***Dorsal venous arch*** Superficial veins on the dorsum of the foot that unite to form the small and great saphenous veins. The great saphenous vein is the longest vein of the body. The location of the great saphenous vein is about 2.5 cm (1 in.) anterior to the medial malleolus of the tibia and is fairly constant. Knowledge of this

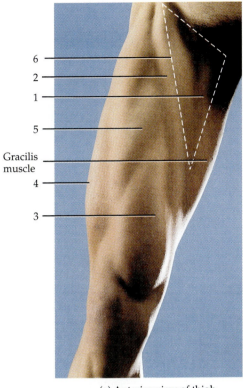

6
2
1
5
Gracilis
muscle
4
3

(a) Anterior view of thigh

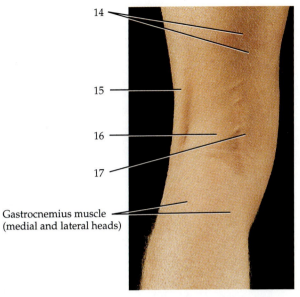

14
15
16
17

Gastrocnemius muscle
(medial and lateral heads)

(c) Posterior view of back of knee

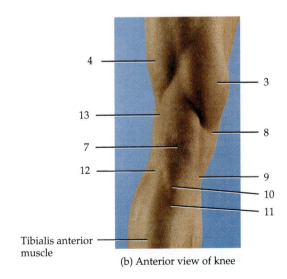

4
13
7
12
3
8
9
10
11

Tibialis anterior
muscle

(b) Anterior view of knee

___ Adductor longus muscle
___ Femoral triangle
___ Lateral condyle of femur
___ Lateral condyle of tibia
___ Medial condyle of femur
___ Medial condyle of tibia
___ Patella
___ Patellar ligament
___ Popliteal fossa
___ Rectus femoris muscle
___ Sartorius muscle
___ Site of
    semitendinosus and
    semimembranosus
    muscles
___ Tendon of biceps
    femoris muscle
___ Tendon of semitendinosus
    muscle
___ Tibial tuberosity
___ Vastus lateralis muscle
___ Vastus medialis muscle

FIGURE 11.11   Surface anatomy of the thigh and knee.

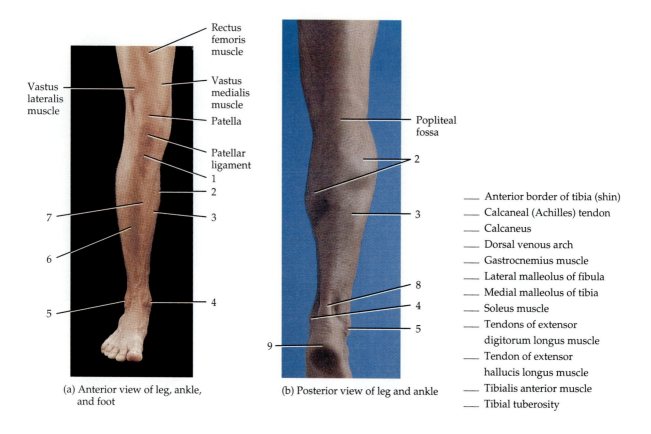

Rectus femoris muscle

Vastus lateralis muscle

Vastus medialis muscle

Patella

Patellar ligament

1

2

7

3

6

5

4

(a) Anterior view of leg, ankle, and foot

Popliteal fossa

2

3

8

4

5

9

(b) Posterior view of leg and ankle

____ Anterior border of tibia (shin)
____ Calcaneal (Achilles) tendon
____ Calcaneus
____ Dorsal venous arch
____ Gastrocnemius muscle
____ Lateral malleolus of fibula
____ Medial malleolus of tibia
____ Soleus muscle
____ Tendons of extensor digitorum longus muscle
____ Tendon of extensor hallucis longus muscle
____ Tibialis anterior muscle
____ Tibial tuberosity

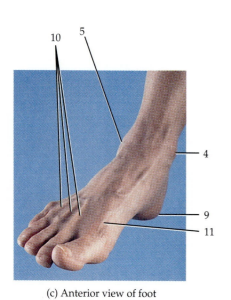

10   5

4

9

11

(c) Anterior view of foot

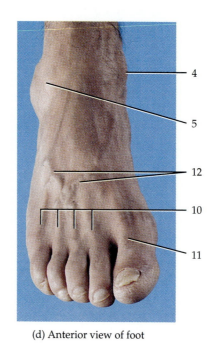

4

5

12

10

11

(d) Anterior view of foot

FIGURE 11.12   Surface anatomy of the leg, ankle, and foot.

location may be lifesaving when an urgent transfusion is needed in obese or collapsed patients when other veins cannot be detected.

11. *Tendons of extensor digitorum longus muscle* Visible in line with phalanges II through V (or 2–5).

12. *Tendon of extensor hallucis longus muscle* Visible in line with phalanx I (great toe). Pulsations in the dorsalis pedis artery may be felt in most people just lateral to this tendon when the blood vessel passes over the navicular and cuneiform bones of the tarsus.

ANSWER THE LABORATORY REPORT QUESTIONS AT THE END OF THE EXERCISE.

Surface Anatomy 11

Student _____ Date _____

Laboratory Section _____ Score/Grade _____

PART 1. Multiple Choice

_____ 1. The term used to refer to the crown of the skull is (a) occipital (b) mental (c) parietal (d) zygomatic

_____ 2. The laryngeal cartilage in the midline of the anterior cervical region known as the Adam's apple is the (a) cricoid (b) epiglottis (c) arytenoid (d) thyroid

_____ 3. Inflammation of which muscle is associated with "stiff neck"? (a) teres major (b) trapezius (c) deltoid (d) cervicalis

_____ 4. The skeletal muscle located directly on either side of the vertebral column is the (a) serratus anterior (b) infraspinatus (c) teres major (d) erector spinae

_____ 5. The suprasternal notch and xiphoid process are associated with the (a) sternum (b) scapula (c) clavicle (d) ribs

_____ 6. The expanded end of the spine of the scapula is the (a) acromion (b) linea alba (c) olecranon (d) superior angle

_____ 7. Which nerve can be palpated as a rounded cord in a groove posterior to the medial epicondyle? (a) median (b) radial (c) ulnar (d) brachial

_____ 8. Which carpal bone can be palpated as a projection distal to the styloid process of the ulna? (a) trapezoid (b) trapezium (c) hamate (d) pisiform

_____ 9. The lateral rounded contour on the anterior surface of the hand (at the base of the thumb) formed by the muscles of the thumb is the (a) "anatomical snuffbox" (b) thenar eminence (c) hypothenar eminence (d) dorsal venous arch

_____ 10. The superior margin of the hipbone is the (a) pubic symphysis (b) iliac spine (c) acetabulum (d) iliac crest

_____ 11. Which bony structure bears the weight of the body when a person is seated? (a) greater trochanter (b) iliac crest (c) ischial tuberosity (d) gluteal fold

_____ 12. Which muscle is *not* a component of the quadriceps femoris group? (a) biceps femoris (b) vastus lateralis (c) vastus medialis (d) rectus femoris

_____ 13. The diamond-shaped space on the posterior aspect of the knee is the (a) cubital fossa (b) posterior triangle (c) popliteal fossa (d) nuchal groove

_____ 14. The projection of the distal end of the tibia that forms the prominence on one side of the ankle is the (a) medial condyle (b) medial malleolus (c) lateral condyle (d) lateral malleolus

_____ 15. The tendon that can be seen in line with the great toe belongs to which muscle? (a) extensor digiti minimi (b) extensor digitorum longus (c) extensor hallucis longus (d) extensor carpi radialis

PART 2. Completion

16. The laryngeal cartilage that connects the larynx to the trachea is the _____ cartilage.

17. The _____ triangle is bordered by the mandible, sternum, cervical midline, and sternocleidomastoid muscle.

18. The depression on the superior surface of the sternum between the medial ends of the clavicles is the

    _____.

19. The principal superficial chest muscle is the _____.

20. Tendinous intersections are associated with the _____ muscle.

21. The muscle that forms the rounded prominence of the shoulder is the _____ muscle.

22. The triangular space in the anterior aspect of the elbow is the _____.

23. The "anatomical snuffbox" is bordered by the tendons of the extensor pollicis brevis muscle and the

    _____ muscle.

24. The dorsal aspects of the distal ends of metacarpals II through V are commonly referred to as

    _____.

25. The tendon of the _____ muscle is in line with phalanx V.

26. The dimple that forms about 4 cm lateral to the midline just above the buttocks lies superficial to the

    _____.

27. The femoral projection that can be palpated about 20 cm (8 in.) inferior to the iliac crest is the

    _____.

28. The continuation of the quadriceps femoris tendon inferior to the patella is the

    _____.

29. The tendon of insertion for the gastrocnemius and soleus muscles is the _____ tendon.

30. Superficial veins on the dorsum of the foot that unite to form the small and great saphenous veins

    belong to the _____.

31. The prominent veins along the lateral cervical regions are the _____ veins.

32. The pronounced vertebral spine of C7 is the _____.

33. The most reliable surface anatomy feature of the chest is the _____.

34. A slight groove extending from the xiphoid process to the pubic symphysis is the

    _____.

35. The vein that crosses the cubital fossa and is frequently used to remove blood is the

    _____ vein.

36. Respiratory sounds are clearly heard in the triangle of _____.

37. The heart lies on the diaphragm deep to the _____ junction.

38. A horizontal line drawn across the highest point of each iliac crest is called the

_____ line.

39. The anatomical landmark related to appendectomy and appendicitis is _____.

40. The lateral edge of the rectus abdominis muscle is called the _____.

PART **3.** MATCHING

_____	**41.** Mental region	A. Inferior edges of costal cartilages of ribs 7 through 10
_____	**42.** Xiphoid process	B. Forms elbow
_____	**43.** Arm	C. Inferior portion of sternum
_____	**44.** Nucha	D. Manus
_____	**45.** Shoulder	E. Crus
_____	**46.** Gluteus maximus muscle	F. Brachium
_____	**47.** Costal margin	G. Tarsus
_____	**48.** Olecranon	H. Anterior part of mandible
_____	**49.** Wrist	I. Brain case
_____	**50.** Auricular region	J. Component of hamstrings
_____	**51.** Semitendinosus muscle	K. Forms main part of prominence of buttock
_____	**52.** Leg	L. Carpus
_____	**53.** Cranium	M. Posterior neck region
_____	**54.** Ankle	N. Acromial
_____	**55.** Hand	O. Ear

Nervous Tissue

The ***nervous system*** has three basic functions: (1) sensory, (2) integrative, and (3) motor.

1. ***Sensory function*** It *senses* certain changes (stimuli), both within your body (the internal environment), such as stretching of your stomach or an increase in blood acidity, and outside your body (the external environment), such as a raindrop landing on your arm or the aroma of a rose.
2. ***Integrative function*** It *analyzes* the sensory information, *stores* some aspects, and *makes decisions* regarding appropriate behaviors.
3. ***Motor function*** It may *respond* to stimuli by initiating muscular contractions or glandular secretions.

The branch of medical science that deals with the normal functioning and disorders of the nervous system is called ***neurology*** (noo-ROL-ō-jē; *neuro* = nerve or nervous system; *logos* = study of).

A. NERVOUS SYSTEM DIVISIONS

The two principal divisions of the nervous system are the ***central nervous system (CNS)*** and the ***peripheral*** (pe-RIF-er-al) ***nervous system (PNS).*** The CNS consists of the ***brain*** and ***spinal cord.*** Within the CNS, various sorts of incoming sensory information are integrated and correlated, thoughts and emotions are generated, and memories are formed and stored. Most nerve impulses that stimulate muscles to contract and glands to secrete originate in the CNS.

The CNS is connected to sensory receptors, muscles, and glands in peripheral parts of the body by the PNS. The PNS consists of ***cranial nerves*** that arise from the brain and ***spinal nerves*** that emerge from the spinal cord. Portions of these nerves carry nerve impulses into the CNS while other portions carry impulses out of the CNS.

The input component of the PNS consists of nerve cells called ***sensory*** or ***afferent*** (AF-er-ent; *ad* = toward; *ferre* = to carry) ***neurons.*** They conduct nerve impulses from sensory receptors located in various parts of the body *to the CNS* and end within the CNS. The output component consists of nerve cells called ***motor*** or ***efferent*** (EF-er-ent; *ex* = away from; *ferre* = to carry) ***neurons.*** They originate within the CNS and conduct nerve impulses *from the CNS* to muscles and glands.

The PNS may be subdivided further into a ***somatic*** (*soma* = body) ***nervous system (SNS)*** and an ***autonomic*** (*auto* = self; *nomos* = law) ***nervous system (ANS).*** The major difference between the two is the ***effector,*** the tissue that receives stimulation or inhibition from the nervous system (skeletal muscle, smooth muscle, cardiac muscle, or a gland). The SNS consists of sensory neurons that convey information from cutaneous and special sense receptors primarily in the head, body wall, and limbs to the CNS and motor neurons from the CNS that conduct impulses to *skeletal muscles* only. Because these motor responses can be consciously controlled, this portion of the SNS is *voluntary.*

The ANS consists of sensory neurons that convey information from receptors primarily in the viscera to the CNS and motor neurons from the CNS that conduct impulses to smooth muscle, cardiac muscle, and glands. Since its motor responses are not normally under conscious control, the ANS is *involuntary.*

The motor portion of the ANS consists of two branches, the ***sympathetic division*** and the ***parasympathetic division.*** With few exceptions, the viscera receive instructions from both. Usually, the two divisions have opposing actions. For example, sympathetic neurons speed the heartbeat while parasympathetic neurons slow it down. Processes promoted by sympathetic neurons often involve expenditure of energy while those promoted by parasympathetic neurons restore and conserve body energy.

In this exercise you will identify the parts of a neuron and the components of a reflex arc.

B. HISTOLOGY OF NERVOUS TISSUE

Despite its complexity, the nervous system consists of only two principal kinds of cells: neurons and neuroglia. *Neurons* (nerve cells) constitute the nervous tissue and are highly specialized for nerve impulse (nerve action potential) conduction. Mature neurons have only limited capacity for replacement or repair. *Neuroglia* (noo-ROG-lē-a; *neuro* = nerve; *glia* = glue) can divide and multiply and support, nurture, and protect neurons and maintain homeostasis of the fluid that bathes neurons. They do not transmit nerve impulses. Brain tumors are commonly derived from neuroglia. Such tumors, called *gliomas*, are highly malignant and rapidly enlarging.

A neuron consists of the following parts:

1. *Cell body* Contains a nucleus, cytoplasm, lysosomes, mitochondria, Golgi complex, chromatophilic substance (Nissl bodies), and neurofibrils.
2. *Dendrites* (*dendro* = tree) Usually short, highly branched extensions of the cell body that conduct nerve impulses toward the cell body.
3. *Axon* (*axon* = axis) Single, usually relatively long process that conducts nerve impulses away from the cell body to another neuron, muscle fiber, or gland cell. An axon, in turn, consists of the following:
   a. *Axon hillock* (*hilloc* = small hill) The origin of an axon from the cell body represented as a small cone-shaped region.
   b. *Initial segment* First portion of an axon. Except in sensory neurons, nerve impulses arise at the junction of the axon hillock and initial segment, a region called the *trigger zone.*
   c. *Axoplasm* Cytoplasm of an axon.
   d. *Axolemma* (*lemma* = sheath or husk) Plasma membrane around the axoplasm.
   e. *Axon collateral* Side branch of an axon.
   f. *Axon terminals* Fine, branching processes of an axon or axon collateral.
   g. *Synaptic end bulbs* Bulblike structures at distal end of axon terminals that contain *synaptic vesicles* for neurotransmitters.
   h. *Myelin sheath* Multilayered, lipid and protein segmented covering of many axons, especially large peripheral ones; the myelin sheath is produced by peripheral nervous system neuroglia called *neurolemmocytes (Schwann cells)* and central nervous system neuroglia called *oligodendrocytes* (described shortly).
   i. *Neurolemma (sheath of Schwann)* Peripheral, nucleated cytoplasmic layer of the neurolemmocyte (Schwann cell) that encloses the myelin sheath. It is found only around axons in the peripheral nervous system.
   j. *Neurofibral node (node of Ranvier;* pronounced RON-vē-ā) Unmyelinated gap between segments of the myelin sheath.

Using your textbook and models of neurons as a guide, label Figure 12.1.

Now obtain a prepared slide of an ox spinal cord (transverse section), human spinal cord (transverse and longitudinal sections), nerve endings in skeletal muscle, and a nerve trunk (transverse and longitudinal sections). Examine each under high power and identify as many parts of the neuron as you can.

Neurons may be classified on the basis of structure and function. Structural classification is based on the number of processes extending from the cell body. Structurally, neurons are *multipolar* (several dendrites and one axon), *bipolar* (one main dendrite and one axon), and *unipolar* (a single process that branches into an axon and a dendrite). Functional classification is based on the type of information carried and the direction in which the information is carried. Functionally, neurons are classified as *sensory (afferent)*, which carry nerve impulses from receptors toward the central nervous system; *motor (efferent)*, which carry nerve impulses away from the central nervous system to effector (muscles or glands); and *association (connecting* or *interneurons)*, neurons that are located in the central nervous system and carry nerve impulses between sensory and motor neurons. The neuron you have already labeled in Figure 12.1 is a motor neuron.

Using your textbook and models of neurons as a guide, label Figure 12.2.

C. HISTOLOGY OF NEUROGLIA

Among the types of neuroglial cells are the following:

1. *Astrocytes* (AS-trō-sīts; *astro* = star; *cyte* = cell) star-shaped cells with many processes. Participate in the metabolism of neurotransmitters (glutamate and γ-aminobutyric acid) and maintain the proper balance of potassium ions (K^+) for generation of nerve impulses by CNS neurons; participate in brain development by assisting migration of neurons; help form the

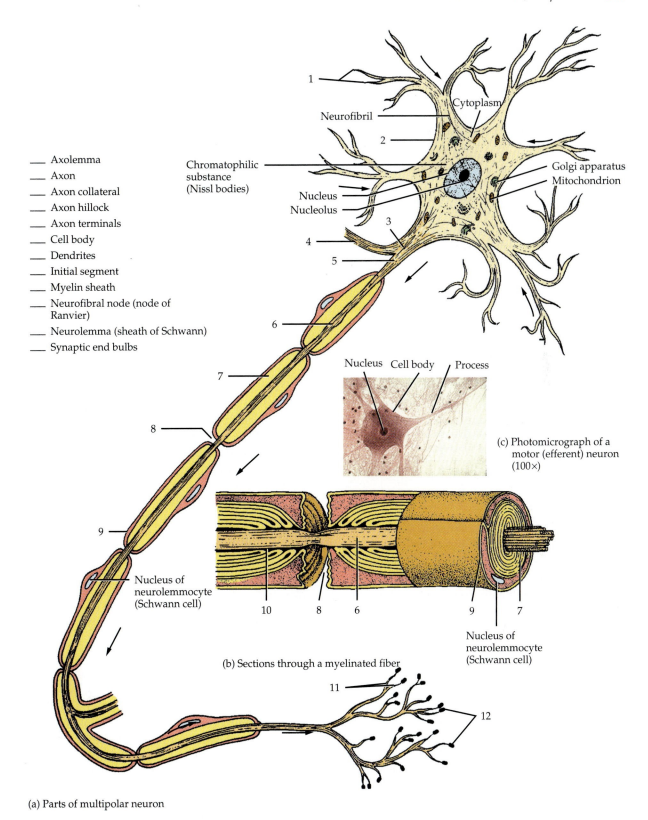

__ Axolemma
__ Axon
__ Axon collateral
__ Axon hillock
__ Axon terminals
__ Cell body
__ Dendrites
__ Initial segment
__ Myelin sheath
__ Neurofibral node (node of Ranvier)
__ Neurolemma (sheath of Schwann)
__ Synaptic end bulbs

Cytoplasm
Neurofibril
Chromatophilic substance (Nissl bodies)
Golgi apparatus
Mitochondrion
Nucleus
Nucleolus

Nucleus of neurolemmocyte (Schwann cell)

Nucleus   Cell body   Process

(c) Photomicrograph of a motor (efferent) neuron (100×)

Nucleus of neurolemmocyte (Schwann cell)

(b) Sections through a myelinated fiber

(a) Parts of multipolar neuron

FIGURE 12.1   Structure of a neuron. In (a), arrows indicate the direction in which the nerve impulse travels.

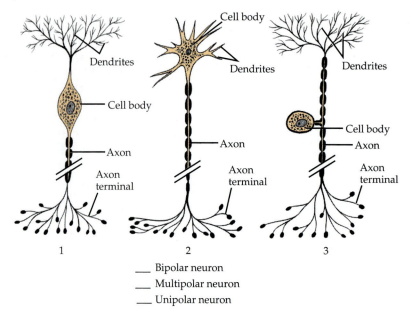

___ Bipolar neuron

___ Multipolar neuron

___ Unipolar neuron

Figure 12.2   Structural classification of neurons.

blood-brain barrier, which regulates the passage of substances into the brain; twine around neurons to form a supporting network; and provide a link between neurons and blood vessels.

   a. *Protoplasmic astrocytes* are found in the gray matter of the central nervous system.

   b. *Fibrous astrocytes* are found in the white matter of the central nervous system.

2. *Oligodendrocytes* (ol'-i-gō-DEN-drō-sīts; *oligo* = few; *dendro* = tree) Resemble astrocytes but with fewer and shorter processes; they produce a myelin sheath on axons of neurons of the CNS.

3. *Microglia* (mī-KROG-lē-a; *micro* = small) Small cells derived from monocytes with few processes; although normally stationary, they can migrate to damaged nervous tissue and there carry on phagocytosis; they are also called *brain macrophages.*

4. *Ependymal* (e-PEN-di-mal; *epi* = above; *dyma* = garment) *cells* Epithelial cells arranged in a single layer that range from squamous to columnar in shape; many are ciliated; they form a continuous epithelial lining for the central canal of the spinal cord and for the ventricles of the brain, spaces that contain networks of capillaries that form cerebrospinal fluid; ependymal cells probably assist in circulating cerebrospinal fluid in these areas.

5. *Neurolemmocytes* (noo'-rō-LE-mō-sīts) *(Schwann cells)* Flattened cells arranged around axons.

Produce a phospholipid myelin sheath around axons and dendrites of neurons of PNS.

6. *Satellite* (SAT-i-līt) *cells* Flattened cells arranged around the cell bodies of ganglia (collections of neuron cell bodies outside the CNS).

Obtain prepared slides of astrocytes (protoplasmic and fibrous), oligodendrocytes, microglia, ependymal cells, neurolemmocytes, and satellite cells. Using your textbook, Figure 12.3, and models of neuroglia as a guide, identify the various kinds of cells. In the spaces provided, draw each of the cells.

D. NEURONAL CIRCUITS

The CNS contains billions of neurons organized into complicated patterns called *neuronal pools.* Each pool differs from all others and has its own role in regulating homeostasis. A neuronal pool may contain thousands or even millions of neurons.

   The functional contact between two neurons or between a neuron and an effector is called a *synapse.* At a synapse, the neuron sending the signal is called a *presynaptic neuron,* and the neuron receiving the message is called a *postsynaptic neuron.*

   Neuronal pools in the CNS are arranged in patterns called *circuits* over which the nerve impulses are conducted. In *simple series circuits* a presynaptic neuron stimulates only a single neu-

Protoplasmic astrocyte

Fibrous astrocyte

Oligodendrocyte

Microglial cell

Ependymal cell

Neurolemmocyte

Satellite cell

ample, a small number of neurons in the brain that govern a particular body movement stimulate a much larger number of neurons in the spinal cord. Sensory signals also feed into diverging circuits and are often relayed to several regions of the brain.

In another arrangement, called *convergence*, several presynaptic neurons synapse with a single postsynaptic neuron. This arrangement permits more effective stimulation or inhibition of the postsynaptic neuron. In one type of *converging circuit*, the postsynaptic neuron receives nerve impulses from several different sources. For example, a single motor neuron that synapses with skeletal muscle fibers at neuromuscular junctions receives input from several pathways that originate in different brain regions.

Some circuits in your body are constructed so that once the presynaptic cell is stimulated, it will cause the postsynaptic cell to transmit a series of nerve impulses. One such circuit is called a *reverberating (oscillatory) circuit*. In this pattern, the incoming impulse stimulates the first neuron, which stimulates the second, which stimulates the third, and so on. Branches from later neurons synapse with earlier ones, however, sending the impulse back through the circuit again and again. The output signal may last from a few seconds to many hours, depending on the number of synapses and the arrangement of neurons in the circuit. Inhibitory neurons may turn off a reverberating circuit after a period of time. Among the body responses thought to be the result of output signals from reverberating circuits are breathing, coordinated muscular activities, waking up, sleeping (when reverberation stops), and short-term memory. One form of epilepsy (grand mal) is probably caused by abnormal reverberating circuits.

A fourth type of circuit is the *parallel after-discharge circuit.* In this circuit, a single presynaptic cell stimulates a group of neurons, each of which synapses with a common postsynaptic cell. If the input is excitatory, the postsynaptic neuron then can send out a stream of impulses in quick succession. It is thought that parallel after-discharge circuits may be employed for precise activities such as mathematical calculations.

Using your textbook as a guide, label Figure 12.4.

E. REFLEX ARC

A *reflex* is a fast, predictable, automatic response to a change in the environment (stimulus). The stimulus is applied to the periphery and conducted either

ron in a pool. The single neuron then stimulates another, and so on. Most circuits, however, are more complex.

A single presynaptic neuron may synapse with several postsynaptic neurons. Such an arrangement, called *divergence*, permits one presynaptic neuron to influence several postsynaptic neurons or several muscle fibers or gland cells at the same time. In a *diverging circuit*, the nerve impulse from a single presynaptic neuron causes the stimulation of increasing numbers of cells along the circuit. For ex-

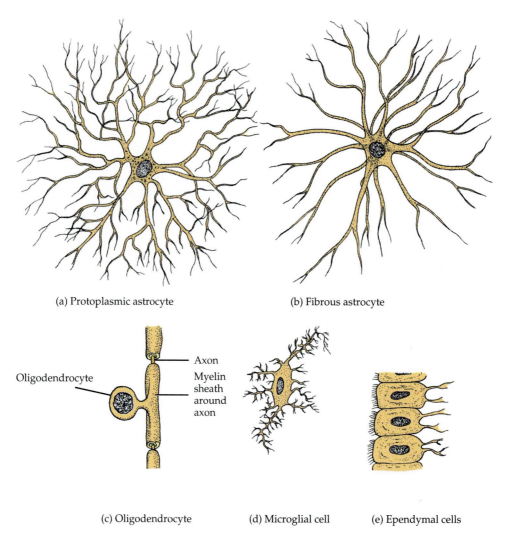

(a) Protoplasmic astrocyte

(b) Fibrous astrocyte

Oligodendrocyte

Axon

Myelin
sheath
around
axon

(c) Oligodendrocyte

(d) Microglial cell

(e) Ependymal cells

FIGURE 12.3   Neuroglia.

to the brain or spinal cord. Reflexes serve to restore functions to homeostasis. For a reflex to occur, a stimulus must elicit a response in one or more structural units within the body termed *reflex arcs.* A reflex arc consists of the following components:

1. *Receptor* The distal end of a sensory neuron (dendrite) or an associated sensory structure that serves as a receptor. A receptor converts a stimulus from its particular form of energy (such as temperature, pressure, or stretch) into the electrical energy utilized by neurons. If this localized depolarization is of threshold value, an action potential (nerve impulse) will be initiated in a sensory neuron.

2. *Sensory (afferent) neuron* A nerve cell that carries the action potential from the receptor to the central nervous system (CNS).

3. *Integrating center* Region in the CNS where the sensory neuron makes a functional connection with one or more neurons. This CNS synapse may occur with an association neuron or a motor neuron. It is here at the CNS synapse where the initial processing of the sensory information occurs: the more complex the reflex involved, the greater the number of CNS synapses involved in a reflex arc.

4. *Motor (efferent) neuron* A nerve cell that transmits the action potential generated by the sensory neuron or an association neuron away from the CNS to the effector.

5. *Effector* The part of the body, either a muscle or a gland, that responds to the motor neuron impulse and thus the stimulus.

Label the components of a reflex arc in Figure 12.5.

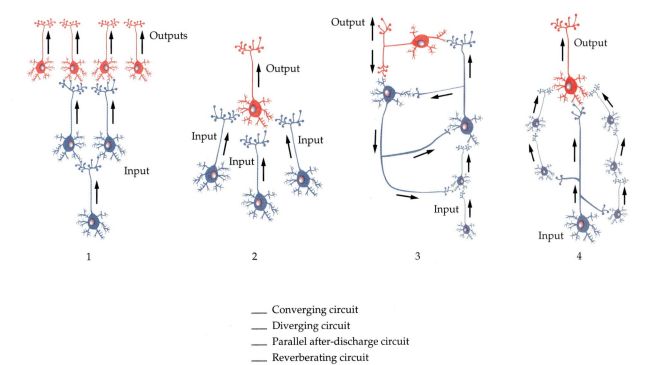

_____ Converging circuit
_____ Diverging circuit
_____ Parallel after-discharge circuit
_____ Reverberating circuit

FIGURE 12.4   Neuronal circuit.

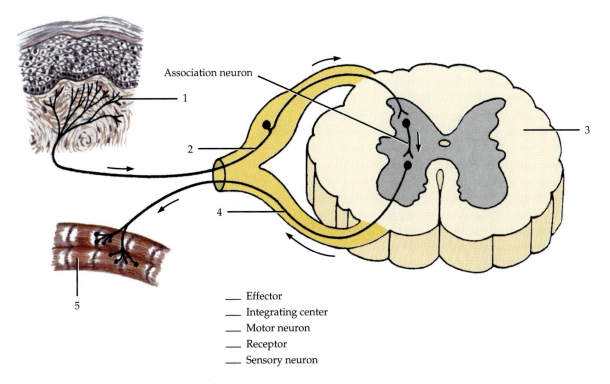

_____ Effector
_____ Integrating center
_____ Motor neuron
_____ Receptor
_____ Sensory neuron

FIGURE 12.5   Components of a reflex arc.

F. DEMONSTRATION OF REFLEX ARC

FLEXICOMP™ is a computer hardware and software system designed to demonstrate characteristics of the reflex arc. It consists of a computer-interfaced goniometer and percussion mallet that permit quantification of the reflex response for various joints of the body and provide visual representation of the reflex arc. The program is available from INTELITOOL® (1-800-227-3805).

ANSWER THE LABORATORY REPORT QUESTIONS AT THE END OF THE EXERCISE.

Nervous Tissue 12

Student _____   Date _____

Laboratory Section _____   Score/Grade _____

PART 1. Multiple Choice

_____ 1. The portion of a neuron that conducts nerve impulses away from the cell body is the (a) dendrite (b) axon (c) receptor (d) effector

_____ 2. The fine branching filaments of an axon are called (a) myelin sheaths (b) axolemmas (c) axon terminals (d) axon hillocks

_____ 3. The component of a reflex arc that responds to a motor impulse is the (a) integrating center (b) receptor (c) sensory neuron (d) effector

_____ 4. Which type of neuron conducts nerve impulses toward the central nervous system? (a) sensory (b) association (c) connecting (d) motor

_____ 5. In a reflex arc, the nerve impulse is transmitted directly to the effector by the (a) sensory neuron (b) motor neuron (c) integrating center (d) receptor

_____ 6. Bulblike structures at the distal ends of axon terminals that contain storage sacs for neurotransmitters are called (a) dendrites (b) synaptic end bulbs (c) axon collaterals (d) neurofi-brils

_____ 7. Which neuroglial cell is phagocytic? (a) oligodendrocyte (b) protoplasmic astrocyte (c) microglial cell (d) fibrous astrocyte

PART 2. Completion

8. A neuron that contains several dendrites and one axon is classified as _____.

9. The portion of a neuron that contains the nucleus and cytoplasm is the _____.

10. The lipid and protein covering around many peripheral axons is called the _____.

11. The two types of cells that compose the nervous system are neurons and _____.

12. The peripheral, nucleated layer of the neurolemmocyte (Schwann cell) that encloses the myelin sheath

    is the _____.

13. The side branch of an axon is referred to as the _____.

14. The part of a neuron that conducts nerve impulses toward the cell body is the

    _____.

15. Neurons that carry nerve impulses between sensory neurons and motor neurons are called

_____ neurons.

16. Neurons with one dendrite and one axon are classified as _____.

17. Unmyelinated gaps between segments of the myelin sheath are known as _____.

18. The neuroglial cell that produces a myelin sheath around axons of neurons of the central nervous

system is called a(n) _____.

19. In a reflex arc, the muscle or gland that responds to a motor impulse is called the

_____.

20. The functional contact between two neurons or between a neuron and an effector is called a(n)

_____.

21. The circuit that probably plays a role in breathing, waking up, sleeping, and short-term memory is

a(n) _____ circuit.

Nervous System

13

In this exercise, you will examine the principal structural features of the spinal cord and spinal nerves, perform several experiments on reflexes. Next, you will identify the principal structural features of the brain, trace the course of cerebrospinal fluid, identify the cranial nerves, perform several experiments designed to test for cranial nerve function, and examine the structure and function of the autonomic nervous system. You will also dissect and study the sheep brain.

A. SPINAL CORD AND SPINAL NERVES

1. Meninges

The *meninges* (me-NIN-jēz) are connective tissue coverings that run continuously around the spinal cord and brain (*meninx*, pronounced MĒ-ninks, is singular). They protect the central nervous system. The spinal meninges are:

a. *Dura mater* (DYOO-ra MĀ-ter; *dura* = tough; *mater* = mother) The most superficial meninx composed of dense irregular connective tissue. Between the wall of the vertebral canal and the dura mater is the *epidural space*, which is filled with fat, connective tissue, and blood vessels.
b. *Arachnoid* (a-RAK-noyd; *arachne* = spider) The middle meninx is an avascular covering composed of very delicate collagen and elastic fibers. Between the arachnoid and the dura mater is a space called the *subdural space*, which contains interstitial fluid.
c. *Pia mater* (PĒ-a MĀ-ter; *pia* = delicate) The deep meninx is a thin, transparent connective tissue layer that adheres to the surface of the brain and spinal cord. It consists of interlacing collagen and a few elastic fibers and contains blood vessels. Between the pia mater and the arachnoid is a space called the *subarachnoid*

space where cerebrospinal fluid circulates. Extensions of the pia mater called *denticulate* (den-TIK-yoo-lāt; *denticulus* = a small tooth) *ligaments* are attached laterally to the dura mater along the length of the spinal cord and suspend the spinal cord and afford protection against shock and sudden displacement.

Label the meninges, subarachnoid space, and denticulate ligament in Figure 13.1.

2. General Features

Obtain a model or preserved specimen of the spinal cord and identify the following general features:

a. *Cervical enlargement* Between vertebrae C4 and T1; origin of nerves to upper limbs.
b. *Lumbar enlargement* Between vertebrae T9 and T12; origin of nerves to lower limbs.
c. *Conus medullaris* (KŌ-nus med-yoo-LAR-is; *konos* = cone). Tapered conical portion of spinal cord that ends at the intervertebral disc between L1 and L2.
d. *Filum terminale* (FĪ-lum ter-mi-NAL-ē; *filum* = filament; *terminale* = terminal) Nonnervous fibrous tissue that arises from the conus medullaris and anchors the spinal cord to the coccyx; consists mostly of pia mater.
e. *Cauda equina* (KAW-da ē-KWĪ-na; meaning "horse's tail") Spinal nerves that angle inferiorly in the vertebral canal giving the appearance of wisps of coarse hair.
f. *Anterior median fissure* Deep, wide groove on the anterior surface of the spinal cord.
g. *Posterior median sulcus* Shallow, narrow groove on the posterior surface of the spinal cord.

After you have located the parts on a model or preserved specimen of the spinal cord, label the spinal cord in Figure 13.2.

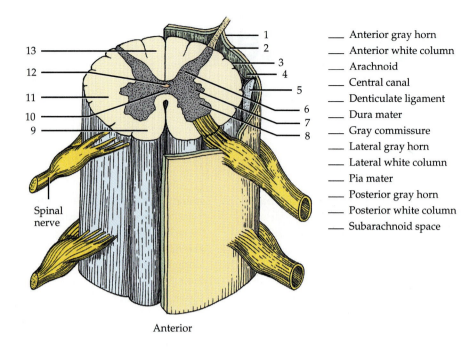

Labels on the right side of figure:

- ___ Anterior gray horn
- ___ Anterior white column
- ___ Arachnoid
- ___ Central canal
- ___ Denticulate ligament
- ___ Dura mater
- ___ Gray commissure
- ___ Lateral gray horn
- ___ Lateral white column
- ___ Pia mater
- ___ Posterior gray horn
- ___ Posterior white column
- ___ Subarachnoid space

Spinal nerve

Anterior

FIGURE 13.1　Transverse (cross) section of spinal cord showing meninges on right side of figure.

3. Transverse Section of Spinal Cord

In a freshly dissected section of the brain or spinal cord, some regions look white and glistening whereas others appear gray. *White matter* refers to aggregations of myelinated processes from many neurons. The whitish color of myelin gives white matter its name. The *gray matter* of the nervous system contains either nerve cell bodies, dendrites, and axon terminals or bundles of unmyelinated axons and neuroglia. They look grayish, rather than white, because there is no myelin in these areas.

Obtain a model or specimen of the spinal cord in transverse section and note the gray matter, shaped like a letter H or a butterfly. Identify the following parts:

a. *Gray commissure* (KOM-mi-shur)　Cross bar of the letter H.
b. *Central canal*　Small space in the center of the gray commissure that contains cerebrospinal fluid.
c. *Anterior gray horn*　Anterior region of the upright portion of the H.
d. *Posterior gray horn*　Posterior region of the upright portion of the H.
e. *Lateral gray horn*　Intermediate region between the anterior and posterior gray horns present in the thoracic, upper lumbar, and sacral segments of the spinal cord.

f. *Anterior white column*　Anterior region of white matter.
g. *Posterior white column*　Posterior region of white matter.
h. *Lateral white column*　Intermediate region of white matter between the anterior and posterior white columns.

Label these parts of the spinal cord in transverse section in Figure 13.1.

Examine a prepared slide of a spinal cord in transverse section and see how many structures you can identify.

4. Spinal Nerve Attachments

Spinal nerves are the paths of communication between the spinal cord tracts and most of the body. The 31 pairs of spinal nerves are named and numbered according to the region of the spinal cord from which they emerge. The first cervical pair emerges between the atlas and occipital bone; all other spinal nerves leave the vertebral column from intervertebral foramina between adjoining vertebrae. There are 8 pairs of cervical nerves, 12 pairs of thoracic nerves, 5 pairs of lumbar nerves, 5 pairs of sacral nerves, and 1 pair of coccygeal nerves. Label the spinal nerves in Figure 13.2.

Each pair of spinal nerves is connected to the spinal cord by two points of attachment called roots.

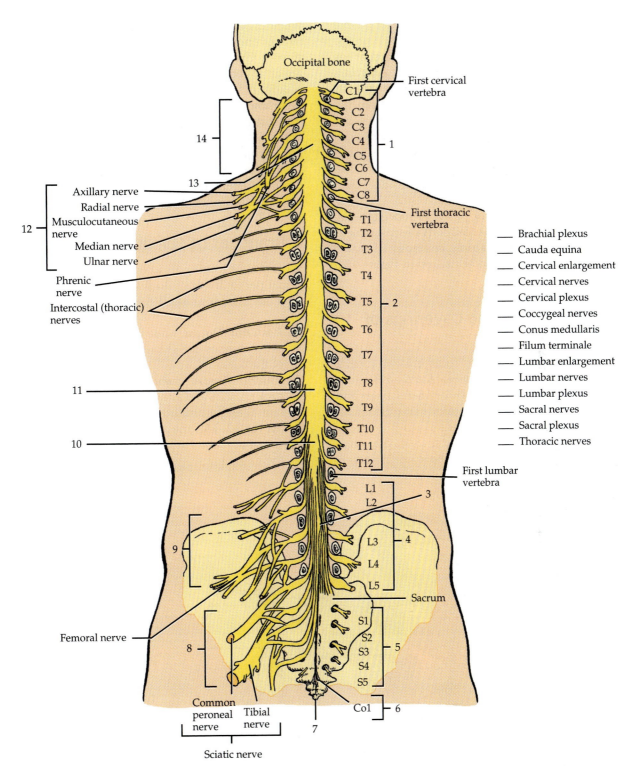

Occipital bone

First cervical vertebra

C1
C2
C3
C4
C5
C6
C7
C8

14

13

Axillary nerve
Radial nerve
Musculocutaneous nerve
Median nerve
Ulnar nerve

12

Phrenic nerve

Intercostal (thoracic) nerves

First thoracic vertebra

T1
T2
T3
T4
T5
T6
T7
T8
T9
T10
T11
T12

1

2

11

10

First lumbar vertebra

L1
L2
L3
L4
L5

3
4

9

Sacrum

S1
S2
S3
S4
S5

5

Femoral nerve

8

Common peroneal nerve   Tibial nerve

Co1   6

7

Sciatic nerve

___ Brachial plexus
___ Cauda equina
___ Cervical enlargement
___ Cervical nerves
___ Cervical plexus
___ Coccygeal nerves
___ Conus medullaris
___ Filum terminale
___ Lumbar enlargement
___ Lumbar nerves
___ Lumbar plexus
___ Sacral nerves
___ Sacral plexus
___ Thoracic nerves

Posterior view

FIGURE 13.2   Spinal cord.

The *posterior (sensory) root* contains sensory nerve fibers only and conducts nerve impulses from the periphery to the spinal cord. Each posterior root has a swelling, the *posterior (sensory) root ganglion*, which contains the cell bodies of the sensory neurons from the periphery. Fibers extend from the ganglion into the posterior gray horn. The other point of attachment, the *anterior (motor) root*, contains motor nerve fibers only and conducts nerve impulses from the spinal cord to the periphery. The cell bodies of the motor neurons are located in lateral or anterior gray horns.

Label the posterior root, posterior root ganglion, anterior root, spinal nerve, cell body of sensory neuron, axon of sensory neuron, cell body of motor neuron, and axon of motor neuron in Figure 13.3 on page 228.

5. Components and Coverings of Spinal Nerves

The posterior and anterior roots unite to form a spinal nerve at the intervertebral foramen. Because the posterior root contains sensory nerve fibers and the anterior root contains motor nerve fibers, all spinal nerves are *mixed nerves.*

Spinal nerves are covered by several connective tissue layers. Individual nerve fibers within a nerve, whether myelinated or unmyelinated, are wrapped in a covering called the *endoneurium* (en'-dō-NOO-rē-um). Groups of fibers with their endoneurium are arranged in bundles called *fascicles*, and each bundle is wrapped in a covering called the *perineurium* (per'-i-NOO-rē-um). All the fascicles, in turn, are wrapped in a covering called the *epineurium* (ep'-i-NOO-rē-um). This is the outermost covering around the entire nerve.

Obtain a prepared slide of a nerve in transverse section and identify the fibers, endoneurium, perineurium, epineurium, and fascicles. Now label Figure 13.4.

6. Branches of Spinal Nerves

Shortly after leaving its intervertebral foramen, a spinal nerve divides into several branches called *rami* (singular is *ramus* [RĀ-mus]):

a. *Posterior (Dorsal) ramus* Innervates (supplies) deep muscles and skin of the posterior surface of the back.

b. *Anterior (Ventral ramus)* Innervates superficial back muscles and all structures of the limbs and lateral and anterior trunk; except for thoracic nerves T2–T11, the anterior rami of the other spinal nerves form plexuses before innervating their structures.

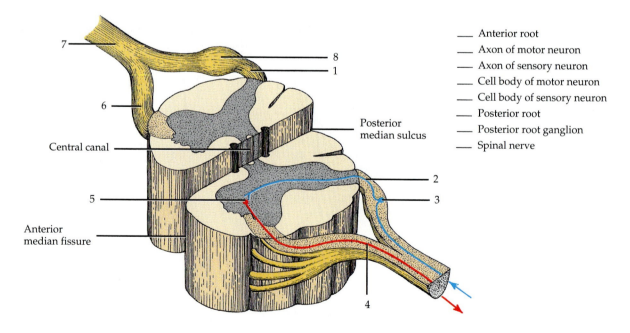

_____ Anterior root
_____ Axon of motor neuron
_____ Axon of sensory neuron
_____ Cell body of motor neuron
_____ Cell body of sensory neuron
_____ Posterior root
_____ Posterior root ganglion
_____ Spinal nerve

Posterior median sulcus

Central canal

Anterior median fissure

Sections through the thoracic spinal cord

FIGURE 13.3   Spinal nerve attachments.

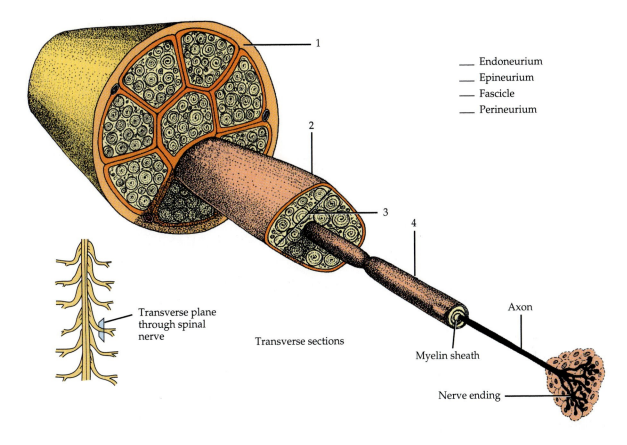

Endoneurium
Epineurium
Fascicle
Perineurium

Transverse plane through spinal nerve

Transverse sections

Axon

Myelin sheath

Nerve ending

FIGURE 13.4   Coverings of a spinal nerve.

c. *Meningeal branch* Innervates vertebrae, vertebral ligaments, blood vessels of the spinal cord, and meninges.

d. *Rami communicantes* (RĀ-mē ko-myoo-nē-KAN-tēz) Gray and white rami communicantes are components of the autonomic nervous system; they connect the anterior rami with sympathetic trunk ganglia.

7. Plexuses

The anterior rami of spinal nerves, except for T2–T11, do not go directly to body structures they supply. Instead, they join with adjacent nerves on either side of the body to form networks called *plexuses* (PLEK-sus-ēz; *plexus* = braid).

a. *Cervical plexus* (SER-vi-kul PLEK-sus) Formed by the anterior rami of the first four cervical nerves (C1–C4) with contributions from C5; one is located on each side of the neck alongside the first four cervical vertebrae; the plexus supplies the skin and muscles of the head, neck, and upper part of shoulders.

Using your textbook as a guide, label the nerves of the cervical plexus in Figure 13.5.

b. *Brachial* (BRĀ-kē-al) *plexus* Formed by the anterior rami of spinal nerves C5–C8 and T1 with contributions from C4 and T2; each is located on either side of the last four cervical and first thoracic vertebrae and extends inferiorly and laterally, superior to the first rib behind the clavicle, and into the axilla; the plexus constitutes the entire nerve supply for the upper limbs and shoulder region.

Using your textbook as a guide, label the nerves of the brachial plexus in Figure 13.6.

c. *Lumbar* (LUM-bar) *plexus* Formed by the anterior rami of spinal nerves L1–L4; each is located on either side of the first four lumbar vertebrae posterior to the psoas major muscle and anterior to the quadratus lumborum muscle; the plexus supplies the anterolateral abdominal wall, external genitals, and part of the lower limbs.

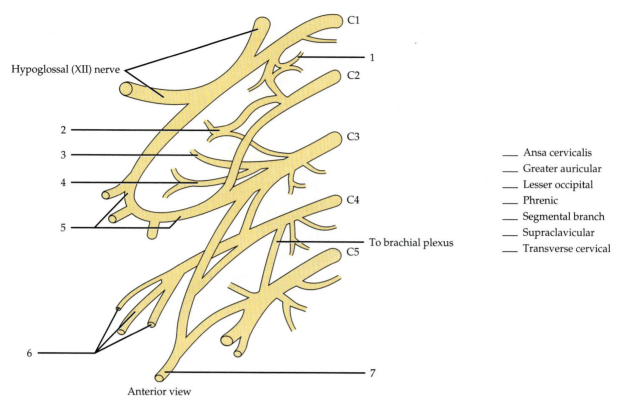

Hypoglossal (XII) nerve

C1
1
C2
2
3
4
C3
5
C4
To brachial plexus
C5
6
7

Anterior view

___ Ansa cervicalis
___ Greater auricular
___ Lesser occipital
___ Phrenic
___ Segmental branch
___ Supraclavicular
___ Transverse cervical

FIGURE 13.5   Origin of cervical plexus.

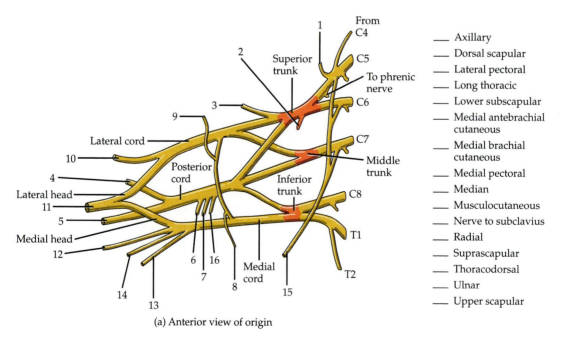

1
From
C4
2
Superior
trunk
C5
To phrenic
nerve
3
C6
9
Lateral cord
C7
10
Posterior
cord
Middle
trunk
4
Inferior
trunk
Lateral head
C8
11
5
Medial head
12
T1
6 16
7
Medial
cord
8
T2
14
15
13

(a) Anterior view of origin

___ Axillary
___ Dorsal scapular
___ Lateral pectoral
___ Long thoracic
___ Lower subscapular
___ Medial antebrachial
       cutaneous
___ Medial brachial
       cutaneous
___ Medial pectoral
___ Median
___ Musculocutaneous
___ Nerve to subclavius
___ Radial
___ Suprascapular
___ Thoracodorsal
___ Ulnar
___ Upper scapular

FIGURE 13.6   Origin of brachial plexus.

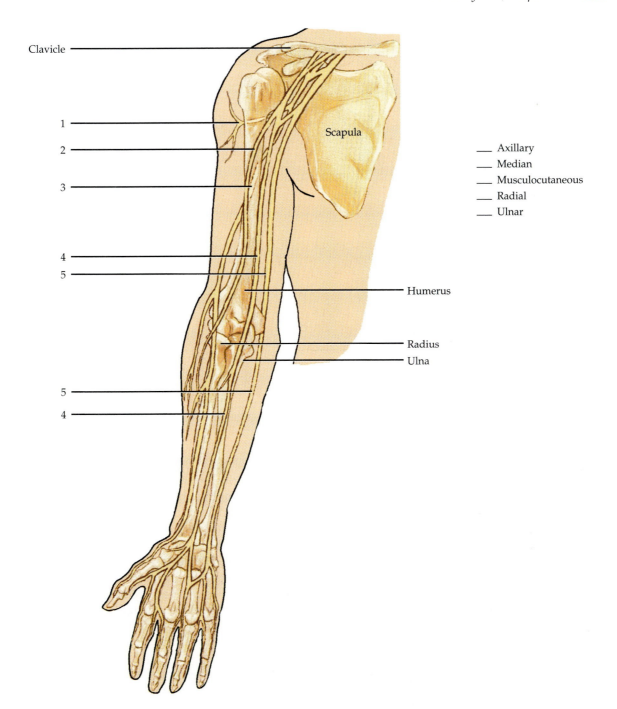

Clavicle

1

2

3

4

5

5

4

Scapula

Humerus

Radius

Ulna

___ Axillary
___ Median
___ Musculocutaneous
___ Radial
___ Ulnar

(b) Anterior view of distribution

FIGURE 13.6 (Continued)   Origin and distribution of brachial plexus.

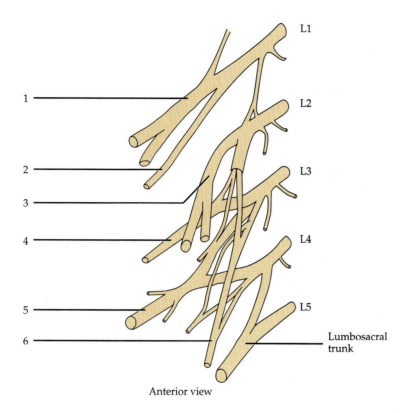

L1
L2
L3
L4
L5

Lumbosacral
trunk

Anterior view

_____ Femoral
_____ Genitofemoral
_____ Iliohypogastric
_____ Ilioinguinal
_____ Lateral femoral cutaneous
_____ Obturator

FIGURE 13.7   Origin of lumbar plexus.

Using your textbook as a guide, label the nerves of the lumbar plexus in Figure 13.7.

d. *Sacral* (SĀ-kral) *plexus* Formed by the anterior rami of spinal nerves L4–L5 and S1–S4; each is located largely anterior to the sacrum; the plexus supplies the buttocks, perineum, and lower limbs.

Using your textbook as a guide, label the nerves of the sacral plexus in Figure 13.8.

Label the cervical, brachial, lumbar, and sacral plexuses in Figure 13.2. Also note the names of some of the major peripheral nerves that arise from the plexuses.

For each nerve listed in Table 13.1, indicate the plexus to which it belongs and the structure(s) it innervates.

8. Spinal Cord Tracts

The vital function of conveying sensory and motor information to and from the brain is carried out by sensory (ascending) and motor (descending) tracts and pathways in the spinal cord. The names of the tracts and pathways indicate the white column in which the tract travels, where the cell bodies of the tract originate, and where the axons of the tract terminate. For example, the anterior spinothalamic

tract is located in the *anterior* white column, it originates in the *spinal cord*, and it terminates in the *thalamus* of the brain. Since it conveys nerve impulses from the spinal cord upward to the brain, it is a sensory (ascending) tract.

a. SOMATIC SENSORY PATHWAYS

Somatic sensory pathways from receptors to the cerebral cortex involve three-neuron sets. Axon collaterals (branches) of somatic sensory neurons simultaneously carry signals into the cerebellum and the reticular formation of the brain stem.

1. *First-order neurons* Carry signals from the somatic receptors into either the brain stem or spinal cord. From the face, mouth, teeth, and eyes, somatic sensory impulses propagate along *cranial nerves* into the brain stem. From the back of the head, neck, and body, somatic sensory impulses propagate along *spinal nerves* into the spinal cord.

2. *Second-order neurons* Carry signals from the spinal cord and brain stem to the thalamus. Axons of second-order neurons cross over (decussate) to the opposite side in the spinal cord or brain stem before ascending to the thalamus.

3. *Third-order neurons* Project from the thalamus to the primary somatosensory area of the cor-

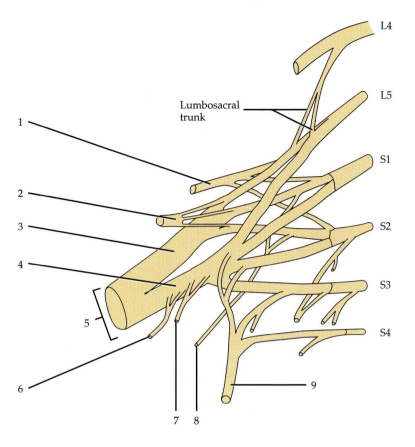

L4

L5

S1

S2

S3

S4

Lumbosacral trunk

1

2

3

4

5

6

7   8

9

___ Common peroneal
___ Inferior gluteal
___ Nerve to obturator internus and superior gemellus
___ Nerve to quadratus femoris and inferior gemellus
___ Posterior femoral cutaneous
___ Pudendal
___ Sciatic
___ Superior gluteal
___ Tibial

(a) Origin in anterior view

FIGURE 13.8 Origin of sacral plexus.

tex (postcentral gyrus; see Figure 13.14 on page 245), where conscious perception of the sensations results.

There are two general pathways by which somatic sensory signals entering the spinal cord ascend to the cerebral cortex: the *posterior column–medial lemniscus pathway* and the *anterolateral (spinothalamic) pathways.*

1. Posterior Column–Medial Lemniscus Pathway to the Cerebral Cortex Nerve impulses for conscious proprioception and most tactile sensations ascend to the cerebral cortex along a common pathway formed by three-neuron sets. First-order neurons extend from sensory receptors into the spinal cord and up to the medulla oblongata on the same side of the body. The cell bodies of these first-order neurons are in the posterior (dorsal) root ganglia of spinal nerves. Their axons form the *posterior column (fasciculus gracilis;* fa-SIK-yoo-lus gras-I-lus and *fasciculus cuneatus;* kyoo-nē-AT-us) in the spinal cord. The axon terminals form synapses with second-order neurons in the medulla. The cell body

of a second-order neuron is located in the nucleus cuneatus (which receives input conducted along axons in the fasciculus cuneatus from the neck, upper limbs, and upper chest) or nucleus gracilis (which receives input conducted along axons in the fasciculus gracilis from the trunk and lower limbs). The axon of the second-order neuron crosses to the opposite side of the medulla and enters the *medial lemniscus,* a projection tract that extends from the medulla to the thalamus. In the thalamus, the axon terminals of second-order neurons synapse with third-order neurons, which project their axons to the somatosensory area of the cerebral cortex.

Impulses conducted along the posterior column-medial lemniscus pathway give rise to several highly evolved and refined sensations. These are:

Discriminative touch The ability to recognize the exact location of a light touch and to make two-point discriminations.

Stereognosis The ability to recognize by feel the size, shape, and texture of an object. Examples are identifying (with closed eyes) a paper clip put into your hand or reading braille.

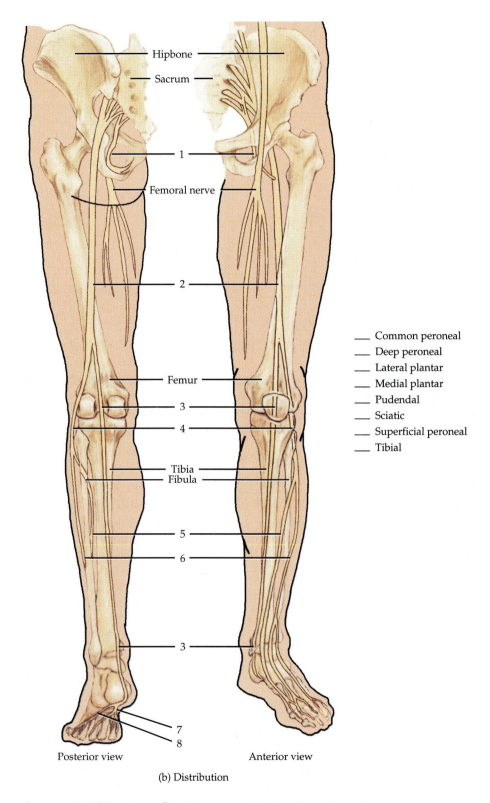

Hipbone

Sacrum

1

Femoral nerve

2

___ Common peroneal
___ Deep peroneal
___ Lateral plantar
___ Medial plantar
___ Pudendal
___ Sciatic
___ Superficial peroneal
___ Tibial

Femur

3

4

Tibia
Fibula

5

6

3

7
8

Posterior view   Anterior view

(b) Distribution

FIGURE 13.8 (*Continued*)   Origin and distribution of sacral plexus.

TABLE 13.1
Nerves of Plexuses and Innervations

Nerve	Plexus	Innervation
Musculocutaneous (mus'-kyoo-lō-kyoo-TAN-ē-us)		
Femoral (FEM-or-al)		
Phrenic (FREN-ik)		
Pudendal (pyoo-DEN-dal)		
Axillary (AK-si-lar-ē)		
Transverse cervical (SER-vi-kul)		
Radial (RĀ-dē-al)		
Obturator (OB-too-rā-tor)		
Tibial (TIB-ē-al)		
Thoracodorsal (thō-RA-kō-dor-sal)		
Perforating cutaneous (PER-fō-rā-ting kyoo'-TA-nē-us)		
Ulnar (UL-nar)		
Long thoracic (thō-RAS-ik)		
Median (MĒ-dē-an)		
Iliohypogastric (il'-ē-ō-hī-pō-GAS-trik)		
Deep peroneal (per'-ō-NĒ-al)		
Ansa cervicalis (AN-sa ser-vi-KAL-is)		
Sciatic (sī-AT-ik)		

Proprioception The awareness of the precise position of body parts, and *kinesthesia,* the awareness of directions of movement.

Weight discrimination The ability to assess the weight of an object.

Vibratory sensations The ability to sense rapidly fluctuating touch.

2. Anterolateral (Spinothalamic) Pathways to the Cerebral Cortex The *anterolateral* (*spinothalamic;* spī-nō-THAL-am-ik) *pathways* carry mainly pain and temperature impulses. In addition, they relay the sensations of tickle and itch and some tactile impulses, which give rise to a very crude, not well-localized touch or pressure sensation. Like the posterior column–medial lemniscus pathway, the anterolateral pathways are also composed of three-neuron sets. The first-order neuron connects a receptor of the neck, trunk, or limbs with the spinal cord. The cell body of the first-order neuron is in the posterior root ganglion. The axon of the first-order neuron synapses with the second-order neuron, which is located in the posterior gray horn of the spinal cord. The axon of the second-order neuron continues to the opposite side of the spinal cord and passes superiorly to the brain stem in either the *lateral spinothalamic tract* or the *anterior spinothalamic tract.* The axon from the second-order neuron ends in the thalamus. There, it synapses with the third-order neuron. The axon of the third-order neuron projects to the somatosensory area of the cerebral cortex. The lateral spinothalamic tract conveys sensory impulses for pain and temperature whereas the anterior spinothalamic tract conveys impulses for tickle, itch, crude touch, and pressure.

3. Somatic Sensory Pathways to the Cerebellum Two tracts in the spinal cord, the *posterior spinocerebellar* (spī'-nō-ser-e-BEL-ar) *tract* and the *anterior spinocerebellar tract,* are major routes for the subconscious proprioceptive input to reach the cerebellum. Sensory input conveyed to the cerebellum along these two pathways is critical for posture, balance, and coordination of skilled movements.

b. SOMATIC MOTOR PATHWAYS

The most direct somatic motor pathways extend from the cerebral cortex to skeletal muscles. Other pathways are less direct and include synapses in the basal ganglia, thalamus, reticular formation, and cerebellum.

1. Direct Pathways Voluntary motor impulses are propagated from the motor cortex to voluntary motor neurons (somatic motor neurons) that innervate skeletal muscles via the *direct* or *pyramidal* (pi-RAM-i-dal) *pathways.* The simplest of these pathways consists of sets of two neurons, upper motor neurons and lower motor neurons. About one million pyramidal-shaped cell bodies of direct pathway *upper motor neurons* (*UMNs*) are in the cortex. Their axons descend through the internal capsule of the cerebrum. In the medulla oblongata, the axon bundles form the ventral bulges known as the *pyramids.* About 90% of these axons also cross (decussate) to the opposite side in the medulla oblongata. They terminate in nuclei of cranial nerves or in the anterior gray horn of the spinal cord. *Lower motor neurons* (*LMNs*) extend from the motor nuclei of nine cranial nerves to muscles of

the face and head and from the anterior horn of each spinal cord segment to skeletal muscle fibers of the trunk and limbs. Close to their termination point, most upper motor neurons synapse with an association neuron, which, in turn, synapses with a lower motor neuron. A few upper motor neurons synapse directly with lower motor neurons.

The direct pathways convey impulses from the cortex that result in precise, voluntary movements. The main parts of the body governed by the direct pathways are the face, vocal cords (for speech), and hands and feet of the limbs. They channel nerve impulses into three tracts:

a. *Lateral corticospinal* (kor'-ti-kō-SPĪ-nal) *tracts* These pathways begin in the right and left motor cortex and descend through the *internal capsule* of the cerebrum and through the cerebral peduncle of the midbrain and the pons on the same side. About 90% of the axons of upper motor neurons cross over to the opposite side in the medulla oblongata. These axons form the lateral corticospinal tracts in the right and left lateral white columns of the spinal cord. Thus the motor cortex of the right side of the brain controls muscles on the left side of the body, and vice versa. The lower motor neurons then receive input from both upper motor neurons and association neurons. Axons of lower motor neurons (somatic motor neurons) exit all levels of the spinal cord via the anterior roots of spinal nerves and terminate in skeletal muscles. These motor neurons control skilled movements of the hands and feet.

b. *Corticobulbar* (kor'-ti-kō-BUL-bar) *tracts* The axons of upper motor neurons of these tracts accompany the corticospinal tracts from the motor cortex through the internal capsule to the brain stem. There some cross whereas others remain uncrossed. They terminate in the nuclei of nine pairs of cranial nerves in the pons and medulla: the oculomotor (III), trochlear (IV), trigeminal (V), abducens (VI), facial (VII), glossopharyngeal (IX), vagus (X), accessory (XI), and hypoglossal (XII). The lower motor neurons of cranial nerves convey impulses that control voluntary movements of the eyes, tongue, and neck; chewing; facial expression; and speech.

c. *Anterior corticospinal tracts* About 10% of the axons of upper motor neurons do not cross in the medulla oblongata. They pass through the medulla oblongata, descend on the same side, and form the anterior corticospinal tracts in the right and left anterior white columns. At several spinal cord levels, some of the axons of these upper motor neurons cross. After crossing to the opposite side, they synapse with association or lower motor neurons in the anterior gray horn of the spinal cord. Axons of these lower motor neurons exit the cervical and upper thoracic segments of the cord via the anterior roots of spinal nerves. They terminate in skeletal muscles that control movements of the neck and part of the trunk, thus coordinating movements of the axial skeleton.

2. Indirect Pathways The *indirect* (*extrapyramidal*) *pathways* include all descending (motor) tracts other than the corticospinal and corticobulbar tracts. Nerve impulses conducted along the indirect pathways follow complex, polysynaptic circuits that involve the motor cortex, basal ganglia, limbic system, thalamus, cerebellum, reticular formation, and nuclei in the brain stem. Axons of upper motor neurons that carry motor signals from the indirect pathways descend from various nuclei of the brain stem into five major tracts of the spinal cord and terminate on association neurons or lower motor neurons.

Lower motor neurons receive both excitatory and inhibitory input from many presynaptic neurons in both direct and indirect pathways, an example of convergence. For this reason, lower motor neurons are also called the *final common pathway.* Most nerve impulses from the brain are conveyed to association neurons before being received by lower motor neurons. The sum total of the input from upper motor neurons and association neurons determines the final response of the lower motor neu-

ron. It is not just a simple matter of the brain sending an impulse and the muscle always contracting.

The five major tracts of the indirect pathways are the *rubrospinal* (ROO-brō-spī-nal), *tectospinal* (TEK-tō-spī-nal), *vestibulospinal* (ves-TIB-yoo-lō-spī-nal), *lateral reticulospinal* (re-TIK-yoo-lō-spī-nal), and *medial reticulospinal.*

Label the sensory (ascending) and motor (descending) tracts shown in Figure 13.9.

Indicate the function of each sensory and motor tract listed in Table 13.2.

9. Reflex Experiments

In this section, you will determine the responses obtained in various common human reflexes. While performing the procedures it is important for you to consider the various components of a reflex arc, and to try to determine the receptors and effector organs involved.

a. Patellar reflex Have your partner sit on a table so that the knee is off of the table and the leg hangs freely. Strike the quadriceps (patellar) tendon just inferior to the kneecap with the reflex hammer. What is the response?

_____

_____

_____

_____

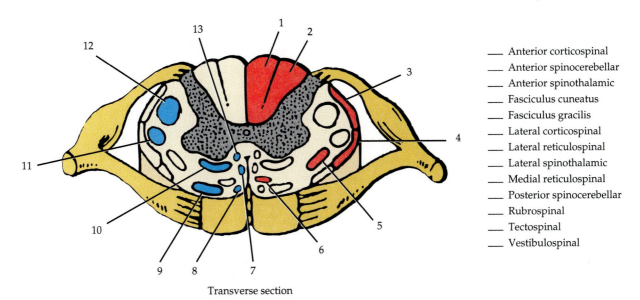

— Anterior corticospinal
— Anterior spinocerebellar
— Anterior spinothalamic
— Fasciculus cuneatus
— Fasciculus gracilis
— Lateral corticospinal
— Lateral reticulospinal
— Lateral spinothalamic
— Medial reticulospinal
— Posterior spinocerebellar
— Rubrospinal
— Tectospinal
— Vestibulospinal

Transverse section

FIGURE 13.9 Selected sensory (ascending) and motor (descending) tracts of the spinal cord. The sensory tracts are indicated in red, and the motor tracts are indicated in blue.

TABLE 13.2
Sensory (Ascending) and Motor (Descending) Tracts and Their Functions

Sensory (ascending) tracts	Function
Posterior column	
Lateral spinothalamic	
Anterior spinothalamic	
Posterior spinocerebellar	
Anterior spinocerebellar	
Motor (descending) tracts	**Function**
Direct	
Lateral corticospinal	
Anterior corticospinal	
Corticobulbar	
Indirect	
Rubrospinal	
Tectospinal	
Vestibulospinal	
Lateral reticulospinal	
Medial reticulospinal	

Test the subject while he or she interlocks fingers and pulls one hand against the other. Is there any change in the level of activity (sensitivity) of the reflex?

_____

_____

_____

Formulate a hypothesis regarding the purpose of the patellar reflex.

_____

_____

_____

b. *Achilles reflex* Have a subject kneel on a chair and let the feet hang freely over the edge of the chair. Bend one foot to increase the tension on the gastrocnemius muscle. Tap the calcaneal (Achilles) tendon with the reflex hammer.

What is the result?

_____

_____

_____

_____

c. *Plantar reflex* Scratch the sole of your partner's foot by moving a blunt object along the sole toward the toes.
What is the result?

_____

_____

_____

_____

What is the Babinski sign?

_____

_____

_____

Why is it normal in children under the age of 18 months?

_____

_____

_____

_____

What does the Babinski sign indicate in an adult?

_____

_____

_____

_____

B. BRAIN

1. Parts

The brain may be divided into four principal parts: (1) *brain stem*, which consists of the medulla oblongata, pons, and midbrain; (2) *diencephalon* (dī-en-SEF-a-lon; *dia* = through; *enkephalos* = brain), which consists primarily of the thalamus and hypothalamus; (3) *cerebellum*, (ser'-e-BEL-um) which is posterior to the brain stem; and (4) *cerebrum*, (se-RĒ-brum) which is superior to the brain stem and comprises about seven-eighths of the total weight of the brain.

Examine a model and preserved specimen of the brain and identify the parts just described. Then refer to Figure 13.10 and label the parts of the brain.

2. Meninges

As in the spinal cord, the brain is protected by *meninges*. The cranial meninges are continuous with the spinal meninges. The cranial meninges are the superficial *dura mater*, the middle *arachnoid*, and the deep *pia mater*. The cranial dura mater consists of two layers called the *periosteal layer* (adheres to the cranial bones and serves as a periosteum) and the *meningeal layer* (thinner, inner layer that corresponds to the spinal dura mater).

Refer to Figure 13.11 on page 241 and label all of the meninges.

3. Cerebrospinal Fluid (CSF)

The central nervous system is nourished and protected by *cerebrospinal fluid (CSF)*. The fluid circulates through the subarachnoid space around the brain and spinal cord and through the *ventricles* (VEN-tri-kuls; *ventriculus* = little belly or cavity) of the brain. The ventricles are cavities in the brain that communicate with each other, with the central canal of the spinal cord, and with the subarachnoid space.

Cerebrospinal fluid is formed primarily by filtration and secretion from networks of capillaries and ependymal cells in the ventricles, called *choroid* (KŌ-royd; *chorion* = membrane) *plexuses* (see Figure 13.11). Each of the two *lateral ventricles* is located within a hemisphere (side) of the cerebrum under the corpus callosum. Anteriorly, the lateral ventricles are separated by a thin membrane called the *septum pellucidum* (pe-LOO-si-dum; *pellucidus* = allowing passage of light). The fluid formed in the choroid plexuses of the lateral ventricles circulates through an opening called the *interventricular foramen* into the third ventricle. The *third ventricle* is a slit between and inferior to the right and left halves of the thalamus and between the lateral ventricles. More fluid is added by the choroid plexus of the third ventricle. Then the fluid circulates through a canal-like structure called the *cerebral aqueduct* (AK-we-dukt) into the fourth ventricle. The *fourth ventricle* lies between the inferior brain

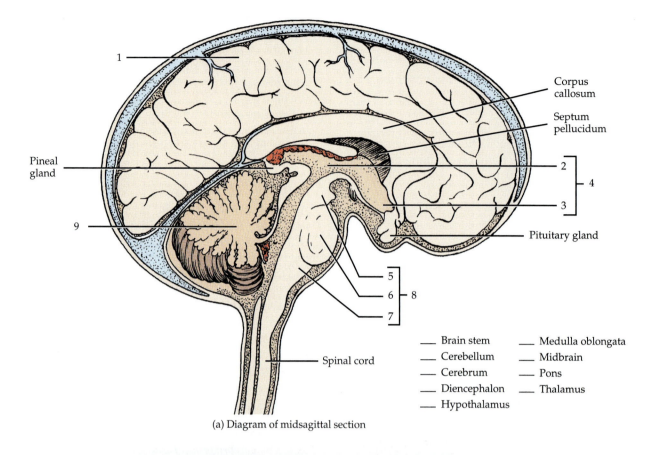

Corpus callosum

Septum pellucidum

Pineal gland

Pituitary gland

Spinal cord

____ Brain stem   ____ Medulla oblongata
____ Cerebellum   ____ Midbrain
____ Cerebrum   ____ Pons
____ Diencephalon ____ Thalamus
____ Hypothalamus

(a) Diagram of midsagittal section

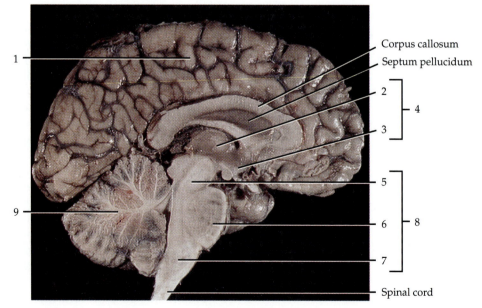

Corpus callosum

Septum pellucidum

Spinal cord

(b) Photograph of midsagittal section

FIGURE 13.10   Principal parts of the brain.

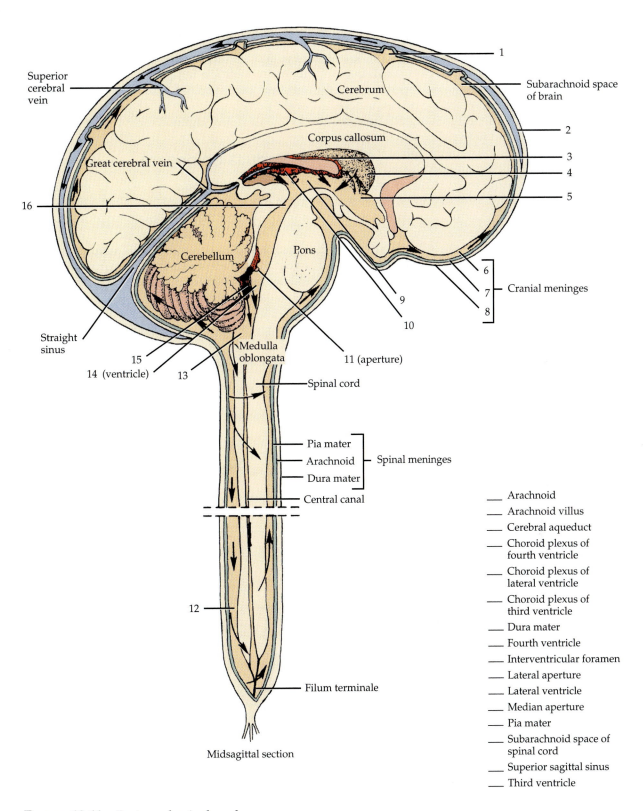

Superior cerebral vein

Cerebrum

Subarachnoid space of brain

1

2

Corpus callosum

3

Great cerebral vein

4

5

16

Cerebellum

Pons

6

7   Cranial meninges

8

9

10

Straight sinus

Medulla oblongata

11 (aperture)

15

14 (ventricle)   13

Spinal cord

Pia mater

Arachnoid   Spinal meninges

Dura mater

Central canal

12

Filum terminale

Midsagittal section

___ Arachnoid

___ Arachnoid villus

___ Cerebral aqueduct

___ Choroid plexus of fourth ventricle

___ Choroid plexus of lateral ventricle

___ Choroid plexus of third ventricle

___ Dura mater

___ Fourth ventricle

___ Interventricular foramen

___ Lateral aperture

___ Lateral ventricle

___ Median aperture

___ Pia mater

___ Subarachnoid space of spinal cord

___ Superior sagittal sinus

___ Third ventricle

FIGURE 13.11   Brain and spinal cord.

stem and the cerebellum. More fluid is added by the choroid plexus of the fourth ventricle. The roof of the fourth ventricle has three openings: one *median aperture* (AP-er-chur) and two *lateral apertures*. The fluid circulates through the apertures into the subarachnoid space around the posterior portion of the brain and inferiorly through the central canal of the spinal cord to the subarachnoid space around the posterior surface of the spinal cord, up the anterior surface of the spinal cord, and around the anterior part of the brain. Most of the cerebrospinal fluid is absorbed into a vein called the superior sagittal sinus through its arachnoid villi. Normally, cerebrospinal fluid is absorbed as rapidly as it is formed.

Refer to Figure 13.11 and label the choroid plexus of the lateral ventricle, lateral ventricle, interventricular foramen, choroid plexus of third ventricle, third ventricle, cerebral aqueduct, choroid plexus of fourth ventricle, fourth ventricle, median aperture, lateral aperture, subarachnoid space of spinal cord, superior sagittal sinus, and arachnoid villus.

Note the arrows in Figure 13.11, which indicate the path taken by cerebrospinal fluid. With the aid of your textbook, starting at the choroid plexus of the lateral ventricle and ending at the superior sagittal sinus, see if you can follow the remaining path of the fluid.

Now complete Figure 13.12.

4. Medulla Oblongata

The *medulla oblongata* (me-DULL-la ob'-long-GA-ta), or just simply *medulla*, is a continuation of the superior part of the spinal cord and forms the inferior part of the brain stem. The medulla contains all sensory and motor tracts that communicate between the spinal cord and various parts of the brain. On the ventral side of the medulla are two roughly triangular bulges called *pyramids*. They contain the largest motor tracks that run from the cerebral cortex to the spinal cord. Most fibers in the left pyramid cross to the right side of the spinal cord and most fibers in the right pyramid cross to the left side of

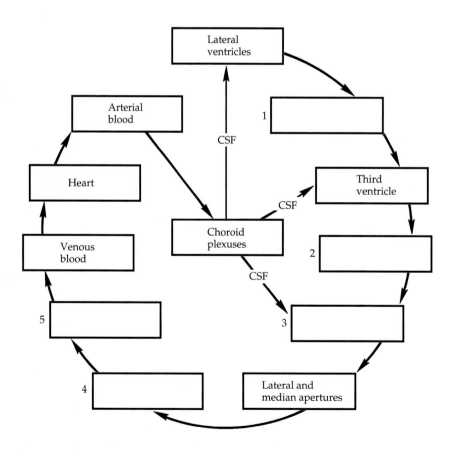

FIGURE 13.12   Formation, circulation, and absorption of cerebrospinal fluid (CSF).

the spinal cord. This crossing is called the ***decussation*** (dē'-ku-SĀ-shun) ***of pyramids*** and explains why motor areas of one side of the cerebral cortex control muscular movements on the opposite side of the body. The medulla contains vital reflex centers called the ***cardiovascular center,*** (regulates heart rate and blood vessel diameter) and ***medullary rhythmicity center*** (adjusts basic rhythm of breathing). Other centers in the medulla coordinate swallowing, vomiting, coughing, sneezing, and hiccuping. The posterior side of the medulla contains two pairs of prominent nuclei: ***nucleus gracilis*** and ***nucleus cuneatus.*** Most sensory impulses initiated on one side of the body cross to the thalamus on the opposite side either in the spinal cord or in these medullary nuclei. The medulla also contains the nuclei of origin of several cranial nerves. These are the cochlear and vestibular branches of the vestibulocochlear (VIII) nerves, glossopharyngeal (IX) nerves, vagus (X) nerves, spinal portions of the accessory (XI) nerves, and hypoglossal (XII) nerves. Examine a model or specimen of the brain and identify the parts of the medulla. Then refer to Figure 13.10 and locate the medulla.

5. Pons

The ***pons*** (pons = bridge) lies directly superior the medulla oblongata and anterior to the cerebellum. As the name implies, the pons is a bridge connecting the spinal cord with the brain, and other parts of the brain to each other. These connections are provided by axons that extend in two principle directions. The transverse axons are within paired tracts that connect the right and left sides of the cerebellum. They form the ***middle cerebellar peduncles.*** The longitudinal axons of the pons belong to the motor and sensory tracts that connect the medulla to the midbrain. The pons contains the nuclei or origin of the following pairs of cranial nerves: trigeminal (V) nerves, abducens (VI) nerves, facial (VII) nerves, and vestibular branch of the vestibulocochlear (VIII) nerves. Other important nuclei in the pons are the ***pneumotaxic*** (noo-mō-TAK-sik) ***area*** and ***apneustic*** (ap-NOO-stik) ***area*** that help control respiration.

Identify the pons on a model or specimen of the brain. Locate the pons in Figure 13.10.

6. Midbrain

The ***midbrain*** extends from the pons to the inferior portion of the cerebrum. The ventral portion of the midbrain contains the paired ***cerebral peduncles*** (pe-DUNG-kulz; *pedunculus* = stemlike portion), which connect the upper parts of the brain to inferior parts of the brain and spinal cord. The posterior part of the midbrain contains four rounded elevations called the ***corpora quadrigemina*** (KOR-por-ra kwad-ri-JEM-in-a; *corpus* = body; *quadrigeminus* = group of four). Two of the elevations, the ***superior colliculi*** (ko-LIK-yoo-lī) serve as reflex centers for movements of the eyeballs and the head in response to visual and other stimuli. ***Colliculus,*** which is singular, means small mound. The other two elevations, the ***inferior colliculi***, serve as reflex centers for movements of the head and trunk in response to auditory stimuli. The midbrain contains the nuclei of origin for two pairs of cranial nerves: oculomotor (III) and trochlear (IV).

Identify the parts of the midbrain on a model or specimen of the brain. Locate the midbrain in Figure 13.10.

7. Thalamus

The ***thalamus*** (THAL-a-mus; *thalamos* = inner chamber) is a large oval structure that comprises about 80% of the diencephalon. It is located superior to the midbrain and consists of two masses of gray matter organized into nuclei and covered by a layer of white matter. The two masses are joined by a bridge of gray matter called the ***intermediate mass.*** The thalamus contains numerous nuclei that serve as relay stations for all sensory impulses. The most prominent are the ***medial geniculate*** (je-NIK-yoo-lāt) ***nuclei*** (hearing), ***lateral geniculate nuclei*** (vision), ***ventral posterior nuclei*** (general sensations and taste), ***ventral lateral nuclei*** (voluntary motor actions), and the ***ventral anterior nuclei*** (voluntary motor actions and arousal). The thalamus also allows crude appreciation of some sensations, such as pain, temperature, and pressure. Precise localization of such sensations depends on nerve impulses being relayed from the thalamus to the cerebral cortex.

Identify the thalamic nuclei on a model or specimen of the brain. Then refer to Figure 13.13 and label the nuclei.

8. Hypothalamus

The ***hypothalamus*** (*hypo* = under) is located inferior to the thalamus and forms the floor and part of the wall of the third ventricle. Among the functions served by the hypothalamus are the control and integration of the activities of the autonomic nervous

Midsagittal section

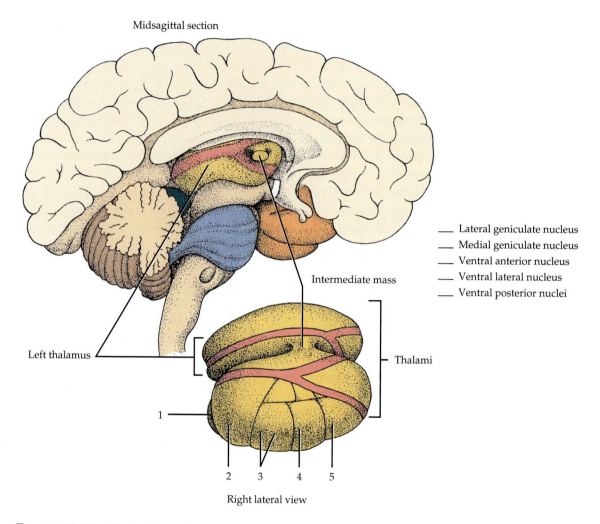

Lateral geniculate nucleus
Medial geniculate nucleus
Ventral anterior nucleus
Ventral lateral nucleus
Ventral posterior nuclei

Intermediate mass

Left thalamus

Thalami

1

2   3   4   5

Right lateral view

FIGURE 13.13   Thalamic nuclei.

system and parts of the endocrine system (pituitary gland) and the control of body temperature. The hypothalamus also assumes a role in feelings of rage and aggression, food intake, thirst, and the waking state and sleep patterns.

Identify the hypothalamus on a model or specimen of the brain. Locate the hypothalamus in Figure 13.10.

9. Cerebrum

The *cerebrum* is the largest portion of the brain and is supported on the brain stem. Its outer surface consists of gray matter and is called the *cerebral cortex* (*cortex* = rind or bark). Beneath the cerebral cortex is the cerebral white matter. The upfolds of the cerebral cortex are termed *gyri* (JĪ-rī) or *convolutions,* the deep downfolds are termed *fissures,* and the shallow downfolds are termed *sulci* (SUL-sī).

The *longitudinal fissure* separates the cerebrum into right and left halves called *hemispheres.* Each hemisphere is further divided into lobes by sulci or fissures. The *central sulcus* (SUL-kus) separates the *frontal lobe* from the *parietal lobe.* The *lateral cerebral sulcus* separates the frontal lobe from the *temporal lobe.* The *parieto-occipital sulcus* separates the *parietal lobe* from the *occipital lobe.* Another prominent fissure, the *transverse fissure,* separates the cerebrum from the cerebellum. Another lobe of the cerebrum, the *insula,* lies deep within the lateral cerebral fissure under the parietal, frontal, and temporal lobes. It cannot be seen in external view. Two important gyri on either side of the *central sulcus* are the *precentral gyrus* and the *postcentral gyrus.* The olfactory (I) and optic (II) cranial nerves are associated with the cerebrum.

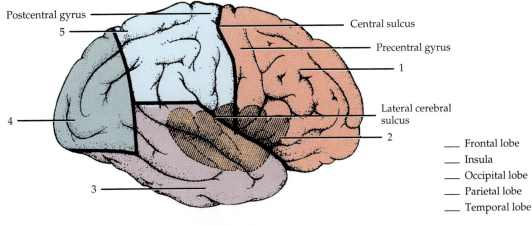

(a) Diagram of right lateral view

___ Frontal lobe
___ Insula
___ Occipital lobe
___ Parietal lobe
___ Temporal lobe

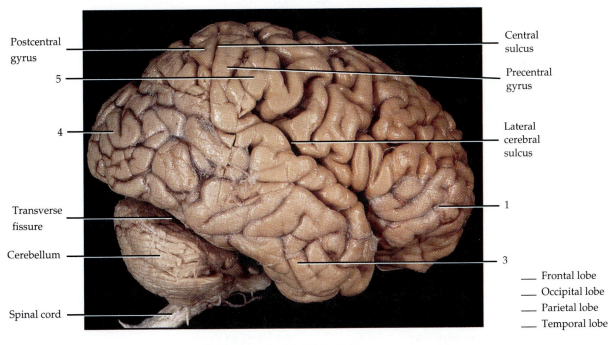

(b) Photograph of right lateral side

___ Frontal lobe
___ Occipital lobe
___ Parietal lobe
___ Temporal lobe

FIGURE 13.14   Lobes of cerebrum.

Examine a model of the brain and identify the parts of the cerebrum just described. Refer to Figure 13.14 and label the lobes.

10. Functional Areas of Cerebral Cortex

The functions of the cerebrum are numerous and complex. In a general way, the cerebral cortex can be divided into sensory, motor, and association areas.

The *sensory areas* receive and interpret sensory impulses, the *motor areas* control muscular movement, and the *associated areas* are concerned with emotional and intellectual processes.

The principal sensory and motor areas of the cerebral cortex are indicated by numbers based on K. Brodmann's map of the cerebral cortex. His map, first published in 1909, attempts to correlate structure and function. Refer to Figure 13.15 and match the name of the sensory or motor area next to the appropriate number.

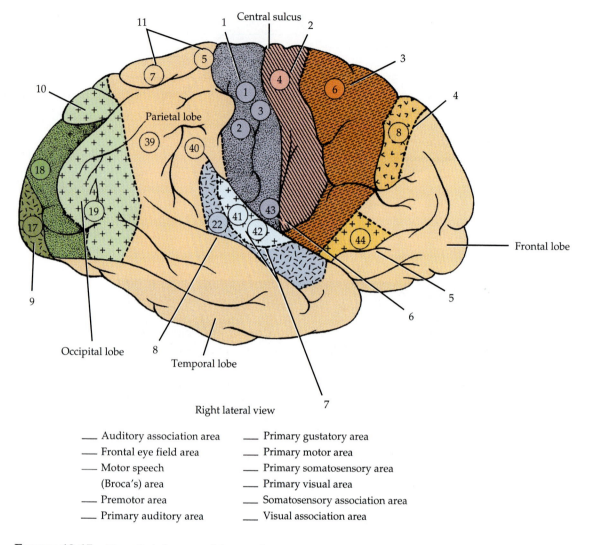

FIGURE 13.15　Functional areas of the cerebrum.

---

11. Cerebral White Matter

The white matter underlying the cerebral cortex consists of myelinated axons organized into tracts and extending in three principal directions.

a. *Association fibers* connect and transmit nerve impulses between gyri in the same hemisphere.
b. *Commissural fibers* transmit impulses from the gyri in one cerebral hemisphere to the corresponding gyri in the opposite cerebral hemisphere. Three important groups of commissural fibers are the *corpus callosum, anterior commissure,* and *posterior commissure.*
c. *Projection fibers* form motor and sensory tracts that transmit impulses from the cerebrum and other parts of the brain to the spinal cord or from the spinal cord to the brain.

12. Basal Ganglia

Basal ganglia (GANG-lē-a) or *cerebral nuclei* are groups of gray matter in the cerebral hemispheres. The largest of the basal ganglia is the *corpus striatum* (strī-Ā-tum; *corpus* = body; *striatum* = striped), which consists of the *caudate* (*cauda* = tail) *nucleus* and the *lentiform* (*lenticala* = shaped like a lentil or lens) or *lenticular nucleus*. The lentiform nucleus, in turn, is subdivided into a lateral *putamen* (pu-TĀ-men; *putamen* = shell) and a medial *globus pallidus* (*globus* = ball; *pallid* = pale).

　The basal ganglia are interconnected by many nerve fibers. They also receive input from and provide output to the cerebral cortex, thalamus, and hypothalamus. The caudate nucleus and the putamen control large automatic movements of skeletal muscles, such as swinging the arms while walking. The globus pallidus is concerned with the regulation of muscle tone required for specific body movements.

Examine a model or specimen of the brain and identify the basal ganglia described. Now refer to Figure 13.16 and label the basal ganglia shown.

13. Cerebellum

The **cerebellum** is inferior to the posterior portion of the cerebrum and separated from it by the **transverse fissure** and by an extension of the cranial dura mater called the **tentorium** (*tentorium* = tent) **cerebelli**. The central constricted area of the cerebellum is called the **vermis** (meaning worm-shaped) and the lateral portions are referred to as **cerebellar hemispheres**. The surface of the cerebellum, called the **cerebellar cortex**, consists of gray matter thrown into a series of slender parallel ridges called **folia** (*folia* = leaf). Deep to the gray matter are **white matter tracts** called **arbor vitae** (*arbor* = tree; *vita* = life) that resemble branches of a tree.

The cerebellum is attached to the brain stem by three paired bundles of fibers called **cerebellar peduncles**. The **inferior cerebellar peduncles** connect the cerebellum with the medulla oblongata and the spinal cord. These peduncles contain both sensory and motor axons that carry information into and out of the cerebellum. The **middle cerebellar peduncles** contain only sensory axons, which conduct input from the pons into the cerebellum. The **superior cerebellar peduncles** contain mostly motor axons that conduct output from the cerebellum into the midbrain.

The cerebellum compares the intended movement programmed by motor areas in the cerebrum with what is actually happening. It constantly receives sensory input from proprioceptors in muscles, tendons, and joints, receptors for equilibrium, and visual receptors of the eyes. If the intent of the cerebral motor areas is not being attained by skeletal muscles, the cerebellum detects the variation and sends feedback signals to the motor areas to either stimulate or inhibit the activity of skeletal muscles. This interaction helps to smooth and coordinate complex sequences of skeletal muscle contractions. Besides coordinating skilled movements, the cerebellum is the main brain region that regulates posture and balance. These aspects of cerebellar function make possible all skilled motor activities, from catching a baseball to dancing.

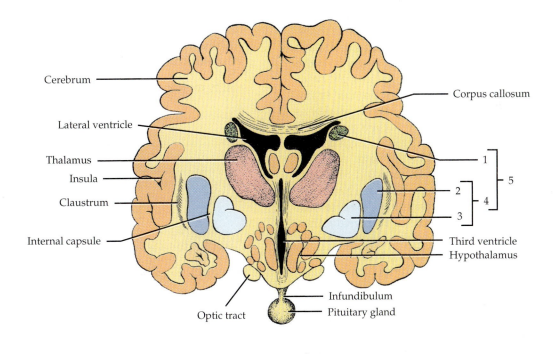

Frontal section of cerebrum

___ Caudate nucleus   ___ Lentiform nucleus

___ Corpus striatum   ___ Putamen

___ Globus pallidus

FIGURE 13.16   Basal ganglia.

Examine a model or specimen of the brain and locate the parts of the cerebellum. Identify the cerebellum in Figure 13.10.

14. Determination of Cerebellar Function

Working with your lab partner, perform the following tests to determine cerebellar function:

1. With your upper limbs down straight at your side, walk heel to toe for 20 ft without losing your balance.
2. Stand away from any supporting object and place your feet together and upper limbs down straight at your side; look straight ahead. Now close your eyes and stand for 2 to 3 min. Your lab partner should stand off to one side and then immediately in front of his or her partner and observe any and all body movements.
3. Stand away from any supporting object and place your feet together and upper limbs down straight at your side; look straight ahead. Now, with your eyes open raise one foot off the ground and run the heel of that foot down the shin of the other leg. Your lab partner should observe the smoothness of the movement and whether your heel remains in contact with the shin at all times.
4. Place your left palm in the supinated (palm up) position with the elbow at a 90° angle from the body. Now place the palm of your right hand into the left palm. Alternately pronate (palm down) and supinate the right hand as quickly as possible, while contacting the left palm only when the movement is completed (i.e., supinated right hand gently slaps left palm, and then pronated right hand gently slaps left palm). Do not let the medial surface of the right hand contact the left palm at any time. Note the ease with which your partner completes this activity. Then have your partner repeat the actions with the right and left hands switching roles.
5. Stand away from any supporting object and place your feet together and upper limbs abducted (away from midline) and palms facing anteriorly; look straight ahead. Now, gently close your eyes and touch the tip of your nose with your right index finger and return the right upper limb to its abducted position. Now touch the tip of your nose with your left index finger, and return the left upper limb to its abducted position. Slowly increase the rapidity of the activity.
6. With your eyes closed, touch your nose with the index finger of each hand.

C. CRANIAL NERVES

Of the 12 pairs of *cranial nerves*, 10 originate from the brain stem, but all pass through foramina in the base of the skull. The cranial nerves are designated by Roman numerals and names. The Roman numerals indicate the order in which the nerves arise from the brain, from front to back. The names indicate the distribution or function of the nerves.

Obtain a model of the brain and using your text and any other aids available identify the 12 pairs of cranial nerves. Now refer to Figure 13.17 and label the cranial nerves.

D. TESTS OF CRANIAL NERVE FUNCTION

The following simple tests may be performed to determine cranial nerve function. Although they provide only superficial information, they will help you to understand how the various cranial nerves function. Perform each of the tests with your partner.

1. *Olfactory (I) nerve* Have your partner smell several familiar substances such as spices, first using one nostril and then the other. Your partner should be able to distinguish the odors equally with each nostril.
   What structures could be malfunctioning if the
   ability to detect odors is lost? _____
   _____
   _____
   _____

2. *Optic (II) nerve* Have your partner read a portion of a printed page using each eye. Do the same while using a Snellen chart at a distance of 6.10 m (20 ft).
   Describe the visual pathway from the optic
   nerve to the cerebral cortex. _____
   _____
   _____
   _____

3. *Oculomotor (III), trochlear* (TROK-lē-ar) *(IV),* and *abducens* (ab-DOO-sens) *(VI) nerves* To test the motor responses of these nerves, have your partner follow your finger with his or her eyes without moving the head. Move your finger up, down, medially, and laterally.

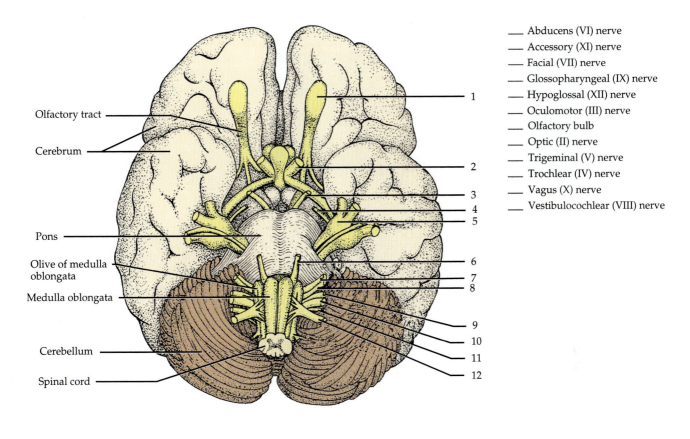

Olfactory tract

Cerebrum

Pons

Olive of medulla oblongata

Medulla oblongata

Cerebellum

Spinal cord

1
2
3
4
5
6
7
8
9
10
11
12

___ Abducens (VI) nerve
___ Accessory (XI) nerve
___ Facial (VII) nerve
___ Glossopharyngeal (IX) nerve
___ Hypoglossal (XII) nerve
___ Oculomotor (III) nerve
___ Olfactory bulb
___ Optic (II) nerve
___ Trigeminal (V) nerve
___ Trochlear (IV) nerve
___ Vagus (X) nerve
___ Vestibulocochlear (VIII) nerve

(a) Diagram of inferior surface

FIGURE 13.17   Cranial nerves of human brain.

Which nerves control which movements of the eyeball? _____

_____

Look for signs of ptosis (drooping of one or both eyelids).
Which cranial nerve innervates the upper eye lid? _____

CAUTION! *Do not make contact with the eyes.*

To test for the pupillary light reflex, shine a small flashlight into each eye from the side. Observe the pupil.

What cranial nerve controls this reflex? _____

4. **Trigeminal** (trī-JEM-i-nal) **(V) nerve** To test the motor responses of this nerve, have your partner close his or her jaws tightly.

What muscles are used? _____

_____

Now while holding your hand under your partner's lower jaw to provide resistance, ask your partner to open his or her mouth. To test the sensory responses of this nerve, have your partner close his or her eyes and lightly whisk a piece of dry cotton over the mandibular, maxillary, and ophthalmic areas on each side of the face. Do the same with cotton that has been moistened with cold water.
What cranial nerves bring about this response?

_____

5. **Facial (VII) nerve** To test the motor responses of this nerve, ask your partner to bring the corners of the mouth straight back, smile while showing his or her teeth, whistle, puff his or her cheeks, frown, raise his or her eyebrows, and wrinkle his or her forehead.

Define Bell's palsy. _____

_____

To test the sensory responses of this nerve, touch the tip of your partner's tongue with an

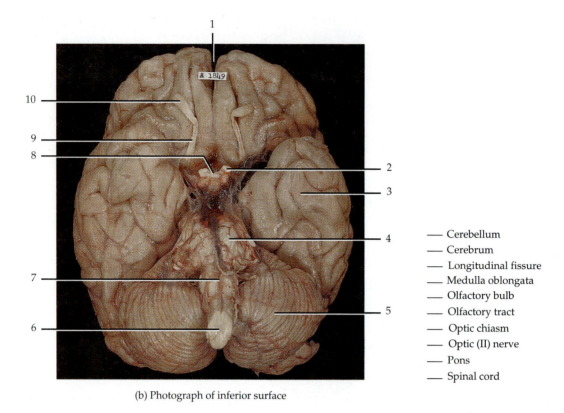

(b) Photograph of inferior surface

FIGURE 13.17 *(Continued)*   Cranial nerves of human brain.

— Cerebellum
— Cerebrum
— Longitudinal fissure
— Medulla oblongata
— Olfactory bulb
— Olfactory tract
— Optic chiasm
— Optic (II) nerve
— Pons
— Spinal cord

applicator that has been dipped into a salt solution. Rinse the mouth out with water and now place an applicator dipped into a sugar solution along the anterior surface of the tongue. What are the four basic tastes that can be distinguished? _____

_____

6. *Vestibulocochlear* (ves-tib'-yoo-lō-KŌK-lē-ar) *(VIII) nerve*
   a. To test for the functioning of the vestibular portion of this nerve, have your partner sit on a swivel stool with his or her head slightly bent forward toward the chest. Now have a student (the operator) turn your partner *slowly and very carefully* to the right about 10 times (about one turn every 2 sec). Stop the stool suddenly and look for nystagmus (nis-TAG-mus), which is rapid movement or quivering of the eyeballs. Note the direction of the nystagmus.

CAUTION! *Do not allow the subject to stand or walk until dizziness is completely gone.*

   b. When the nystagmus movements have ceased have the subject sit on the chair and, with his or her eyes open, reach forward and touch the operator's index finger with his or her finger. Now spin the subject to the right about 10 times (as before) and then suddenly stop the stool. *Immediately* after stopping the stool have the subject again, with his or her eyes open, try to reach forward and touch the operator's index finger with his or her finger. Note the degree of ease or difficulty with which this is accomplished.
   c. Again, when the nystagmus movements have ceased, repeat procedure b. This time, however, *immediately* after stopping the chair have the operator extend an index finger and have the subject open his or her eyes, locate the finger visually, and then *close his or her eyes and try to touch the operator's index finger with his or her eyes closed.* Note the degree of ease or difficulty with which this is accomplished, as well as the direction (right or left) in which the subject missed the operator's finger.

d. Now repeat procedure a with the subject resting his or her head on their right shoulder. Note any changes that might occur with the nystagmus response. _____

_____

Formulate a hypothesis as to the differences in results obtained in procedures a through d.

_____

_____

_____

_____

To test the functioning of the cochlear portion of the nerve, have your subject sit on a chair with his or her eyes closed with six members of the class standing in a wide circle around the individual. Now, one at a time, and in a random order, have the students click two coins together and have the subject point in the direction from which the sound originated. Then, with the students still standing in a circle of equal radius around the chair, and with the subject's eyes still closed, hand a ticking watch to one of the individuals in the circle and have him or her *slowly* move the watch closer to the subject until the subject first hears the sound and can correctly indicate from which direction the sound is originating. Repeat this procedure until the distance at which the sound is first heard and the direction correctly identified has been determined for all six directions. Measure the distances between the subject and each encircling student. Are all six distances

equal? _____

_____

7. **Glossopharyngeal** (glos-ō-fa-RIN-jē-al) **(IX)** and **Vagus (X) nerves** The palatal (gag) reflex can be used to test the functioning of both of these nerves. Using a cotton-tipped applicator, *very slowly and gently* touch your partner's uvula. Develop a hypothesis concerning the purpose

of this reflex. _____

_____

_____

To test the sensory responses of these nerves, have your partner swallow. Does swallowing occur easily? Now *gently* hold your partner's tongue down with a tongue depressor and ask him or her to say "ah." Does the uvula move? Are the movements on both sides of the soft palate the same? The sensory function of the glossopharyngeal nerve may be tested by *lightly* applying a cotton-tipped applicator dipped in quinine to the top, sides, and back of the tongue.

In which area of the tongue was the quinine

tasted? _____

What taste sensation is located there? _____

_____

8. **Accessory (XI) nerve** The strength and muscle tone of the sternocleidomastoid and trapezius muscles indicate the proper functioning of the accessory nerve. To ascertain the strength of the sternocleidomastoid muscle, have your partner turn his or her head from side to side against *slight* resistance that you supply by placing your hands on either side of your partner's head. To ascertain the strength of the trapezius muscle, place your hands on your partner's shoulders and while *gently* pressing down firmly ask him or her to shrug his or her shoulders.

Do both muscles appear to be reasonably

strong? _____

9. **Hypoglossal (XII) nerve** Have your partner protrude his or her tongue. It should protrude without deviation. Now have your partner protrude his or her tongue and move it from side to side while you attempt to *gently* resist the movements with a tongue depressor.

E. DISSECTION OF SHEEP BRAIN

CAUTION! *Please reread Section D, "Precautions Related to Dissection" at the beginning of the laboratory manual on page xiii before you begin your dissection.*

The brains of the fetal pig, sheep, and human show many similarities. They possess the protective membranes called the **meninges**, which can easily be seen as you proceed with the dissections. The outermost layer is the **dura mater**. It is the toughest, protective one, and may be missing in the preserved sheep brain. The middle membrane is the **arachnoid**, and the inner one containing blood vessels and adhering closely to the surface of the brain itself is the **pia mater**.

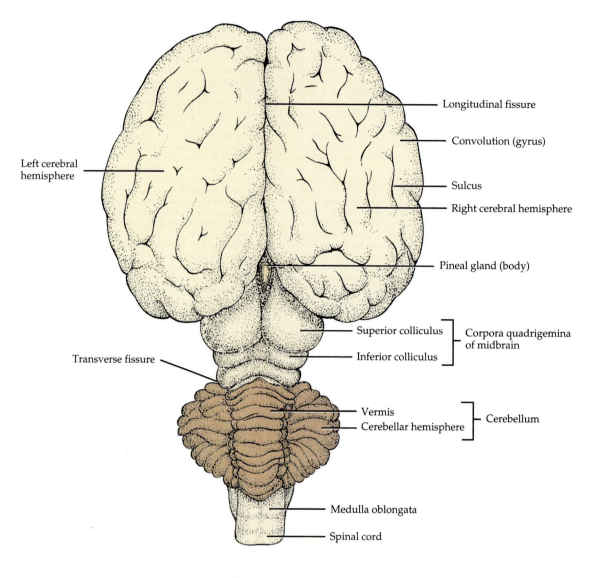

Dorsal view

FIGURE 13.18 Sheep brain in which cerebellum has been spread apart from the cerebrum.

PROCEDURE

1. Use Figures 13.18 through 13.21 as references for this dissection.
2. The most prominent external parts of the brain are the pair of large *cerebral hemispheres* and the posterior *cerebellum* on the dorsal surface. These large hemispheres are separated from each other by the *longitudinal fissure*. The *transverse fissure* separates the hemispheres from the cerebellum. The surfaces of these hemispheres form many *gyri*, or raised ridges, that are separated by grooves, or *sulci*.

3. If you spread the hemispheres gently apart you can see, deep in the longitudinal fissure, thick bundles of white transverse fibers. These bundles form the *corpus callosum*, which connects the hemispheres.
4. Most of the following structures can be identified by examining a midsagittal section of the sheep brain, or by cutting an intact brain along the longitudinal fissure completely through the corpus callosum.
5. If you break through the thin ventral wall, the *septum pellucidum* of the corpus callosum, you

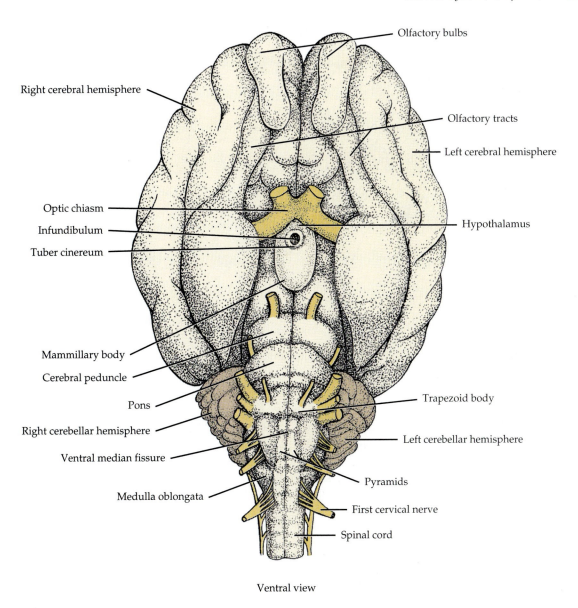

Olfactory bulbs

Right cerebral hemisphere

Olfactory tracts

Left cerebral hemisphere

Optic chiasm

Infundibulum

Tuber cinereum

Hypothalamus

Mammillary body

Cerebral peduncle

Pons

Right cerebellar hemisphere

Ventral median fissure

Medulla oblongata

Trapezoid body

Left cerebellar hemisphere

Pyramids

First cervical nerve

Spinal cord

Ventral view

FIGURE 13.19　Sheep brain.

can see part of a large chamber, the *lateral ventricle*, inside the hemisphere.

6. Each hemisphere has one of these ventricles. Ventral to the septum pellucidum, locate a smaller band of white fibers called the *fornix*. Close by where the fornix disappears is a small, round bundle of fibers called the *anterior commissure*.

7. The *third ventricle* and the *thalamus* are located ventral to the fornix. The third ventricle is outlined by its shiny epithelial lining, and the thalamus forms the lateral walls of this ventricle. This ventricle is crossed by a large circular mass of tissue, the *intermediate mass*, which connects the two sides of the thalamus. Each lateral ventricle communicates with the third ventricle through an opening, the *interventricular foramen*, which lies in a depression anterior to the intermediate mass and can be located with a dull probe.

8. Spreading the cerebral hemispheres and the cerebellum apart reveals the roof of the midbrain (mesencephalon), which is seen as two pairs of round swellings collectively called the *corpora quadrigemina*. The larger, more anterior pair are the *superior colliculi*. The smaller

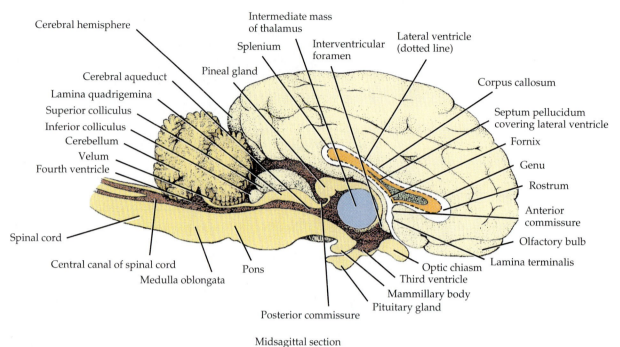

Cerebral hemisphere
Intermediate mass of thalamus
Splenium
Interventricular foramen
Lateral ventricle (dotted line)
Pineal gland
Cerebral aqueduct
Lamina quadrigemina
Superior colliculus
Inferior colliculus
Cerebellum
Velum
Fourth ventricle
Corpus callosum
Septum pellucidum covering lateral ventricle
Fornix
Genu
Rostrum
Anterior commissure
Spinal cord
Olfactory bulb
Lamina terminalis
Central canal of spinal cord
Pons
Medulla oblongata
Optic chiasm
Third ventricle
Mammillary body
Pituitary gland
Posterior commissure

Midsagittal section

FIGURE 13.20   Sheep brain.

posterior pair are the *inferior colliculi*. The *pineal gland (body)* is seen directly between the superior colliculi. Just posterior to the inferior colliculi, appearing as a thin white strand, is the *trochlear (IV) nerve*.

9. The cerebellum is connected to the brain stem by three prominent fiber tracts called *peduncles*. The *superior cerebellar peduncle* connects the cerebellum with the midbrain, the *inferior cerebellar peduncle* connects the cerebellum with the medulla, and the *cerebellar peduncle* connects the cerebellum with the pons.

10. Most of the following parts can be located on the ventral surface of the intact brain.

11. Just beneath the cerebral hemispheres are two *olfactory bulbs*, which continue posteriorly as two *olfactory tracts*. Posterior to these tracts, the *optic (II) nerves* undergo a crossing (decussation) known as the *optic chiasma*.

12. Locate the *pituitary gland (hypophysis)* just posterior to the chiasma. This gland is connected to the *hypothalamus* portion of the diencephalon by a stalk called the *infundibulum*. The *mammillary body* appears immediately posterior to the infundibulum.

13. Just posterior to this body are the paired *cerebral peduncles*, from which arise the large *oculomotor (III) nerves*. They may be partially covered by the pituitary gland.

14. The *pons* is a posterior extension of the hypothalamus and the *medulla oblongata* is a posterior extension of the pons.

15. The *cerebral aqueduct* dorsal to the peduncles runs posteriorly and connects the third ventricle with the *fourth ventricle*, which is located dorsal to the medulla and ventral to the cerebellum.

16. The medulla merges with the *spinal cord*, and is separated by the *ventral median fissure*. The *pyramids* are the longitudinal bands of tissue on either side of this fissure.

17. Identify the remaining *cranial nerves* on the ventral surface of the brain. They are the trigeminal (V), abducens (VI), facial (VII), vestibulocochlear (VIII), glossopharyngeal (IX), vagus (X), accessory (XI), and hypoglossal (XII). The previously identified cranial nerves are the olfactory (I), optic (II), oculomotor (III), and trochlear (IV), for a total of twelve.

18. A transverse section through a cerebral hemisphere reveals *gray matter* near the surface of the *cerebral cortex* and *white matter* beneath this layer.

19. A transverse section through the spinal cord reveals the *central canal*, which is connected to the fourth ventricle and contains *cerebrospinal fluid.*

20. A midsagittal section through the cerebellum reveals a treelike arrangement of gray and white matter called the *arbor vitae* (tree of life).

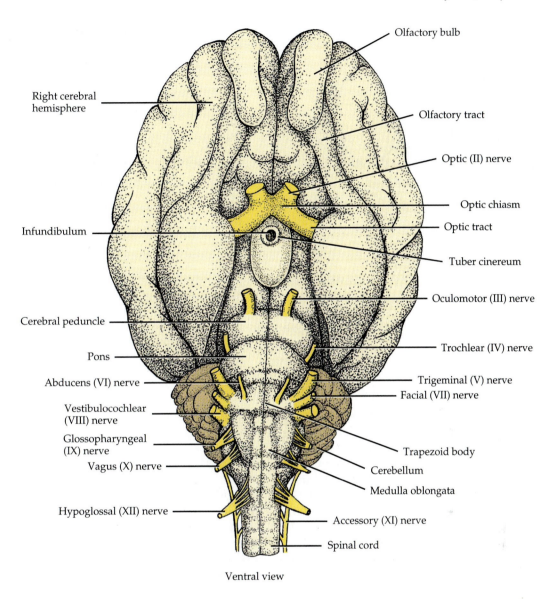

Olfactory bulb

Right cerebral hemisphere

Olfactory tract

Optic (II) nerve

Optic chiasm

Optic tract

Infundibulum

Tuber cinereum

Oculomotor (III) nerve

Cerebral peduncle

Pons

Trochlear (IV) nerve

Abducens (VI) nerve

Trigeminal (V) nerve

Facial (VII) nerve

Vestibulocochlear (VIII) nerve

Glossopharyngeal (IX) nerve

Trapezoid body

Vagus (X) nerve

Cerebellum

Medulla oblongata

Hypoglossal (XII) nerve

Accessory (XI) nerve

Spinal cord

Ventral view

FIGURE 13.21　Cranial nerves of a sheep brain.

F. AUTONOMIC NERVOUS SYSTEM

The *autonomic nervous system (ANS)* regulates the activities of smooth muscle, cardiac muscle, and certain glands, usually involuntarily. In the somatic nervous system (SNS), which is voluntary, the cell bodies of the motor neurons are in the CNS, and their axons extend all the way to skeletal muscles in spinal nerves. The ANS always has two motor neurons in the pathway. The first motor neuron, the *preganglionic neuron*, has its cell body in the CNS. Its axon leaves the CNS and synapses in an autonomic ganglion with the second neuron called the *postganglionic neuron*. The cell body of the postganglionic

neuron is inside an autonomic ganglion, and its axon terminates in a *visceral effector* (muscle or gland).

Label the components of the autonomic pathway shown in Figure 13.22.

The ANS consists of two divisions: sympathetic and parasympathetic (Figure 13.23 on page 257). Most viscera are innervated by both divisions. In general, nerve impulses from one division stimulate a structure, whereas nerve impulses from the other division decrease its activity.

In the *sympathetic division*, the cell bodies of the preganglionic neurons are located in the lateral gray horns of the spinal cord in the thoracic and first two lumbar segments. The axons of preganglionic neurons are myelinated and leave the spinal cord

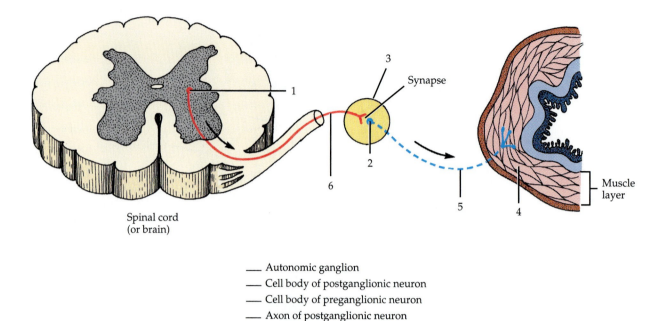

Spinal cord
(or brain)

_____ Autonomic ganglion
_____ Cell body of postganglionic neuron
_____ Cell body of preganglionic neuron
_____ Axon of postganglionic neuron
_____ Axon of preganglionic neuron
_____ Visceral effector

FIGURE 13.22   Components of an autonomic pathway.

through the ventral (anterior) root of a spinal nerve. Each axon travels briefly in a ventral ramus and then through a small branch called a ***white ramus communicans*** to enter a sympathetic trunk ganglion. These ganglia lie in a vertical row, on either side of the vertebral column, from the base of the skull to the coccyx. In the ganglion, the axon may synapse with a postganglionic neuron, travel upward or downward through the sympathetic trunk ganglia to synapse with postganglionic neurons at different levels, or pass through the ganglion without synapsing to form part of the splanchnic nerves. If the preganglionic axon synapses in a sympathetic trunk ganglion, it reenters the anterior or posterior ramus of a spinal nerve via a small branch called a ***gray ramus communicans***. If the preganglionic axon forms part of the splanchnic nerves, it passes through the sympathetic trunk ganglion but synapses with a postganglionic neuron in a prevertebral (collateral) ganglion. These ganglia are anterior to the vertebral column close to large abdominal arteries from which their names are derived (celiac, superior mesenteric, and inferior mesenteric).

In the ***parasympathetic division***, the cell bodies of the preganglionic neurons are located in nuclei in the brain stem and lateral gray horn of the second through fourth sacral segments of the spinal cord. The axons emerge as part of cranial or spinal nerves.

The preganglionic axons synapse with postganglionic neurons in terminal ganglia, near or within visceral effectors.

The sympathetic division is primarily concerned with processes that expend energy. During stress, the sympathetic division sets into operation a series of reactions collectively called the ***fight-or-flight response***, designed to help the body counteract the stress and return to homeostasis. During the fight-or-flight response, the heart and breathing rates increase and the blood sugar level rises, among other things.

The parasympathetic division is primarily concerned with activities that restore and conserve energy. It is thus called the ***rest-repose system***. Under normal conditions, the parasympathetic division dominates the sympathetic division in order to maintain homeostasis.

Autonomic fibers, like other axons of the nervous system, release neurotransmitters at synapses as well as at points of contact with visceral effectors ***(neuroeffector junctions)***. On the basis of the neurotransmitter produced, autonomic fibers may be classified as either cholinergic or adrenergic. ***Cholinergic*** (kō-lin-ER-jik) ***fibers*** release ***acetylcholine (ACh)*** and include the following: (1) all sympathetic and parasympathetic preganglionic axons, (2) all parasympathetic postganglionic axons, and (3) a few sympathetic postganglionic axons.

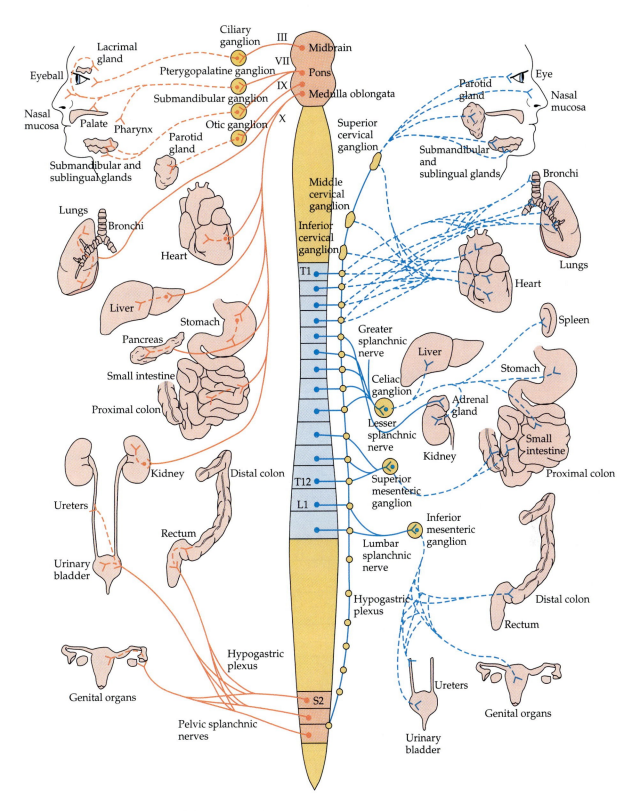

FIGURE 13.23   Structure of the autonomic nervous system.

Adrenergic (ad'-ren-ER-jik) *fibers* produce *norepinephrine (NE)* or *epinephrine (adrenalin)* in the modified postganglionic sympathetic fibers of the adrenal medulla. Most sympathetic postganglionic axons are adrenergic.

The actual effects produced by ACh are determined by the type of receptor with which it interacts. The two types of cholinergic receptors are known as nicotinic receptors and muscarinic receptors. *Nicotinic receptors* are found on both sympathetic and parasympathetic postganglionic neurons. These receptors are so named because the actions of ACh on them are similar to those produced by nicotine. *Muscarinic receptors* are found on effectors (muscles and glands) innervated by parasympathetic postganglionic axons. These receptors are so named because the actions of ACh on them are similar to those produced by muscarine, a toxin produced by a mushroom. The ef-

fects of NE and epinephrine, like those of ACh, are also determined by the type of receptor with which they interact. Such receptors are found on visceral effectors innervated by most sympathetic postganglionic axons and are referred to as *alpha receptors* and *beta receptors.* In general, alpha receptors are excitatory. Some beta receptors are excitatory, while others are inhibitory. Although cells of most effectors contain either alpha or beta receptors, some effector cells contain both. NE, in general, stimulates alpha receptors to a greater extent than beta receptors, and epinephrine, in general, stimulates both alpha and beta receptors about equally.

Using your textbook as a reference, write in the effects of sympathetic and parasympathetic stimulation for the visceral effectors listed in Table 13.3.

ANSWER THE LABORATORY REPORT QUESTIONS AT THE END OF THE EXERCISE.

TABLE 13.3
Activities of the Autonomic Nervous System

Visceral effector	Effect of sympathetic stimulation	Effect of parasympathetic stimulation
GLANDS		
Sweat		
Lacrimal (tear)		
Adrenal medulla		
Liver		
Kidney (juxtaglomerular cells)		
Pancreas		
SMOOTH MUSCLE		
Iris, radial muscle		
Iris, circular muscle		
Ciliary muscle of eye		
Salivary gland arterioles		
Gastric gland arterioles		
Intestinal gland arterioles		

TABLE 13.3 *(Continued)*

Visceral effector	Effect of sympathetic stimulation	Effect of parasympathetic stimulation
Lungs, bronchial muscle		
Heart arterioles		
Skin and mucosal arterioles		
Skeletal muscle arterioles		
Abdominal viscera arterioles		
Brain arterioles		
Systemic veins		
Gallbladder and ducts		
Stomach and intestines		
Kidney		
Ureter		
Spleen		
Urinary bladder		
Uterus		
Sex organs		
Hair follicles (arrector pili muscle)		

CARDIAC MUSCLE (HEART)

Nervous System 13

Student _____   Date _____

Laboratory Section _____   Score/Grade _____

PART 1. Multiple Choice

_____ 1. The tapered, conical portion of the spinal cord is the (a) filum terminale (b) conus medullaris (c) cauda equina (d) lumbar enlargement

_____ 2. The superficial meninx composed of dense fibrous connective tissue is the (a) pia mater (b) arachnoid (c) dura mater (d) denticulate

_____ 3. The portion of a spinal nerve that contains motor nerve fibers only is the (a) posterior root (b) posterior root ganglion (c) lateral root (d) anterior root

_____ 4. The connective tissue covering around individual nerve fibers is the (a) endoneurium (b) epineurium (c) perineurium (d) ectoneurium

_____ 5. On the basis of organization, which does *not* belong with the others? (a) pons (b) medulla oblongata (c) thalamus (d) midbrain

_____ 6. The lateral ventricles are connected to the third ventricle by the (a) interventricular foramen (b) cerebral aqueduct (c) median aperture (d) lateral aperture

_____ 7. The vital centers for heartbeat, respiration, and blood vessel diameter regulation are found in the (a) pons (b) cerebrum (c) cerebellum (d) medulla oblongata

_____ 8. The reflex centers for movements of the head and trunk in response to auditory stimuli are located in the (a) inferior colliculi (b) medial geniculate nucleus (c) superior colliculi (d) ventral posterior nucleus

_____ 9. Which thalamic nucleus controls general sensations and taste? (a) medial geniculate (b) ventral posterior (c) ventral lateral (d) ventral anterior

_____ 10. Integration of the autonomic nervous system, control of body temperature, and the regulation of food intake and thirst are functions of the (a) pons (b) thalamus (c) cerebrum (d) hypothalamus

_____ 11. The left and right cerebral hemispheres are separated from each other by the (a) central sulcus (b) transverse fissure (c) longitudinal fissure (d) insula

_____ 12. Which structure does *not* belong with the others? (a) putamen (b) caudate nucleus (c) insula (d) globus pallidus

_____ 13. Which peduncles connect the cerebellum with the midbrain? (a) superior (b) inferior (c) middle (d) lateral

_____ 14. Which cranial nerve has the most anterior origin? (a) XI (b) IX (c) VII (d) IV

_____ 15. Extensions of the pia mater that suspend the spinal cord and protect against shock are the (a) choroid plexuses (b) pyramids (c) denticulate ligaments (d) superior colliculi

_____ **16.** Which branch of a spinal nerve enters into formation of plexuses? (a) meningeal (b) dorsal (c) rami communicantes (d) ventral

_____ **17.** Which plexus innervates the upper limbs and shoulders? (a) sacral (b) brachial (c) lumbar (d) cervical

_____ **18.** How many pairs of thoracic spinal nerves are there? (a) 1 (b) 5 (c) 7 (d) 12

PART 2. Completion

19. The narrow, shallow groove on the posterior surface of the spinal cord is the _____.

20. The space between the dura mater and wall of the vertebral canal is called the _____.

21. In a spinal nerve, the cell bodies of sensory neurons are found in the _____.

22. The superficial connective tissue covering around a spinal nerve is the _____.

23. The middle meninx is referred to as the _____.

24. The nuclei of origin for cranial nerves IX, X, XI, and XII are found in the _____.

25. The portion of the brain containing the cerebral peduncles is the _____.

26. Cranial nerves V, VI, VII, and VIII have their nuclei of origin in the _____.

27. A shallow downfold of the cerebral cortex is called a(n) _____.

28. The _____ separates the frontal lobe of the cerebrum from the parietal lobe.

29. White matter tracts of the cerebellum are called _____.

30. The space between the dura mater and the arachnoid is referred to as the _____.

31. Together, the thalamus and hypothalamus constitute the _____.

32. Cerebrospinal fluid passes from the third ventricle into the fourth ventricle through the

_____.

33. The cerebrum is separated from the cerebellum by the _____ fissure.

34. The branches of a spinal nerve that are components of the autonomic nervous system are known as

_____.

35. The plexus that innervates the buttocks, perineum, and lower limbs is the _____ plexus.

36. There are _____ pairs of spinal nerves.

37. The part of the brain that coordinates subconscious movements in skeletal muscles is the

_____.

38. The cell bodies of _____ neurons of the ANS are found inside autonomic ganglia.

39. The portion of the ANS concerned with the fight-or-flight response is the _____ division.

40. The autonomic ganglia that are anterior to the vertebral column and close to large abdominal

arteries are called _____ ganglia.

41. The _____ nerve innervates the diaphragm.

42. The _____ tract conveys nerve impulses for touch and pressure.

43. The flexor muscles of the thigh and extensor muscles of the leg are innervated by the

_____ nerve.

44. The _____ tract conveys nerve impulses related to muscle tone and posture.

45. The _____ nerve supplies the extensor muscles of the arm and forearm.

46. The _____ area of the cerebral cortex receives sensations from cutaneous, muscular, and visceral receptors in various parts of the body.

47. The portion of the cerebral cortex that translates thoughts into speech is the _____ area.

48. The term _____ refers to aggregations of myelinated processes from many neurons.

49. A(n) _____ spinal nerve contains preganglionic autonomic nervous system neurons.

50. _____-order neurons carry sensory information from the spinal cord and brain stem to the thalamus.

51. The part of the brain that allows crude appreciation of some sensations, such as pain, temperature, and pressure, is the _____.

52. Autonomic fibers are classified as either cholinergic or _____.

PART 3. Matching

Using B for brachial, C for cervical, L for lumbar, and S for sacral, indicate to which plexus the following nerves belong.

_____ **53.** Sciatic

_____ **54.** Femoral

_____ **55.** Radial

_____ **56.** Ansa cervicalis

_____ **57.** Median

_____ **58.** Obturator

_____ **59.** Pudendal

_____ **60.** Tibial

_____ **61.** Ilioinguinal

_____ **62.** Ulnar

_____ **63.** Phrenic

PART **4.** **Matching**

_____ **64.** Posterior column-medial meniscus pathway

_____ **65.** Spinocerebellar tracts

_____ **66.** Lateral corticospinal tracts

_____ **67.** Lateral spinothalamic tract

_____ **68.** Anterior corticospinal tracts

_____ **69.** Tectospinal tract

_____ **70.** Lateral reticulospinal tract

A. Conduct subconscious proprioceptive input to the cerebellum

B. Convey impulses that control movements of the neck and part of the trunk

C. Conducts impulses for conscious proprioception and most tactile sensations

D. Conveys impulses that facilitate flexor reflexes, inhibit extensor reflexes, and decrease muscle tone in muscles of the axial skeleton and proximal limbs

E. Control skilled movements of the distal limbs

F. Conduct mainly pain and temperature impulses

G. Conveys impulses that move the head and eyes in response to visual stimuli

General Senses and Sensory and Motor Pathways

14

Now that you have studied the nervous system, you will study sensations. In its broadest sense, *sensation* is the conscious or subconscious awareness of external or internal stimuli. If the stimulus is strong enough, one or more nerve impulses arise in sensory nerve fibers. After the nerve impulses propagate to a region of the spinal cord or brain, they are translated into a sensation. The nature of the sensation and the type of reaction generated vary with the level of the central nervous system at which the sensation is translated.

A. CHARACTERISTICS OF SENSATIONS

Conscious sensations may be characterized into three types. *Superficial sensations* include touch, temperature, two-point discrimination, and pain. *Deep sensations* include muscle and joint pain, vibratory sense, and proprioception (muscle and joint position or location). *Combined sensations* are involved in the process termed *stereognosis,* which is the process of recognizing objects by the sense of touch while the eyes are closed, and *topognosis,* which is the ability to localize cutaneous sensation.

For a sensation to be detected (either consciously or subconsciously) four events must occur:

1. *Stimulation* A *stimulus,* or change in the environment, capable of activating certain sensory neurons must be present.
2. *Transduction* A *sensory receptor* or *sense organ* must receive the stimulus and *transduce* (convert) it to a receptor potential. A sensory receptor or sense organ is a specialized peripheral structure or a specialized type of neuron that is sensitive to certain types of stimuli.
3. *Impulse generation and conduction* The generator potential elicits nerve impulses that are conducted along a sensory neural pathway to

the CNS. The sensory neuron that conveys such impulses to the CNS is termed a *first-order neuron.* Additional neurons (termed *second-order neurons* and *third-order neurons*) carry the action potential to a particular region within the brain. The number of neurons (two or three) involved in the transmission of information to the brain varies depending upon the sensory tract utilized to transmit the information to the brain.

4. *Integration* A region of the CNS must receive and integrate the information carried via the action potential into a sensation. Most conscious sensations or perceptions occur in the cerebral cortex of the brain after passing through the thalamus. In other words, you see, hear, and feel in the brain. You seem to see with your eyes, hear with your ears, and feel pain in an injured part of your body because sensory impulses from each part of the body arrive in a specific region in the cerebral cortex, which interprets the sensation as coming from the stimulated sensory receptors.

A *sensory unit* consists of a single peripheral neuron and its terminal ending. The neuron's cell body is typically located in either a posterior spinal ganglion or within a cranial nerve ganglion. The peripheral processes of this neuron may terminate either as free nerve endings or in association with a sensory receptor. Sensory receptors may be either very simple or quite complex, containing highly specialized neurons, epithelium, and connective tissue components. All sensory receptors contain the terminal processes of a neuron, exhibit a high degree of excitability, and possess a specific threshold for a particular type of stimulus, which is termed a *sensory modality.* In addition, a sensory unit possesses a peripheral *receptive field,* which is an area within which a stimulus of appropriate quality and strength will cause a sensory neuron to initiate an action potential. The majority of sensory impulses are conducted to the sensory areas of the cerebral cortex, for this is the region of the brain that initiates the processing of sensory information that ultimately results in the conscious feeling of the sensory stimulus. Different sets of sensory nerve fibers, when activated, will elicit different sensations by virtue of their unique central nervous system connections. Therefore, a particular sensory neuron will provoke an identical sensation *regardless of how it is excited*. This interpretation of sensory perception is termed *Muller's Doctrine of Specific Energies.*

One characteristic of sensations, that of *projection,* describes the process by which the brain refers sensations to their point of *learned origin* of the stimulation. This process accounts for a phenomenon termed *phantom pain* that is quitecommon in amputees. When phantom pain is experienced the individual feels pain (or one of many other sensations, such as itching or tickling) in the part of the body that was amputated because of the irritation of a nerve ending in the healing wound surface of the amputation. The action potential is carried to the brain and projected to the portion of the limb that is no longer intact.

A second characteristic of many sensations is *adaptation,* that is, a change in sensitivity, usually a decrease, even though a stimulus is still being applied. For example, when you first get into a tub of hot water, you might feel an intense burning sensation. However, after a brief period of time the sensation decreases to one of comfortable warmth, even though the stimulus (hot water) is still present and has not diminished in intensity.

A third characteristic is that of *afterimages,* that is, the persistence of a sensation after the stimulus has been removed. One common example of an afterimage occurs when you look at a bright light and then look away. You will still see the light for several seconds afterward.

A fourth characteristic of sensations is that of *modality.* Modality is the possession of distinct properties by which one sensation may be distinguished from another. For example, pain, pressure, touch, body position, equilibrium, hearing, vision, smell, and taste are all distinctive because the brain perceives each differently.

B. CLASSIFICATION OF RECEPTORS

Receptors vary in their complexity. The *somatic (general) senses* include touch, pressure, vibration, itch, tickle, warmth, cold, and pain plus proprioception (detection of body positions and move-

ments). Somatic receptors are relatively simple and utilize relatively simple neural pathways.

Somatic sensations arise in receptors located in the skin (cutaneous sensations) or in muscles, tendons, joints, and the inner ear (proprioceptive sensations). *Cutaneous* (kyoo-TĀ-nē-us; *cutis* = skin) *sensations* include:

   a. *Tactile* (TAK-fīl; *tact* = touch) *sensations* (touch, pressure, vibration, itch, and tickle).
   b. *Thermal sensations* (cold and warmth).
   c. *Pain sensations.*

Proprioceptive (prō-prē-ō-SEP-tiv) *sensations* provide us with an awareness of the activities of muscles, tendons, and joints and equilibrium.

The receptors for the *special senses*—smell, taste, vision, hearing, and equilibrium—are located in sense organs such as the eye and ear. Special sense receptors are relatively complex and utilize relatively complex neural pathways.

Special sensations arise in receptors located in organs in different parts of the body. The special sensations are as follows:

   a. *Olfactory* (ōl-FAK-tō-rē) *sensations* (smell).
   b. *Gustatory* (GUS-ta-tō-rē) *sensations* (taste).
   c. *Visual sensations* (sight).
   d. *Auditory sensations* (hearing).
   e. *Equilibrium* (ē'-kwi-LIB-rē-um) *sensations* (orientation of the body).

Two widely used classifications of receptors are based on the location of the receptors and the type of stimuli they detect. The classification according to location is as follows:

1. *Exteroceptors* (EKS'-ter-ō-sep'-tors), located at or near the body surface, provide information about the *external* environment. They transmit sensations such as hearing, sight, smell, taste, touch, pressure, temperature, and pain.
2. *Interoceptors* or *visceroceptors* (VIS-er-ō-sep'-tors), located in blood vessels and viscera, provide information about the *internal* environment. Sensations from these receptors often do not reach conscious perception but may be perceived as pain or pressure.
3. *Proprioceptors* (PRŌ-prē-ō-sep'-tors; *proprio* = one's own) or stretch receptors, located in muscles, tendons, joints, and the internal ear, are stimulated by stretching or movement. They provide information about body position, muscle tension, and the position of our joints and equilibrium.

The classification of sensations is based on the type of stimuli they detect is as follows:

1. *Mechanoreceptors* detect mechanical pressure or stretching. They provide sensationsof touch, pressure, vibration, proprioception (awareness of the location and movement of body parts), hearing, equilibrium, and blood pressure.
2. *Thermoreceptors* detect changes in temperature.
3. *Nociceptors* (NŌ-sē-sep'-tors) detect pain.
4. *Photoreceptors* detect light that strikes the retina of the eye.
5. *Chemoreceptors* detect chemicals in the mouth (taste), nose (smell), and chemicals in body fluids such as water, oxygen, carbon dioxide, hydrogen ions, certain other electrolytes, hormones, and glucose.

C. RECEPTORS FOR GENERAL SENSES

1. Tactile Receptors

Although touch, pressure, and vibration are classified as separate sensations, all are detected by mechanoreceptors.

Touch sensations generally result from stimulation of tactile receptors in the skin or in tissues immediately deep to the skin. *Crude touch* refers to the ability to perceive that something has touched the skin, although the size or texture cannot be determined. *Discriminative touch* refers to the ability to recognize exactly what point on the body is touched.

Touch receptors include corpuscles of touch, hair root plexuses, and type I and type II cutaneous mechanoreceptors. *Corpuscles of touch* or *Meissner's* (MĪS-ners) *corpuscles* are receptors for discriminative touch that are found in dermal papillae. They have already been discussed in Exercise 5. *Hair root plexuses* are dendrites arranged in networks around hair follicles that detect movement when hairs are moved. *Type I cutaneous mechanoreceptors,* also called *tactile* or *Merkel* (MER-kel) *discs,* are the flattened portions of dendrites of sensory neurons that make contact with epidermal cells of the stratum basale called Merkel cells. They are distributed in many of the same locations as corpuscles of touch and also function in discriminative touch. *Type II cutaneous mechanoreceptors,* or *end organs of Ruffini,* are embedded deeply in the dermis and in deeper tissues of the body. They detect heavy and continuous touch sensations.

Receptors for touch, like other cutaneous receptors, are not randomly distributed; some areas contain many receptors, others contain few. Such a clustering of receptors is called *punctate distribution.* Touch receptors are most numerous in the fingertips, palms, and soles. They are also abundant in the eyelids, tip of the tongue, lips, nipples, clitoris, and tip of the penis.

Examine prepared slides of corpuscles of touch (Meissner's corpuscles), hair root plexuses, type I cutaneous mechanoreceptors (tactile or Merkel discs), and type II cutaneous mechanoreceptors (end organs of Ruffini). With the aid of your textbook, draw each of the receptors in the spaces that follow.

Corpuscles of touch (Meissner's corpuscles)

Hair root plexuses

Type I cutaneous mechanoreceptors
(tactile or Merkel discs)

Type II cutaneous mechanoreceptors
(end organs of Ruffini)

Pressure sensations generally result from stimulation of tactile receptors in deeper tissues. *Pressure* is a sustained sensation that is felt over a larger area than touch. Receptors for pressure sensations include lamellated corpuscles and type II cutaneous mechanoreceptors. *Lamellated (Pacinian) corpuscles* are found in the subcutaneous layer and have already been discussed in Exercise 5.

Pressure receptors are found in the subcutaneous tissue under the skin, in the deep subcutaneous tissues that lie under mucous membranes, around joints and tendons, in the perimysium of muscles, in the mammary glands, in the external genitals of both sexes, and in some viscera such as urinary bladder and pancreas.

Examine a prepared slide of lamellated (Pacinian) corpuscles. With the aid of your textbook, draw the receptors in the spaces that follow.

Lamellated (Pacinian) corpuscles

Vibration sensations result from rapidly repetitive sensory signals from tactile receptors. Receptors for vibration include corpuscles of touch (Meissner's corpuscles) that detect low-frequency vibration and lamellated (Pacinian) corpuscles that detect high-frequency vibration.

The *itch sensation* results from stimulation of free nerve endings by certain chemicals, such as bradykinin, often as a result of a local inflammatory response. Free nerve endings also are thought to mediate the *tickle sensation.*

2. Thermoreceptors

The cutaneous receptors for the sensation of warmth and coolness are free (naked) nerve endings that are widely distributed in the dermis and subcutaneous connective tissue. They are also located in the cornea of the eye, tip of the tongue, and external genitals. Separate thermoreceptors respond to warm and cold stimuli.

3. Pain Receptors

Receptors for *pain*, called *nociceptors* (NŌ-sē-sep'-tors; *noci* = harmful), are free (naked) nerve endings (see Figure 5.1). Pain receptors are found in practically every tissue of the body and adapt only

slightly or not at all. They may be excited by any type of stimulus. Excessive stimulation of any sense organ causes pain. For example, when stimuli for other sensations such as touch, pressure, heat, and cold reach a certain threshold, they stimulate pain receptors as well. Pain receptors, because of their sensitivity to all stimuli, have a general protective function of informing us of changes that could be potentially dangerous to health or life. Adaptation to pain does not readily occur. This low level of adaptation is important, because pain indicates disorder or disease. If we became used to it and ignored it, irreparable damage could result.

4. Proprioceptive Receptors

An awareness of the activities of muscles, tendons, and joints and equilibrium is provided by the *proprioceptive (kinesthetic) sense.* It informs us of the degree to which tendons are tensed and muscles are contracted. The proprioceptive sense enables us to recognize the location and rate of movement of one part of the body in relation to other parts. It also allows us to estimate weight and to determine the muscular effort necessary to perform a task. With the proprioceptive sense, we can judge the position and movements of our limbs without using our eyes when we walk, type, play a musical instrument, or dress in the dark.

Proprioceptors, the receptors for proprioception, are located in skeletal muscles and tendons, in and around joints, and in the internal ear. Proprioceptors adapt only slightly. This slight adaptation is beneficial because the brain must be apprised of the status of different parts of the body at all times so that adjustments can be made to ensure coordination.

Receptors for proprioception are as follows. The *joint kinesthetic* (kin'-es-THET-ik) *receptors* are located in the articular capsules of joints and ligaments about joints. These receptors provide feedback information on the degree and rate of angulation (change of position) of a joint. *Muscle spindles* are specialized muscle fibers (cells) that consist of endings of sensory neurons. They are located in nearly all skeletal muscles and are more numerous in the muscles of the limbs. Muscle spindles provide feedback information on the degree of muscle stretch. This information is relayed to the central nervous system to assist in the coordination and efficiency of muscle contraction. *Tendon organs (Golgi tendon organs)* are located at the junction of a skeletal muscle and tendon. They function by sensing the tension applied to a tendon and the force of contraction of associated muscles. The information is translated by the CNS.

Proprioceptors in the internal ear are the maculae and cristae that function in equilibrium. These are discussed at the end of the exercise.

Examine prepared slides of joint kinestheticreceptors, muscle spindles, and tendon organs (Golgi tendon organs). With the aid of your textbook, draw the receptors in the spaces that follow.

Joint kinesthetic receptors

Muscle spindles

Tendon organs (Golgi tendon organs)

D. TESTS FOR GENERAL SENSES

Areas of the body that have few cutaneous receptors are relatively insensitive, whereas those regions that contain large numbers of cutaneous receptors are quite sensitive. This difference can be demonstrated by the ***two-point discrimination test*** for touch. In the following tests, students can work in pairs, with one acting as subject and the other as experimenter. The subject will keep his or her eyes closed during the experiments.

1. Two-Point Discrimination Test

In this test, the two points of a measuring compass are applied to the skin and the distance in millimeters (mm) between the two points is varied. The subject indicates when he or she feels two points and when he or she feels only one.

PROCEDURE

CAUTION! *Wash the compass in a fresh bleach solution before using it on another subject to prevent transfer of saliva.*

1. *Very gently*, place the compass on the tip of the tongue, an area where receptors are very densely packed.
2. Narrow the distance between the two points to 1.4 mm. At this distance, the points are able to stimulate two different receptors, and the subject feels that he or she is being touched by two objects.
3. Decrease the distance to less than 1.4 mm. The subject feels only one point, even though both points are touching the tongue, because the points are so close together that they reach only one receptor.
4. Now *gently* place the compass on the back of the neck, where receptors are relatively few and far between. Here the subject feels two distinctly different points only if the distance between them is 36.2 mm or more.
5. The two-point discrimination test shows that the more sensitive the area, the closer the compass points can be placed and still be felt separately.
6. The following order, from greatest to least sensitivity, has been established from the test: tip of tongue, tip of finger, side of nose, back of hand, and back of neck.
7. Test the tip of finger, side of nose, and back of hand and record your results in Section D.1 of the LABORATORY REPORT RESULTS at the end of the exercise.

2. Identifying Touch Receptors

PROCEDURE

1. Using a water-soluble colored felt marking pen, draw a 1-in. square on the back of the forearm and divide the square into 16 smaller squares.
2. With the subject's eyes closed, press a Von Frey hair or bristle against the skin, just enough to cause the hair to bend, once in each of the 16 squares. The pressure should be applied in the same manner each time.
3. The subject should indicate when he or she experiences the sensation of touch, and the experimenter should make dots in square 1 at the places corresponding to the points at which the subject feels the sensations.

4. The subject and the experimenter should switch roles and repeat the test.

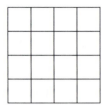

The pair of students working as a team should examine their 1-in. squares after the test is done. They should compare the number of positive and negative responses in each of the 16 small squares, and see how uniformly the touch receptors are distributed throughout the entire 1-in. square. Other general areas used for locating touch receptors are the arm and the back of the hand.

3. Identifying Pressure Receptors

PROCEDURE

1. The experimenter touches the skin of the subject (whose eyes are closed) with the point of a piece of colored chalk.
2. With eyes still closed, the subject then tries to touch the same spot with a piece of differently colored chalk. The distance between the two points is then measured.
3. Proceed using various parts of the body, such as the palm, arm, forearm, and back of neck.
4. Record your results in Section D.3 of the LABORATORY REPORT RESULTS at the end of the exercise.

4. Identifying Thermoreceptors

PROCEDURE

1. Draw a 1-in. square on the back of the wrist.
2. Place a forceps or other metal probe in ice-cold water for a minute, dry it quickly, and, with the *dull* point, explore the area in the square for the presence of cold spots.
3. Keep the probe cold and, using ink, mark the position of each spot that you find.
4. Mark each corresponding place in square 2 with the letter "c."
5. Immerse the forceps in hot water so that it will give a sensation of warmth when removed and applied to the skin, but *avoid having it so hot that it causes pain.*
6. Proceeding as before, locate the position of the warm spots in the same area of the skin.

7. Mark these spots with ink of a different color, and then mark each corresponding place in square 2 with the letter "h."
8. Repeat the entire procedure, using both cold forceps and warm forceps on the back of the hand and the palm, respectively, and mark squares 3 and 4 as you did square 2.

9. In order to demonstrate adaptation of thermoreceptors, fill three 1-liter beakers with 700 ml of (1) ice water, (2) water at room temperature, and (3) water at 45° C. Immerse your left hand in the ice water and your right hand in the water at 45° C for 1 min. Now move your left hand to the beaker with water at room temperature and record the sensation.

_____

Move your right hand to the beaker with the water at room temperature and record the sensation._____
How do you explain the experienced difference in the temperature of the water in the beaker with the water at room temperature?

_____

_____

5. Identifying Pain Receptors

PROCEDURE

1. Using the same 1-in. square of the forearm that was previously used for the touch test in Section D.2, perform the following experiment.
2. Apply a piece of absorbent cotton soaked with water to the area of the forearm for 5 min to soften the skin.
3. Add water to the cotton as needed.
4. Place the blunt end of a probe to the surface of the skin and press enough to produce a sensation of pain. Explore the marked area systematically.
5. Using dots, mark the places in square 5 that correspond to the points that give pain sensation when stimulated.

6. Distinguish between sensations of pain and touch. Are the areas for touch and pain identical?

_____

7. At the end of the test, compare your squares as you did in Section D.2.

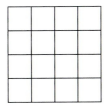

Perform the following test to demonstrate the phenomenon of *referred pain.* Place your elbow in a large shallow pan of ice water, and note the progression of sensation that you experience. At first, you will feel some discomfort in the region of the elbow. Later, pain sensations will be felt elsewhere.

Where do you feel the referred pain? _____

_____

Label the areas of referred pain indicated in Figure 14.1.

6. Identifying Proprioceptors

PROCEDURE

1. Face a blackboard close enough so that you can easily reach to mark it. Mark a small X on the board in front of you and keep the chalk on the X for a moment. Now close your eyes, raise your right hand above your head and then, with your eyes still closed, mark a dot as near as possible to the X. Repeat the procedure by placing your chalk on the X, closing your eyes, raising your arm above your head, and then marking another dot as close as possible to the X. Repeat the procedure a third time. Record your results by estimating or measuring how far you missed the X for each trial in Section D.6 of the LABORATORY REPORT RESULTS at the end of the exercise.

2. Write the word "physiology" on the left line that follows. Now, with your *eyes closed*, write the same word immediately to the right. How do the samples of writing compare?

_____

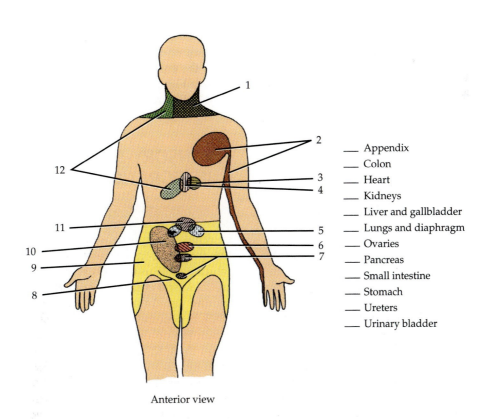

Appendix, Colon, Heart, Kidneys, Liver and gallbladder, Lungs and diaphragm, Ovaries, Pancreas, Small intestine, Stomach, Ureters, Urinary bladder

Anterior view

FIGURE 14.1   Referred pain.

Explain your results. _____

_____

_____

_____

3. The following experiments demonstrate that kinesthetic sensations facilitate repetition of certain acts involving muscular coordination.

   Students should work in pairs for these experiments.

   a. The experimenter asks the subject to carry out certain movements with his or her eyes closed, for example, point to the middle finger of the subject's left hand with the index finger of the subject's right hand.
   b. With his or her eyes closed, the subject extends the right arm as far as possible behind the body, and then brings the index finger quickly to the tip of his or her nose. How accurate is the subject in doing this?

   _____

   c. Ask the subject, with eyes shut, to touch the named fingers of one hand with the index finger of the other hand.
      How well does the subject carry out the directions? _____

   _____

   _____

   _____

E. SOMATIC SENSORY PATHWAYS

Most input from somatic receptors on one side of the body crosses over to the opposite side in the spinal cord or brain stem before ascending to the thalamus. It then projects from the thalamus tothe somatosensory area of the cerebral cortex, where conscious sensations result. Axon collaterals (branches) of somatic sensory neurons also carry signals into the cerebellum and the reticular formation of the brain stem. Two general pathways lead from sensory receptors to the cortex: the posterior column-medial lemniscus pathway and the anterolateral (spinothalamic) pathway.

1. Posterior Column-Medial Lemniscus Pathway

Nerve impulses for conscious proprioception and most tactile sensations ascend to the cerebral cortex along a common pathway formed by three-neuron sets.

Based on the description provided on page 273, label the components of the posterior column-medial lemniscus pathway in Figure 14.2.

2. Anterolateral (Spinothalamic) Pathways

The *anterolateral (spinothalamic) pathways* carry mainly pain and temperature impulses. In addition, they relay tickle, itch, and some tactile impulses, which give rise to a very crude, not well-localized touch or pressure sensation. Like the posterior column-medial lemniscus pathway, the anterolateral pathways are also composed of three-neuron sets. The axon of the second-order neuron extends to the opposite side of the spinal cord and passes superiorly to the brain stem in either the *lateral spinothalamic tract* or *anterior spinothalamic tract.* The lateral spinothalamic tract conveys sensory impulses for pain and temperature, whereas the anterior spinothalamic tract conveys tickle, itch, and crude touch and pressure impulses.

Based on the descriptions provided on page 274, label the components of the lateral spinothalamic tract in Figure 14.3 and the anterior spinothalamic tract in Figure 14.4 on page 275.

F. SENSORY-MOTOR INTEGRATION

Sensory systems provide the input that keeps the central nervous system informed of changes in the external and internal environment. Responses to this information are conveyed to motor systems, which enable us to move about, alter glandular secretions, and change our relationship to the world around us. As sensory information reaches the CNS, it becomes part of a large pool of sensory input. We do not actively respond to every bit of input the CNS receives. Rather, the incoming information is integrated with other information arriving from all other operating sensory receptors. The integration process occurs not just once, butat many stations along the pathways of the CNS and

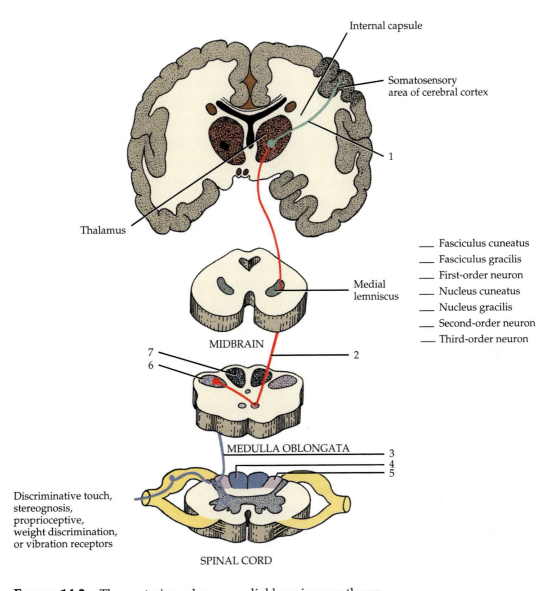

Internal capsule

Somatosensory
area of cerebral cortex

1

Thalamus

____ Fasciculus cuneatus
____ Fasciculus gracilis
____ First-order neuron
____ Nucleus cuneatus
____ Nucleus gracilis
____ Second-order neuron
____ Third-order neuron

Medial
lemniscus

MIDBRAIN

7
6

2

MEDULLA OBLONGATA

3
4
5

Discriminative touch,
stereognosis,
proprioceptive,
weight discrimination,
or vibration receptors

SPINAL CORD

FIGURE 14.2   The posterior column–medial lemniscus pathway.

at both conscious and subconscious levels. It occurs within the spinal cord, brain stem, cerebellum, basal ganglia, and cerebral cortex. As a result, a motor response to make a muscle contract or a gland secrete can be modified at any of these levels. Motor portions of the cerebral motor cortex play the major role in initiating and controlling precise, discrete muscular movements. The basal ganglia largely integrate semivoluntary, automatic movements like walking, swimming, and laughing. The cerebellum assists the motor cortex and basal ganglia by making body movements smooth and coordinated and by contributing significantly to maintaining normal posture and balance.

G. SOMATIC MOTOR PATHWAYS

After receiving and interpreting sensory information, the CNS generates nerve impulses to direct responses to that sensory input. The nerve impulses are sent down the spinal cord in two major motor pathways: the direct pathways and the indirect pathways.

1. Direct Pathways

Voluntary motor impulses propagate from the motor cortex to voluntary motor neurons (somatic motor neurons) that innervate skeletal muscles via

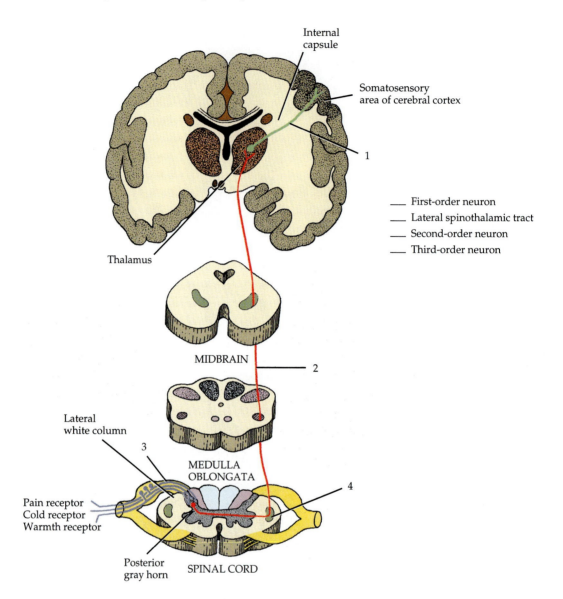

Internal capsule

Somatosensory area of cerebral cortex

1

___ First-order neuron
___ Lateral spinothalamic tract
___ Second-order neuron
___ Third-order neuron

Thalamus

MIDBRAIN   2

Lateral white column

3

MEDULLA OBLONGATA   4

Pain receptor
Cold receptor
Warmth receptor

Posterior gray horn   SPINAL CORD

FIGURE 14.3   The lateral spinothalamic pathway.

the *direct* or *pyramidal* (pi-RAM-i-dal) *pathways.* The direct pathways convey impulses from the cortex that result in precise, voluntary movements and include three tracts: lateral corticospinal (control muscles in the distal portions of the limbs), anterior corticospinal (control muscles of the neck and trunk), and cortico-bulbar (control muscles of the eyes, tongue, and neck; chewing, facial expression, and speech) (see page 236).

Based on the description provided on page 236, label the components of the lateral corticospinal tracts in Figure 14.5.

2. Indirect Pathways

The *indirect pathways* include all motor tracts other than the corticospinal and corticobulbar tracts (see page 237). These are the rubrospinal, tectospinal, vestibulospinal, and lateral and medial reticulospinal.

ANSWER THE LABORATORY REPORT QUESTIONS AT THE END OF THE EXERCISE.

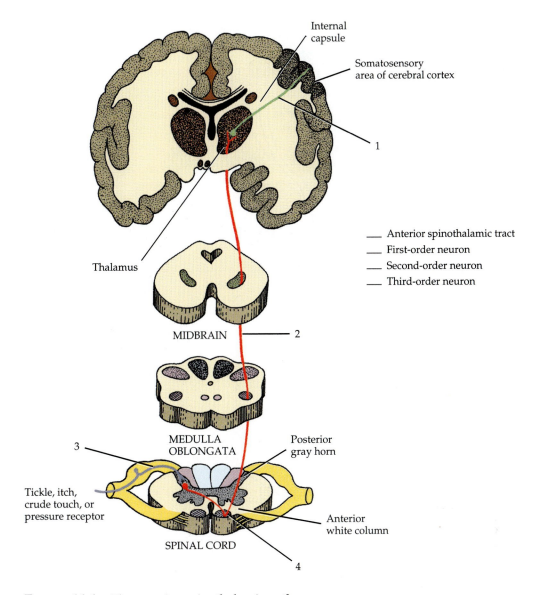

Internal capsule

Somatosensory area of cerebral cortex

1

Thalamus

___ Anterior spinothalamic tract
___ First-order neuron
___ Second-order neuron
___ Third-order neuron

MIDBRAIN — 2

MEDULLA OBLONGATA

Posterior gray horn

3

Tickle, itch, crude touch, or pressure receptor

Anterior white column

SPINAL CORD

4

FIGURE 14.4   The anterior spinothalamic pathway.

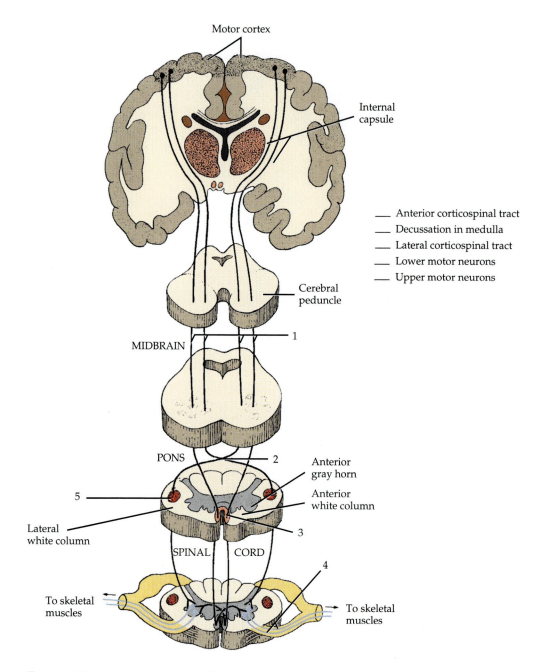

FIGURE 14.5   Direct (pyramidal) pathways: lateral and anterior corticiospinal tracts.

General Senses and Sensory and Motor Pathways 14

Student _____   Date _____

Laboratory Section _____   Score/Grade _____

SECTION D. TESTS FOR GENERAL SENSES

1. Two-Point Discrimination Test

Part of body	Least distance at which two points can be detected
Tip of tongue	1.4 mm
Tip of finger	_____
Side of nose	_____
Back of hand	_____
Back of neck	36.2 mm

3. Identifying Pressure Receptors

Part of body	Distances between points touched by chalk
Palm	_____
Arm	_____
Forearm	_____
Back of neck	_____

6. Identifying Proprioceptors

First trial	_____
Second trial	_____
Third trial	_____

General Senses and Sensory and Motor Pathways 14

Student _____ Date _____

Laboratory Section _____ Score/Grade _____

PART 1. Multiple Choice

_____ 1. The process by which the brain refers sensations to their point of stimulation is referred to as (a) modality (b) projection (c) accommodation (d) convergence

_____ 2. An awareness of the activities of muscles, tendons, and joints is known as (a) referred pain (b) adaptation (c) refraction (d) proprioception

_____ 3. The inability to feel a sensation consciously even though a stimulus is still being applied is called (a) modality (b) projection (c) adaptation (d) afterimage formation

_____ 4. Which receptor does *not* belong with the others? (a) muscle spindle (b) tendon organ (Golgi tendon organ) (c) joint kinesthetic receptor (d) lamellated (Pacinian) corpuscle

_____ 5. A characteristic of sensations by which one sensation may be distinguished from another is called (a) modality (b) projection (c) adaptation (d) afterimage formation

_____ 6. Which are *not* cutaneous receptors? (a) tactile (Merkel) discs (b) muscle spindles (c) lamellated (Pacinian) corpuscles (d) corpuscles of touch (Meissner's corpuscles)

_____ 7. The lateral spinothalamic tract conveys information related to (a) pain (b) touch (c) pressure (d) tickle

_____ 8. Which is an indirect (extrapyramidal) pathway? (a) lateral corticospinal (b) anterior corticospinal (c) tectospinal (d) corticobulbar

_____ 9. The presence of a sensation after a stimulus has been removed is called (a) adaptation (b) afterimage (c) translation (d) learned origin

_____ 10. Receptors that provide information about the external environment are called (a) visceroreceptors (b) proprioceptors (c) baroreceptors (d) exteroceptors

PART 2. Completion

11. Receptors found in blood vessels and viscera are classified as _____.

12. Tactile sensations include touch, pressure, and _____.

13. Receptors for pressure are free (naked) nerve endings, type II cutaneous mechanoreceptors (end organs of Ruffini), and _____ corpuscles.

14. Proprioceptive receptors that provide information about the degree and rate of angulations of joints are _____.

15. The _____ tract conveys sensory impulses for pain and temperature.

16. The _____ conveys motor impulses for precise contraction of muscles in the distal limbs.

17. Pain receptors are called _____.

18. The ability to recognize exactly which point of the body is touched is called _____.

19. _____ neurons extend from cranial nerve motor nuclei or spinal cord anterior horns to skeletal muscle fibers.

20. _____ tracts convey impulses that control voluntary movements of the head and neck.

Special Senses

15

In this exercise, we will consider *special senses* that involve complex receptors and neural pathways. These include *olfactory* (ōl-FAK-tō-rē) *sensations* (smell), *gustatory* (GUS-ta-tō-rē) *sensations* (taste), *visual sensations* (sight), *auditory sensations* (hearing), and *equilibrium* (ē'-kwi-LIB-rē-um) *sensations* (orientation of the body).

A. OLFACTORY SENSATIONS

1. Olfactory Receptors

The receptors for the *olfactory* (ol-FAK-tō-rē; *olfactus* = smell) *sense* are found in the nasal epithelium in the superior portion of the nasal cavity on either side of the nasal septum. The nasal epithelium consists of three principal kinds of cells: olfactory receptors, supporting cells, and basal cells. The *olfactory receptors (cells)* are bipolar neurons. Their cell bodies lie between the supporting cells. The distal (free) end of each olfactory cell contains a knob-shaped dendrite from which six to eight cilia, called *olfactory hairs,* protrude. The *supporting (sustentacular) cells* are columnar epithelial cells of the mucous membrane that lines the nose. *Basal cells* lie between the bases of the supporting cells and produce new olfactory receptors. Within the connective tissue deep to the olfactory epithelium are *olfactory (Bowman's) glands* that secrete mucus.

The unmyelinated axons of the olfactory receptors unite to form the *olfactory (I) nerves,* which pass through foramina in the cribriform plate of the ethmoid bone. The olfactory nerves terminate in paired masses of gray matter, the *olfactory bulbs,* which lie inferior to the frontal lobes of the cerebrum on either side of the crista galli of the ethmoid bone. The first synapse of the olfactory neural pathway occurs in the olfactory bulbs between the axons of the olfactory (I) nerves and the dendrites of neurons inside the olfactory bulbs. Axons of these neurons run posteriorly to form the *olfactory tract.* From here, nerve impulses are conveyed to the primary olfactory area in the temporal lobe of the cerebral cortex. In the cortex, the nerve impulses are interpreted as odor and give rise to the sensation of smell.

Adaptation happens quickly, especially adaptation to odors. For this reason, we become accustomed to some odors and are also able to endure unpleasant ones. Rapid adaptation also accounts for the failure of a person to detect gas that accumulates slowly in a room.

Label the structures associated with olfaction in Figure 15.1.

Now examine a slide of the olfactory epithelium under high power. Identify the olfactory receptors and supporting cells and label Figure 15.2 on page 283.

2. Olfactory Adaptation

PROCEDURE

1. The subject should close his or her eyes after plugging one nostril with cotton.
2. Hold a bottle of oil of cloves, or other substance having a distinct odor, under the open nostril.
3. The subject breathes in through the open nostril, and exhales through the mouth. Note the time required for the odor to disappear, and repeat with the other nostril.
4. As soon as olfactory adaptation has occurred, test an entirely different substance.
5. Compare results for the various materials tested.

Olfactory stimuli such as pepper, onions, ammonia, ether, and chloroform are irritating and may cause tearing because they stimulate the receptors of the trigeminal (V) nerve as well as the olfactory neurons.

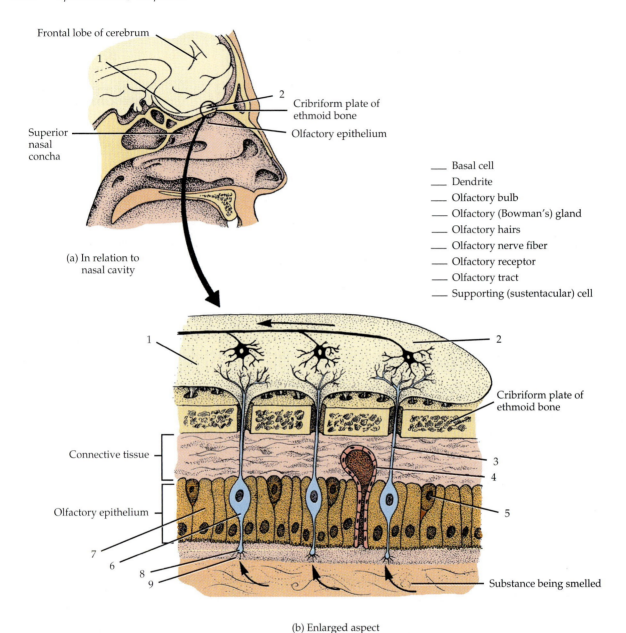

Frontal lobe of cerebrum

Superior
nasal
concha

Cribriform plate of
ethmoid bone

Olfactory epithelium

(a) In relation to
nasal cavity

___ Basal cell
___ Dendrite
___ Olfactory bulb
___ Olfactory (Bowman's) gland
___ Olfactory hairs
___ Olfactory nerve fiber
___ Olfactory receptor
___ Olfactory tract
___ Supporting (sustentacular) cell

Cribriform plate of
ethmoid bone

Connective tissue

Olfactory epithelium

Substance being smelled

(b) Enlarged aspect

FIGURE 15.1   Olfactory receptors.

B. GUSTATORY SENSATIONS

1. Gustatory Receptors

The receptors for *gustatory* (GUS-ta-tō-rē; *gusto* =
taste) *sensations,* or sensations of taste, are located
in the taste buds. Although taste buds are most
numerous on the tongue, they are also found on
the soft palate, pharynx, and larynx. *Taste buds* are
oval bodies consisting of three kinds of cells: supporting cells, gustatory receptors, and basal cells.
The *supporting (sustentacular) cells* are specialized epithelial cells that form a capsule. Inside each
capsule are about 50 *gustatory receptors (cells).*
Each gustatory receptor contains a hairlike process
(gustatory hair) that projects to the surface through
an opening in the taste bud called the *taste pore.*
Gustatory cells make contact with taste stimuli

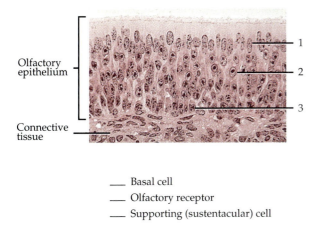

Olfactory epithelium

Connective tissue

1

2

3

___ Basal cell

___ Olfactory receptor

___ Supporting (sustentacular) cell

FIGURE 15.2 Photomicrograph of the olfactory epithelium.

through the taste pore. *Basal cells* are found at the periphery of the taste bud and produce new supporting cells, which then develop into gustatory receptors.

Examine a slide of taste buds and label the structures associated with gustation in Figure 15.3.

Taste buds are located in some connective tissue elevations on the tongue called *papillae* (pa-PIL-lē). They give the upper surface of the tongue its rough texture and appearance. *Circumvallate* (ser-kum-VAL-āt) *papillae* are circular and form an inverted V-shaped row at the posterior portion of the tongue. *Fungiform* (FUN-ji-form) *papillae* are knoblike elevations scattered over the entire surface of the tongue. All circumvallate and most fungiform papillae contain taste buds. *Filiform* (FIL-i-form) *papillae* are threadlike structures that are also distributed over the entire surface of the tongue.

Have your partner protrude his or her tongue and examine its surface with a hand lens to identify the shape and position of the papillae.

2. Identifying Taste Zones

For gustatory cells to be stimulated, substances tasted must be in solution in the saliva in order to enter the taste pores in the taste buds. Despite the many substances tasted, there are basically only four taste sensations: sour, salty, bitter, and sweet. Each taste is due to a different response to different chemicals. Some regions of the tongue react more strongly than others to particular taste sensations.

To identify the taste zones for the four taste sensations, perform the following steps and record the results in Section B.2 of the LABORATORY REPORT RESULTS at the end of the exercise by inserting a plus sign (taste detected) or a minus sign (taste not detected) where appropriate.

PROCEDURE

1. The subject thoroughly dries his or her tongue (use a clean paper towel). The experimenter places some granulated sugar on the tip of the tongue and notes the time. The subject indicates when he or she tastes sugar by raising his or her hand. The experimenter notes the time again and records how long it takes for the subject to taste the sugar.

2. Repeat the experiment, but this time use a drop of sugar solution. Again record how long it takes for the subject to taste the sugar. How do you explain the difference in time periods?

3. The subject rinses his or her mouth again. The experiment is then repeated using the quinine solution (bitter taste), and then the salt solution.

4. After rinsing yet again, the experiment is repeated using the acetic acid solution or vinegar (sour) placed on the tip and *sides* of the tongue.

3. Taste and Inheritance

Taste for certain substances is inherited, and geneticists for many years have been using the chemical *phenylthiocarbamide (PTC)* to test taste. To some individuals this substance tastes bitter; to others it is sweet; and some cannot taste it at all.

PROCEDURE

1. Place a few crystals of PTC on the subject's tongue. Does he or she taste it? If so, describe the taste.

2. Special paper that is flavored with this chemical may be chewed and mixed with saliva and tested in the same manner.

3. Record on the blackboard your response to the PTC test. Usually about 70% of the people tested can taste this compound; 30% cannot. Compare this percentage with the class results.

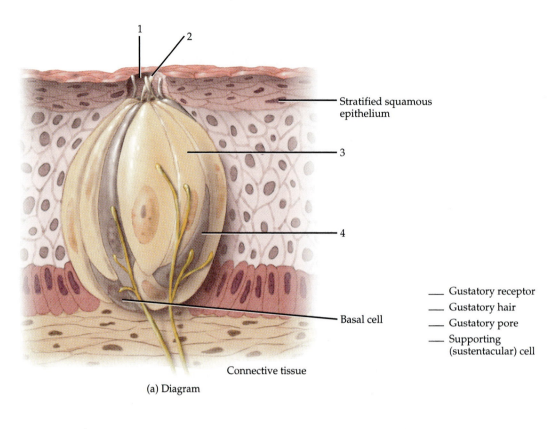

1
2

Stratified squamous
epithelium

3

4

Basal cell

Connective tissue

___ Gustatory receptor
___ Gustatory hair
___ Gustatory pore
___ Supporting
(sustentacular) cell

(a) Diagram

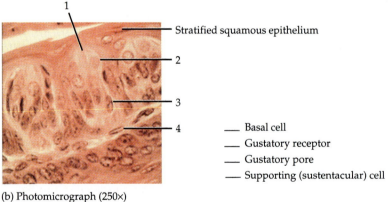

1

Stratified squamous epithelium

2

3

4

___ Basal cell
___ Gustatory receptor
___ Gustatory pore
___ Supporting (sustentacular) cell

(b) Photomicrograph (250×)

FIGURE 15.3   Structure of a taste bud.

4. Taste and Smell

This test combines the effect of smell on the sense of taste.

PROCEDURE

1. Obtain small cubes of carrot, onion, potato, and apple.

2. The subject dries the tongue with a clean paper towel, closes the eyes, and pinches the nostrils shut. The experimenter places the cubes, one by one, on the subject's tongue.

3. The subject attempts to identify each cube in the following sequences: (1) immediately, (2) after chewing (nostrils closed), and (3) after opening the nostrils.

4. Record your results in Section B.4 of the LAB-
ORATORY REPORT RESULTS at the end of the
exercise.

C. VISUAL SENSATIONS

Structures related to *vision* are the eyeball (which
is the receptor organ for visual sensations), optic
(II) nerve, brain, and accessory structures. The extrinsic muscles of the eyeball may be reviewed in
Figure 10.4.

1. Accessory Structures

Among the *accessory structures* are the eyebrows,
eyelids, eyelashes, lacrimal (tearing) apparatus, and
extrinsic eye muscles. *Eyebrows* protect the eyeball
from falling objects, prevent perspiration from getting into the eye, and shade the eye from the direct
rays of the sun. *Eyelids,* or *palpebrae* (PAL-pe-brē),
consist primarily of skeletal muscle covered externally by skin. The underside of the muscle is lined
by a mucous membrane called the *palpebral conjunctiva* (kon-junk-TĪ-va). The *bulbar (ocular) conjunctiva* covers the surface of the eyeball. Also
within eyelids are *tarsal (Meibomian) glands,* modified sebaceous glands whose oily secretion keeps
the eyelids from adhering to each other. Infection
of these glands produces a *chalazion* (cyst) in the
eyelid. Eyelids shade the eyes during sleep, protect
the eyes from light rays and foreign objects, and
spread lubricating secretions over the surface of the
eyeballs. Projecting from the border of each eyelid
is a row of short, thick hairs, the *eyelashes.* Sebaceous glands at the base of the hair follicles of the
eyelashes, called *sebaceous ciliary glands (glands of
Zeis),* pour a lubricating fluid into the follicles. An
infection of these glands is called a *sty.*

The *lacrimal* (LAK-ri-mal; *lacrima* = tear) *apparatus* consists of a group of structures that manufacture and drain tears. Each *lacrimal gland* is
located at the superior lateral portion of both orbits. Leading from the lacrimal glands are 6 to 12
excretory lacrimal ducts that empty tears onto the
surface of the conjunctiva of the upper lid. From
here, the tears pass medially and enter two small
openings called *lacrimal puncta* that appear as two
small pores, one in each papilla of the eyelid, at the
medial commissure of the eye. The tears then pass
into two ducts, the *lacrimal canals,* and are next
conveyed into the lacrimal sac. The *lacrimal sac* is
the superior expanded portion of the *nasolacrimal
duct,* a canal that carries the tears into the nasal cav-

ity. Tears clean, lubricate, and moisten the external
surface of the eyeball.

Label the parts of the lacrimal apparatus in Figure 15.4.

2. Structure of the Eyeball

The eyeball can be divided into three principal layers: (1) fibrous tunic, (2) vascular tunic, and (3) retina (nervous tunic). See Figure 15.5 on page 287.

a. FIBROUS TUNIC

The *fibrous tunic* is the outer coat of the eyeball. It
is divided into the posterior sclera and the anterior
cornea. The *sclera* (SKLE-ra; *skleros* = hard), called
the "white of the eye," is a coat of dense connective
tissue that covers all the eyeball except the most anterior portion (cornea). The sclera gives shape to the
eyeball and protects its inner parts. The anterior
portion of the fibrous tunic is known as the *cornea*
(KOR-nē-a). This nonvascular, transparent coat covers the colored iris. Because it is curved, the cornea
helps focus light. The outer surface of the cornea
contains epithelium that is continuous with the epithelium of the bulbar conjunctiva. At the junction
of the sclera and cornea is the *scleral venous sinus
(canal of Schlemm).*

b. VASCULAR TUNIC

The *vascular tunic* is the middle layer of the eyeball
and consists of three portions: choroid, ciliary body,
and iris. The *choroid* (KŌ-royd) is the posterior portion of the vascular tunic. It is a thin, dark brown
membrane that lines most of the internal surface of
the sclera and contains blood vessels and melanin.
The choroid absorbs light rays so they are not reflected back out of the eyeball and maintains the nutrition of the retina. The anterior portion of the
choroid is the *ciliary* (SIL-ē-ar'-ē) *body,* the thickest
portion of the vascular tunic. It extends from the *ora
serrata* (Ō-ra ser-RĀ-ta) of the retina (inner tunic) to
a point just behind the sclerocorneal junction. The ora
serrata is the jagged margin of the retina. The ciliary
body consists of the *ciliary processes* (folds of the ciliary body that secrete aqueous humor) and the *ciliary muscle* (a circular band of smooth muscle that
alters the shape of the lens for near or far vision). The
iris (*irid* = colored circle), the third portion of the vascular tunic, is the colored portion of the eyeball and
consists of circular and radial smooth-muscle fibers
arranged to form a doughnut-shaped structure. The
hole in the center of the iris is the *pupil,* through
which light enters the eyeball. One function of the iris
is to regulate the amount of light entering the eyeball.

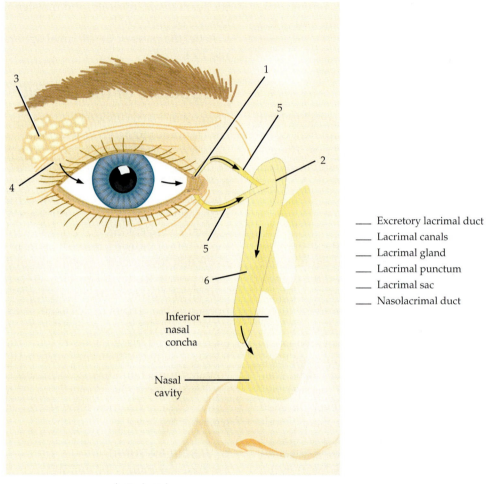

Anterior view

____ Excretory lacrimal duct
____ Lacrimal canals
____ Lacrimal gland
____ Lacrimal punctum
____ Lacrimal sac
____ Nasolacrimal duct

Inferior
nasal
concha

Nasal
cavity

FIGURE 15.4   Lacrimal apparatus.

C. RETINA

The third and inner coat of the eyeball, the *retina (nervous tunic),* is found only in the posterior portion of the eye. Its primary function is image formation. It consists of a pigment epithelium and a neural portion. The outer *pigment epithelium* (nonvisual portion) consists of melanin-containing epithelial cells in contact with the choroid. The inner *neural portion* (visual portion) is composed of three zones of neurons. Named in the order in which they conduct nerve impulses, these are the *photoreceptor layer, bipolar cell layer,* and *ganglion cell layer.* Structurally, the photoreceptor layer is just internal to the pigment epithelium, which lies adjacent to the choroid. The ganglion cell layer is the innermost zone of the neural portion.

The two types of photoreceptors are called rods and cones because of their respective shapes. *Rods* are specialized for vision in dim light. In addition, they allow discrimination between different shades

of dark and light and permit discernment of shapes and movement. *Cones* are specialized for color vision and for sharpness of vision, that is, *visual acuity.* Cones are stimulated only by bright light and are most densely concentrated in the *central fovea,* a small depression in the center of the macula lutea. The *macula lutea* (MAK-yoo-la LOO-tē-a), or yellow spot, is situated in the exact center of the posterior portion of the retina and corresponds to the visual axis of the eye. The fovea is the area of sharpest vision because of the high concentration of cones. Rods are absent from the fovea and macula but increase in density toward the periphery of the retina.

When light stimulates photoreceptors, impulses are conducted across synapses to the bipolar neurons in the intermediate zone of the neural portion of the retina. From there, the impulses pass to the ganglion cell layer. Axons of the ganglion neurons extend posteriorly to a small area of the retina called

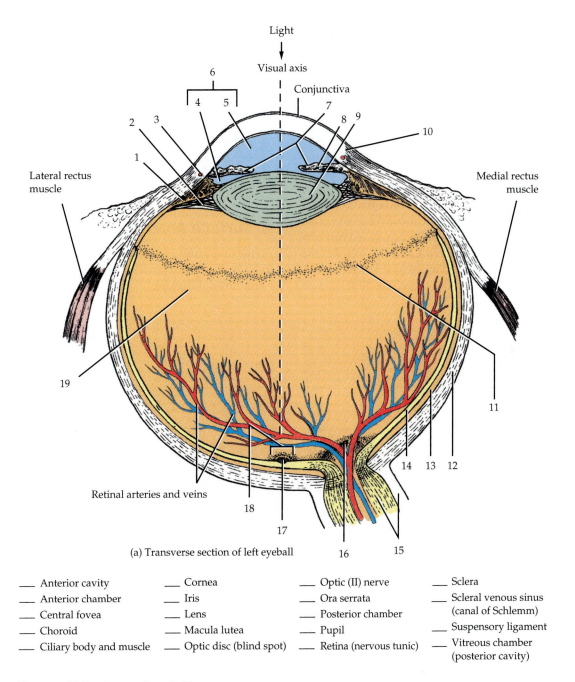

(a) Transverse section of left eyeball

FIGURE 15.5   Parts of eyeball.

____ Anterior cavity   ____ Cornea   ____ Optic (II) nerve   ____ Sclera

____ Anterior chamber   ____ Iris   ____ Ora serrata   ____ Scleral venous sinus (canal of Schlemm)

____ Central fovea   ____ Lens   ____ Posterior chamber

____ Choroid   ____ Macula lutea   ____ Pupil   ____ Suspensory ligament

____ Ciliary body and muscle   ____ Optic disc (blind spot)   ____ Retina (nervous tunic)   ____ Vitreous chamber (posterior cavity)

the *optic disc (blind spot).* This region contains openings through which fibers of the ganglion neurons exit as the *optic (II) nerve.* Because this area contains neither rods nor cones, and only nerve fibers, no image is formed on it. For this reason it is called the blind spot.

d. LENS

The eyeball itself also contains the lens, just behind the pupil and iris. The *lens* is constructed of numerous layers of protein fibers arranged like the layers of an onion. Normally, the lens is perfectly transparent and is enclosed by a clear capsule and held in position by the *suspensory ligaments.* A loss of transparency of the lens is called a *cataract.*

e. INTERIOR

The interior of the eyeball contains a large cavity divided into two smaller cavities. These are called the anterior cavity and the vitreous chamber (posterior cavity) and are separated from each other by the lens. The *anterior cavity,* in turn, has two subdivisions known as the anterior chamber and the posterior chamber. The *anterior chamber* lies posterior to the cornea and anterior to the iris. The *posterior chamber* lies posterior to the iris and anterior to the suspensory ligaments and lens. The anterior cavity is filled with a clear, watery fluid known as the *aqueous* (*aqua* = water) *humor,* which is secreted by the ciliary processes posterior to the iris. From the posterior chamber, the fluid permeates the posterior cavity and then passes anteriorly between the iris and the lens, through the pupil into the anterior chamber. From the anterior chamber, the aqueous humor is drained off into the scleral venous sinus and passes into the blood. Pressure in the eye, called *intraocular pressure (IOP),* is produced mainly by the aqueous humor. Intraocular pressure keeps the retina smoothly applied to the choroid so that the reti-

na may form clear images. Abnormal elevation of intraocular pressure, called *glaucoma* (glaw-KŌ-ma), results in degeneration of the retina and blindness.

The second, larger cavity of the eyeball is the *vitreous chamber (posterior cavity).* It is located between the lens and retina and contains a soft, jellylike substance called the *vitreous body.* This substance contributes to intraocular pressure, helps to prevent the eyeball from collapsing, and holds the retina flush against the internal portions of the eyeball.

Label the parts of the eyeball in Figure 15.5.

3. Surface Anatomy

Refer to Figure 15.6 for a summary of several surface anatomy features of the eyeball and accessory structures of the eye.

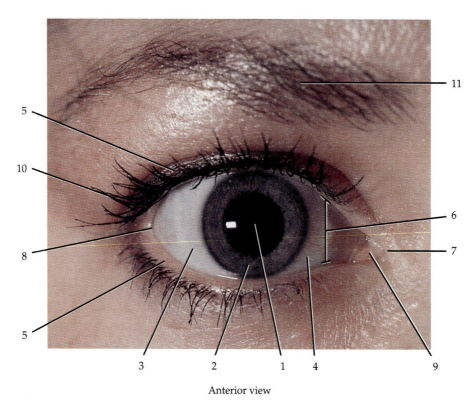

Anterior view

1. **Pupil.** Opening of center of iris of eyeball for light transmission.
2. **Iris.** Circular pigmented muscular membrane behind cornea.
3. **Sclera.** "White" of eye, a coat of fibrous tissue that covers entire eyeball except for cornea.
4. **Conjunctiva.** Membrane that covers exposed surface of eyeball and lines eyelids.
5. **Palpebrae (eyelids).** Folds of skin and muscle lined by conjunctiva.
6. **Palpebral fissure.** Space between eyelids when they are open.

7. **Medial commissure.** Site of union of upper and lower eyelids near nose.
8. **Lateral commissure.** Site of union of upper and lower eyelids away from nose.
9. **Lacrimal caruncle.** Fleshy, yellowish projection of medial commissure that contains modified sweat and sebaceous glands.
10. **Eyelashes.** Hairs on margins of eyelids, usually arranged in two or three rows.
11. **Eyebrows.** Several rows of hair superior to upper eyelids.

FIGURE 15.6   Surface anatomy of eyeball and accessory structures.

4. Dissection of Vertebrate Eye (Beef or Sheep)

CAUTION! *Please reread Section D, "Precautions Related to Dissection" at the beginning of the laboratory manual on page xiii before you begin your dissection.*

a. EXTERNAL EXAMINATION

PROCEDURE

1. Note any *fat* on the surface of the eyeball that protects the eyeball from shock in the orbit. Remove the fat.
2. Locate the *sclera,* the tough external white coat, and the *conjunctiva,* a delicate membrane that covers the anterior surface of the eyeball and is attached near the edge of the cornea. The *cornea* is the anterior, transparent portion of the sclera. It is probably opaque in your specimen due to the preservative.
3. Locate the *optic (II) nerve,* a solid, white cord of nerve fibers on the posterior surface of the eyeball.
4. If possible, identify six *extrinsic eye muscles* that appear as flat bands near the posterior part of the eyeball.

b. INTERNAL EXAMINATION

PROCEDURE

1. With a sharp scalpel, make an incision about 0.6 cm (¼ in.) lateral to the cornea (Figure 15.7 on page 289).
2. Insert scissors into the incision and carefully and slowly cut all the way around the corneal region. The eyeball contains fluid, so take care that it does not squirt out when you make your first incision. Examine the inside of the anterior part of the eyeball.
3. The *lens* is held in position by *suspensory ligaments,* which are delicate fibers. Around the outer margin of the lens, with a pleated appearance, is the black *ciliary body,* which also functions to hold the lens in place. Free the lens and notice how hard it is.
4. The *iris* can be seen just anterior to the lens and is also heavily pigmented or black.
5. The *pupil* is the circular opening in the center of the iris.
6. Examine the inside of the posterior part of the eyeball, identifying the thick *vitreous humor* that fills the space between the lens and retina.
7. The *retina* is the white inner coat beneath the choroid coat and is easily separated from it.

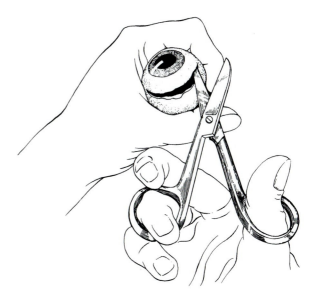

FIGURE 15.7   Procedure for dissecting a vertebrate eye.

8. The *choroid coat* is a dark, iridescent-colored tissue that gets its iridescence from a special structure called the *tapetum lucidum.* The tapetum lucidum, which is not present in the human eye, functions to reflect some light back onto the retina.
9. Finally, identify the *blind spot,* the point at which the retina is attached to the back of the eyeball.

5. Testing for Visual Acuity

The acuteness of vision may be tested by means of a *Snellen Chart.* It consists of letters of different sizes which are read at a distance normally designated at 20 ft. If the subject reads to the line that is marked "50," he or she is said to possess 20/50 vision in that eye, meaning that he or she is reading at 20 ft what a person who has normal vision can read at 50 ft. If he or she reads to the line marked "20," he or she has 20/20 vision in that eye. The normal eye can sufficiently refract light rays from an object 20 ft away to focus a clear object on the retina. Therefore, if you have 20/20 vision, your eyes are perfectly normal. The higher the bottom number, the larger the letter must be for you to see it clearly, and of course the worse or weaker are your eyes.

PROCEDURE

1. Have the subject stand 20 ft from the Snellen Chart and cover the right eye with a 3″ × 5″ (3-in.-by-5-in.) card.

2. Instruct the subject to slowly read down the chart until he or she can no longer focus the letters.

3. Record the number of the last line (20/20, 20/30, or whichever) that can be successfully read.

4. Repeat this procedure covering the left eye.

5. Now the subject should read the chart using both eyes.

6. Record your results in Section C.5 of the LABORATORY REPORT RESULTS at the end of the exercise and change places.

6. Testing for Astigmatism

The eye, with normal ability to refract light, is referred to as an *emmetropic* (em'-e-TROP-ik) eye. It can sufficiently refract light rays from an object 20 ft away to focus a clear object on the retina. If the lens is normal, objects as far away as the horizon and as close as about 20 ft will form images on the sensitive part of the retina. When objects are closer than 20 ft, however, the lens has to sharpen its focus by using the ciliary muscles. Many individuals, however, have abnormalities related to improper refraction. Among these are *myopia* (mī-Ō-pē-a) (nearsightedness), *hypermetropia* (hī'-per-mē-TRŌ-pē-a) (farsightedness), and *astigmatism* (a-STIG-ma-tizm) (irregularities in the surface of the lens or cornea).

Why do you think nearsightedness can be corrected with glasses containing biconcave lenses?

_____

_____

_____

How would you correct farsightedness?_____

_____

_____

PROCEDURE

1. In order to determine the presence of astigmatism, remove any corrective lenses if you are wearing them and look at the center of the following astigmatism test chart, first with one eye, then the other:

2. If all the radiating lines appear equally sharp and equally black, there is no astigmatism.

3. If some of the lines are blurred or less dark than others, astigmatism is present.

4. If you wear corrective lenses, try the test with them on.

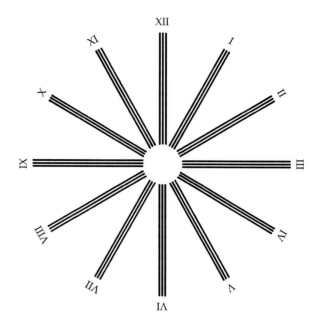

7. Testing for the Blind Spot

PROCEDURE

1. Hold this page about 20 in. from your face with the cross shown below directly in front of your right eye. You should be able to see the cross and the circle when you close your left eye.

2. Now, keeping the left eye closed, slowly bring the page closer to your face while fixing the right eye on the cross.

3. At a certain distance the circle will disappear from your field of vision because its image falls on the blind spot.

8. Image Formation

Formation of an image on the retina requires four basic processes, all concerned with focusing light rays. These are (1) refraction of light rays, (2) accommodation of the lens, (3) constriction of the pupil, and (4) convergence of the eyes.

When light rays traveling through a transparent medium (such as air) pass into a second transparent medium with a different density (such as water), the rays bend at the surface of the two media. This is called *refraction.* As light rays enter the eye, they are refracted at the anterior and posterior surfaces of the cornea. Both surfaces of the lens of the eye further refract the light rays so that they come into exact focus on the retina.

The lens of the eye has the unique ability to change the focusing power of the eye by becoming moderately curved at one moment and greatly curved the next. When the eye is focusing on a close object, the lens curves greatly in order to bend the rays toward the central fovea of the eye. This increase in the curvature of the lens is called *accommodation.*

a. TESTING FOR NEAR-POINT ACCOMMODATION

The following test determines your *near-point accommodation:*

PROCEDURE

1. Using any card that has a letter printed on it, close one eye and focus on the letter.
2. Measure the distance of the card from the eye using a ruler or a meter stick.
3. Now *slowly* bring the card as close as possible to your open eye, and stop when you no longer see a clear, detailed letter.
4. Measure and record this distance. This value is your near-point accommodation.
5. Repeat this procedure three times and then test your other eye.
6. Check Table 15.1 to see whether the near point for your eyes corresponds with that recorded for your age group. (**Note:** *Use a letter that is the size of typical newsprint.*)

b. TESTING FOR CONSTRICTION OF THE PUPIL

PROCEDURE

1. Place a 3″ × 5″ (3-in.-by-5-in.) card on the side of the nose so that a light shining on one side of the face will not affect the eye on the other side.
2. Shine the light from a lamp or a flashlight on one eye, 6 in. away (approximately 15 cm), for about 5 sec. Note the change in the size of the pupil of this eye.

TABLE 15.1
Correlation of Age and Near-Point Accommodation

Age	Inches	Centimeters
10	2.95	7.5
20	3.54	9.0
30	4.53	11.5
40	6.77	17.2
50	20.67	52.5
60	32.80	83.3

3. Remove the light, wait about 3 min, and repeat, but this time observe the pupil of the opposite eye.
4. Wait a few minutes, and repeat the test, observing the pupils of both eyes.

9. Testing for Convergence

In humans, both eyes focus on only one set of objects—a characteristic called *single binocular vision.* The term *convergence* refers to a medial movement of the two eyeballs so that they are both directed toward the object being viewed. The nearer the object, the greater the degree of convergence necessary to maintain single binocular vision.

PROCEDURE

1. Hold a pencil or pen about 2 ft from your nose and focus on its point. Now slowly bring the pencil toward your nose.
2. At some moment you should suddenly see two pencil points, or a blurring of the point.
3. Observe your partner's eyes when he or she does this test.

Images are actually focused upside down on the retina. They also undergo mirror reversal. That is, light reflected from the right side of an object hits the left side of the retina and vice versa. Reflected light from the top of the object crosses light from the bottom of the object and strikes the retina below the central fovea. Reflected light from the bottom of the object crosses light from the top of the object and strikes the retina above the central fovea.

The reason why we do not see a topsy-turvy world is that the brain learns early in life to coordinate visual images with the exact location of objects. The brain stores memories of reaching and touching objects and automatically turns visual images right-side up and right-side around.

10. Testing for Binocular Vision, Depth Perception, Diplopia, and Dominance

Humans are endowed with binocular vision. Each eye sees a different view, although there is a large degree of overlap. The cerebral cortex uses these discrepancies to produce *stereopsis,* or three-dimensional vision, an orientation of the object in space that allows us to perceive its distance from us *(depth perception).*

Even though there are slight differences in the views from each eye, we see a single image because the cerebral cortex integrates the images from each eye into a single perception. If this integration is not present, *diplopia* (dip-LŌ-pē-a), or double vision, results. We do not perceive an "average" or "mean" of both left and right views; rather the view from one of our eyes is *dominant,* i.e., the view we always perceive with both eyes open.

PROCEDURE

1. Place a book at arm's length on a table. Place your left hand at your side and touch the nearest corner of the book with your right index finger. Close the left eye and repeat; close the right eye and repeat. Note the accuracy of your attempts. How accurate was your attempt to touch a

   point with one eye closed? _____
   If the right eye was closed, did you have error

   to the left or the right? _____
2. Use a depth perception tester with both eyes open and with each eye closed. Attempt to align the two arrows and measure monocular perception errors using the scale on the base.

   Length of depth perception error _____
3. Focus on a discrete object in the distance and press *very gently* on your left eyelid. Note how

   many objects you see: _____
4. Stab a pencil through a piece of paper and remove the scrap. Hold the paper at arm's length and focus on a small discrete object which just fills the hole. (You may want to use an X on the blackboard.) *Without moving the paper,* close the left eye and note whether the object is still present. Repeat with right eye

   closed. Note your dominant eye: _____

11. Testing for Afterimages

The rods and cones of the retina are photoreceptors; that is, they contain light-sensitive pigments that absorb light of different wavelengths. *Rhodopsin,* the light-sensitive pigment in rods, consists of opsin and retinal. When light strikes a molecule of rhodopsin, the retinal changes from a curved to a straight shape and breaks away from the opsin. The energy released as a result of the exchange initiates the nerve impulse that causes us to perceive light.

In bright light, all the rhodopsin is decomposed, so that the rods are nonfunctional. Cones function in bright light, in much the same way as rods, but they absorb light of specific wavelengths (red, blue, or green). The particular color perceived depends on which combinations of cones are stimulated. If all three are stimulated, we see white; if none is stimulated, we see black.

If photoreceptors are stimulated by staring at a bright object for a long period of time, they will continue firing briefly after the stimulus has been removed *(positive afterimage).* However, after prolonged firing, these photoreceptors become "bleached," or fatigued, so that they can no longer fire, resulting in a reversal of the image *(negative afterimage).*

PROCEDURE

1. Stare at a light source, *but not a bright light,* for 30 sec. Close the eyes briefly and note the positive afterimage. Open the eyes and focus on a piece of white paper. Note the negative afterimage.
2. Draw a red cross on paper about 1 in. high, with lines ½ in. thick. Stare at the cross without moving the eyes for at least 30 sec. Close the eyes very briefly; open them and stare at a piece of white paper. Note the positive and negative afterimages. Record your observations in Section C.11 of the LABORATORY REPORT RESULTS at the end of the exercise.

12. Testing for Color Blindness

Color blindness is an inherited disorder in which certain cones are absent. It is a sex-linked disorder. About 0.5% of all females and about 8% of all males are affected.

PROCEDURE

1. View Ishihara plates in bright light and compare your responses to those in the plate book or use Holmgren's test for matching colored threads.

2. Note the accuracy of your responses: _____

   _____

13. Visual Pathway

From the rods and cones, impulses are transmitted through bipolar cells to ganglion cells. The cell bodies of the ganglion cells lie in the retina and their axons leave the eye via the *optic (II) nerve.* The axons pass through the *optic chiasm* (kī-AZ-em), a crossing point of the optic nerves. Fibers from the medial retina cross to the opposite side. Fibers from the lateral retina remain uncrossed. Upon passing through the optic chiasm, the fibers, now part of the *optic tract*, enter the brain and terminate in the lateral geniculate nucleus of the thalamus. Here the fibers synapse with the neurons whose axons pass to the visual centers located in the occipital lobes of the cerebral cortex. Label the visual pathway in Figure 15.8.

D. AUDITORY SENSATIONS AND EQUILIBRIUM

In addition to containing receptors for sound waves, the *ear* also contains receptors for equilibrium. The ear is subdivided into three principal regions: (1) external (outer) ear, (2) middle ear, and (3) internal (inner) ear.

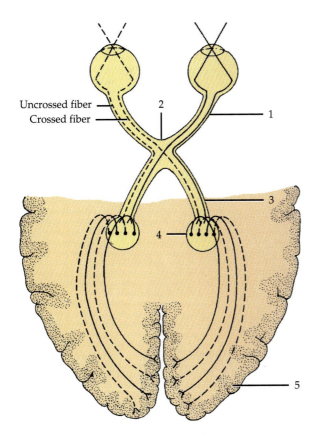

Uncrossed fiber
Crossed fiber

___ Occipital lobe of cerebral cortex
___ Optic chiasm
___ Optic (II) nerve
___ Optic tract
___ Thalamus

FIGURE 15.8 Visual pathway.

1. Structure of Ear

a. EXTERNAL (OUTER) EAR

The *external (outer) ear* collects sound waves and directs them inward. Its structure consists of the auricle, external auditory canal, and tympanic membrane. The *auricle (pinna)* is a trumpet-shaped flap of elastic cartilage covered by thick skin. The rim of the auricle is called the *helix,* and the inferior portion is referred to as the *lobule.* The auricle is attached to the head by ligaments and muscles. The *external auditory canal (meatus)* is a tube, about 2.5 cm (1 in.) in length that leads from the auricle to the eardrum. The walls of the canal consist of bone lined with cartilage that is continuous with the cartilage of the auricle. Near the exterior opening, the canal contains a few hairs and specialized sebaceous glands called *ceruminous* (se-ROO-me-nus) *glands,* which secrete *cerumen* (earwax). The combination of hairs and cerumen prevents foreign objects from entering the ear. The *eardrum,* or *tympanic* (tim-PAN-ik) *membrane,* is a thin, semitransparent partition of fibrous connective tissue located between the external auditory canal and middle ear.

Examine a model or charts and label the parts of the external ear in Figure 15.9.

b. MIDDLE EAR

The *middle ear (tympanic cavity)* is a small, epithelium-lined, air-filled cavity hollowed out of the temporal bone. The area is separated from the external ear by the eardrum and from the internal ear by a very thin bony partition that contains two small membrane-covered openings, called the *oval window* and the *round window.* The posterior wall of the middle ear communicates with the mastoid cells of the temporal bone through a chamber called the *tympanic antrum.*

The anterior wall of the middle ear contains an opening that leads into the *auditory (Eustachian) tube.* The auditory tube connects the middle ear with the nose and nasopharynx. The function of the tube is to equalize air pressure on both sides of the eardrum. Any sudden pressure changes against the eardrum may be equalized by deliberately swallowing.

Extending across the middle ear are three exceedingly small bones called *auditory ossicles* (OS-si-kuls). These are known as the malleus, incus, and stapes. Based on their shape, they are commonly named the hammer, anvil, and stirrup, respectively.

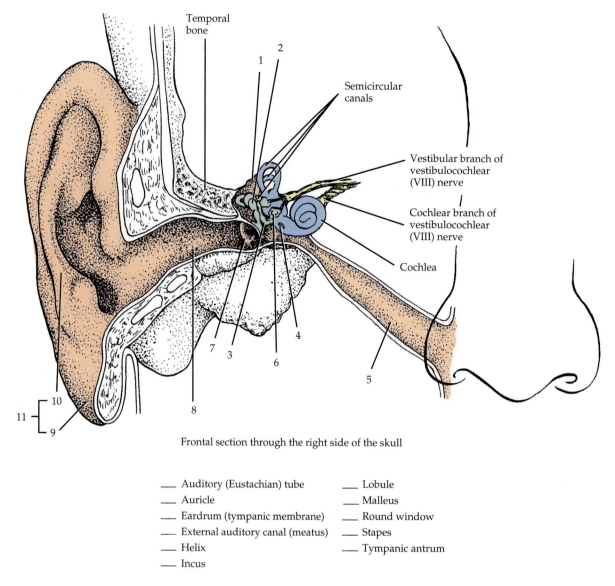

Frontal section through the right side of the skull

___ Auditory (Eustachian) tube   ___ Lobule
___ Auricle   ___ Malleus
___ Eardrum (tympanic membrane)   ___ Round window
___ External auditory canal (meatus) ___ Stapes
___ Helix   ___ Tympanic antrum
___ Incus

FIGURE 15.9   Principal subdivisions of ear.

The "handle" of the *malleus* is attached to the internal surface of the eardrum. Its head articulates with the base of the *incus,* the intermediate bone in the series, which articulates with the stapes. The base of the *stapes* fits into a small opening between the middle and inner ear called the *oval window.* Directly inferior to the oval window is another opening, the *round window.* This opening, which separates the middle and inner ears, is enclosed by a membrane called the *secondary tympanic membrane.*

Examine a model or charts and label the parts of the middle ear in Figures 15.9 and 15.10.

C. INTERNAL (INNER) EAR

The *internal (inner) ear* is also known as the *labyrinth* (LAB-i-rinth). Structurally, it consists of two main divisions: (1) an outer bony labyrinth and (2) an inner membranous labyrinth that fits within the bony labyrinth. The *bony labyrinth* is a series of cavities within the petrous portion of the temporal bone that can be divided into three regions, named on the basis of shape: vestibule, cochlea, and semicircular canals. The bony labyrinth is lined with periosteum and contains a fluid called the *perilymph.* This fluid surrounds the *membranous labyrinth,* a series of sacs and tubes lying inside and having the

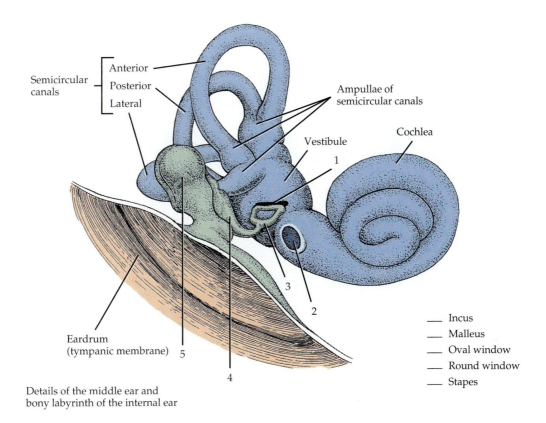

Details of the middle ear and
bony labyrinth of the internal ear

___ Incus
___ Malleus
___ Oval window
___ Round window
___ Stapes

FIGURE 15.10   Ossicles of middle ear.

same general form as the bony labyrinth. Epithelium lines the membranous labyrinth, which is filled with a fluid called the *endolymph.*

The *vestibule* is the oval, central portion of the bony labyrinth (see Figure 15.10). The membranous labyrinth within the vestibule consists of two sacs called the *utricle* (YOO-tri-kul) and *saccule* (SAK-yool). These sacs are connected to each other by a small duct.

Projecting superiorly and posteriorly from the vestibule are the three bony *semicircular canals* (see Figure 15.10). Each is arranged at approximately right angles to the other two. They are called the anterior, posterior, and lateral canals. One end of each canal enlarges into a swelling called the *ampulla* (am-POOL-la; = little jar). Inside the bony semicircular canals lie portions of the membranous labyrinth, the *semicircular ducts (membranous semicircular canals).* These structures communicate with the utricle of the vestibule. Label these structures in Figure 15.11.

Anterior to the vestibule is the *cochlea* (KOK-lē-a; = snail's shell) (label it in Figure 15.11). The cochlea consists of a bony spiral canal that makes about three turns around a central bony core called the *modiolus.* A transverse section through the cochlea shows that the canal is divided by partitions into three separate channels resembling the letter Y lying on its side. The stem of the Y is a bony shelf that protrudes into the canal. The wings of the Y are composed of the vestibular and basilar membranes. The channel above the partition is called the *scala vestibuli.* The channel below is known as the *scala tympani.* The cochlea adjoins the wall of the vestibule, into which the scala vestibuli opens. The scala tympani terminates at the round window. The perilymph of the vestibule is continuous with that of the scala vestibuli. The third channel (between the wings of the Y) is the membranous labyrinth, the *cochlear duct (scala media).* This duct is separated from the scala vestibuli by the *vestibular membrane* and from the scala tympani by the *basilar membrane.* Resting on the basilar membrane is the *spiral organ (organ of Corti),* the organ of hearing. Label these structures in Figure 15.12.

The spiral organ (organ of Corti) is a coiled sheet of epithelial cells on the inner surface of the basilar membrane. This structure is composed of supporting cells and hair cells, which are receptors for auditory sensations. The hair cells have long hairlike processes at their free ends that extend into the endolymph of the cochlear duct. The basal ends of

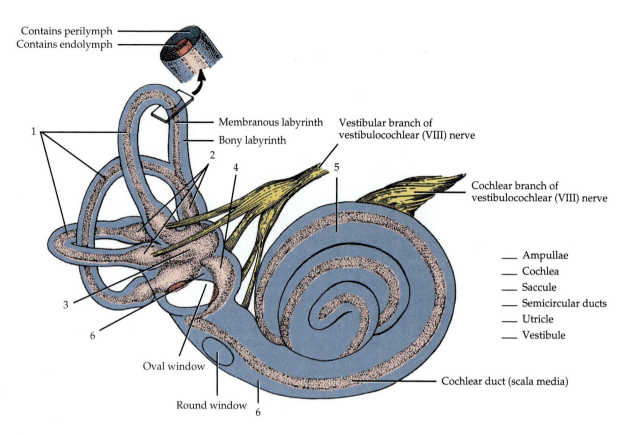

Contains perilymph
Contains endolymph

Membranous labyrinth
Bony labyrinth

Vestibular branch of
vestibulocochlear (VIII) nerve

Cochlear branch of
vestibulocochlear (VIII) nerve

1

2

4

5

3

6

Oval window

Round window

6

___ Ampullae
___ Cochlea
___ Saccule
___ Semicircular ducts
___ Utricle
___ Vestibule

Cochlear duct (scala media)

FIGURE 15.11   Details of inner ear.

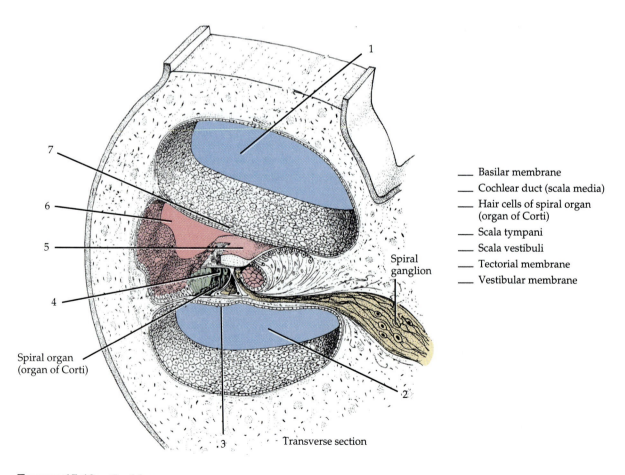

1

7

6

5

4

Spiral organ
(organ of Corti)

Spiral
ganglion

2

3   Transverse section

___ Basilar membrane
___ Cochlear duct (scala media)
___ Hair cells of spiral organ
    (organ of Corti)
___ Scala tympani
___ Scala vestibuli
___ Tectorial membrane
___ Vestibular membrane

FIGURE 15.12   Cochlea.

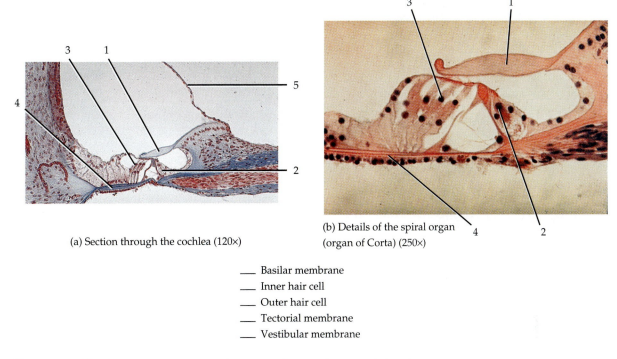

(a) Section through the cochlea (120×)

(b) Details of the spiral organ
(organ of Corta) (250×)

___ Basilar membrane
___ Inner hair cell
___ Outer hair cell
___ Tectorial membrane
___ Vestibular membrane

FIGURE 15.13   Photomicrographs of the spiral organ (organ of Corti).

the hair cells are in contact with fibers of the cochlear branch of the vestibulocochlear (VIII) nerve. Projecting over and in contact with the hair cells of the spiral organ is the *tectorial (tectum = cover) membrane,* a very delicate and flexible gelatinous membrane. Label the tectorial membrane in Figure 15.12.

Obtain a prepared microscope slide of the spiral organ (organ of Corti) and examine under high power. Now label Figure 15.13.

2. Surface Anatomy

Refer to Figure 15.14 for a summary of several surface anatomy features of the ear.

3. Tests for Auditory Acuity

Sound waves result from the alternate compression and decompression of air. The waves have both a frequency and an amplitude. The *frequency* is the distance between crests of a sound wave. It is measured in *Hertz (Hz),* or *cycles per second,* and is perceived as *pitch.* The higher the frequency of a sound, the higher its pitch. Different frequencies displace different areas of the basilar membrane. The *amplitude* of a sound wave is perceived as *intensity (loudness).* Intensity is measured in *decibels (dB).* Population norms have been calculated

that measure the intensity of a sound that can just be heard, i.e., the *threshold* of sound. These levels are said to have an intensity of 0 dB. Each 10 dB reflects a tenfold increase in intensity; thus a sound is 10 times louder than threshold at 10 dB, 100 times greater at 20 dB 1 million times greater at 60 dB, and so on.

a. SOUND LOCALIZATION TEST

The cerebral cortex localizes the source of a sound by evaluating the time lag between the entry of sound into each ear. Without turning the head, it is impossible to distinguish between the origin of sounds that arise at 30° ahead of the right ear and 30° behind the right ear.

PROCEDURE

1. In a quiet room, have a subject occlude one ear with a cotton plug and close the eyes.
2. Move a ticking watch to various points 6 in. away from the ear (side, front, top, back) and ask the subject to point to the location of the sound.
3. Note in which positions the sound is best localized. In what regions was sound localization

most accurate? _____

_____

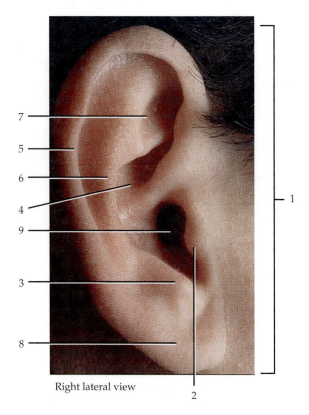

Right lateral view

2

1. **Auricle.** Portion of external ear not contained in head, also called the pinna.
2. **Tragus.** Cartilaginous projection.
3. **Antitragus.** Cartilaginous projection opposite tragus.
4. **Concha.** Hollow of auricle.
5. **Helix.** Superior and posterior free margin of auricle.
6. **Antihelix.** Semicircular ridge posterior and superior to concha.
7. **Triangular fossa.** Depression in superior portion of antihelix.
8. **Lobule.** Inferior portion of auricle devoid of cartilage.
9. **External auditory canal (meatus).** Canal extending from external ear to eardrum.

FIGURE 15.14 Surface anatomy of ear.

b. WEBER'S TEST

Weber's test is diagnostic for both conduction and sensorineural deafness. *Conduction deafness* results from interference with sound waves reaching the inner ear (plugged external auditory canal, fusion of the ossicles, etc.); *sensorineural deafness* is caused by damage to the nerve pathway between the cochlea and the brain. This test should be performed in a quiet room.

PROCEDURE

1. Strike a tuning fork with a mallet and place the tip of the handle on the median line of a subject's forehead.
2. Determine if the tone is equally loud in both ears. _____
3. Have the subject occlude the left ear with cotton and repeat the exercise; occlude the right ear and repeat.

   Describe your results: _____
   _____

If the sound is equally loud in both unoccluded ears, hearing is normal. If the sound is louder in the occluded ear, hearing is normal. This situation parallels conduction deafness, when the cochlea is not receiving environmental background noise but only the vibrations of the tuning fork. If there is sensorineural deafness, the sound is louder in the normal (or nonoccluded) ear, because sensorineural loss produces a deficit in transmission to the cerebral cortex.

4. Equilibrium Apparatus

The term *equilibrium* (balance) has two meanings. One kind of equilibrium, called *static equilibrium,* refers to the position of the body (mainly the head) relative to the force of gravity. The second kind of equilibrium, called *dynamic equilibrium,* is the maintenance of the position of the body (mainly the head) in response to sudden movements (rotation, acceleration, and deceleration). Collectively, the receptor organs for equilibrium are called the *vestibular apparatus,* which includes the maculae in the saccule and utricle and the cristae in the semicircular ducts.

The *maculae* (MAK-yoo-lē) in the walls of the *utricle* and *saccule* are the receptors concerned mainly with static equilibrium. The maculae are small, flat regions that resemble the spiral organ (organ of Corti) microscopically. Maculae are located in planes perpendicular to each other and contain two kinds of cells: *hair (receptor) cells* and *supporting cells.* The hair cells project *stereocilia* (microvilli) and a *kinocilium* (conventional cilium). The columnar supporting cells are scattered among the hair cells. Floating over the hair cells is a thick, gelatinous glycoprotein layer, the *otolithic membrane.* A layer of calcium carbonate crystals, called *otoliths* (*oto* = ear; *lithos* = stone), extends over the entire surface of the otolithic membrane. When the head is tilted, the membrane slides over the hair cells in the direction determined by the tilt of the head. This sliding causes the membrane to pull on the stereocilia, thus initiating a nerve impulse that is conveyed via the vestibular branch of the vestibulocochlear (VIII) nerve to the brain (cerebellum). The cerebellum sends continuous nerve impulses to the motor areas of the cerebral cortex in response to input from the maculae in the utricle and saccule, causing the motor system to increase or decrease its nerve impulses to specific skeletal muscles to maintain static equilibrium.

Label the parts of the macula in Figure 15.15a.

Now consider the role of the cristae in the semicircular ducts in maintaining dynamic equilibrium. The three semicircular ducts are positioned at right angles to one another in three planes: the two vertical ones are called the **anterior** and **posterior semicircular ducts** and the horizontal one is called the **lateral semicircular duct.** This positioning permits correction of an imbalance in three planes. In the **ampulla,** the dilated portion of each duct, is a small elevation called the **crista.** Each crista is composed of a group of **hair (receptor) cells** and **supporting cells** covered by a mass of gelatinous material called the **cupula.** When the head moves, endolymph in the semicircular ducts flows over the hairs and bends them as water in a stream bends the plant life growing at its bottom. Movement of the hairs stimulates sensory neurons, and nerve impulses pass over the vestibular branch of the vestibulocochlear (VIII) nerve. The nerve impulses follow the same pathway as those involved in static equilibrium and are eventually sent to the muscles that contract to maintain body balance in the new position.

Label the parts of the crista in Figure 15.15b.

5. Tests for Equilibrium

You can test equilibrium by using a few simple procedures.

a. BALANCE TEST

This test is used to evaluate static equilibrium.

PROCEDURE

1. Instruct a subject to stand perfectly still with the arms at the sides and the eyes closed.
2. Observe any swaying movements. These are easier to detect if the subject stands in front of a blackboard with a light in front of the subject.

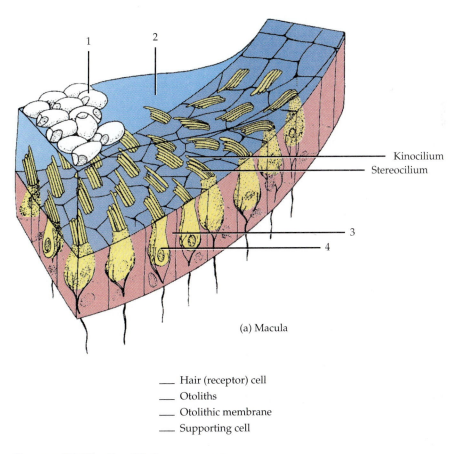

Kinocilium
Stereocilium

(a) Macula

____ Hair (receptor) cell
____ Otoliths
____ Otolithic membrane
____ Supporting cell

FIGURE 15.15   Equilibrium apparatus.

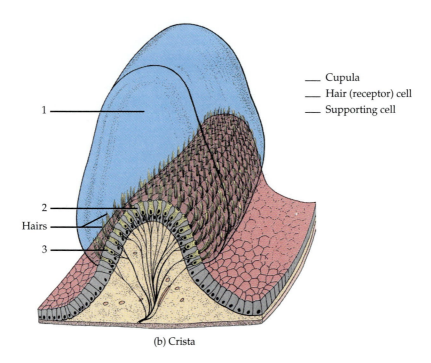

_____ Cupula
_____ Hair (receptor) cell
_____ Supporting cell

1 ——————

2 ——————
Hairs ——————
3 ——————

(b) Crista

FIGURE 15.15 (*Continued*) Equilibrium apparatus.

3. A mark is made at the edge of the shadow of the subject's shoulders and movement of the shadow is then observed.

If the static equilibrium system is dysfunctional the subject will sway or fall, although other proprioreceptors may compensate for the defect.

Record your observations in Section D.5.a of the LABORATORY REPORT RESULTS at the end of the exercise.

b. BARANY TEST

This test evaluates the function of each of the semicircular canals. A subject is rotated, and when rotation is stopped, the momentum causes the endolymph in the lateral (horizontal) canal to continue spinning. The spinning bends the cupula in the direction of rotation. Other sensory input informs us that we are no longer rotating. *Nystagmus,* the reflex movement of the eyes rapidly and then slowly, attempts to compensate for the loss of balance by visual fixation on an object. When the head is tipped toward the shoulder, the anterior semicircular canals are stimulated, so nystagmus is vertical; when the head is tipped slightly forward, the lateral canals are stimulated by the rotation, so nystagmus occurs laterally; when the head is bent onto the chest, the posterior canal receptors are stimulated, so nystagmus is rotational.

PROCEDURE

1. Use a person who is not subject to vertigo (dizziness).
2. The subject sits firmly anchored on a stool, legs up on the stool rung, head tilted onto one shoulder. Then the stool is *very carefully* revolved to the right about 10 times (about one turn every two seconds) and suddenly stopped.
3. Observe the subject's eye movements.

CAUTION! *Watch the subject carefully and be prepared to provide support until the dizziness has passed.*

4. The subject will experience the sensation that the stool is still rotating, which means that the semicircular canals are functioning properly.
5. Repeat on a different subject with the head tipped slightly forward. Use still another subject with the chin touching the chest.
6. Record the direction of nystagmus in all three subjects in Section D.5.b of the LABORATORY REPORT RESULTS at the end of the exercise.

ANSWER THE LABORATORY REPORT QUESTIONS AT THE END OF THE EXERCISE.

Special Senses 15

Student _____   Date _____

Laboratory Section _____   Score/Grade _____

SECTION B. GUSTATORY SENSATIONS
2. Identifying Taste Zones

Areas of Tongue in Which Basic Tastes Are Detected

	Sweet	Bitter	Salty	Sour
Tip of tongue				
Back of tongue				
Sides of tongue				

4. Taste and Smell

	Sensations when placed on dry tongue	Sensations while chewing (nostrils closed)	Sensations while chewing with nostrils opened
Carrot			
Onion			
Potato			
Apple			

SECTION C. VISUAL SENSATIONS

5. Testing for Visual Acuity

Visual activity, left eye   _____

Visual activity, right eye   _____

Visual activity, both eyes   _____

11. Testing for Afterimages

	Appearance of positive afterimage	Appearance of negative afterimage
Bright light		
Red cross		

SECTION D. AUDITORY SENSATIONS AND EQUILIBRIUM

5. Tests For Equilibrium

a. BALANCE TEST

Describe the response of the subject._____

_____

_____

b. BARANY TEST

Subject 1 _____

Subject 2 _____

Subject 3 _____

Special Senses 15

Student _____   Date _____

Laboratory Section _____   Score/Grade _____

PART 1. Multiple Choice

_____   1. The papillae located in an inverted V-shaped row at the posterior portion of the tongue are the (a) circumvallate (b) filiform (c) fungiform (d) gustatory

_____   2. The technical name for the "white of the eye" is the (a) cornea (b) conjunctiva (c) choroid (d) sclera

_____   3. Which is not a component of the vascular tunic? (a) choroid (b) macula lutea (c) iris (d) ciliary body

_____   4. The amount of light entering the eyeball is regulated by the (a) lens (b) iris (c) cornea (d) conjunctiva

_____   5. Which region of the eye is concerned primarily with image formation? (a) retina (b) choroid (c) lens (d) ciliary body

_____   6. The densest concentration of cones is found at the (a) blind spot (b) macula lutea (c) central fovea (d) optic disc

_____   7. Which region of the eyeball contains the vitreous body? (a) anterior chamber (b) vitreous chamber (c) posterior chamber (d) conjunctiva

_____   8. Among the structures found in the middle ear are the (a) vestibule (b) auditory ossicles (c) semicircular canals (d) external auditory canal

_____   9. The receptors for dynamic equilibrium are the (a) saccules (b) utricles (c) cristae in semicircular ducts (d) spiral organs (organs of Corti)

_____   10. The auditory ossicles are attached to the eardrum, to each other, and to the (a) semicircular ducts (b) semicircular canals (c) oval window (d) labyrinth

_____   11. Another name for the internal ear is the (a) labyrinth (b) fenestra (c) cochlea (d) vestibule

_____   12. Orientation of the body relative to the force of gravity is termed (a) the postural reflex (b) the tonal reflex (c) dynamic equilibrium (d) static equilibrium

_____   13. Which sequence best describes the normal flow of tears from the eyes into the nose? (a) lacrimal canals, lacrimal sacs, nasolacrimal ducts (b) lacrimal sacs, lacrimal canals, nasolacrimal ducts (c) nasolacrimal ducts, lacrimal sacs, lacrimal canals (d) lacrimal sacs, nasolacrimal ducts, lacrimal canals

_____   14. A patient whose lens has lost transparency is suffering from (a) glaucoma (b) conjunctivitis (c) cataract (d) trachoma

_____   15. The portion of the eyeball that contains aqueous humor is the (a) anterior cavity (b) lens (c) posterior chamber (d) macula lutea

—————— 16. The organ of hearing located within the inner ear is the (a) vestibule (b) oval window (c) modiolus (d) spiral organ (organ of Corti)

—————— 17. The sense organs of static equilibrium are the (a) semicircular ducts (b) membranous labyrinths (c) maculae in the utricle and saccule (d) pinnae

—————— 18. The membrane that is reflected from the eyelids onto the eyeball is the (a) retina (b) bulbar conjunctiva (c) sclera (d) choroid

—————— 19. Which region of the tongue reacts strongest to bitter tastes? (a) tip (b) center (c) back (d) sides

—————— 20. Which of the following values indicates the best visual acuity? (a) 20/30 (b) 20/40 (c) 20/50 (d) 20/60

—————— 21. Nearsightedness is referred to as (a) emmetropia (b) hypermetropia (c) eumetropia (d) myopia

PART 2. Completion

22. Structures that collectively produce and drain tears are referred to as the ——————————.

23. The three zones of the inner nervous layer of the retina (nervous tunic) are the photoreceptor layer, bipolar cell layer, and ——————————.

24. The small area of the retina where no image is formed is referred to as the ——————————.

25. Abnormal elevation of intraocular pressure (IOP) is called ——————————.

26. In the visual pathway, nerve impulses pass from the optic chiasm to the —————————— before passing to the thalamus.

27. The openings between the middle and inner ears are the oval and —————————— windows.

28. The fluid within the bony labyrinth is called ——————————.

29. The neural pathway for olfaction includes olfactory receptors, olfactory bulbs, ——————————, and cerebral cortex.

30. The posterior wall of the middle ear communicates with the mastoid air cells of the temporal bone through the ——————————.

31. A thin semitransparent partition of fibrous connective tissue that separates the external auditory meatus from the inner ear is the ——————————.

32. The fluid in the membranous labyrinth is called ——————————.

33. The cochlear duct is separated from the scala vestibuli by the ——————————.

34. The gelatinous glycoprotein layer over the hair cells in the maculae is called the——————————.

35. —————————— is blurred vision caused by an irregular curvature of the surface of the cornea or lens.

36. Image formation requires refraction, accommodation, constriction of the pupil, and ——————————.

Endocrine System 16

You have learned how the nervous system controls the body through nerve impulses that are delivered over neurons. Another system of the body, the *endocrine system,* is also involved in controlling bodily functions. The endocrine glands affect bodily activities by releasing chemical messengers, called *hormones,* into the blood stream (the term "hormone" means "to urge on"). The nervous and endocrine systems coordinate their regulatory activities via a complex series of interacting activities. Certain parts of the nervous system stimulate or inhibit the release of hormones. The hormones, in turn, are quite capable of stimulating or inhibiting the flow of particular nerve impulses.

The body contains two different kinds of glands: exocrine and endocrine. *Exocrine glands* secrete their products into ducts. The ducts then carry the secretions into body cavities, into the lumens of various organs, or to the external surface of the body. Examples are sudoriferous (sweat), sebaceous (oil), mucous, and digestive glands. *Endocrine glands,* by contrast, secrete their products (hormones) into the extracellular space around the secretory cells. The secretion then diffuses into blood vessels and the blood. Because they have no ducts, endocrine glands are also called *ductless glands.*

A. ENDOCRINE GLANDS

The endocrine glands are the anterior pituitary gland, thyroid gland, parathyroid glands, adrenal cortex, adrenal medulla, pineal gland, and thymus gland. In addition, several organs of the body contain endocrine tissue, but are not endocrine glands exclusively. They include the pancreas, testes, ovaries, kidneys, hypothalamus, and placenta. The stomach, small intestine, skin, and heart also secrete hormone-like substances and are considered to have endocrine tissue with yet to be completely determined functions. The endocrine glands are organs that together form the *endocrine system.*

Locate the endocrine glands on a torso, and, using your textbook or charts for reference, label Figure 16.1.

All hormones maintain homeostasis by changing the physiological activities of cells. A hormone may stimulate changes in the cells of one or more organs. The cells that respond to the effects of a hormone are called *target cells.*

B. PITUITARY GLAND (HYPOPHYSIS)

The hormones of the *pituitary gland,* also called the *hypophysis* (hī-POF-i-sis), regulate so many body activities that the pituitary gland has been nicknamed the "master gland." This gland lies in the sella turcica of the sphenoid bone and is attached to the hypothalamus via a stalklike structure termed the *infundibulum.* Not only is the hypothalamus of the brain an important regulatory center in the nervous system; it is also a crucial endocrine gland. Cells in the hypothalamus synthesize at least nine different hormones, and the pituitary gland secretes seven more. Together, they play important roles in the regulation of virtually all aspects of growth, development, metabolism, and homeostasis.

The pituitary gland is divided structurally and functionally into an *anterior pituitary gland,* also called the *adenohypophysis* (ad'-i-nō-hī-POF-i-sis), and a *posterior pituitary gland,* also called the *neurohypophysis* (noo'-rō-hī-POF-i-sis). The anterior pituitary gland contains many glandular epitheloid (epithelial-like) cells and forms the glandular part of the pituitary gland. The hypothalamus is connected to the anterior pituitary gland by a series of blood vessels, the *hypothalamic-hypophyseal portal system.* The posterior pituitary gland contains the axon terminals of neurons

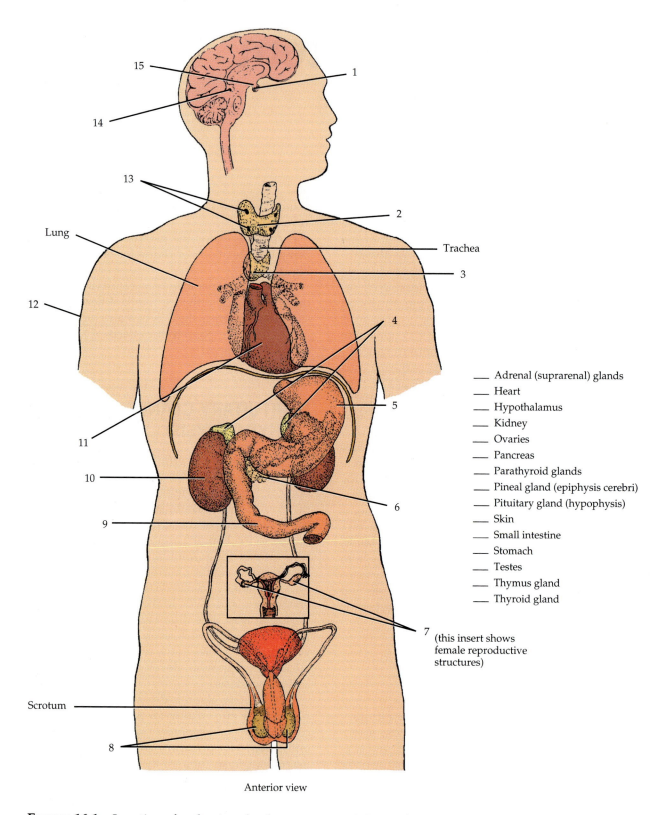

15
1
14
13
2
Lung
Trachea
3
12
4
5
11
10
9
6
8
7

____ Adrenal (suprarenal) glands
____ Heart
____ Hypothalamus
____ Kidney
____ Ovaries
____ Pancreas
____ Parathyroid glands
____ Pineal gland (epiphysis cerebri)
____ Pituitary gland (hypophysis)
____ Skin
____ Small intestine
____ Stomach
____ Testes
____ Thymus gland
____ Thyroid gland

(this insert shows
female reproductive
structures)

Scrotum

Anterior view

FIGURE 16.1   Location of endocrine glands, organs containing endocrine tissue, and associated structures.

whose cell bodies are in the hypothalamus. These terminals form the neural part of the pituitary gland. Between the anterior and posterior pituitary glands is a small, relatively avascular rudimentary zone, the *pars intermedia*, whose role in humans is unknown.

1. Histology of the Pituitary Gland

The anterior pituitary gland releases hormones that regulate a whole range of body activities, from growth to reproduction. However, the release of these hormones is stimulated by *releasing hormones* or inhibited by *inhibiting hormones* that are produced by neurosecretory cells in the hypothalamus of the brain. The hypothalamic hormones reach the anterior pituitary gland through a network of blood vessels.

When the anterior pituitary gland receives proper stimulation from the hypothalamus via releasing or inhibiting hormones, its glandular cells increase or decrease the secretion of any one of seven hormones. They are as follows:

1. *Somatotrophs* (*soma* = body; *tropos* = changing) produce *human growth hormone (hGH)*, which stimulates general body growth and certain aspects of metabolism.
2. *Lactotrophs* (*lact* = milk) synthesize *prolactin (PRL)*, which initiates milk production by suitably prepared mammary glands.
3. *Corticolipotrophs* (*cortex* = rind or bark) synthesize *adrenocorticotropic* (ad-rē-nō-kor'-ti-kō-TRŌ-pik) *hormone (ACTH)*, which stimulates the adrenal cortex to secrete its hormones and *melanocyte-stimulating hormone (MSH)*, which affects skin pigmentation.
4. *Thyrotrophs* (*thyreos* = shield) manufacture *thyroid-stimulating hormone (TSH)*, which controls the thyroid gland.
5. *Gonadotrophs* (*gonas* = seed) produce *follicle-stimulating hormone (FSH)* and *luteinizing hormone (LH)*. Together these hormones stimulate secretion of estrogens and progesterone by the ovaries and maturation of oocytes and secretion of testosterone and production of sperm in the testes.

Except for human growth hormone (hGH), melanocyte-stimulating hormone (MSH), and prolactin (PRL), all the secretions are referred to as *tropins* or *tropic hormones,* which means that their target organs are other endocrine glands.

Follicle-stimulating hormone (FSH) and luteinizing hormone (LH) are also called *gonadotropic* (gō-nad-ō-TRŌ-pik) *hormones* because they regulate the functions of the gonads (ovaries and testes). The gonads are the endocrine glands that produce sex hormones.

Examine a prepared slide of the anterior pituitary gland and identify as many types of cells as possible with the aid of your textbook or a histology textbook.

The posterior pituitary gland is not really an endocrine gland. Instead of synthesizing hormones, it stores and releases hormones synthesized by cells of the hypothalamus. The posterior pituitary gland consists of (1) cells called *pituicytes* (pi-TOO-i-sītz), which are similar in appearance to the neuroglia of the nervous system, and (2) axon terminations of secretory nerve cells of the hypothalamus. The cell bodies of the neurons, called *neurosecretory cells,* originate in nuclei in the hypothalamus. The fibers project from the hypothalamus, form the *supraoptico-hypophyseal* (soo'-pra-op'ti-kō-hī'-pō-FIZ-ē-al) or *hypothalamic-hypophyseal tract,* and terminate on blood capillaries in the posterior pituitary gland. The cell bodies of the neurosecretory cells produce the hormones *oxytocin* (ok'-sē-TŌ-sin), or *OT,* and *antidiuretic hormone (ADH).* These hormones are transported in the neuron fibers into the posterior pituitary gland and are stored in the axon terminals resting on the capillaries. When properly stimulated, the hypothalamus sends impulses over the neurons. The impulses cause release of hormones from the axon terminals into the blood.

Examine a prepared slide of the posterior pituitary gland under high power. Identify the pituicytes and axon terminations of neurosecretory cells with the aid of your textbook.

2. Hormones of the Pituitary Gland

Using your textbook as a reference, give the major functions for the hormones listed below.

a. HORMONES SECRETED BY THE ANTERIOR PITUITARY GLAND

Human growth hormone (hGH)   Also called *somatotropin* and *somatotropic hormone (STH)*

_____

_____

_____

Thyroid-stimulating hormone (TSH)   Also called

thyrotropin _____

_____

_____

_____

Adrenocorticotropic hormone (ACTH) _____

_____

_____

_____

Follicle-stimulating hormone (FSH)

In female: _____

_____

_____

In male: _____

_____

_____

_____

Luteinizing hormone (LH)

In female: _____

_____

_____

_____

In male: _____

_____

_____

Prolactin (PRL)   Also called *lactogenic hormone*

_____

_____

Melanocyte-stimulating hormone (MSH) _____

_____

_____

_____

_____

b. HORMONES STORED AND RELEASED BY THE POSTERIOR PITUITARY GLAND

Oxytocin (OT) _____

_____

_____

_____

Antidiuretic hormone (ADH) _____

_____

_____

_____

C. THYROID GLAND

The *thyroid gland* is a butterfly-shaped endocrine gland located just inferior to the larynx. The right and left *lateral lobes* lie lateral to the trachea and are connected by a mass of tissue called an *isthmus* (IS-mus) that lies anterior to the trachea just inferior to the cricoid cartilage. The *pyramidal lobe,* when present, extends superiorly from the isthmus.

1. Histology of the Thyroid Gland

Histologically, the thyroid gland consists of spherical sacs called *thyroid follicles.* The walls of each follicle consist of epithelial cells that reach the surface of the lumen of the follicle *(follicular cells).* In addition to the follicular cells the thyroid gland also contains *parafollicular cells* (also termed *C cells*). These cells may be within a follicle, where they do not reach the surface of the lumen, or they may be found between follicles in groups of three to four cells. Follicular cells synthesize the hormones *thyroxine* (thī-ROX-sēn) (also termed T_4 or *tetraiodothyronine)* and *triiodothyronine* (trī-ī-ōd-ō-THĪ-rō-nēn) *(T_3).* Together these hormones are referred to as the *thyroid hormones.* Approximately 90% of the hormone secreted by the follicular cells is thyroxine, while the remaining 10% is triiodothyronine. The functions of these two hormones are essentially the same, but they differ in rapidity and intensity of action. T_4 has a significantly longer life within the blood stream, but is also significantly weaker than triiodothyronine. The parafollicular cells produce the hormone *calcitonin* (kal-si-TŌ-nin) *(CT).* Each thyroid follicle is filled with a glycoprotein called *thyroglobulin (TBG),* which is also called *thyroid colloid.*

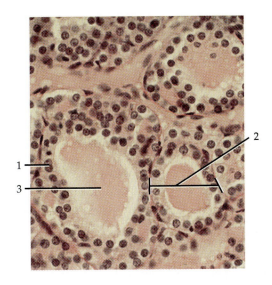

___ Epithelium of follicle

___ Thyroglobulin (TGB)

___ Thyroid follicle

FIGURE 16.2   Histology of thyroid gland (400×).

Examine a prepared slide of the thyroid gland under high power. Identify the thyroid follicles, epithelial cells forming the follicle, and thyroid colloid. Now label the photomicrograph in Figure 16.2.

2. Hormones of the Thyroid Gland

Using your textbook as a reference, give the major functions of the hormones listed below.

Thyroxine (T₄) and *triiodothyronine (T₃)*

_____

_____

_____

Calcitonin (CT) _____

_____

_____

_____

D. PARATHYROID GLANDS

Attached to the posterior surfaces of the lateral lobes of the thyroid gland are small, round masses of tissue called the *parathyroid (para = beside)*

glands. Typically two parathyroid glands, superior and inferior, are attached to each lateral thyroid lobe.

1. Histology of the Parathyroid Glands

Histologically, the parathyroid glands contain two kinds of epithelial cells. The first, a larger, more numerous cell called a *principal (chief) cell,* is believed to be the major synthesizer of *parathyroid hormone (PTH).* The second, a smaller cell called an oxyphil cell, may serve as a reserve cell for hormone synthesis.

Examine a prepared slide of the parathyroid glands under high power. Identify the principal and oxyphil cells. Now label the photomicrograph in Figure 16.3.

2. Hormone of the Parathyroid Glands

Using your textbook as a reference, give the major functions of the hormone listed below:

Parathyroid hormone (PTH)   Also called *para-*

thormone _____

_____

_____

_____

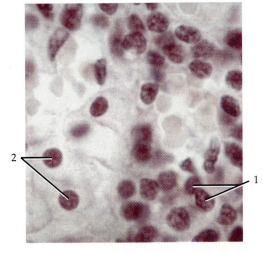

___ Oxyphil cells

___ Principal (chief) cells

FIGURE 16.3   Histology of parathyroid glands (320×).

E. ADRENAL (SUPRARENAL) GLANDS

The two *adrenal (suprarenal) glands* are superior to each kidney, and each is structurally and functionally differentiated into two separate endocrine glands: the superficial *adrenal cortex*, which forms the bulk of the gland in humans, and the deeper *adrenal medulla*. Covering the gland is a thick layer of fatty tissue and an outer thin fibrous connective tissue *capsule*.

1. Histology of the Adrenal Cortex

Histologically, the adrenal cortex is subdivided into three zones, each of which has a different cellular arrangement and secretes different steroid hormones. The outer zone, immediately deep to the capsule, is called the *zona glomerulosa* (*glomerulus* = little ball). Its cells, arranged in arched loops or round balls, primarily secrete a group of hormones called *mineralocorticoids* (min'-er-al-ō-KOR-ti-koyds). The major mineralocorticoid produced by this region is *aldosterone.*

The intermediate zone of the adrenal cortex is the *zona fasciculata* (*fasciculus* = little bundle). This zone, which is the widest of the three, consists of cells arranged in long, relatively straight cords. The zona fasciculata secretes mainly *glucocorticoids* (gloo'-ko-KOR-ti-koyds). The zona fasciculata secretes three glucocorticoids, 95% of which is *cortisol* (also known as *hydrocortisone*). The remaining glucocorticoids synthesized include a small amount of *corticosterone* and a minute amount of *cortisone.*

The deepest zone of the adrenal cortex is termed the *zona reticularis* (*reticular* = net). It is composed of frequently branching cords of cells. This zone synthesizes minute amounts of male hormones (*androgens*).

Examine a prepared slide of the adrenal cortex under high power. Identify the capsule, zona glomerulosa, zona fasciculata, and zona reticularis. Now label Figure 16.4.

2. Hormones of the Adrenal Cortex

Using your textbook as a reference, give the major functions of the hormones listed below.

Mineralocorticoids (mainly *aldosterone*)

_____

_____

_____

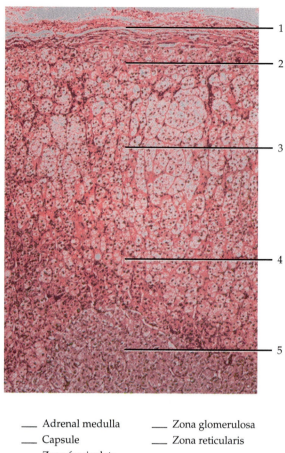

____ Adrenal medulla	____ Zona glomerulosa
____ Capsule	____ Zona reticularis
____ Zona fasciculata	

FIGURE 16.4   Histology of adrenal (suprarenal) glands (10×).

Glucocorticoids (mainly *cortisol*) _____

_____

_____

_____

Androgens

In male: _____

_____

In female: _____

_____

3. Histology of the Adrenal Medulla

The adrenal medulla consists of hormone-producing cells called *chromaffin* (krō-MAF-in; *chroma* = color; *affinia* = affinity for) *cells.* These cells develop from the same embryonic tissue as the postganglionic cells of the sympathetic division of the nervous system. They are directly innervated by preganglionic cells of the sympathetic division of the autonomic nervous system and may be regarded as postganglionic cells that are specialized to secrete hormones. Secretion of hormones from the chromaffin cells is directly controlled by the sympathetic division of the autonomic nervous system, and innervation by the preganglionic fibers allows the gland to respond extremely rapidly to a stimulus. The adrenal medulla secretes the hormones *epinephrine* and *norepinephrine (NE).*

Examine a prepared slide of the adrenal medulla under high power. Identify the chromaffin cells. Now locate the cells in Figure 16.4.

4. Hormones of the Adrenal Medulla

Using your textbook as a reference, give the major functions of the hormones listed below.

Epinephrine and *norepinephrine (NE)* _____

_____

_____

_____

F. PANCREAS

The *pancreas* is classified as both an endocrine and an exocrine gland. Thus, it is referred to as a *heterocrine gland.* We shall discuss only its endocrine functions now. The pancreas is a flattened organ located posterior and slightly inferior to the stomach. The adult pancreas consists of a head, body, and tail.

1. Histology of the Pancreas

The endocrine portion of the pancreas consists of clusters of cells called *pancreatic islets (islets of Langerhans).* They contain four kinds of cells: (1) *alpha cells,* which have more distinguishable plasma membranes, are usually peripheral in the islet, and secrete the hormone *glucagon;* (2) *beta cells,* which generally lie deeper within the islet and secrete the hormone *insulin;* (3) *delta cells,* which secrete *somatostatin;* and (4) *F cells,* which secrete pancreatic polypeptide. The pancreatic islets are surrounded by blood capillaries and by the cells called *acini* that form the exocrine part of the gland.

Examine a prepared slide of the pancreas under high power. Identify the alpha cells, beta cells, and acini (clusters of cells that secrete digestive enzymes) around the pancreatic islets. Now label Figure 16.5.

2. Hormones of the Pancreas

Using your textbook as a reference, give the major functions of the hormones listed on the next page:

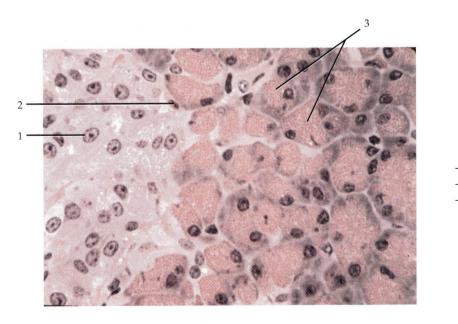

___ Acini

___ Alpha cell

___ Beta cell

FIGURE 16.5   Histology of pancreas (500×).

Glucagon _____

_____

_____

Insulin _____

_____

_____

Somatostatin _____

_____

_____

Pancreatic polypeptide _____

_____

_____

G. TESTES

The *testes* (male gonads) are paired oval glands enclosed by the scrotum. They are partially covered by a serous membrane called the *tunica* (*tunica* = sheath) *vaginalis,* which is derived from the peritoneum and forms during descent of the testes. Internal to the tunica vaginalis is a dense layer of white fibrous tissue, the *tunica albuginea* (al'-byoo-JIN-ē-a; *albus* = white), which extends inward and divides each testis into a series of internal compartments called *lobules.* Each of the 200–300 lobules contains one to three tightly coiled tubules, the convoluted *seminiferous* (*semen* = seed; *ferre* = to carry) *tubules,* which produce sperm by a process called *spermatogenesis* (sper'-ma-tō-JEN-e-sis).

1. Histology of the Testes

Spermatogenic cells are sperm-forming cells in various stages that undergo mitosis and differentiation to eventually produce sperm. Together with supporting cells, they line the seminiferous tubules. The most immature spermatogenic cells are called *spermatogonia* (sper'-ma-tō-GŌ-nē-a; *sperm* = seed; *gonium* = generation or offspring; the singular is *spermatogonium*). They lie next to the basement membrane. Toward the lumen of the tubules are layers of progressively more mature cells. In order of advancing maturity, these are *primary spermatocytes* (SPER-ma-tō-sīts), *secondary spermatocytes,* and *spermatids.* By the time a *sperm cell* or *spermatozoon* (sper'-ma-tō-ZŌ-on; *zoon* = life; the plural is *sperm* or *spermatozoa*), has nearly reached maturity, it is released into the lumen of the tubule and begins to move out of the rete testis.

Embedded among the spermatogenic cells in the tubules are large *sustentacular* (sus'-ten-TAK-yoo-lar; *sustentare* = to support), or *Sertoli, cells*, that extend from the basement membrane to the lumen of the tubule. Sustentacular cells support and protect developing spermatogenic cells; nourish spermatocytes, spermatids, and sperm; phagocytize excess spermatid cytoplasm as development proceeds; and mediate the effects of testosterone and follicle stimulating hormone (FSH). Sustentacular cells also control movements of spermatogenic cells and the release of sperm into the lumen of the seminiferous tubule. They produce fluid for sperm transport and secrete the hormone inhibin, which helps regulate sperm production by inhibiting the secretion of FSH.

In the spaces between adjacent seminiferous tubules are clusters of cells called *interstitial endocrinocytes,* or *Leydig cells.* These cells secrete testosterone, the most important androgen (male sex hormone).

Examine a prepared slide of the testes under high power and identify all of the structures listed. Now label Figure 16.6.

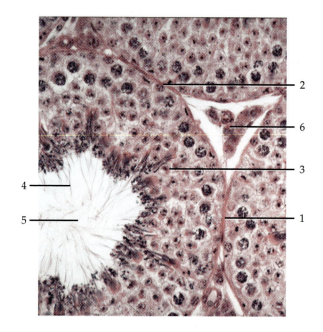

____ Basement membrane

____ Interstitial endocrinocyte
(interstitial cell of Leydig)

____ Lumen of seminiferous tubule

____ Spermatid

____ Spermatogonium

____ Sperm cell

FIGURE 16.6   Histology of testes.

2. Hormones of the Testes

Using your textbook as a reference, give the major functions of the hormones listed below.

Testosterone _____

_____

_____

Inhibin _____

_____

_____

H. OVARIES

The *ovaries* (*ovarium* = egg receptacle), or female gonads, are paired glands resembling unshelled almonds in size and shape. They are positioned in the superior pelvic cavity, one on each side of the uterus, and are maintained in position by a series of ligaments. They are (1) attached to the **broad ligament** of the uterus, which is itself part of the parietal peritoneum, by a fold of peritoneum called the **mesovarium;** (2) anchored to the uterus by the **ovarian ligament;** and (3) attached to the pelvic wall by the **suspensory ligament.** Each ovary also contains a *hilus,* which is the point of entrance for blood vessels and nerves.

1. Histology of the Ovaries

Histologically, each ovary consists of the following parts:

1. *Germinal epithelium* A layer of simple epithelium (low cuboidal or squamous) that covers the free surface of the ovary and is continuous with the mesothelium that covers the mesovarium. The term *germinal epithelium* is a misnomer because it does not give rise to oocytes, although at one time it was believed that it did.
2. *Tunica albuginea* A whitish capsule of dense, irregular connective tissue immediately deep to the germinal epithelium.
3. *Ovarian cortex* A region just deep to the tunica albuginea that consist of dense connective tissue and contains ovarian tissue (described shortly).
4. *Ovarian medulla* A region deep to the ovarian cortex that consists of loose connective tissue and contains blood vessels, lymphatics, and nerves.

5. *Ovarian follicles* (*folliculus* = little bag) Lie in the cortex and consist of *oocytes* (immature ova) in various stages of development and their surrounding cells. When the surrounding cells form a single layer, they are called *follicular cells.* Later in development, when they form several layers, they are referred to as *granulosa cells.* The surrounding cells nourish the developing oocyte and begin to secrete estrogens as the follicle grows larger. Ovarian follicles undergo a series of changes prior to ovulation, progressing through several distinct stages. The most numerous and peripherally arranged follicles are termed *primordial follicles.* If a primordial follicle progresses to ovulation (release of a mature ova), it will sequentially transform into a *primary (preantral) follicle,* then a *secondary (antral) follicle,* and finally a *mature (Graafian) follicle.*
6. *Mature (Graafian) follicle* A large, fluid-filled follicle that soon will rupture and expel a secondary oocyte, a process called *ovulation.*
7. *Corpus luteum* (= yellow body) Contains the remnants of an ovulated mature follicle. The corpus luteum produces progesterone, estrogens, relaxin, and inhibin until it degenerates and turns into fibrous tissue called a *corpus albicans* (= white body).

Examine a prepared slide of the ovary under high power. (You may need to examine more than one slide to see all of the structures listed.) Label Figure 16.7.

2. Hormones of the Ovaries

Using your textbook as a reference, give the major functions of the hormones listed below.

Estrogens _____

_____

_____

_____

Progesterone _____

_____

_____

_____

Relaxin _____

_____

_____

_____

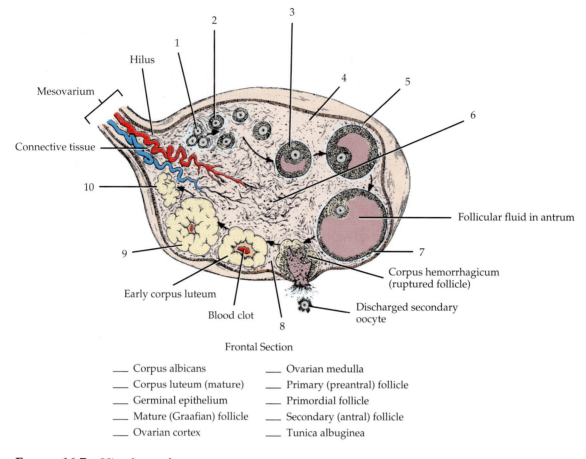

FIGURE 16.7   Histology of ovary.

Labels on diagram: Hilus, Mesovarium, Connective tissue, 10, 9, Early corpus luteum, Blood clot, 8, 1, 2, 3, 4, 5, 6, Follicular fluid in antrum, 7, Corpus hemorrhagicum (ruptured follicle), Discharged secondary oocyte

Frontal Section

___ Corpus albicans
___ Corpus luteum (mature)
___ Germinal epithelium
___ Mature (Graafian) follicle
___ Ovarian cortex

___ Ovarian medulla
___ Primary (preantral) follicle
___ Primordial follicle
___ Secondary (antral) follicle
___ Tunica albuginea

Inhibin _____

_____

_____

_____

I. PINEAL GLAND (EPIPHYSIS CEREBRI)

The cone-shaped endocrine gland attached to the roof of the third ventricle is known as the *pineal* (PĪN-ē-al) *gland (epiphysis cerebri).*

1. Histology of the Pineal Gland

The gland is covered by a capsule formed by the pia mater and consists of masses of neuroglial cells and secretory cells called *pinealocytes* (pin-ē-AL-ō-sīts). Around the cells are scattered preganglionic sympathetic fibers. The pineal gland starts to calcify at about the time of puberty. Such calcium deposits are called *brain sand.* Contrary to a once widely held belief, no evidence suggests that the pineal gland atrophies with age and that the presence of brain sand indicates atrophy. Rather, brain sand may even denote increased secretory activity.

The physiology of the pineal gland is still somewhat obscure. The gland secretes *melatonin.*

2. Hormone of the Pineal Gland

Using your textbook as a reference, give the major functions of the hormone listed below.

Melatonin _____

_____

_____

J. THYMUS GLAND

The *thymus gland* is a bilobed lymphatic gland located in the upper mediastinum posterior to the sternum and between the lungs. The gland is conspicuous in infants and, during puberty, reaches maximum size.

1. Histology of Thymus Gland

After puberty, thymic tissue, which consists primarily of *lymphocytes,* is replaced by fat. By the time a person reaches maturity, the gland has atrophied. Lymphoid tissue of the body consists primarily of lymphocytes that may be distinguished into two kinds: B cells and T cells. Both are derived originally in the embryo from lymphocytic stem cells in red bone marrow. Before migrating to their positions in lymphoid tissue, the descendants of the stem cells follow two distinct pathways. About half of them migrate to the thymus gland, where they are processed to become thymus-dependent lymphocytes, or *T cells.* The thymus gland confers on them the ability to destroy antigens (foreign microbes and substances). The remaining stem cells are processed in some as yet undetermined area of the body, possibly the fetal liver and spleen, and are known as *B cells.* Hormones produced by the thymus gland are *thymosin, thymic humoral factor (THF), thymic factor (TF),* and *thymopoietin.*

2. Hormones of the Thymus Gland

Using your textbook as a reference, write the major functions of the hormones listed below:

Thymosin, thymic humoral factor (THF), thymic

factor (TF), and thymopoietin _____

_____

_____

K. OTHER ENDOCRINE TISSUES

Body tissues other than endocrine glands also secrete hormones. The gastrointestinal tract synthesizes several hormones that regulate digestion in the stomach and small intestine. Among these hormones are *gastrin, secretin, cholecystokinin (CCK),* and *glucose-dependent insulnotropic peptide.*

The placenta produces *human chorionic gonado-tropin (hCG), estrogens, progesterone (PROG), re-laxin,* and *human chorionic somatomammo-tropin (hCS),* all of which are related to pregnancy.

When the kidneys (and liver, to a lesser extent) become hypoxic (subject to below normal levels of oxygen), they release an enzyme called *renal erythropoietic factor.* This is secreted into the blood where it acts on a plasma protein produced in the liver to form a hormone called *erythropoietin* (ē-rith'-rō-POY-ē-tin), or *EPO,* which stimulates red bone marrow to produce more red blood cells and hemoglobin. This ultimately reverses the original stimulus (hypoxia).

Vitamin D, produced by the skin, liver, and kidneys in the presence of sunlight, is converted to its active hormone, *calcitriol,* in the kidneys and liver.

The atria of the heart secrete a hormone called *atrial natriuretic peptide (ANP),* released in response to increased blood volume.

Adipose tissue secretes *leptin,* which suppresses the appetite.

Using your textbook as a reference, write the major functions of the hormones listed below:

Gastrin _____

_____

Secretin _____

_____

Cholecystokinin (CCK) _____

_____

Glucose-dependent insulinotropic peptide _____

_____

Human chorionic gonadotropin (hCG) _____

_____

Human chorionic somatomammotropin (hCS) __

_____

Erythropoietin (EPO) _____

_____

Calcitriol _____

_____

Atrial natriuretic peptide (ANP) _____

_____

Leptin _____

_____

L. PHYSIOLOGY OF THE ENDOCRINE SYSTEM

PROCEDURE USING PHYSIOFINDER

PHYSIOFINDER is an interactive computer program that permits laboratory simulations that do not involve the use of animals or advanced or expensive laboratory equipment. It permits students to perform experiments, analyze data, draw conclusions, and form hypotheses on the basis of collected data. The program is available from Addison-Wesley Longman, Inc., Customer Service (1-800-922-0579).

For activities related to hyperinsulinemia in fish, as well as other activities related to plasma glucose levels, select the appropriate experiment from PHYSIOFINDER Module 11—Control of Plasma Glucose.

ANSWER THE LABORATORY REPORT QUESTIONS AT THE END OF THE EXERCISE.

Endocrine System 16

Student _____   **Date** _____

Laboratory Section _____   **Score/Grade** _____

PART 1. Multiple Choice

_____ 1. Somatotrophs, gonadotrophs, and corticolipotrophs are associated with the (a) thyroid gland (b) anterior pituitary gland (c) parathyroid glands (d) adrenal glands

_____ 2. The posterior pituitary gland is *not* an endocrine gland because it (a) has a rich blood supply (b) is not near the brain (c) does not make hormones (d) contains ducts

_____ 3. Which hormone assumes a role in the development and discharge of a secondary oocyte? (a) hGH (b) TSH (c) LH (d) PRL

_____ 4. The endocrine gland that is probably malfunctioning if a person has a high metabolic rate is the (a) thymus gland (b) posterior pituitary gland (c) anterior pituitary gland (d) thyroid gland

_____ 5. The antagonistic hormones that regulate blood calcium level are (a) hGH-TSH (b) insulin-glucagon (c) aldosterone-cortisone (d) CT-PTH

_____ 6. The endocrine gland that develops from the sympathetic nervous system is the (a) adrenal medulla (b) pancreas (c) thyroid gland (d) anterior pituitary gland

_____ 7. The hormone that aids in sodium conservation and potassium excretion is (a) hydrocortisone (b) CT (c) ADH (d) aldosterone

_____ 8. Which of the following hormones is sympathomimetic? (a) insulin (b) oxytocin (OT) (c) epinephrine (d) testosterone

_____ 9. Which hormone lowers blood sugar level? (a) glucagon (b) melatonin (c) insulin (d) cortisone

_____ 10. The endocrine gland that may assume a role in regulation of the menstrual cycle is the (a) pineal gland (b) thymus gland (c) thyroid gland (d) adrenal gland

PART 2. Completion

11. The pituitary gland is attached to the hypothalamus by a stalklike structure called the

_____.

12. A hormone that acts on another endocrine gland and causes that gland to secrete its own hormones is

called a _____.

13. _____ cells of the anterior pituitary gland synthesize ACTH.

14. The hormone that helps cause contraction of the smooth muscle of the pregnant uterus is

_____.

15. Histologically, the spherical sacs that compose the thyroid gland are called thyroid

_____.

16. The thyroid hormones associated with metabolism are triiodothyronine and _____.

17. Principal and oxyphil cells are associated with the _____ gland.

18. The zona glomerulosa of the adrenal cortex secretes a group of hormones called

_____.

19. The hormones that promote normal metabolism, provide resistance to stress, and function as

anti-inflammatories are _____.

20. The pancreatic hormone that raises blood sugar level is _____.

21. In spermatogenesis, the most immature cells near the basement membrane are called

_____.

22. Cells within the testes that secrete testosterone are known as _____.

23. The ovaries are attached to the uterus by means of the _____ ligament.

24. The female hormones that help cause the development of secondary sex characteristics are called

_____.

25. The endocrine gland that assumes a direct function in the proliferation and maturation of T cells is the

_____.

26. Any hormone that regulates the functions of the gonads is classified as a(n) _____ hormone.

27. The hormone that is stored in the posterior pituitary gland that prevents excessive urine production

is _____.

28. The region of the adrenal cortex that synthesizes androgens is the _____.

29. The hormone-producing cells of the adrenal medulla are called _____ cells.

30. Together, alpha cells, beta cells, delta cells, and F cells constitute the _____.

31. Regulating hormones are produced by the _____ and reach the anterior pituitary gland by a network of blood vessels.

32. FSH and LH are produced by _____ cells of the anterior pituitary gland.

33. The structure in an ovary that produces progesterone, estrogens, relaxin, and inhibin is the

_____.

34. Calcium deposits in the pineal gland are referred to as _____.

35. A gland that is both an exocrine and endocrine gland is known as a _____ gland.

Blood

17

The blood, heart, and blood vessels constitute the *cardiovascular* (*cardio* = heart; *vascular* = blood or blood vessels) *system.* In this exercise you will examine the characteristics of *blood,* a connective tissue also known as *vascular tissue.*

CAUTION! *Please reread Section A, "General Safety Precautions and Procedures," and Section B, "Precautions Related to Working with Blood, Blood Products, or Other Body Fluids," on pages xi–xii, at the beginning of the laboratory manual, before you begin any of the following experiments. Read the experiments before you perform them, to be sure that you understand all the procedures and safety precautions.*

When working with whole blood, wear tight-fitting surgical gloves and safety goggles. Avoid any kind of contact with an open sore, cut, or wound.

After you have completed your experiments, place all glassware in a fresh solution of household bleach or other comparable disinfectant, wash the laboratory tabletop with a fresh solution of household bleach or comparable disinfectant, and dispose of your gloves in the appropriate biohazard container provided by your instructor.

A. COMPONENTS OF BLOOD

Blood is a liquid connective tissue that is composed of two portions: (1) *plasma,* a liquid that contains dissolved substances, and (2) *formed elements,* cells and cell fragments suspended in the plasma. In clinical practice, the most common classification of the formed elements of the blood is the following:

Erythrocytes (e-RITH-rō-sīts), or *red blood cells*
Leukocytes (LOO-kō-sīts), or *white blood cells*
Granular leukocytes (granulocytes)
   Neutrophils
   Eosinophils
   Basophils
Agranular leukocytes (agranulocytes)
   Lymphocytes (T cells, B cells, and natural killer cells)
   Monocytes
Platelets (thrombocytes)

The origin and subsequent development of these formed elements can be seen in Figure 17.1. The process by which the formed elements of blood develop is called *hemopoiesis* (hē-mō-poy-Ē-sis; *hemo* = blood; *poiem* = to make), or *hematopoiesis* (hē-ma-tō-poy-Ē-sis), and the process by which erythrocytes are formed is called *erythropoiesis* (e-rith'-rō-poy-Ē-sis). The immature cells that are eventually capable of developing into mature blood cells are called *hematopoietic stem cells* (see Figure 17.1). Mature blood cells are constantly being replaced, so special cells called *reticuloendothelial cells* (fixed macrophages that line liver sinusoids) have the responsibility of clearing away the dead, disintegrating cell bodies so that small blood vessels are not clogged.

The shapes of the nuclei, staining characteristics, and color of cytoplasmic granules are all useful in differentiation and identification of the various white blood cells. Red blood cells are biconcave discs without nuclei and can be identified easily.

B. PLASMA

When the formed elements are removed from blood, a straw-colored liquid called *plasma* is left. This liquid consists of about 91.5% water and about 8.5% solutes. Among the solutes are proteins (albumins, globulins, and fibrinogen), nonprotein nitrogen (NPN) substances (urea, uric acid, and creatine), foods (amino acids, glucose, fatty acids, glycerides, and glycerol), regulatory substances (enzymes and hormones), gases (oxygen and carbon dioxide), and electrolytes (Na^+, K^+, Ca^{2+}, Mg^{2+}, Cl^-, HCO_3^-, SO_4^{2-}, and HPO_4^{3-}).

1. Physical Characteristics

CAUTION! *Please reread Section A, "General Safety Precautions and Procedures," and Section B, "Precautions Related to Working with Blood, Blood Products, or Other Body Fluids," on pages xi–xii, at*

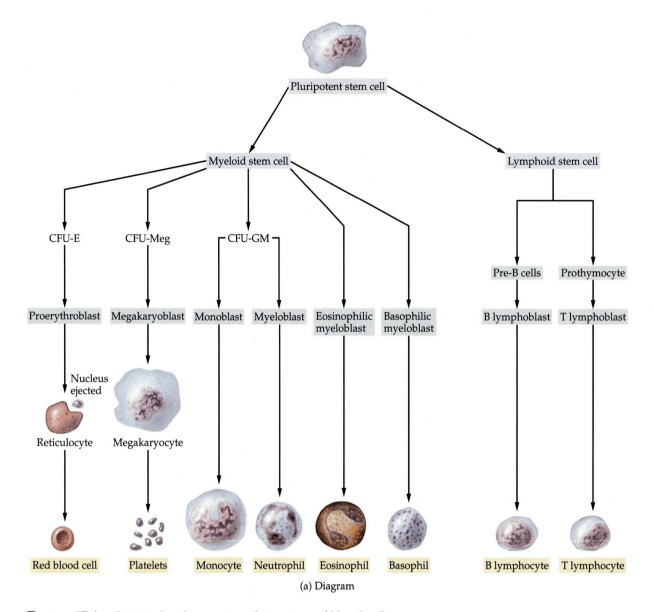

FIGURE 17.1   Origin, development, and structure of blood cells.

the beginning of the laboratory manual before you begin any of the following experiments. Read the experiments before you perform them, to be sure that you understand all the procedures and safety precautions.

When working with whole blood, wear tight-fitting surgical gloves and safety goggles. Avoid any kind of contact with an open sore, cut, or wound.

PROCEDURE

Note: Obtain the plasma or whole blood for this procedure from a clinical laboratory where it has been tested

and certified as noninfectious[1] or from a mammal (other than a human).

1. Obtain about 5 ml of plasma as follows.
2. Centrifuge the whole blood obtained from a clinical laboratory where it has been tested and certified as noninfectious or from a mammal (other than a human). After centrifugation, the

---

[1]Whole blood that has been tested for syphilis; hepatitis A, B, and C; and HIV is available from Carolina Biological Supply Company.

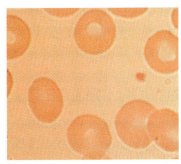

Erythrocytes and platelet (1000×)

Neutrophil (1000×)   Eosinophil (1000×)   Basophil (1000×)

 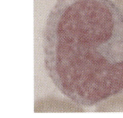

Lymphocyte (1000×)   Monocyte (1000×)

(b) Photomicrographs

FIGURE 17.1 *(Continued)*

bottom of the test tube will contain an anticoagulant (usually EDTA). Right above this will be a layer of erythrocytes, followed by a layer of leukocytes and platelets (buffy coat). The top layer is plasma. Visual inspection of these layers will give an indication of the relative proportions of the various components of blood.

3. Remove 5 ml of plasma using a disposable sterile Pasteur pipette attached to a bulb and place the plasma in a disposable test tube. Note its color. Test the pH of the plasma by dipping the end of a length of pH paper into the plasma and comparing the paper color with the color scale on the pH container. Record the pH in Section B.1 of the LABORATORY REPORT RESULTS at the end of the exercise.

4. Hold the test tube up to a source of natural light and note the color and transparency of the plasma. Record the color and transparency in Section B.1 of the LABORATORY REPORT RESULTS at the end of the exercise.

5. *Dispose of the pipettes and test tubes in the appropriate biohazard container provided by your instructor. Dispose of the pH paper and plasma as per your instructor's directions.*

2. Chemical Constituents

CAUTION! *Please reread Section A, "General Safety Precautions and Procedures," and Section B, "Precautions Related to Working with Blood, Blood Products, or Other Body Fluids," on pages xi–xii, at the beginning of the laboratory manual, before you begin any of the following experiments. Read the experiments before you perform them, to be sure that you understand all the procedures and safety precautions.*

When working with whole blood, wear tight-fitting surgical gloves and safety goggles. Avoid any kind of contact with an open sore, cut, or wound.

PROCEDURE

Note: *Obtain the plasma or whole blood for this procedure from a clinical laboratory where it has been tested and certified as noninfectious or from a mammal (other than a human).*

1. Using a medicine dropper, add 10 ml of distilled water to 5 ml of plasma in a Pyrex test tube.

CAUTION! *Place the test tube in a test-tube rack because it will become too hot to handle.*

2. With forceps, add one Clinitest tablet.

CAUTION! *The concentrated sodium hydroxide in the tablet generates enough heat to make the liquid in the test tube boil. Make sure that the mouth of the test tube is pointed away from you and all other persons in the area.*

3. The color of the solution is graded as follows:

Color	Results
Blue	Negative
Greenish-yellow	1 + (0.5 g/100 ml)
Olive green	2 + (1 g/100 ml)
Orange-yellow	3 + (1.5 g/100 ml)
Brick red (with precipitate)	4 + (more than 2 g/100 ml)

4. Fifteen seconds after boiling has stopped, shake the test tube *gently* and evaluate the color according to the test table on page 321.

CAUTION! *Make sure that the mouth of the test tube is pointed away from you and others.*

A green, yellow, orange, or red color indicates the presence of glucose. Record your results in Section B.2 of the LABORATORY REPORT RESULTS at the end of the exercise.

5. Test the contents of the filter paper for the presence of protein. If protein is present, it will be coagulated by the acetic acid and filtered out of the plasma. Place the contents of the filter paper in a test tube and add 3 ml of water and 3 ml of Biuret reagent. A violet or purple color indicates the presence of protein. Record your results in Section B.2 of the LABORATORY REPORT RESULTS at the end of the exercise.

C. ERYTHROCYTES

Erythrocytes (red blood cells, or RBCs) are biconcave in appearance, have no nucleus, and can neither reproduce nor carry on extensive metabolic activities (Figure 17.1). The interior of the cell contains a red pigment called *hemoglobin,* which is responsible for the red color of blood. The heme portion of hemoglobin combines with oxygen and, to a lesser extent, carbon dioxide and transports them through the blood vessels. An average red blood cell has a life span of about 120 days. A healthy male has about 5.4 million red blood cells per cubic millimeter (mm^3), or per deciliter (dl), of blood; a healthy female, about 4.8 million. Erythropoiesis and red blood cell destruction normally proceed at the same pace. A diagnostic test that informs the physician about the rate of erythropoiesis is the *reticulocyte* (re-TIK-yoo-lō-sīt) *count. Reticulocytes* (see Figure 17.1) are precursor cells of mature red blood cells. Normally, a reticulocyte count is 0.5 to 1.5%. The reticulocyte count is an important diagnostic tool because it is a relatively accurate predictor of the status of red blood cell production in red bone marrow. Normally, red bone marrow replaces about 1% of the adult red blood cells each day. A decreased reticulocyte count (reticulocytopenia) is seen in aplastic anemia and in conditions in which the red bone marrow is not producing red blood cells. An increase in reticulocytes (reticulocytosis) is found in acute and chronic blood loss and certain kinds of anemias such as iron-deficiency anemia.

Note: Although many blood tests in this exercise are sufficient for laboratory demonstration, they are not necessarily clinically accurate.

D. RED BLOOD CELL TESTS

CAUTION! *reread Section A, "General Safety Precautions and Procedures," and Section B, "Precautions Related to Working with Blood, Blood Products, or Other Body Fluids," on pages xi–xii, at the beginning of the laboratory manual before you begin any of the following experiments. Read the experiments before you perform them, to be sure that you understand all the procedures and safety precautions.*

When working with whole blood, wear tight-fitting surgical gloves and safety goggles. Avoid any kind of contact with an open sore, cut, or wound.

1. Source of Blood

Blood should be obtained from a clinical laboratory where it has been tested and certified as noninfectious or from a mammal (other than a human).

Whole blood that has been tested for syphilis; hepatitis A, B, and C; and HIV is available from Carolina Biological Supply Company. This blood can be used for blood grouping and typing, blood glucose, hematocrit, hemoglobin, red blood cell count, white blood cell count, differential white blood cell count, platelet count, sedimentation rate, osmotic fragility studies, and various plasma chemistries. Carolina Biological Supply Company also provides aseptic red blood cells that can be used for blood grouping and typing, blood glucose, hematocrit, hemoglobin, red blood cell count, and osmotic fragility studies. These blood cells are suspended in a modified Alsevere's solution, to which several antibiotics have been added. The cells have been tested for syphilis; hepatitis A, B, and C; and HIV.

2. Filling of Hemocytometer (Counting Chamber)

The procedure for filling the hemocytometer is as follows. Please read it carefully *before* doing 3. Red Blood Cell Count.

PROCEDURE

Note: The blood used for this procedure should be obtained from a clinical laboratory where it has been tested and certified as noninfectious or from a mammal (other than a human).

1. Obtain a hemocytometer and cover slip (Figure 17.2).

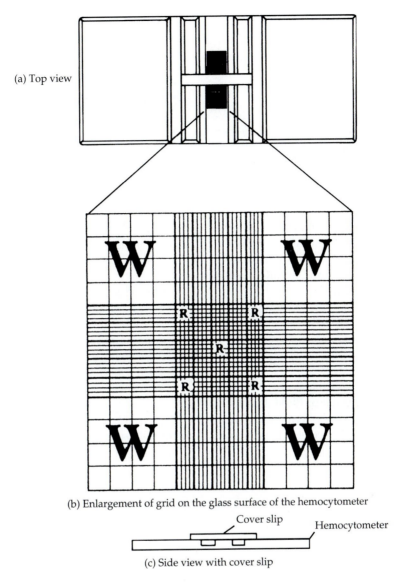

(a) Top view

(b) Enlargement of grid on the glass surface of the hemocytometer

Cover slip   Hemocytometer

(c) Side view with cover slip

FIGURE 17.2 Various parts of a hemocytometer (counting chamber). In (b), the areas used for red blood cell counts are indicated with an R and those for white blood cell counts are indicated with a W.

2. Clean the hemocytometer thoroughly and carefully with alcohol.
3. Place the cover slip on the hemocytometer.
4. Using a Unopette Reservoir System® pipette filled with the proper dilution of blood, place the tip of the pipette on the polished surface of the hemocytometer next to the edge of the cover slip. Although we recommend use of this system, your instructor might wish to use an alternate procedure. *If a different system is used for*

pipetting and diluting, use rubber suction bulbs or pipette pumps. Do not pipette by mouth.

5. Deposit a small drop of diluted blood by squeezing the sides of the reservoir, but do not leave the tip of the pipette in contact with the hemocytometer for more than an instant because this will cause the chamber to overfill. The diluted blood must not overflow the moat; overfilling the moat results in an inaccurate cell count. A properly filled hemocytometer has a

blood specimen only within the space between the cover glass and counting area.

3. Red Blood Cell Count

The purpose of a red blood cell count is to determine the number of circulating red blood cells per cubic millimeter (mm^3), or per deciliter (dl), of blood. Red blood cells carry oxygen to all tissues; thus a drastic reduction in the red cell count will cause immediate reduction in available oxygen.

A decrease in red blood cells can result from a variety of conditions, including impaired cell production, increased cell destruction, and acute blood loss. When the red blood cell count is increased above normal limits, the condition is called ***polycythemia*** (pol'-ē-sī-THĒ-mē-a).

The procedure we will use for determining the number of red blood cells per cubic millimeter (mm^3) of blood is as follows.

PROCEDURE

WARD's Natural Science Establishment, Inc., has made available WARD'S *Simulated Blood*. It is a solution that contains microcomponents that simulate red blood cells, white blood cells, and platelets.

The microcomponents are similar in relative proportion to those found in human blood and can be observed under a microscope without staining.

1. Obtain a Simulated Blood Activity kit.
2. Follow the instructions to perform the red blood cell count.
3. Complete the portion of the Data Table related to red blood cells.
4. Answer the questions related to red blood cells.

ALTERNATE PROCEDURE

Note: The blood used for this procedure should be obtained from a clinical laboratory where it has been tested and certified as noninfectious or from a mammal (other than a human).

1. Obtain a Unopette Reservoir System® for red blood cell determination (the color of the reservoir's bottom surface is red). Identify the reservoir chamber, diluent fluid inside the chamber, pipette, and protective shield for the pipette (Figure 17.3). The diluent fluid contains isotonic saline and sodium azide, an antibacterial agent.

CAUTION! *Do not ingest this fluid; it is for in vitro diagnostic use only.*

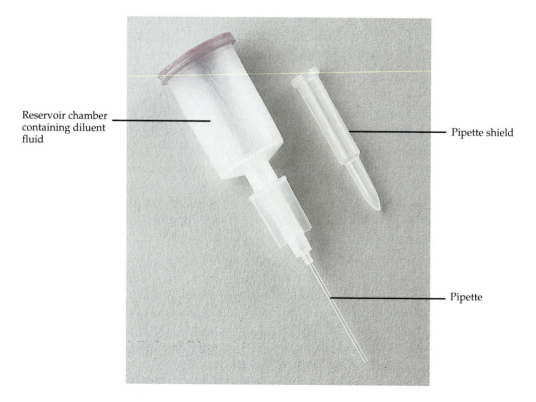

Reservoir chamber containing diluent fluid

Pipette shield

Pipette

FIGURE 17.3   Photograph of the Unopette Reservoir System®.

2. Hold the reservoir on a flat surface in one hand and grasp the pipette assembly in the other hand.

3. Push the tip of the pipette shield firmly through the diaphragm in the neck of the reservoir (Figure 17.4a).

4. Pull out the assembly unit and with a twist remove the protective shield from the pipette assembly.

5. Hold the pipette *almost* horizontally and touch the tip of it to a forming drop of blood that has been placed on a microscope slide (Figure 17.4b). The pipette will fill by capillary action and, when the blood reaches the end of the capillary bore in the neck of the pipette, the filling action will stop.

6. Carefully wipe any excess blood from the tip of the pipette using a Kimwipe or similar wiping tissue, *making certain that no sample is removed from the capillary bore.*

7. Squeeze the reservoir *slightly* to expel a small amount of air.

CAUTION! *Do not expel any liquid. If the reservoir is squeezed too hard, the specimen may be expelled through the overflow chamber, resulting in contamination of the fingers. Reagent contains sodium azide, which is extremely toxic and yields explosive products in metal sinks or pipes. Azide compounds should be diluted with running water before being discarded and disposed of per the directions of your instructor.*

While still maintaining pressure on the reservoir, cover the opening of the overflow chamber of the pipette with your index finger and push the pipette securely into the reservoir neck (Figure 17.4c).

8. Release the pressure on the reservoir and remove your finger from the pipette opening. This will draw the blood into the diluent fluid.

9. Mix the contents of the reservoir chamber by squeezing the reservoir gently several times. It is important to *squeeze gently* so that the diluent fluid is not forced out of the chamber. In

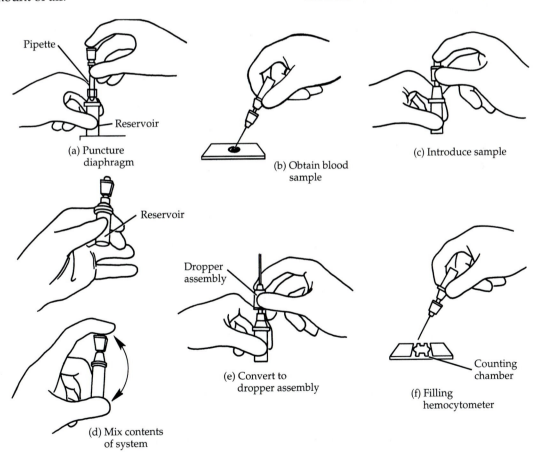

FIGURE 17.4   Preparation of the Unopette Reservoir System®. (Modified from *RBC Determination for Manual Methods* and *WBC Determination for Manual Methods,* Becton-Dickinson, Division of Becton, Dickinson and Company.)

addition to squeezing the reservoir chamber, invert it *gently* several times (Figure 17.4d).

10. Remove the pipette from the reservoir chamber, reverse its position, and replace it on the reservoir chamber (Figure 17.4e). This converts the apparatus into a dropper assembly (Figure 17.4f).

11. Squeeze a few drops out of the reservoir chamber into a disposal container or wipe with Kimwipe or similar wiping tissue. You are now ready to fill the hemocytometer.

12. Follow the procedure outlined in D.2, filling the hemocytometer.

13. Place the hemocytometer on the microscope stage and allow the red blood cells to settle in the hemocytometer (approximately 1–2 min).

14. When counting red blood cells with the hemocytometer, the loaded (filled) chamber should be first located using the low-power objective of the microscope. Upon finding that, switch to the high-power objective via parfocal procedure. Focus the field using the fine adjustment of the microscope. Each red blood cell counted is equal to 10,000, so the degree of error in manual red blood cell counting is quite high, generally in the range of 15 to 20%.

15. Using the high-power lens, count the cells in each of the five squares (E, F, G, H, I) as shown in Figure 17.5. It is suggested that you use a hand counter. As you count, move up and down in a systematic manner. *Note: To avoid overcounting cells at the boundaries, cells that touch the lines on the left and top sides of the hemocytometer should be counted, but not the ones that touch the boundary lines on the right and bottom sides.*

16. Multiply the total number of red blood cells counted in the five squares by 10,000 to obtain the number of red blood cells in 1 mm^3 of blood.

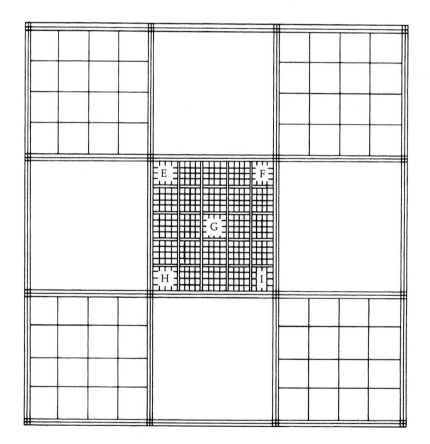

FIGURE 17.5 Hemocytometer (counting chamber). Areas E, F, G, H, and I are all counted in the red blood cell count. The large square containing the smaller squares E, F, G, H, and I is seen in a microscopic field under the 10× magnification. To count the smaller squares (E, F, G, H, and I), the 45× lens is used and therefore E, F, G, H, and I each encompass the entire microscopic field.

Record your value in Section D.1 of the LABO-RATORY REPORT RESULTS at the end of the exercise.

17. The hemocytometer and cover slip should be saved, not discarded.

CAUTION! *Place the hemocytometer and cover slip in a container of fresh bleach solution and dispose of all other materials as directed by your instructor.*

4. Red Blood Cell Volume (Hematocrit)

The percentage of blood volume occupied by the red blood cells is called the ***hematocrit (Hct)*** or ***packed cell volume (PCV).*** It is an important component of a complete blood count and is a standard test for almost all persons having a physical examination or for diagnostic purposes. When a tube of blood is centrifuged, the red blood cells pack into the bottom part of the tube with the plasma on top. The white blood cells and platelets are found in a thin area, the buffy layer, above the red blood cells.

The Readacrit® centrifuge (Figure 17.6) incorporates a built-in hematocrit scale and tube-holding compartments which, when used with special pre-calibrated capillary tubes, permit direct reading of the hematocrit value by measuring the length of the packed red blood cell column. Readacrit® centrifuges present the final hematocrit value without requiring computation by the operator. If the Readacrit® centrifuge or tube reader is not used for

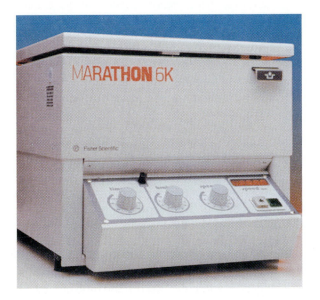

FIGURE 17.6　Centrifuge used for spinning blood to determine hematocrit.

direct reading, measure the total length of blood volume, divide this quantity into length of packed red blood cells, and multiply by 100 for percentage. The calculation is as follows:

$$\frac{\text{length of packed red blood cells}}{\text{total length of blood volume}} \times 100$$

= % volume of whole blood occupied by red blood cells (hematocrit)

In males the normal range is between 40 and 54%, with an average of 47%. In females the normal range is between 38 and 47%, with an average of 42%. Anemic blood may have a hematocrit of 15%; polycythemic blood may have a hematocrit of 65%.

The following procedure for testing red blood cell volume is a micromethod requiring only a drop of blood:

PROCEDURE

Note: The blood used for this procedure should be obtained from a clinical laboratory where it has been tested and certified as noninfectious or from a mammal (other than a human).

1. Place a drop of blood on a microscope slide.
2. Place the unmarked (clear) end of a disposable capillary tube into the drop of blood. Hold the tube slightly below the level of the blood and do not move the tip from the blood or an air bubble will enter the column.

CAUTION! *If an air bubble is present, immediately dispose of the capillary tube in the appropriate biohazard container provided by your instructor and begin again.*

3. Allow blood to flow about two-thirds of the way into the tube. The tube will fill easily if you hold the open end level with or below the blood source.
4. Place your finger over the marked (red) end of the tube and keep it in that position as you seal the blood end of the tube with Seal-ease® or clay or other similar sealing material.
5. According to the directions of your laboratory instructor, place the tube into the centrifuge, making sure that the sealed end is against the rubber ring at the circumference of the centrifuge rotor. Have your laboratory instructor make sure that the tubes are properly balanced. Write the number that indicates the location of your tube. _____

6. When all the tubes from your laboratory section are in place, secure the inside cover of the centrifuge and close the outside cover. Centrifuge on high speed for 4 min.

7. Remove your tube and determine the hematocrit value by reading the length of packed red blood cells directly on the centrifuge scale, or by placing the tube in the tube reader (Figure 17.7) and following instructions on the reader, or by calculation using the equation mentioned previously. When using the tube reader, place the base of the Seal-ease® or clay on the base line and move tube until the top of the plasma is at the line marked 100.

CAUTION! *Immediately dispose of your capillary tube in the appropriate biohazard container provided by your instructor.*

Record your results in Section D.2 of the LABORATORY REPORT RESULTS at the end of the exercise.

What effect would dehydration have on hematocrit? _____

_____

5. Sedimentation Rate

If blood is allowed to stand vertically in a tube, the red blood cells fall to the bottom of the tube and leave clear plasma above them. The distance that the cells fall in 1 hr can be measured and is called

the *sedimentation rate.* It is a function of the amount of fibrinogen and gamma globulin present in plasma and the tendency of the red blood cells to adhere to one another. This rate is greater than normal during menstruation, pregnancy, malignancy, and most infections. Sedimentation rate may be decreased in liver disease. A high rate may indicate tissue destruction in some part of the body; therefore, sedimentation rate is considered a valuable nonspecific diagnostic tool. The normal rate for adults is 0 to 6 mm per hour, for children 0 to 8 mm per hour.

The following method for determining sedimentation rate requires only one drop of blood and is called the Landau micromethod.

PROCEDURE

Note: The blood used for this procedure should be obtained from a clinical laboratory where it has been tested and certified as noninfectious or from a mammal (other than a human).

1. Place a drop of blood on a microscope slide.
2. Using the mechanical suction device, draw the sodium citrate up to the first line that encircles the pipette.
3. Draw the blood up into a disposable sedimentation pipette to the second line. Do not remove the tip of the pipette from the blood or air bubbles will enter the column.

CAUTION! *If air bubbles are drawn into the blood, carefully expel the mixture onto a clean microscope slide, dispose of the slide and pipette in the appropriate biohazard container provided by your instructor, and begin again.*

4. Draw the mixture of fluids up into the bulb and mix by expelling it into the lumen of the tube. Draw and mix the fluids six times. If any air bubbles appear, use the procedure described in step 3.
5. Adjust the level of the blood as close to zero as possible. (Exactly zero is very difficult to get.)
6. Remove the suction device by placing the lower end of the pipette on the *gloved* index finger of the left hand before removing the device from the other end. The blood will leave the pipette if the end is not held closed.
7. Place the lower end of the pipette on the base of the pipette rack (Figure 17.8) and the opposite end at the top of the rack. The tube must be exactly perpendicular. Record the time at which it is put in the rack.

Time _____.

FIGURE 17.7   Microhematocrit tube reader.

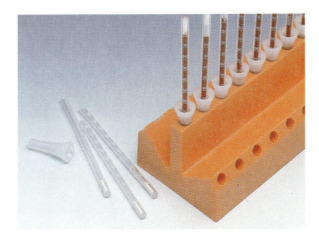

Figure 17.8   One type of sedimentation tube rack.

8. One hour later, measure the distance from the top of the plasma to the top of the red blood cell layer with an accurate millimeter scale, and record your results in Section D.3 of the LABORATORY REPORT RESULTS at the end of the exercise.

CAUTION! *Dispose of the pipette in the appropriate biohazard container provided by your instructor. Using a medicine dropper, rinse the base of the pipette rack with fresh bleach solution.*

6. Hemoglobin Determination

It is possible for one to have anemia even if one's red blood cell count is normal. For example, the red blood cells present may be deficient in hemoglobin or they may be smaller than normal. Thus, the amount of hemoglobin per unit volume of blood, and not necessarily the number of red blood cells, is the determining factor for anemia. The hemoglobin content of blood is expressed in grams (g) per 100 ml of blood, and the normal ranges vary with the technique used. An average range is 12 to 15 g/100 ml in females and 13 to 16 g/100 ml in males.

One procedure that can be used for determining hemoglobin uses an apparatus called a ***hemoglobinometer*** (see Figure 17.9a). This instrument compares the absorption of light by hemoglobin in a blood sample of known depth to that of a standardized glass plate.[2]

[2]If Unopette tests are available, follow the procedure outlined in Unopette test no. 5857, "Cyanmethemoglobin Determination for Manual Methods," for hemoglobin estimation.

PROCEDURE

Note: *The blood used for this procedure should be obtained from a clinical laboratory where it has been tested and certified as noninfectious or from a mammal (other than a human).*

1. Obtain a hemoglobinometer and examine its parts (Figure 17.9a). Note the four scales on the side of the instrument. The top scale gives hemoglobin content in g/100 ml of blood; the other three scales give comparative readings, that is, the amount of hemoglobin expressed as a percent of 15.6, 14.5, or 13.8 g/100 ml of blood.
2. Remove the two pieces of glass from the blood chamber assembly (Figure 17.9b) and clean them with alcohol and wipe them with lens paper. The piece of glass that contains the moat will receive the blood sample; the other piece of glass serves as a cover glass.
3. Replace the glass pieces half-way onto the clip with the moat plate on the bottom.
4. Obtain a sample of blood.
5. Apply the blood to the moat and completely cover the raised surface of the plate with blood.
6. Hemolyze the blood by agitating it with the tip of a hemolysis applicator for about 30 to 45 sec. Hemolysis is complete when the appearance of the blood changes from cloudy to transparent. This change reflects the lysis of the red blood cell membranes and the liberation of hemoglobin.
7. Push the specimen chamber into the clip and push the clip into the hemocytometer.
8. Holding the hemocytometer in your left hand, look through the eyepiece while depressing the light switch button. You will see a split field.
9. Move the slide button back and forth with the right index finger until the two sides of the field match in color and shading.
10. Read the value indicated on the scale marked 15.6. Determine the grams of hemoglobin per 100 ml of the blood sample by reading the number above the index mark on the hemocytometer. Record your results in Section D.4 of the LABORATORY REPORT RESULTS at the end of the exercise.

CAUTION! *Remove the glass pieces from the chamber and wash in fresh bleach solution. Dry the glass pieces with lens paper.*

11. What type of anemia would reveal a normal or slightly below normal hematocrit, yet a significantly reduced hemoglobin level?

_____

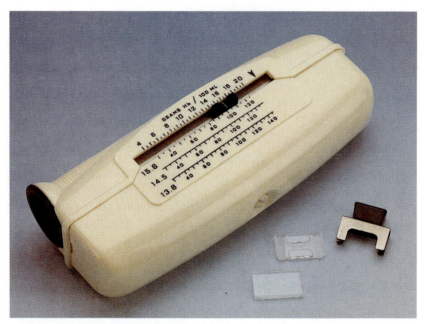

(a) Cambridge Instruments Hemoglobin-Meter

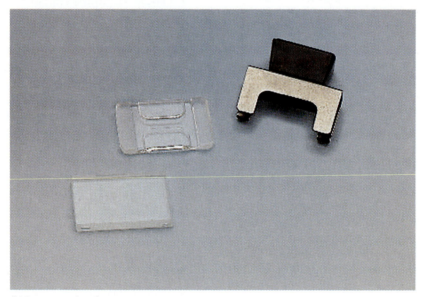

(b) Specimen chamber

FIGURE 17.9   Hemoglobinometer.

ALTERNATE PROCEDURE

The **Tallquist measurement** of hemoglobin is an old and somewhat inaccurate technique, although it is quick and inexpensive. The hemoglobinometer gives more accurate (15%) results.

Note: The blood used for this procedure should be obtained from a clinical laboratory where it has been test-

ed and certified as noninfectious or from a mammal (other than a human).

1. Obtain a sample of blood.
2. Place one drop of blood on the Tallquist® paper.
3. As soon as the blood no longer appears shiny, match its color with the scale provided. Record

your observations in Section D.4 of the LABO-RATORY REPORT RESULTS at the end of the exercise.

7. Oxyhemoglobin Saturation

The ability of blood to carry oxygen is determined by proper respiratory system function, hematocrit, and the ability of hemoglobin to bind reversibly with oxygen within the alveoli (air sacs) of the lungs and release oxygen at the tissue level. Hemoglobin's interaction with oxygen is measured by an *oxygen-hemoglobin dissociation curve*. Such a curve demonstrates that the number of available oxygen-binding sites on hemoglobin that actually bind to oxygen is proportional to the partial pressure of oxygen (pO_2). Arterial blood usually has a pO_2 of 95, causing 97% of the available hemoglobin to be saturated with oxygen. Venous blood, however, has a pO_2 of 40, so only 75% of the hemoglobin is saturated with oxygen; 25% is given off to tissues. The determination of percent saturation for hemoglobin is a very sensitive test of pulmonary function. However, such a test needs to be interpreted carefully, as an individual with a normally functioning pulmonary system may exhibit abnormally low oxyhemoglobin saturation due to a variety of possible causes, such as *methemoglobinemia* (transformation of normal oxyhemoglobin due to the reduction of normal Fe^{2+} to Fe^{3+} within the hemoglobin mole-cule) or *carbon monoxide (CO) poisoning.* In order to determine the percent hemoglobin saturation in an unknown sample of blood one would need to compare the absorption spectrum of the blood sample to that obtained from a pure sample of *oxyhemoglobin (HbO₂), carboxyhemoglobin (HbCO),* and *reduced hemoglobin (Hb)* (carboxyhemoglobin is included due to the normal presence of carbon monoxide in today's polluted atmosphere). This is shown in Figure 17.10. The spectrum obtained with the unknown sample of blood would be some combination of these three, since it would contain a certain percentage of each form of hemoglobin. Such a test is possible due to the fact that all three of these types of hemoglobin compounds are different colors, and would therefore absorb different portions of the light spectrum. Through a complex determination of the relative contribution of each type of hemoglobin to the resulting spectrum for the unknown sample, the percentage of each hemoglobin form may be determined.

8. Spectrum of Oxyhemoglobin and Reduced Hemoglobin

The following procedure details a simplified, although less accurate method for determining the absorption spectra for fully oxygenated and deoxygenated hemoglobin. Exposing blood to air saturates it with oxygen, while sodium dithionate reduces it by removing oxygen.

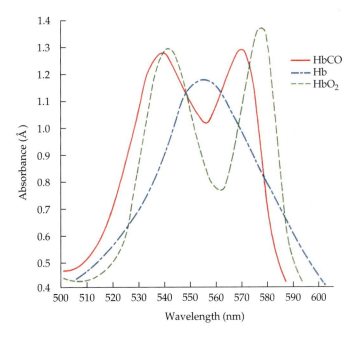

FIGURE 17.10   Absorption spectra of carboxyhemoglobin (HbCO), reduced hemoglobin (Hb), and oxyhemoglobin (HbO₂).

PROCEDURE

Note: The blood used for this procedure should be obtained from a clinical laboratory where it has been tested and certified as noninfectious or from a mammal (other than a human).

1. Turn on the spectrophotometer and let it warm up for several minutes.
2. After the instrument has warmed up, set the wavelength at 500 nm and adjust the meter needle to 0% transmittance (absorbance at infinity) by using the zero control knob.
3. Place a cuvette containing distilled water into the cuvette holder and adjust the needle to 100% transmittance by using the light control knob.
4. Place 8 ml of distilled water in a clean test tube.
5. Obtain several drops of blood.
6. Mix the distilled water with the blood by placing a stopper in the mouth of the test tube and then inverting the test tube, thereby adding the blood to the distilled water.
7. Transfer half of the test tube contents (4 ml) into a second clean test tube.
8. Add 0.2 ml of 1% sodium dithionite solution to the second test tube and mix thoroughly. *(Note: In order for the procedure to work properly, the sodium dithionite solution must be fresh and made up just prior to use, and a spectral determination must be completed within 5 min of addition of sodium dithionite to the second tube.)*
9. Record the absorbances of solutions 1 (oxyhemoglobin) and 2 (reduced hemoglobin) at 500 nm wavelength.
10. Standardize the spectrophotometer at 510 nm utilizing the cuvette containing only distilled water as outlined in steps 2 and 3 above.
11. Determine the absorbances for solutions 1 and 2 at 510 nm.
12. Repeat procedures 2, 3, and 9 at each of the following wavelengths: 520, 530, 540, 560, 570, 580, 590, 600 nm.
13. Record and graph your results in Section D.5 of the LABORATORY REPORT RESULTS at the end of the exercise. Be sure to compare and give physiological reasons for the observed differences in the two spectra.

E. LEUKOCYTES

Leukocytes (white blood cells, or *WBCs)* are different from red blood cells in that they have nuclei and do not contain hemoglobin (see Figure 17.1). They are less numerous than red blood cells, ranging from 5000 to 10,000 cells per cubic millimeter (mm^3) or per deciliter (dl) of blood. The ratio, therefore, of red blood cells to white blood cells is about 700:1.

As Figure 17.1 shows, leukocytes can be differentiated by their appearance. They are divided into two major groups, granular leukocytes and agranular leukocytes. *Granular leukocytes,* which are formed from red bone marrow, have large granules in the cytoplasm that can be seen under a microscope and possess lobed nuclei. The three types of granular leukocytes are *neutrophils, eosinophils,* and *basophils. Agranular leukocytes,* which are also formed from red bone marrow, contain small granules that cannot be seen under a light microscope, and usually have spherical nuclei. The two types of agranular leukocytes are *lymphocytes* and *monocytes.*

Leukocytes as a group function in phagocytosis, producing antibodies, and combating allergies. The life span of a leukocyte usually ranges from a few hours to a few months. Some lymphocytes, called T and B memory cells, can live throughout one's life once they are formed.

F. WHITE BLOOD CELL TESTS

CAUTION! *Please reread Section A, "General Safety Precautions and Procedures," and Section B, "Precautions Related to Working with Blood, Blood Products, or Other Body Fluids," on pages xi–xii at the beginning of the laboratory manual, before you begin any of the following experiments. Read the experiments before you perform them, to be sure that you understand all the procedures and safety precautions.*

When working with whole blood, wear tight-fitting surgical gloves and safety goggles. Avoid any kind of contact with an open sore, cut, or wound.

1. White Blood Cell Count

This procedure determines the number of circulating white blood cells in the body. Because white blood cells are a vital part of the body's immune defense system, any abnormalities in the white blood cell count must be carefully noted.

An increase in number *(leukocytosis)* may result from such conditions as bacterial or viral infection, metabolic disorders, chemical and drug poisoning, and acute hemorrhage. A decrease in number *(leukopenia)* may result from typhoid infection, measles, infectious hepatitis, tuberculosis, or cirrhosis of the liver.

PROCEDURE

WARD'S Natural Science Establishment, Inc., has made available WARD'S *Simulated Blood*. It is a solution that contains microcomponents that simulate red blood cells, white blood cells, and platelets. The microcomponents are similar in relative proportion to those found in human blood and can be observed under a microscope without staining.

1. Obtain a Simulated Blood Activity kit.
2. Follow the instructions to perform the white blood cell count.
3. Complete the portion of the Data Table related to counted white blood cells.
4. Answer the questions related to white blood cells.

ALTERNATE PROCEDURE

Note: *The blood used for this procedure should be obtained from a clinical laboratory where it has been tested and certified as noninfectious or from a mammal (other than a human).*

   Whole blood that has been tested for syphilis; hepatitis A, B, and C; and HIV is available from Carolina Biological Supply Company. This blood can be used for white blood cell count and differential white blood cell count.

1. Follow steps 1–14 of D.3, "Red Blood Cell Count," but use the Unopette Reservoir System® for white blood cell determination (the color of the reservoir's bottom surface is white).
2. Using the low-power objective (10×), count the cells in each of the four corner squares (A, B, C, D in Figure 17.11) of the hemocytometer. The

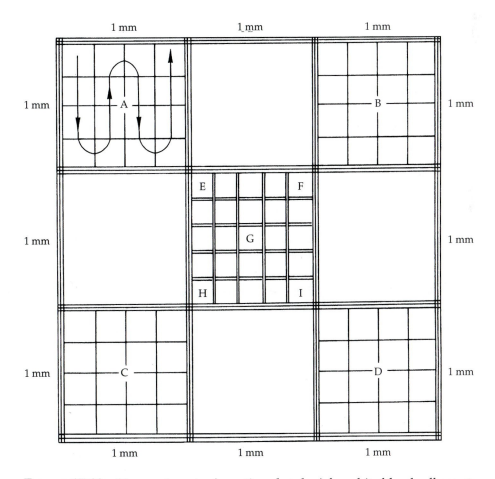

FIGURE 17.11   Hemocytometer (counting chamber) for white blood cell count (as seen through scanning lens). A, B, C, and D are areas counted in the white blood cell count, (when viewed through 10× lens, 1 sq mm is seen in the microscopic field). Areas A, B, C, and D each equal 1 sq. mm; therefore, a total of 4 sq mm is counted in the white blood cell count.

direction to follow when counting white blood cells is indicated in square A in Figure 17.11. *Note: To avoid overcounting of cells at the boundaries, the cells that touch the lines on the left and top sides of the hemocytometer should be counted, but not the ones that touch the boundary lines on the right and bottom sides. It is suggested that you use a hand counter.*

3. Multiply the results by 50 to obtain the amount of circulating white blood cells per mm^3 of blood and record your results in Section F.6 of the LABORATORY REPORT RESULTS at the end of the exercise.

4. The factor of 50 is the dilution factor, or $2.5 \times 20 = 50$. The volume correction factor of 2.5 is arrived at in this manner: each of the corner areas (A, B, C, and D) is exactly 1 mm^3 by 0.1 mm deep (Figure 17.11). Therefore, the volume of each of these corner areas is 0.1 mm^3. Because four of them are counted, the total volume of diluted blood examined is 0.4 mm^3. However, because we want to know the number of cells in 1 mm^3 instead of 0.4 mm^3, we must multiply our count by 2.5 $(0.4 \times 2.5 = 1.0)$.

CAUTION! *Place the hemocytometer and cover slip in a container of fresh bleach solution and dispose of all other materials as directed by your instructor.*

5. The hemocytometer should be saved, not discarded.

2. Differential White Blood Cell Count

The purpose of a differential white blood cell count is to determine the relative *percentages* of each of the five normally circulating types of white blood cells in a total count of 100 white blood cells. A normal differential white blood cell count might appear as follows:

Type of white blood cell	Normal
Neutrophils	60–70%
Eosinophils	2–4%
Basophils	0.5–1%
Lymphocytes	20–25%
Monocytes	3–8%

Significant elevations of different types of white blood cells usually indicate specific pathological conditions. For example, a high neutrophil count may indicate acute appendicitis or infection. Lymphocytes predominate in antigen-antibody reactions, specific

leukemias, and in infectious mononucleosis. An increase in eosinophils may be seen in allergic reactions and parasitic infections. An elevated percentage of monocytes may result from chronic infections. An increase in basophils is rare and denotes a specific type of leukemia and allergic reactions.

The procedure for making a differential white blood cell count is as follows:

PROCEDURE

Note: The blood used for this procedure should be obtained from a clinical laboratory where it has been tested and certified as noninfectious or from a mammal (other than a human).

1. Obtain a drop of blood.
2. Use a second slide as a spreader (Figure 17.12).
3. Draw the spreader toward the drop of blood (in this direction →) until it touches the drop. The blood should fan out to the edges of the spreader slide.
4. Keeping the spreader at a 25° angle, press the edge of the spreader firmly against the slide and push the spreader rapidly over the entire length of the slide (in this direction ←). The drop of blood will thin out toward the end of the slide.
5. Let the smear dry.
6. To stain the slide follow this procedure:
   a. Place the slide on a staining rack.
   b. Cover the entire slide with Wright's stain.
   c. Let stain stand for 1 min.
   d. Add 25 drops of buffer solution and mix completely with Wright's stain.
   e. Let the mixture stand for 8 min.
   f. Wash the mixture off completely with distilled water.
7. Let the slide dry completely before counting.
8. To proceed with counting the cells, use an area of slide where blood is thinnest (one cell thick). This is called the *feathering edge*. Count cells under an oil-immersion lens. Count a total of 100 white cells.
9. The actual counting may be done by moving the slide either up and down or from side to side as follows:

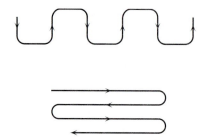

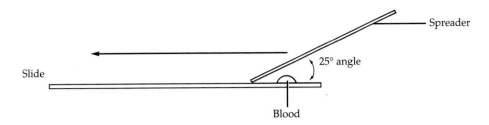

FIGURE 17.12   Procedure to prepare a blood smear.

10. Record each white blood cell you observe by making a mark in the following chart until you have recorded 100 cells.

ALTERNATE PROCEDURE

1. Obtain a prepared slide of stained blood.
2. Repeat steps 9 and 10 in the preceding procedure.

G. PLATELETS

Platelets (thrombocytes) are formed from fragments of the cytoplasm of megakaryocytes (see Figure 17.1). The fragments become enclosed in pieces of cell membrane from the megakaryocytes and develop into platelets. Platelets are very small, disc-shaped cell fragments without nuclei. Between 250,000 and 400,000 are found in each cubic millimeter or deciliter (dl) of blood. They function to prevent fluid loss by starting a chain of reactions that results in blood clotting. They have a short life span, probably only one week, because they are expended in clotting and are just too simple to carry on extensive metabolic activity.

PROCEDURE

WARD'S Natural Science Establishment, Inc., has made available WARD'S *Simulated Blood.* It is a solution that contains microcomponents that simulate red blood cells, white blood cells, and platelets. The microcomponents are similar in relative proportion to those found in human blood and can be observed under a microscope without staining.

1. Obtain a Simulated Blood Activity kit.
2. Follow the instructions to perform the platelet count.
3. Complete the portion of the Data Table related to platelets.
4. Answer the questions related to platelets.

H. DRAWINGS OF BLOOD CELLS

Examine a prepared slide of a stained smear of blood cells using the oil-immersion objective. With the aid of your textbook, identify red blood cells, neutrophils, basophils, eosinophils, lymphocytes, monocytes, and platelets. Using colored crayons or pencil crayons, draw and label all of the different

Type of white blood cell	Number observed	Percentage
Neutrophils		
Eosinophils		
Basophils		
Lymphocytes		
Monocytes		

blood cells in Section H of the LABORATORY RE-
PORT RESULTS at the end of the exercise. Be very
accurate in drawing and coloring the granules and
the nuclear shapes, because both of these are used
to identify the various types of cells.

I. BLOOD GROUPING (TYPING)

The plasma membranes of red blood cells contain
genetically determined antigens called *isoantigens.*
The plasma of blood contains genetically determined antibodies called *isoantibodies.* The antibodies cause the *agglutination (clumping)* of the
red blood cells carrying the corresponding antigen.

These proteins, antigens, and antibodies are responsible for the two major classifications of blood
groups: the ABO group and the Rh system. In addition to the ABO group and the Rh system, other
human blood groups include: MNSs, P, Lutheran,
Kell, Lewis, Duffy, Kidd, Diego, and Sutter. Fortu-
nately these different antigenic factors do not exhibit extreme degrees of antigenicity and, therefore,
usually cause very weak transfusion reactions or
no reaction at all.

1. ABO Group

This major blood grouping is based on two antigens symbolized as *A* and *B* (Figure 17.13). Individuals whose red blood cells produce only antigen
A have blood type A. Individuals who produce only
antigen *B* have blood type B. If both *A* and *B* antigens are produced, the result is type AB, whereas
the absence of both *A* and *B* antigens results in the
so-called type O.

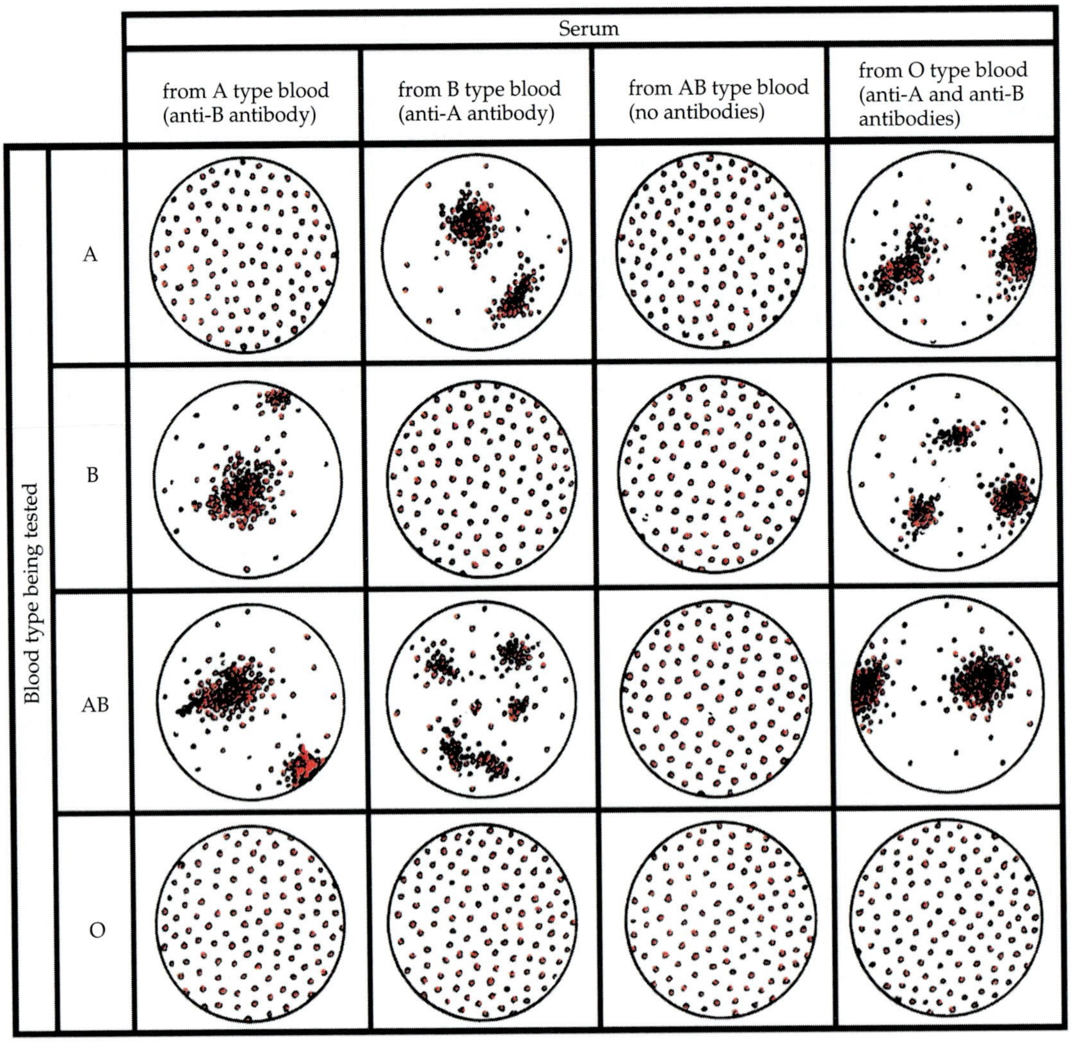

FIGURE 17.13   ABO blood groupings.

The antibodies in blood plasma are anti-*A* antibody, which attacks antigen *A*, and anti-*B* antibody, which attacks antigen *B*. The antigens and antibodies formed by each of the four blood types and the various agglutination reactions that occur when whole blood samples are mixed with serum are shown in Figure 17.13. You do not have antibodies that attack the antigens of your own red blood cells. For example, a type A person has antigen *A* but not antibody anti-*A*.

Antigens and antibodies are of critical importance in blood transfusions. In an incompatible blood transfusion, the donated red blood cells are attacked by the recipient antibodies, causing the blood cells to agglutinate. Agglutinated cells become lodged in small capillaries throughout the body and, over a period of hours, the cells swell, rupture, and release hemoglobin into the blood. Such a reaction as it relates to red blood cells is called **hemolysis** (*lysis* = dissolve). The degree of agglutination depends on the titer (strength or amount) of antibody in the blood. Agglutinated cells can block blood vessels and may lead to kidney or brain damage and death, and the liberated hemoglobin may also cause kidney damage.

Because cells of type O blood contain neither of the two antigens (*A* or *B*), moderate amounts of this blood can be transfused into a recipient of any ABO blood type without *immediate* agglutination. For this reason, type O blood is referred to as the **universal donor.** However, transfusing large amounts of type O blood into a recipient of type A, B, or AB can cause *delayed* agglutination of the recipient's red blood cells, because the transfused antibodies (type O blood contains anti-*A* and anti-*B* antibodies) are not sufficiently diluted to prevent the reaction.

Persons with type AB blood are sometimes called **universal recipients** because their blood contains no antibodies to agglutinate donor red blood cells. *Small quantities* of blood from all other ABO types can be transfused into type AB blood without adverse reaction. However, again, if *large quantities* of type A, B, or O blood are transfused into type AB, the antibodies in the donor blood might accumulate in sufficient quantity to clump the recipient's red blood cells.

Table 17.1 summarizes the various interactions of the four blood types. Table 17.2 lists the incidence of these different blood types in the United States, comparing some races.

2. ABO Blood Grouping Test

CAUTION! *Please reread Section A, "General Safety Precautions and Procedures" and Section B, "Precautions Related to Working with Blood, Blood Products, or Other Body Fluids" on pages xi–xii at the beginning of the laboratory manual before you begin any of the following experiments. You should also read the experiments before you perform them to be sure that you understand all the procedures and safety precautions.*

When working with whole blood, wear tight-fitting surgical gloves and safety goggles. Avoid any kind of contact with an open sore, cut, or wound.

Blood should be obtained from a clinical laboratory where it has been tested and certified as noninfectious or from a mammal (other than a human).

TABLE 17.1
Summary of ABO System Interactions

Blood Type	A	B	AB	O
Antigen on red blood cell	*A*	*B*	*A* and *B*	Neither *A* nor *B*
Antibody in plasma	anti-*B*	anti-*A*	Neither anti-*A* nor anti-*B*	*a* and *b*
Compatible donor blood types	A, O	B, O	A, B, AB, O	O
Incompatible donor blood types	B, AB	A, AB	—	A, B, AB
Genotype (genetic make-up)	*AO* and *AA*	*BO* and *BB*	*AB*	*OO*

TABLE 17.2
Incidence of Human Blood Groups in the United States

	Blood Groups (percentages)				
	O	**A**	**B**	**AB**	**Rh$^+$**
Whites	45	40	11	4	85
Blacks	49	27	20	4	95
Japanese	31	38	21	10	~ 100
Chinese	42	27	25	6	~ 100
Native Indians	79	16	4	1	~ 100
Korean	32	28	30	10	~ 100

Whole blood that has been tested for syphilis; he-patitis A, B, and C; and HIV is available from Carolina Biological Supply Company. This blood can be used for blood grouping and typing. Carolina Biological Supply Company also provides aseptic red blood cells that can be used for blood grouping and typing. These blood cells are suspended in a modified alse-vere's solution to which several antibiotics have been added. The cells have been tested for syphilis; hepatitis A, B, and C; and HIV.

Also available for this exercise is WARD'S Simulated ABO and Rh Blood Typing Activity.

The procedure for ABO sampling is as follows:

PROCEDURE

1. Using a wax pencil, divide a glass slide in half and label the left side A and the right side B.
2. On the left side, place one large drop of anti-A serum, and on the right side place one large drop of anti-B serum.
3. Next to the drops of antisera, place one drop of blood, being careful not to mix samples on the left and right sides.
4. Using a mixing stick or a toothpick, mix the blood on the left side with the anti-A serum, and then, using a *different stick or different toothpick,* mix the blood on the right side with the anti-*B.*

CAUTION! *Immediately dispose of the stick or toothpick in the appropriate biohazard container provided by your instructor.*

5. *Gently* tilt the slide back and forth and observe it for 1 minute.
6. Record your results in Section I.1 of the LABO-RATORY REPORT RESULTS at the end of the exercise, using "+" for clumping (agglutination) and "−" for no clumping.

CAUTION! *Dispose of your slide in the appropriate biohazard container provided by your instructor.*

3. Rh System

The Rh factor is the other major classification of blood grouping. This group was designated Rh because the blood of the rhesus monkey was used in the first research and development. As is the ABO grouping, this classification is based on antigens that lie on the surfaces of red blood cells. The designation Rh$^+$ (positive) is given to those that have the antigen, and Rh$^-$ (negative) is for those that lack the antigen. The estimation is that 85% of whites and 95% of blacks in the United States are Rh$^+$, whereas 15% of whites and 5% of blacks are Rh$^-$ (see Table 17.2).

The Rh factor is extremely important in pregnancy and childbirth. Under normal circumstances, human plasma does not contain anti-Rh antibodies. If, however, a woman who is Rh$^-$ becomes pregnant with an Rh$^+$ child, her blood may produce antibodies that will react with the blood of a subsequent child.[3] The first child is unaffected because the mother's body has not yet produced these antibodies. This is a serious reaction and hemolysis may occur in the fetal blood.

The hemolysis produced by this fetal-maternal incompatibility is called *hemolytic disease of newborn,* or *HDN (erythroblastosis fetalis),* and could

---

[3]Be sure to note the important difference in the production of antibodies in the ABO and Rh systems. An Rh$^-$ person can *produce antibodies in response to the stimulus of invading Rh antigen;* by contrast, any antibodies of the ABO system that exist in the blood of a person *occur naturally and are present regardless of whether or not ABO antigens are introduced.*

be fatal for the newborn infant. A drug called Rho-GAM, given to Rh⁻ mothers immediately after delivery or abortion, prevents the production of antibodies by the mother so that the fetus of the next pregnancy is protected.

4. Rh Blood Grouping Test

CAUTION! *Please reread Section A, "General Safety Precautions and Procedures," and Section B, "Precautions Related to Working with Blood, Blood Products, or Other Body Fluids," on pages xi–xii, at the beginning of the laboratory manual, before you begin any of the following experiments. Read the experiments before you perform them, to be sure that you understand all the procedures and safety precautions.*

When working with whole blood, wear tight-fitting surgical gloves and safety goggles. Avoid any kind of contact with an open sore, cut, or wound.

Note: Remember that this procedure is sufficient for laboratory demonstration, but it should not be considered clinically accurate, since the anti-Rh (anti-D) serum deteriorates rapidly at room temperature and the anti-Rh antibodies are far less potent in their agglutinizing capability than the anti-A or anti-B antibodies. This test is less accurate than the ABO determination.

Blood should be obtained from a clinical laboratory where it has been tested and certified as noninfectious or from a mammal (other than a human).

Whole blood that has been tested for syphilis; hepatitis A, B, and C; and HIV is available from Carolina Biological Supply Company. This blood can be used for blood grouping and typing. Carolina Biological Supply Company also provides aseptic red blood cells that can be used for blood grouping and typing. These blood cells are suspended in a modified alsevere's solution to which several antibiotics have been added. The cells have been tested for syphilis; hepatitis A, B, and C; and HIV.

Also available for this exercise is WARD'S Simulated ABO and Rh Blood Typing Activity.

The procedure for Rh grouping is as follows:

PROCEDURE

1. Place one large drop of anti-Rh (anti-D) serum on a glass slide.[4]
2. Add one drop of blood and mix using a mixing stick or toothpick.

CAUTION! *Immediately dispose of the stick or toothpick in the appropriate biohazard container provided by your instructor.*

3. Place the slide on a preheated warming box, and gently rock the box back and forth for 2 minutes. (Unlike ABO typing, Rh typing is better done on a heated warming box.)
4. Record your results, using "+" for clumping and "−" for no clumping. Record whether you are Rh⁺ or Rh⁻ in Section I.4 of the LABORATORY REPORT RESULTS at the end of the exercise.

CAUTION! *Dispose of your slide in the appropriate biohazard container provided by your instructor.*

ANSWER THE LABORATORY REPORT QUESTIONS AT THE END OF THE EXERCISE.

---

[4]The Rh antigen is more specifically termed the D antigen after the Fisher-Race nomenclature, which is based on genetic concepts or theories of inheritance.

Blood 17

Student _____   Date _____

Laboratory Section _____   Score/Grade _____

SECTION B. PLASMA

1. *Physical Characteristics*

pH _____

Color _____

Transparency (clear, translucent, opaque) _____

2. *Chemical Constituents*

Is glucose present? _____

Is protein present? _____

SECTION D. RED BLOOD CELL TESTS

1. Red blood cell count results: _____ RBCs per mm^3.

2. Red blood cell volume (hematocrit) results: _____%.

3. Sedimentation rate results: _____ mm per hour.

4. Hemoglobin determination results: hemoglobinometer, _____ g per 100 ml.

Tallquist® paper, _____ g per 100 ml.

5. Complete the following table:

Wavelength

	500	510	520	530	540	550	560	570	580	590	600
Solution 1											
Solution 2											

Graph your results here.

SECTION F. WHITE BLOOD CELL TESTS

6. White blood cell count results: _____ WBCs per mm^3.

SECTION H. DRAWINGS OF BLOOD CELLS

SECTION I. BLOOD GROUPING (TYPING)

1. In determining the ABO blood grouping, did you observe clumping when the blood was mixed with

_____ anti-A serum only

_____ anti-B serum only

_____ both anti-A and anti-B serums

_____ neither anti-A nor anti-B serum

2. Based on your observations, what is the ABO grouping?

_____ A   _____ B   _____ AB   _____ O

3. Based on your observations, briefly explain why you identify your ABO blood grouping as you do.

_____

_____

4. Record the results of the ABO blood grouping test done by your class.

Type	Anti-A (present or absent)	Anti-B (present or absent)	Number of individuals	Class percentage
A				
B				
AB				
O				

5. Based on your observations, are you Rh$^+$ or Rh$^-$?

_____ Rh$^+$ _____ Rh$^-$

6. Record the results of the Rh test done by your class.

Type	Number of individuals	Class percentage
Rh$^+$		
Rh$^-$		

Blood 17

Student _____   Date _____

Laboratory Section _____   Score/Grade _____

PART 1. Multiple Choice

_____ 1. The process by which all blood cells are formed is called (a) hemocytoblastosis (b) erythro-poiesis (c) hemopoiesis (d) leukocytosis

_____ 2. An inability of body cells to receive adequate amounts of oxygen may indicate a malfunc-tion of (a) neutrophils (b) leukocytes (c) lymphocytes (d) erythrocytes

_____ 3. Special cells of the body that have the responsibility of clearing away dead, disintegrating bodies of red and white blood cells are called (a) agranular leukocytes (b) reticuloendothe-lial cells (c) erythrocytes (d) thrombocytes

_____ 4. The name of the test procedure that informs the physician about the rate of erythropoiesis is called the (a) reticulocyte count (b) sedimentation rate (c) hemoglobin count (d) differ-ential white blood cell count

_____ 5. The normal red blood cell count per cubic millimeter in a male is about (a) 5.4 million (b) 7 million (c) 4 million (d) more than 9 million

_____ 6. The normal number of leukocytes per cubic millimeter is (a) 5000 to 10,000 (b) 8000 to 12,000 (c) 2000 to 4000 (d) over 15,000

_____ 7. Under the microscope, red blood cells appear as (a) circular discs with centrally located nuclei (b) circular discs with lobed nuclei (c) oval discs with many nuclei (d) biconcave discs without nuclei

_____ 8. An increase in the number of white blood cells is called (a) leukopenia (b) hematocrit (c) polycythemia (d) leukocytosis

_____ 9. Platelets are formed from a special large cell that breaks up into small fragments. This cell is called a(n) (a) eosinophil (b) hemocytoblast (c) megakaryocyte (d) platelet

_____ 10. The blood type showing the highest incidence in Caucasians in the United States is (a) A (b) O (c) AB (d) B

PART 2. Completion

11. Another name for red blood cells is _____.

12. Blood gets its red color from the presence of _____.

13. The life span of a red blood cell is approximately _____.

14. A good method for routine testing for anemia is _____.

15. The normal sedimentation rate value for adults is _____.

16. The normal ratio of red blood cells to white blood cells is about _____.

17. The granular leukocytes are formed from _____ tissue.

18. The number of platelets per cubic millimeter (mm^3) or deciliter (dl) found normally in blood is

_____.

19. The function of platelets is to prevent blood loss by starting a chain of reactions resulting in

_____.

20. In the ABO blood grouping system, the genetically determined structures on the surface of red blood

cells are called _____.

21. The hemolysis produced by fetal-maternal incompatibility of blood cells is called

_____.

22. The part of the cell where agglutinogens are located is _____.

23. Cells derived from mesenchyme that give rise to all formed elements in blood are called

_____.

PART 3. Matching

_____ **24.** Hematocytoblast   A. Response to tissue by invading bacteria

_____ **25.** Polycythemia   B. An increase in the normal red blood cell count

_____ **26.** A high neutrophil count   C. Leukemia and infectious mononucleosis

_____ **27.** A high monocyte count   D. Immature cells that develop into mature blood cells

_____ **28.** Leukopenia   E. Chronic infections

_____ **29.** A high lymphocyte count   F. Liquid portion of blood without the formed elements
                                              and clotting substances
_____ **30.** Plasma

_____ **31.** A high eosinophil count   G. An allergic reaction

_____ **32.** Serum   H. A decrease in the normal white blood cell count

                                           I. Liquid portion of blood without the formed elements

Heart

18

As noted in Chapter 17, the cardiovascular system consists of the blood, heart, and blood vessels. Having considered the origin, structural features, and functions of blood, we now examine the heart, the center of the cardiovascular system. The *heart* is a hollow, muscular organ that pumps blood through miles and miles of blood vessels. The heart rests on the diaphragm, near the midline of the thoracic cavity in the *mediastinum.* About two thirds of the heart's mass lying to the left of the body's midline. Its pointed end, the *apex,* projects inferiorly to the left, and its broad end, the *base,* projects superiorly to the right. The main parts of the heart and associated structures to be discussed here are the pericardium, wall, chambers, great vessels, and valves.

A. PERICARDIUM

The *pericardium* (*peri* = around; *cardio* = heart), is a triple-layered sac that surrounds and protects the heart. It consists of two principal portions: the fibrous pericardium and the serous pericardium (Figure 18.1). The superficial *fibrous pericardium* is a tough, inelastic, dense irregular connective tissue membrane. The fibrous pericardium prevents over-

stretching of the heart, provides protection, and anchors the heart in the mediastinum. The deeper *serous pericardium,* is a thinner, delicate membrane that forms a double layer around the heart. The superficial *parietal layer* of the serous pericardium is fused with the fibrous pericardium. The deeper *visceral layer* of the serous pericardium is also called the *epicardium.* Between the parietal and visceral layers of the serous pericardium is a thin film of serous fluid known as *pericardial fluid.* It is a slippery secretion of pericardial cells that reduces friction between the membranes as the heart moves. The potential space that houses the pericardial fluid is called the *pericardial cavity.* Identify these structures using a specimen, model, or chart of a heart and label Figure 18.1.

B. HEART WALL

Three layers of tissue compose the heart: the epicardium (external layer), the myocardium (middle layer), and the endocardium (inner layer). The *epicardium* (*epi* = above), which is also the visceral layer of the serous pericardium, is the thin, transparent superficial layer of the heart wall. The middle

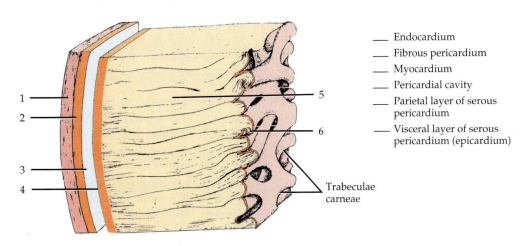

1
2
3
4
5
6

___ Endocardium
___ Fibrous pericardium
___ Myocardium
___ Pericardial cavity
___ Parietal layer of serous pericardium
___ Visceral layer of serous pericardium (epicardium)

Trabeculae carneae

FIGURE 18.1   Structure of pericardium and heart wall.

myocardium (*myo* = muscle), which is composed of cardiac muscle tissue, forms the bulk of the heart and is responsible for contraction. The *endocardium* (*endo* = within) is a thin, deep layer of endothelium and areolar connective tissue that lines the inside of the myocardium and covers the heart valves and the tendons that hold them open. Label the layers of the heart in Figure 18.1.

C. CHAMBERS AND GREAT VESSELS OF HEART

The heart contains four cavities called *chambers.* Each of the two superior chambers is called an *atrium* (*atrium* = court). The *right atrium* collects deoxygenated blood from systemic circulation, and the *left atrium* collects oxygenated blood from pulmonary circulation. The two inferior chambers are called *ventricles* (*ventricle* = little belly). The *right ventricle* receives deoxygenated blood from the right atrium and pumps it to the lungs to become oxygenated. The *left ventricle* receives oxygenated blood from the left atrium and pumps it to all body cells, except the air sacs of the lungs.

During one phase of a heartbeat, the atria empty of blood and become smaller. This produces the appearance of a wrinkled pouch on the anterior surface of each atrium called an *auricle* (OR-i-kul; *auris* = ear), so named because of its resemblance to a dog's ear. Each auricle slightly increases the surface area of an atrium so that it can hold a greater volume of blood. On the surface of the heart are a series of grooves, each called a *sulcus* (SUL-kus), which contain coronary blood vessels and a variable amount of fat. The sulci (SUL-kē) also mark the external boundaries between chambers of the heart. The deep *coronary* (*corona* = crown) *sulcus* encircles most of the heart and marks the boundary between the superior atria and inferior ventricles. The *anterior interventricular sulcus* is a shallow groove on the anterior surface of the heart that marks the boundary between the right and left ventricles. This sulcus continues around to the posterior surface of the heart as the *posterior interventricular sulcus,* which marks the boundary between the ventricles of the posterior aspect of the heart.

Label these structures in Figures 18.2 and 18.3.

The *right atrium* forms the right border of the heart. It receives deoxygenated blood from systemic circulation through three veins: superior vena cava, inferior vena cava, and coronary sinus.

The anterior and posterior walls of the right atrium differ considerably. Whereas the posterior wall is smooth, the anterior wall is rough due to the presence of muscular ridges called *pectinate* (*pectin* = comb) *muscles.* Between the right atrium and left atrium is a connective tissue partition called the *interatrial septum.* A prominent feature of this septum is an oval depression called the *fossa ovalis.* It is the remnant of the foramen ovale, an opening in the interatrial septum of the fetal heart that normally closes soon after birth (see Figure 18.3a).

The *right ventricle* forms most of the anterior surface of the heart. Deoxygenated blood passes from the right atrium into the right ventricle through a valve called the *tricuspid valve.* The cusps of the valve are connected to tendonlike cords, the *chordae tendineae* (KOR-dē ten-DIN-ē-ē; *chorda* = cord; *tendo* = tendon), which, in turn, are connected to cone-shaped muscular columns in the inner surface of the ventricle called *papillary* (*papilla* = nipple) *muscles.* (Details about the tricuspid and other valves mentioned in this discussion of the heart chambers will follow shortly.) Another prominent feature of the right ventricle is a series of ridges and folds called *trabeculae carneae* (tra-BEK-yoo-lē KAR-nē-ē; *trabecula* = little beam; *carneous* = fleshy). The right ventricle is separated from the left ventricle by a partition called the *interventricular septum.*

Deoxygenated blood passes from the right ventricle through a valve called the *pulmonary semilunar valve* into the *pulmonary* (*pulmo* = lung) *trunk* to enter pulmonary circulation. The pulmonary trunk divides into a right and a left *pulmonary artery,* each of which carries blood to their respective lungs. In the lungs, the deoxygenated blood releases carbon dioxide and takes on oxygen. The blood is then oxygenated blood.

The *left atrium* forms most of the base of the heart. It receives oxygenated blood from pulmonary circulation in the lungs through four *pulmonary veins.* Like the right atrium, the left atrium is characterized by a smooth posterior wall. Pectinate muscles, however, are found only in the right atrium.

The *left ventricle* forms the apex of the heart. Oxygenated blood passes from the left atrium into the left ventricle through a valve called the *bicuspid (mitral) valve.* Like the tricuspid valve, the bicuspid valve has chordae tendineae attached to papillary muscles. However, as the name indicates, the bicuspid valve has only two cusps (not three as in the tricuspid valve). The left ventricle also contains trabeculae carneae.

Oxygenated blood passes into systemic circulation from the left ventricle through a valve called

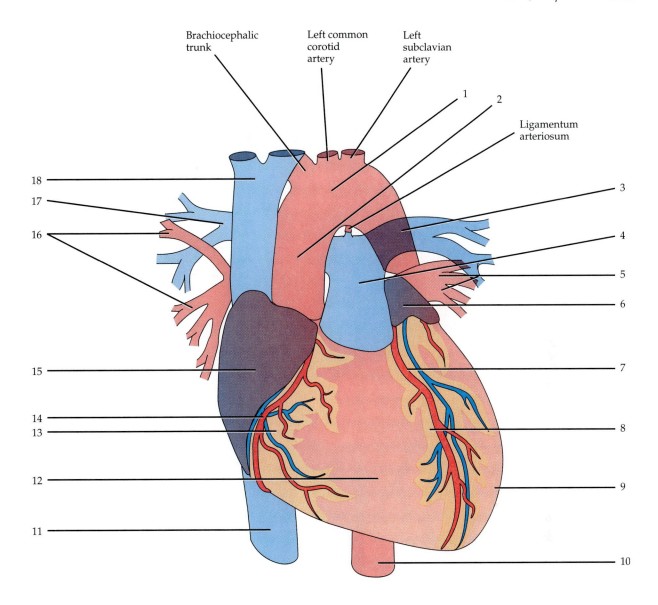

Brachiocephalic trunk

Left common corotid artery

Left subclavian artery

Ligamentum arteriosum

1
2
3
4
5
6
7
8
9
10
11
12
13
14
15
16
17
18

(a) Diagram of anterior external view

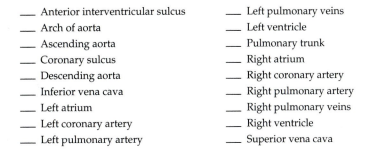

___ Anterior interventricular sulcus
___ Arch of aorta
___ Ascending aorta
___ Coronary sulcus
___ Descending aorta
___ Inferior vena cava
___ Left atrium
___ Left coronary artery
___ Left pulmonary artery

___ Left pulmonary veins
___ Left ventricle
___ Pulmonary trunk
___ Right atrium
___ Right coronary artery
___ Right pulmonary artery
___ Right pulmonary veins
___ Right ventricle
___ Superior vena cava

FIGURE 18.2   External surface of the human heart. Red-colored vessels carry oxygenated blood; blue-colored vessels carry deoxygenated blood.

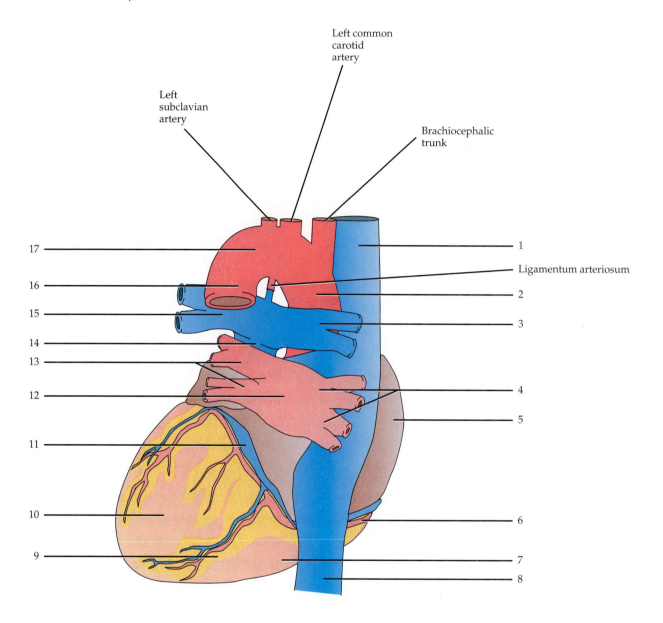

Left common
carotid
artery

Left
subclavian
artery

Brachiocephalic
trunk

17

16

15

14

13

12

11

10

9

1

Ligamentum arteriosum

2

3

4

5

6

7

8

(b) Diagram of posterior external view

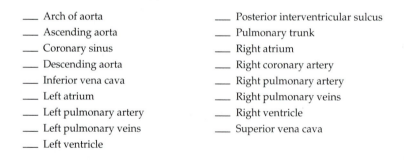

___ Arch of aorta
___ Ascending aorta
___ Coronary sinus
___ Descending aorta
___ Inferior vena cava
___ Left atrium
___ Left pulmonary artery
___ Left pulmonary veins
___ Left ventricle

___ Posterior interventricular sulcus
___ Pulmonary trunk
___ Right atrium
___ Right coronary artery
___ Right pulmonary artery
___ Right pulmonary veins
___ Right ventricle
___ Superior vena cava

FIGURE 18.2 (Continued)   External surface of the human heart.

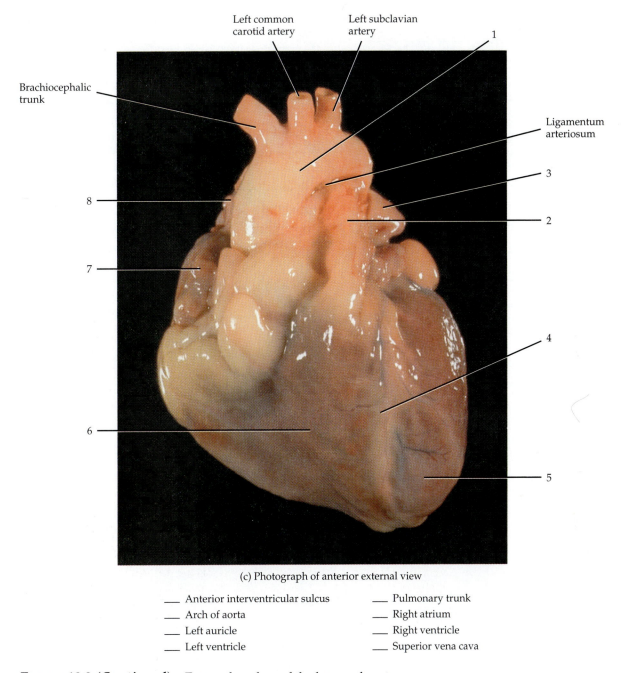

(c) Photograph of anterior external view

___ Anterior interventricular sulcus   ___ Pulmonary trunk
___ Arch of aorta   ___ Right atrium
___ Left auricle   ___ Right ventricle
___ Left ventricle   ___ Superior vena cava

FIGURE 18.2 *(Continued)*   External surface of the human heart.

the *aortic semilunar valve* into the **ascending aorta.** From here some of the blood flows into the left and right **coronary arteries,** which branch from the ascending aorta and carry the blood to the heart wall. The remainder of the blood passes into the **arch of the aorta** and **descending aorta** (**thoracic aorta** and **abdominal aorta**). Branches of the arch of the aorta and descending aorta carry the blood throughout the systemic circulation.

During fetal life, a temporary blood vessel, called the ductus arteriosus, connects the pulmonary trunk with the aorta. It redirects blood so that only a small amount enters the nonfunctioning fetal lungs. The ductus arteriosus normally closes shortly after birth, leaving a remnant known as the **ligamentum arteriosum,** which interconnects the arch of the aorta and pulmonary trunk.

Label these structures in Figures 18.2 and 18.3.

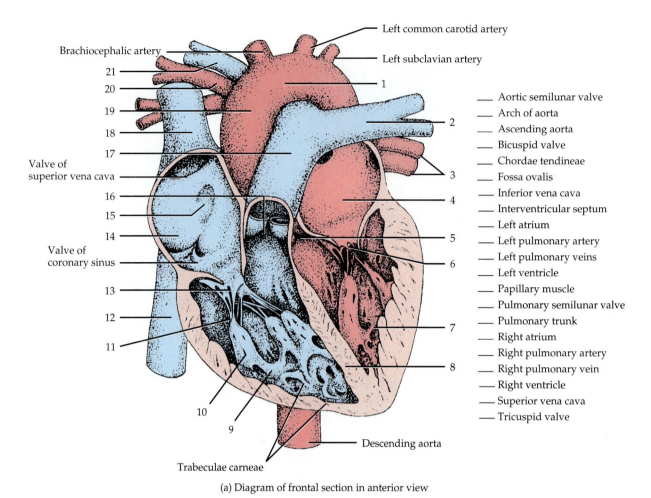

Left common carotid artery

Brachiocephalic artery

Left subclavian artery

21

20

19

18

17

Valve of
superior vena cava

16

15

14

Valve of
coronary sinus

13

12

11

1

2

3

4

5

6

7

8

___ Aortic semilunar valve
___ Arch of aorta
___ Ascending aorta
___ Bicuspid valve
___ Chordae tendineae
___ Fossa ovalis
___ Inferior vena cava
___ Interventricular septum
___ Left atrium
___ Left pulmonary artery
___ Left pulmonary veins
___ Left ventricle
___ Papillary muscle
___ Pulmonary semilunar valve
___ Pulmonary trunk
___ Right atrium
___ Right pulmonary artery
___ Right pulmonary vein
___ Right ventricle
___ Superior vena cava
___ Tricuspid valve

10

9

Descending aorta

Trabeculae carneae

(a) Diagram of frontal section in anterior view

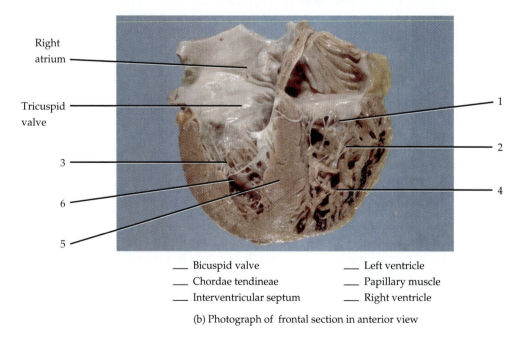

Right
atrium

Tricuspid
valve

3

6

5

1

2

4

___ Bicuspid valve   ___ Left ventricle
___ Chordae tendineae   ___ Papillary muscle
___ Interventricular septum ___ Right ventricle

(b) Photograph of frontal section in anterior view

FIGURE 18.3   Structure of the human heart.

D. VALVES OF HEART

As each chamber of the heart contracts, it pushes a volume of blood into a ventricle or out of the heart through an artery. To prevent backflow of blood, the heart has *valves.* These structures are composed of dense connective tissue covered by endocardium. Valves open and close in response to pressure changes as the heart contracts and relaxes.

Atrioventricular (AV) valves lie between the atria and ventricles. The right AV valve between the right atrium and right ventricle is also called the *tricuspid* (trī-KUS-pid) *valve* because it consists of three cusps (flaps). The left AV valve between the left atrium and left ventricle has two cusps and is called the *bicuspid (mitral) valve.* When an AV valve is open, the pointed ends of the cusps project into a ventricle. Tendonlike cords called *chordae tendineae* connect the pointed ends and undersurfaces to *papillary muscles* (muscular columns) that are located on the inner surface of the ventricles.

Near the origin of both arteries that emerge from the heart, there are heart valves that allow ejection of blood from the heart but prevent blood from flowing backward into the heart. These are the *semilunar (SL) valves.* The *pulmonary semilunar valve* lies in the opening where the pulmonary trunk leaves the right ventricle. The *aortic semilunar valve* is situated at the opening between the left ventricle and the aorta.

Both valves consist of three semilunar (half-moon, or crescent-shaped) cusps. Each cusp is attached by its convex margin to the artery wall. The free borders of the cusps curve outward and project into the opening inside the blood vessel. When blood starts to flow backward toward the heart as the ventricles relax, it fills the cusps and tightly closes the semilunar valves.

Label the atrioventricular and semilunar valves in Figures 18.2 and 18.3.

E. BLOOD SUPPLY OF HEART

Nutrients could not possibly diffuse from the chambers of the heart through all the layers of cells that make up the heart tissue. For this reason, the wall of the heart has its own blood vessels. The flow of blood through the many vessels that pierce the myocardium is called the *coronary (cardiac) circulation.* The arteries of the heart encircle it like a crown encircles the head (*corona* = crown). While it is contracting, the heart receives little flow of oxygenated blood by way of the *coronary arteries,* which branch from the ascending aorta. When the heart relaxes, however, the high pressure of blood in the aorta propels blood through the coronary arteries, into capillaries, and then into *coronary veins.*

Two coronary arteries, the right and left coronary arteries, branch from the ascending aorta and supply oxygenated blood to the myocardium. The *left coronary artery* passes inferior to the left auricle and divides into the anterior interventricular and circumflex branches. The *anterior interventricular branch* or *left anterior descending (LAD) artery* is in the anterior interventricular sulcus and supplies oxygenated blood to the walls of both ventricles. The *circumflex branch* lies in the coronary sulcus and distributes oxygenated blood to the walls of the left ventricle and left atrium.

The *right coronary artery* supplies small branches (atrial branches) to the right atrium. It continues inferior to the right auricle and divides into the posterior interventricular and marginal branches. The *posterior interventricular branch* follows the posterior interventricular sulcus, supplying walls of the two ventricles with oxygenated blood. The *marginal branch* in the coronary sulcus transports oxygenated blood to the myocardium of the right ventricle.

Label the arteries in Figure 18.4a on page 354.

After blood passes through the arteries of coronary circulation, where it delivers oxygen and nutrients, it passes into veins, where it collects carbon dioxide and wastes. The deoxygenated blood then drains into a large vascular sinus on the posterior surface of the heart, called the *coronary sinus,* which empties into the right atrium. A vascular sinus is a vein with a thin wall that has no smooth muscle to alter its diameter. The principal tributaries carrying blood into the coronary sinus are the *great cardiac vein,* which drains the anterior aspect of the heart, and the *middle cardiac vein,* which drains the posterior aspect of the heart.

Label the veins in Figure 18.4b on page 354.

F. DISSECTION OF SHEEP HEART

The anatomy of the sheep heart closely resembles that of the human heart. Use Figures 18.5 and 18.6 on pages 355–356 as references for this dissection. In addition, models of human hearts can also be used as references.

CAUTION! *Please reread Section D, "Precautions Related to Dissection" at the beginning of the laboratory manual on page xiii before you begin your dissection.*

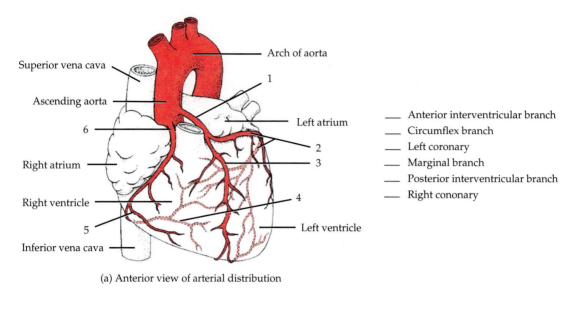

Superior vena cava

Ascending aorta

6

Right atrium

Right ventricle

5

Inferior vena cava

Arch of aorta

1

Left atrium

2

3

4

Left ventricle

___ Anterior interventricular branch
___ Circumflex branch
___ Left coronary
___ Marginal branch
___ Posterior interventricular branch
___ Right cononary

(a) Anterior view of arterial distribution

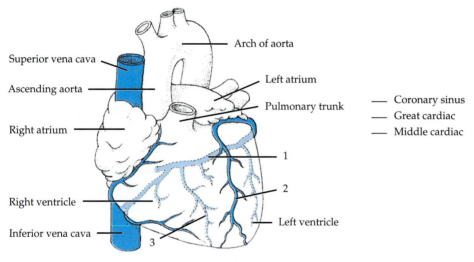

Superior vena cava

Ascending aorta

Right atrium

Right ventricle

Inferior vena cava

Arch of aorta

Left atrium

Pulmonary trunk

1

2

Left ventricle

3

___ Coronary sinus
___ Great cardiac
___ Middle cardiac

(b) Anterior view of venous drainage

FIGURE 18.4 Coronary (cardiac) circulation.

First examine the **pericardium**, a fibroserous membrane that encloses the heart, which may have already been removed in preparing the sheep heart for preservation. The **myocardium** is the middle layer and constitutes the main muscle portion of the heart. The **endocardium** (the third layer) is the inner lining of the heart. Use the figures to determine which is the ventral surface of the heart and then identify the **pulmonary trunk** emerging from the anterior ventral surface, near the midline, and medial to the **left auricle**. A longitudinal depression on the ventral surface, called the **anterior longitudinal sulcus**, separates the right ventricle from the left ventricle. Locate the **coronary blood vessels** lying in this sulcus.

PROCEDURE

1. Remove any fat or pulmonary tissue that is present.
2. In cutting the sheep heart open to examine the chambers, valves, and vessels, the anterior longitudinal sulcus is used as a guide.
3. Carefully make a shallow incision through the ventral wall of the pulmonary trunk and the right ventricle, trying not to cut the dorsal surface of either structure.
4. The incision is best made *less than an inch to the right of, and parallel to,* the previously mentioned anterior longitudinal sulcus.
5. If necessary, the incision can be continued to where the pulmonary trunk branches into a **right**

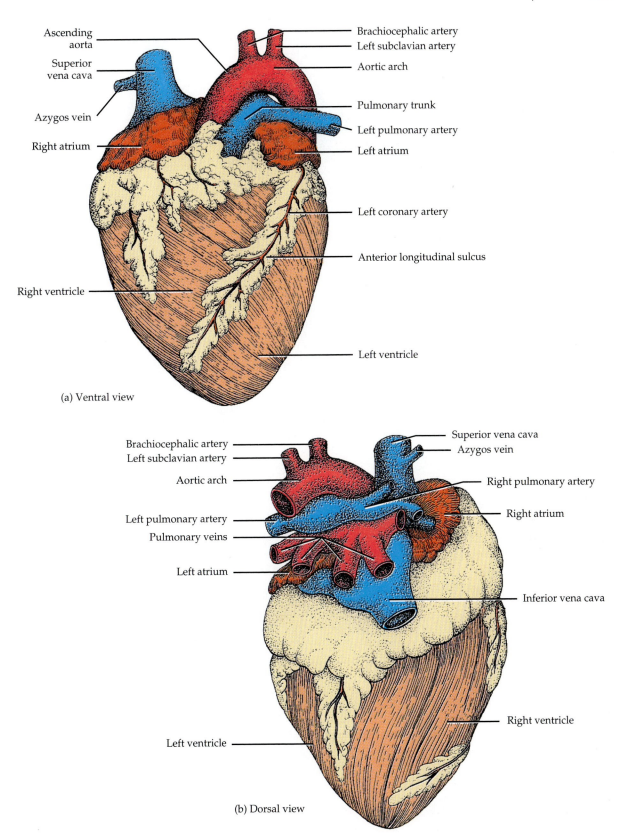

Ascending aorta

Superior vena cava

Azygos vein

Right atrium

Brachiocephalic artery

Left subclavian artery

Aortic arch

Pulmonary trunk

Left pulmonary artery

Left atrium

Left coronary artery

Anterior longitudinal sulcus

Right ventricle

Left ventricle

(a) Ventral view

Brachiocephalic artery

Left subclavian artery

Aortic arch

Left pulmonary artery

Pulmonary veins

Left atrium

Superior vena cava

Azygos vein

Right pulmonary artery

Right atrium

Inferior vena cava

Right ventricle

Left ventricle

(b) Dorsal view

FIGURE 18.5   External structure of a cat or sheep heart.

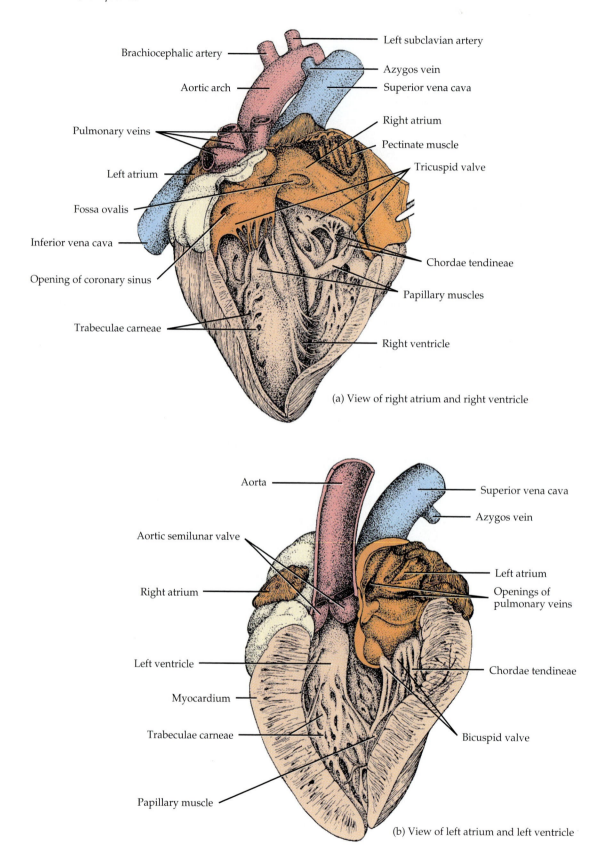

Left subclavian artery

Brachiocephalic artery

Aortic arch

Azygos vein

Superior vena cava

Pulmonary veins

Right atrium

Pectinate muscle

Tricuspid valve

Left atrium

Fossa ovalis

Inferior vena cava

Opening of coronary sinus

Chordae tendineae

Papillary muscles

Trabeculae carneae

Right ventricle

(a) View of right atrium and right ventricle

Aorta

Superior vena cava

Azygos vein

Aortic semilunar valve

Right atrium

Left atrium

Openings of pulmonary veins

Left ventricle

Myocardium

Chordae tendineae

Trabeculae carneae

Bicuspid valve

Papillary muscle

(b) View of left atrium and left ventricle

FIGURE 18.6   Internal structure of a cat or sheep heart.

pulmonary artery, which goes to the right lung, and a *left pulmonary artery*, which goes to the left lung. The *pulmonary semilunar valve* of the pulmonary artery can be clearly seen upon opening it. In any of these internal dissections of the heart, any coagulated blood or latex should be immediately removed so that all important structures can be located and identified.

6. Keeping the cut still parallel to the sulcus, extend the incision around and through the dorsal ventricular wall until you reach the *interventricular septum.*

7. Now examine the dorsal surface of the heart and locate the thin-walled *superior vena cava* directly above the *right auricle*. This vein proceeds posteriorly straight into the right atrium.

8. Make a second longitudinal cut, this time through the superior vena cava (dorsal wall).

9. Extend the cut posteriorly through the right atrium on the left of the right auricle. Proceed posteriorly to the dorsal right ventricular wall and join your first incision.

10. The entire internal right side of the heart should now be clearly seen when carefully spread apart. The interior of the superior vena cava, right atrium, and right ventricle will now be examined. Start with the right auricle and locate the *pectinate muscle*, the large opening of the *inferior vena cava* on the left side of the right atrium, and the opening of the *coronary sinus* just below the opening of the inferior vena cava. By using a dull probe and gentle pressure, most of the vessels can be traced to the dorsal surface of the heart.

11. Now find the wall that separates the two atria, the *interatrial septum*. Also find the *fossa ovalis*, an oval-shaped depression ventral to the entrance of the inferior vena cava.

12. The *tricuspid valve* between the right atrium and the right ventricle should be examined to locate the three cusps, as its name indicates. From the cusps of the valve itself, and tracing posteriorly, the *chordae tendineae*, which hold the valve in place, should be identified. Still tracing posteriorly, the chordae are seen to originate from the *papillary muscles*, which themselves originate from the wall of the right ventricle itself.

13. Look carefully again at the dorsal surface of the left atrium and locate as many *pulmonary veins* (normally, four) as possible.

14. Make your third longitudinal cut through the most lateral of the pulmonary veins that you have located.

15. Continue posteriorly through the left atrial wall and the left ventricle to the *apex* of the heart.

16. Compare the difference in the thickness of the wall between the right and left ventricles. Explain your answer.

17. Examine the *bicuspid (mitral) valve*, again counting the cusps. Determine if the left side of the heart has basically the same structures as studied on the right side.

18. Probe from the left ventricle to the *aorta* as it emerges from the heart, examining the *aortic semilunar valve*. Find the openings of the right and left main coronary arteries.

19. Locate now the *brachiocephalic artery*, which is one of the first branches from the arch of the aorta. This artery continues branching and terminates by supplying the arms and head as its name indicates.

20. Connecting the aorta with the pulmonary artery is the remnant of the *ductus arteriosus*, called the *ligamentum arteriosum*. It may not be present in your sheep heart.

ANSWER THE LABORATORY REPORT QUESTIONS AT THE END OF THE EXERCISE.

Heart 18

Student _____   Date _____

Laboratory Section _____   Score/Grade _____

PART 1. Multiple Choice

_____ 1. Which of the following veins drain the blood from most of the vessels supplying the heart wall? (a) vasa vasorum (b) superior vena cava (c) coronary sinus (d) inferior vena cava

_____ 2. The atrioventricular valve on the same side of the heart as the origin of the aorta is the (a) aortic semilunar (b) tricuspid (c) bicuspid (d) pulmonary semilunar

_____ 3. Which valve does the blood go through just before entering the pulmonary trunk on the way to the lungs? (a) tricuspid (b) pulmonary semilunar (c) aortic semilunar (d) bicuspid

_____ 4. The pointed end of the heart that projects inferiorly and to the left is the (a) costal surface (b) base (c) apex (d) coronary sulcus

_____ 5. Which of these structures is more internal? (a) fibrous pericardium (b) visceral layer of serous pericardium (c) parietal layer of serous pericardium (d) myocardium

_____ 6. The musculature of the heart is referred to as the (a) endocardium (b) myocardium (c) epicardium (d) pericardium

_____ 7. The depression in the interatrial septum corresponding to the foramen ovale of fetal circulation is the (a) interventricular sulcus (b) pectinate muscle (c) chordae tendineae (d) fossa ovalis

PART 2. Completion

8. Malfunction of the _____ valve would interfere with the flow of blood from the right atrium to the right ventricle.

9. Deoxygenated blood is sent to the lungs through the _____.

10. The triple-layered sac that encloses the heart is called the _____.

11. The two inferior chambers of the heart are separated by the _____.

12. The earlike flap of tissue on each atrium is called a(n) _____.

13. The large vein that drains blood from most parts of the body superior to the heart and empties into

the right atrium is the _____.

14. The cusps of atrioventricular valves are prevented from inverting by the presence of cords called

_____, which are attached to papillary muscle.

15. A groove on the surface of the heart that houses blood vessels and a variable amount of fat is called

    a(n) _____.

16. The branch of the left coronary artery that distributes blood to the left atrium and left ventricle is the

    _____.

17. The ridges and folds in the ventricles are called _____.

18. The _____ vein drains the anterior aspect of the heart.

PART 3. Special Exercise

Draw a model of the heart and carefully label the four chambers, the four valves in their proper places, and the major blood vessels entering and exiting from the heart.

Blood Vessels

Blood vessels are networks of tubes that carry blood throughout the body. They are called arteries, arterioles, capillaries, venules, or veins. In this exercise, you will study the histology of blood vessels and identify the principal arteries and veins of the human cardiovascular system.

A. ARTERIES AND ARTERIOLES

Arteries (AR-ter-ēs; *aer* = air; *tereo* = to carry) are blood vessels that carry blood *away* from the heart to body tissues. Arteries are constructed of three coats of tissue called *tunics* and a hollow core, called a *lumen,* through which blood flows (Figure 19.1). The deep coat is called the *tunica interna* and consists of a lining of endothelium in contact with the blood and a layer of elastic tissue called the *internal elastic lamina.* The middle coat, or *tunica media,* is usually the thickest layer and consists of elastic fibers and smooth muscle fibers. This tunic is responsible for two major properties of arteries: *elasticity* and *contractility.* The superficial coat, or *tunica externa,* is composed principally of elastic and collagen fibers. An *external elastic lamina* may separate the tunica externa from the tunica media.

Obtain a prepared slide of a transverse section of an artery and identify the tunics using Figure 19.1 as a guide.

As arteries approach various tissues of the body, they become smaller and are known as *arterioles* (ar-TER-rē-ōls; *arteriola* = small artery). Arterioles play a key role in regulating blood flow from arteries into capillaries. When arterioles enter a tissue, they branch into countless microscopic blood vessels called capillaries.

B. CAPILLARIES

Capillaries (KAP-i-lar'-ēs; *capillaris* = hairlike) are microscopic blood vessels that connect arterioles and venules. Their function is to permit the exchange of nutrients and wastes between blood and body tissues. This function is related to the fact that capillaries consist of only a single layer of endothelium.

C. VENULES AND VEINS

When several capillaries unite, they form small veins called *venules* (VEN-yools; *venula* = little vein). They collect blood from capillaries and drain it into veins.

Veins (VĀNS) are composed of the same three tunics as arteries, but there are variations in their relative thicknesses. The tunica interna of veins is thinner than that of their accompanying arteries. In addition, the tunica media of veins is much thinner than that of accompanying arteries with relatively little smooth muscle or elastic fibers. The tunica externa is the thickest layer consisting of collagen and elastic fibers (Figure 19.1). Functionally, veins return blood from tissues *to* the heart.

Obtain a prepared slide of a transverse section of an artery and its accompanying vein and compare them, using Figure 19.1 as a guide.

D. CIRCULATORY ROUTES

The two basic postnatal (after birth) circulatory routes are systemic and pulmonary circulation (Figure 19.2). Some other circulatory routes, which are all subdivisions of systemic circulation, include hepatic portal circulation, coronary (cardiac) circulation, fetal circulation, and the cerebral arterial circle (circle of Willis). The latter is found at the base of the brain (see Table 19.2).

1. Systemic Circulation

The largest route is the *systemic circulation* (see Figures 19.3 through 19.11 and Tables 19.1 through 19.11). This route includes the flow of blood from

Lumen

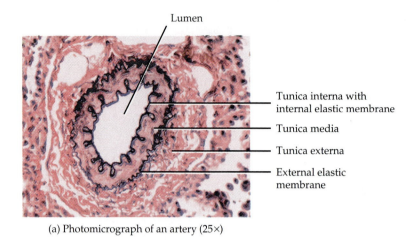

Tunica interna with
internal elastic membrane

Tunica media

Tunica externa

External elastic
membrane

(a) Photomicrograph of an artery (25×)

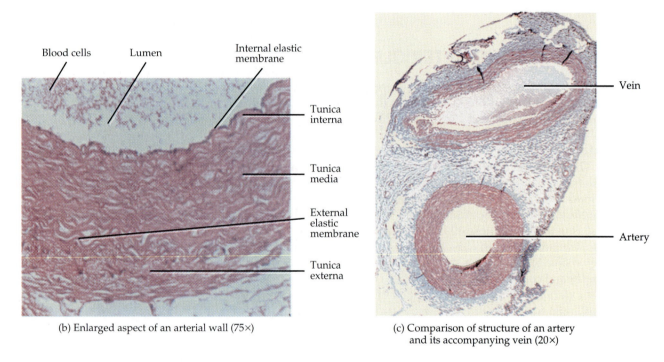

Blood cells   Lumen   Internal elastic
membrane

Tunica
interna

Tunica
media

External
elastic
membrane

Tunica
externa

(b) Enlarged aspect of an arterial wall (75×)

Vein

Artery

(c) Comparison of structure of an artery
and its accompanying vein (20×)

FIGURE 19.1   Histology of blood vessels.

the left ventricle to all parts of the body. The function of the systemic circulation is to carry oxygen and nutrients to body tissues and to remove carbon dioxide and other wastes from them. All systemic arteries branch from the *aorta.* As the aorta emerges from the left ventricle, it passes superiorly and posteriorly to the pulmonary trunk. At this point, it is called the *ascending aorta.* The ascending aorta gives off two coronary branches (right and left coronary arteries) to the heart muscle. Then it turns to the left, forming the *arch of the aorta* before descending to the level of the intervertebral disc between the fourth and the fifth thoracic vertebrae as the *descending aorta.* The descending aorta lies close to the vertebral bodies, passes through the diaphragm, and divides at the level of the fourth lumbar vertebra into two *common iliac arteries,* which carry blood to the lower limbs. The section of the descending aorta between the arch of the aorta and the diaphragm is referred to as the *thoracic aorta.*

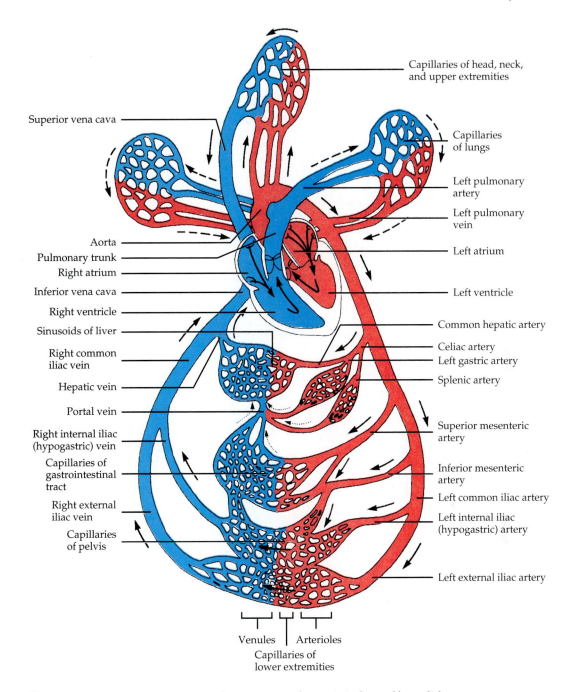

Superior vena cava

Aorta
Pulmonary trunk
Right atrium
Inferior vena cava
Right ventricle
Sinusoids of liver
Right common
iliac vein
Hepatic vein
Portal vein
Right internal iliac
(hypogastric) vein
Capillaries of
gastrointestinal
tract
Right external
iliac vein
Capillaries
of pelvis

Capillaries of head, neck,
and upper extremities

Capillaries
of lungs
Left pulmonary
artery
Left pulmonary
vein
Left atrium

Left ventricle

Common hepatic artery
Celiac artery
Left gastric artery
Splenic artery

Superior mesenteric
artery

Inferior mesenteric
artery
Left common iliac artery
Left internal iliac
(hypogastric) artery

Left external iliac artery

Venules | Arterioles
Capillaries of
lower extremities

FIGURE 19.2   Circulatory routes. Systemic circulation is indicated by solid arrows;
pulmonary circulation by broken arrows; and hepatic portal circulation by dotted
arrows.

The section between the diaphragm and the common iliac arteries is termed the *abdominal aorta.* Each section of the aorta gives off arteries that continue to branch into distributing arteries leading to organs and finally into the arterioles and capillaries that service the systemic tissues (except the air sacs of the lungs).

Deoxygenated blood is returned to the heart through the systemic veins. All the veins of the systemic circulation flow into either the *superior vena cava, inferior vena cava,* or *coronary sinus.* They in turn empty into the right atrium.

Refer to Tables 19.1 through 19.11 and Figures 19.3 through 19.12.

TABLE 19.1
Aorta and Its Branches (Figure 19.3)

OVERVIEW: The *aorta* (ā-OR-ta) is the largest artery of the body, about 2 to 3 cm (0.8 to 1.2 in.) in diameter. It begins at the left ventricle and contains a valve at its origin, called the aortic semilunar valve (see Figure 18.3a), which prevents backflow of blood into the left ventricle during its diastole (relaxation). The principal divisions of the aorta are the ascending aorta, arch of the aorta, thoracic aorta, and abdominal aorta.

Division of aorta	Arterial branch		Region supplied
Ascending aorta (ā-OR-ta)	Right and left coronary		Heart
Arch of aorta	Brachiocephalic (brā'-kē-ō-se FAL-ik) trunk	Right common carotid (ka-ROT-id)	Right side of head and neck
		Right subclavian (sub KLĀ-vē-an)	Right upper limb
	Left common carotid		Left side of head and neck
	Left subclavian		Left upper limb
Thoracic (thō-RAS-ik) *aorta*	Intercostals (in'-ter-KOS-tals)		Intercostal and chest muscles, pleurae
	Superior phrenics (FREN-iks)		Posterior and superior surfaces of diaphragm
	Bronchials (BRONG-kē-als)		Bronchi of lungs
	Esophageals (e-sof'-a-JĒ-als)		Esophagus
Abdominal (ab-DOM-i-nal) *aorta*	Inferior phrenics (FREN-iks)		Inferior surface of diaphragm
	Celiac	Common hepatic (he-PAT-ik)	Liver
		Left gastric (GAS-trik)	Stomach and esophagus
		Splenic (SPLĒN-ik)	Spleen, pancreas, stomach
	Superior mesenteric (MES-en-ter'-ik)		Small intestine, cecum, ascending and transverse colons, and pancreas
	Suprarenals (soo'-pra-RĒ-nals)		Adrenal (suprarenal) glands
	Renals (RĒ-nals)		Kidneys
	Gonadals (gō-NAD-als)	Testiculars (tes-TIK-yoo-lars) or	Testes
		Ovarians (ō-VA-rē-ans)	Ovaries
	Inferior mesenteric (MES-en-ter'-ik)		Transverse, descending, sigmoid colons and rectum
	Common iliacs (IL-ē-aks)	External iliacs	Lower limbs
		Internal iliacs (hypogastrics)	Uterus, prostate gland, muscles of buttocks, and urinary bladder

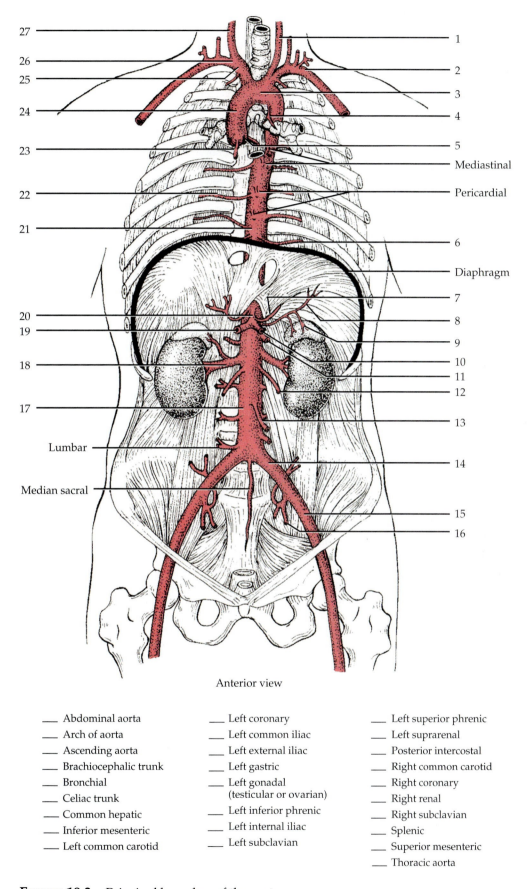

27 ___
26 ___
25 ___
24 ___
23 ___
22 ___
21 ___
20 ___
19 ___
18 ___
17 ___
Lumbar ___
Median sacral ___

1 ___
2 ___
3 ___
4 ___
5 ___
Mediastinal
Pericardial
6 ___
Diaphragm
7 ___
8 ___
9 ___
10 ___
11 ___
12 ___
13 ___
14 ___
15 ___
16 ___

Anterior view

___ Abdominal aorta
___ Arch of aorta
___ Ascending aorta
___ Brachiocephalic trunk
___ Bronchial
___ Celiac trunk
___ Common hepatic
___ Inferior mesenteric
___ Left common carotid

___ Left coronary
___ Left common iliac
___ Left external iliac
___ Left gastric
___ Left gonadal
   (testicular or ovarian)
___ Left inferior phrenic
___ Left internal iliac
___ Left subclavian

___ Left superior phrenic
___ Left suprarenal
___ Posterior intercostal
___ Right common carotid
___ Right coronary
___ Right renal
___ Right subclavian
___ Splenic
___ Superior mesenteric
___ Thoracic aorta

FIGURE 19.3 Principal branches of the aorta.

TABLE 19.2
Arch of Aorta (Figures 19.4, 19.5, and 19.6)

OVERVIEW: The *arch of the aorta* is about 4½ cm (almost 2 in.) in length and is the continuation of the ascending aorta. It emerges from the pericardium posterior to the sternum at the level of the sternal angle. Initially, the arch is directed superiorly, posteriorly and to the left, and then inferiorly on the left side of the body of the fourth thoracic vertebra. Actually, the arch is directed not only from right to left, but from anterior to posterior as well. The arch of the aorta ends at the level of the intervertebral disc between the fourth and fifth thoracic vertebrae, where it becomes the thoracic aorta. The thymus gland lies anterior of the arch of the aorta, while the trachea lies posterior.

Three major arteries branch from the superior aspect of the arch of the aorta. In order of their origination, they are the brachiocephalic trunk, left common carotid artery, and left subclavian artery.

Branch	Description and region
Brachiocephalic (brā'-kē-ō-se-FAL-ik; *brachium* = arm; *cephalic* = head) *trunk*	The *brachiocephalic trunk,* which is found only on the right side, is the first and largest branch off the arch of the aorta. There is no left brachiocephalic artery. It bifurcates (divides) at the right sternoclavicular joint to form the right subclavian artery and right common carotid artery. The *right subclavian* (sub-KLĀ-vē-an) *artery* extends from the brachiocephalic to the first rib and then passes into the armpit (axilla) and supplies the arm, forearm, and hand. Continuation of the right subclavian into the axilla is called the *axillary* (AK-si-ler'-ē) *artery.* From here, it continues into the arm as the *brachial* (BRĀ-kē-al) *artery.* At the bend of the elbow, the brachial artery divides into the medial *ulnar* (UL-nar) and lateral *radial* (RĀ-dē-al) *arteries.* These vessels pass down to the palm, one on each side of the forearm. In the palm, branches of the two arteries anastomose to form two palmar arches—the *superficial palmar* (PAL-mar) *arch* and the *deep palmar arch.* From these arches arise the *digital* (DIJ-i-tal) *arteries,* which supply the fingers and thumb (Figure 19.4).
	Before passing into the axilla, the right subclavian gives off a major branch to the brain called the *right vertebral* (VER-te-bral) *artery.* The right vertebral artery passes through the foramina of transverse processes of the cervical vertebrae and enters the skull through the foramen magnum to reach the inferior surface of the brain. Here it unites with the left vertebral artery to form the *basilar* (BAS-i-lar) *artery* (Figures 19.5 and 19.6).
	The *right common carotid artery* passes upward in the neck. At the upper level of the larynx, it divides into the *right external* and *right internal carotid* (ka-ROT-id) *arteries.* The external carotid supplies the right side of the thyroid gland, tongue, throat, face, ear, scalp, and dura mater. The internal carotid supplies the brain, right eye, and right sides of the forehead and nose (Figure 19.5).
	Inside the cranium, anastomoses of the left and right internal carotids along with the basilar artery form an arrangement of blood vessels at the base of the brain near the sella turcica called the *cerebral* (se-RĒ-bral) *arterial circle (circle of Willis).* From this circle arise arteries supplying most of the brain. Essentially the cerebral arterial circle is formed by the union of the *anterior cerebral arteries* (branches of internal carotids) and *posterior cerebral arteries* (branches of basilar artery). Posterior cerebral arteries are connected with internal carotids by the *posterior communicating* (ko-MYOO-ni-kā'-ting) *arteries.* The anterior cerebral arteries are connected by the *anterior communicating arteries.* The *internal carotid* (ka-ROT-id) *arteries* are also considered part of the cerebral arterial circle. The function of the cerebral arterial circle is to equalize blood pressure to the brain and provide alternate routes for blood to the brain, should the arteries become damaged.
Left common carotid (ka-ROT-id)	The *left common carotid* is the second branch off the arch of the aorta. Corresponding to the right common carotid, it divides into basically the same branches with the same names, except that the arteries are now labeled "left" instead of "right."
Left subclavian (sub-KLĀ-vē-an)	The *left subclavian artery* is the third branch off the arch of the aorta. It distributes blood to the left vertebral artery and vessels of the left upper limb. Arteries branching from the left subclavian are named like those of the right subclavian.

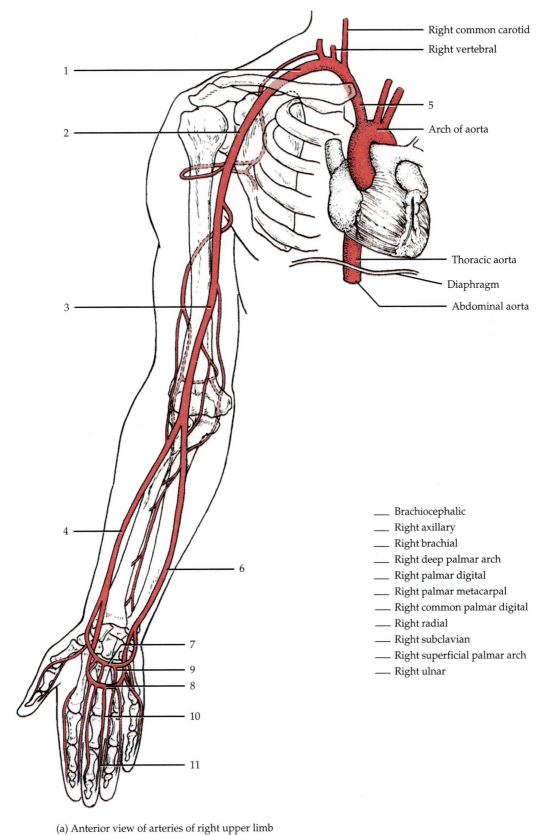

Right common carotid

Right vertebral

5

Arch of aorta

Thoracic aorta

Diaphragm

Abdominal aorta

___ Brachiocephalic
___ Right axillary
___ Right brachial
___ Right deep palmar arch
___ Right palmar digital
___ Right palmar metacarpal
___ Right common palmar digital
___ Right radial
___ Right subclavian
___ Right superficial palmar arch
___ Right ulnar

(a) Anterior view of arteries of right upper limb

FIGURE 19.4   Branches of the arch of the aorta.

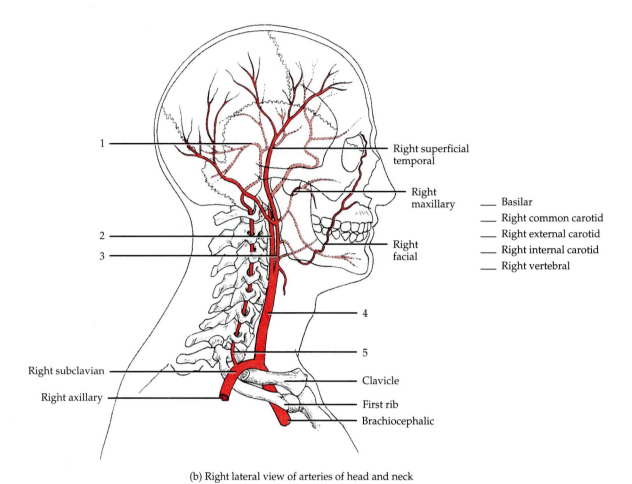

Basilar
Right common carotid
Right external carotid
Right internal carotid
Right vertebral

Right superficial temporal

Right maxillary

Right facial

Right subclavian

Right axillary

Clavicle

First rib

Brachiocephalic

(b) Right lateral view of arteries of head and neck

FIGURE 19.5   Arteries of neck and head.

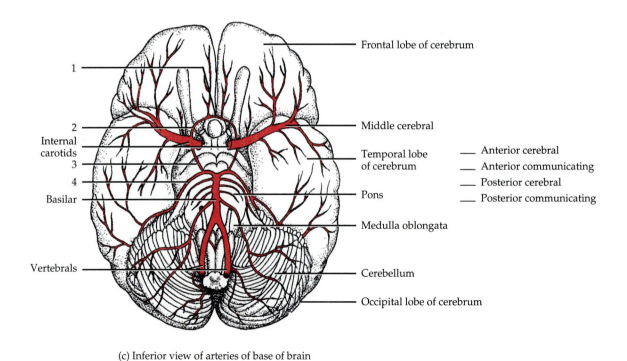

Frontal lobe of cerebrum

Internal carotids

Basilar

Vertebrals

Middle cerebral

Temporal lobe of cerebrum

Pons

Medulla oblongata

Cerebellum

Occipital lobe of cerebrum

Anterior cerebral
Anterior communicating
Posterior cerebral
Posterior communicating

(c) Inferior view of arteries of base of brain

FIGURE 19.6   Branches of the arch of the aorta.

Write the names of the missing arteries in the following scheme of circulation. Be sure to indicate left or right where applicable.

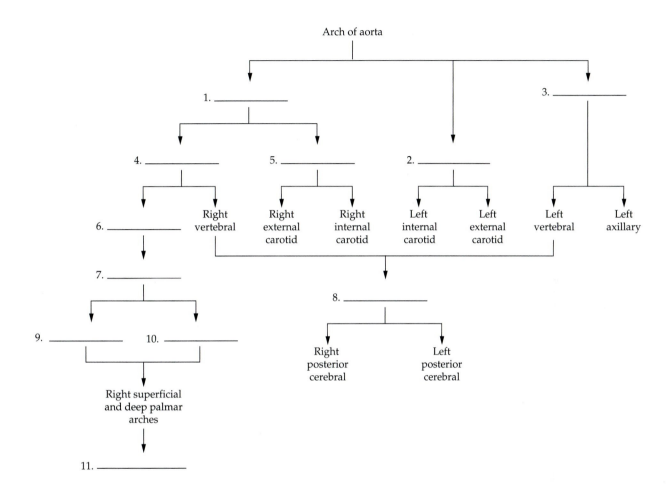

TABLE 19.3
Thoracic Aorta (Figure 19.3)

OVERVIEW: The *thoracic aorta* is about 20 cm (8 in.) long and is a continuation of the arch of the aorta. It begins at the level of the intervertebral disc between the fourth and fifth thoracic vertebrae, where it lies to the left of the vertebral column. As it descends, it moves closer to the midline and terminates at an opening in the diaphragm (aortic hiatus) in front of the vertebral column at the level of the intervertebral disc between the twelfth thoracic and first lumbar vertebrae.

Along its course, the thoracic aorta sends off numerous small arteries to the viscera *(visceral branches)* and body wall structures *(parietal branches).*

Branch	Description and region supplied
VISCERAL	
Pericardial (per'-i-KAR-dē-al; *peri* = around; *cardia* = heart) *arteries*	Two or three minute *pericardial arteries* supply blood to the pericardium.
Bronchial (BRONG-kē-al; *bronchus* = windpipe) *arteries*	One right and two left *bronchial arteries* supply the bronchial tubes, visceral pleurae, bronchial lymph nodes, and esophagus. (Whereas the right bronchial artery arises from the third posterior intercostal artery, the two left bronchial arteries arise from the thoracic aorta.)
Esophageal (e-sof'-a-JĒ-al) *arteries*	Four or five *esophageal arteries* supply the esophagus.
Mediastinal (mē'-dē-as-TĪ-nal) *arteries*	Numerous small *mediastinal arteries* supply blood to structures in the posterior mediastinum.
PARIETAL	
Posterior intercostal (in'-ter-KOS-tal; *inter* = between; *costa* = rib) *arteries*	Nine pairs of *posterior intercostal arteries* supply the intercostal, pectoral major and minor, and serratus anterior muscles; overlying subcutaneous tissue and skin; mammary glands; and vertebral canal and its contents.
Subcostal (SUB-kos-tal; *sub* = under) *arteries*	The left and right *subcostal arteries* have a distribution similar to that of the posterior intercostals.
Superior phrenic (FREN-ik; *phren* = diaphragm) *arteries*	Small *superior phrenic arteries* supply the posterior and superior surfaces of the diaphragm.

TABLE 19.4
Abdominal Aorta (Figure 19.7)

OVERVIEW: The *abdominal aorta* is the continuation of the thoracic aorta. It begins at the aortic hiatus in the diaphragm and ends at about the level of the fourth lumbar vertebra, where it divides into right and left common iliac arteries. The abdominal aorta lies in front of the vertebral column.

As with the thoracic aorta, the abdominal aorta gives off *visceral* and *parietal branches*. The unpaired visceral branches arise from the anterior surface of the aorta and include the *celiac, superior mesenteric,* and *inferior mesenteric arteries.* The paired visceral branches arise from the lateral surfaces of the aorta and include the *suprarenal, renal,* and *gonadal arteries.* The paired parietal branches arise from the posterolateral surfaces of the aorta and include the *inferior phrenic* and *lumbar arteries.* The unpaired parietal artery is the *median sacral.*

Branch	Description and region supplied
VISCERAL	
Celiac (SĒ-lē-ak; *koilia* = abdominal cavity) *trunk*	The *celiac artery (trunk)* is the first visceral aortic branch below the diaphragm. It has three branches: (1) *common hepatic* (he-PAT-ik) *artery,* (2) *left gastric* (GAS-trik) *artery,* and (3) *splenic* (SPLĒN-ik) *artery.* The common hepatic artery has three main branches: (1) *hepatic artery proper,* a continuation of the common hepatic artery, which supplies the liver, gallbladder and stomach; (2) *right gastric artery,* which supplies the stomach; and (3) *gastroduodenal* (gas'-trō-doo'-ō-DE-nal) *artery,* which supplies the stomach, duodenum, pancreas, and greater omentum. The *left gastric artery* supplies the stomach and the esophagus. The *splenic artery* supplies the spleen and has three main branches: (1) *pancreatic* (pan'-krē-AT-ik) *artery,* which supplies the pancreas; (2) *left gastroepiploic* (gas'-trō-ep'-i-PLO-ik) *artery,* which supplies the stomach and greater omentum; and (3) *short gastric* (GAS-trik) *artery,* which supplies the stomach.
Superior mesenteric (MES-en-ter'-ik; *meso* = middle; *enteron* = intestine) *artery*	The *superior mesenteric artery* anastomoses extensively and has several principal branches: (1) *inferior pancreaticoduodenal* (pan'-krē-at'-i-kō-doo'-ō-DE-nal) *artery,* which supplies the pancreas and duodenum; (2) *jejunal* (je-JOO-nal) and *ileal* (IL-ē-al) *arteries,* which supply the jejunum and ileum, respectively; (3) *ileocolic* (il'-ē-ō-KOL-ik) *artery,* which supplies the ileum and ascending colon of the large intestine; (4) *right colic* (KOL-ik) *artery,* which supplies the ascending colon; and (5) *middle colic artery,* which supplies the transverse colon of the large intestine.
Suprarenals (soo'-pra-RĒ-nals; *supra* = above; *renal* = kidney) *arteries*	Right and left *suprarenal arteries* supply blood to the adrenal (suprarenal) glands. The glands are also supplied by branches of the renal and inferior phrenic arteries.
Renals (RĒ-nals; *renal* = kidney) *arteries*	Right and left *renal arteries* carry blood to the kidneys, adrenal (suprarenal) glands, and ureters.
Gonadals (gō-NAD-als) [*testiculars* (tes-TIK-yoo-lars) or *ovarians* (ō-VA-rē-ans)] *arteries*	Right and left *testicular arteries* extend into the scrotum and supply the testes, epididymis, and ureters; right and left *ovarian arteries* are distributed to the ovaries, uterine (Fallopian) tubes, and ureters.
Inferior mesenteric (MES-en-ter'-ik) *arteries*	The principal branches of the *inferior mesenteric artery,* which also anastomoses, are the (1) *left colic* (KOL-ik) *artery,* which supplies the transverse and descending colons; (2) *sigmoid* (SIG-moyd) *arteries,* which supply the descending and sigmoid colons; and (3) *superior rectal* (REK-tal) *artery,* which supplies the rectum.
PARIETAL	
Inferior phrenics (FREN-iks) *arteries*	The *inferior phrenic arteries* are distributed to the inferior surface of the diaphragm and adrenal (suprarenal) glands.
Lumbars (LUM-bars; *lumbar* = loin) *arteries*	The *lumbar arteries* supply the spinal cord and its meninges and the muscles and skin of the lumbar region of the back.
Median sacral (SĀ-kral)	The *median sacral artery* supplies the sacrum, coccyx.

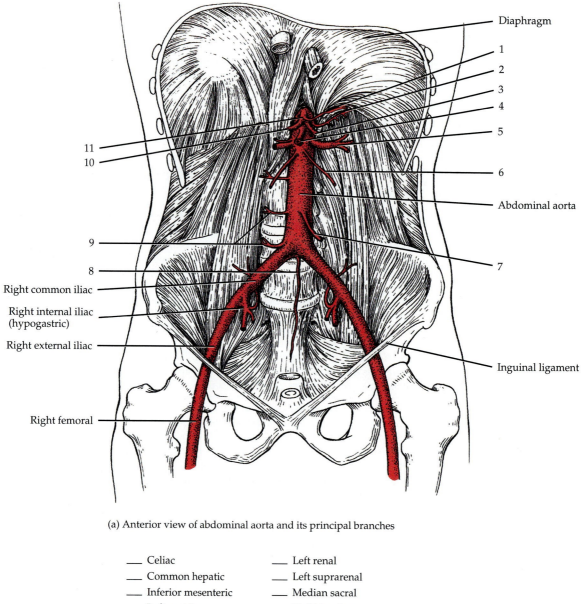

(a) Anterior view of abdominal aorta and its principal branches

____ Celiac   ____ Left renal

____ Common hepatic   ____ Left suprarenal

____ Inferior mesenteric   ____ Median sacral

____ Left gastric   ____ Right lumbars

____ Left gonadal   ____ Splenic
      (testicular or ovarian)   ____ Superior mesenteric

FIGURE 19.7   Abdominal arteries.

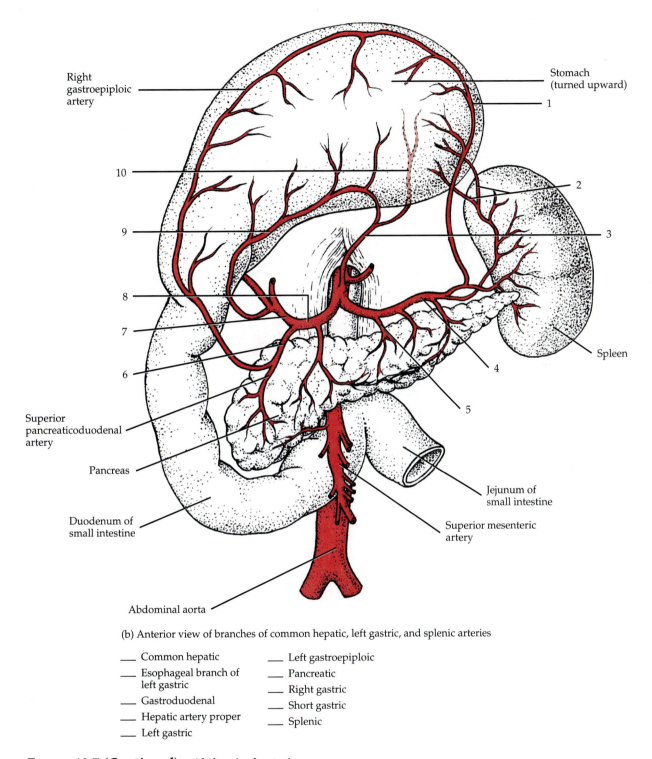

Right gastroepiploic artery

Stomach (turned upward)

1

10

2

9

3

8

7

6

Superior pancreaticoduodenal artery

Pancreas

Duodenum of small intestine

Spleen

4

5

Jejunum of small intestine

Superior mesenteric artery

Abdominal aorta

(b) Anterior view of branches of common hepatic, left gastric, and splenic arteries

___ Common hepatic

___ Esophageal branch of left gastric

___ Gastroduodenal

___ Hepatic artery proper

___ Left gastric

___ Left gastroepiploic

___ Pancreatic

___ Right gastric

___ Short gastric

___ Splenic

FIGURE 19.7 *(Continued)*   Abdominal arteries.

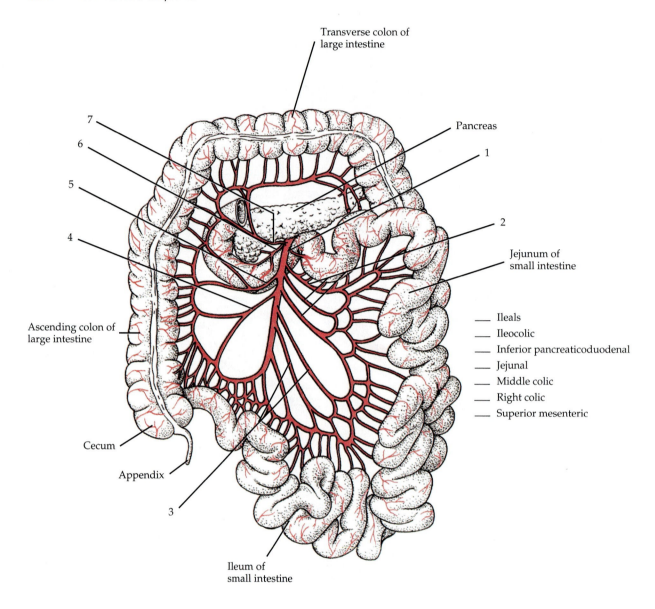

Transverse colon of
large intestine

Pancreas

1

2

Jejunum of
small intestine

___ Ileals

___ Ileocolic

___ Inferior pancreaticoduodenal

___ Jejunal

___ Middle colic

___ Right colic

___ Superior mesenteric

7

6

5

4

Ascending colon of
large intestine

Cecum

Appendix

3

Ileum of
small intestine

(c) Anterior view of branches of superior mesenteric artery

FIGURE 19.7 (*Continued*)   Abdominal arteries.

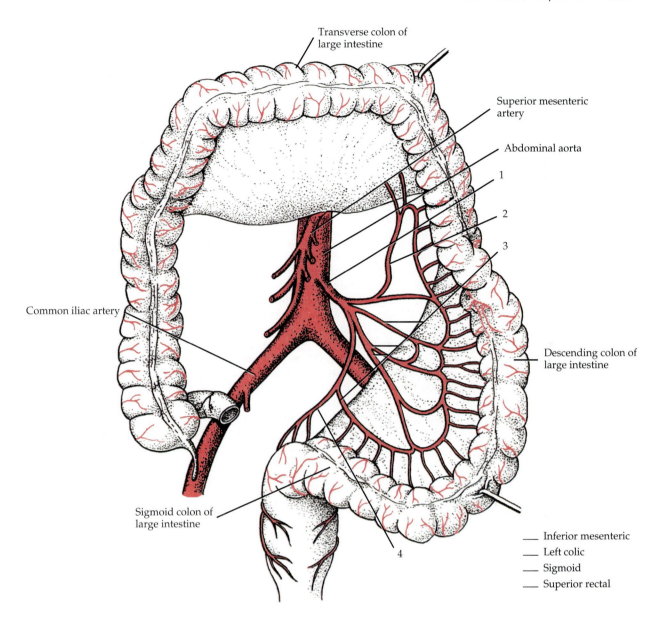

Transverse colon of large intestine

Superior mesenteric artery

Abdominal aorta

1

2

3

Common iliac artery

Descending colon of large intestine

Sigmoid colon of large intestine

4

___ Inferior mesenteric
___ Left colic
___ Sigmoid
___ Superior rectal

(d) Anterior view of branches of inferior mesenteric artery

FIGURE 19.7 (Continued)   Abdominal arteries.

TABLE 19.5
Arteries of Pelvis and Lower Limbs (Extremities) (Figure 19.8)

OVERVIEW: The abdominal aorta terminates by dividing into the right and left *common iliac arteries.* These, in turn, divide into the *internal* and *external iliac arteries.* Upon entering the thighs, the external iliac arteries become the *femoral arteries,* then the *popliteal arteries* posterior to the knee, and then the *anterior* and *posterior tibial arteries* at the knees.

Branch	Description and region supplied
Common iliac (IL-ē-ak; *iliac* = ilium) *arteries*	At about the level of the fourth lumbar vertebra, the abdominal aorta divides into the right and left *common iliac arteries.* Each passes inferiorly about 5 cm (2 in.) and gives rise to two branches: internal iliac and external iliac.
Internal iliacs	The *internal iliac (hypogastric) arteries* form branches that supply the psoas minor, gluteal muscles, quadratus lumborum, medial side of each thigh, urinary bladder, rectum, prostate gland, ductus (vas) deferens, uterus, and vagina.
External iliacs	The *external iliac arteries* diverge through the greater (false) pelvis and enter the thighs to become the right and left *femoral* (FEM-o-ral) *arteries.* Both femorals send branches back up to the genitals and the wall of the abdomen. Other branches run to the muscles of the thigh. The femoral continues down the medial and posterior side of the thigh posterior to the knee joint, where it becomes the *popliteal* (pop'-li-TĒ-al) *artery.* Between the knee and ankle, the popliteal runs down on the posterior aspect of the leg and is called the *posterior tibial* (TIB-ē-al) *artery.* Inferior to the knee, the *peroneal* (per'-ō-NĒ-al) *artery* branches off the posterior tibial to supply structures on the medial side of the fibula and calcaneus. In the calf, the *anterior tibial artery* branches off the popliteal and runs along the anterior surface of the leg. At the ankle, it becomes the *dorsalis pedis* (PED-is) *artery.* At the ankle, the posterior tibial divides into the *medial* and *lateral plantar* (PLAN-tar) *arteries.* The lateral plantar artery and the dorsalis pedis artery unite to form the *plantar arch.* From this arch, *digital arteries* supply the toes.

Write the names of the missing arteries in the following scheme of circulation. Be sure to indicate left or right where applicable.

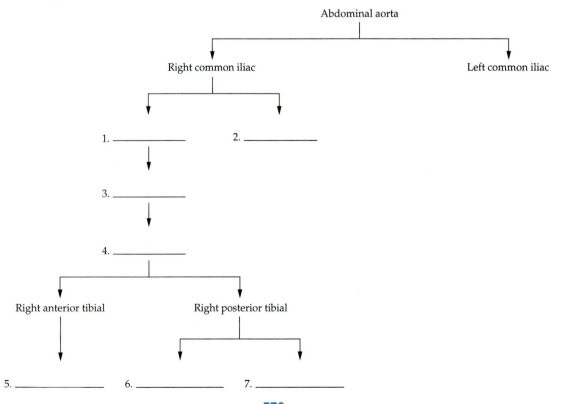

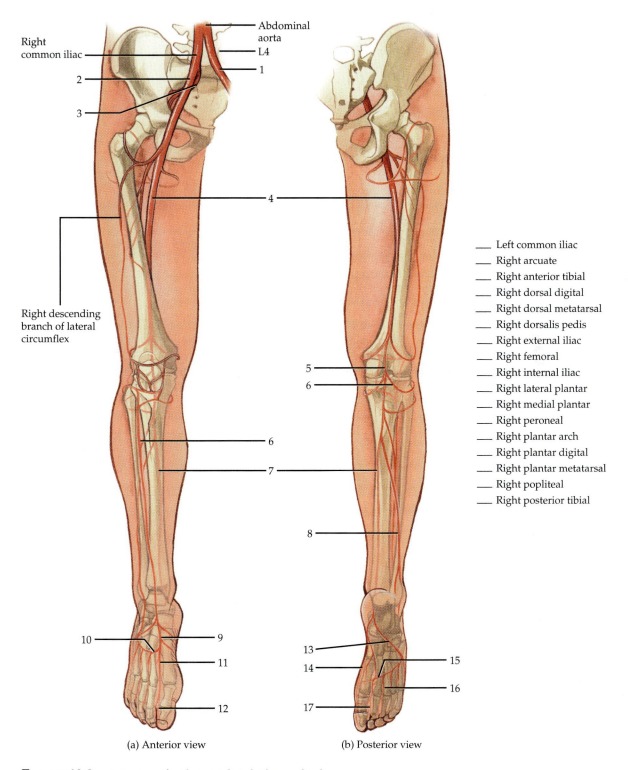

Abdominal aorta

L4

Right common iliac

Right descending branch of lateral circumflex

___ Left common iliac
___ Right arcuate
___ Right anterior tibial
___ Right dorsal digital
___ Right dorsal metatarsal
___ Right dorsalis pedis
___ Right external iliac
___ Right femoral
___ Right internal iliac
___ Right lateral plantar
___ Right medial plantar
___ Right peroneal
___ Right plantar arch
___ Right plantar digital
___ Right plantar metatarsal
___ Right popliteal
___ Right posterior tibial

(a) Anterior view   (b) Posterior view

FIGURE 19.8   Arteries of pelvis and right lower limb.

TABLE 19.6
Veins of Systemic Circulation (See Figure 19.10)

OVERVIEW: Deoxygenated blood returns to the right atrium from three veins: the *coronary sinus, superior vena cava*, and *inferior vena cava.* The coronary sinus receives blood from the cardiac veins; the superior vena cava receives blood from veins superior to the diaphragm, except the air sacs of the lungs. This includes the head, neck, upper limbs, and thoracic wall. The inferior vena cava receives blood from veins inferior to the diaphragm. This includes the lower limbs, most of the abdominal walls, and abdominal viscera.

Vein	Description and region drained
Coronary (KOR-o-nar-ē; *corona* = crown) *sinus*	The *coronary sinus* receives almost all venous blood from the myocardium. It is located in the coronary sulcus (see Fig. 18.4b) and opens into the right atrium between the orifice of the inferior vena cava and the tricuspid valve.
Superior vena cava (VE-na CA-va; *vena* = vein; *cava* = cavelike) *(SVC)*	The *SVC* is about 7½ cm (3 in.) long, 2 cm (1 in.) diameter and empties its blood into the superior part of the right atrium. It begins posterior to the right first costal cartilage by the union of the right and left brachiocephalic veins and ends at the level of the right third costal cartilage where it enters the right atrium. The SVC drains the head, neck, chest, and upper limbs.
Inferior vena cava *(IVC)*	The *IVC* is the largest vein in the body, about 3½ cm (1½ in.) in diameter. It begins anterior to the fifth lumbar vertebra by the union of the common iliac veins, ascends behind the peritoneum to the right of the midline, pierces the costal tendon of the diaphragm at the level of the eighth thoracic vertebra, and enters the inferior part of the right atrium. The IVC drains the abdomen, pelvis, and lower limbs. The IVC is commonly compressed during the later stages of pregnancy owing to the enlargement of the uterus. This produces edema of the ankles and feet and temporary varicose veins.

TABLE 19.7
Veins of Head and Neck (Figure 19.9)

OVERVIEW: The majority of blood draining from the head is pushed into three pairs of veins: *internal jugular, external jugular*, and *vertebral.* Within the brain, all veins drain into dural venous sinuses and then into the internal jugular veins. *Dural venous sinuses* are endothelial-lined venous channels between layers of the cranial dura mater.

Vein	Description and region drained
Internal jugulars (JUG-yoo-lars; *jugular* = throat) *veins*	Right and left *internal jugular veins* receive blood from the face and neck. They arise as a continuation of the *sigmoid* (SIG-moyd) *sinuses* at the base of the skull. Intracranial vascular sinuses are located between layers of the dura mater and receive blood from the brain. Other sinuses that drain into the internal jugulars include the *superior sagittal* (SAJ-i-tal) *sinus, inferior sagittal sinus, straight sinus*, and *transverse (lateral) sinuses.* Internal jugulars descend on either side of the neck and pass behind the clavicles, where they join with the right and left subclavian veins. Unions of the internal jugulars and subclavians form the right and left *brachiocephalic* (brā'-kē-ō-se-FAL-ik) *veins.* From here blood flows into the superior vena cava. The general structures drained by the internal jugular veins are the brain (through the dural venous sinuses), face and neck.
External jugulars	Right and left *external jugular veins* run down the neck along the outside of the internal jugulars. They drain blood from the scalp, and superficial and deep regions of the face.
Vertebrals (VER-te-brals; *vertebra* = vertebrae) *veins*	The right and left *vertebral veins* originate inferior to the occipital condyles. They descend through successive transverse foramina of the first six cervical vertebrae and emerge from the formina of the sixth cervical vertebra to enter the brachiocephalic veins in the root of the neck. The vertebral veins drain deep structures in the neck such as the cervical vertebrae, cervical spinal cord, and some neck muscles.

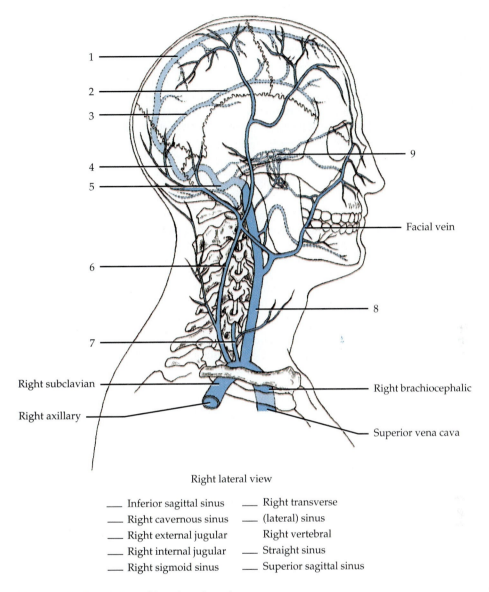

Right lateral view

___ Inferior sagittal sinus	___ Right transverse
___ Right cavernous sinus	___ (lateral) sinus
___ Right external jugular	Right vertebral
___ Right internal jugular	___ Straight sinus
___ Right sigmoid sinus	___ Superior sagittal sinus

FIGURE 19.9   Veins of head and neck.

TABLE 19.8
Veins of Upper Limbs (Extremities) (Figure 19.10)

OVERVIEW: Blood from each upper limb is returned to the heart by superficial and
deep veins. Both sets of veins contain valves. *Superficial veins* are located just below the
skin and are often visible. They anastomose extensively with each other and with deep
veins, and do not accompany arteries. *Deep veins* are located deep in the body. They
usually accompany arteries, and many have the same names as corresponding arteries.

Vein	Description and region drained
SUPERFICIAL	
Cephalic (se-FAL-ik; *ceph* = head) *veins*	The *cephalic vein* of each upper limb begins in the medial part of the *dorsal venous* (VĒ-nus) *arch* and winds upward around the radial border of the forearm. Anterior to the elbow, it is connected to the basilic vein by the *median cubital* (KYOO-bi-tal) *vein.* Just below the elbow, the cephalic vein unites with the *accessory cephalic vein* to form the cephalic vein of the upper limb. Ultimately, the cephalic vein empties into the axillary vein. The cephalic veins drain blood from the lateral aspect of the upper limbs.
Basilic (ba-SIL-ik; *basilikos* = royal) *veins*	The *basilic vein* of each upper limb originates in the ulnar part of the *dorsal venous arch.* It extends along the posterior surface of the ulna to a point near the elbow where it receives the *median cubital vein.* If a vein must be punctured for an injection, transfusion, or removal of a blood sample, the median cubitals are preferred. After receiving the median cubital vein, the basilic continues ascending on the medial side until it reaches the middle of the arm. There it penetrates the tissues deeply and runs alongside the brachial artery until it joins the brachial vein. As the basilic and brachial veins merge in the axillary area, they form the axillary vein.
Median antebrachials (an'-tē-BRĀ-kē-als; *ante* = in front of, before; *brachium* = arm) *veins*	The *median antebrachial veins* drain the *palmar venous plexuses,* ascend on the ulnar side of the anterior forearm, and end in the median cubital veins. They drain the palms and forearms.
DEEP	
Radial (RĀ-dē-al) *veins*	The paired *radial veins* begin at the *deep venous palmar arches.* These arches drain the *palmar metacarpal veins* in the palm.
Ulnar (UL-nar) *veins*	*Ulnar veins* receive tributaries from the *superficial palmar venous arch.* Radial and ulnar veins unite in the bend of the elbow to form the brachial veins.
Brachial (BRĀ-kē-al; *brachium* = arm) *veins*	Located on either side of the brachial arteries, the *brachial veins* join with the basilic veins to form the axillary veins. They drain the forearms, elbow joints, arms, and humerus.
Axillary (AK-si-ler'-ē; *axilla* = armpit) *veins*	The *axillary veins* ascend to the outer borders of the first ribs, where they become the subclavian veins. The axillary veins drain the arms, axillas, and superolateral chest wall.
Subclavians (sub-KLĀ-vē-ans; *sub* = under; *clavicula* = clavicle) *veins*	Right and left *subclavian veins* unite with the internal jugulars to form brachio-cephalic veins. The thoracic duct of the lymphatic system delivers lymph into the left subclavian vein at the junction with the internal jugular. The right lymphatic duct delivers lymph into the right subclavian vein at the corresponding junction.

Write the names of the missing veins in the following scheme of circulation. Be sure to indicate left or right where applicable.

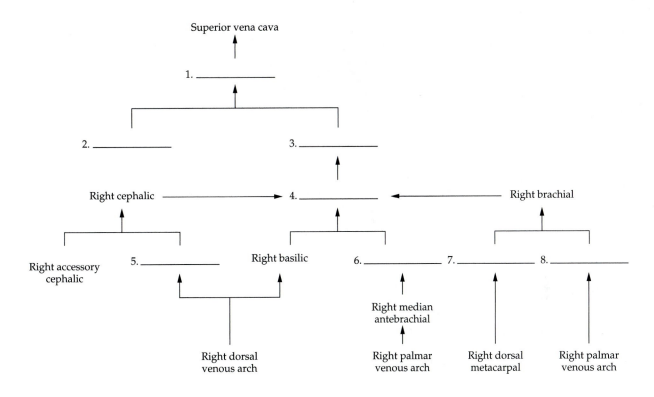

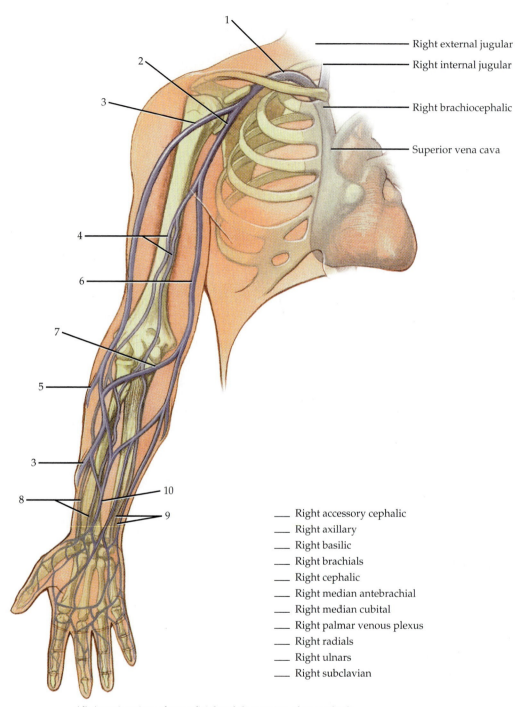

1
2
3

Right external jugular

Right internal jugular

Right brachiocephalic

Superior vena cava

4

6

7

5

3

8

10

9

___ Right accessory cephalic
___ Right axillary
___ Right basilic
___ Right brachials
___ Right cephalic
___ Right median antebrachial
___ Right median cubital
___ Right palmar venous plexus
___ Right radials
___ Right ulnars
___ Right subclavian

(d) Anterior view of superficial and deep veins of upper limb

FIGURE 19.10   Veins of the right upper limb.

TABLE **19.9**
Veins of Thorax (Figure 19.11)

OVERVIEW: Although the brachiocephalic veins drain some portions of the thorax, most
thoracic structures are drained by a network of veins called the *azygos system*. This is a
network of veins on each side of the vertebral column: *azygos, hemiazygos,* and *accessory hemiazygos.* They show considerable variation in origin, course, tributaries, anastomoses, and termination. Ultimately, they empty into the superior vena cava.

Vein	Description and region drained
Brachiocephalic (brā-kē-ō-se-FAL-ik)	Right and left ***brachiocephalic veins,*** formed by the union of the subclavians and internal jugulars, drain blood from the head, neck, upper limbs, mammary glands, and upper thorax. Brachiocephalics unite to form the superior vena cava.
Azygos (az-Ī-gos = unpaired) *system*	The ***azygos system,*** besides collecting blood from the thorax, may serve as a bypass for the inferior vena cava that drains blood from the lower body. Several small veins directly link the azygos system with the inferior vena cava. Large veins that drain the lower limbs and abdomen pass blood into the azygos system. If the inferior vena cava or hepatic portal vein becomes obstructed, the azygos system can return blood from the lower body to the superior vena cava.
Azygos vein	The ***azygos vein*** lies anterior to the vertebral column, slightly to the right of the midline. It begins at the junction of the right ascending lumbar and right subcostal veins near the diaphragm. At the level of the fourth thoracic vertebra, it arches over the root of the right lung to end in the superior vena cava. Generally, the azygos vein drains the right side of the thoracic wall, thoracic viscera, and abdominal wall.
Hemiazygos (HEM-ē-az-ī-gos) *hemi* = half) *vein*	The ***hemiazygos vein*** is anterior to the vertebral column and slightly to the left of the midline. It begins at the junction of the left ascending lumbar and left subcostal veins. It terminates by joining the azygos vein at about the level of the ninth thoracic vertebra. Generally, the hemiazygos vein drains the left side of the thoracic wall, thoracic viscera, and abdominal wall. Specifically, the hemiazygos vein receives blood from the ninth through eleventh ***left intercostal veins, esophageal, mediastinal, pericardial, bronchial,*** and ***left subcostal veins.***
Accessory hemiazygos vein	The ***accessory hemiazygos vein*** is also anterior to the vertebral column and to the left of the midline. It begins at the fourth or fifth intercostal space and descends from the fifth to the eighth thoracic vertebra. It terminates by joining the azygos vein at about the level of the eighth thoracic vertebra. The accessory hemiazygos vein drains the left side of the thoracic wall. It receives blood from the fourth through eighth ***left intercostal veins.*** The first through third left intercostal nerves open into the left brachiocephalic vein.

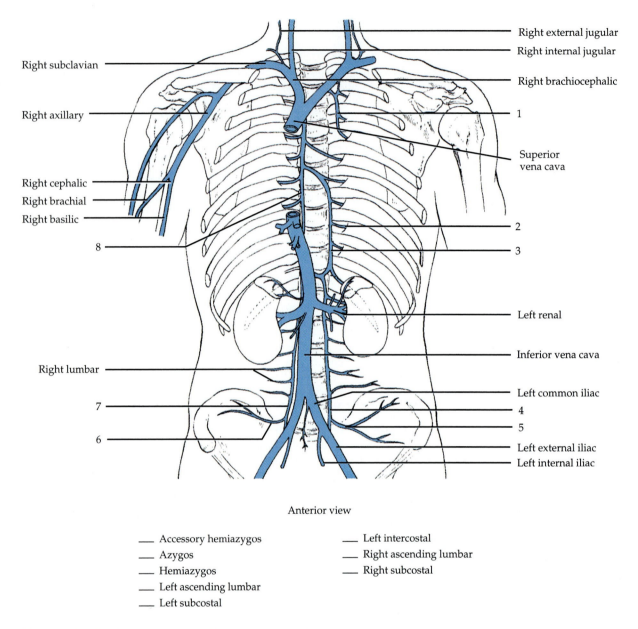

Right external jugular

Right internal jugular

Right subclavian

Right brachiocephalic

Right axillary

1

Superior
vena cava

Right cephalic

Right brachial

Right basilic

8

2

3

Left renal

Inferior vena cava

Right lumbar

Left common iliac

7

4

6

5

Left external iliac

Left internal iliac

Anterior view

___ Accessory hemiazygos   ___ Left intercostal

___ Azygos   ___ Right ascending lumbar

___ Hemiazygos   ___ Right subcostal

___ Left ascending lumbar

___ Left subcostal

FIGURE 19.11   Veins of thorax, abdomen, and pelvis.

TABLE 19.10
Veins of Abdomen and Pelvis (Figure 19.11)

OVERVIEW: Blood from the abdominopelvic viscera and abdominal wall returns to the
heart via the *inferior vena cava*. Many small veins enter the inferior vena cava. Most
carry return flow from parietal branches of the abdominal aorta and their names correspond to the names of the arteries. The inferior vena cava does not receive veins directly
from the gastrointestinal tract, spleen, pancreas, and gallbladder. These organs pass
their blood into a common vein, the *hepatic portal vein,* which delivers the blood to the
liver. The hepatic portal vein is formed by the union of the superior mesenteric and
splenic veins. This special flow of venous blood is called *hepatic portal circulation,*
which is described shortly.

Vein	Description and region drained
Inferior vena cava (VĒ-na CĀ-va)	The *inferior vena cava* is formed by the union of two common iliac veins that drain the lower limbs and abdomen. The inferior vena cava extends superiorly through the abdomen and thorax to the right atrium.
Common iliac (IL-ē-ak; *iliac* = pertaining to the ilium) *veins*	The *common iliac veins* are formed by the union of the internal (hypogastric) and external iliac veins and represent the distal continuation of the inferior vena cava at its bifurcation (branching).
Internal iliac veins	Tributaries of the *internal iliac veins* basically correspond to branches of the internal iliac arteries. Generally, the veins drain the thigh, buttocks, external genitals, and pelvis.
External iliac veins	The *external iliac veins* are a continuation of the femoral veins and receive blood from the lower limbs and inferior part of the anterior abdominal wall. The external iliac veins drain the lower limbs, cremasteric muscle in males, and the abdominal wall.
Renal (RĒ-nal; *renal* = kidney) *veins*	The *renal veins* are large and pass anterior to the renal arteries. The *renal veins* drain the kidneys.
Gonadal (gō-NAD-al) [*testicular* (tes-TIK-yoo-lor) or *ovarian* (ō-VA-rē-an)] *veins*	The *testicular veins* drain the testes (left testicular vein empties into the left renal vein and right testicular drains into the inferior vena cava); the *ovarian veins* drain the ovaries (left ovarian vein empties into the left renal vein and right ovarian drains into the inferior vena cava).
Suprarenal (soo'-pra-RĒ-nal; *supra* = above) *veins*	The *suprarenal veins* drain the adrenal (suprarenal) glands (left suprarenal vein empties into the left renal vein and right suprarenal vein empties into the superior vena cava).
Inferior phrenic (FREN-ik; *phrenic* = diaphragm) *veins*	The *inferior phrenic veins* drain the diaphragm (left interior phrenic vein sends a tributary to the left renal vein and right inferior phrenic vein empties into the superior vena cava).
Hepatic (he-PAT-ik; *hepatic* = liver) *veins*	The *hepatic veins* drain the liver.
Lumbars (LUM-bars)	A series of parallel *lumbar veins* drain blood from both sides of the posterior abdominal wall, vertebral canal, spinal cord, and meninges. The lumbars connect at right angles with the right and left *ascending lumbar veins,* which form the origin of the corresponding azygos or hemiazygos vein. The lumbars drain blood into the ascending lumbars and then run to the inferior vena cava, where they release the remainder of the flow.

TABLE 19.11
Veins of Lower Limbs (Extremities) (Figure 19.12)

OVERVIEW: As with the upper limbs, blood from each lower limb is drained by *superficial* and *deep veins*. The superficial veins often anastomose with each other and with deep veins along their length. Deep veins, for the most part, have the same names as their accompanying arteries. All veins of the lower limbs have valves, which are more numerous than in veins of the upper limbs.

Vein	Description and region drained
SUPERFICIAL VEINS	
Great saphenous (sa-FĒ-nus; *saphenes* = clearly visible) *veins*	The *great saphenous vein,* the longest vein in the body, begins at the medial end of the *dorsal venous arch* of the foot. It passes anterior to the medial malleolus and then upward along the medial aspect of the leg and thigh just deep to the skin. It receives tributaries from superficial tissues and connects with the deep veins as well. It empties into the femoral vein in the groin. The great saphenous vein is frequently used for prolonged administration of intravenous fluids. This is particularly important in very young babies and in patients of any age who are in shock and whose veins are collapsed. It and the small saphenous vein are subject to varicosity. The great saphenous veins are also used as a source of vascular grafts, especially for coronary bypass surgery.
Small saphenous veins	The *small saphenous veins* begin at the lateral end of the dorsal venous arch of the foot. They pass behind the lateral malleolus and ascends under the skin along the posterior aspect of the leg. It receives blood from the foot and posterior portion of the leg. It empties into the popliteal vein posterior to the knee.
DEEP VEINS	
Posterior tibial (TIB-ē-al) *veins*	The *posterior tibial vein* is formed by the union of the *medial* and *lateral plantar* (PLAN-tar) *veins* behind the medial malleolus. It ascends deep in the muscle at the back of the leg, receives blood from the *peroneal* (per'-ō-NĒ-al) *veins,* which drain the lateral and posterior leg muscles.
Anterior tibial veins	The *anterior tibial veins* arise in the dorsal venous arch. They run between the tibia and fibula and unites with the posterior tibial to form the popliteal vein. The anterior tibial veins drain the ankle joint, knee joint, tibiofibular joint, and anterior portion of leg.
Popliteal (pop'-li-TĒ-al; *popliteus* = hollow behind knee) *veins*	The *popliteal vein,* just behind the knee, receives blood from the anterior and posterior tibials and the small saphenous vein. The popliteal veins drain the knee joint and the skin, muscle, and bones of portions of the calf and thigh around the knee joint.
Femoral (FEM-o-ral) *veins*	The *femoral vein* is the upward continuation of the popliteal just above the knee. The femorals run up the posterior surface of the thighs and drain the thighs, femurs, external genitals, and superficial lymph nodes. After receiving the great saphenous veins in the groin, they continue as the *external iliac veins.*

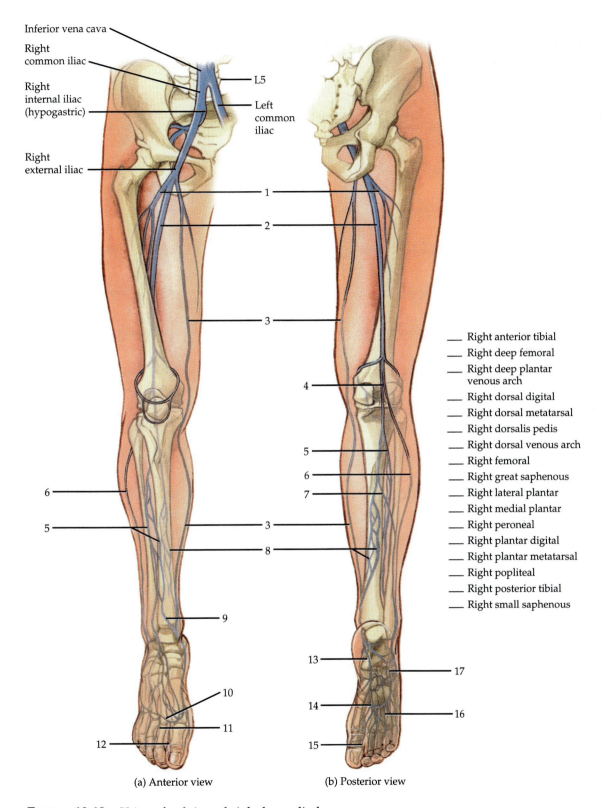

Inferior vena cava

Right common iliac

Right internal iliac (hypogastric)

Right external iliac

L5

Left common iliac

1

2

3

4

5

6

7

3

8

9

10

11

12

13

14

15

16

17

5

6

___ Right anterior tibial
___ Right deep femoral
___ Right deep plantar venous arch
___ Right dorsal digital
___ Right dorsal metatarsal
___ Right dorsalis pedis
___ Right dorsal venous arch
___ Right femoral
___ Right great saphenous
___ Right lateral plantar
___ Right medial plantar
___ Right peroneal
___ Right plantar digital
___ Right plantar metatarsal
___ Right popliteal
___ Right posterior tibial
___ Right small saphenous

(a) Anterior view

(b) Posterior view

FIGURE 19.12　Veins of pelvis and right lower limb.

Write the names of the missing veins in the following scheme of circulation. Be sure to indicate left or right where applicable.

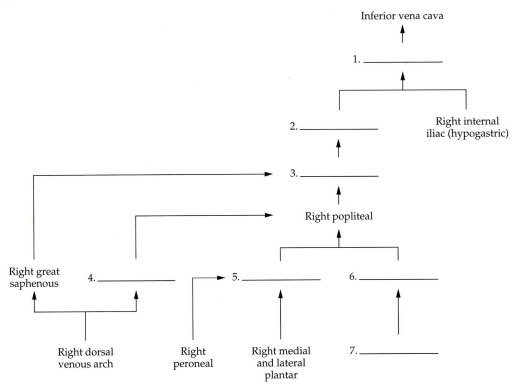

2. Hepatic Portal Circulation

The **hepatic** (*hepato* = liver) **portal circulation** detours venous blood from the gastrointestinal organs and spleen through the liver before it returns to the heart (Figure 19.13). A *portal system* carries blood between two capillary networks, from one location in the body to another without passing through the heart, in this case from capillaries of the gastrointestinal tract to sinusoids of the liver. After a meal, hepatic portal blood is rich with absorbed substances. The liver stores some and modifies others before they pass into the general circulation. For example, the liver converts glucose into glycogen for storage. It also modifies other digested substances so they may be used by cells, detoxifies harmful substances that have been absorbed by the gastrointestinal tract, and destroys bacteria by phagocytosis.

The **hepatic portal vein** is formed by the union of the (1) superior mesenteric and (2) splenic veins. The **superior mesenteric vein** drains blood from the small intestine, portions of the large intestine, stomach, and pancreas through the *jejunal, ileal, ileocolic, right colic, middle colic, pancreaticoduodenal,* and *right*

gastroepiploic veins. The **splenic vein** drains blood from the stomach, pancreas, and portions of the large intestine through the *superior rectal, sigmoidal,* and *left colic veins.* The right and left gastric veins, which open directly into the hepatic portal vein, drain the stomach. The *cystic vein,* which also opens into the hepatic portal vein, drains the gallbladder.

At the same time the liver receives deoxygenated blood via the hepatic portal system, it also receives oxygenated blood from the systemic circulation via the hepatic artery. Ultimately, all blood leaves the liver through the **hepatic veins,** which drain into the inferior vena cava.

Using your textbook, charts, or models for reference, label Figure 19.13.

3. Pulmonary Circulation

The **pulmonary** (*pulmo* = lung) **circulation** carries deoxygenated blood from the right ventricle to the air sacs within the lungs and returns oxygenated blood from the air sacs within the lungs to the left atrium (Figure 19.14 on page 390). The **pulmonary trunk** emerges from the right ventricle and passes

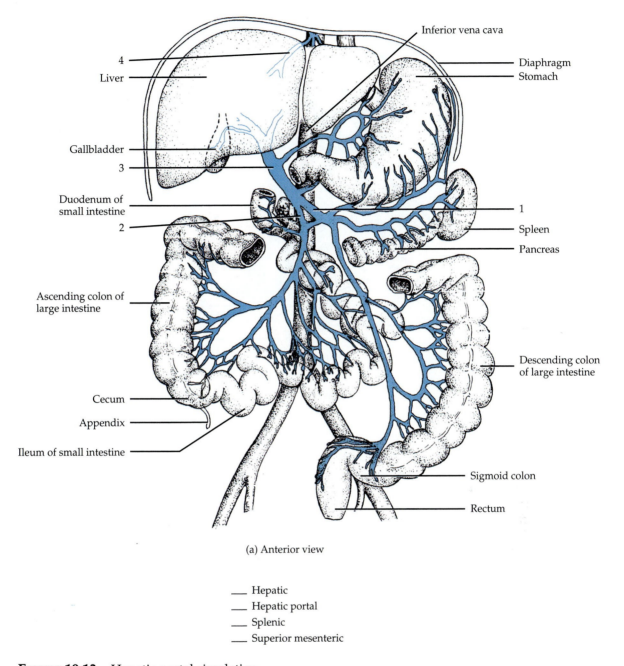

(a) Anterior view

___ Hepatic
___ Hepatic portal
___ Splenic
___ Superior mesenteric

FIGURE 19.13   Hepatic portal circulation.

superiorly, posteriorly, and to the left. It then divides into two branches: the *right pulmonary artery* extends to the right lung; the *left pulmonary artery* goes to the left lung. The pulmonary arteries are the only postnatal (after birth) arteries that carry deoxygenated blood. On entering the lungs, the branches divide and subdivide until finally they form capillaries around the air sacs within the lungs. CO_2 passes from the blood into these air sacs and is exhaled. Inhaled O_2 passes from the air sacs into the blood. The pulmonary capillaries unite, form venules and veins, and eventually two *pulmonary veins* exit from each lung and transport the oxygenated blood to the left atrium. The pulmonary veins are the only postnatal veins that carry oxygenated blood. Contractions of the left ventricle then send the blood into the systemic circulation.

Study a chart or model of the pulmonary circulation, trace the path of blood through it, and label Figure 19.14.

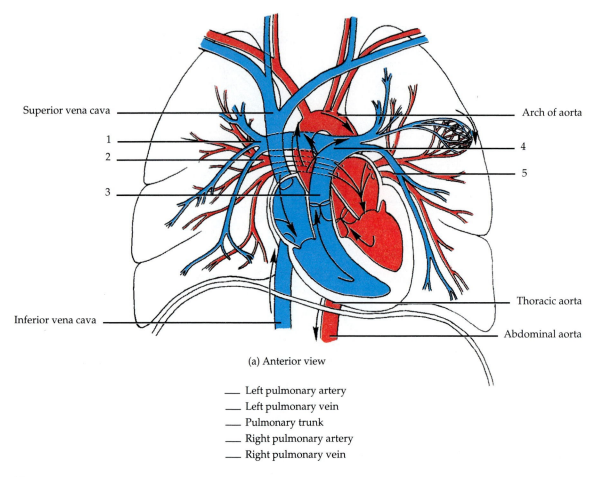

(a) Anterior view

____ Left pulmonary artery
____ Left pulmonary vein
____ Pulmonary trunk
____ Right pulmonary artery
____ Right pulmonary vein

FIGURE 19.14   Pulmonary circulation.

4. Fetal Circulation

The circulatory system of a fetus, called *fetal circulation,* differs from the postnatal circulation because the lungs, kidneys, and gastrointestinal tract begin to function at birth. The fetus obtains its O_2 and nutrients by diffusion from the maternal blood and eliminates its CO_2 and wastes by diffusion into the maternal blood (Fig. 19.15).

The exchange of materials between fetal and maternal circulation occurs through a structure called the *placenta* (pla-SEN-ta). It is attached to the umbilicus (navel) of the fetus by the umbilical (um-BIL-i-kal) cord, and it communicates with the mother through countless small blood vessels that emerge from the uterine wall. The umbilical cord contains blood vessels that branch into capillaries in the placenta. Wastes from the fetal blood diffuse out of the capillaries, into spaces containing maternal blood (intervillous spaces) in the placenta, and finally into the mother's uterine blood vessels. Nutrients travel the opposite route—from the maternal

blood vessels to the intervillous spaces to the fetal capillaries. Normally, there is no direct mixing of maternal and fetal blood since all exchanges occur by diffusion through capillary walls.

Blood passes from the fetus to the placenta via two *umbilical arteries.* These branches of the internal iliac (hypogastric) arteries are within the umbilical cord. At the placenta, fetal blood picks up O_2 and nutrients and eliminates CO_2 and wastes. The oxygenated blood returns from the placenta via a single *umbilical vein.* This vein ascends to the liver of the fetus, where it divides into two branches. Some blood flows through the branch that joins the hepatic portal vein and enters the liver. Most of the blood flows into the second branch, the *ductus venosus* (DUK-tus ve-NŌ-sus), which drains into the inferior vena cava. Thus a good portion of the O_2 and nutrients in the blood bypasses the fetal liver and is delivered to the developing fetal brain.

Circulation through other portions of the fetus is similar to postnatal circulation. Deoxygenated blood returning from the lower regions mingles

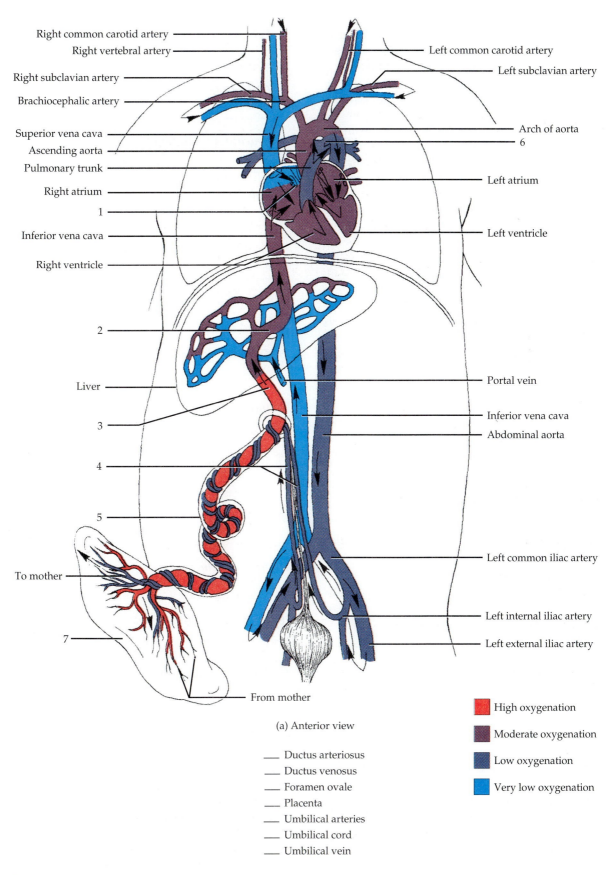

Right common carotid artery

Right vertebral artery

Right subclavian artery

Brachiocephalic artery

Superior vena cava

Ascending aorta

Pulmonary trunk

Right atrium

1

Inferior vena cava

Right ventricle

2

Liver

3

4

5

To mother

7

From mother

Left common carotid artery

Left subclavian artery

Arch of aorta

6

Left atrium

Left ventricle

Portal vein

Inferior vena cava

Abdominal aorta

Left common iliac artery

Left internal iliac artery

Left external iliac artery

(a) Anterior view

___ Ductus arteriosus
___ Ductus venosus
___ Foramen ovale
___ Placenta
___ Umbilical arteries
___ Umbilical cord
___ Umbilical vein

High oxygenation

Moderate oxygenation

Low oxygenation

Very low oxygenation

FIGURE 19.15 Fetal circulation.

with oxygenated blood from the ductus venosus in the inferior vena cava. This mixed blood then enters the right atrium. Deoxygenated blood returning from the upper regions of the fetus enters the superior vena cava and passes into the right atrium.

Most of the fetal blood does not pass from the right ventricle to the lungs, as it does in postnatal circulation, since the fetal lungs do not operate. In the fetus, an opening called the *foramen ovale* (fō-RĀ-men ō-VAL-ē) exists in the septum between the right and left atria. About one-third of the blood passes through the foramen ovale directly into the systemic circulation. The blood that does pass into the right ventricle is pumped into the pulmonary trunk, but little of this blood reaches the lungs. Most is sent through the *ductus arteriosus* (ar-tē-rē-Ō-sus). This vessel connects the pulmonary trunk with the aorta and allows most blood to bypass the fetal lungs. The blood in the aorta is carried to all parts of the fetus through the systemic circulation. When the common iliac arteries branch into the external and internal iliacs, part of the blood flows into the internal iliacs. It then goes to the umbilical arteries and back to the placenta for another exchange of materials. The only fetal vessel that carries fully oxygenated blood is the umbilical vein.

Label Figure 19.15.

E. BLOOD VESSEL EXERCISE

For each vessel listed, indicate the region supplied (if an artery) or the region drained (if a vein):

1. *Coronary artery* _____

   _____

2. *Internal iliac veins* _____

   _____

3. *Lumbar arteries* _____

   _____

4. *Renal artery* _____

   _____

5. *Left gastric artery* _____

   _____

6. *External jugular vein* _____

   _____

7. *Left subclavian artery* _____

   _____

8. *Axillary vein* _____

   _____

9. *Brachiocephalic veins* _____

   _____

10. *Transverse sinuses* _____

    _____

11. *Hepatic artery* _____

    _____

12. *Inferior mesenteric artery* _____

    _____

13. *Suprarenal artery* _____

    _____

14. *Inferior phrenic artery* _____

    _____

15. *Great saphenous vein* _____

    _____

16. *Popliteal vein* _____

    _____

17. *Azygos vein* _____

    _____

18. *Internal iliac (hypogastric) artery* _____

    _____

19. *Internal carotid artery* _____

    _____

20. *Cephalic vein* _____

    _____

ANSWER THE LABORATORY REPORT QUESTIONS AT THE END OF THE EXERCISE.

Blood Vessels 19

Student _____   Date _____

Laboratory Section _____   Score/Grade _____

PART 1. Multiple Choice

_____ 1. The largest of the circulatory routes is (a) systemic (b) pulmonary (c) coronary (d) hepatic portal

_____ 2. All arteries of systemic circulation branch from the (a) superior vena cava (b) aorta (c) pulmonary artery (d) coronary artery

_____ 3. The arterial system that supplies the brain with blood is the (a) hepatic portal system (b) pulmonary system (c) cerebral arterial circle (circle of Willis) (d) carotid system

_____ 4. An obstruction in the inferior vena cava would hamper the return of blood from the (a) head and neck (b) upper limbs (c) thorax (d) abdomen and pelvis

_____ 5. Which statement best describes arteries? (a) all carry oxygenated blood to the heart (b) all contain valves to prevent the backflow of blood (c) all carry blood away from the heart (d) only large arteries are lined with endothelium

_____ 6. Which statement is *not* true of veins? (a) they have less elastic tissue and smooth muscle than arteries (b) their tunica externa is the thickest coat (c) most veins in the limbs have valves (d) they always carry deoxygenated blood

_____ 7. A thrombus in the first branch of the arch of the aorta would affect the flow of blood to the (a) left side of the head and neck (b) myocardium of the heart (c) right side of the head and neck and right upper limb (d) left upper limb

_____ 8. If a vein must be punctured for an injection, transfusion, or removal of a blood sample, the likely site would be the (a) median cubital (b) subclavian (c) hemiazygos (d) anterior tibial

_____ 9. In hepatic portal circulation, blood is eventually returned to the inferior vena cava through the (a) superior mesenteric vein (b) hepatic portal vein (c) hepatic artery (d) hepatic veins

_____ 10. Which of the following are involved in pulmonary circulation? (a) superior vena cava, right atrium, and left ventricle (b) inferior vena cava, right atrium, and left ventricle (c) right ventricle, pulmonary artery, and left atrium (d) left ventricle, aorta, and inferior vena cava

_____ 11. If a thrombus in the left common iliac vein dislodged, into which arteriole system would it first find its way? (a) brain (b) kidneys (c) lungs (d) left arm

_____ 12. In fetal circulation, the blood containing the highest amount of oxygen is found in the (a) umbilical arteries (b) ductus venosus (c) aorta (d) umbilical vein

_____ 13. The greatest amount of elastic tissue found in the arteries is located in which coat? (a) tunica interna (b) tunica media (c) tunica externa (d) tunica adventitia

_____ 14. Which coat of an artery contains endothelium? (a) tunica interna (b) tunica media (c) tunica externa (d) tunica adventitia

_____ 15. Permitting the exchange of nutrients and gases between the blood and tissue cells is the primary function of (a) capillaries (b) arteries (c) veins (d) arterioles

_____ 16. The circulatory route that runs from the gastrointestinal tract to the liver is called (a) coronary circulation (b) pulmonary circulation (c) hepatic portal circulation (d) cerebral circulation

_____ 17. Which of the following statements about systemic circulation is *not* correct? (a) its purpose is to carry oxygen and nutrients to body tissues and to remove carbon dioxide (b) all systemic arteries branch from the aorta (c) it involves the flow of blood from the left ventricle to all parts of the body except the lungs (d) it involves the flow of blood from the body to the left atrium

_____ 18. The opening in the septum between the right and left atria of a fetus is called the (a) foramen ovale (b) ductus venosus (c) foramen rotundum (d) foramen spinosum

_____ 19. The branch of the umbilical vein in the fetus that connects with the inferior vena cava, bypassing the liver, is the (a) foramen ovale (b) ductus venosus (c) ductus arteriosus (d) patent ductus

_____ 20. Which of the vessels does *not* belong with the others? (a) brachiocephalic artery (b) left common carotid artery (c) celiac artery (d) left subclavian artery

PART 2. Matching

_____ 21. Aortic branch that supplies the head and associated structures

_____ 22. Artery that distributes blood to the small intestine and part of the large intestine

_____ 23. Vessel into which veins of the head and neck, upper limbs, and thorax enter

_____ 24. Vessel into which veins of the abdomen, pelvis, and lower limbs enter

_____ 25. Vein that drains the head and associated structures

_____ 26. Longest vein in the body

_____ 27. Vein just behind the knee

_____ 28. Artery that supplies a major part of the large intestine and rectum

_____ 29. First branch off of the arch of the aorta

_____ 30. Arteries supplying the heart

A. Inferior vena cava
B. Superior vena cava
C. Superior mesenteric
D. Common carotid
E. Jugular
F. Brachiocephalic
G. Coronary
H. Inferior mesenteric
I. Popliteal
J. Great saphenous

Cardiovascular Physiology

20

A. CARDIAC CONDUCTION SYSTEM AND ELECTROCARDIOGRAM (ECG OR EKG)

The heart is innervated by the autonomic nervous system (ANS), which modifies, but does not initiate, the cardiac cycle. The heart can continue to contract if separated from the ANS. This is possible because the heart has an *intrinsic pacemaker* termed the *sinoatrial (SA) node.* The SA node is called an intrinsic pacemaker because of its ability to rhythmically, spontaneously depolarize, thereby reaching threshold value and initiating an action potential (nerve impulse). This ability to spontaneously depolarize is due to the SA node's permeability characteristics to sodium, potassium, and calcium ions. The SA node is connected to a series of specialized muscle cells that comprise the *cardiac conduction system.* This system distributes the action potentials that stimulate the cardiac muscle fibers to contract.

Embedded in the posterior wall of the right atrium, the *sinoatrial (SA) node* propagates an action potential throughout both atria via gap junctions. The net effect of the action potential is the contraction of the atria. As the action potential propagates along the cardiac muscle fibers of the atria, it reaches the *atrioventriculuar (AV) node,* which is located in the septal wall between the two atria. The AV node depolarizes, but in such a way as to slow the conduction velocity of the action potential. The spread of the action potential continues as it enters the *atrioventricular (AV) bundle* (sometimes referred to as the *bundle of His*). This constitutes the only electrical connection between the atria and ventricles. The action potential is then conducted along the AV bundle and enters the left and right bundle branches, which run through the interventricular septum toward the apex of the heart. The bundle branches then spread out as a network of large diameter *conduction myofibers (Purkinje fibers)* which conduct the action potential, to the apex of the heart

first and then to the remainder of the myocardium. The wave of depolarization that occurs, at this period of time, initiates ventricular contraction.

Label the components of the cardiac conduction system shown in Figure 20.1.

The electrocardiogram (ECG) provides a record of the electrical events happening within the heart. The ECG is obtained by placing recording electrodes on the surface of the body. These recording electrodes detect the changing potential differences of the heart on the body's surfaces, and a time-dependence plot of these differences is called an *electrocardiogram (ECG).* An *electrocardiograph* is an instrument used to record these changes. As the electrical impulses are transmitted throughout the cardiac conduction system and myocardial cells, a different electrical impulse is generated. These impulses are transmitted from the electrodes to a recording needle that graphs the impulses as a series of up-and-down (vertical) waves called *deflection waves* (Figure 20.2).

1. Electrocardiographic Recordings

In a clinical setting, the recording of an ECG requires the placement of electrodes on the arms and legs, and at six different positions on the chest. The electrocardiograph then amplifies the electrical activity of the heart rendering 12 tracings based upon different combinations of the arm/leg and chest leads. Each tracing is different due the varying positions relative to the heart. Cardiologists can compare these recording with those of "normal" recordings to determine: conduction system abnormalities, cardiac damage, and enlargement of the heart.

A typical lead II ECG record (Figure 20.2) produces three recognizable waves, or series of deflections and intervals.

The first wave is called the *P wave,* it is a small upward deflection representing the *depolarization of the atria.* The *QRS complex* is the second wave, which begins as a downward deflection, followed by an upright deflection and another downward

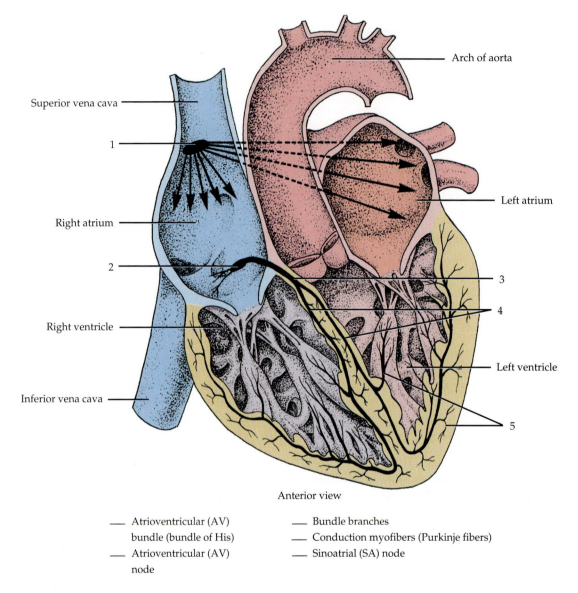

Anterior view

—— Atrioventricular (AV)   —— Bundle branches
     bundle (bundle of His)   —— Conduction myofibers (Purkinje fibers)
—— Atrioventricular (AV)   —— Sinoatrial (SA) node
     node

FIGURE 20.1   Conduction system of heart.

wave. This complex represents both *ventricular de-plorization* and *atria repolarization.* The third wave, the ***T wave*** which is large than the P wave, and smaller and wider than the QRS complex, indicates *ventricular repolarization.* It occurs just prior to the ventricles relaxing.

In order to analyze an ECG, the period between the waves, which are called *intervals* or *segments* must be examined. The ***P-Q interval*** represents the conduction time from the start of atrial excitation to the beginning of ventricular excitation. A lengthened P-Q interval could be indicative of scar tissue in the heart, bicuspid valve stenosis, coronary artery disease and rheumatic fever. The ***S-T segment*** represents the period when all of the ventricular muscle

is fully depolarized. This is the plateau phase of the actin potential. An elongated S-T segment may be indicative of acute myocardial infarction or ischemia (insufficient oxygen) of the cardiac musculature.

2. Recording an ECG

Note: Depending on the type of laboratory equipment available, select from the following procedures related to electrocardiography.

PROCEDURE USING PHYSIOFINDER

PHYSIOFINDER is an interactive computer program that permits laboratory simulations that do not involve the use of animals or advanced or

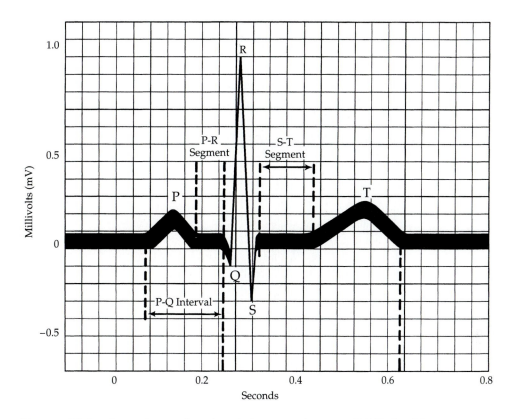

FIGURE 20.2   Recordings of a normal electrocardiogram (ECG), lead II.

expensive laboratory equipment. It permits students to perform experiments, analyze data, draw conclusions, and form hypotheses on the basis of collected data. The program is available from Addison-Wesley Longman, Inc., Customer Service (1-800-922-0579).

For activities related to electrocardiography, select the appropriate experiments from PHYSIOFINDER Module 4—Control of the Heart.

PROCEDURE USING CARDIOCOMPTM

CARDIOCOMP™ is a computer hardware and software system designed to acquire data related to electrocardiography by permitting students to analyze their cardiac biopotentials. The program is available from INTELITOOL® (1-800-227-3805).

PROCEDURE USING A PHYSIOGRAPH-TYPE POLYGRAPH, ELECTROCARDIOGRAPH, OR OSCILLOSCOPE

In the following procedure a physiograph-type polygraph is used, but an electrocardiograph or oscilloscope may also be used (Figure 20.3a).

In recording the ECG, we will utilize only the standard three limb leads. Usually only two electrodes will be in actual use at any one time. The third electrode will be automatically switched off by the *lead-selector switch* of the polygraph.

For most general purposes three leads are usually used. However, electrocardiologists use additional leads, including several chest wall electrodes (Figure 20.3b). These leads are positioned around the chest, and encircle the heart so that there are six intersecting lines on a horizontal plane through the atrioventricular node.

The procedure for recording electrocardiograms using the physiograph is as follows:

PROCEDURE

1. Either you or your laboratory partner should lie on a table or cot, roll down long stockings or socks to the ankles, and *remove all metal jewelry*.
2. Swab the skin with alcohol where electrodes will be applied. Electrode cream, jelly, or saline paste is applied to the skin only where the electrode will make contact, such as to the inner forearm just above the wrists and to the inside of the legs just above the ankles.
3. The cream is then applied on the concave surface of the electrode plates, which are then fastened securely to the limb area surfaces with

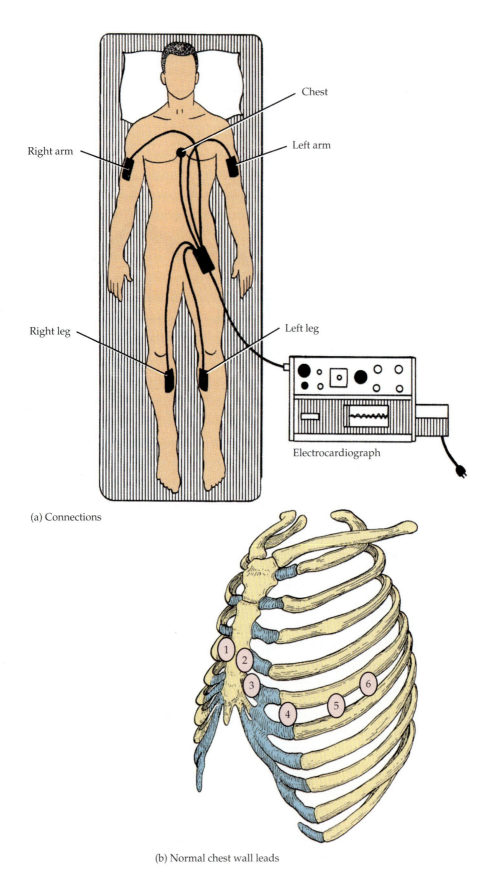

(a) Connections

(b) Normal chest wall leads

FIGURE 20.3 Electrocardiograph.

rubber straps or adhesive tape (rings). Adjust the straps to fit snugly, but *not so tight as to impede blood flow*.

4. Firmly connect the proper end of the electrode cables to the electrode plates, using the standard limb leads previously described, and connect the other end to a cardiac preamplifier or to a self-contained electrocardiograph.

5. After you apply the electrodes and connect the instrument to the subject, ascertain the following before actually recording:
   a. Recording power switch is on.
   b. Paper is sufficient for the entire recording, and the pen is centered properly on the paper with ink flowing freely.
   c. Preamplifier sensitivity has been set at 10 mm per millivolt.
   d. Subject is lying quietly and is relaxed.
   Note: Usually five or six ECG complex recordings from each lead are sufficient. All electrocardiographers have chosen a standard paper speed of 25 mm per second.

6. Turn the instrument on and record for 30 seconds from each lead. Obtain directions from your laboratory instructor on the procedure to be followed when switching from lead to lead. When the recording leads are switched, note the change by marking it on the electrocardiogram. The subject must remain very still and relax completely during the recording period.

7. When the recording is finished, turn off the reading instrument and disconnect the leads from the subject and remove the electrodes, cleaning off the excess cream or paste. The skin can be washed with water to remove electrode cream.

8. Identify and letter the P, QRS, and T waves, and compare the waves with those shown in Figure 20.2.

9. Calculate the duration of the waves and the P-Q interval. Attach this recording to Section A of the LABORATORY REPORT RESULTS at the end of the exercise.

B. CARDIAC CYCLE

In a normal *cardiac cycle* (heartbeat), the two atria contract while the two ventricles relax. Then, while the two ventricles contract, the two atria relax. The term *systole* (SIS-tō-lē; *systole* = contraction) refers to the phase of contraction; the phase of relaxation is *diastole* (dī-AS-tō-lē; *diastole* = expansion). A cardiac cycle consists of a systole and diastole of both atria plus a systole and diastole of both ventricles.

For the purposes of our discussion, we divide the cardiac cycle of a resting adult into three main phases (Figure 20.4).

1. *Isovolumic relaxation.* Repolarization of the ventricular muscle fibers (T wave in the ECG) initiates relaxation. As the ventricles relax, pressure within the chambers drops, and blood starts to flow from the aorta back toward the left ventricle. As this blood becomes trapped in the cusps of the aortic semilunar valve, the valve closes. Rebound of blood off the closed cusps produces a bump called the *dicrotic wave* on the aortic pressure curve. Pressure within the left ventricle, however, remains higher than that within the right atria, so the bicuspid valve remains closed.

   With the closure of the aortic semilunar valve, there is a brief interval when ventricular blood volume remains the same because the aortic semilunar and bicuspid valves are closed. This period is called *isovolumic relaxation.* As the ventricle continues to relax, the pressure falls quickly. When ventricular pressure drops below atrial pressure, the bicuspid valve opens and ventricular filling begins.

2. *Ventricular filling.* The major part of ventricular filling occurs just after the AV valves open. Blood that has been flowing into the atria and building up while the ventricles were contracting now rushes into the ventricles. The first third of ventricular filling time thus is known as the period of *rapid ventricular filling.* During the middle third, called *diastasis*, a much smaller volume of blood flows into the ventricles.

   Firing of the SA node results in atrial depolarization, noted as the P wave on the ECG. Atrial systole (contraction) follows the P wave. Atrial systole occurs in the last third of the ventricular filling period and accounts for the final 15–20% of the blood that fills the ventricles. This period of ventricular filling is termed the period of atrial systole. At the end of ventricular diastole, there are about 130 ml in each ventricle. This volume of blood is called *end-diastolic volume (EDV)*. Atrial contraction is not absolutely necessary for adequate blood flow at normal heart rates (exercise, for example). Throughout the period of ventricular filling, the AV valves are open and the semilunar valves are closed.

3. *Ventricular systole.* Near the end of atrial systole, the action potential from the SA node has

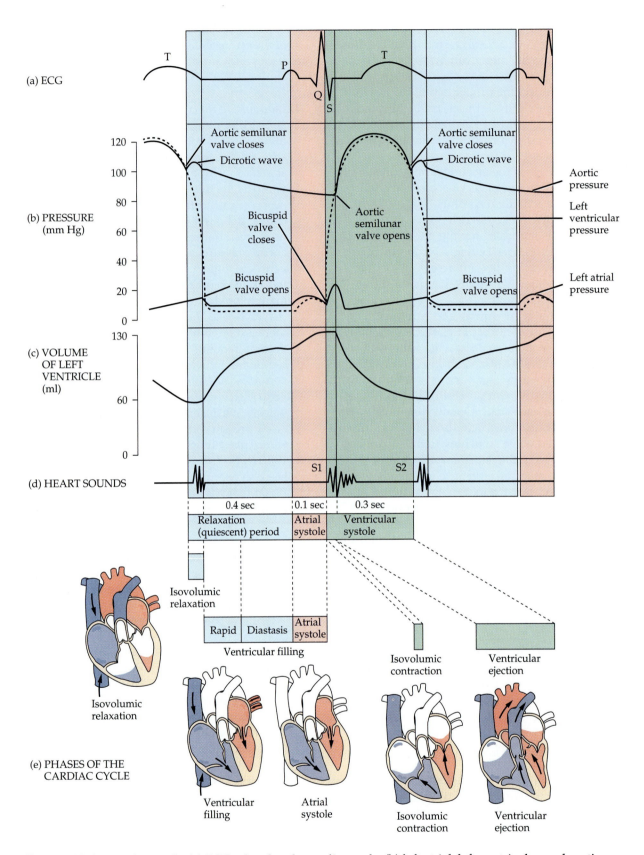

(a) ECG

(b) PRESSURE (mm Hg)

Aortic semilunar valve closes
Dicrotic wave
Aortic pressure

Bicuspid valve closes
Aortic semilunar valve opens
Aortic semilunar valve closes
Dicrotic wave
Left ventricular pressure

Bicuspid valve opens
Bicuspid valve opens
Left atrial pressure

(c) VOLUME OF LEFT VENTRICLE (ml)

(d) HEART SOUNDS

S1   S2

0.4 sec   0.1 sec   0.3 sec

Relaxation (quiescent) period   Atrial systole   Ventricular systole

Isovolumic relaxation

Rapid | Diastasis | Atrial systole
Ventricular filling

Isovolumic contraction

Ventricular ejection

Isovolumic relaxation

Ventricular filling   Atrial systole   Isovolumic contraction   Ventricular ejection

(e) PHASES OF THE CARDIAC CYCLE

FIGURE 20.4 Cardiac cycle. (a) ECG related to the cardiac cycle; (b) left atrial, left ventricular, and aortic pressure changes along with the opening and closing of the valves during cardiac cycle; (c) left ventricular volume during the cardiac cycle; (d) heart sounds related to the cardiac cycle; (e) phases of the cardiac cycle.

passed through the AV node and into the ventricles, causing them to depolarize. Onset of ventricular depolarization is indicated by the QRS complex in the ECG. Then, *ventricular systole* begins, and blood is pushed up against the AV valves, forcing them shut. This period of the cardiac cycle, when all valves are closed and pressure continues to climb within the left ventricle is termed the period of *isovolumetric contraction.* During this time cardiac muscle fibers are contracting and exerting force, but are not yet shortening because it is very difficult to compress any liquid, including blood. Thus the muscle contraction is isometric (same length). Moreover, since there is no escape route for the blood, ventricular volume remains the same (isovolumetric).

As ventricular contraction continues, pressure inside the chambers rises sharply. When left ventricular pressure surpasses aortic pressure the semilunar valves open, and ejection of blood from the heart begins. This period is termed the *period of rapid ventricular ejection* and accounts for approximately one third of the period of ventricular systole. Then, the semilunar valves close and another relaxation period begins. The volume of blood still left in a ventricle following its systole is called *end-systolic volume (ESV).* At rest, ESV is about 60 ml. *Stroke volume,* the volume of blood ejected per beat from the left ventricle, equals end-diastolic volume minus end-systolic volume ($SV = EDV - ESV$).

Since resting heart rate (HR) is about 75 beats/min, each cardiac cycle lasts about 0.8 sec (Figure 20.4). The duration of the various phases of the cardiac cycle are as follows: the total time for ventricular diastole = 0.4 sec, with isovolumic contraction accounting for approximately 40 milliseconds (msec) and atrial contraction accounting for 0.1 sec of that time period; total time for ventricular systole = 0.3 sec, with rapid ventricular ejection accounting for between 50–100 msec of that period.

C. CARDIAC CYCLE EXPERIMENTS

PROCEDURE USING PHYSIOFINDER

PHYSIOFINDER is an interactive computer program that permits laboratory simulations that do not involve the use of animals or advanced or expensive laboratory equipment. It permits students to perform experiments, analyze data, draw conclusions, and form hypotheses on the basis of collected data. The program is available from Addison-Wesley Longman, Inc., Customer Service (1-800-922-0579).

For activities related to the cardiac cycle, select the appropriate experiments from PHYSIOFINDER Module 4—Control of the Heart—and Module 5—Cardiovascular Responses to Exercise.

D. HEART SOUNDS

The beating of a human heart usually produces four sounds, but only two may be detected with a *stethoscope.* (The remaining two sounds may be heard if adequately amplified electronically.) The first sound is created by blood turbulence associated with the closure of the atrioventricular valves soon after ventricular systole begins (See Figure 20.4). This sound is the loudest and longest of the two sounds, and may be best heard over the apex of the heart. The sound produced by the tricuspid valve is best heard in the fifth intercostal space just lateral to the left border of the sternum, while the mitral valve sound is best heard in the fifth intercostal space at the apex of the heart (Figure 20.5).

The second heart sound is created by blood turbulence associated with closure of the semilunar valves at the beginning of ventricular diastole. This sound is of shorter duration and lower intensity, and has a more snapping sound as compared to the quality of the first heart sound. The second heart sound produced by the aortic semilunar valve closing is best heard in the second intercostal space to the right of the sternum. The second sound produced by the pulmonary semilunar valve is best heard in the second intercostal space just to the left of the sternum (Figure 20.5).

Heart sounds provide valuable information about the valves. Abnormal sounds made by the valves are termed *murmurs.* Some murmurs are caused by the noise made by blood flowing back into a chamber of the heart because of improper closure (incompetence) of a valve. Another murmur may be produced by the improper opening (incompetence) of an AV or semilunar valve, which restricts the flow of blood out of a cardiac chamber.

Murmurs do not always indicate that the valves are not functioning properly. Many individuals possess a *functional murmur* that has no clinical significance at all.

Listening to sounds of the body is called *auscultation* (aws-kul-TĀ-shun).

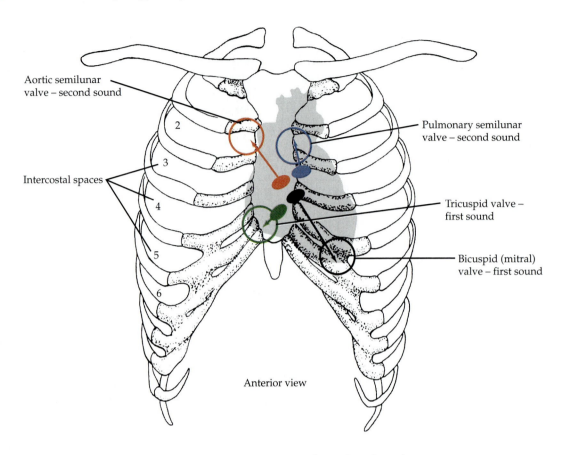

Aortic semilunar
valve – second sound

Pulmonary semilunar
valve – second sound

Intercostal spaces

2

3

4

5

6

Tricuspid valve –
first sound

Bicuspid (mitral)
valve – first sound

Anterior view

FIGURE 20.5   Surface areas where heart valve sounds are best heard.

1. Use of Stethoscope

PROCEDURE

1. The stethoscope should be used in a quiet room.
2. The earpieces of the stethoscope *should be cleaned with alcohol* just before using and should also be pointed slightly forward when placed in the ears. They will be more comfortable in this position, and it will be easier to hear through them.
3. Listen to the heart sounds of your laboratory partner by placing the diaphragm of the stethoscope next to the skin at several positions on the chest wall illustrated in Figure 20.5.
4. The first sound is best heard at the apex of the heart.
5. The second sound is best heard in the second intercostal space.
6. **CAUTION!** *Assuming that your partner has no known or apparent cardiac or other health problems and is capable of such an activity, ask her/him to run in place about 25 steps. Listen to the heart sounds again.*
7. Answer all questions pertaining to Section D of the LABORATORY REPORT RESULTS at the end of the exercise.

E. PULSE RATE

A wave of pressure, called *pulse*, is produced in the arteries due to ventricular contraction. Pulse rate and heart rate are essentially the same. Pulse can be felt readily where an artery is near the surface of the skin and over the surface of a bone. Pulse rates vary considerably in individuals because of time of day, temperature, emotions, stress, and other factors. The normal adult pulse rate of the heart at rest is within a range of 72 to 80 beats per minute. With practice you can learn to take accurate pulse rates by counting the beats per 15 sec and multiplying by 4 to obtain beats per minute. The term *tachycardia* (tak'-e-KAR-dē-a) is applied to a rapid heart rate or pulse rate (over 100 beats/minute). *Bradycardia* (brād-e-KAR-dē-a) indicates a slow heart rate or pulse rate (below 50 beats/minute).

Although the pulse may be detected in most surface arteries, the pulse rate is usually determined on the *radial artery* of the wrist.

1. Radial Pulse

PROCEDURE

1. Using your index and middle finger palpate your laboratory partner's radial artery.
2. The thumb should never be used because it has its own prominent pulse.
3. Palpate the area behind your partner's thumb just inside the bony prominence on the lateral aspect of the wrist.
4. **CAUTION!** *Do not apply too much pressure.*
5. Count the pulse, change positions, and record your results in Section E of the LABORATORY REPORT RESULTS at the end of the exercise.
6. Calculate the *median pulse rate* of the entire class.

2. Carotid Pulse

PROCEDURE

1. Using the same fingers that you used for the radial pulse, place them on either side of your partner's larynx (voice box).
2. *Gently* press downward and toward the back until you feel the pulse. You must feel the pulse clearly with at least two fingers, so adjust your hand accordingly.
3. The radial and carotid pulse can be compared under the following conditions: (a) sitting quietly, (b) standing quietly, (c) right after walking 60 steps, (d) right after running in place 60 steps, *assuming that your partner has no known or apparent cardiac or other health problems and is capable of such an activity*. Notice how long it takes the pulse to return to normal after the walking and running exercises.
4. Record your results in Section E of the LABORATORY REPORT RESULTS at the end of the exercise.
5. Compare your radial and carotid pulse in the table provided.

F. BLOOD PRESSURE (AUSCULTATION METHOD)

In physiological terms, the term *blood pressure* actually refers to the interaction of several different pressures (the pressure only within the arteries, or pressure within the veins, or venous pressure; the pressure within the pulmonary system, or pulmonary pressure; and the pressure within all of the other vascular beds, termed systemic pressure). Clinically, however, the term *blood pressure* only refers to the pressure within the large arteries.

Arterial pressure can be measured either directly, by the insertion of a needle or catheter directly into the artery in such a way that the needle is pointing "upstream," or against the flow of blood within the vessel, or indirectly, by an instrument called a *sphygmomanometer* (sfig'-mō-ma-NOM-e-ter; *sphygmo* = pulse). Regardless of which method is utilized, blood pressure is recorded in millimeters of mercury (mm Hg) and is normally taken in the brachial artery.

A commonly used sphygmomanometer consists of an inflatable rubber cuff attached by a rubber tube to a compressible hand pump or bulb. Another tube attaches to the cuff and to a mercury column marked off in millimeters or an anaeroid gauge that measures the pressure in millimeters of mercury (mm Hg). The highest pressure in the artery, occurring during ventricular systole, is termed *systolic blood pressure.* The lowest pressure, occurring during ventricular diastole, is termed *diastolic blood pressure.* Blood pressures are usually expressed as a ratio of systolic to diastolic pressure. The average blood pressure of a young adult is about 120 mm Hg systolic and 80 mm Hg diastolic, abbreviated to 120/80. The difference between systolic and diastolic pressure is called *pulse pressure.* Pulse pressure may be used clinically to indicate several physiological and pathological parameters, and usually averages 40 mm Hg in a healthy individual. The ratio of systolic pressure to diastolic pressure to pulse pressure may also be utilized clinically as a diagnostic tool, and is usually 3:2:1.

This exercise employs the indirect *auscultatory method*, in which the sounds of blood flow are heard with a stethoscope. Blood flow in an artery is impeded by increasing pressure within a sphygmomanometer. When the cuff of the sphygmomanometer applies sufficient pressure to completely occlude blood flow, no sounds can be heard distal to the cuff because no blood can flow through the artery. When cuff pressure drops below the maximal (systolic) pressure in the artery, blood is heard passing through the vessel. When cuff pressure drops below the lowest (diastolic) pressure in the vessel, the sound becomes muffled and usually disappears. The sounds heard through the stethoscope via this procedure are termed *Korotkoff* (kō-ROT-kof) *sounds*.

Indirect blood pressure can be taken in any artery that can be occluded easily. The brachial artery has the advantage of being at approximately the same level as the heart, so brachial pressure closely reflects aortic pressure.

The procedure for determining blood pressure using the sphygmomanometer is as follows:

PROCEDURE

1. Either you or your laboratory partner should be comfortably seated, at ease, with your arm bared, slightly flexed, abducted, and perfectly relaxed. You may, for convenience, rest the forearm on a table in the supinated position.
2. Wrap the deflated cuff of the sphygmomanometer around the arm with the lower edge about 2.54 cm (1 in.) above the antecubital space. Close the valve on the neck of the rubber bulb.
3. Clean the earpieces of the stethoscope with alcohol before using it. Using the diaphragm of the stethoscope, find the pulse in the brachial artery just above the bend of the elbow, on the inner margin of the biceps brachii muscle.
4. Inflate the cuff by squeezing the bulb until the air pressure within it just exceeds 170 mm Hg. At this point the wall of the brachial artery is compressed tightly, and no blood should be able to flow through.
5. Place the diaphragm of the stethoscope firmly over the brachial artery and while watching the pressure gauge, slowly turn the valve, releasing air from the cuff. Listen carefully for Korotkoff's sounds as you watch the pressure fall. The first loud, rapping sound you hear will be the *systolic pressure*.
6. Continue listening as the pressure falls. The pressure recorded on the mercury column when the sounds become faint or disappear is the *diastolic pressure* reading. It measures the force of blood in the arteries during ventricular re-

laxation and specifically reflects the peripheral resistance of the arteries.

7. Repeat this procedure for both readings two or three times to see if you get consistent results. *Allow a few minutes between readings*. Record all results in Section F of the LABORATORY REPORT RESULTS at the end of the exercise.
8. Have a partner stand and record his or her blood pressure several times for each arm. Record all results in the table provided in Section F of the LABORATORY REPORT RESULTS at the end of the exercise.
9. *Now, assuming that your partner has no known or apparent cardiac or other health problems, and is capable of such an activity* have your partner do some exercise, such as running in place 40 or 50 steps, and measure the blood pressure again *immediately after the completion of the exercise*. Record the pulse pressure in the table provided in Section F of the LABORATORY REPORT RESULTS at the end of the exercise.

G. OBSERVING BLOOD FLOW

PROCEDURE USING PHYSIOFINDER

PHYSIOFINDER is an interactive computer program that permits laboratory simulations that do not involve the use of animals or advanced or expensive laboratory equipment. It permits students to perform experiments, analyze data, draw conclusions, and form hypotheses on the basis of collected data. The program is available from Addison-Wesley Longman, Inc., Customer Service (1-800-922-0579).

For activities related to observing blood flow, select the appropriate experiments from PHYSIOFINDER Module 6—Control of Microcirculation.

ANSWER THE LABORATORY REPORT QUESTIONS AT THE END OF THE EXERCISE.

Cardiovascular Physiology 20

Student _____   **Date** _____

Laboratory Section _____   **Score/Grade** _____

SECTION A. CARDIAC CONDUCTION SYSTEM
AND ELECTROCARDIOGRAM (ECG OR EKG)

Attach examples of the electrocardiogram strips you obtained.

LEAD I

Electrocardiogram Strip

LEAD II

Electrocardiogram Strip

LEAD III

Electrocardiogram Strip

SECTION D. HEART SOUNDS

1. Which heart sound is the loudest? _____

2. Did you hear a third sound? _____

3. Where does the first sound originate? _____

4. Where does the second sound originate? _____

5. After you exercised, how did the heart sounds differ from before? _____

6. Did they differ in rate and intensity? _____

7. Did the first or second increase in loudness? _____

SECTION E. PULSE RATE

1. Radial pulse rate count results: _____ pulses per minute.

2. Have all radial pulse rates put on the blackboard, arranging them from the highest to the lowest. The median pulse rate is found exactly half-way down from the top.

What is the median radial pulse rate of the class? _____

What was the highest rate? _____

What was the lowest rate? _____

3. Compare your radial and carotid pulse rates by filling in the following table.

	Radial pulse rate	Carotid pulse rate
Sitting quietly		
Standing quietly		
After walking		
After running in place		

SECTION F. BLOOD PRESSURE (AUSCULTATION METHOD)

Record your systolic and diastolic blood pressures in the following table.

	Systolic pressure		Diastolic pressure		Pulse pressure
	Left arm	Right arm	Left arm	Right arm	
Sitting					
Standing					
After running					

Cardiovascular Physiology 20

Student _____   Date _____

Laboratory Section _____   Score/Grade _____

PART 1. Multiple Choice

_____ 1. When the semilunar valves are open during a cardiac cycle, which of the following occur?
I—atrioventricular valves are closed;
II—ventricles are in systole;
III—ventricles are in diastole;
IV—blood enters the aorta;
V—blood enters the pulmonary trunk;
VI—atrial contraction.
(a) I, II, IV, and V (b) I, II, and VI (c) II, IV, and V (d) I, III, IV, and VI

_____ 2. When ventricular pressure drops below atrial pressure, which phase of the cardiac cycle occurs? (a) ventricular filling (b) ventricular systole (c) isovolumetric contraction (d) isovolumetric relaxation

_____ 3. During which phase of the cardiac cycle are all four chambers in diastole? (a) ventricular filling (b) relaxation (c) isovolumetric contraction (d) isovolumetric relaxation

_____ 4. Which of the following statements is *not true*? (a) Pulse pressure is the difference between systolic and diastolic blood pressures. (b) Both systolic and diastolic blood pressures can be obtained via the pulse method. (c) Systolic blood pressure is obtained at the first loud, rapping sound you hear when you are measuring blood pressure via the auscultatory method. (d) Diastolic blood pressure is obtained when the sound heard through the stethoscope when you are measuring blood pressure via the ausculatory method becomes muffled and usually disappears.

_____ 5. The two distinct heart sounds, described phonetically as lubb and dupp, represent (a) contraction of the ventricles and relaxation of the atria (b) contraction of the atria and relaxation of the ventricles (c) blood turbulence associated with closing of the atrioventricular and semilunar valves (d) surging of blood into the pulmonary artery and aorta.

PART 2. Completion

6. Systole and diastole of both atria plus systole and diastole of both ventricles is called

_____ .

7. Blood flow through the heart is controlled by speed of the cardiac cycle, venous return to the heart,

opening and closing of the valves, and _____ .

8. Heart sounds provide valuable information about the _____.

9. Abnormal or peculiar heart sounds are called _____.

10. Although the pulse may be detected in most surface arteries, pulse rate is usually determined on the

    _____.

11. The heart has an intrinsic regulating system called the cardiac _____ system.

12. Electrical impulses accompanying the cardiac cycle are recorded by the _____.

13. The typical ECG produces three clearly recognizable waves. The first wave, which indicates depolar-

    ization of the atria, is called the _____.

14. Various up-and-down impulses produced by an ECG are called _____.

15. The instrument normally used to measure blood pressure is called a(n) _____.

16. The artery that is normally used to evaluate blood pressure is the _____.

17. Rapping or thumping sounds heard clinically when blood pressure is being taken are called

    _____ sounds.

18. The difference between systolic and diastolic pressure is called _____.

19. An average blood pressure value for an adult is _____.

20. An average pulse pressure is _____.

Lymphatic System

21

The *lymphatic* (lim-FAT-ik) *system* consists of a fluid called lymph flowing within lymphatic vessels (lymphatics), several structures and organs that contain lymphatic tissue, and red bone marrow, which houses stem cells that develop into lymphocytes (Fig. 21.1). Lymphatic tissue is a specialized form of reticular connective tissue that contains large numbers of lymphocytes.

A. LYMPHATIC VESSELS

Lymphatic vessels originate as microscopic *lymphatic capillaries* in spaces between cells. They are found in most parts of the body; they are absent in avascular tissue, the central nervous system, splenic pulp, and bone marrow. They are slightly larger than blood capillaries and have a unique structure that permits interstitial fluid to flow into them but not out. Lymphatic capillaries also differ from blood capillaries in that they end blindly; blood capillaries have an arterial and a venous end. In addition, lymphatic capillaries are structurally adapted to ensure the return of proteins to the cardiovascular system when they leak out of blood capillaries.

Just as blood capillaries converge to form venules and veins, lymphatic capillaries unite to form larger and larger lymph vessels called *lymphatic vessels* (Figure 21.2). Lymphatic vessels resemble veins in structure, but have thinner walls and more valves, and contain lymph nodes at various intervals along their length (Figure 21.2). Ultimately, lymphatic vessels deliver lymph into two main channels—the thoracic duct and the right lymphatic duct. These will be described shortly.

Lymphangiography (lim-fan'-jē-OG-ra-fē) is the x-ray examination of lymphatic vessels and lymph organs after they are filled with a radiopaque substance. Such an x-ray is called a *lymphangiogram* (lim-FAN-jē-ō-gram). Lymphangiograms are useful in detecting edema and carcinomas, and in locating lymph nodes for surgical and radiotherapeutic treatment.

B. LYMPHATIC TISSUE

1. Thymus Gland

Usually a bilobed lymphatic organ, the *thymus gland* is located in the mediastinum, posterior to the sternum and between the lungs (Figure 21.1). Its role in immunity is to produce thymic hormones, which are thought to aid in the maturation of T-cells.

2. Lymph Nodes

The oval or bean-shaped structures located along the length of lymphatic vessels are called *lymph nodes*. A lymph node contains a slight depression on one side called a *hilus* (HĪ-lus), where blood vessels and efferent lymphatic vessels leave the node. Each node is covered by a *capsule* of dense connective tissue that extends into the node. The capsular extensions are called *trabeculae*. Internal to the capsule is a supporting network of reticular fibers and fibroblasts. The capsule, trabeculae, and reticular fibers and fibroblasts constitute the stroma (framework) of a lymph node. The parenchyma (functioning part) of a lymph node is specialized into two regions: cortex and medulla. The outer *cortex* contains many *follicles,* which are regions of densely packed lymphocytes arranged in masses called *lymphatic nodules*. The outer rim of each follicle contains *T cells (T-lymphocytes)* plus *macrophages* and *follicular dendritic cells,* which participate in activation of T cells. The central area of each follicle, called the *germinal center*, is the site where *B cells (B-lymphocytes)* proliferate into antibody-secreting plasma cells. The inner region of a lymph node is called the *medulla*. In the medulla, the lymphocytes are tightly packed in strands called *medullary cords*. These cords also contain macrophages and plasma cells.

Lymph flows through a node in one direction. It enters through *afferent* (*ad* = to; *ferre* = to carry) *lymphatic vessels* that enter the convex surface of the node at several points. They contain valves that open toward the node so that the lymph is directed *inward*.

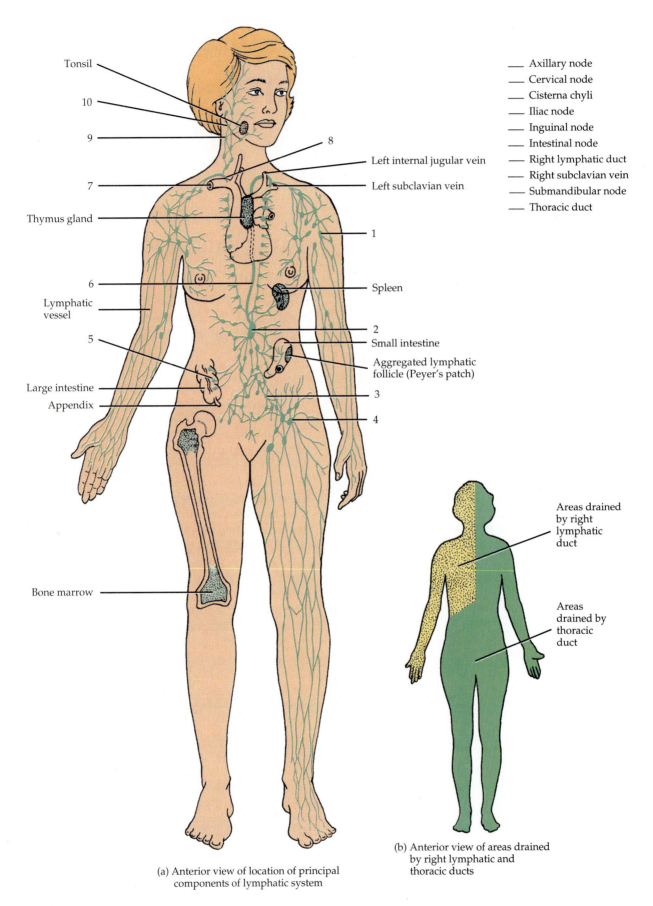

Tonsil

10

9

7

Thymus gland

Lymphatic
vessel

6

5

Large intestine

Appendix

Bone marrow

8

Left internal jugular vein

Left subclavian vein

1

Spleen

2

Small intestine

Aggregated lymphatic
follicle (Peyer's patch)

3

4

___ Axillary node
___ Cervical node
___ Cisterna chyli
___ Iliac node
___ Inguinal node
___ Intestinal node
___ Right lymphatic duct
___ Right subclavian vein
___ Submandibular node
___ Thoracic duct

Areas drained
by right
lymphatic
duct

Areas
drained by
thoracic
duct

(a) Anterior view of location of principal
components of lymphatic system

(b) Anterior view of areas drained
by right lymphatic and
thoracic ducts

FIGURE 21.1 Lymphatic system.

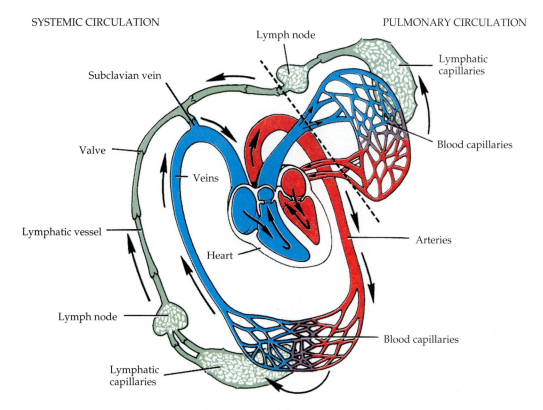

SYSTEMIC CIRCULATION PULMONARY CIRCULATION

Lymph node

Subclavian vein

Lymphatic capillaries

Valve

Blood capillaries

Veins

Arteries

Heart

Lymphatic vessel

Lymph node

Blood capillaries

Lymphatic capillaries

Arrows show direction of flow of lymph and blood

FIGURE 21.2 Relationship of lymphatic system to cardiovascular system.

Inside the node, the lymph enters the *sinuses,* which are a series of irregular channels between the medullary cords. Lymph flows through sinuses in the cortex *(cortical sinuses)* and then in the medulla *(medullary sinuses)* and exit the lymph node via one or two *efferent* (*ex* = away) *lymphatic vessels.* Efferent lymphatic vessels are wider than afferent vessels and contain valves that open away from the node to convey lymph *out* of the node. Efferent lymphatic vessels emerge from one side of the lymph node at a slight depression called a *hilus* (HĪ-lus). Blood vessels also enter and leave the node at the hilus.

The lymph nodes filter foreign substances from the lymph as it passes back toward the bloodstream. These substances are trapped by the reticular fibers within the node. Then, macrophages destroy some foreign substances by phagocytosis and lymphocytes bring about destruction of others by immune responses. Plasma cells and T cells that proliferate within lymph nodes can circulate to other parts of the body.

Label the parts of a lymph node in Figure 21.3 and the various groups of lymph nodes in Figure 21.1.

3. Spleen

The oval *spleen* is the largest mass of lymphatic tissue in the body (Figure 21.1). It is situated in the left hypochondriac region between the stomach and diaphragm.

The splenic artery and vein and the efferent lymphatic vessels pass through the hilus. Since the spleen has no afferent lymphatic vessels or lymph sinuses, it does not filter lymph. One key splenic function related to immunity is the production of B cells, which develop into antibody-producing plasma cells. The spleen also phagocytizes bacteria and worn-out and damaged red blood cells and platelets. During early fetal development, the spleen participates in blood cell formation.

C. LYMPH CIRCULATION

When plasma is filtered by blood capillaries, it passes into the interstitial spaces; it is then known as interstitial fluid. When this fluid passes from interstitial spaces into lymph capillaries, it is called

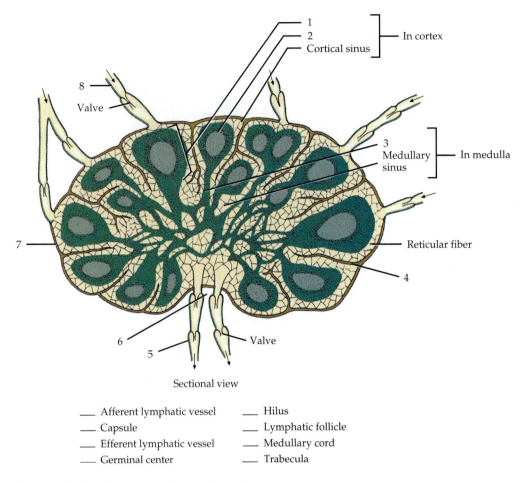

Sectional view

___ Afferent lymphatic vessel   ___ Hilus
___ Capsule   ___ Lymphatic follicle
___ Efferent lymphatic vessel   ___ Medullary cord
___ Germinal center   ___ Trabecula

FIGURE 21.3   Structure of a lymph node.

lymph (*lympha* = clear water). Lymph from lymph capillaries flows into lymphatic vessels that run toward lymph nodes. At the nodes, afferent vessels penetrate the capsules at numerous points, and the lymph passes through the sinuses of the nodes. Efferent vessels from the nodes unite to form *lymph trunks*.

The principal trunks pass their lymph into two main channels, the thoracic duct and the right lymphatic duct. The ***thoracic (left lymphatic) duct*** begins as a dilation in front of the second lumbar vertebra called the ***cisterna chyli*** (sis-TER-na KĪ-lē). The thoracic duct is the main collecting duct of the lymphatic system and receives lymph from the left side of the head, neck, and chest, the left upper limb, and the entire body inferior to the ribs (Figure 21.1).

The ***right lymphatic duct*** drains lymph from the upper right side of the body (Figure 21.1). Ultimately, the thoracic duct empties all of its lymph into the junction of the left internal jugular vein and left subclavian vein, and the right lymphatic duct empties all of its lymph into the junction of the right internal jugular vein and right subclavian vein. Thus, lymph is drained back into the blood and the cycle repeats itself continuously.

Edema, an excessive accumulation of interstitial fluid in tissue spaces, may be caused by an obstruction, such as an infected node or a blockage of vessels, in the pathway between the lymphatic capillaries and the subclavian veins. Another cause is excessive lymph formation and increased permeability of blood capillary walls. A rise in capillary blood pressure, in which interstitial fluid is formed faster than it is passed into lymphatic vessels, also may result in edema.

ANSWER THE LABORATORY REPORT QUESTIONS AT THE END OF THE EXERCISE.

Lymphatic System 21

Student _____ **Date** _____

Laboratory Section _____ **Score/Grade** _____

PART 1. Completion

1. Small masses of lymphatic tissue located along the length of the lymphatic vessels are called

   _____.

2. Lymphatic vessels have thinner walls than veins, but resemble veins in that they also have

   _____.

3. All lymphatic vessels converge, get larger, and eventually merge into two main channels, the thoracic

   duct and the _____.

4. Cells in the lymph nodes that carry on phagocytosis are _____.

5. Lymph is conveyed out of lymph nodes in _____ vessels.

6. B cells in lymph nodes produce certain cells that are responsible for the production of antibodies.

   These cells are called _____.

7. The x-ray examination of lymphatic vessels and lymph organs after they are filled with a radiopaque

   substance is called _____.

8. This x-ray examination is useful in detecting edema and _____.

9. The largest mass of lymphatic tissue is the _____.

10. The main collecting duct of the lymphatic system is the _____ duct.

11. _____ are the areas within a lymph node that contain T and B cells.

12. The _____ tonsils are commonly removed by a tonsillectomy.

Respiratory System 22

Cells continually use oxygen (O_2) for the metabolic reactions that release energy from nutrient molecules and produce ATP. At the same time, these reactions release carbon dioxide (CO_2). Since an excessive amount of CO_2 produces acidity that is toxic to cells, the excess CO_2 must be eliminated quickly and efficiently. The two systems that cooperate to supply O_2 and eliminate CO_2 are the cardiovascular system and the respiratory system. The respiratory system provides for gas exchange, intake of O_2, and elimination of CO_2, whereas the cardiovascular system transports the gases in the blood between the lungs and the cells. Failure of either system has the same effect on the body: disruption of homeostasis and rapid death of cells from oxygen starvation and buildup of waste products. In addition to functioning in gas exchange, the respiratory system also contains receptors for the sense of smell, filters inspired air, produces sounds, and helps eliminate wastes.

Respiration is the exchange of gases between the atmosphere, blood, and cells. It takes place in three basic steps:

1. *Pulmonary ventilation* The first process, *pulmonary* (*pulmo* = lung) *ventilation,* or breathing, is the inspiration (inflow) and expiration (outflow) of air between the atmosphere and the lungs.
2. *External (pulmonary) respiration* This is the exchange of gases between the air spaces of the lungs and blood in pulmonary capillaries. The blood gains O_2 and loses CO_2.
3. *Internal (tissue) respiration* The exchange of gases between blood in systemic capillaries and tissue cells is known as internal (tissue) respiration. The blood loses O_2 and gains CO_2. Within cells, the metabolic reactions that consume O_2 and produce CO_2 during production of ATP are termed *cellular respiration.*

The *respiratory system* consists of the nose, pharynx (throat), larynx (voice box), trachea (windpipe), bronchi, and lungs (Figure 22.1). Structurally, the respiratory system consists of two portions. (1) The

term *upper respiratory system* refers to the nose, pharynx, and associated structures. (2) The *lower respiratory system* refers to the larynx, trachea, bronchi, and lungs. Functionally, the respiratory system also consists of two portions. (1) The *conducting portion* consists of a series of interconnecting cavities and tubes—nose, pharynx, larynx, trachea, bronchi, and terminal bronchioles—that conduct air into the lungs. (2) The *respiratory portion* consists of those portions of the respiratory system where the exchange of gases occurs—respiratory bronchioles, alveolar ducts, alveolar sacs, and alveoli.

Using your textbook, charts, or models for reference, label Figure 22.1.

A. ORGANS OF THE RESPIRATORY SYSTEM

1. Nose

The *nose* has an external portion and an internal portion inside the skull. The external portion consists of a supporting framework of bone and hyaline cartilage covered with muscle and skin and lined by mucous membrane. The bridge of the nose is formed by the nasal bones, which hold it in a fixed position. Because it has a framework of pliable hyaline cartilage, the rest of the external nose is somewhat flexible. On the undersurface of the external nose are two openings called the *external nares* (NA-rēz; singular is *naris*), or *nostrils.* The interior structures of the nose are specialized for three functions: (1) incoming air is warmed, moistened, and filtered; (2) olfactory stimuli are received; and (3) large, hollow resonating chambers modify speech sounds.

The internal portion of the nose is a large cavity in the skull that lies inferior to the anterior cranium and superior to the mouth. Anteriorly, the internal nose merges with the external nose, and posteriorly it communicates with the pharynx through two openings called the *internal nares (choanae).* Ducts from the paranasal sinuses

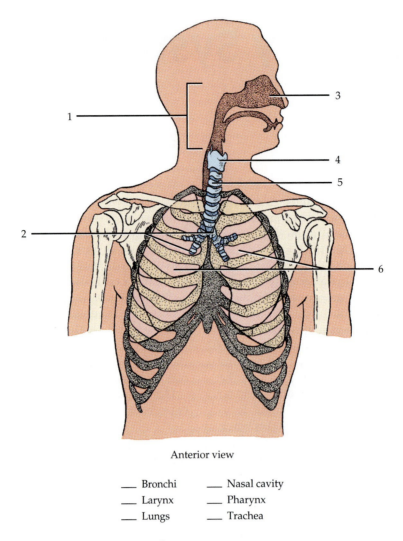

Anterior view

___ Bronchi	___ Nasal cavity
___ Larynx	___ Pharynx
___ Lungs	___ Trachea

FIGURE 22.1   Organs of respiratory system.

(frontal, sphenoidal, maxillary, and ethmoidal) and the nasolacrimal ducts also open into the internal nose. The lateral walls of the internal nose are formed by the ethmoid, maxillae, lacrimal, palatine, and inferior nasal conchae bones. The ethmoid also forms the roof. The floor of the internal nose is formed mostly by the palatine bones and palatine processes of the maxillae, which together comprise the hard palate.

The inside of both the external and internal nose is called the *nasal cavity.* It is divided into right and left sides by a vertical partition called the *nasal septum.* The anterior portion of the septum consists primarily of hyaline cartilage. The remainder is formed by the vomer, perpendicular plate of the ethmoid, maxillae, and palatine bones (see Figure 22.2). The anterior portion of the nasal cavity, just inside the nostrils, is called the *vestibule* and is sur-

rounded by cartilage. The superior nasal cavity is surrounded by bone.

When air enters the nostrils, it passes first through the vestibule. The vestibule is lined by skin containing coarse hairs that filter out large dust particles. The air then passes into the superior nasal cavity. Three shelves formed by projections of the superior, middle, and inferior nasal conchae extend out of each lateral wall of the cavity. The conchae, almost reaching the septum, subdivide each side of the nasal cavity into a series of groovelike passageways—the *superior, middle,* and *inferior meatuses* (mē-Ā-tes-ez; *meatus* = passage; singular is *meatus*). Mucous membrane lines the cavity and its shelves.

The olfactory receptors lie in the membrane lining the superior nasal conchae and adjacent septum. This region is called the **olfactory epithelium.** Inferior to the olfactory epithelium, the mucous

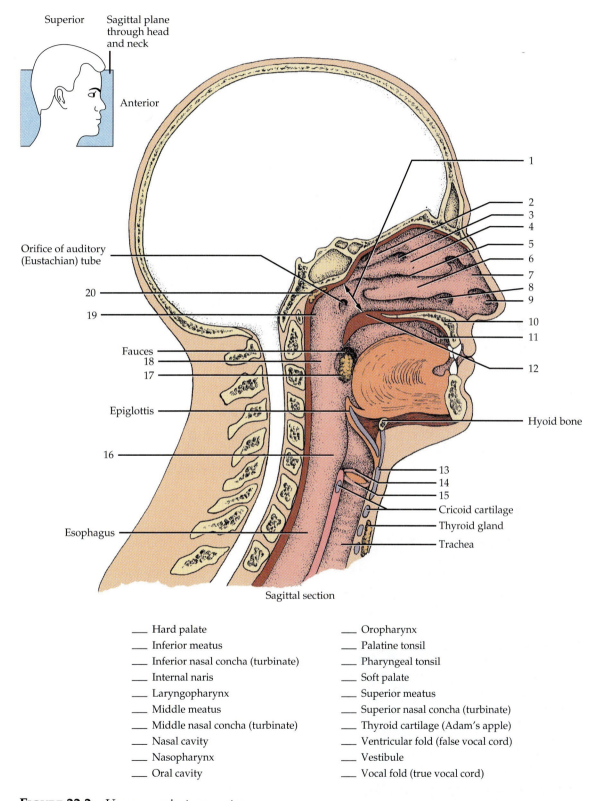

Superior   Sagittal plane
through head
and neck

Anterior

Orifice of auditory
(Eustachian) tube

1

2
3
4
5
6
7
8
9

10
11

12

20

19

Fauces
18
17

Epiglottis

Hyoid bone

16

13
14
15

Cricoid cartilage

Thyroid gland

Esophagus

Trachea

Sagittal section

___ Hard palate
___ Inferior meatus
___ Inferior nasal concha (turbinate)
___ Internal naris
___ Laryngopharynx
___ Middle meatus
___ Middle nasal concha (turbinate)
___ Nasal cavity
___ Nasopharynx
___ Oral cavity

___ Oropharynx
___ Palatine tonsil
___ Pharyngeal tonsil
___ Soft palate
___ Superior meatus
___ Superior nasal concha (turbinate)
___ Thyroid cartilage (Adam's apple)
___ Ventricular fold (false vocal cord)
___ Vestibule
___ Vocal fold (true vocal cord)

FIGURE 22.2   Upper respiratory system.

membrane contains capillaries and pseudostratified ciliated columnar epithelium with many goblet cells. As the air whirls around the conchae and meatuses, it is warmed by blood in the capillaries. Mucus secreted by the goblet cells moistens the air and traps dust particles. Drainage from the nasolacrimal ducts and perhaps secretions from the paranasal sinuses also help moisten the air. The cilia move the mucus–dust packages toward the pharynx so that they can be eliminated from the respiratory tract by swallowing or expectoration (spitting). Substances in cigarette smoke inhibit movement of cilia. When this happens, only coughing can remove mucus–dust particles from the airways. This is one reason that smokers cough often.

Label the hard palate, inferior meatus, inferior nasal concha, internal naris, middle meatus, middle nasal concha, nasal cavity, oral cavity, soft palate, superior meatus, superior nasal concha, and vestibule in Figure 22.2.

The surface anatomy of the nose is shown in Figure 22.3.

2. Pharynx

The *pharynx* (FAR-inks) (throat) is a somewhat funnel-shaped tube about 13 cm (5 in.) long that starts at the internal nares and extends to the level of the cricoid cartilage, the most inferior cartilage of the larynx (voice box). Lying posterior to the nasal and oral cavities and just anterior to the cervical vertebrae, the pharynx is a passageway for air and food, a resonating chamber for speech sounds, and a housing for tonsils.

The pharynx is composed of a superior portion, called the *nasopharynx,* an intermediate portion, the *oropharynx,* and an inferior portion, the *laryngopharynx* (la-rin'-gō-FAR-inks) or *hypopharynx.* The nasopharynx consists of *pseudostratified ciliated columnar epithelium* and has four openings in its wall: two *internal nares* plus two openings into the *auditory (Eustachian) tubes.* The nasopharynx also contains the *pharyngeal tonsil (adenoid).* The oropharynx is lined by *nonkeratinized stratified squamous epithelium* and receives one opening: the *fauces* (FAW-sēz). The oropharynx contains the *palatine* and *lingual tonsils.* The laryngopharynx is also lined by *nonkeratinized stratified squamous epithelium* and becomes continuous with the esophagus posteriorly and the larynx anteriorly.

Label the laryngopharynx, nasopharynx, oropharynx, palatine tonsil, and pharyngeal tonsil in Figure 22.2.

3. Larynx

The *larynx* (LAIR-inks), or voice box, is a short passageway connecting the laryngopharynx with the trachea. Its wall is composed of nine pieces of cartilage.

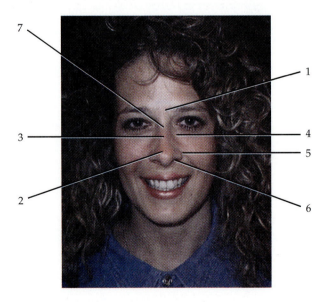

Anterior view

1. **Root.** Superior attachment of nose at forehead located between eyes.
2. **Apex.** Tip of nose.
3. **Dorsum nasi.** Rounded anterior border connecting root and apex; in profile, may be straight, convex, concave, or wavy.
4. **Nasofacial angle.** Point at which side of nose blends with tissues of face.
5. **Ala.** Convex flared portion of inferior lateral surface.
6. **External nares.** External openings into nose (nostrils).
7. **Bridge.** Superior portion of dorsum nasi, superficial to nasal bones.

FIGURE 22.3   Surface anatomy of nose.

a. *Thyroid cartilage (Adam's apple)*　Large anterior piece that gives larynx its triangular shape.

b. *Epiglottis* (*epi* = above; *glotta* = tongue)　Leaf-shaped cartilage on top of larynx that closes off the larynx so that foods and liquids are routed into the esophagus and kept out of the respiratory system. When small particles such as dust, smoke, food, or liquids pass into the larynx, a cough reflex occurs to expel the material.

c. *Cricoid* (KRĪ-koyd; *krikos* = ring) *cartilage*　Ring of cartilage forming the inferior portion of the larynx that is attached to the first ring of tracheal cartilage. The cricoid cartilage is the landmark for making an emergency airway, a procedure called a *tracheostomy* (trā-kē-OS-to-mē).

d. *Arytenoid* (ar-i-TĒ-noyd; *arytaina* = ladle) *cartilages*　Paired, pyramid-shaped cartilages at superior border of cricoid cartilage that attach vocal folds to the intrinsic pharyngeal muscles.

e. *Corniculate* (kor-NIK-yoo-lāt; *corniculate* = shaped like a small horn) *cartilages*　Paired, horn-shaped cartilages at apex of arytenoid cartilages.

f. *Cuneiform* (kyoo-NĒ-i-form; *cuneus* = wedge) *cartilages*　Paired, club-shaped cartilages anterior to the corniculate cartilages that support the vocal folds and epiglottis.

With the aid of your textbook, label the laryngeal cartilages shown in Figure 22.4. Also label the thyroid cartilage in Figure 22.2.

The mucous membrane of the larynx is arranged into two pairs of folds, a superior pair called the *ventricular folds (false vocal cords)* and an inferior pair called the *vocal folds (true vocal cords)*. The space between the vocal folds when they are apart is called the *rima glottidis*. Together, the vocal folds and rima glottidis are referred to as the *glottis*. Movement of

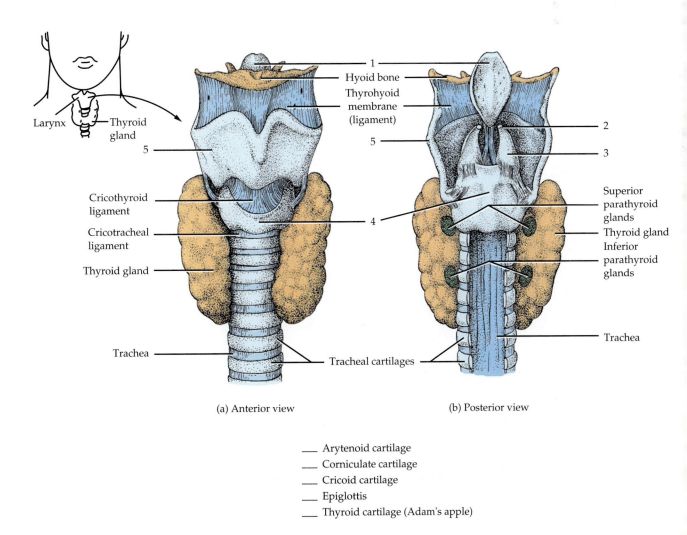

(a) Anterior view　　　(b) Posterior view

____ Arytenoid cartilage
____ Corniculate cartilage
____ Cricoid cartilage
____ Epiglottis
____ Thyroid cartilage (Adam's apple)

FIGURE 22.4　Larynx.

the vocal folds produces sounds; variations in pitch result from (1) varying degrees of tension and (2) varying lengths in males and females.

With the aid of your textbook, label the ventricular folds, vocal folds, and rima glottidis in Figure 22.5. Also label the ventricular and vocal folds in Figure 22.2.

4. Trachea

The *trachea* (TRĀ-kē-a) (windpipe) is a tubular air passageway about 12 cm (4½ in.) in length and 2.5 cm (1 in.) in diameter. It lies anterior to the esophagus and, at its inferior end (fifth thoracic vertebra), divides into right and left primary bronchi (Figure 22.6). The epithelium of the trachea consists of *pseudostratified ciliated columnar epithelium.* This epithelium consists of ciliated columnar cells, goblet cells, and basal cells. The epithelium offers the same protection against dust as the membrane lining the nasal cavity and larynx.

Obtain a prepared slide of pseudostratified ciliated columnar epithelium from the trachea, and, with the aid of your textbook, label the ciliated columnar cells, cilia, goblet cells, and basal cells in Figure 22.7 on page 422.

The trachea consists of smooth muscle, elastic connective tissue, and incomplete *rings of cartilage* (hyaline) shaped like a series of letter Cs. The open ends of the Cs are held together by the *trachealis muscle.* The cartilage provides a rigid support so that the tracheal wall does not collapse inward and obstruct the air passageway, and, because the open

parts of the Cs face the esophagus, the latter can expand into the trachea during swallowing. If the trachea should become obstructed, a *tracheostomy* (trā-kē-OS-tō-mē) may be performed. Another method of opening the air passageway is called *intubation*, in which a tube is passed into the mouth and down through the larynx and the trachea.

5. Bronchi

The trachea terminates by dividing into a *right primary bronchus* (BRONG-kus), going to the right lung, and a *left primary bronchus,* going to the left lung. They continue dividing in the lungs into smaller bronchi, the *secondary (lobar) bronchi* (BRONG-kē), one for each lobe of the lung. These bronchi, in turn, continue dividing into still smaller bronchi called *tertiary (segmental) bronchi,* which divide into *bronchioles.* The next division is into even smaller tubes called *terminal bronchioles.* This entire branching structure of the trachea is commonly referred to as the *bronchial tree.*

Label Figure 22.6.

Bronchography (brong-KOG-ra-fē) is a technique for examining the bronchial tree. With this procedure, an intratracheal catheter is passed transorally or transnasally through the rima glottidis into the trachea. Then an opaque contrast medium is introduced into the trachea and distributed through the bronchial branches. Radiographs of the chest in various positions are taken and the developed film, called *bronchogram* (BRONG-kō-gram), provides a picture of the bronchial tree.

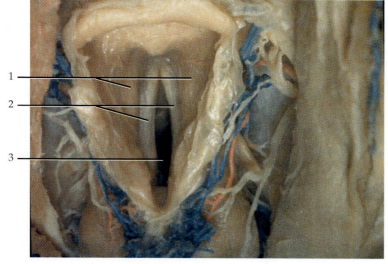

1

2

3

_____ Rima glottidis

_____ Ventricular folds (false vocal cords)

_____ Vocal folds (true vocal cords)

Superior view

FIGURE 22.5   Photograph of larynx.

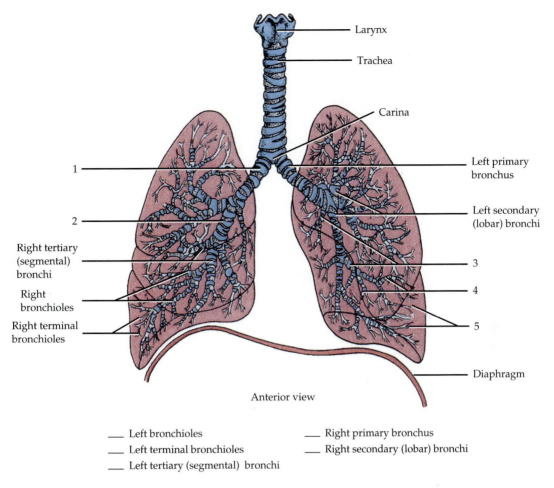

Larynx

Trachea

Carina

1

2

Right tertiary
(segmental)
bronchi

Right
bronchioles

Right terminal
bronchioles

Left primary
bronchus

Left secondary
(lobar) bronchi

3

4

5

Diaphragm

Anterior view

____ Left bronchioles   ____ Right primary bronchus
____ Left terminal bronchioles   ____ Right secondary (lobar) bronchi
____ Left tertiary (segmental) bronchi

FIGURE 22.6   Air passageways to the lungs. Shown is the bronchial tree in relationship to lungs.

6. Lungs

The *lungs* (*lunge* = light, since the lungs float) are paired, cone-shaped organs lying in the thoracic cavity (see Figure 22.1). The *pleural* (*pleura* = side) *membrane* encloses and protects each lung. Whereas the *superficial parietal pleura* lines the wall of the thoracic cavity, the *deep visceral pleura* covers the lungs; the potential space between parietal and visceral pleurae, the *pleural cavity,* contains a lubricating fluid to reduce friction as the lungs expand and recoil.

Major surface features of the lungs include

a. *Base* Broad inferior portion resting on diaphragm.
b. *Apex* Narrow superior portion just above clavicles.
c. *Costal surface* Surface lying against ribs.
d. *Mediastinal surface* Medial surface.

e. *Hilus* Region in mediastinal surface through which bronchial tubes, blood vessels, lymphatic vessels, and nerves enter and exit the lung.
f. *Cardiac notch* Medial concavity in left lung in which heart lies.

Each lung is divided into *lobes* by one or more *fissures*. The right lung has three lobes, *superior, middle,* and *inferior;* the left lung has two lobes, *superior* and *inferior.* The *horizontal fissure* separates the superior lobe from the middle lobe in the right lung; an *oblique fissure* separates the middle lobe from the inferior lobe in the right lung and the superior lobe from the inferior lobe in the left lung. Each lobe receives its own secondary bronchus.

Using your textbook as a reference, label Figure 22.8a on page 423.

Each lobe of a lung is divided into regions called *bronchopulmonary segments,* each supplied by a tertiary bronchus. Each bronchopulmonary

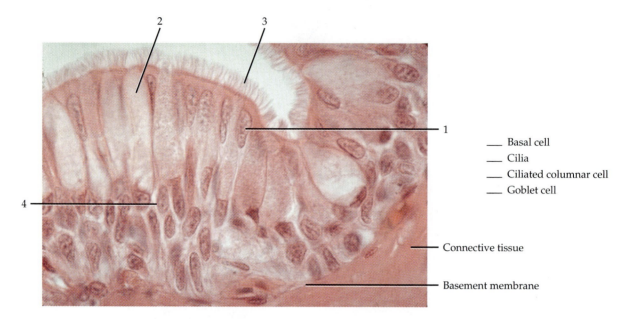

FIGURE 22.7   Histology of trachea (250×).

___ Basal cell
___ Cilia
___ Ciliated columnar cell
___ Goblet cell

Connective tissue

Basement membrane

segment is composed of many smaller compartments called *lobules.* Each lobule is wrapped in elastic connective tissue and contains a lymphatic vessel, arteriole, venule, and branch from a terminal bronchiole. Terminal bronchioles divide into *respiratory bronchioles,* which, in turn, divide into several *alveolar* (al-VĒ-ō-lar) *ducts.* Around the circumference of alveolar ducts are numerous alveoli and alveolar sacs. *Alveoli* (al-VĒ-ō-lī) are cup-shaped outpouchings lined by epithelium and supported by a thin elastic membrane. The singular is *alveolus. Alveolar sacs* are two or more alveoli that share a common opening. Over the alveoli, an arteriole and venule disperse into a network of capillaries. Gas is exchanged between the lungs and blood by diffusion across the alveolar and the capillary walls.

Using your textbook as a reference, label Figure 22.8b.

Each alveolus consists of

a. *Type I alveolar (squamous pulmonary epithelial) cells* Cells that form a continuous lining of the alveolar wall, except for occasional type II alveolar (septal) cells.

b. *Type II alveolar (septal) cells* Cuboidal cells dispersed among type I alveolar cells that secrete a phospholipid substance called *surfactant* (sur-FAK-tant), a surface tension-lowering agent.

c. *Alveolar macrophages (dust cells)* Phagocytic cells that remove fine dust particles and other debris from the alveolar spaces.

Obtain a slide of normal lung tissue and examine it under high power. Using your textbook as a reference, see if you can identify a terminal bronchiole, respiratory bronchiole, alveolar duct, alveolar sac, and alveoli.

If available, examine several pathological slides of lung tissue, such as slides that show emphysema and lung cancer. Compare your observations to the normal lung tissue.

The exchange of respiratory gases between the lungs and blood takes place by diffusion across the alveolar and capillary walls. This membrane, through which the respiratory gases move, is collectively known as the *alveolar-capillary (respiratory) membrane* (Figure 22.9 on page 424). It consists of

1. A layer of type I alveolar (squamous pulmonary epithelial) cells with type II alveolar (septal) cells and alveolar macrophages (dust cells) that constitute the alveolar (epithelial) wall.

2. An epithelial basement membrane underneath the alveolar wall.

3. A capillary basement membrane that is often fused to the epithelial basement membrane.

4. The endothelial cells of the capillary.

Label the components of the alveolar-capillary (respiratory) membrane in Figure 22.9.

Examine the scanning electron micrograph of the cells of the alveolar wall in Figure 22.10 on page 425.

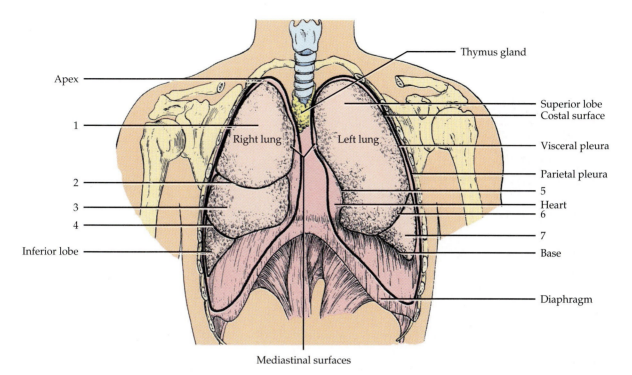

Thymus gland

Apex

1

Right lung   Left lung

Superior lobe
Costal surface

Visceral pleura

Parietal pleura

2

3

4

5

Heart

6

7

Inferior lobe

Base

Diaphragm

Mediastinal surfaces

(a) Coverings and external anatomy in anterior view

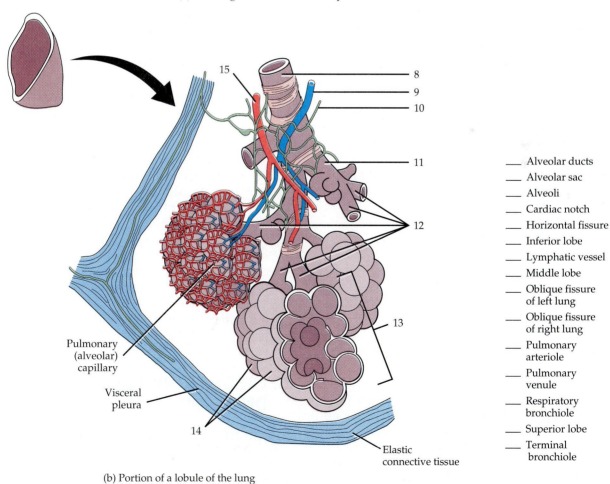

15

8
9
10

11

12

13

__ Alveolar ducts

__ Alveolar sac

__ Alveoli

__ Cardiac notch

__ Horizontal fissure

__ Inferior lobe

__ Lymphatic vessel

__ Middle lobe

__ Oblique fissure
of left lung

__ Oblique fissure
of right lung

__ Pulmonary
arteriole

__ Pulmonary
venule

__ Respiratory
bronchiole

__ Superior lobe

__ Terminal
bronchiole

Pulmonary
(alveolar)
capillary

Visceral
pleura

14

Elastic
connective tissue

(b) Portion of a lobule of the lung

FIGURE 22.8   Lungs.

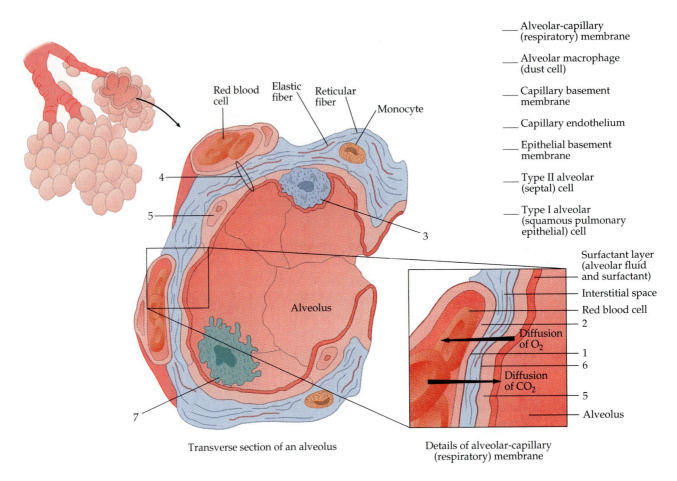

_____ Alveolar-capillary
(respiratory) membrane

_____ Alveolar macrophage
(dust cell)

_____ Capillary basement
membrane

_____ Capillary endothelium

_____ Epithelial basement
membrane

_____ Type II alveolar
(septal) cell

_____ Type I alveolar
(squamous pulmonary
epithelial) cell

Transverse section of an alveolus

Details of alveolar-capillary
(respiratory) membrane

FIGURE 22.9   Alveolar-capillary (respiratory) membrane.

B. DISSECTION OF SHEEP PLUCK

CAUTION! *Please reread Section D, "Precautions Related to Dissection" at the beginning of the laboratory manual on page xiii before you begin your dissection.*

PROCEDURE

1. Preserved sheep pluck may or may not be available for dissection.
2. Pluck consists mainly of a sheep *trachea, bronchi, lungs, heart,* and *great vessels,* and a small portion of the *diaphragm.* It is a good demonstration because it is large and shows the close anatomical correlation between these structures and the systems to which they belong, namely, the respiratory and the cardio-vascular systems.
3. The heart and its great blood vessels have been described in detail in Exercise 18.
4. Pluck can also be used to examine in great detail the trachea and its relationship to the de-

velopment of the bronchi until they branch into each lung. In addition, this specimen is sufficiently large that the bronchial tree can be exposed by careful dissection.
5. This dissection is done by starting at the primary and secondary bronchi and slowly and carefully removing lung tissue as the trees form even smaller branches into lungs.
6. These specimens should be used primarily by the instructor for demonstration, but you can help expose the bronchial tree.

C. LABORATORY TESTS ON RESPIRATION

1. Mechanics of Pulmonary Ventilation (Breathing)

Pulmonary ventilation (breathing) is the process by which gases are exchanged between the atmosphere and the alveoli. Air moves throughout the

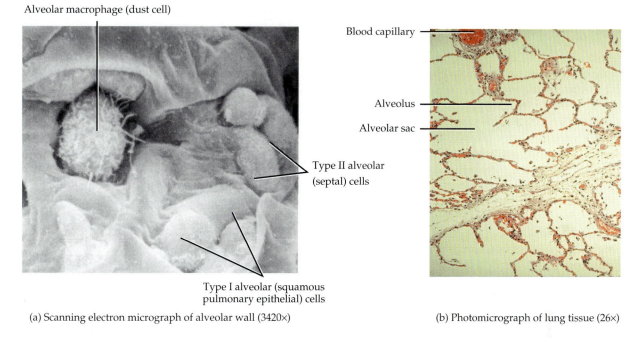

Alveolar macrophage (dust cell)

Type II alveolar (septal) cells

Type I alveolar (squamous pulmonary epithelial) cells

(a) Scanning electron micrograph of alveolar wall (3420×)

Blood capillary

Alveolus

Alveolar sac

(b) Photomicrograph of lung tissue (26×)

FIGURE 22.10 Histology of lung tissue.

respiratory system as a result of pressure gradients (differences) between the external environment and the respiratory system. Likewise, oxygen and carbon dioxide then move between the respiratory alveoli and the pulmonary capillaries of the cardiovascular system, and the peripheral capillaries of the cardiovascular system and the tissues of the body as a result of diffusion gradients. We breathe in (inhale) when the pressure inside the thoracic cavity and lungs is less than the air pressure in the atmosphere; similarly we breathe out (exhale) when the pressure inside the lungs and thoracic cavity is greater than the pressure in the atmosphere.

a. INSPIRATION

Breathing in is called *inspiration (inhalation).* When the thoracic cavity is at rest, and no air movement is occurring, the pressure inside the lungs equals the pressure of the atmosphere, which is approximately 760 mm Hg, or 1 atmosphere (atm), at sea level. For air to flow into the lungs, something must happen to reduce the pressure within the lungs to a value lower than atmospheric pressure. This condition is achieved by increasing the volume of the thoracic cavity.

In order for inspiration to occur, the thoracic cavity must be expanded. This increases lung volume and thus decreases pressure in the lungs. The first step toward increasing lung volume (size) involves contraction of the principal inspiratory muscles—the diaphragm and external intercostals.

Contraction of the diaphragm causes it to flatten. This increases the vertical length of the thoracic cavity and accounts for about 75% of the air that enters the lungs during inspiration. At the same time the diaphragm contracts, the external intercostals contract. As a result, the ribs are pulled superiorly and outward, increasing the anterior-posterior diameter of the thoracic cavity. This movement of the ribs by the external intercostals is much like the movement of a bucket handle when a bucket is placed on its side and the handle is moved from a vertical to a more horizontal position. Because the thoracic cavity is a closed system, this increase in the vertical length and anterior-posterior diameter of the thoracic cavity causes the volume of the thoracic cavity to increase, and the pressure within the cavity to decrease. As a result of this decreased pressure within the thoracic cavity, pressure within the lungs decreases, thereby causing air to move into the lungs.

During normal breathing, the pressure between two pleural layers, called *intrapleural (intrathoracic) pressure,* is always subatmospheric. (It may become temporarily positive, but only during modified respiratory movements such as coughing or straining during childbirth or defecation.) When the respiratory system is at rest no air movement in or out of the respiratory tree is occurring, and intrapleural pressure is approximately 4 mm Hg lower than atmospheric pressure, or approximately 756 mm Hg. The overall increase in the size of the thoracic cavity causes intrapleural pressure to fall to approximately 754 mm Hg. The parietal and visceral pleural membranes are normally strongly attached to each other due to surface tension created by a very thin layer of water between the membranes. Therefore, as the walls of the thoracic cavity expand, the parietal pleural lining the thoracic cavity is pulled in all directions, and the visceral pleura is pulled along with it. Consequently the size of the lungs increases, thereby decreasing the pressure within the lungs (called *alveolar (intrapulmonic) pressure*). This decrease in alveolar pressure causes a pressure gradient between the alveoli of the lungs and the external environment. Because the respiratory tree is open to the external environment via the oral and nasal cavities, air moves down the pressure gradient from the atmosphere into the pulmonary alveoli. Air continues to move into the lungs until alveolar pressure equals atmospheric pressure.

b. EXPIRATION

Breathing out, called *expiration (exhalation),* is also achieved by a pressure gradient, but in this case the gradient is reversed so that the pressure in the lungs is greater than the pressure of the atmosphere. Normal quiet expiration, unlike inspiration, is a passive process, in that it does not involve the active contraction of muscles. When inspiration is completed, the diaphragm and external intercostal muscles relax. As the diaphragm returns to its domelike position, the vertical length of the thoracic cavity decreases, returning to its original dimension. Likewise, when the external intercostals relax, the ribs move posteriorly and inferiorly, decreasing the anterior-posterior diameter of the thoracic cavity. These movements return intrapleural pressure to its normal resting value of 756 mm Hg, or 4 mm Hg below atmospheric pressure.

As intrapleural pressure returns to its preinspiration level, the walls of the lungs are no longer pulled outward by the parietal pleura. The elastic recoil of the connective tissue within the lungs and respiratory tree allows the lungs to return to their resting shape and volume, thereby causing alveolar pressure to become slightly greater than atmospheric pressure (or approximately 762 mm Hg). Now air again moves down its pressure gradient, moving from the alveoli through the respiratory tree and out the oral and nasal openings.

In order to demonstrate pulmonary ventilation, a model lung (a bell-jar demonstrator) will be used. This apparatus is basically an artificial thoracic cavity that mimics the organs of the respiratory system and allows the pressure/volume ratios to be manipulated. A diagram of such a device is shown in Figure 22.11. Some of the models will not have a stopcock (rima glottidis); some will not have a tube opening into the pleural space.

PROCEDURE

1. Label the diagram in Figure 22.11 by writing the name of the anatomical structure that corresponds to the following parts of the apparatus: rima glottidis, trachea, primary bronchus, lungs, alveolar pressure, pleural space, and diaphragm.

2. Pull on the rubber membrane (diaphragm) to simulate inspiration. What changes occur in

   intrapleural pressure?_____

   _____

   _____

   In alveolar pressure?_____

   _____

3. Push the rubber membrane (diaphragm) upward to simulate expiration. What changes occur in intrapleural pressure? _____

   _____

   In alveolar pressure?_____

   _____

   Under what normal processes would such changes in intrapleural and alveolar pressures

   be observed?_____

   _____

4. Open the clamp on the tube leading to the intrapleural space. This simulates pneumothorax (air in the pleural cavity). Try to cause inspiration and expiration.

   Explain your results. _____

   _____

5. If present, close the stopcock (rima glottidis) and simulate expiration. This procedure mimics the Valsalva maneuver (forced expiration against a closed rima glottidis as during periods of straining). What changes occur in intrapleural and

   alveolar pressure?_____

   _____

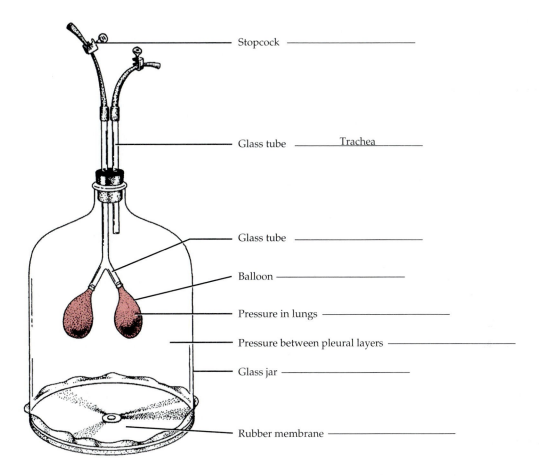

Stopcock _____

Glass tube _____ Trachea _____

Glass tube _____

Balloon _____

Pressure in lungs _____

Pressure between pleural layers _____

Glass jar _____

Rubber membrane _____

FIGURE 22.11   Model lung.

2. Measurement of Chest and Abdomen in Respiration

When the diaphragm contracts, the dome shape of this muscle flattens; the flattening pushes against the abdominal viscera and decreases thoracic volume, which, in turn, decreases alveolar pressure. The volume changes in the abdomen and chest can easily be measured.

PROCEDURE

1. Place a tape measure at the level of the fifth rib (about armpit level) and determine the size of the chest immediately after a
   a. Normal inspiration
   b. Normal expiration
   c. Maximal inspiration
   d. Maximal expiration
2. Repeat step 1 with the tape measure at the level of the waist to determine abdominal size.
3. Record your results in Section C.1 of the LAB-ORATORY REPORT RESULTS at the end of the exercise.

Which measurement increased during inspiration? _____

_____

Which increased during expiration? _____

_____

3. Respiratory Sounds

Air flowing through the respiratory tree creates characteristic sounds that can be detected through the use of a stethoscope. Normal breathing sounds include a soft, breezy sound caused by air filling the lungs during inspiration. As the air exits the lungs during expiration a short, lower-pitched sound may be heard. Perform the following exercises.

PROCEDURE

1. *Clean the earplugs of a stethoscope with alcohol.*
2. Place the diaphragm of the stethoscope just below the larynx and listen for bronchial sounds during both inspiration and expiration.

3. Move the stethoscope slowly downward toward the bronchial tubes until the sounds are no longer heard.

4. Place the stethoscope under the scapula, under the clavicle, over different intercostal spaces (the spaces between the ribs) on the chest, and listen for any sound during inspiration and expiration.

4. Use of a Pneumograph

The *pneumograph* is an instrument that measures variations in breathing patterns caused by various physical or chemical factors. The chest pneumograph, which is attached to a polygraph recorder via electrical leads, consists of a rubber bellows that fits around the chest just below the rib cage. As the subject breathes, chest movements cause changes in the air pressure within the pneumograph that are transmitted to the recorder. Normal inspiration and expiration can thus be recorded, and the effects of a wide range of physical and chemical factors on these movements can be studied.

PROCEDURE

Perform the following exercises and record your values in Section C.2 of the LABORATORY REPORT RESULTS at the end of the exercise. Label and save all recordings and attach them to Section C.2 of the LABORATORY REPORT RESULTS. In addition, label the inspiratory and expiratory phases of each recording, determine their duration, and then calculate the respiratory rate.

1. Place a respiratory pneumograph around the chest at the level of the sixth rib and attach it at the back. Connect the electrical lead to the recorder and adjust the instrument's centering and sensitivity so that the needle deflects as the subject breathes. If it does not, adjust the pneumograph bellows up or down on the chest, or loosen the degree of the tightness around the subject's chest.

2. Seat the subject so that the pneumograph recording is not visible to the subject.

3. Set the polygraph at a slow speed (approximately 1 cm/sec) and record normal, quiet breathing (eupnea) for approximately 30 sec.

4. Have the subject inhale deeply and hold his or her breath for as long as possible. Record the breathing pattern during the breath holding and 30 to 60 sec after the resumption of breathing at the end of breath holding.

5. Record the respiratory movements of a subject in Section C.2 of the LABORATORY REPORT RESULTS at the end of the exercise during the following activities:
   a. Reading
   b. Swallowing water
   c. Laughing
   d. Yawning
   e. Coughing
   f. Sniffing
   g. Doing a rather difficult long-division calculation

5. Measurement of Respiratory Volumes

In clinical practice, *respiration* refers to one complete respiratory cycle, that is, one inspiration and one expiration. A normal adult has 14 to 18 respirations in a minute, during which the lungs exchange specific volumes of air with the atmosphere.

As the following respiratory volumes and capacities are discussed, keep in mind that the values given vary with age, height, sex, and physiological state. The volume of air expired under normal, quiet inspiration is approximately 500 ml. The volume of air expired under normal, quiet breathing conditions is equal to that of inspiration, and this volume of air is called *tidal volume* (Figure 22.12). Of this volume of air, only approximately 350 ml reaches the alveoli. The other 150 ml of air does not reach the alveoli because it remains in the spaces not designed for air exchange (*anatomical dead space*) (nose, pharynx, larynx, trachea, and bronchi) as well as those areas designed for air exchange but not currently being utilized by the lungs (*physiological dead space*).

If we take a very deep breath, we can inspire much more than the 500-ml tidal volume taken in during normal, quiet respiration. The additional inhaled air, called the *inspiratory reserve volume*, averages 3100 ml above the tidal volume of quiet respiration. If we inspire *normally* and then expire *as forcibly as possible*, a normal individual would be able to exhale the 500 ml of air taken in during a normal, quiet tidal volume, as well as an additional 1200 ml of air, which is termed *expiratory reserve volume*. Even after a forceful expiration there is still some air remaining in the lungs, which is termed *residual volume* (1200 ml). This volume of air ensures gas exchange between the lungs and the cardiovascular system during brief time periods between respiratory cycles, or during extended periods of no respiratory activity (apnea).

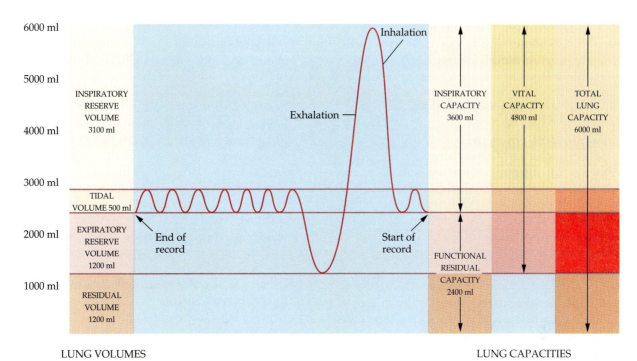

FIGURE 22.12   Spirogram of lung volumes and capacities.

Opening the thoracic cavity allows the intrapleural pressure to equal the atmospheric pressure, forcing out some of the residual volume. The air remaining is called the *minimal volume.* Minimal volume provides a medical and legal tool for determining whether a baby was born dead or died after birth. The presence of minimal volume can be demonstrated by placing a piece of lung in water and watching it float. Fetal lungs contain no air, and so the lung of a stillborn will not float in water.

Lung capacities are combinations of specific lung volumes. *Inspiratory capacity,* the total inspiratory ability of the lungs, is the sum of tidal volume plus inspiratory reserve volume (3600 ml). *Functional residual capacity* is the sum of residual volume plus expiratory reserve volume (2400 ml). *Vital capacity* is the sum of the inspiratory reserve volume, tidal volume, and expiratory reserve volume (4800 ml). Finally, *total lung capacity* is the sum of all volumes (6000 ml).

In a normal person the volumes and capacities of air in the lungs depends upon body size and build, as well as body position and the physiological state of the pulmonary system. Pulmonary volumes and capacities may change when a person lies down due to the movement of abdominal contents upon assuming a supine position. When the supine position is assumed the abdominal contents tend to press superiorly upon the diaphragm, thereby decreasing the volume of the thoracic cavity. In addition, an individual assuming the supine position increases the volume of blood within the pulmonary circulation, thereby decreasing the space for pulmonary air.

Note: Depending on the type of laboratory equipment available, select from the following procedures related to measurement of respiratory volumes.

PROCEDURE USING PHYSIOFINDER

PHYSIOFINDER is an interactive computer program that permits laboratory simulations that do not involve the use of animals or advanced or expensive laboratory equipment. It permits students to perform experiments, analyze data, draw conclusions, and form hypotheses on the basis of collected data. The program is available from Addison-Wesley Longman, Inc., Customer Service (1-800-922-0579).

For activities related to respiratory volumes, select the appropriate experiments from PHYSIOFINDER module 10—Control of Breathing.

PROCEDURE USING SPIROCOMP™

SPIROCOMP™ is a computerized spirometry system consisting of hardware and software designed to allow for quick and easy measurement of standard lung volumes. A stacked bar graph display of the data clearly depicts the relationship between the various volumes. The built-in group analysis features also allows you to view gender-specific data. The program is available from INTELITOOL® (1-800-227-3805).

PROCEDURE USING HANDHELD RESPIROMETER

A *respirometer (spirometer)* is an instrument used to measure volumes of air exchanged in breathing. Of the several different respirometers available, we will make use of two: a handheld respirometer and the Collins respirometer. One type of handheld respirometer is the Pulmometer (Figure 22.13). It measures and provides a direct digital display of certain respiratory volumes and capacities.

PROCEDURE

1. Set the needle or the digital indicator to zero.
2. Place a *clean* mouthpiece in the spirometer tube and hold the tube in your hand while you breathe normally for a few respirations. (*Note: Inhale through the nose and exhale through the mouth.*)
3. Place the mouthpiece in your mouth and inhale; exhale three normal breaths through the mouthpiece.

FIGURE 22.13   Handheld respirometer.

4. Divide the total volume of air expired by three to determine your average tidal volume. Record in Section C.3 of the LABORATORY REPORT RESULTS at the end of the exercise.
5. Return the scale indicator to zero.
6. Breathe normally for a few respirations. Following a normal expiration of tidal volume, forcibly exhale as much air as possible into the mouthpiece.
7. Repeat step 6 twice. Divide the total amount of air expired by three to determine your average expiratory reserve volume. Record this value in Section C.3 of the LABORATORY REPORT RESULTS at the end of the exercise.
8. To determine vital capacity, breathe deeply a few times and inhale as much air as possible. Exhale as fully and as steadily as possible into the spirometer tube.
9. Repeat step 8 twice. Divide the total volume of expired air by three to determine your average vital capacity. Record this value in Section C.3 of the LABORATORY REPORT RESULTS at the end of the exercise.
10. Calculate your inspiratory reserve volume by substituting the known volumes in this equation: Vital capacity = Inspiratory reserve volume + tidal volume + expiratory reserve volume. Record this volume in Section C.3 of the LABORATORY REPORT RESULTS at the end of the exercise.
11. Compare your vital capacity with the normal values shown in Tables 22.1 and 22.2.
12. Sit quietly and count your respirations per minute. Calculate your minute volume of respirations (MVR) by multiplying the breaths per minute by your tidal volume. Record this value in Section C.3 of the LABORATORY REPORT RESULTS at the end of the exercise.

$$\frac{\text{\# breaths}}{\text{minute}} \times \frac{\text{\# ml}}{\text{breath}} = \frac{\text{\# ml}}{\text{minute}}$$

13. Repeat the above procedures while lying in the supine position. Compare the various lung volumes and capacities obtained while in the supine position with those obtained in the upright position.

PROCEDURE USING COLLINS RESPIROMETER

The Collins respirometer, shown in Figure 22.14, is a closed system that allows the measurement of volumes of air inhaled and exhaled. The instrument consists of a weighted drum, containing air, inverted

TABLE 22.1
Predicted Vital Capacities for Females

								Height in centimeters and inches											
	cm	152	154	156	158	160	162	164	166	168	170	172	174	176	178	180	182	184	186	188
Age	in.	59.8	60.6	61.4	62.2	63.0	63.7	64.6	65.4	66.1	66.9	67.7	68.5	69.3	70.1	70.9	71.7	72.4	73.2	74.0
16		3070	3110	3150	3190	3230	3270	3310	3350	3390	3430	3470	3510	3550	3590	3630	3670	3715	3755	3800
17		3055	3095	3135	3175	3215	3255	3295	3335	3375	3415	3455	3495	3535	3575	3615	3655	3695	3740	3780
18		3040	3080	3120	3160	3200	3240	3280	3320	3360	3400	3440	3480	3520	3560	3600	3640	3680	3720	3760
20		3010	3050	3090	3130	3170	3210	3250	3290	3330	3370	3410	3450	3490	3525	3565	3605	3645	3695	3720
22		2980	3020	3060	3095	3135	3175	3215	3255	3290	3330	3370	3410	3450	3490	3530	3570	3610	3650	3685
24		2950	2985	3025	3065	3100	3140	3180	3200	3260	3300	3335	3375	3415	3455	3490	3530	3570	3610	3650
26		2920	2960	3000	3035	3070	3110	3150	3190	3230	3265	3300	3340	3380	3420	3455	3495	3530	3570	3610
28		2890	2930	2965	3000	3040	3070	3115	3155	3190	3230	3270	3305	3345	3380	3420	3460	3495	3535	3570
30		2860	2895	2935	2970	3010	3045	3085	3120	3160	3195	3235	3270	3310	3345	3385	3420	3460	3495	3535
32		2825	2865	2900	2940	2975	3015	3050	3090	3125	3160	3200	3235	3275	3310	3350	3385	3425	3460	3495
34		2795	2835	2870	2910	2945	2980	3020	3055	3090	3130	3165	3200	3240	3275	3310	3350	3385	3425	3460
36		2765	2805	2840	2875	2910	2950	2985	3020	3060	3095	3130	3165	3205	3240	3275	3310	3350	3385	3420
38		2735	2770	2810	2845	2880	2915	2950	2990	3025	3060	3095	3130	3170	3205	3240	3275	3310	3350	3385
40		2705	2740	2775	2810	2850	2885	2920	2955	2990	3025	3060	3095	3135	3170	3205	3240	3275	3310	3345
42		2675	2710	2745	2780	2815	2850	2885	2920	2955	2990	3025	3060	3100	3135	3170	3205	3240	3275	3310
44		2645	2680	2715	2750	2785	2820	2855	2890	2925	2960	2995	3030	3060	3095	3130	3165	3200	3235	3270
46		2615	2650	2685	2715	2750	2785	2820	2855	2890	2925	2960	2995	3030	3060	3095	3130	3165	3200	3235
48		2585	2620	2650	2685	2715	2750	2785	2820	2855	2890	2925	2960	2995	3030	3060	3095	3130	3160	3195
50		2555	2590	2625	2655	2690	2720	2755	2785	2820	2855	2890	2925	2955	2990	3025	3060	3090	3125	3155
52		2525	2555	2590	2625	2655	2690	2720	2755	2790	2820	2855	2890	2925	2955	2990	3020	3055	3090	3125
54		2495	2530	2560	2590	2625	2655	2690	2720	2755	2790	2820	2855	2885	2920	2950	2985	3020	3050	3085
56		2460	2495	2525	2560	2590	2625	2655	2690	2720	2755	2790	2820	2855	2885	2920	2950	2980	3015	3045
58		2430	2460	2495	2525	2560	2590	2625	2655	2690	2720	2750	2785	2815	2850	2880	2920	2945	2975	3010
60		2400	2430	2460	2495	2525	2560	2590	2625	2655	2685	2720	2750	2780	2810	2845	2875	2915	2940	2970
62		2370	2405	2435	2465	2495	2525	2560	2590	2620	2655	2685	2715	2745	2775	2810	2840	2870	2900	2935
64		2340	2370	2400	2430	2465	2495	2525	2555	2585	2620	2650	2680	2710	2740	2770	2805	2835	2865	2895
66		2310	2340	2370	2400	2430	2460	2495	2525	2555	2585	2615	2645	2675	2705	2735	2765	2800	2825	2860
68		2280	2310	2340	2370	2400	2430	2460	2490	2520	2550	2580	2610	2640	2670	2700	2730	2760	2795	2820
70		2250	2280	2310	2340	2370	2400	2425	2455	2485	2515	2545	2575	2605	2635	2665	2695	2725	2755	2780
72		2220	2250	2280	2310	2335	2365	2395	2425	2455	2480	2510	2540	2570	2600	2630	2660	2685	2715	2745
74		2190	2220	2245	2275	2305	2335	2360	2390	2420	2450	2475	2505	2535	2565	2590	2620	2650	2680	2710

Source: E. A. Gaensler and G. W. Wright, *Archives of Environmental Health,* 12 (Feb.): 146–189 (1966).

over a chamber of water. The air-filled chamber is connected to the subject's mouth by a tube. When the subject inspires, air is removed from the chamber, causing the drum to sink and producing an upward deflection. This deflection is recorded by the stylus on the graph paper on the kymograph (rotating drum). When the subject expires, air is added, causing the drum to rise and producing a downward deflection. These deflections are recorded as a *spirogram* (see Figure 22.12). The horizontal (*x*) axis of a spirogram is graduated in millimeters (mm) and records elapsed time; the vertical (*y*) axis is gradu-

ated in milliliters (ml) and records air volumes. Spirometric studies measure lung capacities and rates and depths of ventilation for diagnostic purposes. Spirometry is indicated for individuals with labored breathing, and is used to diagnose respiratory disorders such as bronchial asthma and emphysema.

The air in the body is at a different temperature than the air contained in the respirometer; body air is also saturated with water vapor. To make the correction for these differences, one measures the temperature of the respirometer and uses Table 22.3 to

TABLE 22.2
Predicted Vital Capacities for Males

							Height in centimeters and inches											
Age	**cm** **in.**	**152** **59.8**	**154** **60.6**	**156** **61.4**	**158** **62.2**	**160** **63.0**	**162** **63.7**	**164** **64.6**	**166** **65.4**	**168** **66.1**	**170** **66.9**	**172** **67.7**	**174** **68.5**	**176** **69.3**	**178** **70.1**	**180** **70.9**	**182** **71.7**	**184** **72.4**	**186** **73.2**	**188** **74.0**
16		3920	3975	4025	4075	4130	4180	4230	4285	4335	4385	4440	4490	4540	4590	4645	4695	4745	4800	4850
18		3890	3940	3995	4045	4095	4145	4200	4250	4300	4350	4405	4455	4505	4555	4610	4660	4710	4760	4815
20		3860	3910	3960	4015	4065	4115	4165	4215	4265	4320	4370	4420	4470	4520	4570	4625	4675	4725	4775
22		3830	3880	3930	3980	4030	4080	4135	4185	4235	4285	4335	4385	4435	4485	4535	4585	4635	4685	4735
24		3785	3835	3885	3935	3985	4035	4085	4135	4185	4235	4285	4330	4380	4430	4480	4530	4580	4630	4680
26		3755	3805	3855	3905	3955	4000	4050	4100	4150	4200	4250	4300	4350	4395	4445	4495	4545	4595	4645
28		3725	3775	3820	3870	3920	3970	4020	4070	4115	4165	4215	4265	4310	4360	4410	4460	4510	4555	4605
30		3695	3740	3790	3840	3890	3935	3985	4035	4080	4130	4180	4230	4275	4325	4375	4425	4470	4520	4570
32		3665	3710	3760	3810	3855	3905	3950	4000	4050	4095	4145	4195	4240	4290	4340	4385	4435	4485	4530
34		3620	3665	3715	3760	3810	3855	3905	3950	4000	4045	4095	4140	4190	4225	4285	4330	4380	4425	4475
36		3585	3635	3680	3730	3775	3825	3870	3920	3965	4010	4060	4105	4155	4200	4250	4295	4340	4390	4435
38		3555	3605	3650	3695	3745	3790	3840	3885	3930	3980	4025	4070	4120	4165	4210	4260	4305	4350	4400
40		3525	3575	3620	3665	3710	3760	3805	3850	3900	3945	3990	4035	4085	4130	4175	4220	4270	4315	4360
42		3495	3540	3590	3635	3680	3725	3770	3820	3865	3910	3955	4000	4050	4095	4140	4185	4230	4280	4325
44		3450	3495	3540	3585	3630	3675	3725	3770	3815	3860	3905	3950	3995	4040	4085	4130	4175	4220	4270
46		3420	3465	3510	3555	3600	3645	3690	3735	3780	3825	3870	3915	3960	4005	4050	4095	4140	4185	4230
48		3390	3435	3480	3525	3570	3615	3655	3700	3745	3790	3835	3880	3925	3970	4015	4060	4105	4150	4190
50		3345	3390	3430	3475	3520	3565	3610	3650	3695	3740	3785	3830	3870	3915	3960	4005	4050	4090	4135
52		3315	3353	3400	3445	3490	3530	3575	3620	3660	3705	3750	3795	3835	3880	3925	3970	4010	4055	4100
54		3285	3325	3370	3415	3455	3500	3540	3585	3630	3670	3715	3760	3800	3845	3890	3930	3975	4020	4060
56		3255	3295	3340	3380	3425	3465	3510	3550	3595	3640	3680	3725	3765	3810	3850	3895	3940	3980	4025
58		3210	3250	3290	3335	3375	3420	3460	3500	3545	3585	3630	3670	3715	3755	3800	3840	3880	3925	3965
60		3175	3220	3260	3300	3345	3385	3430	3470	3500	3555	3595	3635	3680	3720	3760	3805	3845	3885	3930
62		3150	3190	3230	3270	3310	3350	3390	3440	3480	3520	3560	3600	3640	3680	3730	3770	3810	3850	3890
64		3120	3160	3200	3240	3280	3320	3360	3400	3440	3490	3530	3570	3610	3650	3690	3730	3770	3810	3850
66		3070	3110	3150	3190	3230	3270	3310	3350	3390	3430	3470	3510	3550	3600	3640	3680	3720	3760	3800
68		3040	3080	3120	3160	3200	3240	3280	3320	3360	3400	3440	3480	3520	3560	3600	3640	3680	3720	3760
70		3010	3050	3090	3130	3170	3210	3250	3290	3330	3370	3410	3450	3480	3520	3560	3600	3640	3680	3720
72		2980	3020	3060	3100	3140	3180	3210	3250	3290	3330	3370	3410	3450	3490	3530	3570	3610	3650	3680
74		2930	2970	3010	3050	3090	3130	3170	3200	3240	3280	3320	3360	3400	3440	3470	3510	3550	3590	3630

Source: E. A. Gaensler and G. W. Wright, *Archives of Environmental Health,* 12 (Feb.): 146–189 (1966).

obtain a conversion factor: ***body temperature, pressure (atmospheric), saturated*** (with water vapor) ***(BTPS).*** All measured volumes are then multiplied by this factor.

PROCEDURE

1. *When using the Collins respirometer, the disposable mouthpiece is discarded after use by each subject, and the hose is then detached and rinsed with 70% alcohol.* CAUTION! *This procedure must be repeated with every student using the equipment.*

2. Students should work in pairs, with one student operating the respirometer while the other student is tested.

3. Before starting the recording, a little practice may be necessary to learn to inhale and exhale only through your mouth and into the hose. A nose clip may be used to prevent leakage from the nose.

4. Make sure that the respirometer is functioning. Your instructor may have already prepared the instrument for use.
   a. The leveling screws must be raised or lowered so that the bell does not rub on the metal body.
   b. Paper must be fed from a roll onto the kymograph *or* taped onto the kymograph in single sheets.

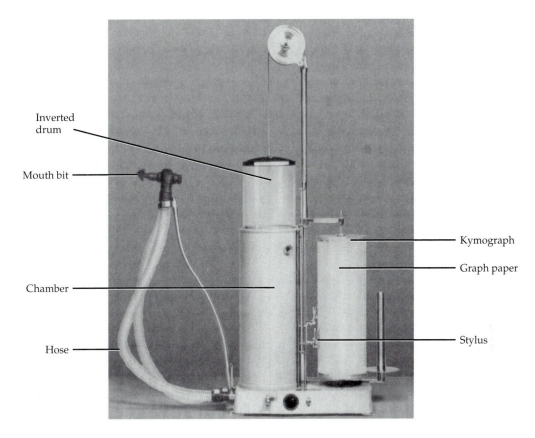

FIGURE 22.14   Collins respirometer. This type is commonly used in college biology laboratories.

c. The soda-lime that absorbs carbon dioxide must be pink (fresh) and not purple (saturated with CO_2).

d. Pen reservoirs must be full and primed.

5. Raise and lower the drum several times to flush the respirometer of stale air and position the ventilometer pen (the upper pen, usually with black ink) so that it will begin writing in the center of the spirogram paper.

6. Set the free-breathing valve so that the opening is seen through the side of the valve.

7. Place a *sterile or disposable mouthpiece in your mouth* and close your nose with a clamp or with the fingers.

8. Breathe several times to become accustomed to the use of this instrument.

9. Have your lab partner close the free-breathing valve (turn it completely to the opposite direction so that no opening is seen) and turn on the respirometer to the slow speed. At this setting, the kymograph moves 32 mm (the distance between two vertical lines) each minute.

10. Perform the following exercises and record your values in Section C.4 of the LABORATORY REPORT RESULTS at the end of the exercise. In each case, * indicates the placement of the mouth on the mouthpiece.

11. Inspire normally, then * exhale normally three times. This volume is your tidal volume. Repeat two more times and average and record the values.

12. Expire normally, then * exhale as much air as possible three times, recording this volume. This value is your expiratory reserve volume. Repeat two more times and record the values.

13. After taking a deep breath, * exhale as much air as possible three times. This volume is your vital capacity. Repeat two more times and record the values. Compare your vital capacity to the normal value shown in Tables 22.1 and 22.2.

14. Because your vital capacity consists of tidal volume, inspiratory reserve volume, and expiratory reserve volume, and because you have already measured tidal volume in step 11 and expiratory reserve volume in step 12, you can calculate your inspiratory reserve volume by subtracting tidal volume and expiratory reserve volume from vital capacity. Inspiratory reserve volume = (vital capacity) − (tidal volume + expiratory reserve volume). Record this value.

TABLE 22.3
Temperature Variation Conversion Factors

Temperature °C (°F)	Conversion factor
20 (68.0)	1.102
21 (69.8)	1.096
22 (71.6)	1.091
23 (73.4)	1.085
24 (75.2)	1.080
25 (77.0)	1.075
26 (78.8)	1.068
27 (80.6)	1.063
28 (82.4)	1.057
29 (84.2)	1.051
30 (86.0)	1.045
31 (87.8)	1.039
32 (89.6)	1.032
33 (91.4)	1.026
34 (93.2)	1.020
35 (95.0)	1.014
36 (96.8)	1.007
37 (98.6)	1.000

In some cases, an individual with a pulmonary disorder has a nearly normal vital capacity. If the rate of expiration is timed, however, the extent of the pulmonary disorder becomes apparent. In order to do this, an individual expels air into a Collins respirometer as fast as possible and the expired volume is measured per unit of time. Such a test is called *forced expiratory volume (FEV$_T$)*. The T indicates that the volume of air is timed. FEV_1 is the volume of air forcefully expired in 1 sec, FEV_2 is the volume expired in 2 sec, and so on. A normal individual should be able to expel 83% of the total capacity during the first second, 94% in 2 sec, and 97% in 3 sec. For individuals with disorders such as emphysema and asthma, the percentage can be considerably lower, depending on the extent of the problem.

FEV_1 is determined according to the following procedure.

PROCEDURE

1. Apply a noseclip to prevent leakage of air through the nose.
2. Turn on the kymograph.
3. Before placing the mouthpiece in your mouth, inhale as deeply as possible.
4. Expel all the air you can into the mouthpiece.
5. Turn off the kymograph.
6. Draw a vertical line on the spirogram at the starting point of exhalation. Mark this A.

7. Using the Collins VC timed interval ruler, draw a vertical line to the left of A and label it line B. The time between lines A and B is 1 sec.
8. The FEV_1 is the point where the spirogram tracing crosses line B.
9. Record your value here _____
10. Now read your vital capacity from the spirogram and record the value here _____
11. In order to adjust for differences in temperature in the respirometer, use Table 22.3 as a guide. Determine the temperature in the respirometer, find the appropriate conversion factor and multiply the conversion factor by your vital capacity.
12. If, for example, the temperature of the respirometer is 75.2°F (24°C), the conversion factor is 1.080. And, if your vital capacity is 5600 ml, then

$$1.080 \times 5600 \text{ ml} = 6048 \text{ ml}$$

13. Use the same conversion factor and multiply it by your FEV_1. If your FEV_1 is 4000 ml, then

$$1.080 \times 4000 \text{ ml} = 4320 \text{ ml}$$

14. To calculate FEV_1, divide 4320 by 6048.

$$FEV_1 = \frac{4320}{6048} = 71\%$$

15. Repeat the procedure three times and record your FEV_1 in Section C.4 of the LABORATORY REPORT RESULTS at the end of the exercise.

Compare your results using the Collins respirometer with those using the handheld respirometer.

D. LABORATORY TESTS COMBINING RESPIRATORY AND CARDIOVASCULAR INTERACTIONS

1. *Experimental setup* Review earlier explanations on recording the following:
   a. Blood pressure with sphygmomanometer and stethoscope (Section E in Exercise 20).
   b. Radial pulse (Section D.1 in Exercise 20).

CAUTION! *Assuming that your partner has no known or apparent cardiac or other health problems and is capable of such an activity, ask her/him to perform the following activities after attaching the various pieces of apparatus in order to record respiratory movements and blood pressure.*

2. *Experimental procedure* Some form of regulated exercise is necessary for this experiment. Choose one of the following forms of exercise to have your subject participate in:

   a. *Riding exercise cycle* If this form is chosen, set the resistance to be felt while riding the cycle, but *not so high* that the subject cannot complete the 4-min exercise period without difficulty.

   b. *Harvard Step Test* In this form of exercise the subject is to step up onto a 20-in. platform (a chair will substitute quite well) with one foot at a time. The subject is to bring *both* feet up onto the platform before stepping back down to the floor. The subject is also to remain erect at all times, and to do 30 complete cycles (up onto the platform and back down) per minute.

3. *Experimental protocol* Record respiratory movements, blood pressure, and pulse during each of the procedures listed. Record your data in the table provided in Section D of the LABORATORY REPORT RESULTS at the end of the exercise. All data should be recorded *simultaneously*, thereby requiring participation of all members of the experimental group.

   a. *Basal readings* Have the subject sit erect and quiet for 3 min. Obtain readings for

      Respiratory rate and depth

      Systolic and diastolic blood pressure

      Pulse rate

When the basal readings have been recorded, obtain additional readings after (1) sitting quietly on the exercise cycle for 3 min, if this form of exercise is to be utilized, or (2) standing quietly for 3 min in front of the platform that is to be utilized for the Harvard Step Test.

   b. *Readings after 1 min of exercise* After the subject has exercised for 1 min, obtain additional readings.

   c. *Readings after 2 and 3 min of exercise* Again, obtain additional readings after the subject has completed 2 and 3 min of exercise.

   d. *Readings upon completion of exercise* When the subject has completed 4 min of exercise, obtain readings for respiratory rate, respiratory depth, heart rate, and systolic and diastolic blood pressure immediately upon completion *while the subject remains seated on the exercise cycle or stands erect on the floor*, depending upon the type of exercise utilized.

   e. *Readings 1, 2, 3, and 5 min after completion of exercise* With the subject *still sitting on the exercise cycle or still standing* erect on the floor obtain additional readings at the above time intervals after completion of the exercise period.

ANSWER THE LABORATORY REPORT QUESTIONS AT THE END OF THE EXERCISE.

Respiratory System 22

Student _____   **Date** _____

Laboratory Section _____   **Score/Grade** _____

SECTION C. LABORATORY TESTS ON RESPIRATION

1. Measurement of Chest and Abdomen in Respiration

Record the results of your chest measurements in inches, in the following table.

Size after a	Chest	Abdomen
Normal inspiration		
Normal expiration		
Maximal inspiration		
Maximal expiration		

2. Use of Pneumograph

Attach a sample of any one of the following activities.

Reading
Swallowing water
Laughing
Yawning
Coughing
Sniffing

3. Use of Handheld Respirometer

Tidal volume _____

Expiratory reserve volume _____

Vital capacity _____

Inspiratory reserve volume _____

Minute volume of respiration (MVR) _____

4. Use of Collins Respirometer

Record the results of your exercises using the Collins respirometer in the following table.

	Tidal volume (1)	Expiratory reserve (2)	Vital capacity (3)	Inspiratory reserve (4)	FEV$_1$ (5)
First time	ml	ml	ml	ml	ml
Second time	ml	ml	ml	ml	ml
Third time	ml	ml	ml	ml	ml
Your average	ml	ml	ml	ml	ml
Normal value	ml	ml	ml	ml	ml

SECTION D. COMBINED RESPIRATORY AND CARDIOVASCULAR INTERACTIONS

Record the results of your exercises in the following table.

	Respiratory rate and depth	Systolic and diastolic pressure	Pulse rate
Basal readings while sitting erect and quiet for 3 min			
Basal readings while standing on cycle or in front of platform for 3 min			
Readings after 1 min of exercise			
Readings after 2 min of exercise			
Readings after 3 min of exercise			
Readings after 4 min of exercise			
Readings after 1 min following completion of exercise			
Readings after 2 min following completion of exercise			
Readings after 3 min following completion of exercise			
Readings after 5 min following completion of exercise			

Respiratory System 22

Student _____   Date _____

Laboratory Section _____   Score/Grade _____

PART 1. Multiple Choice

_____ 1. The overall exchange of gases between the atmosphere, blood, and cells is called (a) inspiration (b) respiration (c) expiration (d) none of these

_____ 2. The lateral walls of the internal nose are formed by the ethmoid bone, maxillae, lacrimal, inferior conchae, and the (a) hyoid bone (b) nasal bone (c) palatine bone (d) occipital bone

_____ 3. The portion of the pharynx that contains the pharyngeal tonsils is the (a) oropharynx (b) laryngopharynx (c) nasopharynx (d) pharyngeal orifice

_____ 4. The Adam's apple is a common term for the (a) thyroid cartilage (b) cricoid cartilage (c) epiglottis (d) none of these

_____ 5. The C-shaped rings of cartilage of the trachea not only prevent the trachea from collapsing but also aid in the process of (a) lubrication (b) removing foreign particles (c) gas exchange (d) swallowing

_____ 6. Of the following structures, the smallest in diameter is the (a) left primary bronchus (b) bronchioles (c) secondary bronchi (d) alveolar ducts

_____ 7. The structures of the lung that actually contain the alveoli are the (a) respiratory bronchioles (b) fissures (c) lobules (d) terminal bronchioles

_____ 8. From superficial to deep, the structure(s) that you would encounter first among the following is (are) the (a) bronchi (b) parietal pleura (c) pleural cavity (d) secondary bronchi

PART 2. Completion

9. An advantage of nasal breathing is that the air is warmed, moistened, and _____.

10. Improper fusion of the palatine and maxillary bones results in a condition called

_____.

11. The protective lid of cartilage that prevents food from entering the trachea is the

_____.

12. After removal of the _____ an individual would be unable to speak.

13. The upper respiratory tract is able to trap and remove dust because of its lining of

_____.

14. The passage of a tube into the mouth and down through the larynx and trachea to bypass an obstruction is called _____.

15. A radiograph of the bronchial tree after administration of an iodinated medium is called a

    _____.

16. The sequence of respiratory tubes from largest to smallest is trachea, primary bronchi, secondary

    bronchi, bronchioles, _____ , respiratory bronchioles, and alveolar ducts.

17. Both the external and internal nose are divided internally by a vertical partition called the

    _____.

18. The undersurface of the external nose contains two openings called the nostrils or

    _____.

19. The functions of the pharynx are to serve as a passageway for air and food and to provide a resonat-

    ing chamber for _____.

20. An inflammation of the membrane that encloses and protects the lungs is called

    _____.

21. The anterior portion of the nasal cavity just inside the nostrils is called the _____.

22. Groovelike passageways in the nasal cavity formed by the conchae are called

    _____.

23. The portion of the pharynx that contains the palatine and lingual tonsils is the

    _____.

24. Each bronchopulmonary segment of a lung is subdivided into many compartments called

    _____.

25. A(n) _____ is an outpouching lined by epithelium and supported by a thin elastic
    membrane.

26. The _____ cartilage attaches the larynx to the trachea.

27. The portion of a lung that rests on the diaphragm is the _____.

28. Phagocytic cells in the alveolar wall are called _____.

29. The surface of a lung lying against the ribs is called the _____ surface.

30. The _____ is a structure in the medial surface of a lung through which bronchi,
    blood vessels, lymphatic vessels, and nerves pass.

31. The nose, pharynx, and associated structures comprise the _____ respiratory sys-
    tem.

32. The _____ is the rounded, anterior border of the nose that connects the root and
    apex.

PART 3. Matching

_____ **33.** Tidal volume

_____ **34.** Inspiratory reserve volume

_____ **35.** Inspiratory capacity

_____ **36.** Expiratory reserve volume

_____ **37.** Vital capacity

_____ **38.** Total lung capacity

A. 1200 ml of air

B. 3600 ml of air

C. 4800 ml of air

D. 500 ml of air

E. 3100 ml of air

F. 6000 ml of air

Digestive System

23

Digestion occurs basically as two events—mechanical digestion and chemical digestion. *Mechanical digestion* consists of various movements of the gastrointestinal tract that help chemical digestion. These movements include physical breakdown of food by the teeth and complete churning and mixing of this food with enzymes by the smooth muscles of the stomach and small intestine. *Chemical digestion* consists of a series of catabolic (hydrolysis) reactions that break down the large nutrient molecules that we eat, such as carbohydrates, lipids, and proteins, into much smaller molecules that can be absorbed and used by body cells.

A. GENERAL ORGANIZATION OF DIGESTIVE SYSTEM

Digestive organs are usually divided into two main groups. The first is the *gastrointestinal (GI) tract*, or *alimentary* (*alimentum* = nourishment) *canal*, a continuous tube running from the mouth to the anus, and measuring about 9 m (30 ft) in length in a cadaver. This tract is composed of the mouth, pharynx, esophagus, stomach, small intestine, and large intestine. The small intestine has three regions: duodenum, jejunum, and ileum. The large intestine has four regions: cecum, colon, rectum, and anal canal. The colon is divided into ascending colon, transverse colon, descending colon, and sigmoid colon.

The second group of organs composing the digestive system consists of the *accessory structures* such as the teeth, tongue, salivary glands, liver, gallbladder, and pancreas (see Figure 23.1).

Using your textbook, charts, or models for reference, label Figure 23.1.

The wall of the gastrointestinal tract, especially from the stomach to the anal canal, has the same basic arrangement of tissues. The four layers (tunics) of the tract, from deep to superficial, are the *mucosa, submucosa, muscularis*, and *serosa* (see Figure 23.9).

Inferior to the diaphragm, the serosa is also called the *peritoneum* (per'-i-tō-NĒ-um; *peri* = around; *tonos* = tension). The peritoneum is composed of a layer of simple squamous epithelium (called mesothelium) and an underlying layer of connective tissue. The *parietal peritoneum* lines the wall of the abdominopelvic cavity, and the *visceral peritoneum* covers some of the organs in the cavity. The potential space between the parietal and visceral portions of the peritoneum is called the *peritoneal cavity*. Unlike the two other serous membranes of the body, the pericardium and the pleura, which smoothly cover the heart and lungs, the peritoneum contains large folds that weave in between the viscera. The important extensions of the peritoneum are the *mesentery* (MEZ-en-ter'-ē; *meso* = middle; *enteron* = intestine) *mesocolon, falciform* (FAL-si-form) *ligament, lesser omentum* (ō-MENT-um), and *greater omentum.*

Inflammation of the peritoneum, called *peritonitis*, is a serious condition because the peritoneal membranes are continuous with one another, enabling the infection to spread to all the organs in the cavity.

B. ORGANS OF DIGESTIVE SYSTEM

1. Mouth (Oral Cavity)

The *mouth*, also called the *oral*, or *buccal* (BUK-al; *bucca* = cheeks), *cavity*, is formed by the cheeks, hard and soft palates, and tongue. The *hard palate* forms the anterior portion of the roof of the mouth and the *soft palate* forms the posterior portion. The *tongue* forms the floor of the oral cavity and is composed of skeletal muscle covered by mucous membrane. Partial digestion of carbohydrates and triglycerides occurs in the mouth.

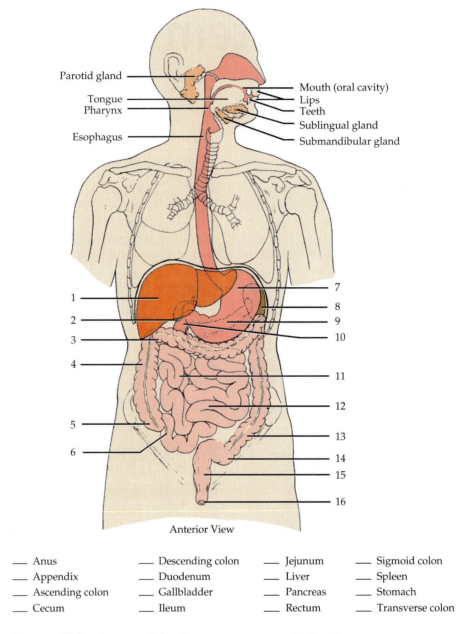

Parotid gland

Tongue
Pharynx

Esophagus

Mouth (oral cavity)
Lips
Teeth
Sublingual gland
Submandibular gland

1
2
3
4
5
6

7
8
9
10
11
12
13
14
15
16

Anterior View

— Anus	— Descending colon	— Jejunum	— Sigmoid colon
— Appendix	— Duodenum	— Liver	— Spleen
— Ascending colon	— Gallbladder	— Pancreas	— Stomach
— Cecum	— Ileum	— Rectum	— Transverse colon

FIGURE 23.1   Organs of the digestive system and related structures.

a. ***Cheeks*** Lateral walls of oral cavity. Muscular structures covered by skin and lined by nonkeratinized stratified squamous epithelium; anterior portions terminate in the ***superior*** and ***inferior labia*** (lips).

b. ***Vermilion*** (ver-MIL-yon) Transition zone of lips where outer skin and inner mucous membranes meet.

c. ***Labial frenulum*** (LĀ-bē-al FREN-yoo-lum; *labium* = fleshy border; *frenulum* = small bridle) Midline fold of mucous membrane that attaches the inner surface of each lip to its corresponding gum.

d. ***Vestibule*** (= entrance to a canal) Space bounded externally by cheeks and lips and internally by gums and teeth.

e. ***Oral cavity proper*** Space extending from the gums and teeth to the ***fauces*** (FAW-sēz; *fauces* = passages), opening of oral cavity proper into pharynx. Area is enclosed by the dental arches.

f. ***Hard palate*** Formed by maxillae and palatine bones and covered by mucous membrane.

g. *Soft palate* Arch-shaped muscular partition between oropharynx and nasopharynx lined by mucous membrane. Hanging from free border of soft palate is a muscular projection, the *uvula* (YOU-vyoo-la = little grape).

h. *Palatoglossal arch (anterior pillar)* Muscular fold that extends inferiorly, laterally, and anteriorly to the side of the base of tongue.

i. *Palatopharyngeal* (PAL-a-tō-fa-rin'-jē-al) *arch (posterior pillar)* Muscular fold that extends inferiorly, laterally, and posteriorly to the side of pharynx. *Palatine tonsils* are between arches and *lingual tonsil* is at base of tongue.

j. *Tongue* Movable, muscular organ on floor of oral cavity. *Extrinsic muscles* originate outside tongue (to bones in the area), insert into connective tissues, and move tongue from side to side and in and out to maneuver food for chewing and swallowing; *intrinsic muscles* originate and insert into connective tissues within the tongue and alter shape and size of tongue for speech and swallowing.

k. *Lingual* (*lingua* = tongue) *frenulum* Midline fold of mucous membrane on undersurface of tongue that helps restrict its movement posteriorly.

l. *Papillae* (pa-PIL-ē = nipple-shaped projections) Projections of lamina propria on surface of tongue covered with epithelium; *filiform* (= threadlike) *papillae* are conical projections in parallel rows over anterior two-thirds of tongue; *fungiform* (= shaped like a mushroom) *papillae* are mushroomlike elevations distributed among filiform papillae and more numerous near tip of tongue (appear as red dots and most contain taste buds); *circumvallate* (*circum* = around; *vallare* = to wall) *papillae* are arranged in the form of an inverted V on the posterior surface of tongue (all contain taste buds).

Using a mirror, examine your mouth and locate as many of the structures (a through l) as you can. Label Figure 23.2.

2. Salivary Glands

Most saliva is secreted by the *salivary glands*, which lie outside the mouth and pour their contents into ducts that empty into the oral cavity. The carbohydrate-digesting enzyme in saliva is salivary amylase. The three pairs of salivary glands are the *parotid* (*para* = near; *otia* = ear) *glands* (anterior and inferior to the ears), which secrete into the oral cavity vestibule through *parotid (Stensen's) ducts; submandibular glands* (deep to the base of the tongue in the posterior part of the floor of the mouth), which secrete on either side of the lingual frenulum in the floor of the oral cavity through *submandibular (Wharton's) ducts;* and *sublingual glands* (superior to the submandibular glands), which secrete into the floor of the oral cavity through *lesser sublingual (Rivinus') ducts.*

Label Figure 23.3.

The parotid glands are compound tubuloacinar glands, whereas the submandibulars and sublinguals are compound acinar glands (see Figure 23.4).

Examine prepared slides of the three different types of salivary glands and compare your observations to Figure 23.4.

3. Teeth

Teeth (dentes) are located in the sockets of the alveolar processes of the mandible and maxillae. The alveolar processes are covered by *gingivae* (jin-JĪ-vē) or gums, which extend slightly into each socket. The sockets are lined by a dense fibrous connective tissue called a *periodontal* (*peri* = around; *odous* = tooth) *ligament*, which anchors the teeth in position and acts as a shock absorber during chewing.

Following are the parts of a tooth:

a. *Crown* Exposed portion above level of gums.

b. *Root* One to three projections embedded in socket.

c. *Neck* Constricted junction line of the crown and root near the gum line.

d. *Dentin* Calcified connective tissue that gives teeth their basic shape and rigidity.

e. *Pulp cavity* Enlarged part of cavity in crown within dentin.

f. *Pulp* Connective tissue containing blood vessels, lymphatic vessels, and nerves.

g. *Root canal* Narrow extension of pulp cavity in root.

h. *Apical foramen* Opening in base of root canal through which blood vessels, lymphatic vessels, and nerves enter tooth.

i. *Enamel* Covering of crown that consists primarily of calcium phosphate and calcium carbonate.

j. *Cementum* Bonelike substance that covers and attaches root to periodontal ligament.

With the aid of your textbook, label the parts of a tooth shown in Figure 23.5.

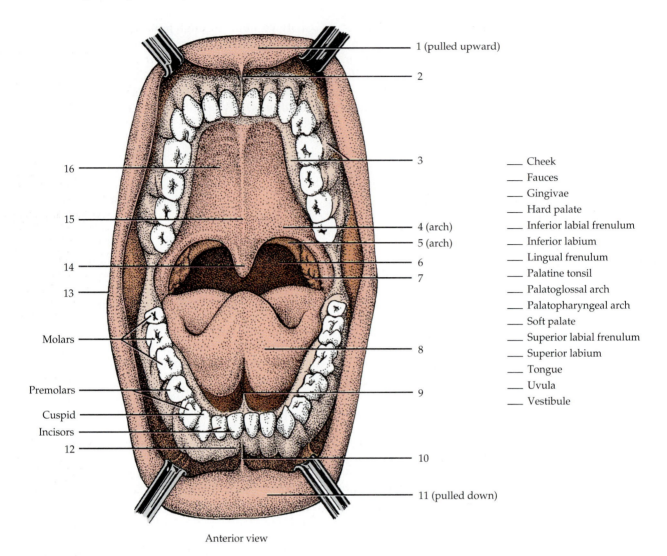

Anterior view

FIGURE 23.2   Mouth (oral cavity).

Labels on the diagram:

1 (pulled upward)
2
3
4 (arch)
5 (arch)
6
7
8
9
10
11 (pulled down)
16
15
14
13
Molars
Premolars
Cuspid
Incisors
12

___ Cheek
___ Fauces
___ Gingivae
___ Hard palate
___ Inferior labial frenulum
___ Inferior labium
___ Lingual frenulum
___ Palatine tonsil
___ Palatoglossal arch
___ Palatopharyngeal arch
___ Soft palate
___ Superior labial frenulum
___ Superior labium
___ Tongue
___ Uvula
___ Vestibule

4. Dentitions

Dentitions (sets of teeth) are of two types: *deciduous* (baby) and *permanent*. Deciduous teeth begin to erupt at about 6 months of age, and one pair appears at about each month thereafter until all 20 are present. The deciduous teeth are as follows:

a. *Incisors* Central incisors closest to midline, with lateral incisors on either side. Incisors are chisel-shaped, adapted for cutting into food, have only one root.
b. *Cuspids (canines)* Posterior to incisors. Cuspids have pointed surfaces (cusps) for tearing and shredding food, have only one root.
c. *Molars* First and second molars posterior to canines. Molars crush and grind food. Upper molars have four cusps and three roots, lower molars have four cusps and two roots.

All deciduous teeth are usually lost between 6 and 12 years of age and replaced by permanent dentition consisting of 32 teeth that appear between age 6 and adulthood. The permanent teeth are:

a. *Incisors* Central incisors and lateral incisors replace those of deciduous dentition.
b. *Cuspids (canines)* These replace those of deciduous dentition.
c. *Premolars (bicuspids)* First and second premolars replace deciduous molars. Premolars crush and grind food, have two cusps and one root (upper first premolars have two roots).
d. *Molars* These erupt behind premolars as jaw grows to accommodate them and do not replace any deciduous teeth. First molars erupt at age 6, second at age 12, and third (wisdom teeth) after age 18.

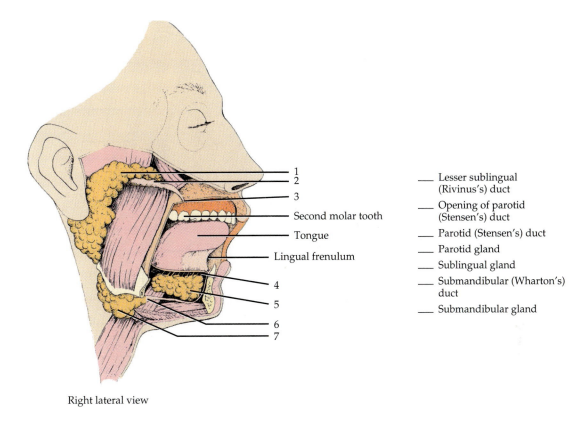

Right lateral view

FIGURE 23.3 Location of salivary glands.

___ Lesser sublingual
   (Rivinus's) duct

___ Opening of parotid
   (Stensen's) duct

___ Parotid (Stensen's) duct

___ Parotid gland

___ Sublingual gland

___ Submandibular (Wharton's)
   duct

___ Submandibular gland

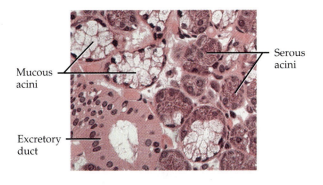

FIGURE 23.4 Histology of salivary glands (100×).

Using a mirror, examine your mouth and locate as many teeth of the permanent dentition as you can.

With the aid of your textbook, label the deciduous and permanent dentitions in Figure 23.6.

5. Esophagus

The *esophagus* (e-SOF-a-gus; *oisein* = to carry; *phagema* = food) is a muscular, collapsible tube posterior to the trachea. The structure is 23 to 25 cm (10 in.) long and extends from the laryngopharynx through the mediastinum and esoph-ageal hiatus in the diaphragm and terminates in the superior portion of the stomach. The esophagus conveys food from the pharynx to the stomach by peristalsis.

Histologically, the esophagus consists of a *mucosa* (nonkeratinized stratified squamous epi-the-lium, lamina propria, muscularis mucosae), *submucosa* (areolar connective tissue, blood vessels, mucous glands), *muscularis* (superior third striated, middle third striated and smooth, inferior third smooth), and *adventitia* (ad-ven-TISH-ya). The esophagus is not covered by a serosa.

Examine a prepared slide of a transverse section of the esophagus that shows its various coats. With the aid of your textbook, label Figure 23.7.

6. Stomach

The *stomach* is a J-shaped enlargement of the gastrointestinal tract inferior to the diaphragm (see Figure 23.1). It is in the epigastric, umbilical, and left hypochondriac regions of the abdomen. The superior part is connected to the esophagus; the inferior part empties into the duodenum, the first portion of the small intestine. The stomach is divided into four main areas: cardia, fundus, body, and pylorus.

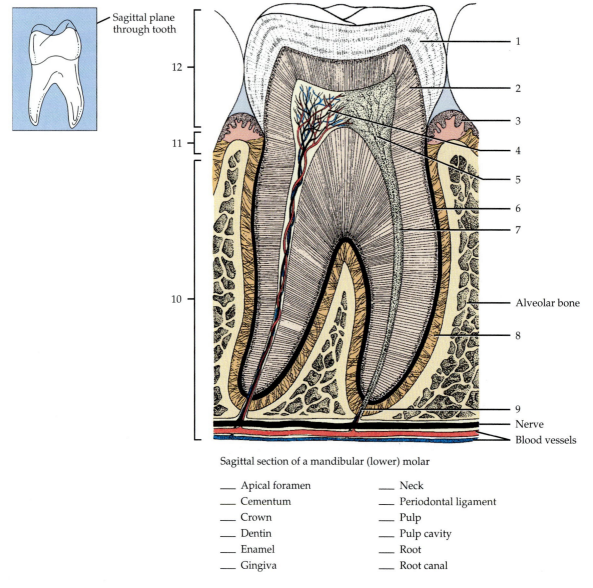

Sagittal plane through tooth

12

11

10

1
2
3
4
5
6
7

Alveolar bone

8

9
Nerve
Blood vessels

Sagittal section of a mandibular (lower) molar

___ Apical foramen   ___ Neck
___ Cementum   ___ Periodontal ligament
___ Crown   ___ Pulp
___ Dentin   ___ Pulp cavity
___ Enamel   ___ Root
___ Gingiva   ___ Root canal

FIGURE 23.5   Parts of a tooth.

The **cardia** (CAR-dē-a) surrounds the lower esophageal sphincter, a physiological sphincter in the esophagus just superior to the diaphragm. The rounded portion superior to and to the left of the cardia is the **fundus** (FUN-dus). Inferior to the fundus, the large central portion of the stomach is called the **body**. The narrow, inferior region is the **pylorus** (pī-LOR-us; *pyle* = gate; *ouros* = guard). The pylorus consists of a **pyloric antrum** (AN-trum = cave), which is closer to the body of the stomach, and a **pyloric canal**, which is closer to the duodenum. The concave medial border of the stomach is called the **lesser curvature**, and the convex lateral border is the **greater curvature**. The pylorus communicates with the duodenum of the small intestine via a sphincter called the **pyloric sphincter (valve)**. The main chemical activity of the stomach is to begin the digestion of proteins.

Label Figure 23.8.

The surface of the **mucosa** is a layer of simple columnar epithelial cells called **mucous surface cells** (Figure 23.9). The mucosa contains a **lamina propria** (areolar connective tissue) and a **muscularis mucosae** (smooth muscle). Epithelial cells extend down into the lamina propria, forming many narrow channels called **gastric pits** and columns of secretory

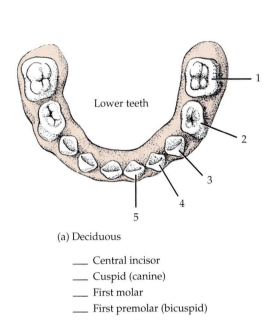

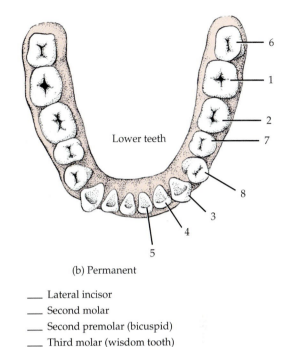

(a) Deciduous

___ Central incisor

___ Cuspid (canine)

___ First molar

___ First premolar (bicuspid)

FIGURE 23.6   Dentitions.

(b) Permanent

___ Lateral incisor

___ Second molar

___ Second premolar (bicuspid)

___ Third molar (wisdom tooth)

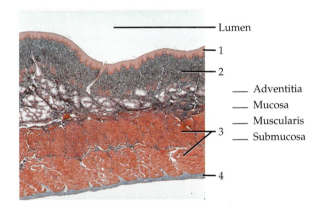

Lumen

___ Adventitia

___ Mucosa

___ Muscularis

___ Submucosa

FIGURE 23.7   Histology of esophagus (10×).

cells called *gastric glands.* Secretions from several gastric glands flow into each gastric pit and then into the lumen of the stomach. The gastric glands include three types of *exocrine gland* cells that secrete their products into the stomach lumen: mucous neck cells, chief cells, and parietal cells. Both mucous surface cells and *mucous neck cells* secrete mucus. The *chief (zymogenic) cells* secrete pepsinogen and gastric lipase. *Parietal (oxyntic) cells* produce hydrochloric acid. The secretions of the mucous, chief, and parietal cells are collectively called *gastric juice.* In addition, gastric glands include one type of enteroendocrine cell. An *enteroendocrine (enteron =*

intestine) *cell* is a hormone-producing cell in the gastrointestinal mucosa. One such cell is a *G cell*, located mainly in the pyloric antrum, that secretes the hormone gastrin into the bloodstream.

Examine a prepared slide of a section of the stomach that shows its various layers. With the aid of your textbook, label Figure 23.9 on page 451.

7. Pancreas

The *pancreas (pan = all; kreas = flesh)* is a retroperitoneal gland posterior to the greater curvature of the stomach (see Figure 23.1). The gland consists of a *head* (expanded portion near duodenum), *body* (central portion), and *tail* (terminal tapering portion).

Histologically, the pancreas consists of *pancreatic islets (islets of Langerhans)* that contain (1) glucagon-producing *alpha cells*, (2) insulin-producing *beta cells*, (3) somatostatin-producing *delta cells*, and pancreatic polypeptide-producing *F cells* (see Figure 16.5). The pancreas also consists of *acini* that produce pancreatic juice (see Figure 16.5). Pancreatic juice contains enzymes that assist in the chemical breakdown of carbohydrates, proteins, triglycerides, and nucleic acids.

Pancreatic juice is delivered from the pancreas to the duodenum by a large main tube, the *pancreatic duct (duct of Wirsung)*. This duct unites with the

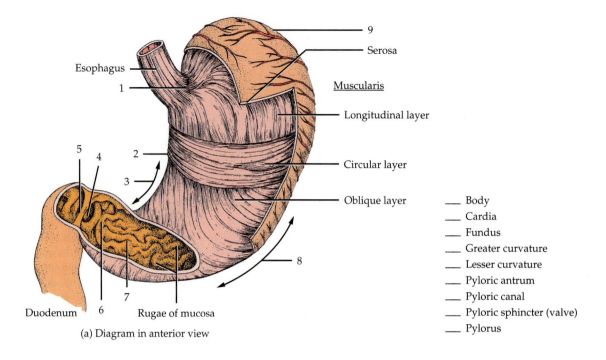

Esophagus

Muscularis

9
Serosa
Longitudinal layer
Circular layer
Oblique layer

___ Body
___ Cardia
___ Fundus
___ Greater curvature
___ Lesser curvature
___ Pyloric antrum
___ Pyloric canal
___ Pyloric sphincter (valve)
___ Pylorus

Duodenum

Rugae of mucosa

(a) Diagram in anterior view

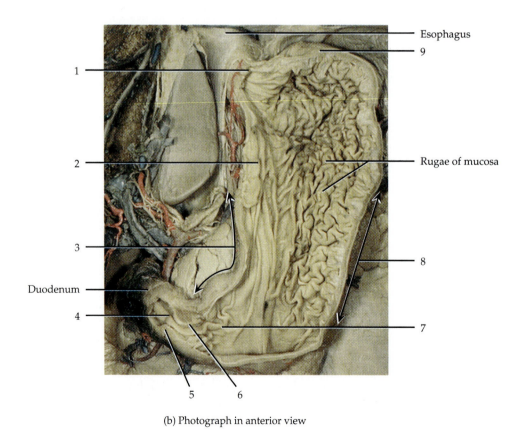

Esophagus
9

Rugae of mucosa

Duodenum

(b) Photograph in anterior view

FIGURE 23.8   Stomach. External and internal anatomy.

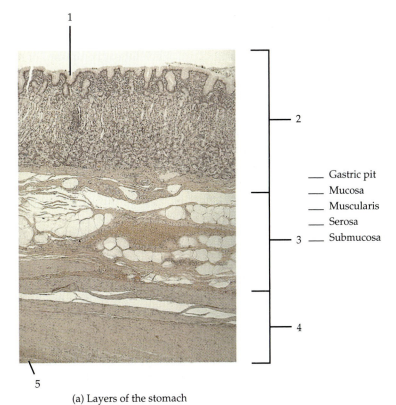

1

2

___ Gastric pit
___ Mucosa
___ Muscularis
___ Serosa
3 ___ Submucosa

4

5

(a) Layers of the stomach

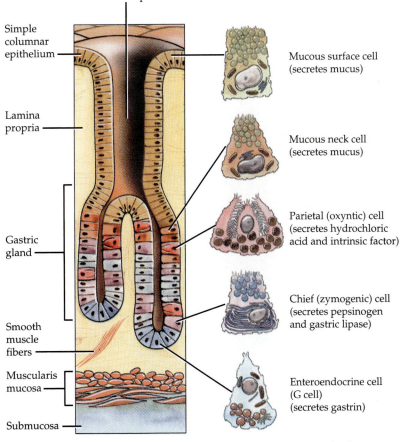

Gastric pit

Simple
columnar
epithelium

Lamina
propria

Gastric
gland

Smooth
muscle
fibers

Muscularis
mucosa

Submucosa

Mucous surface cell
(secretes mucus)

Mucous neck cell
(secretes mucus)

Parietal (oxyntic) cell
(secretes hydrochloric
acid and intrinsic factor)

Chief (zymogenic) cell
(secretes pepsinogen
and gastric lipase)

Enteroendocrine cell
(G cell)
(secretes gastrin)

(b) Sectional view of the stomach mucosa showing gastric glands

FIGURE 23.9   Histology of stomach. (17×)

common bile duct from the liver and pancreas and enters the duodenum in a common duct called the **hepatopancreatic ampulla (ampulla of Vater)**. The ampulla opens on an elevation of the duodenal mucosa, the **duodenal papilla**. An **accessory pancreatic duct (duct of Santorini)** may also lead from the pancreas and empty into the duodenum about 2.5 cm (1 in.) superior to the hepato-pancreatic ampulla. With the aid of your textbook, label the structures associated with the pancreas in Figure 23.10.

8. Liver

The **liver** is located inferior to the diaphragm (see Figure 23.1). It occupies most of the right hypochondriac and part of the epigastric regions of the abdomen. The gland is divided into two principal lobes, the **right lobe** and **left lobe**, separated by the **falciform ligament**. The falciform ligament attaches the liver to the anterior abdominal wall and diaphragm. The right lobe consists of an inferior **quadrate lobe** and a posterior **caudate lobe**.

Each lobe is composed of microscopic functional units called **lobules**. Among the structures in a lobule are cords of **hepatocytes (liver cells)** arranged in a radial pattern around a **central vein**; **sinusoids**, endothelial lined spaces between hepatocytes through which blood flows; and **stellate reticuloendothelial (Kupffer) cells** that destroy bacteria and worn-out blood cells by phagocytosis.

Examine a prepared slide of several liver lobules. Compare your observations with Figure 23.11 on page 453.

Bile is manufactured by hepatocytes and functions in the emulsification of triglycerides in the small intestine. The liquid is passed to the small intestine as follows: Hepatocytes secrete bile into **bile canaliculi** (kan'-a-LIK-yoo-lī = small canals) that empty into small ducts. The small ducts merge into larger **right** and **left hepatic ducts**, one in each principal lobe of the liver. The right and left hepatic ducts unite outside the liver to form a single **common hepatic duct**. This duct joins the **cystic** (kystis = bladder) **duct** from the gallbladder to become the

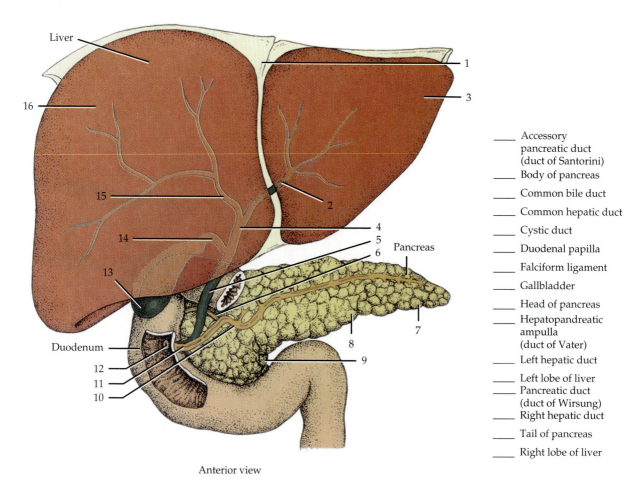

_____ Accessory pancreatic duct (duct of Santorini)

_____ Body of pancreas

_____ Common bile duct

_____ Common hepatic duct

_____ Cystic duct

_____ Duodenal papilla

_____ Falciform ligament

_____ Gallbladder

_____ Head of pancreas

_____ Hepatopandreatic ampulla (duct of Vater)

_____ Left hepatic duct

_____ Left lobe of liver

_____ Pancreatic duct (duct of Wirsung)

_____ Right hepatic duct

_____ Tail of pancreas

_____ Right lobe of liver

Anterior view

FIGURE 23.10   Relations of the liver, gallbladder, duodenum, and pancreas.

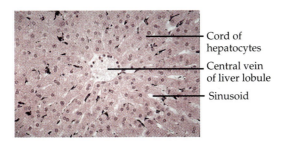

— Cord of
hepatocytes

— Central vein
of liver lobule

— Sinusoid

FIGURE 23.11   Photomicrograph of liver
lobule (100×).

common bile duct, which empties into the duodenum at the hepatopancreatic ampulla (ampulla of
Vater). When triglycerides are not being digested, a
valve around the hepatopancreatic ampulla, the
*sphincter of the hepatopancreatic ampulla (sphinc-
ter of Oddi)*, closes, and bile backs up into the gallbladder via the cystic duct. In the gallbladder, bile
is stored and concentrated.

With the aid of your textbook, label the structures
associated with the liver in Figure 23.10.

9. Gallbladder

The *gallbladder* (*galla* = bile) is a pear-shaped sac
in a fossa along the posterior surface of the liver (see
Figure 23.1). The gallbladder stores and concentrates
bile. The cystic duct of the gallbladder and common
hepatic duct of the liver merge to form the common
bile duct. The *mucosa* of the gallbladder consists of
simple columnar epithelium that contains rugae.
The *muscularis* consists of smooth muscle and the
outer coat consists of visceral peritoneum.

With the aid of your textbook, label the structures
associated with the gallbladder in Figure 23.10.

10. Small Intestine

The bulk of digestion and absorption occurs in the
small intestine, which begins at the pyloric sphincter (valve) of the stomach, coils through the central
and inferior part of the abdomen, and joins the large
intestine at the ileocecal sphincter (see Figure 23.1).
The mesentery attaches the small intestine to the
posterior abdominal wall. The small intestine is
about 6.35 m (21 ft) long and is divided into three
segments: *duodenum* (doo'-ō-DĒ-num), which begins at the stomach; *jejunum* (jē-JOO-num), the middle segment; and *ileum* (IL-ē-um), which terminates
at the large intestine (Figure 23.12 on page 454).

The wall of the small intestine is composed of the
same four coats that make up most of the gastrointestinal tract. However, special features of both the

mucosa and the submucosa facilitate the processes
of digestion and absorption (Figure 23.13). The mucosa forms a series of *villi* (= tuft of hair; *villus* is
singular). These projections are 0.5 to 1 mm long
and give the intestinal mucosa a velvety appearance. The large number of villi (20 to 40 per square
millimeter) vastly increases the surface area of the
epithelium available for absorption and digestion.
Each villus has a core of lamina propria. Embedded
in this connective tissue are an arteriole, a venule,
a capillary network, and a *lacteal* (LAK-tē-al),
which is a lymphatic capillary. Nutrients being absorbed by the epithelial cells covering the villus
pass through the wall of a capillary or a lacteal to
enter blood or lymph, respectively.

The epithelium of the mucosa consists of simple
columnar epithelium and contains absorptive cells,
goblet cells, enteroendocrine cells, and Paneth cells.
The apical (free) membrane of the absorptive cells
features *microvilli* (mī'-krō-VIL-ī), which are
microscopic, fingerlike projections of the plasma
membrane that contain actin filaments. In a photomicrograph taken through a light microscope, the
microvilli are too small to be seen individually. They
form a fuzzy line, called the **brush border,** at the apical surface of the absorptive cells, next to the lumen
of the small intestine. Larger amounts of digested
nutrients can diffuse into the absorptive cells of the
intestinal wall because the microvilli greatly increase the surface area of the plasma membrane. It
is estimated that there are about 200 million microvilli per square millimeter of small intestine.

The mucosa contains many cavities lined with
glandular epithelium. Cells lining the cavities form
the *intestinal glands (crypts of Lieberkühn)* and secrete intestinal juice. *Paneth cells* are found in the
deepest parts of the intestinal glands. They secrete
lysozyme, a bactericidal enzyme, and are also capable of phagocytosis. They may have a role in regulating the microbial population in the intestines.
The submucosa of the duodenum contains *duodenal (Brunner's) glands.* They secrete an alkaline
mucus that helps neutralize gastric acid in the
chyme. Some of the epithelial cells in the mucosa
are goblet cells, which secrete additional mucus.

The lamina propria of the small intestine has an
abundance of mucosa-associated lymphoid tissue
(MALT) in the form of lymphatic nodules, masses
of lymphatic tissue not surrounded by a capsule.
Solitary lymphatic nodules are most numerous in
the lower part of the ileum. Groups of lymphatic
nodules, referred to as *aggregated lymphatic follicles (Peyer's patches),* are numerous in the ileum.
The muscularis mucosae consist of smooth muscle.

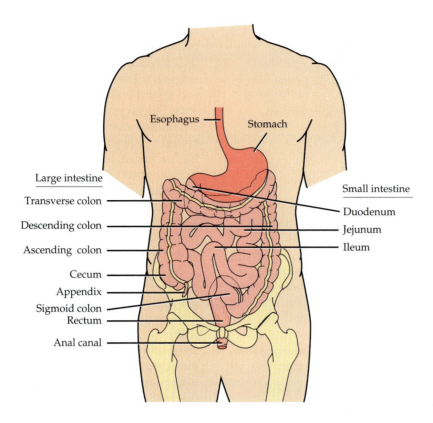

FIGURE 23.12   Intestines. Note the parts of the small intestine on the right side of the illustration and the parts of the large intestine on the left side.

The *muscularis* of the small intestine consists of two layers of smooth muscle. The outer, thinner layer contains longitudinally arranged fibers. The inner, thicker layer contains circularly arranged fibers.

Except for a major portion of the duodenum, the *serosa* (or visceral peritoneum) completely surrounds the small intestine.

Obtain prepared slides of the small intestine (section through its tunics and villi) and identify as many structures as you can, using Figure 23.13 on page 455 and your textbook as references.

11. Large Intestine

The *large intestine* functions in the completion of absorption of water that leads to the formation of feces, and in the expulsion of feces from the body. Bacteria residing in the large intestine manufacture certain vitamins (some B vitamins and vitamin K). The large intestine is about 1½ m (5 ft) long and extends from the ileum to the anus (see Figure 23.12). It is attached to the posterior abdominal wall by an extension of visceral peritoneum called mesocolon.

The large intestine is divided into four principal regions: cecum, colon, rectum, and anal canal.

The opening from the ileum into the large intestine is guarded by a fold of mucous membrane, the *ileocecal sphincter* (*valve*). Hanging below the valve is a blind pouch, the *cecum*, to which is attached the *vermiform appendix* (*vermis* = worm; *appendix* = appendage) by an extension of visceral peritoneum called the *mesoappendix*. Inflammation of the vermiform appendix is called *appendicitis*. The open end of the cecum merges with the *colon* (*kolon* = food passage). The first division of the colon is the *ascending colon*, which ascends on the right side of the abdomen and turns abruptly to the left at the inferior surface of the liver *(right colic [hepatic] flexure)*. The *transverse colon* continues across the abdomen, curves at the inferior surface of the spleen *(left colic [splenic] flexure)*, and passes down the left side of the abdomen as the *descending colon*. The *sigmoid colon* begins near the iliac crest, projects medially toward the midline, and terminates at the rectum at the level of the third sacral vertebra. The *rectum* is the last 20 cm (7 to 8 in.) of the gastrointestinal tract. Its terminal 2 to 3 cm

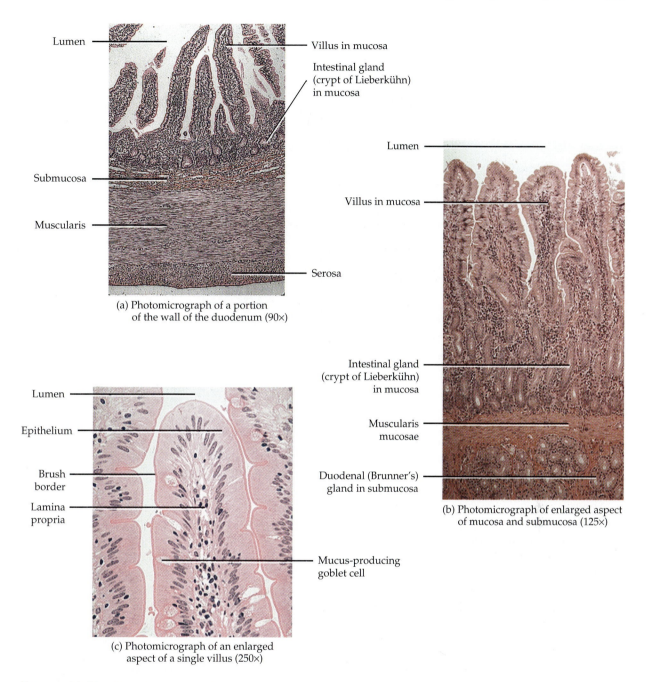

Lumen

Villus in mucosa

Intestinal gland
(crypt of Lieberkühn)
in mucosa

Submucosa

Muscularis

Serosa

(a) Photomicrograph of a portion
of the wall of the duodenum (90×)

Lumen

Villus in mucosa

Intestinal gland
(crypt of Lieberkühn)
in mucosa

Muscularis
mucosae

Duodenal (Brunner's)
gland in submucosa

(b) Photomicrograph of enlarged aspect
of mucosa and submucosa (125×)

Lumen

Epithelium

Brush
border

Lamina
propria

Mucus-producing
goblet cell

(c) Photomicrograph of an enlarged
aspect of a single villus (250×)

FIGURE 23.13   Histology of small intestine.

(1 in.) is known as the *anal canal*. The opening of the anal canal to the exterior is the *anus*.

With the aid of your textbook, label the parts of the large intestine in Figure 23.14.

The wall of the large intestine differs from that of the small intestine in several respects. No villi or permanent circular folds are found in the mucosa. The *mucosa* consists of simple columnar epithelium, lamina propria (areolar connective tissue), and muscularis mucosae (smooth muscle). The epithelium contains mostly absorptive and goblet cells. The absorptive cells function primarily in water absorption. The goblet cells secrete mucus that lubricates the colonic contents as they pass through. Both absorptive and goblet cells are located in long, straight, tubular intestinal glands that extend the full thickness of the mucosa. Solitary lymphatic nodules are also found in the mucosa.

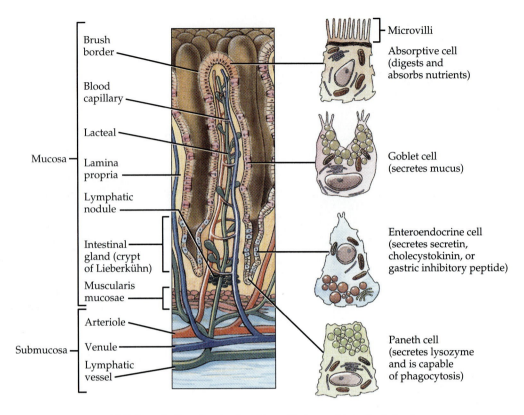

(d) Enlarged villus showing lacteal, capillaries, and intestinal gland

FIGURE 23.13 *(Continued)*   Histology of small intestine.

The *submucosa* of the large intestine is similar to that found in the rest of the gastrointestinal tract.

The *muscularis* consists of an external layer of longitudinal muscles and an internal layer of circular muscles. Unlike other parts of the GI tract, portions of the longitudinal muscles are thickened, forming three conspicuous longitudinal bands called *taeniae coli* (TĒ-nē-ē KŌ-lī; *taenia* = flat band), alternating with a wall section with less or no longitudinal muscle. Each band runs the length of most of the large intestine. Tonic contractions of the bands gather the colon into a series of pouches called *haustra* (HAWS-tra; singular is *haustrum* = shaped like a pouch), which give the colon a puckered appearance. There is a single layer of circular muscle between taeniae coli. The *serosa* of the large intestine is part of the visceral peritoneum. Small pouches of visceral peritoneum filled with fat are attached to taeniae coli and are called *epiploic appendages.*

Examine a prepared slide of the large intestine showing its tunics. Compare your observations to Figure 23.15.

C. DEGLUTITION

Swallowing, or *deglutition* (dē-gloo-TISH-un), is the mechanism that moves food from the mouth through the esophagus to the stomach. It is facilitated by saliva and mucus and involves a complex series of actions by the mouth, pharynx, and esophagus. Swallowing can be initiated voluntarily, but the remainder of the activity is almost completely under reflex control. Swal-lowing is a very orderly sequence of events involving the movement of food from the mouth to the stomach, while simultaneously involving the inhibition of respiratory activity and the prevention of entrance of foreign particles into the trachea.

Swallowing is divided into three phases: (1) the *voluntary phase,* when a *bolus* (soft flexible mass of food) is moved into the oropharynx; (2) the *pharyngeal phase,* involving the movement of the bolus through the pharynx into the esophagus; and (3) the *esophageal phase,* which involves the movement of the bolus from the esophagus, through the lower esophageal (gastroesophageal) sphincter, and

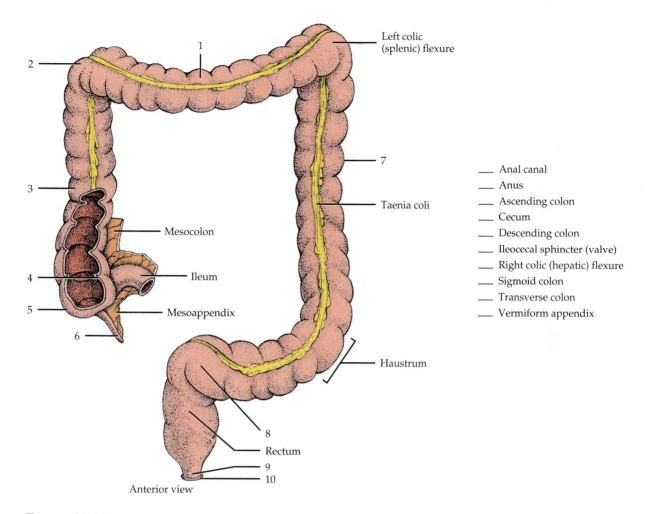

Left colic
(splenic) flexure

7

Taenia coli

Mesocolon

Ileum

Mesoappendix

____ Anal canal
____ Anus
____ Ascending colon
____ Cecum
____ Descending colon
____ Ileocecal sphincter (valve)
____ Right colic (hepatic) flexure
____ Sigmoid colon
____ Transverse colon
____ Vermiform appendix

Haustrum

8
Rectum
9
10

Anterior view

FIGURE 23.14   Large intestine.

into the stomach. Both the pharyngeal and esophageal phases are involuntary. The passage of solid or semisolid food from the mouth to the stomach takes 4 to 8 sec. Very soft foods and liquids pass through in about 1 sec.

PROCEDURE

1. Chew a cracker and note the movements of your tongue as you begin to swallow. Chew another cracker, lean over your chair, placing your mouth below the level of your pharynx, and attempt to swallow again.
2. Note the movements of your laboratory partner's larynx during the pharyngeal stage of deglutition.
3. *Clean the earpieces of a stethoscope with alcohol.* Using the stethoscope, listen to the sounds that occur when your partner swallows water. Place the stethoscope just to the left of the midline at the level of the sixth rib. You should hear the

sound of water as it makes contact with the lower esophageal sphincter (valve), followed by the sound of water passing through the sphincter after the sphincter relaxes. The sphincter is at the junction of the esophagus and stomach.

4. Record your observations in Section C of the LABORATORY REPORT RESULTS at the end of the exercise. At what point can deglutition be

stopped voluntarily? _____

_____

D. CHEMISTRY OF DIGESTION

In order to understand how food is digested, let us examine the various processes that occur between absorption and final utilization of nutrients by the body's cells.

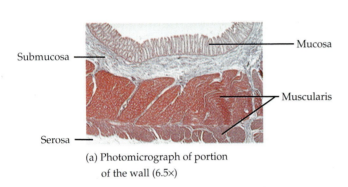

(a) Photomicrograph of portion
of the wall (6.5×)

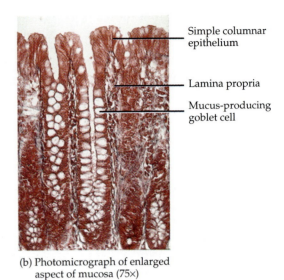

(b) Photomicrograph of enlarged
aspect of mucosa (75×)

FIGURE 23.15   Histology of large intestine.

Nutrients are chemical substances in food that provide energy, act as building blocks to form new body components, or assist body processes. The six major classes of nutrients are carbohydrates, lipids, proteins, minerals, vitamins, and water. Cells break down carbohydrates, lipids, and proteins to release energy, or use them to build new structures and new regulatory substances, such as hormones and enzymes.

Enzymes are produced by living cells to catalyze or speed up many of the reactions in the body. Enzymes are proteins and act as *catalysts* to speed up reactions without being permanently altered by the reaction. Digestive enzymes function as catalysts to speed chemical reactions in the gastrointestinal tract.

Because digestive enzymes function outside the cells that produce them, they are capable of also reacting within a test tube and therefore provide an excellent means of studying enzyme activity.

CAUTION! *Please reread Section A, "General Safety Precautions and Procedures" on page xi and Section C, "Precautions Related to Working with Reagents" on page xii at the beginning of the laboratory manual before you begin any of the following experiments. You should also read the experiments before you perform them to be sure that you understand all the procedures and safety precautions.*

1. Positive Tests for Sugar and Starch

Salivary amylase is an enzyme produced by the salivary glands. This enzyme starts starch digestion and *hydrolyzes* (splits using water) it into maltose (a disaccharide), maltotriose (a trisaccharide) and α-dextrins. We measure the amount of starch and sugar present before and after enzymatic activity. It is expected that the amount of starch should *decrease* and the sugar level should *increase* as a result of salivary amylase activity.

a. TEST FOR SUGAR

Benedict's test is commonly used to detect sugars. Glucose (monosaccharide), maltose (disaccharide), or any other reducing sugars react with *Benedict's solution,* forming insoluble red cuprous oxide. The precipitate of cuprous oxide can usually be seen in the bottom of the tube when standing. Benedict's solution turns green, yellow, orange, or red depending on the amount of reducing sugar present according to the following scale:

blue (−)
green (+)
yellow (++)
orange (+++)
red (++++)

Test for the presence of sugar (maltose) as follows:

PROCEDURE

1. Using separate medicine droppers, place 2 ml of maltose and 2 ml of Benedict's solution in a Pyrex test tube.
2. *Using a test tube holder*, place the test tube in a boiling water bath and heat for 5 minutes. **CAUTION!** *Make sure that the mouth of the test tube is pointed away from you and all other persons in the area.* Note the color change.
3. *Using a test tube holder*, remove the test tube from the water bath.
4. Repeat the same procedure using a starch solution instead of maltose. Notice that the color does not change, because the solution contains no sugar. Now you have a method for detecting sugar.

b. TEST FOR STARCH

Lugol's solution is a brown-colored iodine solution used to test certain polysaccharides, especially starch. Starch, for example, gives a **deep blue** to **black** color with Lugol's solution (the black is really a concentrated blue color). Cellulose, monosaccharides, and disaccharides do not react. A negative test is indicated by a yellow to brown color of the solution itself, or possibly some other color (other than blue to black) resulting from pigments present in the substance being tested.

PROCEDURE

1. Using separate medicine droppers, place a drop of starch solution on a spot plate and then add a drop of Lugol's solution to it. Notice the black color that forms as the starch-iodine complex develops.
2. Repeat the test using a maltose solution in place of the starch solution. Note that there is no color change, because Lugol's solution and maltose do not combine. Now you have a method for detecting starch.

c. DIGESTION OF STARCH

PROCEDURE

1. Using separate medicine droppers, transfer 3 ml of a starch solution to a small beaker, add a fresh enzyme solution consisting of a pinch of amylase powder in 3 ml of water, and mix thoroughly with a glass rod.
2. Wait for 1 min, record the time, remove 1 drop of the mixture with a glass rod to the depres-

sion of a spot plate, and then test for starch with Lugol's solution.
3. At 1-min intervals, test 1-drop samples of the mixture until you no longer note a positive test for starch. Keep the glass rod in the mixture, stirring it from time to time.
4. After the starch test is seen to be negative, test the remaining mixture for the presence of glucose. Do this by adding 2 ml of the mixture with a medicine dropper to 2 ml of the Benedict's solution in a Pyrex test tube and heat in a boiling water bath as per the procedure outlined in a.2, "Test for Sugar." Answer questions a through c in Section D.1 of the LABORATORY REPORT RESULTS at the end of the exercise.

2. Effect of Temperature on Starch Digestion

In the following procedure, you will test starch digestion at five different temperatures to determine how temperature influences enzyme activity. Lugol's solution is again used for presence or absence of starch.

A fresh enzyme solution consisting of a pinch of amylase powder in 3 ml of water should be used as before. Five constant-temperature water baths should be available. Starting with the lowest temperature, these are: 0°C or cooler, 10°C, 40°C, 60°C, and boiling.

PROCEDURE

1. Prepare 10 Pyrex test tubes, 5 containing 1 ml each of enzyme solution and 5 containing 1 ml each of starch solution.
2. Using rubber bands, pair the tubes (i.e., a tube containing enzyme with one containing starch) and use a test tube holder to place one pair into each water bath. **CAUTION!** *Make sure that the mouth of the test tube is pointed away from you and everyone else in the area.*
3. Permit the tubes to adapt to the bath temperatures for about 5 min, then mix the enzyme and starch solutions of each pair together, and *using a test tube holder* place the single test tube in its respective water bath.
4. After 30 sec, *using a test tube holder*, remove the test tubes from the water baths and test all five tubes for starch on a spot plate using Lugol's solution.
5. Repeat every 30 sec until you have determined the time required for the starch to disappear (that is, to be digested).

6. Record and graph your results in Section D.2 of the LABORATORY REPORT RESULTS at the end of the exercise.

3. Effect of pH on Starch Digestion

You can demonstrate the effectiveness of salivary amylase digestion at different pH readings.

PROCEDURE

1. Prepare three buffer solutions as follows:

    Solution A pH 4.0
    Solution B pH 7.0
    Solution C pH 9.0

2. Once again a fresh enzyme solution consisting of a pinch of amylase powder in 3 ml of water should be used.
3. Using medicine droppers, mix 4 ml of a starch solution with 2 ml of buffer solution A in a Pyrex test tube.
4. Repeat this procedure with buffer solutions B and C.
5. You now have three test tubes of a starch-buffer solution, each at a different pH (4.0, 7.0, and 9.0).
6. Using separate medicine droppers, place one drop of starch-buffer solution A on a spot plate and immediately add one drop of the saliva.
7. Test for starch disappearance using Lugol's solution, and record the time when starch first disappears completely.
8. Repeat this test for the other two starch buffers (solutions B and C) and record the time when starch is no longer present at each pH.
9. Record and explain your results in Section D.3 of the LABORATORY REPORT RESULTS at the end of the exercise.

4. Action of Bile on Triglycerides

Bile is important in the process of lipid digestion because of its emulsifying effects (breaking down of large globules to smaller, uniformly distributed particles) on triglycerides (fats) and oils. *Bile does not contain any enzymes.* Emulsification of triglycerides by means of bile salts serves to increase the surface area of the triglyceride that will be exposed to the action of the lipase.

PROCEDURE

1. Place 5 ml of water into one Pyrex test tube and 5 ml of bile solution into a second.
2. Using a medicine dropper, add one drop of vegetable oil that has been colored with a fat-soluble dye, such as Sudan B, into each tube.

3. Place a stopper in both tubes. Shake them *vigorously,* and then let them stand in a test tube rack undisturbed for 10 min. Triglycerides or oils that are broken into sufficiently small droplets will remain suspended in water in the form of an *emulsion.* If emulsification has not occurred, the triglyceride or oil will lie on the surface of the water.
4. Answer questions in Section D.4 of the LABORATORY REPORT RESULTS at the end of the exercise.

5. Digestion of Triglycerides

You can demonstrate the effect of pancreatic juice on triglycerides by the use of pancreatin, which contains all the enzymes present in pancreatic juice. Because the optimum pH of the pancreatic enzymes ranges from 7.0 to 8.8, the pancreatin is prepared in sodium carbonate. The enzyme used in this test is *pancreatic lipase*, which digests triglycerides to fatty acids and glycerol. The fatty acid produced changes the color of *blue* litmus to *red.*

PROCEDURE

1. Using a medicine dropper, place 5 ml of litmus cream (heavy cream to which powdered litmus has been added to give it a blue color) in a Pyrex test tube, and, *using a test tube holder,* place the test tube in a 40°C water bath.
2. Repeat the procedure with another 5-ml portion in a second tube, but put it in an ice bath.
3. When the tubes have adapted to their respective temperatures (in about 5 min), use a medicine dropper to add 5 ml of pancreatin to each tube and, *using a test tube holder,* replace them in their water baths until a color change occurs in one tube.
4. Summarize your results and your explanation in Section D.5 of the LABORATORY REPORT RESULTS at the end of the exercise.

6. Digestion of Protein

Here you demonstrate the effect of pepsin on protein and the factors affecting the rate of action of pepsin. *Pepsin,* a proteolytic enzyme, is secreted in inactive form (pepsinogen) by chief (zymogenic) cells in the lining of the stomach. Pepsin digests proteins (fibrin in this experiment) to peptides. The efficiency of pepsin activity depends on the pH of the solution, the optimum being 1.5 to 2.5. Pepsin is almost completely inactive in neutral or alkaline solutions.

PROCEDURE

1. Prepare and number the following five Pyrex test tubes. For this test, the quantity of each solution must be *measured carefully.*

    Tube 1—5 ml of 0.5% pepsin; 5 ml of 0.8% HCl
    Tube 2—5 ml of pepsin; 5 ml of water
    Tube 3—5 ml of pepsin, boiled for 10 min in a water bath; 5 ml of 0.8% HCl

CAUTION! *Make sure that the mouth of the test tube is pointed away from you and all other persons in the area. Using a test tube holder, remove the test tube from the water bath.*

    Tube 4—5 ml of pepsin; 5 ml of 0.5% NaOH
    Tube 5—5 ml of water; 5 ml of 0.8% HCl

2. First determine the approximate pH of each test tube using Hydrion paper (range 1 to 11) and put the values in the table provided in Section D.6 of the LABORATORY REPORT RESULTS at the end of the exercise.

3. Using forceps, place a small amount of fibrin (the protein) in each test tube. An amount near the size of a pea will be sufficient. *Using a test tube holder,* put the tubes in a 40°C water bath and *carefully* shake occasionally. Maintain 40°C temperature closely. The tubes should remain in the water bath for *at least 1½ hr.*

4. Watch the changes that the fibrin undergoes. The swelling that occurs in some tubes should not be confused with digestion. Digested fibrin becomes transparent and disappears (dissolves) as the protein is digested to soluble peptides.

5. Finish the experiment when the fibrin is digested in one of the five tubes.

6. Record all your observations in the table provided in Section D.6 of the LABORATORY REPORT RESULTS at the end of the exercise.

ANSWER THE LABORATORY REPORT QUESTIONS AT THE END OF THE EXERCISE.

Digestive System 23

Student _____   **Date** _____

Laboratory Section _____   **Score/Grade** _____

SECTION C. DEGLUTITION

1. Tongue movement at beginning of deglutition: _____

_____

2. Description of swallowing with head below pharynx: _____

_____

3. Laryngeal movements during pharyngeal deglutition: _____

_____

4. Time lapse between two sounds of deglutition: _____

_____

SECTION D. CHEMISTRY OF DIGESTION

1. Digestion of Starch

a. How long did it take for the starch to be digested? _____

b. Did your observation indicate the presence of maltose? _____

c. What is the meaning of this result? _____

2. Effect of Temperature on Starch Digestion

d. Record the time required for starch to disappear at the temperatures tested.

_____0°C   _____40°C   _____Boiling

_____10°C   _____60°C

e. What is the optimum temperature for starch digestion? _____

f. What happens to amylase when it is boiled? _____

g. Is the effect of boiling reversible or irreversible? _____

h. Is the effect of freezing reversible or irreversible? _____

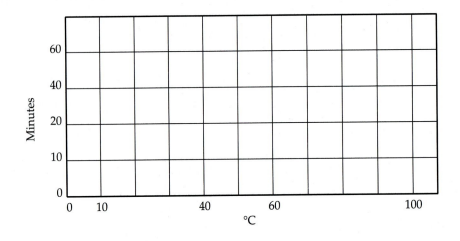

3. Effect of pH on Starch Digestion

i. Record the time required for starch to disappear at the pH readings tested.

Solution A: pH 4.0 _____

Solution B: pH 7.0 _____

Solution C: pH 9.0 _____

j. Explain your results

A: pH 4.0 _____

B: pH 7.0 _____

C: pH 9.0 _____

k. What is the optimum pH for the action of salivary amylase? _____

_____

4. Action of Bile on Triglycerides

l. What difference can you detect in the appearance of the mixtures in the two test tubes? _____

_____

m. How does emulsification of lipids by means of bile aid in the digestion of triglycerides? _____

_____

5. Digestion of Triglycerides

Record the results of the digestion of triglycerides by pancreatic lipase in pancreatic juice in the following table.

Tube no.	Temperature of water bath	Change in pH (color)	Explanation of results
1			
2			

6. Digestion of Protein

Record the results of the digestion of protein by pepsin in the following table.

Tube no.	Tube contents	pH	Digestion observed (yes or no)	Explanation of results
1				
2				
3				
4				
5				

Digestive System 23

Student _____   Date _____

Laboratory Section _____   Score/Grade _____

PART 1. Multiple Choice

_____ 1. If an incision has to be made in the small intestine to remove an obstruction, the first layer of tissue to be cut is the (a) muscularis (b) mucosa (c) serosa (d) submucosa

_____ 2. Mesentery, lesser omentum, and greater omentum are all directly associated with the (a) peritoneum (b) liver (c) esophagus (d) mucosa of the gastrointestinal tract

_____ 3. Chemical digestion of carbohydrates is initiated in the (a) stomach (b) small intestine (c) mouth (d) large intestine

_____ 4. A tumor of the villi and circular folds would interfere most directly with the body's ability to carry on (a) absorption (b) deglutition (c) mastication (d) peristalsis

_____ 5. The main chemical activity of the stomach is to begin the digestion of (a) triglycerides (b) proteins (c) carbohydrates (d) all of the above

_____ 6. Surgical cutting of the lingual frenulum would occur in which part of the body? (a) salivary glands (b) esophagus (c) nasal cavity (d) tongue

_____ 7. To free the small intestine from the posterior abdominal wall, which of the following would have to be cut? (a) mesocolon (b) mesentery (c) lesser omentum (d) falciform ligament

_____ 8. The cells of gastric glands that produce secretions directly involved in chemical digestion are the (a) mucous (b) parietal (c) chief (d) pancreatic islets (islets of Langerhans)

_____ 9. An obstruction in the hepatopancreatic ampulla (ampulla of Vater) would affect the ability to transport (a) bile and pancreatic juice (b) gastric juice (c) salivary amylase (d) intestinal juice

_____ 10. The terminal portion of the small intestine is known as the (a) duodenum (b) ileum (c) jejunum (d) pyloric sphincter (valve)

_____ 11. The portion of the large intestine closest to the liver is the (a) right colic flexure (b) rectum (c) sigmoid colon (d) left colic flexure

_____ 12. The lamina propria is found in which coat? (a) serosa (b) muscularis (c) submucosa (d) mucosa

_____ 13. Which structure attaches the liver to the anterior abdominal wall and diaphragm? (a) lesser omentum (b) greater omentum (c) mesocolon (d) falciform ligament

_____ 14. The opening between the oral cavity and pharynx is called the (a) vermilion border (b) fauces (c) vestibule (d) lingual frenulum

_____ 15. All of the following are parts of a tooth *except* the (a) crown (b) root (c) cervix (d) papilla

_____ **16.** Cells of the liver that destroy worn-out white and red blood cells and bacteria are termed (a) hepatocytes (b) stellate reticuloendothelial (Kupffer) cells (c) alpha cells (d) beta cells

_____ **17.** Bile is manufactured by which cells? (a) alpha (b) beta (c) hepatocytes (d) stellate reticuloendothelial (Kupffer)

_____ **18.** Which part of the small intestine secretes the intestinal digestive enzymes? (a) intestinal glands (b) duodenal (Brunner's) glands (c) lacteals (d) microvilli

_____ **19.** Structures that give the colon a puckered appearance are called (a) taenia coli (b) villi (c) rugae (d) haustra

PART 2. Completion

20. An acute inflammation of the serous membrane lining the abdominal cavity and covering the abdominal viscera is referred to as _____.

21. The _____ is a sphincter (valve) between the ileum and large intestine.

22. The portion of the small intestine that is attached to the stomach is the _____.

23. The portion of the stomach closest to the esophagus is the _____.

24. The _____ forms the floor of the oral cavity and is composed of skeletal muscle covered with mucous membrane.

25. The convex lateral border of the stomach is called the _____.

26. The three special structures found in the wall of the small intestine that increase its efficiency in absorbing nutrients are the villi, circular folds, and _____.

27. The three pairs of salivary glands are the parotids, submandibulars, and _____.

28. The small intestine is divided into three segments: duodenum, ileum, and _____.

29. The large intestine is divided into four main regions: the cecum, colon, rectum, and

_____.

30. The enzyme that is present in saliva is called _____.

31. The transition of the lips where the outer skin and inner mucous membrane meet is called the

_____.

32. The _____ papillae are arranged in the form of an inverted V on the posterior surface of the tongue.

33. The portion of a tooth containing blood vessels, lymphatic vessels, and nerves is the

_____.

34. The teeth present in a permanent dentition, but not in a deciduous dentition, that replace the deciduous molars are the _____.

35. The portion of the gastrointestinal tract that conveys food from the pharynx to the stomach is the

_____.

36. The inferior region of the stomach connected to the small intestine is the _____.

37. The clusters of cells in the pancreas that secrete digestive enzyme are called _____.

38. The caudate lobe, quadrate lobe, and central vein are all associated with the _____.

39. The common bile duct is formed by the union of the common hepatic duct and

_____ duct.

40. The pear-shaped sac that stores bile is the _____.

41. _____ glands of the small intestine secrete an alkaline substance to protect the mucosa from excess acid.

42. The _____ attaches the large intestine to the posterior abdominal wall.

43. The _____ is the last 20 cm (7 to 8 in.) of the gastrointestinal tract.

44. A midline fold of mucous membrane that attaches the inner surface of each lip to its corresponding

gum is the _____.

45. The bonelike substance that gives teeth their basic shape is called _____.

46. The _____ anchors teeth in position and helps to dissipate chewing forces.

47. The portion of the colon that terminates at the rectum is the _____ colon.

48. The palatine tonsils are between the palatoglossal and _____ arches.

49. The teeth closest to the midline are the _____.

50. The vermiform appendix is attached to the _____.

51. A soft, flexible mass of food that is easily swallowed is called a _____.

52. The cells in gastric glands that secrete pepsinogen are called _____.

PART 3. Matching

_____ **53.** Pancreatic lipase
_____ **54.** Benedict's solution
_____ **55.** Lugol's solution
_____ **56.** Salivary amylase
_____ **57.** Pepsin

A. Commonly used solution in the test for starch
B. Capable of digesting starch
C. Capable of digesting protein
D. Commonly used solution for detecting reducing sugars
E. Digests triglycerides into fatty acids and glycerol and changes the color of blue litmus to red

Urinary System

24

The *urinary system* functions to keep the body in homeostasis by controlling the composition and volume of the blood. The system accomplishes these functions by removing and restoring selected amounts of water and various solutes. The kidneys also excrete selected amounts of various wastes, assume a role in erythropoiesis by secreting erythropoietin, help control blood pH, help regulate blood pressure by secreting renin (which activates the renin-angiotensin pathway), participate in the synthesis of vitamin D, and perform gluconeogenesis (synthesis of glucose molecules) during periods of fasting or starvation.

The urinary system consists of two kidneys, two ureters, one urinary bladder, and a single urethra (see Figure 24.1). Other systems that help in waste elimination are the respiratory, integumentary, and digestive systems.

Using your textbook, charts, or models for reference, label Figure 24.1.

A. ORGANS OF URINARY SYSTEM

1. Kidneys

The paired *kidneys,* which resemble kidney beans in shape, are found just superior to the waist between the parietal peritoneum and the posterior wall of the abdomen. Because they are behind the peritoneal lining of the abdominal cavity, they are referred to as *retroperitoneal* (re'-trō-per-i-tō-NĒ-al; *retro* = behind). The kidneys are located between the levels of the last thoracic and third lumbar vertebrae, with the right kidney slightly lower than the left because of the position of the liver.

Three layers of tissue surround each kidney: the *renal* (*renalis* = kidney) *capsule, adipose capsule (perirenal fat),* and *renal fascia.* They function to protect the kidney and hold it firmly in place.

Near the center of the kidney's concave border, which faces the vertebral column, is a vertical fissure called the *renal hilus,* through which the ureter

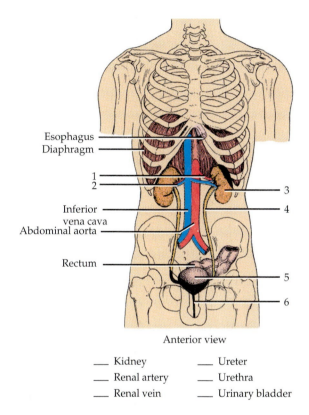

FIGURE 24.1 Organs of male urinary system and associated structures.

Labels on figure:
Esophagus
Diaphragm
1
2
Inferior vena cava
Abdominal aorta
Rectum
3
4
5
6

Anterior view

___ Kidney   ___ Ureter
___ Renal artery   ___ Urethra
___ Renal vein   ___ Urinary bladder

leaves the kidney and through which blood vessels, lymphatics, and nerves enter and exit the kidney. The hilus is the entrance to a cavity in the kidney called the *renal sinus.*

If a frontal section is made through a kidney, the following structures can be seen:

a. *Renal cortex* (*cortex* = rind or bark) Superficial, narrow, reddish area.

b. *Renal medulla* (*medulla* = inner portion) Deep, wide, reddish-brown area.

c. *Renal (medullary) pyramids* Striated triangular structures, 8 to 18 in number, in the renal medulla. The bases of the renal pyramids face the renal cortex, and the apices, called *renal papillae,* are directed toward the center of the kidney.

   d. *Renal columns* Cortical substance between renal pyramids.

   e. *Renal pelvis* Large cavity in the renal sinus, enlarged proximal portion of ureter.

   f. *Major calyces* (KĀ-li-sēz) Consist of 2 or 3 cuplike extensions of the renal pelvis.

   g. *Minor calyces* Consist of 8 to 18 cuplike extensions of the major calyces.

Within the renal cortex and renal pyramids of each kidney are more than 1 million microscopic units called nephrons, the functional units of the kidneys (described shortly). As a result of their activity in regulating the volume and chemistry of the blood, they produce urine. Urine passes from the nephrons to the minor calyces, major calyces, renal pelvis, ureter, urinary bladder, and urethra.

   Examine a specimen, model, or chart of the kidney and with the aid of your textbook, label Figure 24.2.

2. Nephrons

Basically, a *nephron* (NEF-ron) consists of (1) a renal corpuscle (KOR-pus-sul; *corpus* = body; *cle* = tiny) where fluid is filtered and (2) a renal tubule into which the filtered fluid (filtrate) passes. A *renal corpuscle* has two components: a tuft (knot) of capillaries called a *glomerulus* (glō-MER-yoo-lus; *glomus* = ball; *ulus* = small) surrounded by a double-walled epithelial cup, called a *glomerular (Bowman's) capsule,* lying in the renal cortex of the kidney. The inner wall of the capsule, the *visceral layer,* consists of epithelial cells called *podocytes* and surrounds the glomerulus. A space called the *capsular (Bowman's) space* separates the visceral layer from the outer wall of the capsule, the *parietal layer,* which is composed of simple squamous epithelium.

   The visceral layer of the glomerular (Bowman's) capsule and endothelium of the glomerulus form an *endothelial-capsular (filtration) membrane,* a very effective filter. Electron microscopy has determined that the membrane consists of the following

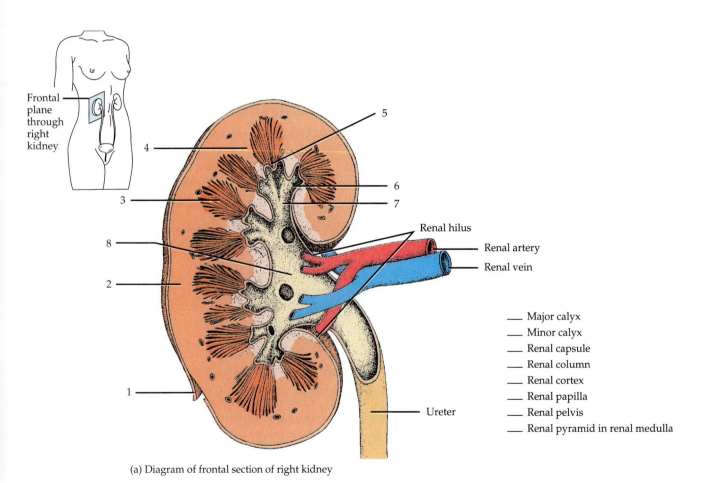

Frontal plane through right kidney

Renal hilus

Renal artery

Renal vein

Ureter

___ Major calyx
___ Minor calyx
___ Renal capsule
___ Renal column
___ Renal cortex
___ Renal papilla
___ Renal pelvis
___ Renal pyramid in renal medulla

(a) Diagram of frontal section of right kidney

FIGURE 24.2   Kidney.

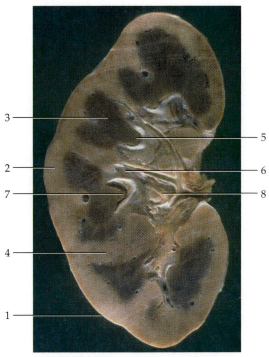

(b) Photograph of frontal section of right kidney

FIGURE 24.2 *(Continued)* Kidney.

components, given in the order in which substances are filtered (Figure 24.3).

  a. *Endothelial fenestrations (pores) of the glomerulus* The single layer of endothelial cells has large fenestrations (pores) that prevent filtration of blood cells but allow all components of blood plasma to pass through.
  b. *Basement membrane (basal lamina) of the glomerulus* This layer of extracellular material lies between the endothelium and the visceral layer of the glomerular capsule. It consists of fibrils in a glycoprotein matrix and prevents filtration of larger proteins.
  c. *Slit membranes between pedicels* The specialized epithelial cells that cover the glomerular capillaries are called *podocytes* (*podos* = foot). Extending from each podocyte are thousands of footlike structures called *pedicels* (PED-i-sels; *pediculus* = little foot). The pedicels cover the basement membrane, except for spaces between them, which are called *filtration slits.* A thin membrane, the *slit membrane*, extends across filtration slits and prevents filtration of medium-sized proteins.

The endothelial-capsular membrane filters blood passing through the kidney. Blood cells and large molecules, such as proteins, are retained by the fil-

ter and eventually are recycled into the blood. The filtered substances pass through the membrane and into the space between the parietal and visceral layers of the glomerular (Bowman's) capsule, and then enter the renal tubule (described shortly).

As the filtered fluid (filtrate) passes through the remaining parts of a nephron, substances are selectively added and removed. The end product of these activities is urine. Nephrons are frequently classified into two kinds. A *cortical nephron* usually has its glomerulus in the superficial renal cortex, and the remainder of the nephron penetrates only into the superficial renal medulla. A *juxtamedullary nephron* usually has its glomerulus close to the corticomedullary junction, and other parts of the nephron penetrate deeply into the renal medulla (see Figure 24.4). The following description of the remaining components of a nephron applies to juxtamedullary nephrons.

After the filtrate leaves the glomerular (Bowman's) capsule, it passes through the following parts of a renal tubule:

  a. *Proximal convoluted tubule (PCT)* Coiled tubule in the renal cortex that originates at the glomerular (Bowman's) capsule; consists of simple cuboidal epithelium with microvilli.

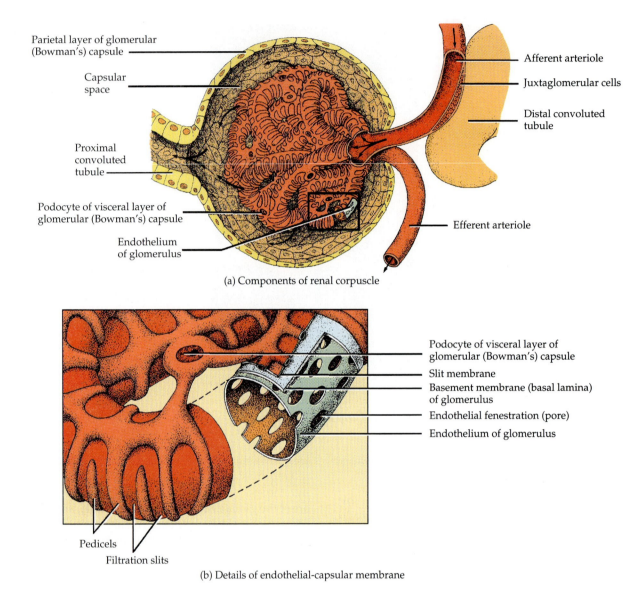

Parietal layer of glomerular
(Bowman's) capsule

Capsular
space

Proximal
convoluted
tubule

Podocyte of visceral layer of
glomerular (Bowman's) capsule

Endothelium
of glomerulus

Afferent arteriole

Juxtaglomerular cells

Distal convoluted
tubule

Efferent arteriole

(a) Components of renal corpuscle

Podocyte of visceral layer of
glomerular (Bowman's) capsule

Slit membrane

Basement membrane (basal lamina)
of glomerulus

Endothelial fenestration (pore)

Endothelium of glomerulus

Pedicels

Filtration slits

(b) Details of endothelial-capsular membrane

FIGURE 24.3   Endothelial-capsular membrane.

b. *Loop of Henle (nephron loop)* U-shaped tubule that connects the proximal and distal convoluted tubules. It consists of a *descending limb of the loop of Henle,* an extension of the proximal convoluted tubule that dips down into the renal medulla and consists of simple squamous epithelium, and an *ascending limb of the loop of Henle* that ascends into the renal medulla and approaches the renal cortex and consists of simple squamous, cuboidal, and columnar epithelium; the ascending limb is wider in diameter than the descending limb.

c. *Distal convoluted tubule (DCT)* Coiled extension of ascending limb in the renal cortex; consists of simple cuboidal epithelium with fewer microvilli than in the proximal convoluted tubule.

Most of the simple cuboidal cells are *principal cells,* which are sensitive to antidiuretic hormone and aldosterone, two hormones that regulate kidney functions. A few of the cells are *intercalated cells,* which can secrete H^+ to rid the body of excess acids.

Distal convoluted tubules terminate by merging with straight *collecting ducts.* In the renal medulla, collecting ducts receive distal convoluted tubules from several nephrons, pass through the renal pyramids, and open at the renal papillae into minor calyces through about 30 large *papillary ducts.* The processed filtrate, called urine, passes from the collecting ducts to papillary ducts, minor calyces, major calyces, renal pelvis, ureter, urinary bladder, and urethra.

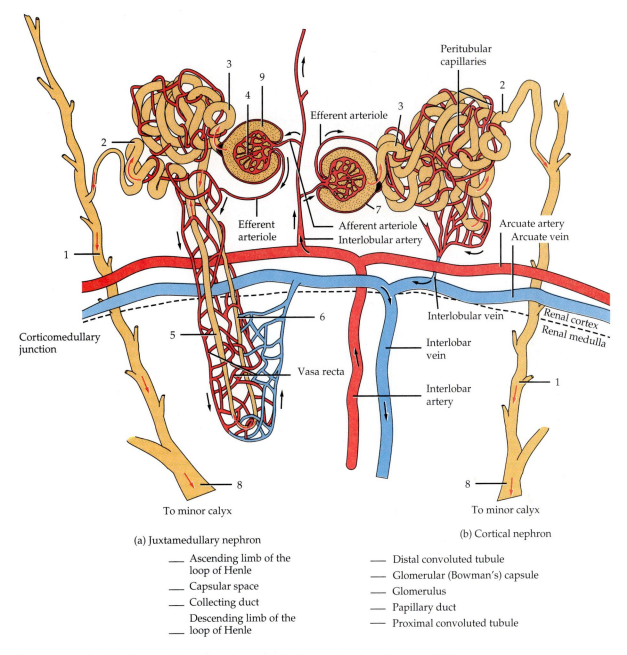

Peritubular
capillaries

Efferent arteriole

Efferent
arteriole

Afferent arteriole

Interlobular artery

Corticomedullary
junction

Vasa recta

Interlobular vein

Interlobar
vein

Interlobar
artery

Arcuate artery
Arcuate vein

Renal cortex
Renal medulla

To minor calyx

To minor calyx

(a) Juxtamedullary nephron

___ Ascending limb of the
    loop of Henle

___ Capsular space

___ Collecting duct

    Descending limb of the
___ loop of Henle

(b) Cortical nephron

___ Distal convoluted tubule

___ Glomerular (Bowman's) capsule

___ Glomerulus

___ Papillary duct

___ Proximal convoluted tubule

FIGURE 24.4 Nephrons. The colored arrows indicate the direction in which the filtrate flows.

With the aid of your textbook, label the parts of a nephron and associated structures in Figure 24.4.

Examine prepared slides of various components of nephrons and compare your observations to Figure 24.5 on page 476.

3. Blood and Nerve Supply

Nephrons are abundantly supplied with blood vessels, and the kidneys actually receive around 20 to 25% of the total cardiac output, or approximately 1200 ml, every minute. The blood supply originates in each kidney with the **renal artery,** which divides into many branches, eventually supplying the nephron and its complete tubule.

Before or immediately after entering the renal hilus, the renal artery divides into a larger anterior branch and a smaller posterior branch. From these branches, five **segmental arteries** originate. Each gives off several branches, the **interlobar arteries,** which pass between the renal pyramids in the renal columns. At the bases of the pyramids, the interlobar

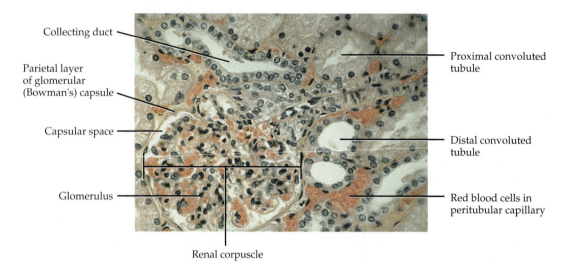

Collecting duct

Parietal layer
of glomerular
(Bowman's) capsule

Capsular space

Glomerulus

Proximal convoluted
tubule

Distal convoluted
tubule

Red blood cells in
peritubular capillary

Renal corpuscle

FIGURE 24.5   Histology of a nephrons. (250×)

arteries arch between the renal medulla and renal cortex and here are known as *arcuate* (*arcuatus* = shaped like a bow) *arteries*. Branches of the arcuate arteries, called *interlobular arteries,* enter the renal cortex. *Afferent* (*ad* = toward; *ferre* = to carry) *arterioles,* branches of the interlobular arteries, are distributed to the *glomeruli.* Blood leaves the glomeruli via *efferent* (*efferens* = to bring out) *arterioles.*

The next sequence of blood vessels depends on the type of nephron. Around convoluted tubules, efferent arterioles of cortical nephrons divide to form capillary networks called *peritubular* (*peri* = around) *capillaries.* Efferent arterioles of juxtamedullary nephrons also form peritubular capillaries and, in addition, form long loops of blood vessels around medullary structures called *vasa recta* (VĀ-sa REK-ta; *vasa* = vessels; *recta* = straight). Peritubular capillaries eventually reunite to form *interlobular veins.* Blood then drains into *arcuate veins, interlobar veins,* and *segmental veins.* Blood leaves the kidneys through the *renal vein* that exits at the hilus. (The vasa recta pass blood into the interlobular veins, arcuate veins, interlobar veins, and renal veins.)

In each nephron, the final portion of the ascending limb of the loop of Henle makes contact with the afferent arteriole serving its own renal corpuscle. The cells of the renal tubule in this region are tall and crowded together. Collectively, they are known as the *macula densa* (*macula* = spot; *densa* = dense). These cells monitor the Na$^+$ and Cl$^-$ concentration of fluid in the tubule lumen. Next to the macula densa, the wall of the afferent arteriole (and sometimes efferent arteriole) contains modified smooth

muscle fibers called *juxtaglomerular (JG) cells.* Together with the macula densa, they constitute the *juxtaglomerular apparatus,* or *JGA.* The JGA helps regulate arterial blood pressure and the rate of blood filtration by the kidneys. The distal convoluted tubule begins a short distance past the macula densa.

Using Figures 24.4 and 24.6 as guides, trace a drop of blood from its entrance into the renal artery to its exit through the renal vein. As you do so, name in sequence each blood vessel through which blood passes for both cortical and juxtaglomerular nephrons.

Label Figure 24.7.

The nerve supply to the kidneys comes from the *renal plexus* of the autonomic system. The nerves are vasomotor because they regulate the circulation of blood in the kidney by regulating the diameters of the arterioles.

4. Ureters

The body has two retroperitoneal *ureters* (YOO-re-ters), one for each kidney; each ureter is a continuation of the renal pelvis and runs to the urinary bladder (see Figure 24.1). Urine is carried through the ureters mostly by peristaltic contractions of the muscular layer of the ureters. Each ureter extends 25 to 30 cm (10 to 12 in.). At the base of the urinary bladder, the ureters turn medially and enter the posterior aspect of the urinary bladder.

Histologically, the ureters consist of an inner *mucosa* of transitional epithelium and an underlying lamina propria (connective tissue), a middle *muscularis* (inner longitudinal and outer circular

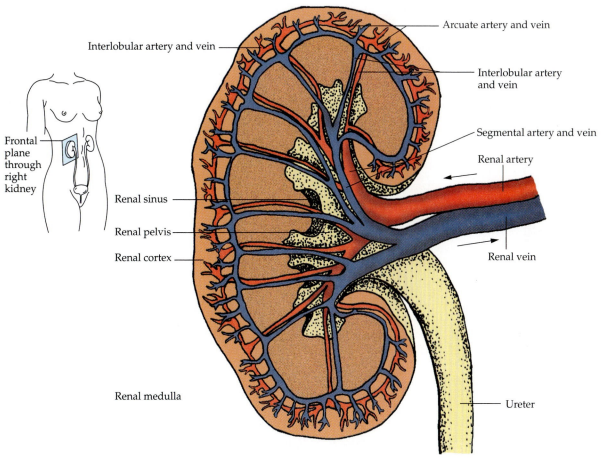

Interlobular artery and vein

Arcuate artery and vein

Interlobular artery and vein

Segmental artery and vein

Renal artery

Frontal plane through right kidney

Renal sinus

Renal pelvis

Renal cortex

Renal vein

Ureter

Renal medulla

Frontal section of right kidney

FIGURE 24.6   Macroscopic blood vessels of the kidney. Macroscopic and microscopic blood vessels are shown in Figure 24.4.

smooth muscle), and an outer *adventitia* (areolar connective tissue).

Examine a prepared slide of the wall of the ureter showing its various layers. With the aid of your textbook, label Figure 24.8.

5. Urinary Bladder

The *urinary bladder* is a hollow muscular organ located in the pelvic cavity posterior to the pubic symphysis (see Figure 24.1). In the male, the bladder is directly anterior to the rectum; in the female, it is anterior to the vagina and inferior to the uterus.

At the base of the interior of the urinary bladder is the *trigone* (TRĪ-gōn; *trigonium* = triangle), a triangular area bounded by the opening into the urethra (internal urethral orifice) and ureteral openings into the bladder. The *mucosa* of the urinary bladder consists of transitional epithelium and lam-

ina propria (connective tissue). Rugae (folds in the mucosa) are also present. The *muscularis,* also called the *detrusor* (de-TROO-ser; *detrudere* = to push down) *muscle,* consists of three layers of smooth muscle: inner longitudinal, middle circular, and outer longitudinal. In the region around the opening to the urethra, the circular muscle fibers form an *internal urethral sphincter.* Inferior to this is the *external urethral sphincter* which is composed of skeletal muscle and is a modification of the urogenital diaphragm. The superficial coat of the urinary bladder is the *adventitia,* a layer of areolar connective tissue that is continuous with that of the ureters. Over the superior surface of the urinary bladder, the adventitia is also covered with peritoneum and the two together constitute the *serosa.*

Urine is expelled from the bladder by an act called *micturition* (mik'-too-RISH-un; *micturire* = to urinate), commonly known as urination, or

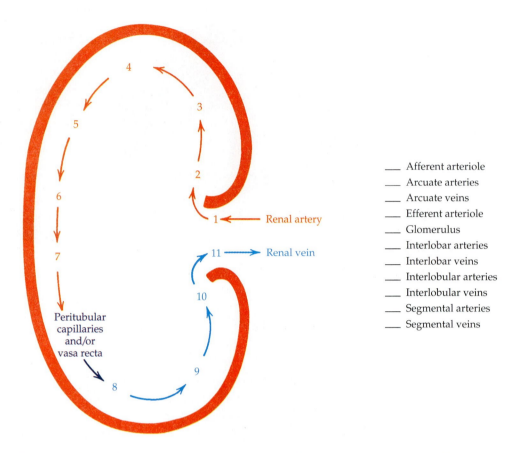

____ Afferent arteriole
____ Arcuate arteries
____ Arcuate veins
____ Efferent arteriole
____ Glomerulus
____ Interlobar arteries
____ Interlobar veins
____ Interlobular arteries
____ Interlobular veins
____ Segmental arteries
____ Segmental veins

FIGURE 24.7   Blood supply of the right kidney. This view is designed to show the *sequence* of blood flow, not the anatomical location of blood vessels, which is shown in Figure 24.6.

voiding. The average capacity of the urinary bladder is 700 to 800 ml.

Using your textbook as a guide, label the external urethral sphincter, internal urethral sphincter, ureteral openings, and ureters in Figure 24.9.

Examine a prepared slide of the wall of the urinary bladder. With the aid of your textbook, label Figure 24.10 on page 481.

6. Urethra

The *urethra* is a small tube leading from the internal urethral orifice in the floor of the urinary bladder to the exterior of the body. In females, this tube is posterior to the pubic symphysis and is embedded in the anterior wall of the vagina; its length is approximately 3.8 cm (1½ in.). The opening of the urethra to the exterior, the ***external urethral orifice,*** is between the clitoris and vaginal orifice. In males, its length is around 20 cm (8 in.), and it follows a route different from that of the female. Immediately inferior to the urinary bladder, the urethra passes

through the prostate gland (prostatic urethra), pierces the urogenital diaphragm (membranous urethra), and traverses the penis (spongy urethra). The urethra is the terminal portion of the urinary system, and serves as the passageway for discharging urine from the body. In addition, in the male, the urethra serves as the duct through which reproductive secretions are discharged from the body.

Label the urethra and urethral orifice in Figure 24.9.

B. DISSECTION OF SHEEP (OR PIG) KIDNEY

The sheep kidney is very similar to both the human and cat kidney. You may use Figure 24.2 as a reference for this dissection.

CAUTION! *Please reread Section D, "Precautions Related to Dissection" at the beginning of the laboratory manual on page xiii before you begin your dissection.*

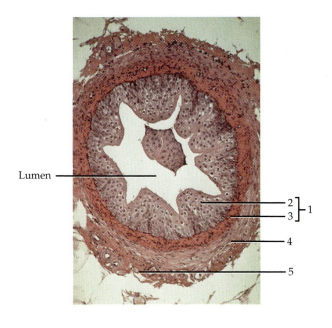

Lumen

2
3 } 1
4
5

___ Adventitia

___ Lamina propria of mucosa

___ Mucosa

___ Muscularis

___ Transitional epithelium of mucosa

FIGURE 24.8   Histology of ureter

PROCEDURE

1. Examine the intact kidney and notice the renal hilus and the fatty tissue that normally surrounds the kidney. Strip away the fat.
2. As you peel the fat off, look carefully for the **adrenal (suprarenal) gland.** This gland is usually found attached to the superior surface of the kidney, as it is in the human. Most preserved kidneys do not have this gland. If it is present, remove it, cut it in half, and note its distinct outer **cortex** and inner **medulla.**
3. Look at the **renal hilus,** which is the concave area of the kidney. From here the **ureter, renal artery,** and **renal vein** enter and exit.
4. Differentiate these blood vessels by examining the thickness of their walls. Which vessel has the thicker wall?
5. With a sharp scalpel *carefully* make a frontal section through the kidney.
6. Identify the **renal capsule** as a thin, tough layer of connective tissue completely surrounding the kidney.
7. Immediately beneath this capsule is an outer light-colored area called the **renal cortex.** The inner dark-colored area is the **renal medulla.**

8. The **renal pelvis** is the large chamber formed by the expansion of the ureter inside the kidney. This renal pelvis divides into many smaller areas called **renal calyces,** each of which has a dark tuft of kidney tissue called a **renal pyramid.**
9. The bases of these pyramids face the cortical area. Their apices, called **renal papillae,** are directed toward the center of the kidney.
10. The calyces collect urine from collecting ducts and drain it into the renal pelvis and out through the ureter.
11. The renal artery divides into several branches that pass between the renal pyramids. These vessels are small and delicate and may be too difficult to dissect and trace through the renal medulla.

C. RENAL PHYSIOLOGY EXPERIMENTS

PROCEDURE USING PHYSIOFINDER

PHYSIOFINDER is an interactive computer program that permits laboratory simulations which do not involve the use of animals or of advanced or expensive laboratory equipment. It permits students to perform experiments, analyze data, draw conclusions, and form hypotheses on the basis of collected data. The program is available from Addison-Wesley Longman, Inc., Customer Service (1-800-922-0579).

For activities related to renal physiology, select the appropriate experiments from PHYSIOFINDER Module 7—Renal Water Regulation, Module 8—Renal Sodium Regulation, and Module 9—Renal Hydrogen Ion Regulation.

D. URINE

The kidneys function to maintain bodily homeostasis. This is accomplished by three processes: (1) filtration of the blood by the glomeruli, (2) tubular reabsorption, and (3) tubular secretion. As a result of these three functions, **urine** is formed and eliminated from the body. Urine contains a high concentration of solutes, and in a healthy person, the volume, pH, and solute concentration of urine will vary with the needs of the internal environment. In certain pathological conditions, the characteristics of urine may change drastically. An analysis of the volume and the physical and chemical properties of urine tells us much about the state of the body.

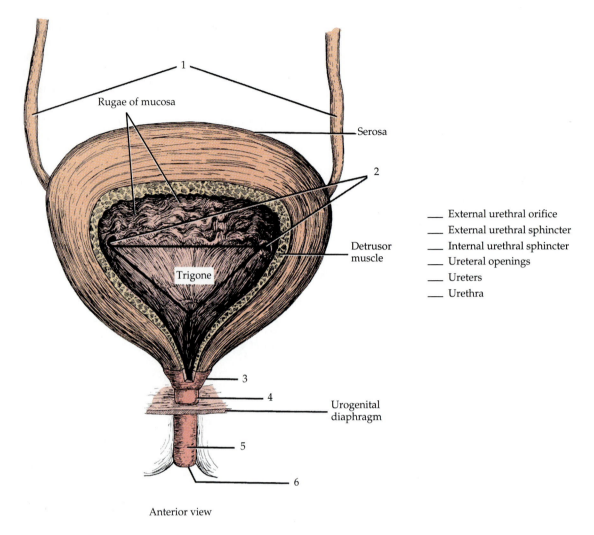

Rugae of mucosa

Serosa

Detrusor
muscle

Trigone

___ External urethral orifice
___ External urethral sphincter
___ Internal urethral sphincter
___ Ureteral openings
___ Ureters
___ Urethra

Urogenital
diaphragm

Anterior view

FIGURE 24.9   Urinary bladder and female urethra.

1. Physical Characteristics

Normal urine usually varies between a straw yellow and an amber transparent color, and possesses a characteristic odor. Urine color varies considerably according to the ratio of solutes to water and according to an individual's diet.

Cloudy urine sometimes reflects the secretion of mucin from the urinary tract lining and is not necessarily an indicator of a pathological condition. The normal pH of urine ranges between 4.6 and 8.0 and averages 6.0. The pH of urine is also strongly affected by diet, with a high-protein diet lowering the pH, and a mostly vegetable diet increasing the pH of the urine.

Specific gravity is the ratio of the weight of a volume of a substance to the weight of an equal volume of distilled water. Water has a specific gravity of 1.000. The specific gravity of urine depends on the amount of solids in solution, and normally ranges from 1.001 to 1.035. The greater the concentration of solutes, the higher the specific gravity. In certain conditions, such as diabetes mellitus, specific gravity is high because of the high glucose content.

2. Abnormal Constituents

When the body's metabolism becomes abnormal, many substances not normally found in urine may appear in varying amounts, while normal constituents may appear in abnormal amounts. *Urinalysis* is the analysis of the physical and chemical properties of urine, and is a vital tool in diagnosing pathological conditions.

  a. *Albumin* Normally absent from a urine sample because the molecules are too large to be filtered out of the blood via the endothelial-capsular membrane. When albumin is found in the urine, the condition is called *albuminuria*.

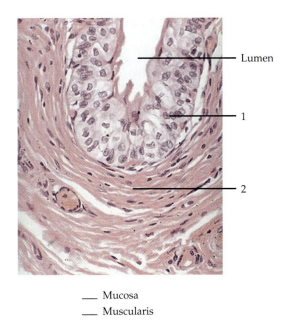

___ Lumen

___ 1

___ 2

___ Mucosa
___ Muscularis

FIGURE 24.10   Histology of urinary bladder (400×).

b. *Glucose* Urine normally contains such small amounts of *glucose* that clinically glucose is considered to be absent from urine samples. Its presence in significant amounts is called *glucosuria*, and the most common cause is a high blood sugar (glucose) level seen in certain diseases such as diabetes mellitus.

c. *Erythrocytes* **Hematuria** is the term utilized to describe the presence of red blood cells in a urine sample. Hematuria usually indicates the presence of a pathological condition within the kidneys.

d. *Leucocytes* **Pyuria** (pī-YOO-rē-a) is the condition that occurs when white blood cells and other components of pus are found in the urine, and this usually indicates a pathological condition.

e. *Ketone bodies* Normal urine contains a small amount of *ketone (acetone) bodies.* Their appearance in large quantities in the urine produces the condition called *ketosis (acetonuria)* and may indicate physiological abnormalities.

f. *Casts* Tiny masses of various substances that have hardened and assumed the shape of the lumens of the nephron tubules. They are microscopic in size and are composed of many different substances.

g. *Calculi* Insoluble *calculi* (stones) are various salts that have solidified in the urinary tract. They may be found anywhere from the kidney tubules to the external opening, and their presence causes considerable pain as they attempt to pass through the various lumens of the urinary system.

E. URINALYSIS

In this exercise you will determine some of the characteristics of urine and perform tests for some abnormal constituents that may be present in urine. Some of these tests may be used in determining unknowns in urine specimens.

CAUTION! *Please reread Section A, "General Safety Precautions and Procedures" on page xi, and Section C, "Precautions Related to Working with Reagents" on page xii at the beginning of the laboratory manual before you begin any of the following experiments. Read the experiments before you perform them, to be sure that you understand all the procedures and safety precautions.*

When working with urine, avoid any kind of contact with an open sore, cut, or wound. Wear tight-fitting surgical gloves and safety goggles.

Work with your own urine only. After you have completed your experiments, place all glassware in a fresh household bleach solution or other comparable disinfectant, wash the laboratory tabletop with a fresh household bleach solution or comparable disinfectant, and dispose of the gloves and Chemstrips® in the appropriate biohazard container provided by your instructor.

1. Urine Collection
PROCEDURE

1. A specimen of urine may be collected at any time for routine tests; urine voided within 3 hr after meals, however, may contain abnormal constituents. For this reason the first voiding in the morning is preferred.

2. Both males and females should collect a midstream sample of urine in a sterile container. A midstream sample is essential to avoid contamination from the external genitalia, and to avoid the presence of pus cells and bacteria that are normally found in the urethra.

3. If not examined immediately, the specimen should be refrigerated to prevent unnecessary bacterial growth.

4. Before testing *always* mix urine by swirling, inverting the container, or stirring with a wooden swab stick.

5. *Keep all containers clean!*

6. Wrap all papers and sticks and put them in the garbage pail.

7. Rinse all test tubes and glass containers carefully with *cold water* after they have cooled.

8. Flush sinks well with cold water.

9. Obtain either your own freshly voided urine sample or a provided sample if one is available.

2. Chemstrip® Testing

Alternate methods can be employed for several of the tests you are about to perform. One alternate method is the use of plastic strips to which are attached paper squares impregnated with various reagents. These strips display a color reaction when dipped into urine with any abnormal constituents.

PROCEDURE

1. At this point, you should take a Chemstrip® and test your urine sample for the following: pH, protein, glucose, ketones, bilirubin, and blood (hemoglobin). Record your results below.
2. Determine which component(s) of urine are measured by the Chemstrip® you are using. If you are using a strip that tests for multiple substances, determine which squares measure which substances. Locate the color chart used to read the results. Note the appropriate time for reading each test.
3. Remove a test strip from the vial and *replace the cap.* Dip the test strip into your urine sample for no longer than 1 sec, being sure that all reagents on the strip are immersed.
4. Remove any excess urine from the strip by drawing the edge of the strip along the rim of the container holding your urine sample.
5. After the appropriate time, as indicated on the Chemstrip® vial, hold the strip close to the color blocks on the vial.
6. Make sure that the strip blocks are properly lined up with the color chart on the vial.

Test	Chemstrip® result
pH	_____
Protein	_____
Glucose	_____
Ketones	_____
Bilirubin	_____
Blood	_____

3. Physical Analysis

a. COLOR

Normal urine varies in color from straw yellow to amber because of the pigment *urochrome*, a byproduct of hemoglobin destruction. Observe the color of your urine sample. Some abnormal colors are as follows:

Color	Possible Cause
Silvery, milky	Pus, bacteria, epithelial cells
Smoky brown, rust	Blood
Orange, green, blue, red	Medications or liver disease

Record the color of your urine in Section E.3.a of the LABORATORY REPORT RESULTS at the end of the exercise.

How would sickle-cell anemia affect urine color?

_____

_____

b. TRANSPARENCY

A fresh urine sample should be clear. Cloudy urine may be due to substances such as mucin, phosphates, urates, fat, pus, mucus, microbes, crystals, and epithelial cells.

PROCEDURE

1. To determine transparency, cover the container and shake your urine sample and observe the degree of cloudiness.
2. Record your observations in Section E.3.b of the LABORATORY REPORT RESULTS at the end of the exercise.

c. pH

The pH of urine varies with several factors already indicated. You can test the pH of your urine by using either a Chemstrip® or pH paper. Because you have already determined the pH of your urine by using a Chemstrip®, you might want to verify the results using pH paper.

PROCEDURE

1. Place a strip of pH paper into your urine sample three consecutive times.
2. Shake off any excess urine.
3. Let the pH paper sit for 1 min and then compare it to the color chart provided.
4. Record your observations in Section E.3.c of the LABORATORY REPORT RESULTS at the end of the exercise.

How would the consumption of antacid (sodium bicarbonate) affect urine pH? _____

_____

_____

d. SPECIFIC GRAVITY

The specific gravity is easily determined using a urinometer (hydrometer). The urinometer is a float with a numbered scale near the top that indicates specific gravity directly.

PROCEDURE

1. Familiarize yourself with the scale on the urinometer neck. Determine the change in specific gravity represented by each calibration.
2. Allow your urine to reach room temperature. Urinometers are calibrated to read the specific gravity at 15°C (69°F). If the temperature differs, add or subtract 0.001 for each 3°C above or below 15°C.
3. Fill the cylinder ¾ full of urine and insert the urinometer. Make sure that it is free-floating; if not, spin the neck gently.
4. Read the scale at the bottom of the meniscus when the urinometer is at rest.
5. Record the specific gravity in Section E.3.d of the LABORATORY REPORT RESULTS at the end of the exercise.
6. Rinse the urinometer and cylinder. *Follow your instructor's directions for cleaning them.*
   How would dehydration affect the specific gravity of urine? _____

_____

_____

4. Chemical Analysis

a. CHLORIDE AND SODIUM CHLORIDE

Most of the sodium chloride (NaCl) present in the renal filtrate is reabsorbed; a small amount re-mains as a normal component of urine. The normal value for chloride (Cl^-) is 476 mg/100 ml; for sodium (Na^+) it is 294 mg/100 ml.

PROCEDURE

1. Using a medicine dropper, place 10 drops of urine into a Pyrex test tube.
2. Using a medicine dropper, add 1 drop of 20% potassium chromate and *gently* agitate the tube. It should be a yellow color.

3. Using a medicine dropper, add 2.9% silver nitrate solution *one drop at a time*, counting the drops and *gently* agitating the test tube during the time the solution is being added.
4. Count the number of drops needed to change the color of the solution from bright yellow to brown.
5. Determine the chloride and sodium chloride concentrations of the sample and record your results in Section E.4.a of the LABORATORY REPORT RESULTS at the end of the exercise. Each drop of silver nitrate added in step 4 is equivalent to (1) 61 mg of Cl^- per 100 ml of urine, and (2) 100 mg of NaCl per 100 ml of urine.

Thus, to determine the chloride concentration of the sample, multiply the number of drops times 61 to obtain the number of mg of Cl^- per 100 ml of urine.

To determine the sodium chloride concentration of the sample, multiply the number of drops times 100 to obtain the number of mg of NaCl per 100 ml of urine.

What would a high NaCl content in the urine indicate? _____

_____

_____

b. GLUCOSE

Glucosuria (the presence of glucose in the urine) occurs in patients with diabetes mellitus or other disorders. Traces of glucose may occur in normal urine, but detection of these small amounts requires special tests.

(1) Benedict's Test Benedict's solution is commonly used to detect reducing sugars in urine and is not specific for just glucose.

PROCEDURE

1. In a Pyrex test tube, combine 10 drops of urine with 5 ml of Benedict's solution. Mix the solution.
2. Using a test tube holder, place the test tube in a boiling water bath for 5 min.
   CAUTION! *Make sure that the mouth of the test tube is pointed away from you and all other persons in the area.*
3. *Using a test tube holder*, remove the test tube from the water bath and read the results according to the following chart.

4. Record your results in Section E.4.b of the LAB-
ORATORY REPORT RESULTS at the end of the
exercise.

Color	Results
Blue	Negative
Greenish yellow	1 + (0.5 g/100 ml)
Olive green	2 + (1 g/100 ml)
Orange-yellow	3 + (1.5 g/100 ml)
Brick red (with precipitate)	4 + (> 2 g/100 ml)

(2) Clinitest® Reagent Method

PROCEDURE

1. Using medicine droppers, place 10 drops of
water and 5 drops of urine in a Pyrex test tube.

CAUTION! *Place the test tube in a test tube rack
because it will become too hot to handle.*

2. With forceps, add one Clinitest® tablet.

CAUTION! *The concentrated sodium hydroxide in
the tablet generates enough heat to make the liquid in
the test tube boil. Make sure that the mouth of the test
tube is pointed away from you and all other persons in
the area.*

3. The color of the solution is graded as in Benedict's test.
4. Fifteen seconds after boiling has stopped, shake
the test tube *gently* and evaluate the color according to the Benedict's test table above. Dis-
regard any color change that occurs after 15 sec.

CAUTION! *Make sure the mouth of the test tube is
pointed away from you and others.*

5. Record your results in Section E.4.b of the LAB-
ORATORY REPORT RESULTS at the end of the
exercise.
How would a high-sugar diet affect the urine?

Explain. _____

_____

_____

(3) Chemstrip® Method   Record your results in
Section E.4.b of the LABORATORY REPORT RE-
SULTS at the end of the exercise.

c. PROTEIN

Normal urine contains traces of proteins that are
hard to detect through regular laboratory procedures. Albumin is the most abundant serum protein and is the one usually detected. Because tests
for albumin are determined by precipitating the
protein either by heat (coagulation) or by adding a
reagent, the urine sample should either be filtered
or centrifuged. The test for protein will be done by
the Albutest® reagent method and the Chemstrip®
method.

(1) Albutest® Reagent Method

PROCEDURE

1. Using forceps, place an Albutest® tablet on a
clean, dry paper towel and add one drop of
urine.
2. After the drop has been absorbed, add two
drops of water and allow these to penetrate before reading.
3. Compare the color (in daylight or fluorescent
light) on top of the tablet with the color chart
provided in lab.
4. If albumin is present in the urine, a *blue-green*
spot will remain on the surface of the tablet
after the water is added. The amount of protein
is indicated by the intensity of the blue-green
color.
5. If the test is negative, the original color of the
tablet will not be changed at the completion of
the test.
6. Record your results in Section E.4.c of the LAB-
ORATORY REPORT RESULTS at the end of the
exercise.
Why is protein not normally found in urine?

_____

_____

(2) Chemstrip® Method   Record your results in
Section E.4.c of the LABORATORY REPORT RE-
SULTS at the end of the exercise.

d. KETONE (ACETONE) BODIES

The presence of ketone (acetone) bodies in urine is
a result of abnormal fat catabolism. Reagents such
as sodium nitroprusside, ammonium sulfate, and
ammonium hydroxide are available in the form of
tablets. Ketones turn purple when added to these
chemicals.

(1) Acetest® Tablet Method

PROCEDURE

1. Using forceps, place an Acetest® tablet on a clean, dry paper towel, and, using a medicine dropper, place one drop of urine on the tablet.
2. If acetone or ketone is present, a *lavender-purple* color develops within 30 sec. If the tablet becomes cream-colored from wetting, the results are negative. Compare results with the color chart that comes with the reagent.
3. Record your results in Section E.4.d of the LABORATORY REPORT RESULTS at the end of the exercise.

Why would starvation cause ketones? _____

_____

_____

(2) Chemstrip® Method   Record your results in Section E.4.d of the LABORATORY REPORT RESULTS at the end of the exercise.

e. BILE PIGMENTS

Bile pigments, biliverdin and bilirubin, are not normally present in urine. The presence of large quantities of bilirubin in the extracellular fluids produces jaundice, a yellowish tint to the body tissues, including yellowness of the skin and deep tissues.

(1) Shaken Tube Test for Bile Pigments

PROCEDURE

1. Fill a Pyrex test tube halfway with urine, stopper the tube, and shake it vigorously, being careful to keep the stopper in the tube.
2. A yellow color of foam indicates the presence of bile pigments.
3. Record your results in Section E.2.e of the LABORATORY REPORT RESULTS at the end of the exercise.

(2) Ictotest® for Bilirubin

PROCEDURE

1. Using a medicine dropper, place a drop of urine on one square of the special mat provided in the Ictotest® kit.
2. Using forceps, place one Ictotest® reagent tablet in the center of the moistened area.

CAUTION! *Do not touch the tablets with your fingers. Recap the bottle.*

3. Add one drop of water directly to the tablet; after 5 sec, add another drop of water to the tablet so that the water runs off onto the mat. Observe the color of the mat around the tablet at 60 sec. The presence of bilirubin will turn the mat *blue* or *purple*. A slight *pink* or *red* color is negative for bilirubin.
4. Record your results in Section E.2.e of the LABORATORY REPORT RESULTS at the end of the exercise.

(3) Chemstrip® Method   Record your results in Section E.2.e of the LABORATORY REPORT RESULTS at the end of the exercise.

f. HEMOGLOBIN

Hemoglobin is not normally found in urine.

(1) Chemstrip® Method   Record your results in Section E.4.f of the LABORATORY REPORT RESULTS at the end of the exercise.

5. Microscopic Analysis

Before you actually examine the sediment of urine microscopically, it will be helpful to read the following description of some of the major components of urinary sediment. For purposes of discussion, the constituents of sediment will be classified into three major groups: (1) cells, (2) casts, and (3) crystals.

Cells in urinary sediment are derived from the lining of the urinary tract as a result of normal wear and tear inflammations, and various disease processes. Using Figure 24.11a as a guide, see how many cells you can identify in your urine sample:

a. **Transitional (urothelial) cells** Cells derived from the transitional epithelium that lines the renal pelvis, ureter, and urinary bladder.
b. **Squamous epthelial cells** Cells derived from the distal portion of the urethra or inflamed areas of the urinary bladder.
c. **Red blood cells** Usually not associated with normal urine. Their prescence in urine (hematuria) may be associated with a variety of diseases.
d. **White blood cells** More than a few indicates pyuria (the presence of white blood cells and other components of pus); indicates an infection.

Casts are hardened masses of material that assume the shape of the lumen of a renal tubule in which they form. They are formed by the precipitation of proteins and agglutination of cells. Casts form as a result of protein in urine passing through renal tubules, highly acidic urine, and highly concentrated urine. Casts are named after the cells or substances that compose them or on the basis of their appearance. Using Figure 24.11b as a guide, see how many casts you can identify in your urine sample:

a. **Hyaline casts** Consist mostly of a microprotein derived from renal tubule epithelial cells.
b. **Epithelial casts** Consist largely of tubular epithelial cells.
c. **Granular casts** Considered one phase in the breakdown of cellular casts. Depending on the degree of breakdown, they may be classified as coarsely granular and finely granular.
d. **Waxy casts** Considered the end stage of the breakdown of cellular casts.
e. **White blood cell casts** Consist of aggregates of white blood cells.
f. **Red blood cell casts** Consist of aggregates of red blood cells.

Crystals form from the end product of tissue metabolism and consumption of excessive amounts of certain foods and drugs. Identification of crystals is based on their shape. Using Figure 24.11c as a guide, see how many crystals you can identify in your urine sample:

a. **Calcium oxalate** Range in shape from oval to dumbbell to octahedron (8-sided) to dodecahedron (12-sided); latter two shapes appear as "envelopes."
b. **Uric acid** Yellow or red-brown rhombic prisms, hexagonal or square plates, or spheres.
c. **Triple phosphate** Appear as six- to eight-sided prisms ("coffin lids") or feathery forms.

If you allow a urine specimen to stand undisturbed for a few hours, many suspended materials will settle to the bottom. A much faster method is to centrifuge a urine sample.

PROCEDURE

1. Place 5 ml of fresh urine in a centrifuge tube. Follow the directions of your instructor to centrifuge the tube for 5 min at a slow speed (1500 revolutions per minute [rpm]).
2. Dispose of the supernatant (the clear urine) and mix the sediment by shaking the test tube.
3. Using a long medicine dropper or a Pasteur pipette with a bulb, place a small drop of sediment on a clean glass slide, add one drop of

Sedi-stain® or methylene blue, and place a cover glass over the specimen.
4. Using low power and reduced light, examine the sediment for any of the microscopic elements pictured in Figure 24.11.
5. Increase the light and use high power or oil immersion to look for crystals.

Draw the results of your observation in Section E.5 of the LABORATORY REPORT RESULTS at the end of the exercise.

Why might some red blood cells in urinary sediment be crenated? _____

_____

Why might yeast be seen in the urine of a person with diabetes mellitus? _____

_____

6. Unknown Specimens

a. UNKNOWNS PREPARED BY INSTRUCTOR

When the composition of a substance has not been defined, it is called an **unknown**. In this exercise, the unknowns will be urine specimens to which the instructor has added glucose, albumin, or any other detectable substance. Each unknown contains only one added substance. Perform the previously outlined tests until you identify the substance in your unknown.

Record your results in Section E.6.a of the LABORATORY REPORT RESULTS at the end of the exercise.

b. UNKNOWNS PREPARED BY CLASS (OPTIONAL)

The class should be divided into two groups. Each group adds certain substances, such as glucose, protein, starch, or fat, to normal, freshly voided urine, keeping accurate records as to what was added to each sample. The two groups then exchange samples, and each group does the basic chemical tests on urine to detect which substances were added. Each student should add one substance to one urine sample and see if another student can detect what was added. Record your results in Section E.6.b of the LABORATORY REPORT RESULTS at the end of the exercise.

ANSWER THE LABORATORY REPORT QUESTIONS AT THE END OF THE EXERCISE.

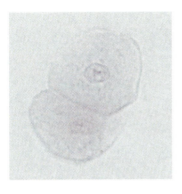

Transitional epithelial cells (400×)

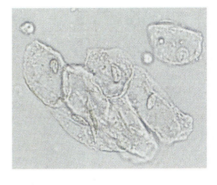

Squamous epithelial cells (400×)

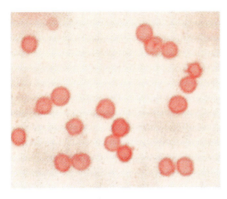

Red blood cells (400×)

Neutrophils (400×)

(a) Cells

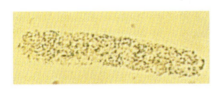

Hyaline cast (400×)

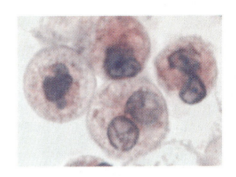

Epithelial cell cast (400×)

Finely granular cast (400×)

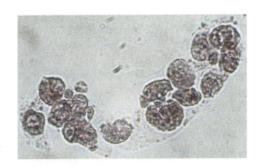

Waxy cast (400×)

(b) Casts

FIGURE 24.11   Photographs of some microscopic components of urinary sediment.

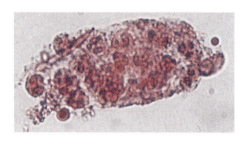

White blood cell cast (400×)

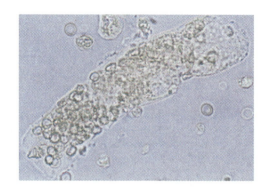

Red blood cell cast (400×)

(b) Casts (continued)

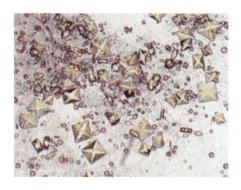

Calcium oxalate crystals (400×)

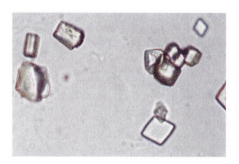

Uric acid crystals (400×)

Triple phosphate crystals (400×)

(c) Crystals

FIGURE 24.11 *(Continued)*

Urinary System 24

Student _____   Date _____

Laboratory Section _____   Score/Grade _____

SECTION E. URINALYSIS

3. Physical Analysis

Characteristic	Normal	Your sample
a. Color	Straw yellow to amber	_____
b. Sediment	None	_____
c. pH	5.0-7.8	_____
d. Specific gravity	1.008-1.030	_____

4. Chemical Analysis

Record your results in the spaces provided.

a. CHLORIDE AND SODIUM CHLORIDE

Chloride concentration _____ mg/100 ml

Sodium chloride concentration _____ mg/100 ml

b. GLUCOSE

Benedict's test _____

Clinitest® tablet _____

Chemstrip® _____

c. PROTEIN

Albutest® reagent tablets _____

Chemstrip® _____

d. KETONE (ACETONE) BODIES

Acetest® tablet _____

Chemstrip® _____

e. BILE PIGMENTS

Shaken tube (bile pigments) _____

Ictotest® (bilirubin) _____

Chemstrip® _____

f. HEMOGLOBIN

Chemstrip® _____

5. Microscopic Analysis

Draw some of the substances (types of cells, types of crystals, or other elements) that you found in the microscopic examination of urinary sediment.

6. Unknown Specimens

a. What substance did you find in the unknown specimen that was prepared by your instructor?

_____

b. What substance did you find in the unknown specimen that was prepared by other students?

_____

Urinary System 24

Student _____   Date _____

Laboratory Section _____   Score/Grade _____

PART 1. Multiple Choice

_____ 1. Beginning at the deepest layer and moving toward the superficial layer, identify the order of tissue layers surrounding the kidney. (a) renal capsule, renal fascia, adipose capsule (b) renal fascia, adipose capsule, renal capsule (c) adipose capsule, renal capsule, renal fascia (d) renal capsule, adipose capsule, renal fascia

_____ 2. The functional unit of the kidney is the (a) nephron (b) ureter (c) urethra (d) hilus

_____ 3. Substances filtered by the kidney must pass through the endothelial-capsular membrane, which is composed of several parts. Which of the following choices lists the correct order of the parts as substances pass through the membrane? (a) epithelium of the visceral layer of glomerular (Bowman's) capsule, endothelium of the glomerulus, basement membrane of the glomerulus (b) endothelium of the glomerulus, basement membrane of the glomerulus, epithelium of the visceral layer of glomerular (Bowman's) capsule (c) basement membrane of the glomerulus, endothelium of the glomerulus, epithelium of the visceral layer of the glomerular (Bowman's) capsule (d) epithelium of the visceral layer of the glomerular (Bowman's) capsule, basement membrane of the glomerulus, endothelium of the glomerulus

_____ 4. In the glomerular (Bowman's) capsule, the afferent arteriole divides into a capillary network called a(n) (a) glomerulus (b) interlobular artery (c) peritubular capillary (d) efferent arteriole

_____ 5. Transport of urine from the renal pelvis into the urinary bladder is the function of the (a) urethra (b) calculi (c) casts (d) ureters

_____ 6. The terminal portion of the urinary system is the (a) urethra (b) urinary bladder (c) ureter (d) nephron

_____ 7. Damage to the renal medulla would interfere first with the functioning of which parts of a juxtamedullary nephron? (a) glomerular (Bowman's) capsule (b) distal convoluted tubule (c) collecting duct (d) proximal convoluted tubule

_____ 8. An obstruction in the glomerulus would affect the flow of blood into the (a) renal artery (b) efferent arteriole (c) afferent arteriole (d) intralobular artery

_____ 9. Urine that leaves the distal convoluted tubule passes through the following structures in which sequence? (a) collecting duct, hilus, calyces, ureter (b) collecting duct, calyces, pelvis, ureter (c) calyces, collecting duct, pelvis, ureter (d) calyces, hilus, pelvis, ureter

_____ 10. The position of the kidneys posterior to the peritoneal lining of the abdominal cavity is described by the term (a) retroperitoneal (b) anteroperitoneal (c) ptosis (d) inferoperitoneal

_____ 11. Of the following structures, the one to receive filtrate *last* as it passes through the nephron is the (a) proximal convoluted tubule (b) ascending limb of the loop of Henle (c) glomerulus (d) collecting duct

_____ 12. Peristalsis of the ureter is a function of the (a) serosa (b) mucosa (c) submucosa (d) muscularis

_____ 13. The trigone and the detrusor muscle are associated with the (a) kidney (b) urinary bladder (c) urethra (d) ureters

_____ 14. The notch on the medial surface of the kidney through which blood vessels enter and exit is called the (a) renal medulla (b) major calyx (c) renal hilus (d) renal column

_____ 15. Blood is drained from the kidneys by the (a) renal arteries (b) interlobar arteries (c) interlobular veins (d) renal veins

_____ 16. The epithelium of the urinary bladder that permits distension is (a) stratified squamous (b) transitional (c) simple squamous (d) pseudostratified

_____ 17. How many times a day is the entire volume of blood in the body filtered by the kidneys? (a) 100 times (b) 5 times (c) 30 times (d) 60 times

_____ 18. The average urine capacity of the urinary bladder is (a) 1000 to 1200 ml (b) 50 to 100 ml (c) 700 to 800 ml (d) 200 to 300 ml

_____ 19. The normal pH of urine is between (a) 4.6 and 8.0 (b) 2.0 and 4.8 (c) 10.0 and 12.0 (d) none of the above

_____ 20. Normal urine has a specific gravity of approximately (a) 1.001 to 1.035 (b) 1.030 to 1.080 (c) 1.100 to 1.200 (d) none of the above

_____ 21. The special chemical that may be used to detect glucose in the urine is (a) sulfosalicylic acid (b) Benedict's solution (c) Lugol's solution (d) nitric acid

PART 2. Completion

22. In addition to the urinary system, other systems that help eliminate wastes are the respiratory, integumentary, and _____ systems.

23. The double-walled cup found in a nephron is called a(n) _____.

24. The special capillary network found inside of this double-walled cup is the _____.

25. The major blood vessel that enters each kidney is the _____.

26. The nerve supply to the kidneys comes from the autonomic nervous system and is called the

_____.

27. Urine is expelled from the urinary bladder by an act called urination, voiding, or

_____.

28. The small tube in the urinary system that leads from the floor of the urinary bladder to the outside is

the _____.

29. The apices of renal pyramids are referred to as renal _____.

30. The cortical substance between renal pyramids is called a renal _____.

31. Cuplike extensions of the renal pelvis, usually two or three in number, are referred to as

_____.

32. Epithelial cells of the visceral layer of the glomerular (Bowman's) capsule are called

_____ .

33. Distal convoluted tubules terminate by merging with _____ .

34. Long loops of blood vessels around the medullary structures of juxtamedullary nephrons are called

_____ .

35. Which blood vessel comes next in this sequence? Interlobar artery, arcuate artery, interlobular artery,

_____ .

36. The abnormal condition in which red blood cells are found in the urine in appreciable amounts is

called _____ .

37. Various salts that solidify in the urinary tract are called _____ .

38. Various substances that have hardened and assumed the shape of the lumens of the nephron tubules

are the _____ .

39. The pH of urine in individuals on high-protein diets tends to be _____ than

normal.

40. The greater the concentration of solutes in urine, the greater will be its _____ .

pH and Acid-Base Balance

25

A. THE CONCEPT OF pH

When molecules of inorganic acids, bases, or salts dissolve in water, they undergo *ionization* (ī'-on-i-ZĀ-shun), or *dissociation* (dis'-sō-sē-Ā-shun); that is, they separate into ions.

An *acid* can be defined as a substance that dissociates into one or more *hydrogen ions (H+)* and one or more *anions* (negative ions). Because H^+ is a single proton with a charge of +1, an acid can also be defined as a proton donor. A *base*, by contrast, dissociates into one or more *hydroxide ions (OH⁻)* and one or more *cations* (positive ions). A base can also be viewed as a proton acceptor. Hydroxide ions have a strong attraction for protons. A *salt*, when dissolved in water, dissociates into cations and anions, neither of which is H^+ or OH^-. Acids and bases react with one another to form salts. Body fluids must constantly contain balanced quantities of acids and bases. In solutions such as those found inside or outside body cells, acids dissociate into hydrogen ions (H^+) and anions. Bases, on the other hand, dissociate into hydroxide ions (OH^-) cations. The more hydrogen ions that exist in a solution, the more acidic the solution; conversely, the more hydroxide ions, the more basic (alkaline) the solution.

Biochemical reactions—those that occur in living systems—are very sensitive to even small changes in acidity or alkalinity. Any departure from the narrow limits of normal H^+ and OH^- concentrations may greatly modify cell functions and disrupt homeostasis. For this reason, the acids and bases that are constantly formed in the body must be kept in balance.

A solution's acidity or alkalinity is expressed on the *pH scale*, which runs from 0 to 14. This scale is based on the concentration of H^+ in a solution. The midpoint of the scale is 7, where the concentrations of H^+ and OH^- are equal. A substance with a pH of 7, such as distilled (pure) water, is neutral. A solution that has more H^+ than OH^- is an *acidic solu-tion* and has a pH below 7. A solution that has more OH^- than H^+ is a *basic (alkaline) solution* and has a pH above 7. A change of one whole number on the pH scale represents a 10-fold change from the previous concentration. A pH of 1 denotes 10 times more H^+ than a pH of 2. A pH of 3 indicates 10 times fewer H^+ than a pH of 2 and 100 times fewer H^+ than a pH of 1.

B. MEASURING pH

1. Using Litmus Paper

PROCEDURE

CAUTION! *Please reread Section A, "General Safety Precautions and Procedures," on page xi, and Section C, "Precautions Related to Working with Reagents," on page xii at the beginning of the laboratory manual, before you begin any of the following experiments. Read the experiments before you perform them, to be sure that you understand all the procedures and safety precautions.*

1. Before you begin, it is important to know that *an acid solution will turn blue litmus paper red* and *a basic (alkaline) solution will turn red litmus paper blue.*
2. Using forceps, dip a strip of red litmus paper into each of the solutions to be tested. Use a *new strip* for each solution. Record your observations in the Table in Section B.1 of the LABORATORY REPORT RESULTS at the end of the exercise.
3. Using forceps, dip a strip of blue litmus paper into each of the solutions to be tested. Use a *new strip* for each solution. Record your observations in the Table in Section B.1 of the LABORATORY REPORT RESULTS at the end of the exercise.
4. Using your textbook as a guide, determine the pH of the following body fluids:

Body Fluid	pH
Bile	
Saliva	
Gastric juice	
Blood	
Pancreatic juice	
Semen	
Urine	

2. Using pH Paper

PROCEDURE

CAUTION! *Please reread Section A, "General Safety Precautions and Procedures," on page xi, and Section C, "Precautions Related to Working with Reagents," on page xii, at the beginning of the laboratory manual, before you begin any of the following experiments. Read the experiments before you perform them, to be sure that you understand all the procedures and safety precautions.*

1. Using forceps, dip a strip of pH paper into each of the solutions to be tested. Use a *new strip* for each solution. Use wide-range pH paper to determine to approximate pH and narrow-range pH paper to determine a more accurate pH.
2. Compare the color of the strip of pH paper to the color chart on the pH paper container.
3. Record your observations in the Table in Section B.2 of the LABORATORY REPORT RESULTS at the end of the exercise.

3. Using a pH Meter

Because there are different types of pH meters, your instructor will demonstrate how to use the pH meter in your laboratory.

PROCEDURE

CAUTION! *Please reread Section A, "General Safety Precautions and Procedures," on page xi, and Section C, "Precautions Related to Working with Reagents," on page xii, at the beginning of the laboratory manual, before you begin any of the following experiments. Read the experiments before you perform them, to be sure that you understand all the procedures and safety precautions.*

1. Examine the pH meter that has been made available to you and identify the following parts: (1) electrodes, (2) pH dial or digital display, (3) temperature control, and (4) calibration control.
2. Plug in the pH meter and turn it on. (*NOTE: Some models take up to one-half hour to warm up.*)
3. Using the temperature control knob, adjust the pH meter for the temperature of the solutions to be tested.
4. To calibrate the pH meter, place the electrode(s) in a beaker that contains a pH 7 buffer solution. The electrode(s) should be immersed at least 1 in. into the solution. Adjust the pH meter with the appropriate controls so that the meter will show a pH value of 7. Now the instrument is calibrated.
5. Depress the standby button and remove the electrode(s) from the buffer solution. *The electrode(s) should not touch anything.* Rinse the electrode(s) with distilled water, using a wash bottle. The rinse water can be collected in an empty beaker.
6. Immerse the electrode(s) into the first solution to be tested. Release the standby button and note the pH. Record the pH in the Table in Section B.3 of the LABORATORY REPORT RESULTS at the end of the exercise.
7. Depress the standby button, remove the electrode(s) from the solution, and rinse with distilled water. Test the pH of the remaining solutions and record each pH in the Table in Section B.3 of the LABORATORY REPORT RESULTS at the end of the exercise.

C. ACID-BASE BALANCE

A very important electrolyte in terms of the body's acid-base balance is the hydrogen ion (H^+). Although some hydrogen ions enter the body in ingested foods, most are produced as a result of the cellular metabolism of substances such as glucose, fatty acids, and amino acids. One of the major challenges to homeostasis is keeping the H^+ concentration at an appropriate level to maintain proper acid-base balance.

The balance of acids and bases is maintained by controlling the H^+ concentration of body fluids, particularly extracellular fluid. In a healthy person, the pH of the extracellular fluid remains between 7.35 and 7.45. Metabolism typically produces a significant excess of H^+. If there were no mechanisms for disposal of acids, the rising concentration of H^+

in body fluids would quickly lead to death. Homeostasis of H^+ concentration within a narrow pH range is essential to survival and depends on three major mechanisms.

1. *Buffer systems* Buffers act quickly to bind H^+ temporarily, which removes excess H^+ from solution but not from the body.
2. *Exhalation of carbon dioxide* By increasing the rate and depth of breathing, more carbon dioxide can be exhaled. This reduces the level of carbonic acid and is effective within minutes.
3. *Kidney excretion* The slowest mechanism, taking hours or days, but the only way to eliminate acids other than carbonic acid is through their passage into urine and their excretion by the kidneys.

In the following experiments, you will note the relationship between buffers and the exhalation of carbon dioxide to pH.

1. Buffers and pH

Most *buffer systems* of the body consist of a weak acid and the salt of that acid, which functions as a weak base. Buffers function to prevent rapid, drastic changes in the pH of a body fluid by changing strong acids and bases into weak acids and bases. Buffers work within fractions of a second. A strong acid dissociates into H^+ more easily than does a weak acid. Strong acids therefore lower pH more than weak ones because strong acids contribute more H^+. Similarly, strong bases dissociate more easily into OH^-. The principal buffer systems of the body fluids are the carbonic acid-bicarbonate system, the phosphate system, and the protein buffer system.

a. CARBONIC ACID-BICARBONATE BUFFER SYSTEM

The *carbonic acid-bicarbonate buffer system* is based on the bicarbonate ion (HCO_3^-), which can act as a weak base, and carbonic acid (H_2CO_3), which can act as a weak acid. Thus, the buffer system can compensate for either an excess or a shortage of H^+. For example, if there is an excess of H^+ (an acid condition), HCO_3^- can function as a weak base and remove the excess H^+ as follows:

$$H^+ + HCO_3^- \rightarrow H_2CO_3 \rightarrow H_2O + CO_2$$

Hydrogen ion · Bicarbonate ion (weak base) · Carbonic acid · Water · Carbon dioxide

On the other hand, if there is a shortage of H^+ ions (an alkaline condition), H_2CO_3 can function as a weak acid and provide H^+ as follows:

$$H_2CO_3 \rightarrow H^+ + HCO_3^-$$

Carbonic acid (weak acid) · Hydrogen ion · Bicarbonate ion

A typical bicarbonate buffer system consists of a mixture of carbonic acid (H_2CO_3) and its salt, sodium bicarbonate ($NaHCO_3$). The carbonic acid-bicarbonate buffer system is an important regulator of blood pH. When a strong acid, such as hydrochloric acid (HCl) is added to a buffer solution containing sodium bicarbonate, which behaves like a weak base, the following reaction occurs:

$$HCl + NaHCO_3 \rightarrow NaCl + H_2CO_3$$

Hydrochloric acid (strong acid) · Sodium bicarbonate (weak base) · Sodium chloride · Carbonic acid (weak acid)

If a strong base, such as sodium hydroxide ($NaOH$), is added to a buffer solution containing a weak acid, such as carbonic acid, the following reaction occurs:

$$NaOH + H_2CO_3 \rightarrow H_2O + NaHCO_3$$

Sodium hydroxide (strong base) · Carbonic acid (weak acid) · Water · Sodium bicarbonate (weak base)

Normal metabolism produces more acids than bases and thus tends to acidify the blood rather than make it more alkaline. Accordingly, the body needs more bicarbonate salt than it needs carbonic acid. Bicarbonate molecules outnumber carbonic acid molecules 20:1.

b. PHOSPHATE BUFFER SYSTEM

The *phosphate buffer system* acts in the same manner as the carbonic acid-bicarbonate buffer system. The components of the phosphate buffer system are the sodium salts of dihydrogen phosphate and sodium monohydrogen phosphate ions. The dihydrogen phosphate ion acts as the weak acid and is capable of buffering strong bases.

$$NaOH + NaH_2PO_4 \rightarrow H_2O + Na_2HPO_4$$

Sodium hydroxide (strong base) · Sodium dihydrogen phosphate (weak acid) · Water · Sodium monohydrogen phosphate (weak base)

The monohydrogen phosphate ion acts as the weak base and is capable of buffering strong acids.

$$HCl + Na_2HPO_4 \rightarrow NaCl + NaH_2PO_4$$

| Hydrochloric acid (strong acid) | Sodium monohydrogen phosphate (weak base) | Sodium chloride (salt) | Sodium dihydrogen phosphate (weak acid) |

Because the phosphate concentration is highest in intracellular fluid, the phosphate buffer system is an important regulator of pH in the cytosol. It also is present at a lower level in extracellular fluids and acts to buffer acids in urine. NaH_2PO_4 is formed when excess H^+ in the kidney tubules combines with Na_2HPO_4. In this reaction, $Na+$ released from Na_2HPO_4 forms sodium bicarbonate ($NaHCO_3$) and passes into the blood. The H^+ that replaces Na^+ becomes part of the NaH_2PO_4 that passes into the urine. This reaction is one of the mechanisms by which the kidneys help maintain pH by the acidification of urine.

C. PROTEIN BUFFER SYSTEM

The *protein buffer system* is the most abundant buffer in body cells and plasma. Inside red blood cells the protein hemoglobin is an especially good buffer. Proteins are composed of amino acids. An amino acid is an organic compound that contains at least one carboxyl group (COOH) and at least one amine group (NH_2). The free carboxyl group at one end of a protein acts like an acid by releasing hydrogen ions (H^+) ions when pH rises and can dissociate in this way:

$$
\begin{array}{ccc}
R & & R \\
| & & | \\
NH_2-C-COOH & \rightarrow & NH_2-C-COO + H^+ \\
| & & | \\
H & & H
\end{array}
$$

The H^+ is then able react with any excess hydroxide ion (OH^-) in the solution to form water.

The free amine group at the other end of a protein can act as a base by combining with hydrogen ions when pH falls as follows:

$$
\begin{array}{ccc}
R & & R \\
| & & | \\
COOH-C-NH_2 + H^+ & \rightarrow & COOH-C-NH_3{}^+ \\
| & & | \\
H & & H
\end{array}
$$

Thus, proteins act as both acidic and basic buffers.

The following exercise will demonstrate how buffers resist changes in pH.

PROCEDURE

CAUTION! *Please reread Section A, "General Safety Precautions and Procedures," on page xi, and Section C, "Precautions Related to Working with Reagents," on page xii, at the beginning of the laboratory manual, before you begin any of the following experiments. Read the experiments before you perform them, to be sure that you understand all the procedures and safety precautions.*

1. Using a pH meter, immerse the electrode(s) in a beaker containing distilled water and determine

   the pH. _____

2. Drop by drop, slowly add 0.05 M hydrochloric acid (HCl) to the distilled water. Gently swirl the beaker after each drop is added. Note how many drops it takes for the pH of the solution to

   change one whole number. _____

   What is the pH of the solution? _____

3. Remove the electrode(s) from the solution and rinse with distilled water.

4. Now immerse the electrode(s) in a pH 7 buffer solution. Drop by drop, slowly add 0.05 M HCl to the solution. Gently swirl the beaker after each drop is added. Note how many drops it takes for the pH of the solution to change one

   whole number. _____
   What conclusion can you draw from this

   observation? _____

   _____

   _____

5. Remove the electrode(s) from the solution and rinse with distilled water.

6. Immerse the electrode(s) in a beaker of fresh

   distilled water and determine the pH. _____

   _____

7. Drop by drop, slowly add 0.05 M sodium hydroxide (NaOH) to the distilled water. Gently swirl the beaker after each drop is added. Note how many drops it takes for the pH of the so-

   lution to change one whole number. _____

   _____

   What is the pH of the solution? _____

8. Remove the electrode(s) from the solution and rinse with distilled water.

9. Now immerse the electrode(s) in a pH 7 buffer solution. Drop by drop, slowly add 0.05 M NaOH to the solution. Gently swirl the beaker after each drop is added. Note how many drops it takes for the pH of the solution to change one whole number. _____

What conclusion can you draw from this observation? _____

2. Respirations and pH

Breathing also plays a role in maintaining the pH of the body. An increase in the carbon dioxide (CO_2) concentration in body fluids increases H^+ concentration and thus lowers the pH (makes it more acidic). This is illustrated by the following reactions:

$$CO_2 + H_2O \rightleftharpoons H_2CO_3 \rightleftharpoons H^+ + HCO_3^-$$

Conversely, a decrease in the CO_2 concentration of body fluids raises the pH (makes it more basic).

The pH of body fluids can be adjusted, usually in 1 to 3 min, by a change in the rate and depth of breathing. If the rate and depth of breathing increase, more CO_2 is exhaled, the reaction just given is driven to the left, H^+ concentration falls, and the blood pH rises. Because carbonic acid can be eliminated by exhaling CO_2, it is called a *volatile acid.* If the rate of respiration slows down, less carbon dioxide is exhaled, and the blood pH falls. Doubling the breathing rate increases the pH by about 0.23, from 7.4 to 7.63. Reducing the breathing rate to one-quarter its normal rate lowers the pH by 0.4, from 7.4 to 7.0. These examples show the powerful effect of alterations in breathing on pH of body fluids.

The pH of body fluids, in turn, affects the rate of breathing. If, for example, the blood becomes more acidic, the increase in hydrogen ions is detected by chemoreceptors that stimulate the inspiratory center in the medulla. As a result, the diaphragm and other muscles of respiration contract more forcefully and frequently—the rate and depth of breathing increase.

The same effect is achieved if the blood level of CO_2 increases. The increased rate and depth of respiration remove more CO_2 from blood to reduce the H^+ concentration, and blood pH increases. On the other hand, if the pH of the blood increases, the respiratory center is inhibited and respirations decrease. A decrease in the CO_2 concentration of blood

has the same effect. The decreased rate and depth of respirations cause CO_2 to accumulate in blood and the H^+ concentration increases. The respiratory mechanism normally can eliminate more acid or base than can all the buffers combined, but it is limited to eliminating only the single volatile acid, carbonic acid.

The following exercise will demonstrate the relationship of exhalation of carbon dioxide to pH.

PROCEDURE

CAUTION! *Please reread Section A, "General Safety Precautions and Procedures," on page xi, and Section C, "Precautions Related to Working with Reagents," on page xii, at the beginning of the laboratory manual, before you begin any of the following experiments. Read the experiments before you perform them, to be sure that you understand all the procedures and safety precautions.*

1. Fill a large beaker with 100 ml of distilled water.
2. Add 5 ml of 0.10 normal sodium hydroxide (NaOH) solution and 5 drops of phenol red.
3. Phenol red is a pH indicator. It remains red in a basic solution, changes to orange in a neutral solution, and changes to yellow in an acidic solution.
4. While at rest, exhale through a straw into the solution. Your partner should determine how long it takes for the solution to change from

   orange to yellow. _____

CAUTION! *Perform step 5 only if you have no known or apparent cardiac or other health problems and are capable of such an activity.*

5. *Run in place for about 100 steps.*
6. Exhale through a straw into a fresh solution as prepared in steps 1 and 2. Your partner should determine how long it takes for the solution to change from orange to yellow.
7. Change places with your partner, and repeat the experiment. Explain the difference in time it took for the solution to change from orange to

   yellow at rest and following exercise. _____

   _____

   _____

   _____

D. ACID-BASE IMBALANCES

The normal blood pH range is 7.35 to 7.45. *Acidosis* (or *acidemia*) is a condition in which blood pH is below 7.35. *Alkalosis* (or *alkalemia*) is a condition in which blood pH is higher than 7.45.

A change in blood pH that leads to acidosis or alkalosis can be compensated to return pH to normal. *Compensation* refers to the physiological response to an acid-base imbalance. If a person has an altered pH due to metabolic causes, respiratory mechanisms (hyperventilation or hypoventilation) can help compensate for the alteration. Respiratory compensation occurs within minutes and is maximized within hours. On the other hand, if a person has an altered pH due to respiratory causes, metabolic mechanisms (kidney excretion) can compensate for the alteration. Metabolic compensation may begin in minutes but takes days to reach a maximum.

1. Physiological Effects

The principal physiological effect of acidosis is depression of the CNS through depression of synaptic transmission. If the blood pH falls below 7, depression of the nervous system is so severe that the individual becomes disoriented and comatose and dies. Patients with severe acidosis usually die in a state of coma. On the other hand, the major physiological effect of alkalosis is overexcitability in both the CNS and peripheral nerves. Nerves conduct impulses repetitively, even when not stimulated by normal stimuli, resulting in nervousness, muscle spasms, and even convulsions and death.

In the discussion that follows, note that both respiratory acidosis and alkalosis are primary disorders of blood pCO_2 (normal range 35 to 45 mm Hg). On the other hand, both metabolic acidosis and alkalosis are primary disorders of bicarbonate (HCO_3^-) concentration (normal range 22 to 26 mEq/liter).

2. Respiratory Acidosis

The hallmark of *respiratory acidosis* is an elevated pCO_2 of arterial blood (above 45 mm Hg). Inadequate exhalation of CO_2 decreases the blood pH. It occurs as a result of any condition that decreases the movement of CO_2 from the blood to the alveoli of the lungs to the atmosphere and therefore causes a buildup of carbon dioxide, carbonic acid, and hydrogen ions. Such conditions include emphy-

sema, pulmonary edema, injury to the respiratory center of the medulla, airway obstruction, or disorders of the muscles involved in breathing. Metabolic compensation involves increased excretion of H^+ and increased reabsorption of HCO_3^- by the kidneys. Treatment of respiratory acidosis aims to increase the exhalation of CO_2. Excessive secretions can be suctioned out of the respiratory tract, and artificial respiration can be given. In addition, intravenous administration of bicarbonate and ventilation therapy to remove excessive carbon dioxide can be used.

3. Respiratory Alkalosis

In *respiratory alkalosis* arterial blood pCO_2 is decreased (below 35 mm Hg). Hyperventilation causes the pH to increase. It occurs in conditions that stimulate the respiratory center. Such conditions include oxygen deficiency due to high altitude or pulmonary disease, cerebrovascular accident (CVA), severe anxiety, and aspirin overdose. The kidneys attempt to compensate by decreasing excretion of H^+ and decreasing reabsorption of HCO_3^-. Treatment of respiratory alkalosis is aimed at increasing the level of CO_2 in the body. One simple measure is to have the person breathe into a paper bag and then rebreathe the exhaled mixture of CO_2 and oxygen from the bag.

4. Metabolic Acidosis

In *metabolic acidosis* there is a decrease in HCO_3^- concentration (below 22 mEq/liter). The decrease in pH is caused by loss of bicarbonate, such as may occur with severe diarrhea or renal dysfunction; accumulation of an acid, other than carbonic acid, as may occur in ketosis; or failure of the kidneys to excrete H^+ derived from metabolism of dietary proteins. Compensation is respiratory by hyperventilation. Treatment of metabolic acidosis consists of intravenous solutions of sodium bicarbonate and correcting the cause of acidosis.

5. Metabolic Alkalosis

In *metabolic alkalosis* HCO_3^- concentration is elevated (above 26 mEq/liter). A nonrespiratory loss of acid by the body or excessive intake of alkaline drugs causes the pH to increase. Excessive vomiting of gastric contents results in a substantial loss of hydrochloric acid and is probably the most frequent cause of metabolic alkalosis. Other causes of metabolic alkalosis include gastric suctioning, use of certain diuretics, endocrine disorders, and administration of alkali. Compensation is respiratory by hypoventilation. Treatment of metabolic alkalosis consists of fluid therapy to replace chloride, potassium, and other electrolyte deficiencies and correcting the cause of alkalosis.

A summary of acidosis and alkalosis is presented in Table 25.1.

Based on an analysis of respiratory gases, you can determine if a person has acidosis or alkalosis and whether the acidosis or alkalosis is respiratory or metabolic, as reflected by the change in pH.

In the following table (bottom of page), note the normal ranges for pH, pCO_2, and HCO_3^-. Also note how values above and below normal relate to acidosis and alkalosis.

If a change in pH is a result of an abnormal pCO_2 value, then the condition is respiratory in nature. If, instead, a change in pH is a result of an abnormal HCO_3^- value, the condition is metabolic in nature.

Problems related to acid-base imbalance are reflected in each of the following conditions. Determine whether each is (1) acidosis or alkalosis and (2) metabolic or respiratory:

a. pH = 7.32
   HCO_3^-= 10 mEq/liter

b. pH = 7.48
   pCO_2 = 32 mmHg

c. pH = 7.52
   HCO_3^-= 28 mEq/liter

d. pH = 7.30
   pCO_2 = 48 mmHg

E. RENAL REGULATION OF HYDROGEN ION CONCENTRATION

PHYSIOFINDER is an interactive computer program that permits laboratory simulations that do not involve the use of animals or expensive laboratory equipment. It permits students to perform experiments, analyze data, draw conclusions, and form hypotheses on the basis of collected data. The program is available from Addison-Wesley Longman, Inc., Customer Service (1-800-922-0579).

For activities related to renal regulation of hydrogen ion concentration, select the appropriate experiments from PHYSIOFINDER Module 9—Renal Hydrogen Ion Regulation.

ANSWER THE LABORATORY REPORT QUESTIONS AT THE END OF THE EXERCISE

	pH	pCO_2	HCO_3^-
Normal range	7.35–7.45	35–45 mmHg	22–26 mEq/liter
Acidosis	Below 7.35	Above 45 mmHg	Below 22 mEq/liter
Alkalosis	Above 7.45	Below 35 mmHg	Above 26 Eq/liter

TABLE 25.1
Summary of Acidosis and Alkalosis

Condition	Definition	Common cause	Compensatory mechanism
Respiratory acidosis	Increased pCO_2 (above 45 mm Hg) and decreased pH (below 7.35) if there is no compensation.	Hypoventilation due to emphysema, pulmonary edema, trauma to respiratory center, airway obstructions, dysfunction of muscles of respiration.	Renal: increased excretion of H^+; increased reabsorption of HCO_3^-. If compensation is complete, pH will be within normal range, but pCO_2 will be high.
Respiratory alkalosis	Decreased pCO_2 (below 35 mm Hg) and increased pH (above 7.45) if there is no compensation.	Hyperventilation due to oxygen deficiency, pulmonary disease, cerebrovascular accident (CVA), anxiety, or aspirin overdose.	Renal: decreased excretion of H^+; decreased reabsorption of HCO_3^-. If compensation is complete, pH will be within normal range, but pCO_2 will be low.
Metabolic acidosis	Decreased bicarbonate (below 22 mEq/liter) and decreased pH (below 7.35) if there is no compensation.	Loss of bicarbonate due to diarrhea, accumulation of acid (ketosis), renal dysfunction.	Respiratory: hyperventilation, which increases loss of CO_2. If compensation is complete, pH will be within normal range but HCO_3^- will be low.
Metabolic alkalosis	Increased bicarbonate (above 26 mEq/liter) and increased pH (above 7.45) if there is no compensation.	Loss of acid or excessive intake of alkaline drugs; due to vomiting, gastric suctioning, use of certain diuretics, and administration of alkali.	Respiratory: hypoventilation, which slows loss of CO_2. If compensation is complete, pH will be within normal range, but HCO_3^- will be high.

pH and Acid-Base Balance 25

Student _____   Date _____

Laboratory Section _____   Score/Grade _____

SECTION B. MEASURING pH

1. Using Litmus Paper

Solution	Color of red litmus paper	Color of blue litmus paper	Is the solution acidic or basic?
Milk of magnesia			
Vinegar			
Coffee			
Carbonated soft drink			
Orange juice			
Distilled water			
Baking soda			
Lemon juice			

2. Using pH Paper

Solution	pH
Milk of magnesia	
Vinegar	
Coffee	
Carbonated soft drink	
Orange juice	
Distilled water	
Baking soda	
Lemon juice	

3. Using a pH Meter

Solution	pH
Milk of magnesia	
Vinegar	
Coffee	
Carbonated soft drink	
Orange juice	
Distilled water	
Baking soda	
Lemon juice	

pH and Acid-Base Balance 25

Student _____   Date _____

Laboratory Section _____   Score/Grade _____

PART 1. Multiple Choice

_____ 1. An acid is a substance that dissociates into (a) OH^- (b) H^+ (c) HCO_3^- (d) Na^+

_____ 2. Which of the following pHs is more acidic? (a) 6.89 (b) 6.91 (c) 7.00 (d) 6.83

_____ 3. The pH of bile is (a) 4.2 (b) 6.35 to 6.85 (c) 7.6 to 8.6 (d) 3.0 to 3.5

_____ 4. Which mechanism is quickest to restore pH? (a) exhalation of CO_2 (b) buffers (c) kidney excretion (d) inhalation of oxygen

_____ 5. In the carbonic acid-bicarbonate buffer system, which substance functions to buffer a strong base? (a) NaOH (b) $NaHCO_3$ (c) H_2CO_3 (d) NaCl

_____ 6. The most abundant buffer in body cells and plasma is the (a) protein buffer (b) phosphate buffer (c) hemoglobin buffer (d) carbonic acid-bicarbonate buffer

_____ 7. Doubling the breathing rate increases pH by about (a) 0.75 (b) 2.21 (c) 1.86 (d) 0.23

PART 2. Completion

8. Bases dissociate into _____ ions and cations.

9. A change of one whole number on the pH scale represents a _____-fold change from the previous concentration.

10. A(n) _____ solution will turn red litmus paper blue.

11. The pH of blood is _____.

12. Most H^+ in the body is produced as a result of _____.

13. In the protein buffer system, the carboxyl group acts as a(n) _____.

14. If the rate and depth of respiration increase, pH will _____.

15. If blood becomes more basic, the rate and depth of respiration will _____.

16. A pH higher than 7.45 is referred to as _____.

17. Metabolic acidosis and alkalosis are disorders of _____ concentration.

18. The principal physiological effect of _____ is depression of the CNS.

19. _____ acidosis is characterized by a pH below 7.35 and an elevated pCO_2.

20. Compensation for metabolic acidosis is _____.

Reproductive Systems

26

Reproduction is the process by which new individuals of a species are produced and the genetic material is passed from generation to generation. This maintains continuation of the species. Cell division in a multicellular organism is necessary for growth as well as repair and it involves passing of genetic material from parent cells to daughter cells. In somatic cell division a parent cell produces two identical daughter cells. This process is involved in replacing cells and growth. In reproductive cell division, sperm and egg cells are produced for continuity of the species.

The organs of the male and female reproductive systems may be grouped by function. (1) The testes and ovaries, also called *gonads* (*gonos* = seed), function in the production of gametes—sperm cells and ova, respectively. The gonads also secrete hormones. (2) The *ducts* of the reproductive systems transport, receive, and store gametes. (3) Still other reproductive organs, called *accessory sex glands,* produce materials that support gametes. (4) Finally, several *supporting structures,* including the penis, have various roles in reproduction. In this exercise, you will study the structure of the male and female reproductive organs and associated structures.

A. ORGANS OF MALE REPRODUCTIVE SYSTEM

The *male reproductive system* includes (1) the testes, or male gonads, which produce sperm and secrete hormones; (2) a system of ducts that transport, receive, or store sperm; (3) accessory glands, whose secretions contribute to semen; and (4) several supporting structures, including the penis.

1. Testes

The *testes,* or *testicles,* are paired oval glands that lie in the pelvic cavity for most of fetal life. They usually begin to enter the scrotum during the latter half of the seventh month of fetal development; full descent is not complete until just before birth. If the testes do not descend, the condition is called *cryptorchidism* (krip-TOR-ki-dizm; *kryptos* = hidden; *orchis* = testis). Cryptorchidism may result in sterility, because the cells that stimulate the initial development of sperm cells are destroyed by the higher temperature of the pelvic cavity. The chance of testicular cancer is 30 to 50% greater in cryptorchid testes.

Each testis is partially covered by a serous membrane called the *tunica* (*tunica* = sheath) *vaginalis,* which is derived from the peritoneum. Internal to the tunica vaginalis is a dense white fibrous capsule, the *tunica albuginea* (al'-byoo-JIN-ē-a; *albus* = white), which extends inward and divides the testis into a series of 200 to 300 internal compartments called *lobules.* Each lobule contains one to three tightly coiled *seminiferous* (*semen* = seed; *ferre* = to carry) *tubules* where sperm production *(spermatogenesis)* occurs.

Label the structures associated with the testes in Figure 26.1.

Spermatogenic cells are sperm-forming cells in various stages that undergo mitosis and differentiation to eventually produce sperm. Together with supporting cells, they line the seminiferous tubules. The most immature spermatogenic cells are called *spermatogonia* (sper'-ma-tō-GŌ-nē-a; *sperm* = seed; *gonium* = generation or offspring; singular is *spermatogonium*). They lie next to the basement membrane. Toward the lumen of the tubule are layers of progressively more mature cells. In order of advancing maturity, these are *primary spermatocytes* (SPER-ma-tō-sīts), *secondary spermatocytes,* and *spermatids.* By the time a *sperm cell,* or *spermatozoon* (sper'-ma-tō-ZŌ-on; *zoon* = life; plural is *sperm,* or *spermatozoa*), has nearly reached maturity, it is released into the lumen of the tubule and begins to move out of the rete testis.

Embedded among the spermatogenic cells in the tubules are large *sustentacular* (sus'-ten-TAK-yoo-lar; *sustentare* = to support), or *Sertoli, cells* that

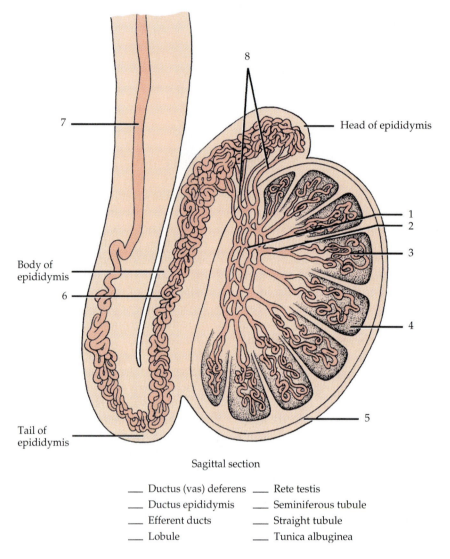

8

7

Head of epididymis

Body of
epididymis

6

1
2

3

4

5

Tail of
epididymis

Sagittal section

___ Ductus (vas) deferens   ___ Rete testis

___ Ductus epididymis   ___ Seminiferous tubule

___ Efferent ducts   ___ Straight tubule

___ Lobule   ___ Tunica albuginea

FIGURE 26.1   Testis showing its system of ducts.

extend from the basement membrane to the lumen of the tubule. Sustentacular cells support and protect developing spermatogenic cells; nourish spermatocytes, spermatids, and sperm; phagocytize excess spermatid cytoplasm as development proceeds; and regulate the effects of testosterone and follicle-stimulating hormone (FSH). Sustentacular cells also control movements of spermatogenic cells and the release of sperm into the lumen of the seminiferous tubule. They produce fluid for sperm transport and secrete the hormone inhibin, which helps regulate sperm production by inhibiting the secretion of FSH. In the spaces between adjacent seminiferous tubules are clusters of cells called *interstitial endocrinocytes,* or *Leydig cells.* These cells

secrete testosterone, the most important androgen (male sex hormone). Because they produce both sperm and hormones, the testes are both exocrine and endocrine glands.

Using your textbook, charts, or models as reference, label Figure 26.2.

Sperm are produced at the rate of about 300 million per day. Once ejaculated, they usually live about 48 hr in the female reproductive tract. The parts of a sperm are as follows:

a. *Head* Contains the *nucleus* (DNA) and *acrosome* (produces hyaluronic acid and proteinases to bring about penetration of secondary oocyte).

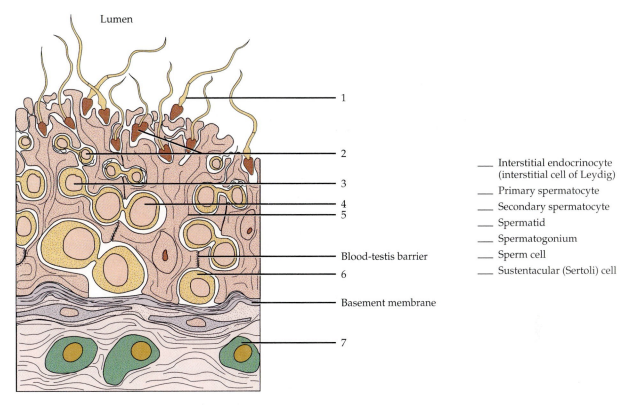

Lumen

—— Interstitial endocrinocyte
    (interstitial cell of Leydig)
—— Primary spermatocyte
—— Secondary spermatocyte
—— Spermatid
—— Spermatogonium
—— Sperm cell
—— Sustentacular (Sertoli) cell

Blood-testis barrier

Basement membrane

(a) Transverse section of a portion of a seminiferous tubule

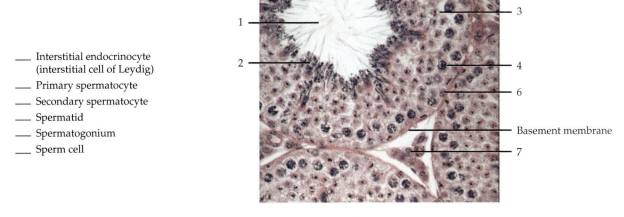

—— Interstitial endocrinocyte
    (interstitial cell of Leydig)
—— Primary spermatocyte
—— Secondary spermatocyte
—— Spermatid
—— Spermatogonium
—— Sperm cell

Basement membrane

(b) Transverse section of a portion of a seminiferous tubule (400x)

FIGURE 26.2   Seminiferous tubules showing various stages of spermatogenesis.
(a) Diagram. (b) Photomicrograph.

b. *Midpiece* Contains numerous mitochondria in which the energy for locomotion is generated.
c. *Tail* Typical flagellum used for locomotion.

With the aid of your textbook, label Figure 26.3 on page 510.

2. Ducts

As sperm cells mature, they are moved through seminiferous tubules into tubes called *straight tubules,* from which they are transported into a network of ducts, the *rete* (RĒ-tē; *rete* = network) *testis.*

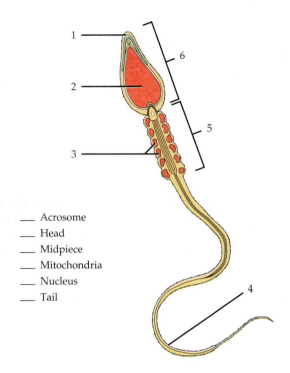

FIGURE 26.3   Parts of a sperm cell.

___ Acrosome
___ Head
___ Midpiece
___ Mitochondria
___ Nucleus
___ Tail

The sperm cells are next transported out of the testes through a series of coiled *efferent ducts* that empty into a single *ductus epididymis* (ep'-i-DID-i-mis; *epi* = above; *didymos* = testis). From there, they are passed into the *ductus (vas) deferens,* which ascends along the posterior border of the testis, penetrates the inguinal canal, enters the pelvic cavity, and loops over the side and down the posterior surface of the urinary bladder. The ductus (vas) deferens and duct from the seminal vesicle (gland) together form the *ejaculatory* (e-JAK-yoo-la-tō'-rē; *ejectus* = to throw out) *duct,* which propels the sperm cells into the *urethra,* the terminal duct of the system. The male urethra is divisible into (1) a *prostatic portion,* which passes through the prostate gland; (2) a *membranous portion,* which passes through the urogenital diaphragm; and (3) a *spongy (penile) portion,* which passes through the corpus spongiosum of the penis (see Figure 26.6 on page 512). The *epididymis* is a comma-shaped organ that is divisible into a head, body, and tail. The head is the superior portion that contains the efferent ducts; the body is the middle portion that contains the ductus epididymis; and the tail is the inferior portion in which the ductus epididymis continues as the ductus (vas) deferens.

Label the various ducts of the male reproductive system in Figure 26.1.

The ductus epididymis is lined with *pseudostratified columnar epithelium.* The free surfaces of the cells contain long, branching microvilli called *stereocilia.* The muscularis deep to the epithelium consists of smooth muscle. Functionally, the ductus epididymis is the site of sperm maturation (increased motility and fertility potential). They require between 10 and 14 days to complete their maturation—that is, to become capable of fertilizing a secondary oocyte. The ductus epididymis also stores sperm cells and propels them toward the urethra during emission by peristaltic contraction of its smooth muscle. Sperm cells may remain in storage in the ductus epididymis up to a month or more. After that, they are expelled from the epididymis or reabsorbed in the epididymis.

Obtain a prepared slide of the ductus epididymis showing its mucosa and muscularis. Compare your observations to Figure 26.4.

Histologically, the *ductus (vas) deferens (seminal duct)* is also lined with *pseudostratified columnar epithelium* and its muscularis consists of three layers of smooth muscle. Peristaltic contractions of the muscularis propel sperm cells toward the urethra during ejaculation. One method of sterilization in males, *vasectomy,* involves removal of a portion of each ductus (vas) deferens.

Obtain a prepared slide of the ductus (vas) deferens showing its mucosa and muscularis. Compare your observations to Figure 26.5.

3. Accessory Sex Glands

Whereas the ducts of the male reproductive system store or transport sperm, a series of *accessory sex glands* secrete most of the liquid portion of *semen.* Semen is a mixture of sperm cells and the secretions of the seminal vesicles, prostate gland, and bulbourethral glands.

The *seminal* (*seminalis* = pertaining to seed) *vesicles* (VES-i-kuls) are paired, convoluted, pouchlike structures posterior to and at the base of the urinary bladder anterior to the rectum. The glands secrete the alkaline viscous component of semen into the ejaculatory duct. The seminal vesicles contribute about 60% of the volume of semen.

The *prostate* (PROS-tāt) *gland,* a single doughnut-shaped gland inferior to the urinary bladder, surrounds the prostatic urethra. The prostate secretes a slightly acidic fluid into the prostatic urethra. The prostatic secretion constitutes about 25% of the total semen produced.

The paired *bulbourethral* (bul'-bō-yoo-RĒ-thral), or *Cowper's, glands,* located inferior to the prostate

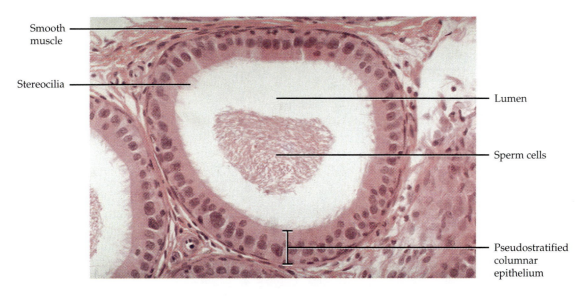

FIGURE 26.4   Histology of ductus epididymis.

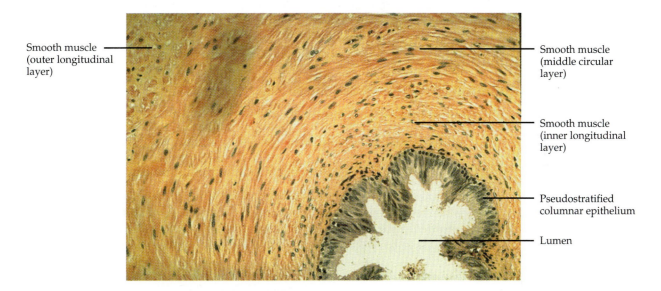

FIGURE 26.5   Histology of ductus (vas) deferens.

on either side of the membranous urethra, are about the size of peas. They secrete an alkaline substance through ducts that open into the spongy (penile) urethra.

With the aid of your textbook, label the accessory glands and associated structures in Figure 26.6.

4. Penis

a. *Body* Main portion of penis composed of three cylindrical masses of tissue, each bound by

fibrous tissue called the ***tunica albuginea.*** The paired dorsolateral masses are called the ***corpora cavernosa penis*** (*corpus* = body; *caverna* = hollow). The smaller midventral mass, the ***corpus spongiosum penis,*** contains the spongy urethra and functions in keeping the spongy urethra open during ejaculation. All three masses are enclosed by fascia and skin and consist of erectile tissue permeated by blood sinuses. With sexual stimulation, the arteries supplying the penis dilate, and large quantities of blood

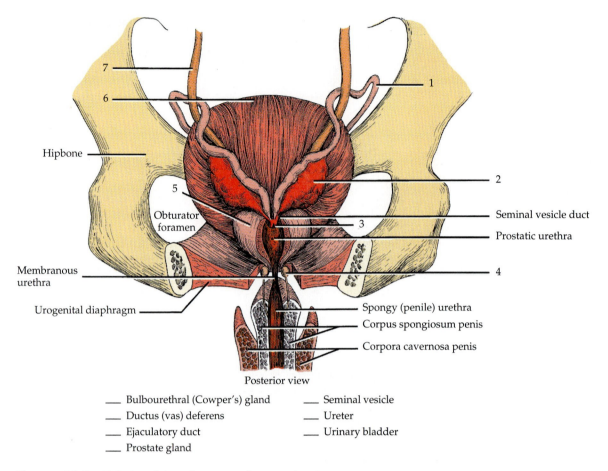

FIGURE 26.6   Relationships of some male reproductive organs.

enter the blood sinuses. Expansion of these spaces compresses the veins draining the penis, so more blood that enters is trapped. These vascular changes result in an *erection,* a parasympathetic reflex. The penis returns to its flaccid state when the arteries constrict and pressure on the veins is relieved.

b. *Root* The attached portion of the penis. It consists of the *bulb of the penis,* the expanded portion of the base of the corpus spongiosum penis, and the *crura* (*crus* = leg; singular is *crus*) *of the penis,* the two separated and tapered portions of the corpora cavernosa penis. The bulb of the penis is attached to the inferior surface of the urogenital diaphragm and enclosed by the bulbospongiosus muscle. Each crus of the penis is attached to the ischial and inferior pubic rami and surrounded by the ischiocavernosus muscle. Contraction of these skeletal muscles aids ejaculation.

c. *Glans* (*glandes* = acorn) *penis* The slightly enlarged distal end of the corpus spongiosum penis. The margin of the glans penis is termed the *corona.* The distal urethra enlarges within the glans penis and forms a terminal slitlike opening, the *external urethral orifice.* Covering the glans in an uncircumsized penis is the loosely fitting *prepuce* (PRĒ-pyoos), or *foreskin.*

With the aid of your textbook, label the parts of the penis in Figure 26.7.

Now that you have completed your study of the organs of the male reproductive system, label Figure 26.8 on page 514.

B. ORGANS OF FEMALE REPRODUCTIVE SYSTEM

The *female reproductive system* includes the female gonads (ovaries), which produce secondary oocytes; uterine (Fallopian) tubes, or oviducts, which transport secondary oocytes and fertilized ova to the uterus; vagina; external organs that compose the vulva; and the mammary glands.

1. Ovaries

The *ovaries* (*ovarium* = egg receptacle) are paired glands that resemble almonds in size and shape. Functionally, the ovaries produce secondary oocytes, discharge them about once a month by a process called ovulation, and secrete female sex hormones (estrogens, progesterone, relaxin, and inhibin). The point of entrance for blood vessels and nerves is the **hilus.** The ovaries are positioned in the superior pelvic cavity, one on each side of the uterus, by a series of ligaments:

a. *Mesovarium* Double-layered fold of pertoneum that attaches ovaries to broad ligaments of uterus.

b. *Ovarian ligament* Anchors ovary to uterus.

c. *Suspensory ligament* Attaches ovary to pelvic wall.

With the aid of your textbook, label the ovarian ligaments in Figure 26.9 on page 515.

Histologically, the ovaries consist of the following parts:

1. *Germinal epithelium* A layer of simple epithelium (low cuboidal or squamous) that covers the surface of the ovary and is continuous with the mesothelium that covers the mesovarium. The term *germinal epithelium* is a misnomer since it does not give rise to oocytes, although at one time it was believed that it did.

2. *Tunica albuginea* A whitish capsule of dense, irregular connective tissue immediately deep to the germinal epithelium.

3. *Ovarian cortex* A region just deep to the tunica albuginea that consists of dense connective tissue and contains ovarian follicles (described shortly).

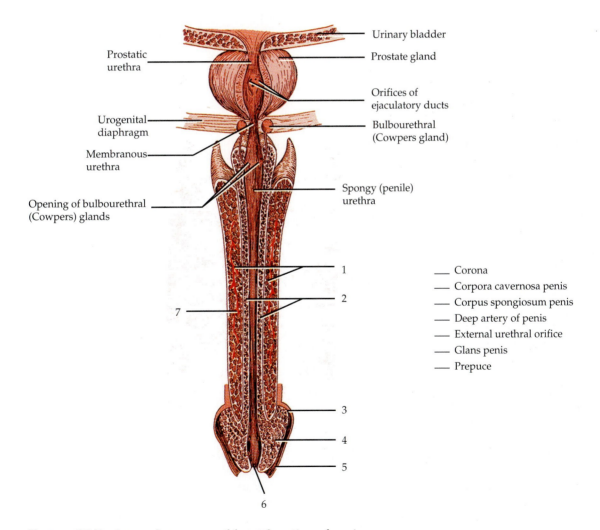

Urinary bladder

Prostate gland

Prostatic urethra

Orifices of ejaculatory ducts

Urogenital diaphragm

Bulbourethral (Cowpers gland)

Membranous urethra

Opening of bulbourethral (Cowpers) glands

Spongy (penile) urethra

1

2

7

___ Corona
___ Corpora cavernosa penis
___ Corpus spongiosum penis
___ Deep artery of penis
___ External urethral orifice
___ Glans penis
___ Prepuce

3

4

5

6

FIGURE 26.7 Internal structure of frontal section of penis.

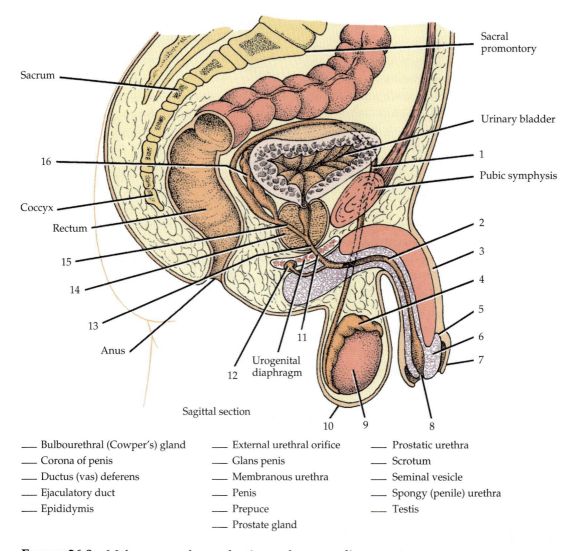

Sacral promontory

Sacrum

Urinary bladder

16

1

Pubic symphysis

Coccyx

2

Rectum

3

15

4

14

5

13

6

Anus

7

Urogenital diaphragm

11

12

10   9   8

Sagittal section

___ Bulbourethral (Cowper's) gland ___ External urethral orifice ___ Prostatic urethra
___ Corona of penis   ___ Glans penis   ___ Scrotum
___ Ductus (vas) deferens   ___ Membranous urethra   ___ Seminal vesicle
___ Ejaculatory duct   ___ Penis   ___ Spongy (penile) urethra
___ Epididymis   ___ Prepuce   ___ Testis
                                    ___ Prostate gland

FIGURE 26.8   Male organs of reproduction and surrounding structures.

4. ***Ovarian medulla*** A region deep to the ovarian cortex that consists of loose connective tissue and contains blood vessels, lymphatics, and nerves.
5. ***Ovarian follicles*** (*folliculus* = little bag) Lie in the cortex and consist of ***oocytes*** in various stages of development and their surrounding cells. When the surrounding cells form a single layer, they are called ***follicular cells.*** Later in development, when they form several layers, they are referred to as ***granulosa cells.*** The surrounding cells nourish the developing oocyte and begin to secrete estrogens as the follicle grows larger. Ovarian follicles undergo a series of changes prior to ovulation, progressing through several distinct stages. The most numerous and peripherally arranged follicles are termed ***primordial follicles.*** If a primordial fol-

licle progresses to ovulation (release of a mature ovum), it will sequentially transform into a ***primary (preantral) follicle,*** then a ***secondary (antral) follicle,*** and finally a ***mature (Graafian) follicle.***
6. ***Mature (Graafian) follicle*** A large, fluid-filled follicle that soon will rupture and expel a secondary oocyte, a process called ***ovulation.***
7. ***Corpus luteum*** (= yellow body) Contains the remnants of an ovulated mature follicle. The corpus luteum produces progesterone, estrogens, relaxin, and inhibin until it degenerates and turns into fibrous tissue called a ***corpus albicans*** (= white body).

With the aid of your textbook, label the parts of an ovary in Figure 26.10 on page 516.

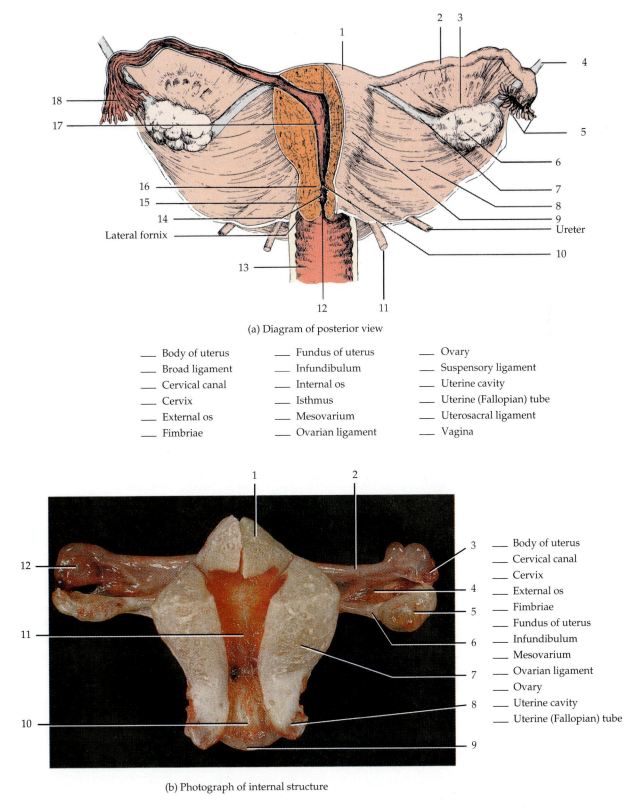

(a) Diagram of posterior view

___ Body of uterus	___ Fundus of uterus	___ Ovary
___ Broad ligament	___ Infundibulum	___ Suspensory ligament
___ Cervical canal	___ Internal os	___ Uterine cavity
___ Cervix	___ Isthmus	___ Uterine (Fallopian) tube
___ External os	___ Mesovarium	___ Uterosacral ligament
___ Fimbriae	___ Ovarian ligament	___ Vagina

___ Body of uterus
___ Cervical canal
___ Cervix
___ External os
___ Fimbriae
___ Fundus of uterus
___ Infundibulum
___ Mesovarium
___ Ovarian ligament
___ Ovary
___ Uterine cavity
___ Uterine (Fallopian) tube

(b) Photograph of internal structure

FIGURE 26.9   Uterus and associated female reproductive structures. The left side of figure (a) has been sectioned to show internal structures.

Obtain prepared slides of the ovary, examine them, and compare your observations to Figure 26.11 on page 517.

2. Uterine (Fallopian) Tubes

The *uterine (Fallopian) tubes,* or *oviducts,* extend laterally from the uterus and transport secondary oocytes from the ovaries to the uterus. Fertilization normally occurs in the uterine tubes. The tubes are positioned between folds of the broad ligaments of the uterus. The funnel-shaped, open distal end of each uterine tube, called the *infundibulum,* is surrounded by a fringe of fingerlike projections called *fimbriae* (FIM-bre-ē; *fimbrae* = fringe). The *ampulla* (am-POOL-la) of the uterine tube is the widest, longest portion, constituting about two-thirds of its length. The *isthmus* (IS-mus) is the short, narrow, thick-walled portion that joins the uterus.

With the aid of your textbook, label the parts of the uterine tubes in Figure 26.9.

Histologically, the mucosa of the uterine tubes consists of ciliated columnar cells and secretory cells. The muscularis is composed of inner circular and outer longitudinal layers of smooth muscle. Wavelike contractions of the muscularis help move the ovum down into the uterus. The serosa is the outer covering.

Examine a prepared slide of the wall of the uterine tube, and compare your observations to Figure 26.12 on page 518.

3. Uterus

The *uterus (womb)* is the site of menstruation, implantation of a fertilized ovum, development of the fetus during pregnancy, and labor. Located between the urinary bladder and the rectum, the organ is shaped like an inverted pear. The uterus is subdivided into the following regions:

a. *Fundus* Dome-shaped portion superior to the uterine tubes.

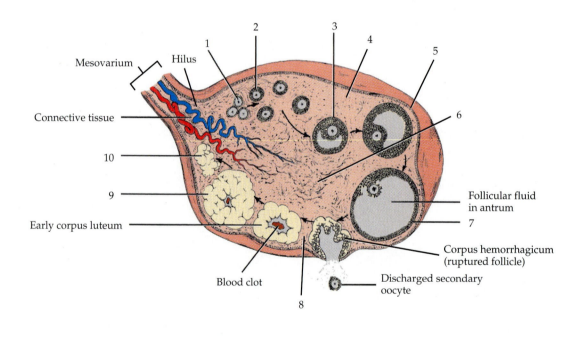

Mesovarium   Hilus

Connective tissue

10

9

Early corpus luteum

Blood clot

Follicular fluid in antrum

7

Corpus hemorrhagicum (ruptured follicle)

Discharged secondary oocyte

___ Corpus albicans
___ Corpus luteum (mature)
___ Cortex of stroma
___ Germinal epithelium
___ Mature (Graafian) follicle

___ Medulla of stroma
___ Primary (preantral) follicle
___ Primordial follicle
___ Secondary (antral) follicle
___ Tunica albuginea

Figure 26.10   Histology of ovary. Arrows indicate sequence of developmental stages that occur as part of ovarian cycle.

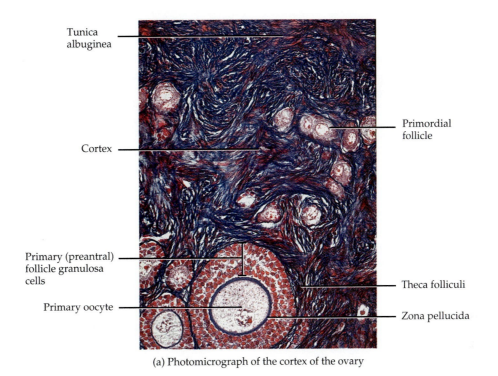

Tunica albuginea

Primordial follicle

Cortex

Primary (preantral) follicle granulosa cells

Theca folliculi

Primary oocyte

Zona pellucida

(a) Photomicrograph of the cortex of the ovary

Theca folliculi

Antrum filled with follicular fluid

Zona pellucida

Primary oocyte

Corona radiata

Secondary (antral) follicle granulosa cells

(b) Photomicrograph of a secondary follicle (248x)

FIGURE 26.11   Histology of ovary.

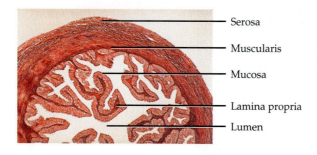

Serosa

Muscularis

Mucosa

Lamina propria

Lumen

FIGURE 26.12   Histology of uterine (Fallopian) tube (50×).

b. *Body* Major, tapering portion.
c. *Cervix* Inferior narrow opening into vagina.
d. *Isthmus* (IS-mus) Constricted region between body and cervix.
e. *Uterine cavity* Interior of the body.
f. *Cervical canal* Interior of the cervix.
g. *Internal os* Site where cervical canal opens into uterine cavity.
h. *External os* Site where cervical canal opens into vagina.

Label these structures in Figure 26.9.

The uterus is maintained in position by the following ligaments:

a. *Broad ligaments* Double folds of parietal peritoneum that anchor the uterus to either side of the pelvic cavity.
b. *Uterosacral ligaments* Parietal peritoneal exten-sions that connect the uterus to the sacrum.
c. *Cardinal (lateral cervical) ligaments* Tissues containing smooth muscle, uterine blood vessels, and nerves. These ligaments extend below the bases of the broad ligaments between the pelvic wall and the cervix and vagina, and are the chief ligaments that maintain the position of the uterus, helping to keep it from dropping into the vagina.
d. *Round ligaments* Extend from uterus to external genitals (labia majora) between folds of broad ligaments.

Label the uterine ligaments in Figure 26.9.

Histologically, the uterus consists of three principal layers: endometrium, myometrium, and perimetrium (serosa). The inner *endometrium* (*endo* = within) is a mucous membrane that consists of simple columnar epithelium, an underlying endometrial stroma composed of connective tissue and

endometrial glands. The endometrium consists of two layers: (1) *stratum functionalis,* the layer closer to the uterine cavity that is shed during menstruation; and (2) *stratum basalis* (ba-SAL-is), the permanent layer that produces a new stratum functionalis after menstruation. The middle *myometrium* (*myo* = muscle) forms the bulk of the uterine wall and consists of three layers of smooth muscle. During labor its coordinated contractions help to expel the fetus. The outer layer is the *perimetrium* (*peri* = around; *metron* = uterus), or *serosa,* part of the visceral peritoneum.

4. Vagina

A muscular, tubular organ lined with a mucous membrane, the *vagina* (*vagina* = sheath) is the passageway for menstrual flow, the receptacle for the penis during copulation, and the inferior portion of the birth canal. The vagina is situated between the urinary bladder and rectum and extends from the cervix of the uterus to the vestibule of the vulva. Recesses called *fornices* (FOR-ni-sēz'; *fornix* = arch or vault) surround the vaginal attachment to the cervix (see Figure 26.9) and make possible the use of contraceptive diaphragms. The opening of the vagina to the exterior, the *vaginal orifice,* may be bordered by a thin fold of vascularized membrane, the *hymen* (*hymen* = membrane).

Label the vagina in Figure 26.9.

Histologically, the mucosa of the vagina consists of nonkeratinized stratified squamous epithelium and connective tissue that lies in a series of transverse folds, the *rugae.* The muscularis is composed of an outer circular and an inner longitudinal layer of smooth muscle.

5. Vulva

The *vulva* (VUL-va; *volvere* = to wrap around), or *pudendum* (pyoo-DEN-dum), is a collective term for the external genitals of the female. It consists of the following parts:

a. *Mons pubis* (MONZ PŪ-bis) Elevation of adipose tissue over the pubic symphysis covered by skin and pubic hair.
b. *Labia majora* (LĀ-bē-a ma-JŌ-ra; *labium* = lip) Two longitudinal folds of skin that extend inferiorly and posteriorly from the mons pubis. Singular is *labium majus.* The folds, covered by pubic hair on their superior lateral surfaces, contain abundant adipose tissue and sebaceous (oil) and sudoriferous (sweat) glands.

c. *Labia minora* (MĪ-nō-ra) Two folds of mucous membrane medial to labia majora. Singular is *labium minus.* The folds have numerous sebaceous glands but few sudoriferous glands and no fat or pubic hair.

d. *Clitoris* (KLI-to-ris) Small cylindrical mass of erectile tissue and nerves at anterior junction of labia minora. The exposed portion is called the *glans* and the covering is called the *prepuce* (foreskin).

e. *Vestibule* Cleft between labia minora containing vaginal orifice, hymen (if present), external urethral orifice, and openings of several ducts.

f. *Vaginal orifice* Opening of vagina to exterior.

g. *Hymen* Thin fold of vascularized membrane that borders vaginal orifice.

h. *External urethral orifice* Opening of urethra to exterior.

i. *Orifices of paraurethral (Skene's) glands* Located on either side of external urethral orifice. The glands secrete mucus.

j. *Orifices of ducts of greater vestibular* (ves-TIB-yoo-lar), or *Bartholin's, glands* Located in a groove between hymen and labia minora. These glands produce a mucoid secretion that supplements lubrication during intercourse.

k. *Orifices of ducts of lesser vestibular glands* Microscopic orifices opening into vestibule.

Using your textbook as an aid, label the parts of the vulva in Figure 26.13.

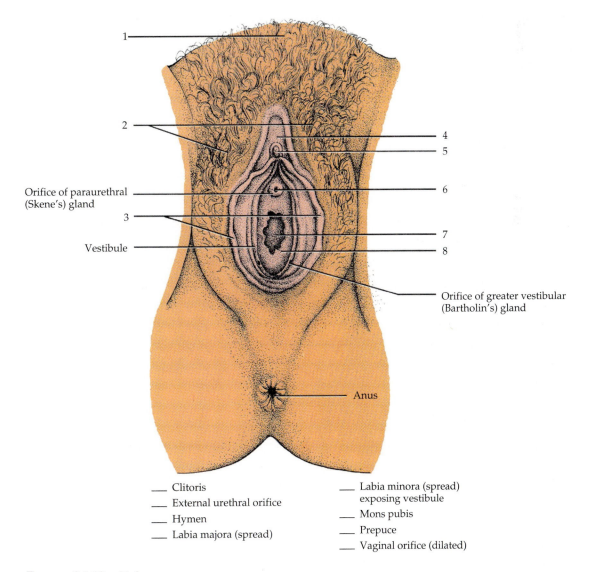

FIGURE 26.13 Vulva.

___ Clitoris
___ External urethral orifice
___ Hymen
___ Labia majora (spread)
___ Labia minora (spread) exposing vestibule
___ Mons pubis
___ Prepuce
___ Vaginal orifice (dilated)

6. Mammary Glands

The *mammary* (*mamma* = breast) *glands* are modified sweat glands that lie over the pectoralis major muscles and are attached to them by a layer of deep fascia. They consist of the following structures:

a. *Lobes* Around 15 to 20 compartments separated by adipose tissue.

b. *Lobules* Smaller compartments in lobes that contain clusters of milk-secreting glands called *alveoli* (*alveolus* = small cavity).

c. *Secondary tubules* Receive milk from alveoli.

d. *Mammary ducts* Receive milk from secondary tubules.

e. *Lactiferous* (*lact* = milk; *ferre* = to carry) *sinuses* Expanded distal portions of mammary ducts that store milk.

f. *Lactiferous ducts* Receive milk from lactiferous sinuses.

g. *Nipple* Projection on anterior surface of mammary gland that contains lactiferous ducts.

h. *Areola* (a-RĒ-ō-la; *areola* = small space) Circular pigmented skin around nipple.

With the aid of your textbook, label the parts of the mammary gland in Figure 26.14.

Examine a prepared slide of alveoli of the mammary gland, and compare your observations to Figure 26.15.

Now that you have completed your study of the organs of the female reproductive system, label Figure 26.16.

7. Female Reproductive Cycle

During their reproductive years, nonpregnant females normally experience a cyclical sequence of changes in the ovaries and uterus. Each cycle takes about a month and involves both oogenesis and preparation of the uterus to receive a fertilized ovum. Hormones secreted by the hypothalamus, anterior pituitary gland, and ovaries control the principal events. The *ovarian cycle* is a series of events associated with the maturation of an oocyte. The *uterine (menstrual) cycle* is a series of changes in the endometrium of the uterus. Each month, the endometrium is prepared for the arrival of a fertilized ovum that will develop in the uterus until birth. If fertilization

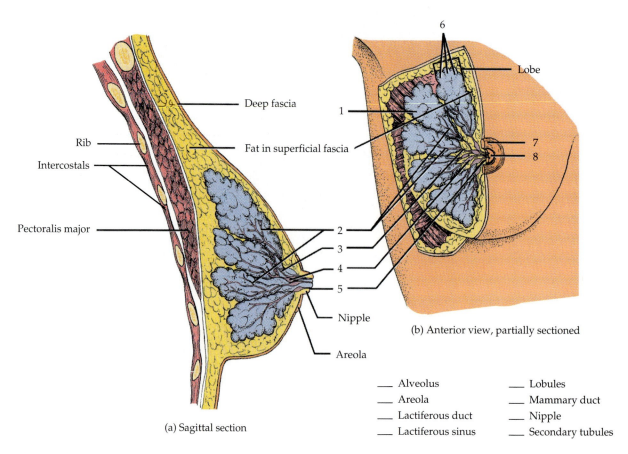

Deep fascia

Rib

Intercostals

Pectoralis major

Fat in superficial fascia

Nipple

Areola

(a) Sagittal section

6

Lobe

1

7

8

2

3

4

5

(b) Anterior view, partially sectioned

___ Alveolus   ___ Lobules

___ Areola   ___ Mammary duct

___ Lactiferous duct   ___ Nipple

___ Lactiferous sinus   ___ Secondary tubules

FIGURE 26.14 Mammary glands.

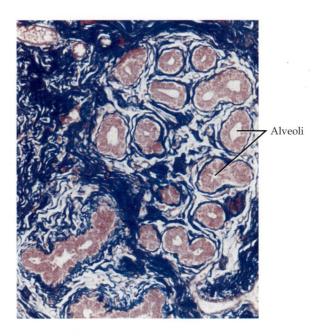

FIGURE 26.15 Histology of mammary gland showing alveoli (180×).

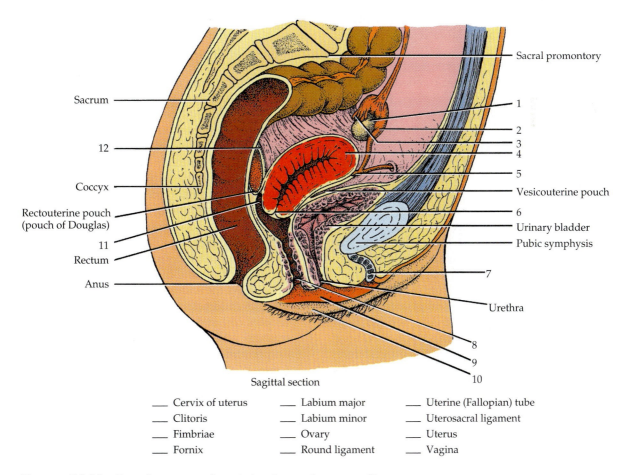

Sagittal section

___ Cervix of uterus	___ Labium major	___ Uterine (Fallopian) tube
___ Clitoris	___ Labium minor	___ Uterosacral ligament
___ Fimbriae	___ Ovary	___ Uterus
___ Fornix	___ Round ligament	___ Vagina

FIGURE 26.16 Female organs of reproduction and surrounding structures.

does not occur, the stratum functionalis portion of the endometrium is shed. The general term *female reproductive cycle* encompasses the ovarian and uterine cycles, the hormonal changes that regulate them, and cyclical changes in the breasts and cervix.

a. HORMONAL REGULATION

The uterine cycle and ovarian cycle are controlled by gonadotropin releasing hormone (GnRH) from the hypothalamus. See Figure 26.17. GnRH stimulates the release of follicle-stimulating hormone (FSH) and luteinizing hormone (LH) from the anterior pituitary gland. FSH stimulates the initial secretion of estrogens by the follicles. LH stimulates the further development of ovarian follicles and their full secretion of estrogens, brings about ovulation, and stimulates the production of estrogens, progesterone, relaxin, and inhibin by the corpus luteum.

b. PHASES OF THE FEMALE REPRODUCTIVE CYCLE

The duration of the female reproductive cycle typically is 24 to 35 days. For this discussion, we shall assume a duration of 28 days, divided into three phases: the menstrual phase, preovulatory phase, and postovulatory phase (Figure 26.17).

(1) Menstrual Phase (Menstruation)   The *menstrual* (MEN-stroo-al) *phase,* also called *menstruation* (men'-stroo-Ā-shun) or *menses* (*mensis* = month), lasts for roughly the first five days of the cycle. (By convention, the first day of menstruation marks the first day of a new cycle.)

Events in the Ovaries During the menstrual phase, 20 or so small secondary (antral) follicles, some in each ovary, begin to enlarge. Follicular fluid, secreted by the granulosa cells and oozing from blood capillaries, accumulates in the enlarging antrum while the oocyte remains near the edge of the follicle (see Figure 26.10).

Events in the Uterus Menstrual flow from the uterus consists of 50 to 150 ml of blood, tissue fluid, mucus, and epithelial cells derived from the endometrium. This discharge occurs because the declining level of estrogens and progesterone causes the uterine spiral arteries to constrict. As a result, the cells they supply become ischemic (deficient in blood) and start to die. Eventually, the entire stratum functionalis sloughs off. At this time the endometrium is very thin because only the stratum basalis remains. The menstrual flow passes from the uterine cavity to the cervix and through the vagina to the exterior.

(2) Preovulatory Phase   The *preovulatory phase,* the second phase of the female reproductive cycle, is the time between menstruation and ovulation. The preovulatory phase of the cycle is more variable in length than the other phases and accounts for most of the difference when cycles are shorter or longer than 28 days. It lasts from days 6 to 13 in a 28-day cycle.

Events in the Ovaries Under the influence of FSH, the group of about 20 secondary follicles continues to grow and begins to secrete estrogens and inhibin. By about day 6, one follicle in one ovary has outgrown all the others and is called the *dominant follicle.* Estrogens and inhibin secreted by the dominant follicle decrease the secretion of FSH, which causes the other less well-developed follicles to stop growing and undergo atresia.

The one dominant follicle becomes the *mature (Graafian) follicle* that continues to enlarge until it is more than 20 mm in diameter and ready for ovulation (see Figure 26.10). This follicle forms a blisterlike bulge on the surface of the ovary. Fraternal (nonidentical) twins may result if two secondary follicles achieve dominance and both ovulate. During the final maturation process, the dominant follicle continues to increase its production of estrogens under the influence of an increasing level of LH. Estrogens are the primary ovarian hormones before ovulation, but small amounts of progesterone are produced by the mature follicle a day or two before ovulation.

With reference to the ovaries, the menstrual phase and preovulatory phase together are termed the *follicular* (fō-LIK-yoo-lar) *phase* because ovarian follicles are growing and developing.

Events in the Uterus Estrogens being liberated into the blood by growing follicles stimulate the repair of the endometrium. Cells of the stratum basalis undergo mitosis and produce a new stratum functionalis. As the endometrium thickens, the short, straight endometrial glands develop and the arterioles coil and lengthen as they penetrate the stratum functionalis. The thickness of the endometrium approximately doubles to about 4 to 6 mm. With reference to the uterus, the preovulatory phase is also termed the *proliferative phase* because the endometrium is proliferating.

(3) Ovulation   The rupture of the mature (Graafian) follicle with release of the secondary oocyte into the pelvic cavity, called *ovulation,* usually occurs on day 14 in a 28-day cycle. During ovulation, the

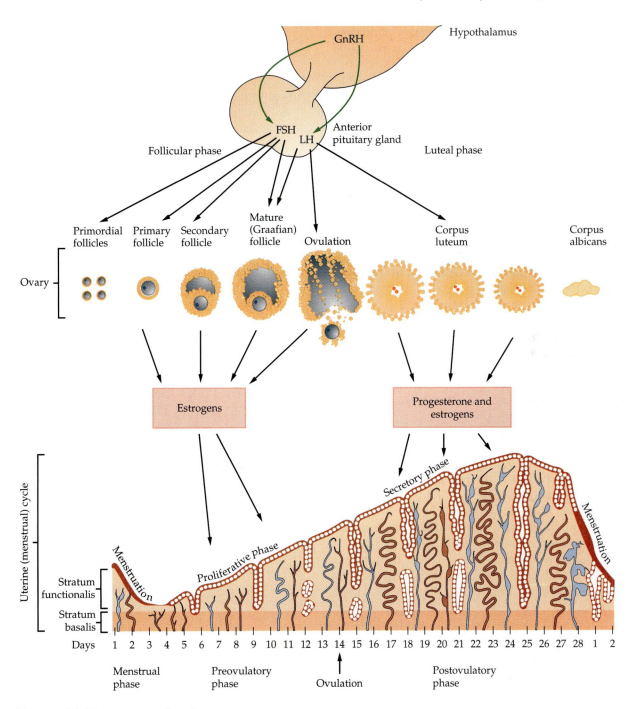

FIGURE 26.17   Menstrual cycle.

secondary oocyte remains surrounded by its zona pellucida and corona radiata. It generally takes a total of about 20 days (spanning the last 6 days of the previous cycle and the first 14 days of the current cycle) for a secondary follicle to develop into a fully mature follicle. During this time the developing ovum completes reduction division (meiosis I) and reaches metaphase of equatorial division (meiosis II). The *high* levels of estrogens during the

last part of the preovulatory phase exert a *positive feedback* effect on both LH and GnRH and cause ovulation.

An over-the-counter home test that detects the LH surge associated with ovulation is available. The test predicts ovulation a day in advance. FSH also increases at this time, but not as dramatically as LH because FSH is stimulated only by the increase in GnRH. The positive feedback effect of estrogens on the hypothalamus and anterior pituitary gland does not occur if progesterone is present at the same time.

After ovulation, the mature follicle collapses and blood within it forms a clot due to minor bleeding during rupture and collapse of the follicle to become the **corpus hemorrhagicum** (*hemo* = blood; *rhegnynai* = to burst forth). (See Figure 26.10.) The clot is eventually absorbed by the remaining follicular cells. In time, the follicular cells enlarge, change character, and form the corpus luteum under the influence of LH. Stimulated by LH, the corpus luteum secretes progesterone, estrogens, relaxin, and inhibin.

(4) Postovulatory Phase   The *postovulatory phase* of the female reproductive cycle is the most constant in duration and lasts for 14 days, from days 15 to 28 in a 28-day cycle. It represents the time between ovulation and the onset of the next menses. After ovulation, LH secretion stimulates the remnants of the mature follicle to develop into the corpus luteum. During its 2-week lifespan, the corpus luteum secretes increasing quantities of progesterone and some estrogens.

Events in One Ovary   If the egg is fertilized and begins to divide, the corpus luteum persists past its normal 2-week lifespan. It is maintained by **human chorionic** (kō-rē-ON-ik) **gonadotropin (hCG),** a hormone produced by the chorion of the embryo as early as 8-12 days after fertilization. The chorion eventually develops into the placenta and the presence of hCG in maternal blood or urine is an indication of pregnancy. As the pregnancy progresses, the placenta itself begins to secrete estrogens to support pregnancy and progesterone to support pregnancy and breast development for lactation. Once the placenta begins its secretion, the role of the corpus luteum becomes minor. With reference to the ovaries, this phase of the cycle is also called the **luteal phase.**

If hCG does not rescue the corpus luteum, after 2 weeks its secretions decline and it degenerates into a corpus albicans (see Figure 26.10). The lack of progesterone and estrogens due to degeneration of the corpus luteum then causes menstruation. In addition, the decreased levels of progesterone, estrogens, and inhibin promote the release of GnRH, FSH, and LH, which stimulate follicular growth, and a new ovarian cycle begins.

Events in the Uterus   Progesterone produced by the corpus luteum is responsible for preparing the endometrium to receive a fertilized ovum. Preparatory activities include growth and coiling of the endometrial glands, which begin to secrete glycogen, vascularization of the superficial endometrium, thickening of the endometrium, and an increase in the amount of tissue fluid. These preparatory changes are maximal about 1 week after ovulation, corresponding to the time of possible arrival of a fertilized ovum. With reference to the uterus, this phase of the cycle is called the **secretory phase** because of the secretory activity of the endometrial glands.

Obtain microscope slides of the endometrium showing the menstrual, preovulatory, and postovulatory phases of the menstrual cycle. See if you can note the differences in thickness of the endometrium, distribution of blood vessels, and distribution and size of endometrial glands.

C. DISSECTION OF FETUS-CONTAINING PIG UTERUS

Examination of the uterus of a pregnant pig reveals that the fetuses are equally spaced in the two uterine horns. Each fetus produces a local enlargement of the horn. The litter size normally ranges from 6 to 12. Your instructor may have you dissect the fetus-containing uterus of a pregnant pig or have one available as a demonstration (Figure 26.18). If you do a dissection, use the following directions. Also examine a chart or model of a human fetus and pregnant uterus if they are available.

CAUTION! *Please reread Section D, "Precautions Related to Dissection" at the beginning of the laboratory manual on page xiii before you begin your dissection.*

PROCEDURE

1. Using a sharp scissors, cut open one of the enlargements of the horn and you will see that each fetus is enclosed together with an elongated, sausage-shaped **chorionic vesicle.**
2. You will also notice many round bumps called **areolae** located over the chorionic surface.
3. The lining of the uterus together with the wall of the chorionic vesicle forms the **placenta.**
4. Carefully cut open the chorionic vesicle, avoiding cutting or breaking the second sac lying within that surrounds the fetus itself.

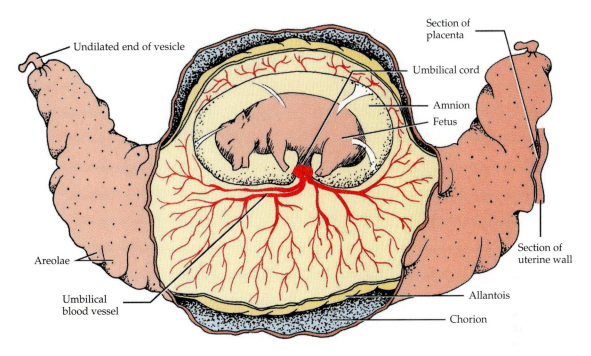

FIGURE 26.18   Fetal pig in opened chorionic vesicle.

5. The vesicle wall is the fusion of two extra-embryonic membranes, the outer *chorion* (KOR-ē-on) and the inner *allantois* (a-LAN-tō-is). The allantois is the large sac growing out from the fetus, and the umbilical cord contains its stalk.

6. The *umbilical blood vessels* are seen in the allantoic wall spreading out in all directions and are also seen entering the *umbilical cord.*

7. A thin-walled nonvascular *amnion* surrounds the fetus. This membrane is filled with *amniotic fluid,* which acts as a protective water cushion and prevents adherence of the fetus and membranes.

ANSWER THE LABORATORY REPORT QUESTIONS AT THE END OF THE EXERCISE.

Reproductive Systems 26

Student _____ Date _____

Laboratory Section _____ Score/Grade _____

PART 1. Multiple Choice

_____ 1. Structures of the male reproductive system responsible for the production of sperm cells are the (a) efferent ducts (b) seminiferous tubules (c) seminal vesicles (d) rete testis

_____ 2. The superior portion of the male urethra is encircled by the (a) epididymis (b) testes (c) prostate gland (d) seminal vesicles

_____ 3. Cryptorchidism is a condition associated with the (a) prostate gland (b) testes (c) seminal vesicles (d) bulbourethral (Cowper's) glands

_____ 4. Weakening of the suspensory ligament would directly affect the position of the (a) mammary glands (b) uterus (c) uterine (Fallopian) tubes (d) ovaries

_____ 5. Organs in the female reproductive system responsible for transporting secondary oocytes and ova from the ovaries to the uterus are the (a) uterine (Fallopian) tubes (b) seminal vesicles (c) inguinal canals (d) none of the above

_____ 6. The name of the process that is responsible for the actual production of sperm cells is called (a) cryptorchidism (b) oogenesis (c) spermatogenesis (d) spermatogonia

_____ 7. Fertilization normally occurs in the (a) uterine (Fallopian) tubes (b) vagina (c) uterus (d) ovaries

_____ 8. The portion of the uterus that assumes an active role during labor is the (a) serosa (b) endometrium (c) peritoneum (d) myometrium

_____ 9. Stereocilia are associated with the (a) ductus (vas) deferens (b) oviduct (c) epididymis (d) rete testis

_____ 10. The major portion of the volume of semen is contributed by the (a) bulbourethral (Cowper's) glands (b) testes (c) prostate gland (d) seminal vesicles

_____ 11. The chief ligament supporting the uterus and keeping it from dropping into the vagina is the (a) cardinal ligament (b) round ligament (c) broad ligament (d) ovarian ligament

_____ 12. Which sequence, from inside to outside, best represents the histology of the uterus? (a) stratum basalis, stratum functionalis, myometrium, perimetrium (b) myometrium, perimetrium, stratum functionalis, stratum basalis (c) stratum functionalis, stratum basalis, myometrium, perimetrium (d) stratum basalis, stratum functionalis, perimetrium, myometrium

_____ 13. The white fibrous capsule that divides the testis into lobules is called the (a) dartos (b) raphe (c) tunica albuginea (d) germinal epithelium

_____ **14.** Which sequence best represents the course taken by sperm cells from their site of origin to the exterior? (a) seminiferous tubules, efferent ducts, epididymis, ductus (vas) deferens, ejaculatory duct, urethra (b) seminiferous tubules, efferent ducts, epididymis, ductus (vas) deferens, urethra, ejaculatory duct (c) seminiferous tubules, efferent ducts, ductus (vas) deferens, epididymis, ejaculatory duct, urethra (d) seminiferous tubules, epididymis, efferent ducts, ductus (vas) deferens, ejaculatory duct, urethra

_____ **15.** The ovaries are anchored to the uterus by the (a) ovarian ligament (b) broad ligament (c) suspensory ligament (d) mesovarium

_____ **16.** The terminal duct for the male reproductive system is the (a) urethra (b) ductus (vas) deferens (c) inguinal canal (d) ejaculatory duct

_____ **17.** The site of sperm cell maturation is the (a) ductus (vas) deferens (b) spermatic cord (c) epididymis (d) testes

_____ **18.** Which of the following is the site of menstruation, implantation of a fertilized ovum, development of the fetus during pregnancy, and labor? (a) uterus (b) uterine (Fallopian) tubes (c) vagina (d) cervix

_____ **19.** Glands lying over the pectoralis major muscles are the (a) lesser vestibular glands (b) adrenal (suprarenal) glands (c) mammary glands (d) greater vestibular (Bartholin's) glands

PART 2. Completion

20. Discharge of a secondary oocyte from the ovary about once each month is a process referred to as _____.

21. The inferior, narrow portion of the uterus that opens into the vagina is the _____.

22. The clusters of milk-secreting cells of the mammary glands are referred to as _____.

23. The distal end of the penis is a slightly enlarged region called the _____.

24. Covering the slightly enlarged region of the penis is a loosely fitting skin called the _____.

25. The circular pigmented area surrounding each nipple of the mammary glands is the _____.

26. After a secondary oocyte leaves the ovary, it enters the open, funnel-shaped distal end of the uterine (Fallopian) tube called the _____.

27. The portion of a sperm cell that contains the nucleus and acrosome is the _____.

28. Vasectomy refers to removal of a portion of the _____.

29. The mass of erectile tissue in the penis that contains the spongy urethra is the _____.

30. Both the mature (Graafian) follicle and _____ of the ovary secrete hormones.

31. The superior dome-shaped portion of the uterus is called the _____.

32. The _____ anchor the uterus to either side of the pelvic cavity.

33. The passageway for menstrual flow and inferior portion of the birth canal is the

_____.

34. Two longitudinal folds of skin that extend inferiorly and posteriorly from the mons pubis and are

covered with pubic hair are the _____.

35. The _____ is a small mass of erectile tissue at the anterior junction of the labia minora.

36. The thin fold of vascularized membrane that borders the vaginal orifice is the

_____.

37. Complete the following sequence for the passage of milk: alveoli, secondary tubules, lactiferous

sinuses, _____, lactiferous ducts, nipple.

38. The layer of simple epithelium covering the free surface of the ovary is the _____.

39. The phase of the menstrual cycle between days 6 and 13 during which endometrial repair occurs is

the _____ phase.

40. During menstruation, the stratum _____ of the endometrium is sloughed off.

41. The most immature spermatogenic cells are called _____.

42. The _____ contains the remnants of an ovulated mature follicle.

43. The hypothalamic hormone that controls the uterine and ovarian cycles is _____.

44. High levels of estrogens exert a positive feedback on LH and GnRH that cause

_____.

Development

27

Developmental anatomy is the study of the sequence of events from fertilization of a secondary oocyte to the formation of a complete organism. Consideration will be given to how reproductive cells are produced and to a few developmental events associated with pregnancy.

A. SPERMATOGENESIS

The process by which the testes produce haploid (*n*) sperm cells involves several phases, including meiosis, and is called *spermatogenesis* (sper'-ma-tō-JEN-e-sis; *spermato* = sperm; *genesis* = to produce). In order to understand spermatogenesis, review the following concepts.

1. In sexual reproduction, a new organism is produced by the union and fusion of sex cells called *gametes* (*gameto* = to marry). Male gametes, produced in the testes, are called sperm cells, and female gametes, produced in the ovaries, are called oocytes.
2. The cell resulting from the union and fusion of gametes, called a *zygote* (*zygosis* = a joining), contains two full sets of chromosomes (DNA), one set from each parent. Through repeated mitotic cell divisions, a zygote develops into a new organism.
3. Gametes differ from all other body cells (somatic cells) in that they contain the *haploid* (one-half) *chromosome number,* symbolized as *n.* In humans, this number is 23, which composes a single set of chromosomes. The nucleus of a somatic cell contains the *diploid chromosome number,* symbolized as 2*n.* In humans, this number is 46, which composes two sets of paired chromosomes. One set of 23 chromosomes comes from the mother and the other set comes from the father.
4. In a diploid cell, two chromosomes that belong to a pair are called *homologous* (*homo* = same) *chromosomes (homologues).* In human diploid

cells, the members of 22 of the 23 pairs of chromosomes are morphologically similar and are called *autosomes.* The other pair, termed X and Y chromosomes, are called the *sex chromosomes* because they determine one's gender. In the female, the homologous pair of sex chromosomes are two similar X chromosomes; in the male, the pair consists of an X and a Y chromosome.
5. If gametes were diploid (2*n*), like somatic cells, the zygote would contain twice the diploid number (4*n*), and with every succeeding generation the chromosome number would continue to double and normal development could not occur.
6. This continual doubling of the chromosome number does not occur because of *meiosis* (*meio* = less), a process by which gametes produced in the testes and ovaries receive the haploid chromosome number. Thus, when haploid (*n*) gametes fuse, the zygote contains the diploid chromosome number (2*n*) and can undergo normal development.

In humans, spermatogenesis takes about 74 days. The seminiferous tubules are lined with immature cells called *spermatogonia* (sper-ma-tō-GŌ-nē-a; *sperm* = seed; *gonium* = generation or offspring), or sperm mother cells (see Figure 27.2 on page 533). Singular is *spermatogonium.* These cells develop from *primordial* (*primordialis* = primitive or early form) *germ cells* that arise from yolk sac endoderm and enter the testes early in development. In the embryonic testes, the primordial germ cells differentiate into spermatogonia but remain dormant until they begin to undergo mitotic proliferation at puberty. Spermatogonia contain the diploid (2*n*) chromosome number. Some spermatogonia remain relatively undifferentiated and capable of extensive mitotic division. Following division, some of the daughter cells remain undifferentiated and serve as a reservoir of precursor cells to prevent depletion of the stem cell population. Such cells remain near the basement membrane. The remainder of the daughter cells differentiate into spermatogonia that lose contact with the basement membrane of the seminiferous tubule, undergo certain developmental changes, and become known as ***primary spermatocytes*** (sper-MAT-ō-sīts'). Primary spermatocytes, like spermatogonia, are diploid (2*n*); that is, they have 46 chromosomes.

1. Meiosis I (Reduction Division)

Each primary spermatocyte enlarges before dividing. Then two nuclear divisions take place as part of meiosis. In the first, called ***meiosis I (reduction division),*** DNA is replicated and 46 chromosomes (each made up of two identical chromatids from the replicated DNA) form and move toward the equatorial plane of the cell. There they line up in homologous pairs so that there are 23 pairs of duplicated chromosomes in the center of the cell. This pairing of homologous chromosomes is called ***synapsis.*** The four chromatids of each homologous pair then become associated with each other to form a ***tetrad.*** In a tetrad, portions of one chromatid may be ex-changed with portions of another. This process, called ***crossing over,*** permits an exchange of genes among maternal and paternal chromosomes (Figure 27.1) that results in the ***recombination*** of genes. Thus, the sperm cells eventually produced are genetically unlike each other and unlike the cell that produced them—one reason for the great variation among humans. Next, the meiotic spindle forms and the kinetochore microtubules organized by the centromeres extend toward the poles of the cell. As the pairs separate, one member of each pair migrates to opposite poles of the dividing cell. The random arrangement of chromosome pairs on the spindle is another reason for variation among humans. The cells formed by the first nuclear division (meiosis I) are called ***secondary spermatocytes.*** Each cell has 23 chromosomes—the haploid number. Each chromosome of the secondary spermatocytes, however, is made up of two chromatids (two copies of the DNA) still attached by a centromere. Moreover, the genes of the chromosomes of secondary spermatocytes may be rearranged as a result of crossing over.

2. Meiosis II (Equatorial Division)

The second nuclear division of meiosis is ***meiosis II (equatorial division).*** There is no replication of DNA. The chromosomes (each composed of two chromatids) line up in single file along the equatorial plane, and the chromatids of each chromosome separate from each other. The cells formed from meiosis II are called ***spermatids.*** Each contains half the original chromosome number, or 23 chromosomes, and is haploid. Each primary spermatocyte therefore produces four spermatids by meiosis I and II. Spermatids lie close to the lumen of the seminiferous tubule.

3. Spermiogenesis

The final stage of spermatogenesis, called ***spermiogenesis*** (sper'-mē-ō-JEN-e-sis), involves the maturation of spermatids into sperm cells. Each spermatid embeds in a sustentacular (Sertoli) cell and develops a head with an acrosome (enzyme-containing granule) and a flagellum (tail). Sustentacular cells extend from the basement membrane to the lumen of the seminiferous tubule, where they nourish the developing spermatids. Since there is no cell division in spermiogenesis, each spermatid develops into a single ***sperm cell (spermatozoon).*** The release of a sperm cell from a sustentacular cell is known as ***spermiation.***

Sperm cells enter the lumen of the seminiferous tubule and migrate to the ductus epididymis, where in 10 to 14 days they complete their maturation and become capable of fertilizing a secondary oocyte. Sperm cells are also stored in the ductus (vas) deferens. Here, they can retain their fertility for up to several months.

With the aid of your textbook, label Figure 27.2.

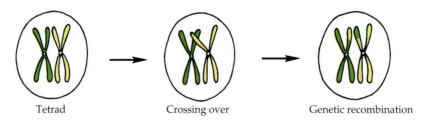

Tetrad   Crossing over   Genetic recombination

FIGURE 27.1   Crossing over within a tetrad, resulting in genetic recombination.

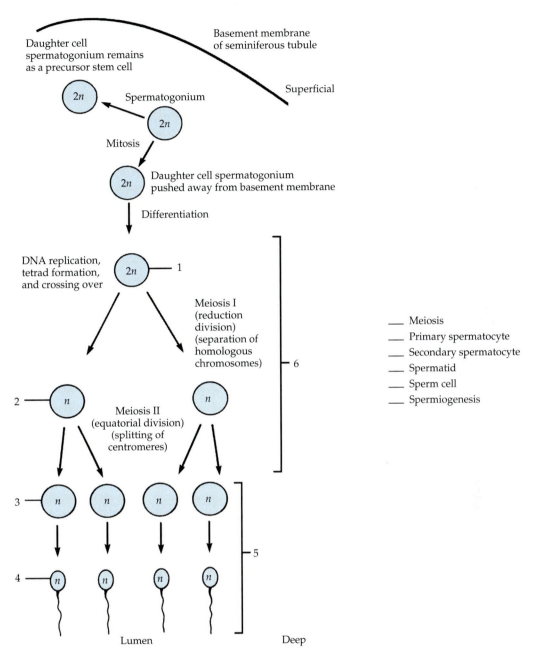

FIGURE 27.2   Spermatogenesis.

___ Meiosis
___ Primary spermatocyte
___ Secondary spermatocyte
___ Spermatid
___ Sperm cell
___ Spermiogenesis

B. OOGENESIS

The formation of haploid (*n*) secondary oocytes in the ovary involves several phases, including meiosis, and is referred to as *oogenesis* (ō'-ō-JEN-e-sis; *oo* = egg; *genesis* = to produce). With some important exceptions, oogenesis occurs in essentially the same manner as spermatogenesis.

1. Meiosis I (Reduction Division)

During early fetal development, primordial (primitive) germ cells migrate from the endoderm of the yolk sac to the ovaries. There, germ cells differentiate within the ovaries into *oogonia* (ō'-ō-GŌ-nē-a; singular is *oogonium* (ō'-ō-GŌ-nē-um). Oogonia are diploid (2*n*) cells that divide mitotically to produce millions of germ cells. Even before birth, many of these germ cells degenerate, a process known as *atresia.* A few develop into larger cells called *primary* (*primus* = first) *oocytes* (Ō'-ō-sītz) that enter prophase of reduction division (meiosis I) during fetal development but do not complete it until puberty. At birth 200,000–2,000,000 oogonia and primary oocytes remain in each ovary. Of these, about 400 will mature and ovulate during a woman's reproductive lifetime; the remaining 99.98% undergo atresia.

Each primary oocyte is surrounded by a single layer of follicular cells, and the entire structure is called a *primordial follicle* (see Figure 26.10). Although the stimulating mechanism is unclear, a few primordial follicles start to grow, even during childhood. They become *primary (preantral) follicles,* which are surrounded first by one layer of cuboidal-shaped follicular cells and then by six to seven layers of cuboidal and low-columnar cells called *granulosa cells.* As a follicle grows, it forms a clear glycoprotein layer, called the *zona pellucida* (pe-LOO-si-da) between the oocyte and the granulosa cells. The innermost layer of granulosa cells becomes firmly attached to the zona pellucida and is called the *corona radiata* (*corona* = crown; *radiata* = radiation). The outermost granulosa cells rest on a basement membrane that separates them from the surrounding ovarian stroma. This outer region is called the *theca folliculi.* As the primary follicle continues to grow, the theca differentiates into two layers: (1) the *theca interna,* a vascularized internal layer of secretory cells, and (2) the *theca externa,* an outer layer of connective tissue cells. The granulosa cells begin to secrete follicular fluid, which builds up in a cavity called the *antrum* in the center of the follicle. The follicle is now termed a *secondary (antral) follicle.* During early childhood, primordial and developing follicles continue to undergo atresia.

After puberty, under the influence of the gonadotropin hormones secreted by the anterior pituitary gland, each month meiosis resumes in one secondary follicle. The diploid primary oocyte completes reduction division (meiosis I) and two haploid cells of unequal size, both with 23 chromosomes (*n*) of two chromatids each, are produced. The follicle in which these events are taking place, termed the *mature (Graafian) follicle* (also called a *vesicular ovarian follicle*) will soon rupture and release its oocyte.

The smaller cell produced by meiosis I, called the *first polar body,* is essentially a packet of discarded nuclear material. The larger cell, known as the *sec-*

ondary oocyte, receives most of the cytoplasm. Once a secondary oocyte is formed, it proceeds to the metaphase of equatorial division (meiosis II) and then stops at this stage.

2. Meiosis II (Equatorial Division)

At ovulation, usually one secondary oocyte (with the first polar body and corona radiata) is expelled into the pelvic cavity. Normally, the cells are swept into the uterine (Fallopian) tube. If fertilization does not occur, the oocyte and other cells degenerate. If sperm cells are present in the uterine tube and one penetrates the secondary oocyte (fertilization), however, equatorial division (meiosis II) resumes. The secondary oocyte splits into two haploid (*n*) cells of unequal size. The larger cell is the *ovum,* or mature egg; the smaller one is the *second polar body.* The nuclei of the sperm cell and the ovum then unite, forming a diploid (2*n*) *zygote.* The first polar body may also undergo another division to produce two polar bodies. If it does, the primary oocyte ultimately gives rise to a single haploid (*n*) ovum and three haploid (*n*) polar bodies, which all degenerate. Thus an oogonium gives rise to a single gamete (ovum), whereas a spermatogonium produces four gametes (sperm).

With the aid of your textbook, label Figure 27.3.

C. EMBRYONIC PERIOD

The *embryonic period* is the first two months of development.

1. Fertilization

During *fertilization* (fer'-til-i-ZĀ-shun; *fertilis* = reproductive) the genetic material from the sperm cell and secondary oocyte merges into a single nucleus (Figure 27.4a). Of the 300–500 million sperm introduced into the vagina, less than 1% reach the secondary oocyte. Fertilization normally occurs in the uterine (Fallopian) tube about 12–24 hours after ovulation. Since ejaculated sperm cells remain viable for about 48 hours and an oocyte is viable for about 24 hours after ovulation, there typically is a 3-day window during which pregnancy can occur—from 2 days before to 1 day after ovulation. Peristaltic contractions and the action of cilia transport the oocyte through the uterine tube. Sperm cells swim up the female tract by whiplike movements of their tail (flagellum). The acrosome of sperm cells produces an

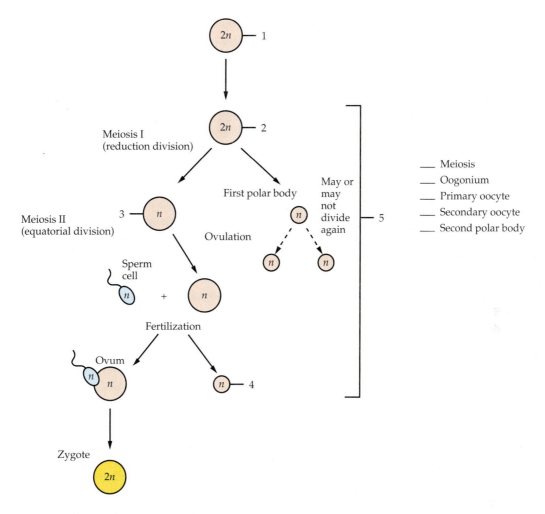

FIGURE 27.3 Oogenesis.

enzyme called **acrosin** that stimulates sperm cell motility and migration within the female reproductive tract. Also, muscular contractions of the uterus, stimulated by prostaglandins in semen, probably aid sperm cell movement toward the uterine tube. Finally, the oocyte is thought to secrete a chemical substance that attracts sperm cells.

Besides contributing to sperm cell movement, the female reproductive tract also confers on sperm cells the capacity to fertilize a secondary oocyte. Although sperm cells undergo maturation in the epididymis, they are still not able to fertilize an oocyte until they have been in the female reproductive tract for several hours.

Capacitation (ka-pas'-i-TĀ-shun) refers to the functional changes that sperm cells undergo in the female reproductive tract that allow them to fertilize a secondary oocyte. During this process, the membrane around the acrosome becomes fragile so that several destructive enzymes—hyaluronidase,

acrosin, and neuraminidase—are secreted by the acrosomes. It requires the collective action of many sperm cells to have just one penetrate the secondary oocyte. The enzymes help penetrate the corona radiata and zona pellucida around the oocyte. Sperm cells bind to receptors in the zona pellucida. Normally only one sperm cell penetrates and enters a secondary oocyte. This event is called **syngamy** (*syn* = together; *gamos* = marriage). Syngamy causes depolarization, which triggers the release of calcium ions inside the cell. Calcium ions stimulate the release of granules by the oocyte that, in turn, promote changes in the zona pellucida to block entry of other sperm cells. This prevents **polyspermy**, fertilization by more than one sperm cell. Once a sperm cell has entered a secondary oocyte, the oocyte completes equatorial division (meiosis II). It divides into a larger ovum (mature egg) and a smaller second polar body that fragments and disintegrates (see Figure 27.3).

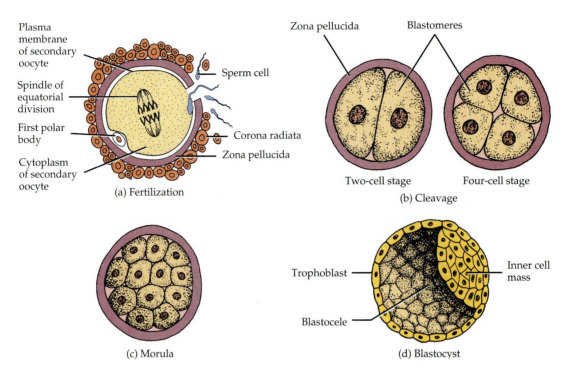

FIGURE 27.4   Fertilization.

When a sperm cell has entered a secondary oocyte, the tail is shed and the nucleus in the head develops into a structure called the *male pronucleus.* The nucleus of the secondary oocyte develops into a *female pronucleus.* After the pronuclei are formed, they fuse to produce a *segmentation nucleus.* The segmentation nucleus is diploid since it contains 23 chromosomes (*n*) from the male pronucleus and 23 chromosomes (*n*) from the female pronucleus. Thus the fusion of the haploid (*n*) pronuclei restores the diploid number (2*n*). The fertilized ovum, consisting of a segmentation nucleus, cytoplasm, and zona pellucida, is called a *zygote* (ZĪ-gōt; *zygosis* = a joining).

2. Formation of the Morula

After fertilization, rapid mitotic cell divisions of the zygote take place. These early divisions of the zygote are called *cleavage* (Figure 27.4b). Although cleavage increases the number of cells, it does not increase the size of the embryo, which is still contained within the zona pellucida.

The first cleavage begins about 24 hours after fertilization and is completed about 30 hours after fertilization, and each succeeding division takes slightly less time. By the second day after fertiliza-

tion, the second cleavage is completed. By the end of the third day, there are 16 cells. The progressively smaller cells produced by cleavage are called *blastomeres* (BLAS-tō-mērz; *blast* = germ, sprout; *meros* = part). Successive cleavages produce a solid sphere of cells, still surrounded by the zona pellucida, called the *morula* (MOR-yoo-la; *morula* = mulberry) (Figure 27.4c). A few days after fertilization, the morula is about the same size as the original zygote.

3. Development of the Blastocyst

By the end of the fourth day, the number of cells in the morula increases and it continues to move through the uterine (Fallopian) tube toward the uterine cavity. At 4½–5 days, the dense cluster of cells has developed into a hollow ball of cells and enters the uterine cavity; it is now called a *blastocyst* (*kystis* = bag) (Figure 27.4d).

The blastocyst has an outer covering of cells called the *trophoblast* (TRŌ-fō-blast; *troph* = nourish), an *inner cell mass (embryoblast),* and an internal fluid-filled cavity called the *blastocele* (BLAS-tō-sēl; *koilos* = hollow). The trophoblast ultimately forms part of the membranes composing the fetal portion of the placenta; part of the inner cell mass develops into the embryo.

PROCEDURE

1. Obtain prepared slides of the embryonic development of the sea urchin. First try to find a zygote. This will appear as a single cell surrounded by an inner fertilization membrane and an outer, jellylike membrane. Draw a zygote in the space provided.
2. Now find several cleavage stages. See if you can isolate two-cell, four-cell, eight-cell, and sixteen-cell stages. Draw the various stages in the spaces provided.
3. Try to find a blastula (called a blastocyst in humans), a hollow ball of cells with a lighter center due to the presence of the blastocele. Draw a blastula in the space provided.

Zygote

2-cell stage

4-cell stage

8-cell stage

16-cell stage

Blastula

4. Implantation

The blastocyst remains free within the cavity of the uterus for a short period of time before it attaches to the uterine wall. During this time, the zona pellucida disintegrates and the blastocyst enlarges. The blastocyst receives nourishment from glycogen-rich secretions of endometrial (uterine) glands, sometimes called uterine milk. About 6 days after fertilization the blastocyst attaches to the endometrium, a process called *implantation* (Figure 27.5). At this time, the endometrium is in its secretory phase.

As the blastocyst implants, usually on the posterior wall of the fundus or body of the uterus, it is oriented so that the inner cell mass is toward the endometrium. The trophoblast develops two layers in the region of contact between the blastocyst and endometrium. These layers are a **syncytiotrophoblast** (sin-sīt'-ē-ō-TRŌF-ō-blast; *syn* = joined; *cyto* = cell) that contains no cell boundaries and a **cytotrophoblast** (sī'-tō-TRŌF-ō-blast) between the inner cell mass and syncytiotrophoblast that is composed of distinct cells (Figure 27.5c). These two layers of trophoblast become part of the chorion (one of the fetal membranes) as they undergo further growth. During implantation, the syncytiotrophoblast secretes enzymes that enable the blastocyst to penetrate the uterine lining. The enzymes digest and liquefy the endometrial cells. The fluid and nutrients further nourish the burrowing blastocyst for about a week after implantation. Eventually, the blastocyst becomes buried in the endometrium. The trophoblast also secretes human chorionic gonadotropin (hCG) that rescues the corpus luteum from degeneration and sustains its secretion of progesterone and estrogens.

5. Primary Germ Layers

After implantation, the first major event of the embryonic period occurs. The inner cell mass of the blastocyst begins to differentiate into the three *primary germ layers:* ectoderm, endoderm, and mesoderm. These are the major embryonic tissues from which all tissues and organs of the body will develop. The process by which the two-layered inner cell mass is converted into a structure composed of the primary germ layers is called *gastrulation* (gas'-troo-LĀ-shun; *gastrula* = little belly).

Within 8 days after fertilization, the cells of the inner cytotrophoblast proliferate and form the amnion (a fetal membrane) and a space, the *amniotic* (am'-nē-OT-ik; *amnion* = lamb) or *amniotic cavity,* over the inner cell mass. The layer of cells of the

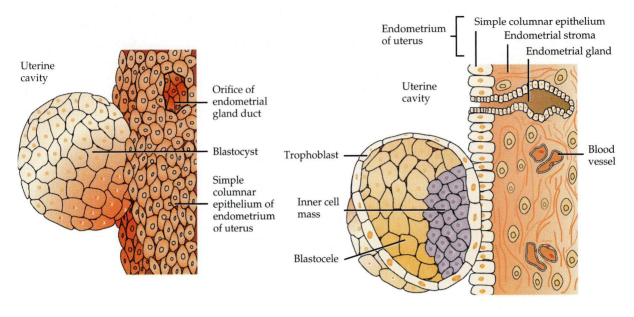

(a) External view, about 6 days after fertilization

(b) Internal view, about 6 days after fertilization

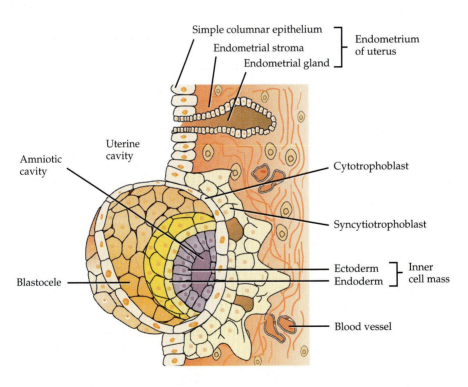

(c) Internal view, about 7 days after fertilization

FIGURE 27.5   Implantation.

inner cell mass that is closer to the amniotic cavity develops into the ***ectoderm*** (*ecto* = outside; *derm* = skin). The layer of the inner cell mass that borders the blastocele develops into the ***endoderm*** (*endo* = inside). As the amniotic cavity forms, the inner cell mass at this stage is called the ***embryonic disc.*** It will form the embryo. At this stage, the embryonic disc contains ectodermal and endodermal cells; the mesodermal cells are scattered external to the disc.

About the 12th day after fertilization, striking changes appear (Figure 27.6a). The cells of the endodermal layer have been dividing rapidly, so that groups of them now extend around in a circle, forming the yolk sac, another fetal membrane

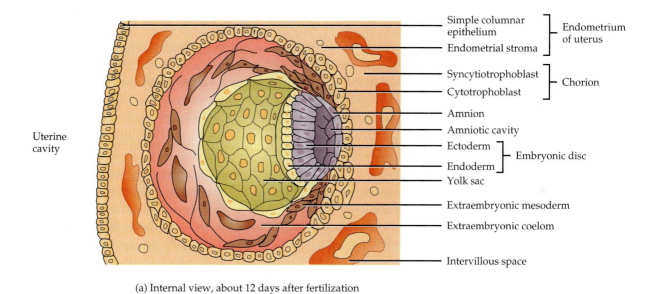

(a) Internal view, about 12 days after fertilization

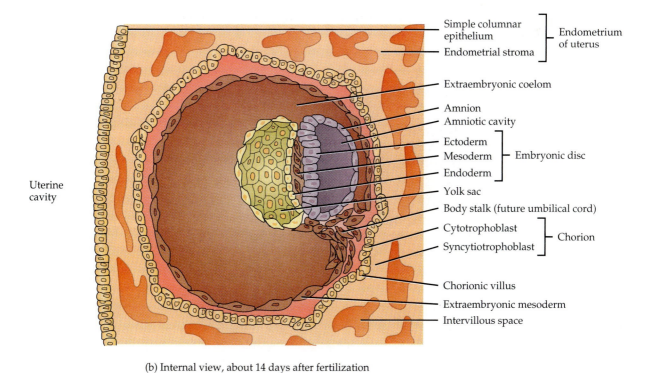

(b) Internal view, about 14 days after fertilization

FIGURE 27.6   Formation of the primary germ layers and associated structures.

(described shortly). Cells of the cytotrophoblast give rise to a loose connective tissue, the *extraembryonic mesoderm* (*meso* = middle). This completely fills the space between the cytotrophoblast and yolk sac. Soon large spaces develop in the extraembryonic mesoderm and come together to form a single, larger cavity called the *extraembryonic coelom* (SĒ-lōm; *koiloma* = cavity), the future ventral body cavity.

About the 14th day, the cells of the embryonic disc differentiate into three distinct layers: the ectoderm, the mesoderm, and the endoderm (Figure 27.6b).

As the embryo develops (Figure 27.6c), the endoderm becomes the epithelial lining of most of the gastrointestinal tract, urinary bladder, gallbladder, liver, pharynx, larynx, trachea, bronchi, lungs, vagina, urethra, and thyroid, parathyroid, and thymus glands, among other structures. The mesoderm develops into muscle; cartilage, bone and other connective tissues; red bone marrow, lymphoid tissue, endothelium of blood and lymphatic vessels, gonads, dermis of the skin, and other structures. The ectoderm develops into the entire nervous system, epidermis of skin, epidermal derivatives of the skin, and portions of the eye and other sense organs.

6. Embryonic Membranes

A second major event that occurs during the embryonic period is the formation of the *embryonic (extraembryonic) membranes* (Figure 27.7). These membranes lie outside the embryo and protect and nourish the embryo and, later, the fetus. The membranes are the yolk sac, amnion, chorion, and allantois.

In many species, the *yolk sac* is a membrane that is the primary source of nourishment for the embryo. A human embryo receives nutrients from the endometrium, however; the yolk sac remains small and functions as an early site of blood formation. The yolk sac also contains cells that migrate into the gonads and differentiate into the primitive germ cells (spermatogonia and oogonia).

The *amnion* is a thin, protective membrane that forms by the eighth day after fertilization and initially overlies the embryonic disc. As the embryo grows, the amnion comes to entirely surround the embryo, creating a cavity that becomes filled with *amniotic fluid.* Most amniotic fluid is initially derived from a filtrate of maternal blood. Later, the fetus makes daily contributions to the fluid by excreting urine into the amniotic cavity. Amniotic fluid serves as a shock absorber for the fetus, helps regulate fetal body temperature, and prevents adhesions between the skin of the fetus and surrounding tissues. Embryonic cells are sloughed off into amniotic fluid; they can be examined in the procedure called *amniocentesis* (am'-nē-ō-sen-TĒ-sis). The amnion usually ruptures just before birth and with its fluid constitutes the "bag of waters."

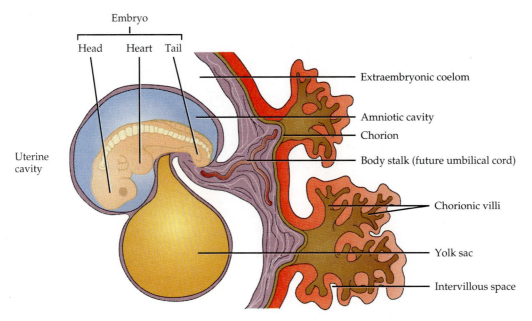

(c) External view, about 25 days after fertilization

FIGURE 27.6 (*Continued*) Formation of the primary germ layers and associated structures.

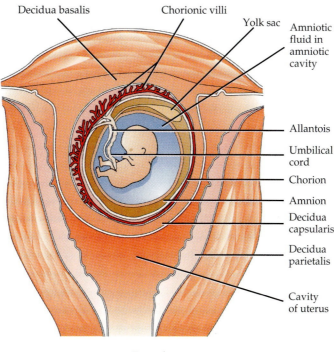

Decidua basalis   Chorionic villi   Yolk sac   Amniotic fluid in amniotic cavity

Allantois

Umbilical cord

Chorion

Amnion

Decidua capsularis

Decidua parietalis

Cavity of uterus

Frontal section

FIGURE 27.7   Embryonic membranes.

The *chorion* (KŌ-rē-on) is derived from the trophoblast of the blastocyst and the mesoderm that lines the trophoblast. It surrounds the embryo and, later, the fetus. Eventually, the chorion becomes the principal embryonic part of the placenta, the structure for exchange of materials between the mother and fetus. It also produces human chorionic gonadotropin (hCG). The amnion, which also surrounds the fetus, eventually fuses to the inner layer of the chorion.

The *allantois* (a-LAN-tō-is; *allas* = sausage) is a small, vascularized outpouching of the yolk sac. It serves as an early site of blood formation. Later its blood vessels serve as the umbilical connection in the placenta between mother and fetus. This connection is the umbilical cord.

7. Placenta and Umbilical Cord

Development of the *placenta* (pla-SEN-ta; *placenta* = flat cake), the third major event of the embryonic period, is accomplished by the third month of pregnancy. The placenta has the shape of a flat cake when fully developed and is formed by the chorion of the embryo and a portion of the endometrium (decidua basalis) of the mother (see Figure 27.7). Functionally, the placenta allows oxygen and nutrients to diffuse into fetal blood from maternal blood. Simul-

taneously, carbon dioxide and wastes diffuse from fetal blood into maternal blood at the placenta.

The placenta also is a protective barrier since most microorganisms cannot cross it. However, certain viruses, such as those that cause AIDS, German measles, chickenpox, measles, encephalitis, and poliomyelitis, may pass through the placenta. The placenta also stores nutrients such as carbohydrates, proteins, calcium, and iron, which are released into fetal circulation as required. Finally, the placenta produces several hormones that are necessary to maintain pregnancy. Almost all drugs, including alcohol, and many substances that can cause birth defects pass freely through the placenta.

If implantation occurs, a portion of the endometrium becomes modified and is known as the *decidua* (dē-SID-yoo-a; *deciduus* = falling off). The decidua includes all but the stratum basalis layer of the endometrium and separates from the endometrium after the fetus is delivered. Different regions of the decidua, which are all areas of the stratum functionalis, are named based on their positions relative to the site of the implanted, fertilized ovum (see Figure 27.7). The *decidua basalis* is the portion of the endometrium between the chorion and the stratum basalis of the uterus. It becomes the maternal part of the placenta. The *decidua capsularis* is the portion of the endometrium that covers the embryo and is located between the embryo and the uterine cavity. The *decidua parietalis* (par-rī-e-TAL-is) is the remaining modified endometrium that lines the noninvolved areas of the entire pregnant uterus. As the embryo and later the fetus enlarge, the decidua capsularis bulges into the uterine cavity and initially fuses with the decidua parietalis, thus obliterating the uterine cavity. By about 27 weeks, the decidua capsularis degenerates and disappears.

During embryonic life, fingerlike projections of the chorion, called *chorionic villi* (kō'-rē-ON-ik VIL-ī), grow into the decidua basalis of the endometrium (see Figure 27.7). These will contain fetal blood vessels of the allantois. They continue growing until they are bathed in maternal blood sinuses called *intervillous* (in-ter-VIL-us) *spaces.* Thus maternal and fetal blood vessels are brought into proximity. It should be noted, however, that maternal and fetal blood do not normally mix. Rather, oxygen and nutrients in the blood of the mother's intervillous spaces diffuse across the cell membranes into the capillaries of the villi while waste products diffuse in the opposite direction. From the capillaries of the villi, nutrients and oxygen enter the fetus through the umbilical vein. Wastes leave the fetus through the umbilical arteries, pass into the capillaries of the villi, and diffuse into the maternal blood. A few materials, such as IgG antibodies, pass from the blood of the mother into the capillaries of the villi.

The *umbilical* (um-BIL-i-kul) *cord* is a vascular connection between mother and fetus. It consists of two umbilical arteries that carry deoxygenated fetal blood to the placenta, one umbilical vein that carries oxygenated blood into the fetus, and supporting mucous connective tissue called Wharton's jelly from the allantois. The entire umbilical cord is surrounded by a layer of amnion (see Figure 27.7).

At delivery, the placenta detaches from the uterus and is termed the *afterbirth.* At this time, the umbilical cord is severed, leaving the baby on its own. The small portion (about an inch) of the cord that remains still attached to the infant begins to wither and falls off, usually within 12–15 days after birth. The area where the cord was attached becomes covered by a thin layer of skin and scar tissue forms. The scar is the *umbilicus (navel).*

Pharmaceutical companies use human placentas to harvest hormones, drugs, and blood. Portions of the placentas are also used for burn coverage. The placenta and umbilical cord veins can also be used in blood vessel grafts.

D. FETAL PERIOD

During the *fetal period,* the months of development after the second month, all the organs of the body grow rapidly from the original primary germ layers, and the organism takes on a human appearance. Some of the principal changes associated with fetal growth are summarized in Table 27.1.

ANSWER THE LABORATORY REPORT QUESTIONS AT THE END OF THE EXERCISE.

TABLE 27.1
Changes Associated with Embryonic and Fetal Growth

End of month	Approximate size and weight	Representative changes
1	0.6 cm (³⁄₁₆ in.)	Eyes, nose, and ears not yet visible. Vertebral column and vertebral canal form. Small buds that will develop into limbs form. Heart forms and starts beating. Body systems begin to form. The central nervous system appears at the start of the third week.
2	3 cm (1¼ in.) 1 g (¹⁄₃₀ oz)	Eyes far apart, eyelids fused, nose flat. Ossification begins. Limbs become distinct and digits are well formed. Major blood vessels form. Many internal organs continue to develop.
3	7½ cm (3 in.) 30 g (1 oz)	Eyes almost fully developed but eyelids still fused, nose develops bridge, and external ears are present. Ossification continues. Limbs are fully formed and nails develop. Heartbeat can be detected. Urine starts to form. Fetus begins to move, but it cannot be felt by mother. Body systems continue to develop.
4	18 cm (6½–7 in.) 100 g (4 oz)	Head large in proportion to rest of body. Face takes on human features and hair appears on head. Skin bright pink. Many bones ossified, and joints begin to form. Rapid development of body systems.
5	25–30 cm (10–12 in.) 200–450 g (½–1 lb)	Head less disproportionate to rest of body. Fine hair (lanugo) covers body. Skin still bright pink. Brown fat forms and is the site of heat production. Fetal movements commonly felt by mother (quickening). Rapid development of body systems.
6	27–35 cm (11–14 in.) 550–800 g (1¼–1½ lb)	Head becomes even less disproportionate to rest of body. Eyelids separate and eyelashes form. Substantial weight gain. Skin wrinkled and pink. Type II alveolar cells begin to produce surfactant.
7	32–42 cm (13–17 in.) 1100–1350 g (2½–3 lb)	Head and body more proportionate. Skin wrinkled and pink. Seven-month fetus (premature baby) is capable of survival. Fetus assumes an upside-down position. Testes descend into scrotum.
8	41–45 cm (16½–18 in.) 2000–2300 g (4½–5 lb)	Subcutaneous fat deposited. Skin less wrinkled. Chances of survival much greater at end of eighth month.
9	50 cm (20 in.) 3200–3400 g (7–7½ lb)	Additional subcutaneous fat accumulates. Lanugo shed. Nails extend to tips of fingers and maybe even beyond.

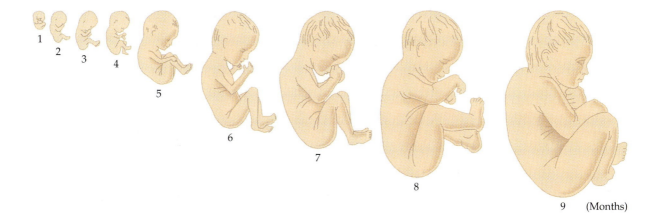

1   2   3   4   5   6   7   8   9   (Months)

Development 27

Student _____   Date _____

Laboratory Section _____   Score/Grade _____

PART 1. Multiple Choice

_____ 1. The basic difference between spermatogenesis and oogenesis is that (a) two more polar bodies are produced in spermatogenesis (b) the secondary oocyte contains the haploid chromosome number, whereas the mature sperm cell contains the diploid number (c) in oogenesis, one secondary oocyte is produced, and in spermatogenesis four mature sperm cells are produced (d) both mitosis and meiosis occur in spermatogenesis, but only meiosis occurs in oogenesis

_____ 2. The union of a sperm cell nucleus and a secondary oocyte nucleus resulting in formation of a zygote is referred to as (a) implantation (b) fertilization (c) gestation (d) parturition

_____ 3. The most advanced stage of development for these stages is the (a) morula (b) zygote (c) ovum (d) blastocyst

_____ 4. Damage to the mesoderm during embryological development would directly affect the formation of (a) muscle tissue (b) the nervous system (c) the epidermis of the skin (d) hair, nails, and skin glands

_____ 5. The placenta, the organ of exchange between mother and fetus, is formed by union of the endometrium with the (a) yolk sac (b) amnion (c) chorion (d) umbilicus

_____ 6. One oogonium produces (a) one ovum and three polar bodies (b) two ova and two polar bodies (c) three ova and one polar body (d) four ova

_____ 7. Implantation is defined as (a) attachment of the blastocyst to the uterine (Fallopian) tube (b) attachment of the blastocyst to the endometrium (c) attachment of the embryo to the endometrium (d) attachment of the morula to the endometrium

_____ 8. Epithelium lining most of the gastrointestinal tract and a number of other organs is derived from (a) ectoderm (b) mesoderm (c) endoderm (d) mesophyll

_____ 9. The nervous system is derived from the (a) ectoderm (b) mesoderm (c) endoderm (d) mesophyll

_____ 10. Which of the following is *not* an embryonic membrane? (a) amnion (b) placenta (c) chorion (d) allantois

PART 2. Completion

11. A normal human sperm cell, as a result of meiosis, contains _____ chromosomes.

12. The process that permits an exchange of genes resulting in their recombination and a part of the variation among humans is called _____.

13. The result of meiosis in spermatogenesis is that each primary spermatocyte produces four

   _____.

14. The stage of spermatogenesis that results in maturation of spermatids into sperm cells is called

   _____.

15. The afterbirth expelled in the final stage of delivery is the _____.

16. After the second month, the developing human is referred to as a(n) _____.

17. Embryonic tissues from which all tissues and organs of the body develop are called the

   _____.

18. The cells of the inner cell mass divide to form two cavities: amniotic cavity and

   _____.

19. Somatic cells that contain two sets of chromosomes are referred to as _____.

20. At the end of the _____ month of development, a heartbeat can be detected.

21. In oogenesis, primordial follicles develop into _____ follicles.

22. The clear, glycoprotein layer between the oocyte and granulosa cells is called the

   _____.

23. _____ refers to the functional changes that sperm cells undergo in the female reproductive tract that allow them to fertilize a secondary oocyte.

24. The _____ of a blastocyst develops into an embryo.

25. The decidua _____ is the portion of the endometrium between the chorion and
   stratum basalis of the uterus.

26. At the end of the _____ month of development, the testes descend into the scrotum.

Genetics

28

Genetics (je-NET-iks) is the branch of biology that studies inheritance. *Inheritance* is the passage of hereditary traits from one generation to another. It is through the passage of hereditary traits that you acquired your characteristics from your parents and will transmit your characteristics to your children. If all individuals were brown-eyed, we could learn nothing of the hereditary basis of eye color. However, because some people are blue-eyed and marry brown-eyed people, we can gain some knowledge of how hereditary traits are transmitted. We constantly analyze the genetic bases of the *differences* between individuals. Some of these differences occur normally, such as differences in eye color, blood groups, or ability to taste PTC (phenylthiocarbamide). Other differences are abnormal, such as physical abnormalities and abnormalities in the processes of metabolism.

A. GENOTYPE AND PHENOTYPE

The vast majority of human cells, except gametes, contain 23 pairs of chromosomes (diploid number) in their nuclei. One chromosome from each pair comes from the mother, and the other comes from the father. The two chromosomes that belong to a pair are called *homologous* (hō-MOL-ō-gus) *chromosomes,* and these homologues contain genes that control the same traits. The homologue of a chromosome that contains a gene for height also contains a gene for height.

The relationship of genes to heredity can be illustrated by the disorder called *phenylketonuria,* or *PKU* (see Figure 28.1). People with PKU are unable to manufacture the enzyme phenylalanine hydroxylase. Current belief is that PKU results from the presence of an abnormal gene symbolized as *p.* The normal gene is symbolized as *P. P* and *p* are said to be alleles. An *allele* is one of many alternative forms of a gene, occupying the same *locus* (position of a gene on a chromosome) in homologous chromosomes. The chromosome that has the gene that di-

rects phenylalanine hydroxylase production will have either *p* or *P* on it. Its homologue will also have either *p* or *P.* Thus every individual will have one of the following genetic makeups, or *genotypes* (JĒ-nō-tīps): *PP, Pp,* or *pp.* Although people with genotypes of *Pp* have the abnormal gene, only those with genotype *pp* suffer from the disorder because the normal gene masks the abnormal one. A gene that masks the expression of its allele is called the *dominant gene,* and the trait expressed is said to be a dominant trait. The homologous gene that is masked is called the *recessive gene.* The trait expressed when two recessive genes are present is called the recessive trait. Several dominant and recessive traits inherited in human beings are listed in Table 28.1.

Traditionally, the dominant gene is symbolized with a capital letter and the recessive one with a lowercase letter. When the same genes appear on homologous chromosomes, as in *PP* or *pp,* the person is said to be *homozygous* for a trait. When the genes on homologous chromosomes are different, however, as in *Pp,* the person is said to be *heterozygous* for the trait. *Phenotype* (FĒ-nō-tīp; *pheno* = showing) refers to how the genetic composition is expressed in the body. An individual with *Pp* has a different genotype from one with *PP,* but both have the same phenotype—which in this case is normal production of phenylalanine hydroxylase.

B. PUNNETT SQUARES

To determine how gametes containing haploid chromosomes unite to form diploid fertilized eggs, special charts called *Punnett squares* are used. The Punnett square is merely a device that helps one visualize all the possible combinations of male and female gametes, and is invaluable as a learning exercise in genetics. Usually, the possible paternal alleles in sperm cells are placed at the side of the chart and the possible maternal alleles in secondary oocytes are placed at the top (Figure 28.1). The spaces in the chart represent the possible genotypes

TABLE 28.1
Selected Hereditary Traits in Humans

Dominant	Recessive
Coarse body hair	Fine body hair
Male pattern baldness	Baldness
Normal skin pigmentation	Albinism
Freckles	Absence of freckles
Astigmatism	Normal vision
Near- or farsightedness	Normal vision
Normal hearing	Deafness
Broad lips	Thin lips
Tongue roller	Inability to roll tongue into a U shape
PTC taster	PTC nontaster
Large eyes	Small eyes
Polydactylism (extra digits)	Normal digits
Brachydactylism (short digits)	Normal digits
Syndactylism (webbed digits)	Normal digits
Feet with normal arches	Flat feet
Hypertension	Normal blood pressure
Diabetes insipidus	Normal excretion
Huntington's chorea	Normal nervous system
Normal mentality	Schizophrenia
Migraine headaches	Normal
Widow's peak	Straight hairline
Curved (hyperextended) thumb	Straight thumb
Normal Cl⁻ transport	Cystic fibrosis
Hypercholesterolemia (familial)	Normal cholesterol level

for that trait in fertilized ova formed by the union of the male and female gametes. Possible combinations are determined simply by dropping the female gamete on the left into the two boxes below it and dropping the female gamete on the right into the two spaces under it. The upper male gamete is then moved across to the two spaces in line with it, and the lower male gamete is moved across to the two spaces in line with it.

C. SEX INHERITANCE

Lining up human chromosomes in pairs reveals that the last pair (the twenty-third pair) differs in males and in females (Figure 28.2a). In females, the pair consists of two rod-shaped chromosomes designated as X chromosomes. One X chromosome is also present in males, but its mate is hook-shaped and called a Y chromosome. The XX pair in the female and the XY pair in the male are called the *sex chromosomes,* and all other pairs of chromosomes are called *autosomes.*

The sex of an individual is determined by the sex chromosomes (Figure 28.2b). When a spermatocyte undergoes meiosis to reduce its chromosome number from diploid to haploid, one daughter cell will contain the X chromosome and the other will contain the Y chromosome. When the secondary oocyte is fertilized by an X-bearing sperm, the offspring normally will be a female (XX). Fertilization by a Y sperm cell normally produces a male (XY).

Sometimes chromosomes fail to move toward opposite poles of a cell in meiotic anaphase. This is called *nondisjunction* and results in one sex cell having two members of a chromosome pair while the other receives none. Thus, eggs can contain two X's or no X (symbolized as 0), and sperm cells may contain both an X and a Y chromosome, two X's or two Y's, or no sex chromosomes at all.

Because the X chromosome contains so many genes unrelated to sex that are necessary for development, a zygote must contain at least one X chromosome to survive. Thus, Y0 and YY zygotes do not develop. However, other zygotes with sex chromosome anomalies do develop. Examples are Turner's syndrome (the presence of only one X chromosome, X0) and Kleinfelter's syndrome (an extra Y chromosome, XYY).

"Extra" X chromosomes (more than two in the female and more than one in the male) have a surprisingly minor effect on the individual, compared with the significant effect of other additional chromosomes. Studies show that only one X chromosome is active in any cell. Any additional X chromosomes are randomly inactivated early in development and do not express the genes contained on them.

These inactivated X chromosomes remain tightly coiled against the cell membrane and can be seen as what are called *Barr bodies* (Figure 28.3 on page 550). Since XY males do not have inactivated X chromosomes, no Barr bodies will be seen in normal male cells.

PROCEDURE

1. Make a buccal smear by *gently* scraping the inside of your cheek with the flat end of a toothpick. Discard the toothpick and *gently* scrape the same area again. This will produce more live cells from a deeper layer of the epithelium of the mucous membrane.

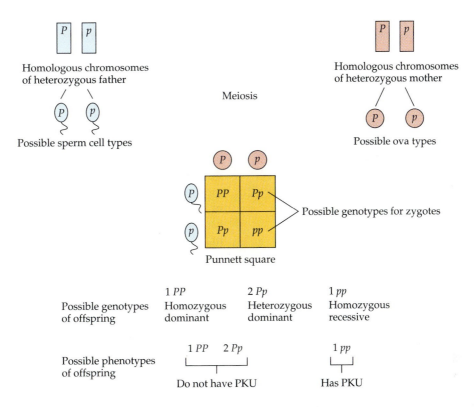

FIGURE 28.1 Inheritance of phenylketonuria (PKU).

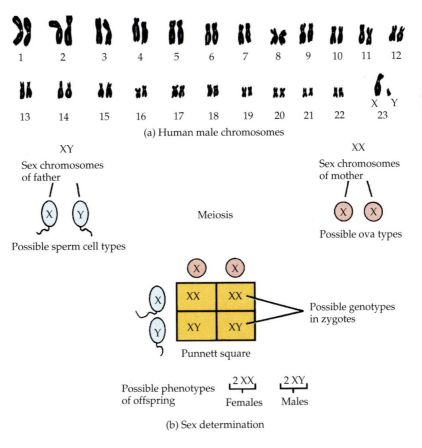

(a) Human male chromosomes

(b) Sex determination

FIGURE 28.2 Inheritance of sex. In (a), note the sex chromosomes, X and Y.

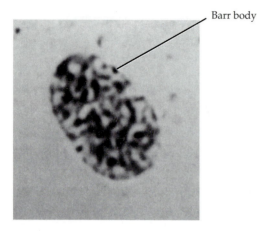

Barr body

FIGURE 28.3   Barr body in cell from buccal mucosa of human female. Feulgen stain; magnification ×300.

2. Spread the material over a clean glass slide.
3. Promptly place the slide in a Coplin jar filled with fixative for 1 min.
4. Remove the slide and wash gently under running tap water.
5. Place the slide in a Coplin jar filled with Giemsa stain for 10 to 20 min. Fresh solutions of stain prepared within 2 hr require 10 min; older stains require a longer time.
6. Wash the slide *gently* under running water and air-dry.
7. Examine the slide under high power and look for interphase nuclei. Identify Barr bodies, small disc-shaped chromatin bodies lying against the nuclear membrane (see Figure 28.3). Depending on the position of the nuclei and the staining technique, Barr bodies should be seen in 30 to 70% of the cells of a normal female.
8. Examine a slide prepared from the buccal epithelium of a class member not of your sex.
9. Draw a cell containing a Barr body in the space provided.

Barr body

D. SEX-LINKED INHERITANCE

As do the other 22 pairs of chromosomes, the sex chromosomes contain genes that are responsible for the transmission of a number of nonsexual traits. Genes for these traits appear on X chromosomes, but many of these genes are absent from Y chromosomes. Traits transmitted by genes on the X chromosome are called *sex-linked traits.* This pattern of heredity is different from the pattern described earlier. About 150 sex-linked traits are known in humans. Examples of sex-linked traits are red–green color blindness and hemophilia.

1. Red–Green Color Blindness

Let us consider the most common type of color blindness, called red–green color blindness. In this condition, there is a deficiency in either red or green cones and red and green are seen as the same color, either red or green, depending on which cone is present. The gene for *red–green color blindness* is a recessive one designated c. Normal color vision, designated C, dominates. The C/c genes are located on the X chromosome. The Y chromosome does not contain these genes. Thus the ability to see colors depends entirely on the X chromosomes. The possible combinations are:

Genotype	Phenotype
$X^C X^C$	Normal female
$X^C X^c$	Normal female (carrying the recessive gene)
$X^c X^c$	Red–green color-blind female
$X^C Y$	Normal male
$X^c Y$	Red–green color-blind male

Only females who have two X^c genes are red–green color-blind. This rare situation can result only from the mating of a color-blind male and a color-blind or carrier female. In $X^C X^c$ females, the trait is masked by the normal, dominant gene. Males, on the other hand, do not have a second X chromosome that would mask the trait. Therefore all males with an X^c gene will be red–green color blind. The inheritance of red–green color blindness is illustrated in Figure 28.4.

2. Hemophilia

Hemophilia is a condition in which the blood fails to clot or clots very slowly after an injury. Hemophilia is a much more serious defect than color blindness because individuals with severe hemophilia can bleed to death from even a small cut.

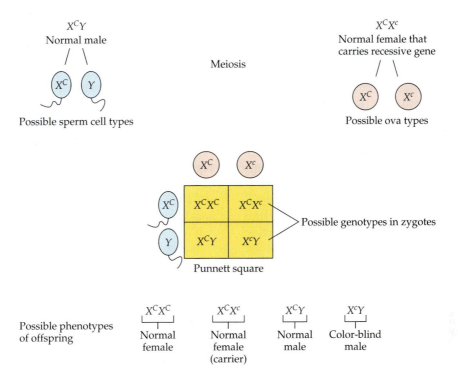

FIGURE 28.4 Inheritance of red–green color blindness.

Hemophilia is caused by a recessive gene as is color blindness. If H represents normal clotting and h represents abnormal clotting, $X^h X^h$ females will be hemophiliacs. Males with $X^H Y$ will be normal and males with $X^h Y$ will be hemophiliacs. Other sex-linked traits in humans are fragile X syndrome, nonfunctional sweat glands, certain forms of diabetes, some types of deafness, uncontrollable rolling of the eyeballs, absence of central incisors, night blindness, one form of cataract, juvenile glaucoma, and juvenile muscular dystrophy.

E. MENDELIAN LAWS

In any genetic cross, all the offspring in the first (that is, the parental, or P_1) generation are symbolized as F_1. The F is from the Latin word *filial,* which means progeny. The second generation is symbolized as F_2, the third as F_3, and continues that way. The recognized "father" of genetics is Gregor Mendel, whose basic experiments were performed on garden peas. As a result of his tests, Gregor Mendel postulated what are now called *Mendelian Laws,* or *Mendelian Principles.* The *First Mendelian Law,* or the *Law of Segregation,* asserts that, in cells of individuals, genes occur in pairs, and that when

those individuals produce germ cells, each germ cell receives only one member of the pair.

This law applies equally to pollen grains (or sperm) and to ova. The genetic cross is represented as follows:

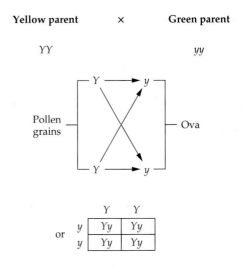

All possible combinations of pollen grains and ova are indicated by the arrows. Notice that all combinations yield the genotype Yy. All these F_1 seeds were yellow, yellow being dominant to green, or, in genetic terms, Y being dominant to y. These F_1

individuals resembled the yellow parent in phenotype (being yellow) but not in genotype (Yy as opposed to YY). Both parents were homozygous. Both members of that pair of alleles were the same. The yellow parent was homozygous for Y and the green parent for y. The F_1 individuals were heterozygous, having one Y and one y.

When the F_1 plants were self-fertilized, the F_2 seeds appeared in the ratio of 3 yellow/1 green. Mendel found similar 3:1 ratios for the other traits he studied, and this type of result has been reported in many species of animals and plants for a variety of traits. Not only does the recessive trait reappear in the F_2, but also in a definite proportion of the individuals, one-fourth of the total. If the sample is small, the ratio may deviate considerably from 3:1, but as the progeny or sampling numbers get larger, the ratio usually comes closer and closer to an exact 3:1 ratio. The reason is that the ratio depends on the random union of gametes. The result is a 3:1 phenotypic ratio, or a 1:2:1 genotypic ratio.

Thus, the four combinations of pollen and ova are expected to occur as follows:

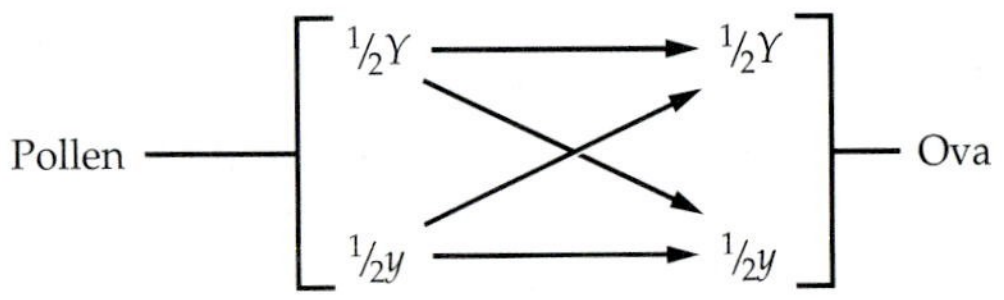

$$\frac{1}{4}YY = \text{yellow}$$
$$\frac{1}{4}Yy = \text{yellow} \quad \left.\right\} \ \tfrac{3}{4}$$
$$\frac{1}{4}Yy = \text{yellow}$$
$$\frac{1}{4}yy = \text{green} \quad \left.\right\} \ \tfrac{1}{4}$$

It is important to realize that these fractions depend on the operation of the laws of probability. A model using coins will emphasize the point. This model consists of two coins, a nickel and a penny, tossed at the same time. The penny may represent the pollen (male parent). At any given toss, the chances are equal that the penny will come up "heads" or that it will come up "tails." Similarly, at any given fertilization the chances are equal that a Y-bearing pollen grain or that a y-bearing one will be transmitted. The nickel represents the ovum. Again, the chances are equal for "heads" or "tails," just as the chances are equal that in any fertilization a Y-bearing or a y-bearing ovum will take part. If we toss the two coins together and do it many times, we will obtain approximately the following:

¼ nickel heads; penny heads	(= YY)
¼ nickel heads; penny tails	(= Yy)
¼ nickel tails; penny heads	(= yY)
¼ nickel tails; penny tails	(= yy)

If we assume "heads" as dominant, we find that three-fourths of the time there is at least one "head" and one-fourth of the time no "heads" (both coins "tails"). Hence this gives us a model of the 3:1 ratio dependent on the laws of probability.

The genetic cross just demonstrated considers only one pair of alleles (yellow versus green or "heads" versus "tails") and is therefore called a **monohybrid cross.** Mendel's second principle applies to genetic crosses in which two traits or two pairs of alleles are considered. These **dihybrid crosses** enabled him to postulate his second principle, the **Principle of Independent Assortment.** This principle states that the segregation of one pair of traits occurs independently of the segregation of a second pair of traits. This is the case only if the traits are caused by genes located on nonhomologous chromosomes.

When Mendel crossed garden peas with round yellow seeds with garden peas with wrinkled green seeds, his F_1 generation showed that yellow and round were dominant. If self-fertilization then occurred, the F_2 generation resulted as follows:

Round yellow	¾ × ¾ = ⁹⁄₁₆
Round green	¾ × ¼ = ³⁄₁₆
Wrinkled yellow	¼ × ¾ = ³⁄₁₆
Wrinkled green	¼ × ¼ = ¹⁄₁₆

Therefore, in a dihybrid cross, the expected phenotypic ratio was 9:3:3:1, with ⁹⁄₁₆ of the F_2 being doubly dominant, and only ¹⁄₁₆ being doubly recessive.

F. MULTIPLE ALLELES

In the genetics examples we have considered to this point, we have discussed only two alleles of each gene. However, many, and possibly all genes, have *multiple alleles;* that is, they exist in more than two allelic forms even though a diploid cell cannot carry more than two alleles.

One example of multiple alleles in humans involves ABO blood groups (Exercise 16). The four basic blood types (phenotypes) of the ABO system are determined by three alleles: I^A, I^B, and i. Alleles I^A and I^B are not dominant over each other. Rather, they are **codominant;** that is, both genes are expressed equally. Both I^A and I^B alleles, however, are dominant over allele i. These three alleles can give rise to six genotypes, as follows:

Genotype	Phenotype (blood type)
$I^A I^A$ or $I^A i$	A
$I^B I^B$ or $I^B i$	B
$I^A I^B$	AB
ii	O

Given this information, is it possible for a child with type O blood to have a mother with type O blood and a father with type AB blood?

Explain _____

_____

_____

If two children in a family have type O blood, the mother has type B blood and the father has type A blood, what is the genotype of the father?

_____

What is the genotype of the mother? _____

G. GENETICS EXERCISES

1. Karyotyping

A group of cytogeneticists meeting in Denver, Colorado, in 1960 adopted a system for classifying and identifying human chromosomes. Chromosome *length* and *centromere position* were the bases for classification. The Denver classification has become a standard for human chromosome studies. By the early 1970s, most human chromosomes could be identified microscopically.

Every chromosome pair could not be identified consistently until chromosome *banding techniques* finally distinguished all 46 human chromosomes. Bands are defined as parts of chromosomes that appear lighter or darker than adjacent regions with particular staining methods.

A *karyotype* is a chart made from a photograph of the chromosomes in metaphase. The chromosomes are cut out and arranged in matched pairs according to length (see Figure 28.2a). Their comparative size, shape, and morphology are then examined to determine if they are normal.

Karyotyping helps scientists to visualize chromosomal abnormalities. For example, individuals with Down syndrome typically have 47 chromosomes, instead of the usual 46, with chromosome 21 being represented three times rather than only twice. The syndrome is characterized by mental retardation, retarded physical development, and distinctive facial features (round head, broad skull, slanting eyes, and large tongue). With chronic myelogenous leukemia, part of the long arm of a chromosome 22 is missing, resulting in the blood disease. The chromosome is referred to as the Philadelphia chromosome, named for the city where it was first detected.

2. PKU Screening

Phenylketonuria (PKU), an inherited metabolic disorder that occurs in approximately 1 in 16,000 births, is transmitted by an autosomal recessive gene (see Figure 28.1). Individuals with this condition do not have the enzyme phenylalanine hydroxylase, which converts the amino acid phenylalanine to tyrosine. As a result, phenylalanine and phenylpyruvic acid accumulate in the blood and urine. These substances are toxic to the central nervous system and can produce irreversible brain damage. Most states in the United States require routine screening for this disorder at birth. The test is accomplished by a simple color change in treated urine.

The procedure for testing for PKU is as follows.

PROCEDURE

1. A Phenistix® test strip is made specifically for testing urine for phenylpyruvic acid. Dip this test strip in freshly voided urine.
2. Compare the color change with the color chart on the Phenistix® bottle. The test is based on the reaction of ferric ions with phenylpyruvic acid to produce a gray-green color.
3. Record your results in Section G.1 of the LABORATORY REPORT RESULTS at the end of the exercise.

3. PTC Inheritance

The ability to taste the chemical compound known as phenylthiocarbamide, commonly called PTC, is inherited. On the average, 7 out of 10 people, on chewing a small piece of paper treated with PTC, detect a definite bitter or sweet taste. Others do not taste anything.

Individuals who can taste something (bitter or sweet) are called "tasters" and have the dominant allele *T*, either as *TT* or *Tt*. A nontaster is a homozygous recessive and is designated as *tt*.

Determine your phenotype for tasting PTC and record your results in Section G.2 of the LABORATORY REPORT RESULTS at the end of the exercise.

Note: If PTC paper is not available, a 0.5% solution of phenylthiourea (PTT) can be substituted because the capacity to taste PTT is also inherited as a dominant.

4. Corn Genetics

Genetic corn may be purchased and used in this exercise. Each ear of corn represents a family of offspring. Mark a starting row with a pin to avoid repetition. Count the kernels (individuals) for each trait (color, wrinkled, or smooth). Record your results in Section G.3 of the LABORATORY REPORT RESULTS at the end of the exercise.

Develop a ratio by using your lowest number as "1" and dividing it into the others to determine what multiples of it they are. See how close you come to Mendel's ratios. Figure out the probable genotype and phenotype of the parent plants if you can. Monohybrid crosses, test crosses, dihybrid crosses, and trihybrid crosses are available.

5. Color Blindness

Using either Stilling or Ishihara test charts, test the entire class for red–green color blindness. Tests for color blindness depend on the person's ability to distinguish various colors from one another and also on his or her ability to judge correctly the degree of contrast between colors.

Of all men, 2% are color-blind to red and 6% to green, so 8% of all men are red–green color-blind. Red–green color blindness is rare in the female, occurring in only 1 of every 250 women. Record your results in Section G.4 of the LABORATORY REPORT RESULTS at the end of the exercise.

6. Mendelian Laws of Inheritance

Follow the procedure outlined in the explanation of the Mendelian Law of Segregation, tossing a nickel and a penny simultaneously to prove the law and determine ratios.

PROCEDURE

1. Toss the nickel and the penny together 10 times to get the genotypes of a family of 10. Repeat this procedure for a total of five times to obtain five families of 10 offspring. Record all of the results on the chart in Section G.5 of the LABORATORY REPORT RESULTS at the end of the exercise.

   Note: Use the following symbols for the following exercises.

   G = gene for yellow
   g = gene for green
   GG = the genotype of an individual pure (homozygous) for yellow
   gg = the genotype of an individual pure (homozygous) for green
   Gg = the genotype of the hybrid (heterozygous) individual, phenotypically yellow
   ♀ = symbol for female
   ♂ = symbol for male

2. Obeying the Mendelian Law of Segregation and using the Punnett square shown in Section G.6 of the LABORATORY REPORT RESULTS at the end of the exercise, cross yellow garden peas with green garden peas (a monohybrid cross). Show the P_1, F_1, and F_2 generations and all the different phenotypes and genotypes.

3. Obeying the Mendelian Law of Independent Assortment and using the Punnett square shown in Section G.8 of the LABORATORY REPORT RESULTS at the end of the exercise, cross the round yellow seeds with the wrinkled green seeds (a dihybrid cross). Show the P_1, F_1, and F_2 generations and all the different phenotypes and genotypes. The F_2 generation can be generated from a Punnett square comparable to that for the monohybrid cross, but with 16 rather than 4 squares.

7. Observing Phenotypes

The pattern of inheritance of many human traits is complex and involves many genes; the inheritance of other traits is controlled by single genes. You are asked to record your phenotype and your genotype, if it can be determined, for several traits controlled by single genes. For example, if you have the phenotypically dominant trait *A*, your genotype will be *A*−, indicating that the allele symbolized as "−"

is not known, since you could be homozygous dominant (*AA*) or heterozygous (*Aa*) for the trait. If you have the recessive trait *a−*, your genotype will be *aa*. Record your phenotype and genotype for the following traits in Section G.10 of the LABORATORY REPORT RESULTS at the end of the exercise.

1. **Attached earlobes** The dominant gene *E* causes earlobes that develop free from the neck; *ee* results in adherent earlobes connected to the cervical skin.
2. **Tongue rolling** The dominant gene *R* causes the development of muscles that allow the tongue to be rolled into a U shape. The *rr* genotype prohibits such rolling.
3. **Hair whorl direction** The dominant gene *W* causes the hair whorl on the cranial surface of the scalp to turn in a clockwise direction; the genotype *ww* determines a counterclockwise whorl.
4. **Little-finger bending** The dominant gene *B* causes the distal segment of the little finger to bend laterally. The genotype *bb* results in a straight distal segment.

5. **Double-jointed thumbs** The dominant gene *J* results in loose ligaments that allow the thumb to be bent out of the constricted orientation caused by the recessive genotype *jj*.
6. **Widow's peak** The dominant gene *W* causes the hairline to extend caudally in the midline of the forehead. The recessive genotype *ww* results in a straight hairline.
7. **Rh factor** A dominant gene *Rh* results in the presence of the Rh antigen on red blood cells. This antigen is not present with the recessive genotype *rhrh*. Use the results obtained in Exercise 16.I.4 to determine your phenotype.

   What additional information would you need to determine your complete genotype if you have the dominant phenotype for these traits?

   _____

   _____

   _____

ANSWER THE LABORATORY REPORT QUESTIONS AT THE END OF THE EXERCISE.

Genetics 28

Student _____ **Date** _____

Laboratory Section _____ **Score/Grade** _____

SECTION G. GENETICS EXERCISES

1. PKU Screening

1. _____ negative—cream color

   _____ 15 mg%—light green

   _____ 40 mg%—medium green

   _____ 100 mg%—dark green

2. PTC Inheritance

2. _____ bitter taste

   _____ sweet taste

   _____ negative (no taste)

3. Corn Genetics

3. Ratio_____ monohybrid cross

   Ratio_____ test cross

   Ratio_____ dihybrid cross

   Ratio_____ trihybrid cross

4. Color Blindness

	Male students	Female students
4. Red color-blind	_____	_____
Green color-blind	_____	_____

5. Mendelian Laws of Inheritance

5. Record the results of tossing a nickel and a penny together 10 times.

	Female (nickel)	Male (penny)	1	2	3	4	5	Total	Class total
Dominant offspring	A Heads	A Heads							
	A Heads	a Tails							
	a Tails	A Heads							
Recessive offspring	a Tails	a Tails							
Ratio dominant to recessive									

6. Complete the following monohybrid cross. Fill in genotypes (within circles) and phenotypes (under circles).

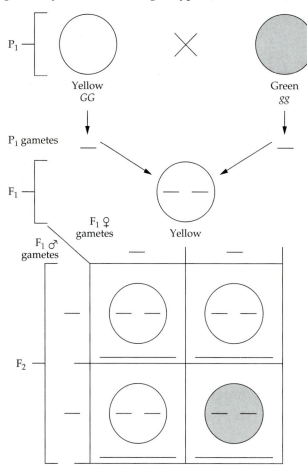

Monohybrid cross in the garden pea *(Pisum sativum)*. G = allele for yellow, *g* = allele for green, P₁ = parental generation, F₁ = first filial generation, F₂ = second filial generation.

7. What is the phenotype ratio of the F₂ generation? _____ yellow/ _____ green.

8. Complete the following dihybrid cross. Fill in genotypes (within circles) and phenotypes (under circles).

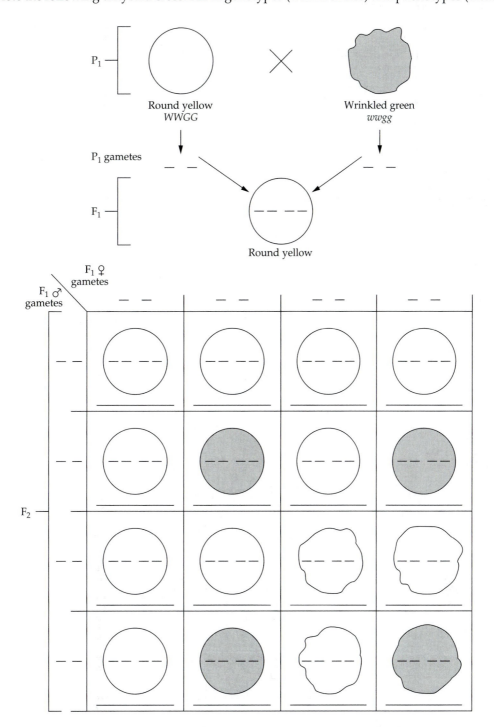

Dihybrid cross in the garden pea *(Pisum sativum).* W = allele for round, *w* = allele for wrinkled, G = allele for yellow, *g* = allele for green.

9. What is the phenotype ratio of the F₂ generation?

_____round yellow/_____ wrinkled yellow/_____ round green/

_____wrinkled green.

10. Observe phenotypes.

Trait	Phenotype	Genotype
Earlobes		
Tongue rolling		
Hair whorl		
Double-jointed thumbs		
Widow's peak		
Rh factor		

Genetics **28**

Student _____   **Date** _____

Laboratory Section _____   **Score/Grade** _____

PART **1.** Multiple Choice

_____   **1.** Using the symbols *Aa* to represent genes, which of the following is true? (a) the trait is homozygous for the dominant characteristic (b) the trait is homozygous for the recessive characteristic (c) the trait is heterozygous (d) sex-linked inheritance is in operation

_____   **2.** Which statement concerning the normal inheritance of sex is correct? (a) all zygotes contain a Y chromosome (b) some ova contain a Y chromosome (c) all ova and all sperm cells contain an X chromosome (d) all ova have an X chromosome, some sperm cells have an X chromosome, and some sperm cells have a Y chromosome

_____   **3.** The genotype that will express characteristics associated with hemophilia (assume that *H* represents the gene for normal blood) is (a) $X^H X^h$ (b) $X^h Y$ (c) $X^H X^H$ (d) $X^H Y$

_____   **4.** The exact position of a gene on a chromosome is called the (a) homologue (b) locus (c) triad (d) allele

PART **2.** Completion

5. When the same genes appear on homologous chromosomes, as in *PP* or *pp*, the individual is said to

be _____ for the trait.

6. If different genes appear on homologous chromosomes, as in *Pp*, the individual is

_____ for the trait.

7. Genetic composition expressed in the body or morphologically is called the body's

_____.

8. The device that helps one visualize all the possible combinations of male and female gametes is called

the _____.

9. The twenty-third pair of human chromosomes are the sex chromosomes. All of the other pairs of

chromosomes are called _____.

10. Red–green color blindness is an inherited trait that is specifically called a(n) _____ trait.

11. The recognized "father" of genetics is _____.

12. A genetic cross that involves only one pair of alleles (or traits) is called a(n) _____ cross.

13. Passage of hereditary traits from one generation to another is called _____.

14. The genetic makeup of an individual is called the person's _____.

15. One of the many alternative forms of a gene is called its _____.

16. Two chromosomes that belong to a pair are called _____ chromosomes.

17. A gene that masks the expression of its allele is called a(n) _____ gene.

18. Failure of chromosomes to move to opposite poles of a cell during meiotic prophase is called

    _____.

19. A normal female who carries the recessive gene for red–green color blindness would have the follow-

    ing genotype: _____.

20. Genes that are expressed equally are said to be _____.

Some Important Units of Measurement

ENGLISH UNITS OF MEASUREMENT

Fundamental or Derived Unit	Units and Equivalents
Length	12 inches (in.) = 1 foot (ft) = 0.333 yard (yd)
	3 ft = 1 yd
	1760 yd = 1 mile (mi)
	5280 ft = 1 mi
Mass	1 ounce (oz) = 28.35 grams (g); 1 g = 0.0353 oz
	1 pound (lb) = 453 g = 16 oz; 1 kilogram (kg) = 2.205 lb
	1 ton = 2000 lb = 907 kg
Time	1 second (sec) = 1/86 400 of a mean solar day
	1 minute (min) = 60 sec
	1 hour (hr) = 60 min = 3600 sec
	1 day = 24 hr = 1440 min = 86 400 sec
Volume	1 fluid dram (fl dr) = 0.125 fluid ounce (fl oz)
	1 fl oz = 8 fl dr = 0.0625 quart (qt) = 0.008 gallon (gal)
	1 qt = 256 fl dr = 32 fl oz = 2 pints (pt) = 0.25 gal
	1 gal = 4 qt = 128 fl oz = 1024 fl dr

METRIC UNITS OF LENGTH AND SOME ENGLISH EQUIVALENTS

Metric Unit	Meaning of Prefix	Metric Equivalent	English Equivalent
1 kilometer (km)	kilo = 1000	1000 m	3280.84 ft or 0.62 mi; 1 mi = 1.61 km
1 hectometer (hm)	hecto = 100	100 m	328 ft
1 dekameter (dam)	deka = 10	10 m	32.8 ft
1 meter (m)	Standard unit of length		39.37 in. or 3.28 ft or 1.09 yd
1 decimeter (dm)	deci = $\frac{1}{10}$	0.1 m	3.94 in.
1 centimeter (cm)	centi = $\frac{1}{100}$	0.01 m	0.394 in.; 1 in. = 2.54 cm
1 millimeter (mm)	milli = $\frac{1}{1000}$	0.001 m = $\frac{1}{10}$ cm	0.0394 in.
1 micrometer (μm) [formerly micron (μ)]	micro = $\frac{1}{1,000,000}$	0.0000001 m = $\frac{1}{10,000}$ cm	3.94×10^{-5} in.
1 nanometer (nm) [formerly millimicron (mμ)]	nano = $\frac{1}{1,000,000,000}$	0.000000001 m = $\frac{1}{10,000,000}$ cm	3.94×10^{-8} in.

TEMPERATURE

Unit	K	°F	°C
1 degree Kelvin (K)	1	$\frac{9}{5}$(K) − 459.7	K + 273.16*
1 degree Fahrenheit (°F)	$\frac{5}{9}$(°F) + 255.4	1	$\frac{5}{9}$(°F − 32)
1 degree Celsius (°C)	°C − 273	$\frac{9}{5}$(°C) + 32	1

* Absolute zero (K) = −273.16°C

VOLUME

Unit	ml	cm³	qt	oz
1 milliliter (ml)	1	1	1.06×10^{-3}	3.392×10^{-2}
1 cubic centimeter (cm³)	1	1	1.06×10^{-3}	3.392×10^{-2}
1 quart (qt)	943	943	1	32
1 fluid ounce (fl oz)	29.5	29.5	3.125×10^{-2}	1

Periodic Table of the Elements

KEY

6	← Atomic Number
C	← Symbol
12.01	← Atomic Weight
Carbon	← Name

1																		2
H																		He
1.0080																		4.003
Hydrogen																		Helium

3	4											5	6	7	8	9	10
Li	Be											B	C	N	O	F	Ne
6.940	9.013											10.82	12.011	14.008	16.000	19.00	20.183
Lithium	Berillium											Boron	Carbon	Nitrogen	Oxygen	Fluorine	Neon

11	12											13	14	15	16	17	18
Na	Mg											Al	Si	P	S	Cl	Ar
22.991	24.32											26.98	28.09	30.975	32.066	35.457	39.944
Sodium	Magnesium											Alminum	Silicon	Phosphorus	Sulfur	Chlorine	Argon

19	20	21	22	23	24	25	26	27	28	29	30	31	32	33	34	35	36
K	Ca	Sc	Ti	V	Cr	Mn	Fe	Co	Ni	Cu	Zn	Ga	Ge	As	Se	Br	Kr
39.100	40.08	44.96	47.90	50.95	52.01	54.94	55.85	58.94	58.71	63.54	65.38	69.72	72.60	74.91	78.96	79.916	83.80
Potassium	Calcium	Scandium	Titanium	Vanadium	Chromium	Manganese	Iron	Cobalt	Nickel	Copper	Zinc	Gallium	Germanium	Arsenic	Selenium	Bromine	Krypton

37	38	39	40	41	42	43	44	45	46	47	48	49	50	51	52	53	54
Rb	Sr	Y	Zr	Nb	Mo	Tc	Ru	Rh	Pd	Ag	Cd	In	Sn	Sb	Te	I	Xe
85.48	87.63	88.92	91.22	92.91	95.95	(99)	101.1	102.91	106.4	107.880	112.41	114.82	118.70	121.76	127.61	126.91	131.30
Rubidium	Strontium	Yttrium	Zirconium	Niobium	Molybdenum	Technetium	Ruthenium	Rhodium	Palladium	Silver	Cadmium	Indium	Tin	Antimony	Tellurium	Iodine	Xenon

55	56	57	72	73	74	75	76	77	78	79	80	81	82	83	84	85	86
Cs	Ba	La	Hf	Ta	W	Re	Os	Ir	Pt	Au	Hg	Tl	Pb	Bi	Po	At	Rn
132.91	137.36	138.92	178.49	180.95	183.86	186.22	190.2	192.2	195.09	197.0	200.61	204.39	207.21	209.00	(210)	(210)	(222)
Cesium	Barium	Lanthanum	Hafnium	Tantalum	Wolfram	Rhenium	Osmium	Iridium	Platinum	Gold	Mercury	Thallium	Lead	Bismuth	Polonium	Astatine	Radon

87	88	89	104	105	106	107	108	109
Fr	Ra	Ac	Unq	Unp	Unh	Uns	Uno	Une
(223)	(226)	(227)	(261)	(262)	(263)	(262)	(265)	(267)
Francium	Radium	Actinium	Unnilquadium	Unnilpentium	Unnilhexium	Unnilseptium	Unniloctium	Unnilennium

58	59	60	61	62	63	64	65	66	67	68	69	70	71
Ce	Pr	Nd	Pm	Sm	Eu	Gd	Tb	Dy	Ho	Er	Tm	Yb	Lu
140.13	140.92	144.27	(147)	150.35	152.0	157.26	158.93	162.51	164.94	167.27	168.94	173.04	174.99
Cerium	Praseodymium	Neodymium	Promethium	Samarium	Europium	Gadolinium	Terbium	Dysprosium	Holmium	Erbium	Thulium	Ytterbium	Lutetium

90	91	92	93	94	95	96	97	98	99	100	101	102	103
Th	Pa	U	Np	Pu	Am	Cm	Bk	Cf	Es	Fm	Md	No	Lw
(232)	(231)	238.07	(237)	(242)	(243)	(247)	(249)	(251)	(254)	(253)	(256)	(253)	257
Thorium	Protactinium	Uranium	Neptunium	Plutonium	Americium	Curium	Berkelium	Californium	Einsteinium	Fermium	Mendelevium	Nobelium	Lawrencium

Eponyms Used in
This Laboratory Manual

APPENDIX

C

An **eponym** is a term that includes reference to a person's name; for example, you may be more familiar with *Achilles tendon* than you are with the technical, but correct term *calcaneal tendon*. Because eponyms remain in frequent use, this glossary indicates which current terms replace eponyms in this manual. In the body of the text eponyms are cited in parentheses, immediately following the current terms where they are used for the first time in a chapter or later in the laboratory manual. In addition, although eponyms are included in the index, they have been cross-referenced to their current terminology.

Eponym	Current Terminology
Achilles tendon	calcaneal tendon
Adam's apple	thyroid cartilage
ampulla of Vater (VA-ter)	hepatopancreatic ampulla
Bartholin's (BAR-tō-linz) gland	greater vestibular gland
Billroth's (BIL-rōtz) cord	splenic cord
Bowman's (BŌ-manz) capsule	glomerular capsule
Bowman's (BŌ-manz) gland	olfactory gland
Broca's (BRŌ-kaz) area	motor speech area
Brunner's (BRUN-erz) gland	duodenal gland
bundle of His (HISS)	atrioventricular (AV) bundle
canal of Schlemm (SHLEM)	scleral venous sinus
circle of Willis (WIL-is)	cerebral arterial circle
Cooper's (KOO-perz) ligament	suspensory ligament of the breast
Cowper's (KOW-perz) gland	bulbourethral gland
crypt of Lieberkühn (LĒ-ber-kyoon)	intestinal gland
duct of Rivinus (ri-VĒ-nus)	lesser sublingual duct
duct of Santorini (san'-tō-RĒ-nē)	accessory duct
duct of Wirsung (VĒR-sung)	pancreatic duct
end organ of Ruffini (roo-FĒ-nē)	type-II cutaneous mechanoreceptor
Eustachian (yoo-STĀ-kē-an) tube	auditory tube
Fallopian (fal-LŌ-pē-an) tube	uterine tube
gland of Zeis (ZĪS)	sebaceous ciliary gland
Golgi (GOL-jē) tendon organ	tendon organ
Graafian (GRAF-ē-an) follicle	mature follicle
Hassall's (HAS-alz) corpuscle	thymic corpuscle
Haversian (ha-VĒR-shun) canal	central canal
Haversian (ha-VĒR-shun) system	osteon
interstitial cell of Leydig (LĪ-dig)	interstitial endocrinocyte
islet of Langerhans (LANG-er-hanz)	pancreatic islet
Kupffer (KOOP-fer) cells	stellate reticuloendothelial cell
loop of Henle (HEN-lē)	loop of the nephron
Malpighian (mal-PIG-ē-an) corpuscle	splenic nodule
Meibomian (mī-BŌ-mē-an) gland	tarsal gland
Meissner's (MĪS-nerz) corpuscle	corpuscle of touch
Merkel (MER-kel) disc	tactile disc
Müller's (MIL-erz) duct	paramesonephric duct
Nissl (NISS-l) bodies	chromatophilic substance
node of Ranvier (ron-VĒ-ā)	neurofibral node
organ of Corti (KOR-tē)	spiral organ
Pacinian (pa-SIN-ē-an) corpuscle	lamellated corpuscle
Peyer's (PI-erz) patches	aggregated lymphatic follicles
plexus of Auerbach (OW-er-bak)	myenteric plexus
plexus of Meissner (MĪS-ner)	submucous plexus
pouch of Douglas	rectouterine pouch
Purkinje (pur-KIN-jē) fiber	conduction myofiber
Rathke's (RATH-kēz) pouch	hypophyseal pouch
Schwann (SCHVON) cell	neurolemmocyte
Sertoli (ser-TŌ-lē) cell	sustentacular cell
Skene's (SKĒNZ) gland	paraurethral gland
sphincter of Oddi (OD-dē)	sphincter of the hepatopancreatic ampulla
Stensen's (STEN-senz) duct	parotid duct
Volkmann's (FŌLK-manz) canal	perforating canal
Wharton's (HWAR-tunz) duct	submandibular duct
Wharton's (HWAR-tunz) jelly	mucous connective tissue
Wormian (WER-mē-an) bone	sutural bone

Figure Credits

1.1 Courtesy of Olympus America, Inc.

3.5 (a)-(f) © Courtesy of Photo Researchers, Inc./Science Source.

4.1 (a) Biophoto Associates/Photo Researchers, Inc. (b) M. I. Walker/Photo Researchers, Inc. (c) Biophoto Associates/Photo Researchers, Inc. (d) Biophoto Associates/Photo Researchers, Inc. (e) Fred Hossler/Visuals Unlimited (f), (g) Ed Reschke.

4.3 (a) Ed Reschke (b) Biophoto Associates/Photo Researchers, Inc. (c) Robert Brons/Biological Photo Service (d) Biophoto Associates/Photo Researchers, Inc. (e) Ed Reschke (f) Bruce Iverson/Visuals Unlimited (g) Fred Hossler/Visuals Unlimited (h) Frederick C. Skvara (i) Chuck Brown/Photo Researchers, Inc.

5.2 Reproduced by permission from Dr. R. G. Kessel and Dr. R. H. Kardon, *Tissues and Organs: A Text Atlas of Scanning Electron Microscopy* H. Freeman.

6.2 Manfred Kage/Peter Arnold, Inc.

6.3 Biophoto Associates/Photo Researchers, Inc.

8.2 (a), (b), (c), (e) Copyright © 1983 by Gerard Tortora, Courtesy of Matt Iacobino and Lynne Borghesi (d) Courtesy of Evan J. Colella (f), (g), (h) Courtesy of Matt Iacobino (i), (j) © 1991 Evan J. Colella.

8.4 Robert A. Chase, M.D/From a Stereoscopic Atlas of Human Anatomy by David L. Bassett.

9.1 Biophoto Associates/Photo Researchers, Inc.

9.3 (a) Alfred Owczarzak/Biological Photo Service (b) Ed Reschke.

9.4 Biophoto Associates/Photo Researchers, Inc.

9.5 Physiograph is a registered trademark of Narco BioSystems Division of International Biomedical, Inc.

9.6 Visuals Unlimited.

9.7 Phototake/Carolina Biological Supply Company.

11.1-11.12 GTS Graphics.

12.1 (c) Ed Reschke.

13.3 John D. Cunningham/Visuals Unlimited.

13.10 Martin M. Rotker/Science Source/Photo Researchers, Inc.

13.14 Fred Hossler/Visuals Unlimited.

13.17 Martin M. Rotker.

13.18 By permission of Tektronix, Inc.

14.6 By Andrew Kuntzman, Wright State University.

14.7 (b) © Carolina Biological Supply/PhotoTake NYC.

14.10 Biomedical Graphics Department, University of Minnesota Hospitals.

14.18 (a) John Cunningham/Visuals Unlimited. (b) Biophoto Associates/Photo Researchers, Inc.

14.19 Copyright © 1983 by Gerard Tortora, Courtesy of James Borghesi.

15.2 R. Calentine/Visuals Unlimited.

15.3 By James R. Smail and Russell A. Whitehead, Macalester College.

15.4 © G. W. Willis/Biological Photo Service.

15.5 © Don W. Fawcett/Visuals Unlimited.

15.6 Frank Awbrey/Visuals Unlimited.

15.13 Manfred Kage/Peter Arnold, Inc.

15.14 Yoav Levy/Phototake NYC.

16.2 Bruce Iverson.

16.3 Project Masters, Inc./The Bergman Collection.

16.4 G. W. Willis, MD/Biological Photo Service.

16.5 R. Calentine/Visuals Unlimited

16.6 Ed Reschke/Peter Arnold, Inc.

16.7 Photograph courtesy of Fisher Scientific.

16.8 Photograph courtesy of Lenni Patti.

16.9 Courtesy of Leica, Inc. Deerfield, IL.

17.1 J. O. Ballard, M.D., Professor of Medicine & Pathology, Penn State University College of Medicine.

17.2 Courtesy of Chihiro Yokochi, MD and Johannes W. Rohen, MD from the book *Photographic Anatomy of the Human Body*, Igaku-Shoin Medical Publishers, New York, NY; 1978.

17.3 Photo courtesy of Becton Dickinson and Company.

18.2 Joseph R. Seibert, Ph.D./Custom Medical Stock Photo, Inc.

18.3 Martin M. Rotker/Photo Researchers, Inc.

19.1 (a) © Martin Rotker/PhotoTake NYC (b) © Carolina Biological Supply/PhotoTake NYC (c) © CNRI / PhotoTake NYC.

22.3 Copyright © Gerard Tortora, Courtesy of Lynne Borghesi.

22.5 Video Surgery/Photo Researchers, Inc.

22.7 Ed Reschke.

22.10 Reproduced by permission from R. G. Kessel and R. H. Kardon, *Tissues and Organs: A Text-Atlas of Scanning Electron Microscopy*, W. H. Freeman, 1979.

22.13 Ward's Natural Science Establishment, Inc.

22.14 Courtesy of Kinetics, Upper Saddle River, NJ.

22.17 Courtesy of Warren E. Collins, Inc. Braintree, MA.

23.3 Courtesy of Lynne M. Borghesi, ©1997.

23.4 Ed Reschke.

23.7 Phototake/Carolina Biological Supply Company.

23.9 John D. Cummingham/Visuals Unlimited.

23.10 (b) G. W. Willis, MD/Biological Photo Service.

23.11 Ed Reschke.

23.13 (a) Ed Reschke/Peter Arnold, Inc. (b) David M. Phillips/Visuals Unlimited, (c) Ed Reschke/Peter Arnold, Inc.

23.15 (a) Bruce Iverson/Visuals Unlimited. (b) Cabisco/Visuals Unlimited.

24.2 (b) Photo Researchers.

24.5 Andrew J. Kuntzman, Ph.D.

24.8 Ed Reschke.

24.9 John D. Cunningham/Visuals Unlimited.

24.10 Bruce Iverson.

24.11 (a1) Courtesy of S. K. Strasinger, *Urinalysis and Body Fluids*, Third Edition, F. A. Davis Co., Philadelphia, 1994. Fig. 10. (a2) Courtesy of S. K. Strasinger, *Urinalysis and Body Fluids*, Third Edition, F. A. Davis Co., Philadelphia, 1994. Fig. 8. (a3) Courtesy of W. Jao, M.D., et. al., *An Atlas of Urinary Sediment*, Abbot Laboratories, Chicago, IL, 1980. Fig. 60, p. 23. (a4) Courtesy of W. Jao, M.D., et. al., *An Atlas of Urinary Sediment*, Abbot Laboratories, Chicago, IL, 1980. Fig. 61, p. 24. (b1) Courtesy of S. K. Strasinger, *Urinalysis and Body Fluids*, Third Edition, F. A. Davis Co., Philadelphia, 1994. Fig. 16. (b2) Courtesy of S. K. Strasinger, *Urinalysis and Body Fluids*, Third Edition, F. A. Davis Co., Philadelphia, 1994. Fig. 21. (b4) Courtesy of W. Jao, M.D., et. al., *An Atlas of Urinary Sediment*, Abbot Laboratories, Chicago, IL, 1980. Fig. 24, p. 11. (b5) Courtesy of S. K. Strasinger, *Urinalysis and Body Fluids*, Third Edition, F. A. Davis Co., Philadelphia, 1994. Fig. 19. (b6) Courtesy of S. K. Strasinger, *Urinalysis and Body Fluids*, Third Edition, F. A. Davis Co., Philadelphia, 1994. Fig. 17. (c1) Courtesy of W. Jao, M.D., et. al., *An Atlas of Urinary Sediment*, Abbot Laboratories, Chicago, IL 1980. Fig. 87, p. 33. (c2) Courtesy of S. K. Strasinger, *Urinalysis and Body Fluids*, 3e, F. A. Davis Co., Philadelphia, 1994. Fig. 29. (c3) Courtesy of W. Jao, M.D., et. al., *An Atlas of Urinary Sediment*, Abbot Laboratories, Chicago, IL, 1980. Fig. 97, p. 36.

24.13 Ed Reschke/Peter Arnold, Inc. (a) David M. Phillips/Visuals Unlimited (b) Ed Reschke/Peter Arnold, Inc.

24.15 (a) Biophoto Associates/Science Source/ Photo Researchers, Inc. (b) Cabisco/Visuals Unlimited.

26.2 Ed Reschke/Peter Arnold, Inc.

26.4 Bruce Iverson.

26.5 Dr. Andrew Kuntzman.

26.9 Martin M. Rotker.

26.11 (a), (b) Biophoto Associates/Photo Researchers, Inc.

26.12 Ed Reschke.

26.15 Science Photo Library/Photo Researchers, Inc.

Index

Little, Brown's Nursing Q&A

Little, Brown's Nursing Q&A
Critical-Thinking Exercises

Sally Lambert Lagerquist, RN, MS

President and Course Coordinator, Review for Nurses, Inc. and RN Tapes Company;
Former Instructor of Undergraduate, Graduate, and Continuing Education in Nursing,
University of California, San Francisco, School of Nursing, San Francisco

1 8 3 7

Little, Brown and Company
Boston   New York   Toronto   London

Library of Congress Cataloging-in-Publication Data

Lagerquist, Sally L.
    Little, Brown's nursing Q&A : critical-thinking exercises / Sally
Lambert Lagerquist.
        p.   cm. — (RN NCLEX review series)
    Includes bibliographical references and index.
    ISBN 0-316-51298-2
    1. Nursing—Examinations, questions, etc.   I. Title.   II. Series.
RT55.L34   1996
610.73'076—dc20   96-10793
                                                        CIP

Notice. The indications for and dosages of all drugs in this book have been recommended in the medical literature and conform to the practices of the general medical community. The medications described do not necessarily have specific approval by the Food and Drug Administration for use in the diseases and dosages for which they are recommended. The package insert for each drug should be consulted for use and dosage as approved by the FDA. Because standards for usage change, it is advisable to keep abreast of revised recommendations, particularly those concerning new drugs.

Printed in the United States of America

MV-NY

Editorial: Evan R. Schnittman, Suzanne Jeans
Production Editor: Katharine S. Mascaro
Copyeditor: Mary Babcock
Production Supervisor: Cate Rickard
Designer: Virginia Pierce
Cover Design: Martucci Design, Inc.

This book is dedicated to:

My family:

A heart full of never-ending gratitude for giving me so much of yourselves all these years, and for making family a priority.

My husband Tom: Your large quantities of loving care and humor gave me renewed energy through all these years of writing nursing review books and lecturing all over the U.S. Thirty years of being with you is not enough! I have a dream . . . to grow older joyfully with you . . . and that will make all the difference!

Our daughter, who is a source of love and joy:
    Ever gentle when it counts
    Loving and caring, expressed in many quiet ways
    Always creative and showing her positive Slavic heritage
    Natural and her "own person"
    And a special person in my life!

Our son, to whom I'm grateful for being:
    Kind and sentimental
    Analytical and challenging of status quo
    Loving and giving, full of surprises
    Energetic and eager to take the "road less travelled"
    Nordic not only in heritage, but in spirit and outlook.

My co-workers, my caring friends:

Lourdes Perez, who almost single-handedly ran our nursing educational company as well as handled most of the interface between me and the outside world. All this made it possible for me to complete this book. You are a true friend, Lourdes, for bolstering my spirits.

Carina Mifuel—thanks for hanging in there when I needed it the most, and for getting office work out in record time.

You both gave me the space and the support I needed to persevere through what seemed like totally never-ending projects (six new books in less than a year!).

Sally Lambert Lagerquist

Contents

Preface

An open letter to all nursing students and graduate nurses who want a practical study guide for hands-on review of nursing knowledge, and who want to apply critical-thinking and diagnostic reasoning skills to real-life clinical situations . . .

Every nursing student needs to be familiar with the most important content in each major clinical area: medical, surgical, obstetric, pediatric, and psychiatric care. *Little, Brown's Nursing Q&A: Critical-Thinking Exercises* is designed to effectively provide the essential content you need in order to master all exams throughout nursing school, as well as achievement tests, and re-entry and challenge exams. This highly practical and concise study guide provides you with important means for developing competency in taking nursing exams while in school, and focuses on *nursing behaviors* (nursing process) applied to *integrated* health care management situations throughout the *life cycle* (childhood and adolescent years, young adulthood, adult years, older adult years).

You'll study 75 cases representing situations that are encountered in all areas of nursing. You'll be able to think through major considerations for each case and review management of patient care in each of the major clinical areas. You'll be able to work with nursing diagnoses (nursing problems) and look for desired patient outcome.

Features of this book:

1. The exam preparation guide (**Part I**) offers tips and strategies for optimum performance.
2. Detailed *case management scenarios*, covering all stages of the life cycle, precede nursing process–oriented critical-thinking exercises. Behavioral and physical components of health care are *integrated*; client needs and nursing problems are selected from the major health care areas and combined within a *life cycle* framework. The detailed case management scenarios are presented for practice and self-evaluation of your ability to *apply the steps of the nursing process.*
3. Reference codes are provided in **Part III** for all questions in this book. The codes represent *five* areas—*nursing process, client needs, cognitive level, human function,* and *critical requirements*—and help you determine which areas you need to concentrate on for further study.
4. The book also includes two additional study aids: an *Index to Nursing Problems* and an *Index to Medical Diagnosis/Case*

Management Scenarios. For each case, questions that cover *nursing problems* are cataloged in the nursing problem index and *medical problems* are cataloged in the case management scenario index. These indexes are indispensable tools for determining which practice questions test similar problems; they focus your review by identifying related content areas.

We compiled this book in direct response to the requests from our students who want a study guide with practice questions and answers to use during nursing school. These requests came from the more than 100,000 students who have attended our Review for Nurses courses held nationwide since 1976, and from subscribers to our taped review series. The enrollees find that the practice tests from our review lectures on each clinical topic, combined with our taped review series, are the most effective and successful ways to review for the nursing exams. We are pleased to offer this edition of the book along with the other NCLEX-RN review study aids available through Little, Brown and Company; Review for Nurses, Inc.*; and Review for Nurses Tapes Company† to help you in your review of nursing.

This study aid is companion to and is coordinated with the following nursing review resources:

- *Little, Brown's NCLEX-RN Examination Review.*
- The NCLEX-RN Review classroom courses offered nationwide and sponsored by *Review for Nurses, Inc.*
- The nursing process–oriented NCLEX-RN Review lecture series on audiocassette and videotape, produced by *Review for Nurses Tapes Company.*
- The special Relaxation Approaches for Nurses, also produced by *Review for Nurses Tapes Company,* on audiocassette.
- Effective Test-taking Techniques, on audiocassette, produced by *Review for Nurses Tapes Company.*
- Review of Diets and Review of Drugs for NCLEX-RN, on audiocassettes, produced by *Review for Nurses Tapes Company.*

The questions included in *Little, Brown's Nursing Q&A: Critical-Thinking Exercises* are field-tested to include relevant trends, latest knowledge, and current practice, and are directly geared to the main clinical nursing content and behaviors currently covered on the NCLEX-RN. Complete answers with detailed rationale are included to help you understand the reasons behind the best choice and to explain why the other choices can be eliminated.

We believe that a question-answer book such as this one should serve several purposes:

1. The primary purpose is to serve as a *tool for evaluation* of knowledge in specific subject areas and application of the nursing process to acute and chronic health problems in children and adults.

---

*Review for Nurses, Inc. is an organization that conducts nationwide review courses.
†Review for Nurses Tapes Company publishes nursing lectures on tapes and printed study guides.

Practice questions in this book can be used for self-evaluation of *baseline content knowledge* and application of nursing process, and to *pinpoint* where further study, review, and practice are needed in any step of the nursing process or subject areas. By taking these practice tests, you will be able to assess your *patterns of difficulties* by determining the types of questions—related to nursing process and human function areas—that pose difficulties for you. Do the questions you answer incorrectly deal with knowledge of nutrition?, elimination?, growth and development?, the process of assessment?, the process of determining priorities of nursing action?, interpersonal relationships?, your own speed and accuracy in reading the case description and the stem and options? The test questions in Part II of this book reflect the type of nursing process–oriented questions that are typical on nursing exams with integrated case scenarios covering the life cycle.

2. The book should also serve as a *learning tool*.

We include detailed answers, with explanations, to *reinforce* what you know, to *fill in gaps* of information, and to *clarify* points as you go through each clinical area in nursing school. Our questions-answers are designed to emphasize how to *apply nursing process* and *transfer your knowledge* of essential concepts and principles to a test question.

The answer sections, with rationale for all the options, help you to look at *reasons* for the correct and incorrect answers and discourage rote memorization of a correct response.

In addition to fulfilling the two main purposes for a study guide in a question-and-answer format, this book includes *special* features that are essential to the process of doing well on exams; namely, **how to study effectively** (how to take notes from lectures and textbooks), **how to prepare yourself emotionally, how to cope with exam anxiety,** and **how to take tests**. We believe that all too often the student is blocked from exceptional performance in exams, not only by lack of basic knowledge, but also by not knowing how to study and not having the test-taking know-how, as well as by a high level of anxiety.

The experience of fear, anxiety, and apprehension about exams is a common pitfall. These feelings can affect how you think, what you remember, and your critical judgment in analyzing questions. Special sections in this book are devoted to test-taking approaches and to anxiety reduction. These sections are aimed at helping you to overcome the feelings of dread and inadequacy, and to gain confidence in what you know, your ability to apply what you know to an unfamiliar situation, and your ability to take tests and handle exam-related anxiety.

This study guide is a tool designed to test nursing students for competency in providing safe nursing care in any health situation in all clinical areas. It is crucial for you to be proficient in taking objective tests. *Little, Brown's Nursing Q&A: Critical-Thinking Exercises* contains critical-thinking exercises with more than 700 objective questions. They will provide you with the opportunity to *master test-taking techniques* and to *apply classroom content* to a test-taking situation, as well as to *evaluate your knowledge base* and identify areas where you may need further study.

Theoretical content has been omitted from this book, as this material is completely reviewed in *Little, Brown's NCLEX-RN Examination Review*, which is co-authored by some of the contributors to this book.

The special experience of our contributing authors in successfully conducting our nationally held NCLEX-RN review courses, plus their expertise in teaching major clinical nursing subjects at leading US schools of nursing, has enabled us to prepare these practice questions and answers for you with a key goal in mind—to assist you in reviewing nursing content thoroughly and effectively, with minimum stress and maximum success and confidence.

We are pleased to offer *Little, Brown's Nursing Q&A: Critical-Thinking Exercises* as a study guide and nursing exam review aid for you. We welcome your comments so that we can continue in our collaborative effort to respond to your needs and to continue to offer effective review study guides that increase your confidence and skills as a nursing practitioner, as well as in test-taking situations. Here's to a successful performance on the exams!

S.L.L.

Acknowledgments

To Evan Schnittman, my editor: Thank you for your vitality and enthusiasm, and above all, your "can do" attitude in implementing our ideas for the "best book ever." You saw an unfilled niche; we made it a reality!

To Suzanne Jeans (Senior Editorial Assistant) and Katie Mascaro (Production Editor): Much thanks for your infinite patience with my inevitable need to add "just one more" thought and idea. Accelerated production wouldn't have happened if it weren't for your commitment and care. You are true gems!

To Bonnie Bergstrom, who spent countless hours and weekends deciphering my handwritten notes in order to wordprocess the manuscript, and who gave valuable feedback in the early editorial and production phase of this book: Bonnie, I couldn't have done it without you!

To the nursing reviewers and special contributors who shared their expertise and made invaluable comments to ensure that this book met the criteria for relevance, usefulness, and excellence as a study aid: Pam Bellefeuille, Dr. Robyn Nelson (Care of the Adult); Dee Gerken, Dr. Mary Parker, Dr. Irene Bobak (Maternity); Kathy Snider, Geraldine Colombraro (Pediatrics); and Marcia Miller (Mental Health).

Contributing Authors

Judith E. Barrett, RN, MSN, MNP
Former Associate Professor, University of San Francisco, School of Nursing, San Francisco

Irene M. Bobak, RN, PhD, FAAN
Professor Emerita, Women's Health and Maternity Nursing, San Francisco State University, San Francisco

Geraldine C. Colombraro, RN, MA, PhD candidate
Assistant Dean, Center for Continuing Education in Nursing and Health Care, Lienhard School of Nursing, Pace University, Pleasantville, New York

Jane Corbett, RN, MS, PhD
Associate Professor, University of San Francisco, School of Nursing, San Francisco

Marlene Farrell, RN, MA
Professor of Nursing, California State University, Los Angeles

Sandra Faux, RN, MN
Former Faculty, University of Illinois College of Nursing, Chicago

Lois A. Fenner Giles, RN, MS
Former Faculty, University of Maryland, School of Nursing, Baltimore, Maryland

Sarah A. Cloud Hardaway, RNP, MSN, PhD candidate
Coordinator and Instructor, Medical Surgical Nursing, Garland County Community College, Hot Springs, Arkansas

Marilyn Brolin Hopkins, RN, MS, DNSc
Professor of Nursing, California State University, Sacramento, California

Sister Mary Brian Kelber, RN, DNSc
Associate Professor, University of San Francisco, School of Nursing, San Francisco

Sally Lambert Lagerquist, RN, MS
President and Course Coordinator, Review for Nurses, Inc. and RN Tapes Company; Former Instructor of Undergraduate, Graduate, and Continuing Education in Nursing, University of California, San Francisco, School of Nursing, San Francisco

Diana R. Lapkin, RN, MS, EdD
Dean, School of Nursing, Salem State College, Salem, Massachusetts

Sister Mary Peter McKusker, RN, MS, MA
Former Instructor, Puno, Peru

Janice Majewski, RN, MS, DNSc
Clinical Research Nurse, Cardiac Rehabilitation, Sequoia Hospital District, Redwood City, California

Karen L. Miller, RN, MSN
Wilmette, Illinois

Robyn M. Nelson, RN, DNSc
Professor of Nursing, California State University, Sacramento, California

Agnes F. Padernal, RN, MS
Former Assistant Clinical Professor, University of California, Los Angeles, School of Nursing, Los Angeles

Frances Ward Quinless, RN, MA, PhD
Chairperson and Associate Professor, University of Medicine and Dentistry of New Jersey, South Orange, New Jersey

Kathy Rose, RN, MS
Former Assistant Professor, University of Illinois School of Nursing, Chicago

Williamina Rose, RN, MS
Former Education Director and Associate Professor, Montana State University, Butte Extended Campus, Butte, Montana

Carole A. Shea, RN, PhD, FAAN
Associate Dean for Academic Affairs and Graduate Director, College of Nursing, Northeastern University, Boston

Connie Perry Simon, RN, MS
Assistant Professor of Nursing, Dillard University, New Orleans, Louisiana

Patricia Sparacino, RN, MS
Associate Clinical Professor and Clinical Nurse Specialist, Cardiovascular Surgery, University of California, San Francisco

Janice Horman Stecchi, RN, EdD
Dean, College of Health Professions, University of Massachusetts Lowell, Lowell, Massachusetts

Janet Jordan Veatch, RN, MN
Clinical Nurse Specialist, Pediatric Oncology, University of California, San Francisco, Department of Nursing, San Francisco

Kathleen Hickel Viger, RN, MS
Director of Surgical Nursing, Maine Medical Center, Portland, Maine

Reviewers

Pam Bellefeuille, RN, MN, CS, CEN
Assistant Professor of Nursing, California State University at Dominguez Hills; Clinical Nurse Specialist and Manager, Emergency Department, Kaiser-Permanente Medical Center, Santa Rosa, California

Dee Gerken, RN, MS, CS
Assistant Professor of Nursing, College of the Desert, Palm Desert, California

Kathleen E. Snider, RN, MS
Professor of Nursing, Department of Health Science, Los Angeles Valley College, Van Nuys, California

Marcia Ann Miller, RN, BSN, MA, MSN
Associate Professor, Nursing Section, Purdue University North Central, Westville, Indiana

How to Use This Book

Student's Use of Case Scenarios

- Read **Part I** first, focusing on the test-preparation guidelines (intellectual, emotional, and physical preparation).
- Read and practice the suggested approaches for anxiety reduction in order to experience self-mastery over anxiety.
- Familiarize yourself with the chapter on *how to take tests*.
- Test your ability to *apply nursing process* to integrated health care situations by answering the questions in **Part II.**
- Set a 1-minute time limit per question.
- Read the case situations and the stems of the questions carefully, paying attention to key words. Be sure not to "read into" the stem.
- Select the *best* answer from the presented options. Circle the letter of the best choice.
- After you have completed a particular exam section (or at the end of the time limit), look up the correct answers.
- Study the rationale given to help you to *understand* the *process* of selecting the best answer and the reasons *why* the *other* options are incorrect. Do not memorize the correct answer, because it is not likely to appear in exactly the same form on any other test you might take; instead, use the reasoning process to help you on other test questions.
- Using the reference codes explained on page 345–346, see if there is a pattern to the type of questions you answered incorrectly: Did you have trouble with questions that deal with sensory-perceptual functions?, fluid and gas? Or did you incorrectly answer questions that require you to analyze presented data, select goals, and intervene with priority measures? If you identify patterns, focus your content review on those areas.
- For more detailed review of specific content areas, refer to *Little, Brown's NCLEX-RN Examination Review*, the references given at the end of this book, and your own class notes or textbooks. Before you can do well on the nursing *process–oriented* questions, you need to know *nursing content*.
- If you answer 80% to 85% of the questions in this book correctly, you can feel confident in your ability to do well on nursing exams.

Instructor's Use of Case Scenarios

Teaching/Learning Objectives that Can Be Met by Using the Case Scenarios

Focus

To separate major nursing problems from secondary ones
To clarify major nursing care problem(s)
To seek a clear statement of the nursing care problem(s)

Language

To define key terms
To practice clear and precise charting

Assumptions

To recognize values, frame of reference, and stereotypes held by patient, family, health professionals, health agency, and society

Assessment/Data Gathering

To determine relevance of assessment data to clinical situation
To take into account the total situation
To categorize data

Analysis

To note patterns and relationships in the case study data
To make appropriate inferences based on assessment data

Evaluation

To develop criteria for expected outcome and unexpected consequences

Overall Benefits

1. Discussing ideas in an organized way, to exchange and explore thoughts and ideas with others
2. Viewing a situation from different perspectives in order to develop in-depth comprehension

Suggestions for Instructor's Use of Case Scenarios

This book has 75 case situations that cover each of the major clinical nursing areas of Pediatrics, Maternity and the Childbearing Family, Acute and Chronic Care of the Adult, Geriatrics, and Mental Health Throughout the Lifespan.

The instructor can select the number of cases that fit his or her class schedule (6-, 12-, or 16-week course). From the 75 cases, the instructor can even select some cases to use one year and then use a completely different set the next year. If problem-based learning and critical thinking is the core component of the nursing program, instructors in *each* clinical rotation and theory class can use all of the cases as part of team teaching during the academic year.

Instructors can use these cases for varying purposes. *One* use is as a substitute for lectures. *Another* is to add them to clinical seminars, with the goal of expanding or reemphasizing aspects of the case that are being discussed in lecture or encountered in the clinical experience at that time.

A *third* possible use of the case scenarios is as supplements in a program that is reduced in scope. A *fourth* possibility is for the instructor to use the cases as a review at the end of each clinical area, or for self-assessment by the students.

The following formats for use in the classroom have proved successful.

A case scenario is assigned for reading; the students meet in small groups to share information based on previous knowledge, and to decide on the answers. They are asked to provide rationale for their choices.

The instructor can then point out relevant data in the vignette, add what further assessment is needed, and introduce new questions related to other conditions.

A "What if . . ." approach can stimulate lively discussions. "If . . . then . . . on the other hand" are useful beginnings for discussions that center on critical thinking; these discussions can have the learning impact of first-hand experience.

Exploration of stereotypes can also come from these cases. Ask the students to discuss their mental picture of a homeless person, an elderly person living alone, a person with a lifestyle dissimilar from their own. The instructor can then propose alternative frames of reference through which students can view a particular situation or set of facts.

This allows for an intensive analysis of the selected case and associated conditions. The students can achieve a depth of knowledge that is useful not only for the short-term (one course), but also for long-term use in their clinical years after graduation.

Sample Study Questions that Can Be Used as Additional Critical-Thinking Exercises

Adult

1. Was the condition related to lifestyle choices?, genetic factors?, environmental factors?
2. Discuss a controversial issue. For example, should CPR be attempted on someone over 85 years of age? Ask students to support their position with reasons.
3. How would your concerns be different if you were seeing the patient for home health care?

Peds

1. Should all children be immunized? Ask the students to support their position.
2. A preschool child is to have a T&A tomorrow morning. What will you do to prepare him so the child is cooperative? How will you present the information if the child does not speak English?, can't hear?, can't see?
3. What is it about the situation that will bring you feelings of satisfaction?, frustration?

Maternity

1. How would your nursing care plan be different if there were complications like diabetes or an infectious disease?
2. What is the significance of the findings?
3. What are your assumptions about bonding?

Mental Health

1. How would you use your senses (sight, hearing, smell, touch) to collect additional data? For example, what visual observations would you make?
2. What are your assumptions about appropriate behaviors?, individual rights?, beliefs?, values?
3. What would you do to validate your initial assumptions about the patient's expectations?

I. *How to Prepare for Examinations*

Introduction

You do not need an inborn gift or a magic secret to do well on tests. You do need to prepare yourself intellectually, emotionally, and physically. This is something you can learn. The person who just whizzes through exams not only has mastered the material, but also has developed a certain test-taking know-how and an effective approach to coping with anxiety.

1. Preparing Yourself Intellectually

The learning process relates and mingles information from many sources. It is impossible to be intellectually prepared without studying the information contained in your textbooks and that conveyed in classroom lectures. This section segregates the components of intellectual preparedness, for simplicity, but all are interrelated.

How to Study

Study necessitates committing certain facts and concepts to memory. If you have had trouble memorizing in the past, or if it has seemed unimportant, we offer some approaches that can increase your success. Give them a try. There is no getting around the fact that training your memory requires effort. But it is an essential step toward your goals—graduation, and your RN License.

Preparation for nursing exams involves memorizing some facts. Some questions directly test your knowledge base. The majority of the questions usually emphasize logical reasoning, interpersonal relationships, and your application of nursing process (i.e., assessing needs, setting priorities, implementing goals, evaluating results).

13 Ways to Aid Your Memory

1. To remember anything, you must believe it is worthwhile. That is the first step. If you do not feel the subject matter itself is worth memorizing, it is hard to feel involved with the content. Make it *important to your self-concept* (e.g., as a nurse, you need to know CPR) that you know the material. Also, recognize that your goal (passing nursing exams) demands this knowledge. If you want to achieve your goal, you will accept the need to memorize certain facts (such as growth and development milestones, lab values, dosages, drug side effects) as an unpleasant but necessary adjunct.

2. In addition, you must begin your study with the intention of remembering the material.

3. Have confidence in yourself and your ability to remember the best answer. Set aside negatives, such as "I didn't study enough." Forget past experiences of not doing well on objective exams. We will talk more about attitude in the chapter on emotional preparation, but you must take a positive approach if you want to study effectively.

4. *Be flexible in your learning and study patterns.* If you are comfortable with the approaches you use and they are effective for you, continue with what is successful. But if you have a mental block against studying or feel you have been unsuccessful in the past, try some new methods. Do not immediately discard an idea because it sounds silly. Give it a chance. If your methods are not meeting the challenge, what is the harm of trying something new?

5. *Schedule time to study* and find out what time frame is effective for you. Perhaps you learn best when you set aside 3 or 4 hours of uninterrupted time and focus totally on study. On the other hand, you may have trouble maintaining your attention for a long period. For you, intense 1-hour periods throughout the day may be more effective.

Many people find it worthwhile to use otherwise wasted time (waiting for and riding public transit, waiting for appointments) to review notes or textbooks. This spreads "study" through the day and makes it less tedious. It takes some planning, but discipline is a necessity regardless of your study method.

Some authorities recommend scheduling study periods in such a way that you will learn material close to the time you intend to use it. If, for example, you have decided to learn yoga and you intend to practice it in the early morning and before dinner, try to find a yoga class that meets at those times.

6. *Eliminate interruptions.* Memory requires your full attention. The mental assimilation process is disrupted by jangling telephones and the competition of television.

7. *Explain and rephrase* the process or concept aloud, first to yourself and then to a friend or classmate. It is impossible to remember something you do not understand. Role playing can be useful. Deliver a mini-lecture to a classmate on some aspect of nursing. Ask your classmate for criticisms and questions. Then switch roles. If you cannot explain something, you may not really know it.

8. *Associate what you want to memorize with something you already know.* If what you are trying to learn disagrees with what you already know, so much the better. Trying to figure out which is right and where the disagreement lies will fix the new information in your mind more quickly.

9. *Assign priorities and establish orders of importance.* Organizing the material so that you can file it in your memory bank requires you to act on the material. This makes remembering easier because you have created a mental outline of what is important, what is subsidiary, and how the whole fits together. If you can remember part of it, often you can then remember the rest. Physically jotting down an outline (add it to your notes) will help fix the material in your mind.

10. *Organize by dividing and grouping.* Associate like or related points. Then place them in the appropriate contexts. Information is best taken in as "little packets."

11. *"Use a word three times and it is yours."* This adage may work for you—not only for words, but for concepts as well. If your task is to learn a number of unfamiliar medical terms, begin *using them in your conversations*—particularly with other nursing students, and even with roommates and friends. Nonnursing students may enjoy the game. En-

courage them to participate by occasionally asking you to spell terms or questioning you on their meaning. Substitute appropriate medical terminology for lay language at every chance.

*12. **Make your mistakes work for you.*** Review test results and any comments from the instructor. Why was your answer incorrect? And what is the correct response? Do some research and find out. You clearly remember embarrassing moments in your life. Apply the same principle to test mistakes—learn from them. It is easy to remember your mistakes. You can then work to eliminate them.

*13. **Finally, use devices and gimmicks as aids.*** Mechanical memory aids are called *mnemonic devices.* You create them yourself. Establish some pattern for remembering otherwise tedious or easily forgotten material (such as the cranial nerves). Such devices are best applied to rote memorization of terms: parts of the body or its systems, for example. Usually the first letter, letters, or syllable of a group of related items are combined to form words—the more bizarre sounding, the better, if it helps jog your memory.

Or perhaps you can identify an existing pattern among the things you must memorize that is reflected in an actual word. For example, the first letters of the terms for the parts of a certain system may spell "carrot." Remember that system as the carrot system; this will jog your memory when you need the specific names.

Mnemonic devices must be kept simple. Creating a more complex system than the one you are attempting to remember is self-defeating.

Individuals who must remember the names of a number of people they meet only briefly often use the association approach. Salesmen employ it in their work, businesswomen make use of it at cocktail parties, and many people apply it to daily situations. When you are introduced to Mr. John Harrington, repeat his name and look for something striking about the man. If you can associate his name with something he is wearing or a physical characteristic, so much the better. John Harrington in the herringbone jacket. If not, focus on the most memorable, preferably outrageous, element of the person's appearance. Long, tall Mr. McFadden. Use rhymes if possible. Electric blue, Lyla LaRue.

This approach can be expanded to facts and ideas. Associate the idea or fact with an illustration—a diagram, a drawing, a map. Visualize that illustration and mentally tack, staple, or glue the fact to it. Picture your class lecturer conveying a certain important piece of information. Visualize his or her actions: noting the number or word on the blackboard, gesturing to indicate it was important, raising the voice to emphasize it.

These are general hints to help you remember more effectively. Let us look now at methods for mastering the lecture and the textbook.

How to Listen and Take Lecture Notes

Knowing how to listen effectively is the key to learning from lectures. You cannot take useful notes until you have developed your listening skills. Listening is not instinctive, nor is effective listening a passive process. You can learn to listen effectively and improve your listening skills with practice.

10 Steps to Effective Listening

*1. **The first step is the appropriate mind-set.*** Figure out why what the lecturer is saying is important to you. You need to have a *reason for*

listening. The subject may interest you, or the topic may be important to you *only* because it will help you to score well on a test. Whatever your reason, motivate yourself to focus on it.

2. *Set up a conducive physical environment for learning.* Eliminate distractions. Choose your seat in the lecture room carefully. The center front section is usually best. Position yourself so that you can see and hear the speaker without glare or craning your neck or shifting constantly in your seat. Avoid sources of noise—fans, open doors, students who are talking. If you find what is going on outside more interesting than what is happening in the lecture, do not sit next to the window. If heat makes you drowsy, avoid the heater or the direct sun.

3. *Get your note-taking supplies together.* A loose-leaf notebook that lies flat works best for most people. You can reorganize and add material later. Wide-lined paper helps keep things in order and will make review easier. Try a nonglare paper—light green or light gray. Have plenty of blank paper in reserve so you do not run out during the lecture.

A pen requires less pressure than a pencil, so it saves energy. Your notes will also be easier to read. Carry two pens; if one runs out of ink, you will have a backup.

Label what you are recording with the name of the course, the date, and the subject matter. Use one large notebook, with dividers for each course, or a series of smaller notebooks, one for each subject.

You have prepared your mind and your materials. You are ready to begin listening.

4. *Do not let the speaker's appearance or mannerisms distract you from the content.* If you disagree with what is being said, pay more attention. We tend to tune out what we do not agree with, but opposing ideas are one of the best sources of new information.

5. *Identify the speaker's organization.* Focus your listening to look for the main ideas. Recognize digressions and figure out what is being supported or explained by them.

6. *Taking notes* helps you to pay attention. Write down all important points, even those you think you know. These notes will remind you of the lecture content later, in your review.

A handy device for improving your listening and concentration abilities is the TQLR approach: *Tune in, Question, Listen,* and *Review.* It summarizes what we have been talking about.

7. *Tune your mind* to the lecture about to begin. Block out distractions. Consolidate and focus your energy on this one task. *Look* at the lecturer to help focus your attention.

8. *Formulate questions* as you focus your mind. "What's the lecturer going to talk about today?" "What should I get out of it?" "I wonder if she'll explain that material I was having trouble with yesterday?" You are becoming interested.

9. *Listen carefully.* Keep your eyes on the speaker and your mind on what is being said. Focus on *main points* and important details, look for the speaker's *organization,* and *relate* one point to another. Be alert to tone, gestures, repetition, illustrations, and useful words such as "remember," "but," "most important," "however," and "rarely."

10. *Review constantly during the lecture.* Think back and see if a pattern is emerging. Clarify anything you are unsure of by asking ques-

tions at the end of the lecture. Look for implications beyond what is being said. Relate the content to your other classes and life experiences.

Ways to practice your listening skills include jotting down one identifying word for each item in a news broadcast. After the broadcast, fill in as much detail as you can remember for each item.

Another practice suggestion is: For 1 minute of each hour of 1 day, give your full listening attention to the person talking. If no one is talking, focus on a sound.

9 Guidelines to Taking Effective Lecture Notes

Taking notes is part of effective listening. Try the following general guidelines if you want to improve your note-taking skills. Many of them are identical with the principles of effective listening.

1. Understanding something does not guarantee remembering it. Write it down to *reinforce* it in your mind.
2. Sort and filter information. Note only the *main points*. Think before you write. You want only information that will be of value later, not a verbatim transcript.
3. Use *key* words or *short* sentences. During digressions, go back and fill in sketchy notations.
4. Develop your own system of *abbreviations, underlining,* and *starring*. Try a skeleton outline format. Show importance by indenting. Leave plenty of space for later additions.
5. Use your own words, but be accurate. *Condense* and *rephrase*. This active involvement helps you understand what is being said.
6. Do not worry about missing a point. *Leave space* and fill it in later.
7. Keep your notes in a *notebook*. Do not use scraps of paper or the backs of envelopes.
8. Set aside time to *rework,* not just recopy, your notes as soon after taking them as possible. Add points. Spell out unclear items. Fill in any gaps. Try summarizing the important points and making up some test questions on the material. Jot them down at the end of your notes. They will be valuable when you are reviewing before exams. Arrange to compare notes with other students.
9. Try to *review* your notes periodically during the course. Do not let this go until the week before exams.

5 R Method

In summary, if the notes you take now do not seem to be of much use, try the 5 R method—*Record, Reduce, Recite, Reflect,* and *Review*.

Make two columns, a 6-inch column at the right and a 2-inch column at the left. This immediately limits the quantity of notes you can *record,* because the lecture notes go in the 6-inch column. Use the 2-inch column to *reduce* your lecture notes. Use "flags"—words such as "important," "memorize," or two- or three-word summaries of the material in the 6-inch column.

Now cover the 6-inch column and *recite,* using the 2-inch column to jog your memory. If you cannot do this, reread your notes and try again.

Reflect on the relationship of one idea to another, the important points, and the interconnections.

And finally, *review* your notes at intervals throughout the course.

How to Use a Textbook

Mastering a textbook requires the same active involvement as learning from a lecture. Think as you read. You may read a novel or a magazine as a diversion from the pressures and anxieties of the day. It is fine to let your mind wander when you are reading for pleasure or to escape. But using your "novel" approach on the textbook will be a disaster.

Remember that *attending the lectures* will make your reading easier and more effective. Often the content of the lecture will parallel the material in the textbook and you will have a better idea of what to look for during your reading. Also, the repetition will help fix the information in your mind.

When you buy a new text, first *look the whole thing over.* Read the *table of contents* to find out what is in the book and how it is organized. Do you know anything about the author—interests, orientation—that will help you understand the material? You may find some of this information in the preface or the introduction. Is the author your lecturer?

Look in the *back of the book.* Find out what is there so you can use it in your study. Does the book have a *glossary?* Take a look at the words in it. Tape a tab or glue a piece of cardboard to the edge of the first page of the glossary to help you find it easily as you study.

Skim through the *index.* See what percentage of the words you recognize. It will help you determine how hard the material will be for you. And remember to use the index to help you review for exams.

See if the book contains an *overview* or a *summary.* If your time is limited, that may be all you read.

Look at a chapter. See how it is constructed. (They will all be organized in the same way.) Is there a summary at the beginning . . . or at the end? Has the author provided *study questions* at the end of each chapter or section or at the back of the book? Know what is there so you can make use of it.

OK4R Method

When there is a choice between "quick and effective" and "slow and effective," it makes sense to choose "quick." There is a quick, effective way to master the textbook. It is simply a matter of focusing your energy. Why spread the process out by plodding along, when with a little more effort you can successfully learn what you need to know in much less time? Use the OK4R method: *Overview, Key ideas, Read, Recall, Reflect,* and *Review.*

When you open the textbook to begin an assignment, do not start to read immediately. First orient yourself. Establish a direction for your mind. Set it on a track.

Overview

To get an *overview,* look at the first and last paragraphs of the material. Read the headings and examine the illustrations. With this approach, you discover the organization and give yourself a basis for selecting and rejecting, to speed your reading. The overview also prods your mind to remember related knowledge, and it arouses your interest, as well as giving you a double exposure.

The point of the overview is to focus your reading. Now you are ready to begin.

Key Ideas

What you are looking for in each paragraph is the *key idea.* You will usually find it in the first sentence. Sort it out from the examples. All these

do is repeat the key idea in a variety of ways, so read them quickly. When you come to the key idea, slow down and read carefully.

A good way to remember key ideas is to *turn them into questions.* This helps you to keep in touch with what you are reading. If you think the key idea is that the dynamics of schizophrenia are interrelated, ask yourself, "How are they interrelated?" Then look for the answers to your question as you read the paragraph or section.

You now have a general idea of the assignment and you are looking for key ideas.

Read

As you *read,* ask yourself, "What main point is being made in this section?" Look at everything you read and relate it to that point. Keep in mind that recognition and learning are not the same thing. You do not necessarily *understand* something simply because it is familiar.

Recall

When you finish a section, look away from the page. You are ready for the second R. Say to yourself, "The main point of this section is _____."
If you can repeat the main point, you have successfully completed the *recall* step. If you cannot remember it, repeat the overview and key idea steps. When you are sure you understand and remember the main point—and *not* before—make notes and mark the textbook.

Jot down your notes from memory, using your own words. Use cue words and short phrases. Making notes from memory minimizes self-deception. Either you can write down the main points or you cannot.

There are many *ways to mark a textbook.* Use whatever works best for you—or a combination of the methods we suggest.

1. Circle (key words) or phrases.
2. Underline major points.
3. Fold corners of pages containing important information or tables and charts to which you refer repeatedly.
4. Use vertical lines in the margin to emphasize important items you have underlined.
5. Place some symbol—an asterisk (*), an X, or something of your own design—in the margin to emphasize the book's key points. Keep these to a minimum. You might have only 10 or 20 marks in an entire book.
6. Note questions, outline major points, summarize complex arguments, or briefly identify important points. Use the top and bottom margins as well as the side ones.

7. Put page numbers in the margins so you can skip explanations and flip quickly from important point to important point when reviewing.
8. If color helps or you enjoy it, use it: yellow highlighter for important words, red lines in the margin, blue for summarizing important points. But do not make your system so complicated that it distracts you from the material.

Reflect

You have read the material and marked the important items in the book. Now *reflect.* This simply means to think. Ask yourself, "How does what I've read relate to what I already know?" Try to link the new with the familiar, to give it more meaning. Then think of examples from your own experience. The best way to remember something is to personalize it.

Review

Finally, *review* the sequence of the main ideas to see how one flows into the next, and the next. Help yourself by scanning the text.

What can you do when you do not have time to read the entire assignment? What you should *not do* is start reading word by word on the first page and see how far you get. Instead, skim. Use the overview and key idea parts of the OK4R approach. Read the first sentence of each paragraph. (They usually contain the main ideas.) Read the headings and look at the illustrations.

What if you work through all the material and still do not understand it? Try to get a simpler book on the same subject and study it first.

If your trouble is vocabulary, sign up for a vocabulary-building course or buy a good vocabulary book and use it.

Remember, the responsibility for learning—mastering the lecture and the textbook material—is yours. Do whatever is necessary to help yourself.

2. *Preparing Yourself Emotionally*

Demystifying Exam Taking

A key to preparing yourself emotionally is *eliminating* as many *potential sources of anxiety* as possible about the test environment. Corrective and adequate information may take care of many of these. This chapter is intended to demystify exam taking and to provide answers to many of your questions.

Identify Your Test-Taking Strengths and Weaknesses

By the time you are ready to graduate from a nursing program, you have been in school for at least 15 years. You have already established a track record for studying and taking exams and have formed certain patterns. Now is the time to assess the patterns you use in test-taking situations. Identify your strengths and your weaknesses. What is your reading speed? What is your reading accuracy? How well do you do with factual questions? with those requiring quantitative analysis? or in application or judgment types of questions? If you have weaknesses in any of these areas, you need to practice and drill to improve them before you take the exams. You cannot prepare yourself emotionally if you are worried about problems in test-taking skills.

Look Forward to the Exam

A positive mental set is essential to emotional preparation. You are your own worst enemy if your attitude is "Oh no, another exam!" Taking a test can be fun—and a challenge. An important step in the challenge is to discern what the writers of the questions had in mind when they prepared the test. Approach it with a *flexible* attitude. You defeat yourself when you think, "This exam better cover what I studied," and you set yourself up for disappointment if it does not cover all those areas. The important thing is not what you think should be there, but what the test writers *have put* there. How can you learn about what the test is like? Focus on the test writers.

Identify with the Test Writers

Who writes your nursing exams, and what is their *perspective?* The test must be broad enough to reflect a consensus by all item writers from all clinical areas as to the solutions to certain nursing problems.

Think "Average"

A highly specialized health science background can be both a blessing and a problem to you. You may know too much detail and specialized content. Remember that the perspective of most nursing exams is basic, general, and noncontroversial. If one of the choices offered in an exam question is a new approach, still under debate, that is not the best answer—even though it may be true in some areas of practice and you may have been exposed to it. *It is best to choose the safest, most conservative answer.*

The thrust of most nursing school exams will be on basic, safe care. What is everyone expected to know?

For example, do not expect too many questions on nursing research and clinical specialization. These areas may be a part of your curriculum in a generic 2-year program, as well as in a 3-year or a 4-year program, but the mandate of most nursing school exams is to test for minimum competency of entry-level practitioners, not researchers and educators. Therefore, exam questions are not meant to test for responses appropriate for specialists in the major clinical areas.

You Can Be "In Charge," Not a Victim

We have answered the first and most important question: Who wrote the exam and what are they looking for? Now let us tackle a common myth about an exam being tricky.

Most nursing exams have no hidden tricks nor a scheme afoot to trap you. If you find yourself wondering "What are they going to pull on us?" change your mental set from passive victim to active strategist. Decide to attack the questions, not to be "done in" by them. Envision yourself a general with a successful confrontational strategy. An attitude of that type provides a tremendous mental uplift.

Allow Yourself to Be "Average"

Do not expect to get a perfect score. That may sound counterproductive, but it is an approach termed *paradoxical intention.* Set as your goal a passing score. Forget that you are at the top of your class or want to be Phi Beta Kappa. This is not a competition for grades. Your goal is to pass each nursing exam and become an RN.

The key is to give yourself permission to score "average." When you permit your mind or body to do what it fears or is resisting, it appears to no longer need to fight it. For example, have you ever tried to fall asleep and the more you tried to sleep, the harder it was? What happened when you opened up a textbook you *had* to read while in bed? How soon before you fell asleep *then?* In the same light, allow yourself to score average. You may find you *will* score much higher when you remove the pressure from yourself and stop thinking you *must* score high.

Know the Purpose of the Test in Order to Make Relevant Choices

Making the wrong choice between two close options is another major reason some students do poorly on exams. They may have mastered the material, they may be excellent clinicians, but they do not select the best answers on a test. They may successfully narrow the four choices down to two, but then they panic and go off on side issues and lose track of what

the test is all about. This is where mental preparedness is important: determining the purpose of a test.

Remember that most nursing exams test for essential nursing behaviors to demonstrate competence. The focus is on *basic, common,* and *general* knowledge. You are being tested to find out if you are a *safe* practitioner, not a clinical specialist. Look for the answer that will let the exam writers know the patient is in safe hands.

Questions on nursing exams usually test for one or more of the following areas: (1) *skills,* (2) *knowledge,* (3) *interpersonal relationships,* and (4) *nursing process.*

Skills

There are "*how to*" questions: How do you irrigate a catheter? How do you apply a sterile dressing?

Knowledge

There are questions on specific facts and knowledge related to categories of human function, pathophysiology, and certain health care trends or statistics.

Certain "pet" subjects appear repeatedly on nursing exams. You might hear about topics from past examinees. These topics, throughout the years, may emerge as patterns of examining, where certain concepts, principles, and procedures are most likely to be tested and thus may be anticipated.

Perhaps on one exam there may have been 40 questions on Addison's disease, 10 on arthritis, and 10 on tetanus. "What kind of an exam is that?" you ask. "I thought the exam would be on common disorders, but I've never even seen a patient with Addison's."

Do not panic. Ask yourself, "What are the basic pathophysiologic principles of Addison's disease?" Acid-base? Corticosteroids? Electrolytes?

In the same light, you may not find one question on commonly seen diabetes, but an uncommon condition that manifests a similar pathophysiology as diabetes, which you *have* studied and clinically observed, may be tested. Apply that same knowledge of underlying pathophysiologic concepts and principles to these unfamiliar situations.

Why have 10 questions on arthritis been included? The key underlying concerns here may be range-of-motion exercises, chronic illness concepts, rehabilitation process, and medication (the corticoids) common to a number of other related disorders.

Why ask about tetanus? Because it is a communicable disease. The examiners want to find out if you know what to report and what not to report, how to prevent it, and how to treat it. Again, it could be representative of a broad spectrum of other diseases.

In a question related to children, you might be asked, "Which of the following diets would you give a child with PKU?" You say to yourself, "The only thing we learned was low phenylalanine, but Lo-Fenolac is not among the choices." Of what is PKU a disturbance or imbalance? Protein metabolism. Knowing that one fact, you choose the diet that is low in protein.

In a question related to a patient on phenothiazines, you see the question "During which patient activity is it important for the nurse to make sure that precautions are taken with the patients?" The choices are (A) going to church on Sunday, (B) going to the museum, (C) going to a concert, and (D) going on a picnic. Let us suppose that you have no idea what

answer is wanted. Look for patterns. Church, concert, museum—all in-
door activities. Picnic—outdoor activity. Explore the choice that differs
from the others (picnic) further. What are some other characteristics of a
picnic? It is usually held outdoors . . . in the afternoon . . . in the sun.
Patients taking phenothiazines sunburn easily. The answer is "picnic."
You could not have memorized the answer to that question. It requires
that you reason it out. Do not be immobilized by such questions. Look for
patterns and similarities throughout the exam.

Interpersonal Relationships

Questions based on interpersonal relationships constitute another portion
of all nursing exams. A caution: In the questions focusing on acute or
chronic health problems of children or adults, do not overlook the possibility that the best answer may be an "emotional coping behavior" type of
answer. You may think the answer is too obvious, but if your first hunch
is "it seems to be a therapeutic communication type of answer," follow that
feeling.

The same principle applies to the questions focusing on emotional disorders. You may find a choice that seems correct, but is oriented to physical
care. Do not ignore that answer if you think it is correct. That test item
could call for a "physical care–oriented" answer.

Nursing Process

Nursing process questions may place heavy emphasis on assessment of
priorities. "What do you do *first?*" "Which is the *most* important?" "What
is the *long-range* goal . . . or the *short-range* goal?" When all the choices
are true, look for the key word—"first," "last," "most," "least."

Aspects of nursing process commonly tested on exams include the following five nursing behaviors, with definitions and examples*:

I. **Assessment**
   A. *Definition:* Ability to establish a database about the patient.
   B. *Examples:*
      1. Can gather objective and subjective data about the patient.
         a. Gathers data from verbal contact with patient, family,
             and/or significant resource persons.
         b. Gathers data from chart, lab reports, progress notes, and
             nursing care plans.
         c. Recognizes signs and symptoms.
         d. Assesses patient's ability to perform activities of daily
             living.
         e. Assesses patient's environment.
         f. Identifies reactions (own, staff) to patient and family.
      2. Can verify information.
         a. Questions information.
         b. Confirms observations.
      3. Can communicate information based on assessment.
II. **Analysis**
   A. *Definition:* Ability to identify actual, potential specific health
       care needs and problems based on assessment.
   B. *Examples:*
      1. Interprets data.

*Adapted from mimeographed sheets distributed at state Student Nurses Association
conventions and to all schools of nursing. Source: State Boards of Nursing.

   2. Collects additional data as indicated.

   3. Identifies and communicates patient's nursing diagnoses.

   4. Determines congruency between patient's needs/problems and health team member's ability to meet patient's needs.

III. Planning

   A. *Definition:* Ability to develop a nursing care plan with goals for meeting patient's needs, and designing strategies to achieve these goals.

   B. *Examples:*

   1. Includes needs based on comfort and priorities.

   2. Includes factors such as age, sex, culture, and ethnicity.

   3. Utilizes resources in the community.

   4. Collaborates and coordinates care with other personnel for delivery of patient's care.

   5. Formulates expected outcomes of nursing interventions.

IV. Implementation

   A. *Definition:* Ability to begin and complete specific actions to meet the set goals.

   B. *Examples:*

   1. Organizes and manages patient's care.

   2. Performs or assists in performing activities of daily living.

   3. Teaches and counsels patient and family and/or health team members.

   4. Provides care and therapeutic measures to achieve established patient goals.

      a. Provides preventive care.

      b. Provides life-saving care.

      c. Prepares for surgery and other procedures.

      d. Uses correct techniques.

   5. Provides care to optimize achievement of patient's health care goals; stimulates growth and maintains optimum health.

      a. Encourages self-care and independence.

      b. Adjusts care to meet needs.

      c. Motivates patient.

      d. Encourages patient to accept treatment.

   6. Supervises, coordinates, and evaluates the delivery of the patient's care provided by other staff.

   7. Records and exchanges information about actions taken.

V. Evaluation

   A. *Definition:* Ability to determine the extent to which goals have been met.

   B. *Examples:*

   1. Determines extent to which goals were appropriate.

   2. Compares actual outcomes with expected outcomes of therapy.

   3. Changes goals and reorders priorities as necessary.

   4. Recognizes effects of measures on patient, family, and staff.

   5. Investigates if measures taken are appropriate.

   6. Evaluates compliance with prescribed and/or proscribed therapy.

   7. Records response to care and/or treatment.

Eliminate Anxiety about the "Unknown"

We have given you general information on the kinds of questions you can expect. Let us look at some specifics that will help decrease anxiety about the "unknowns" and eliminate other reasons for anxiety.

- Find out how long the exam will be.
- Ask how many questions will be on the exam.
- Learn the format of the exam (e.g., multiple choice).
- If multiple choice, check how many options will be offered, and whether there is only one correct answer to each question.
- Note the location of the test and the room in which it will be given.

If you have any doubts about the location of the exam site, or if you are upsetting your mental equilibrium by worrying about what the room will be like, we suggest you make a *"test run."* If you are concerned about the room, look at it. Is it light, dark, cold, overheated?

How the Test Is Scored

You should also know how an exam is scored. Will you receive a pass/fail determination, a letter grade, or a percentage score? Is there a penalty for guessing? Will it be scored "on a curve" or criterion referenced?

Helpful Emotional Attitudes

Forget Anger

Another emotional attitude to work on is a decrease in your level of anger or hostility before you enter the testing room. People who enter an exam furious—with a spouse, friend, parent, child, traffic, or the nursing profession—put themselves at a disadvantage. Anger and hostility distract thought processes, slanting them and upsetting balance. Immediately before an exam, avoid people who incite your anger. You may find this difficult if these people are part of your household, but there are doors and locks, and you can set time limits. As a last resort, consider a "retreat" away, such as a motel room the night before the examination.

Last, Not First

A few final thoughts on emotional preparedness. Decide that you will be the *last* person to leave the exam room, not the first. This will immediately eliminate another area of anxiety, when you notice some people leaving while you are still answering questions or checking your answers. You begin to think, "There must be something wrong. It must be easier than I thought it was." Forget about everyone else and concentrate on using all the time allotted.

Have Tunnel Vision

Prepare yourself, on entering the examination room, to block out everything except the task ahead of you. The psychiatric term for this is *selective inattention;* it is more commonly called *tunnel vision.* Focus only on the test. Block out your worries about sick relatives or the argument you had with a friend. Eliminate from your awareness the person chewing gum next to you, the sneezes and coughs, or the children laughing and the screeching tires outside.

The Test: A "Game" of Skills and Confrontation

In conclusion, we encourage you to "play with" the test. It is a challenge. Walk into the test with an aggressive, confident attitude. Prepare to analyze and dissect and attack and confront the questions. This gives you power. Do not take the passive approach. The correct answer will not magically leap out at you. It is your job to find it. You can put to use your mastery of content, your practical experiences, your memory, your test-taking strategies, and your confidence that it *will* all come together. Why shouldn't it?

3. Preparing Yourself Physically

6 Suggestions

Sleep Versus Cramming

Following all the tips provided earlier on intellectual and emotional preparedness will do little good if you arrive at a nursing exam exhausted. The first principle of being physically prepared is to get adequate sleep. Many people stay up late or all night 2 or 3 days before exams, to cram. We will not say, "Don't do it," because it is your choice. But ask yourself whether the gain in information is worth the sacrifice in effective thinking.

If your nursing exams were simply testing recall of facts, there might be some merit in cramming. Without an adequate background of facts, you have no chance at all in such a situation. But usually only 10–15% of most nursing exam questions test for the recall of facts. The remaining 85% usually test for nursing behaviors and interpersonal relationships, with emphasis on *application* of nursing process. Questions such as these require analysis, critical thought, reasoning, application, and figuring things out. Your brain cells must be rested to allow you to think clearly. So weigh the losses against the gains.

Stress and Medications

Let us also talk about drugs—in lay terms, "uppers" and "downers." You may feel you are so nervous that you cannot think properly. "Perhaps," you think, "medication would calm me down. I think I'll ask my doctor for some." Again, carefully weigh the losses against the gains: the loss of alertness versus the gain in decreased tension. And if you have never before taken tranquilizers, do not take other people's advice in this area— drugs have different effects on different people.

Also consider that there are other effective ways to reduce anxiety and tension without drugs. We will look at some of these in the chapter on anxiety reduction. Give these approaches a try.

Perhaps your problem is depression or lack of energy. You think, "I need something for energy and to keep me awake." Again, weigh the losses against the gains: potential loss of clear, rational thought and powers of concentration versus increased "alertness." You have probably heard many stories about what "uppers" have done for others—how good, alert, and on top of things they can make you feel. But they can have other effects detrimental to successful performance on exams.

So weigh *all* the effects before you make a decision.

Stress and the Gastrointestinal Tract

What you eat before exams is also important. Prepare your body for stress. Gastrointestinal (GI) distress is a normal effect of stress and tension. Treat yourself with care. If you are unaccustomed to spicy foods, do not eat them in the 3 or 4 days preceding the exam. Someone may suggest relaxing at a wonderful Indian restaurant with calming music and great curry. If you do not normally eat curry, have some yogurt or some milk or something bland. Do not increase your chances of GI distress by adding unusual or spicy foods to your diet at this time.

Prepare your GI tract with a diet high in carbohydrates and avoid hypoglycemia. You may want to limit your general food intake. Eating great quantities of food may make you sluggish; you may want to keep the blood supply going to your brain where you need it, not to your digestive tract. These are very simple principles, but things we often overlook when our concentration is elsewhere.

Stress and the Urinary Tract

A normal sign of anxiety is the need for frequent urination. At this time you will be under stress to do well—as you should be. Expect urinary frequency. You can control it somewhat by avoiding dietary sources that act like diuretics—coffee, tea, beer, cigarettes. Be aware of your own body and its reactions to various products.

Stress and Temperature Control

Hot and cold flashes are also signs of anxiety. Prepare for them. Plan to dress in layers. If you are cold, you can keep all your clothing on. As you warm, a natural result of the speeded metabolism caused by anxiety, shed the clothing items one by one. It is terribly frustrating to find yourself in the exam room wearing a wool turtleneck sweater (which was appropriate for the outside temperature) and discover that you are sweltering. You have no alternative but to sit there and feel physically miserable. Plan what you wear.

When choosing your layers, select them from your oldest, most comfortable, most familiar clothing. No one will care what you are wearing. Be comfortable. Pinching, binding clothing is a distraction.

Remember also the effects of stress on hormones. Be prepared for any contingency—regardless of the date. Arm yourself with supplies.

Psychosomatic Distress

Expect that psychosomatic distress will strike! Sometime before an important exam, you may catch cold or develop a sore throat or an earache. You may also find yourself losing your car keys or your wallet, or being forgetful. You may sprain an extremity. What may be happening is a "normal" response to anxiety. It is not uncommon to develop psychosomatic disorders before a major exam. Often, they too will pass once *you* pass this time of stress.

4. Coping with Exam-Related Anxiety

7 Techniques

In the earlier discussion on preparing emotionally for exams, we considered ways of eliminating many of the causes of anxiety. This chapter focuses on mental and physical techniques for coping with and dissipating the natural anxiety associated with the test-taking process.

We highly recommend that you read C. Eugene Walker's book *Learn to Relax: Thirteen Ways to Reduce Tension*. Most of the techniques mentioned below are described in detail by Walker, as are others you may find more effective. You can also find detailed, step-by-step self-mastery exercises on the special audio-cassette tape *Relaxation Approaches for Nurses*, available from this editor.

Cognitive Restructuring, or Turn-Around Thinking

The first approach is borrowed from Albert Ellis's rational-emotive therapy. It is termed *straight thinking*. Ellis claims that we carry with us into certain situations a mental script designed for failure.

A part of the straight-thinking concept is the premise that people think in irrational ways. Read the statements below. If you recognize any of these thoughts, now is the time to begin changing them.

- **Perfectionism:** "Making a mistake is terrible. I've got to get a perfect score."
  Instead, say to yourself, "I don't need or want a perfect score."
- **Powerlessness:** "My emotions can't be controlled." Or, "Self-discipline is too difficult. I'll never be able to crack a book for these exams. And there is no way in the world I'll be able to sit through a 5-day review session."
  Turn it around and say, "Of course I can and will when I make small, realistic goals, such as to review one subject at a time."
- **Comparisons:** "I have to do better than my mother . . . or my sister . . . or my roommate did." Or, "I've got to do better than I did on the last test."
  Restructure that to say, "I just want to do OK, not the best."
- **Putdowns:** "Healthy people don't get upset. I must be crazy." Or, "I'm inferior. I never do well on exams. In fact, nothing ever works

out well for me." "Giving up is the best policy. If I don't know the answer right away, I'll just forget about it. If I don't know the first three or four, I'll just close the booklet and leave. Why torture myself?"

New script to try out: "Test-taking anxiety is normal, not crazy." "I've passed tests before. This is not that different."

- **Magic cures:** "I'll put a tape recorder under my pillow; when I wake up, I'll know all the answers."
  Tell yourself: "Not knowledge by magic, but by knowing *how to figure out* the best answers."
- **Evading responsibility:** "If I don't pass this exam, it's the fault of my nursing school instructors. They didn't teach me the right things."
  Instead, you go back to the books to fill in *your* gaps of knowledge for *yourself.*
- **Punishment:** "If I can't pass this test, I'll show them . . . I'm going to give up nursing."
  Better to view taking each exam as a quiz, not *the* test to end all tests. . . . It's an *episode* in your life, not your whole existence.
- **Unrealistic thinking:** "I'm going to smile and act like I'm happy and carefree, even though I'm scared half out of my mind. I'm not going to yell at my children, or my spouse, or my roommates and let them know how I'm feeling." Or, "Anxiety is always dangerous. I'd better not let anyone know how anxious I feel."
  Instead, let it out!

What should you do if you recognize any of these irrational thought patterns? Begin by accepting responsibility for yourself and your actions. Practice turning around some of your "lines."

Assertive Rights—for Success

Ellis's straight-thinking approach is based on the assumption that each human being possesses 10 assertive rights. These are the attitudes you must begin to develop within yourself to rewrite a mental script that fosters "success" instead of failure. A few of these rights, as they relate to examinations, are:

1. The right to offer no reasons or excuses to justify your behavior. If doing cartwheels on the sidewalk makes you feel better and releases your anxiety, there is no need to justify it—to yourself or to anyone else. Your first commitment is to make yourself feel good so you can focus your attention where you want it to be.
2. The right to change your mind. Go back and change answers if it feels right to do so.
3. The right to make mistakes and to be responsible for them. You have the right *not* to get the "right" answer.
4. The right to say "I don't know why this is the correct answer, but it is," rather than "This is the right answer, and I should know why."
5. The right to go with what feels correct to you, even if you did not select it in a logical manner. You have the right to be illogical in making decisions.
6. The right to say, "I don't understand."
7. The right to say, "I don't care."

That is a capsule summary of Albert Ellis's straight thinking.

Life Structuring

Another approach that may help you to prepare for exams and cope with the accompanying anxiety is *life structuring* or *engineering*. This involves using "games" or rewards to help you get what you want.

For example, you might say, "I'll go home now and review schizophrenia, and when I'm finished, I'll go out with some friends." You are structuring your life around what you want to do and the reward you are going to give yourself for doing it. It is a little game and it is OK. People need rewards.

But be careful how you structure yourself. Do not decide to cover 50 pages or study for an hour. You set yourself up to waste that hour or to learn little from those 50 pages. *Give yourself a specific goal*—something you intend to master. If you complete the task quickly, you can go out earlier. If it takes you a long time, you go out later. In either case you *do* go out—and you *can* enjoy yourself because you know you have accomplished something. You need not feel guilty because you worked for an hour or read the 50 pages but did not learn anything.

Be sure your goals are realistic. You are courting failure if you try to do in 3 hours something you know will take you 6.

There are a number of rewards you can employ between now and the date of the exam to structure your life. Begin to think of some.

Progressive Relaxation

Jacobsen's progressive relaxation approach is also effective in alleviating anxiety. In essence, it involves relaxing the parts of your body one at a time. Close your eyes, focus on a muscle, and relax it. Move on to the next body part, and the next, until your entire body is relaxed. To help yourself learn this technique, record the material that appears in the orientation chapter of *Little, Brown's NCLEX-RN Examination Review* by Sally Lagerquist. Get into a comfortable position in a darkened room, close your eyes, and listen to yourself on the tape. It takes practice. Saying, "That's simple. When I get tense during exams, I'll do it" will not work. Practice the technique until it becomes second nature; then you can call on it automatically when you feel yourself becoming tense.

All the techniques described in this section work for a limited time and for a specific goal. We are not talking about therapy; these are self-help techniques.

Tension and Relaxation

Tightening and letting go is a very simple method of releasing tension. It is fun, it is easy, and it works. If your neck and shoulders are tense, increase and exaggerate that tension, hold it, and suddenly let go. You might say, "I can't do that in the examination room. I'll look ridiculous." Who cares? Give yourself permission to do whatever works for you.

Psychiatric nurses used to ask patients, "Why do you bang your head?" The nonjoking response was, "It feels good when I stop."

Perhaps when you are nervous your hands shake. Exaggerate the shakiness. Shake them as hard as possible for 10 seconds, and suddenly let them go limp. If you do not want to do that in public, find an unobtrusive part of your body to which you can apply the same approach.

Imagine that you are about to have an attack of explosive diarrhea. Tighten your gluteus muscles as hard as you can and stop that attack. Or clench your fists as hard as you can and hold them. Then let go. No one will notice. If you are really self-conscious, wear closed-toed shoes and

contract your toes. Focusing this tension-relaxing technique on another part of your body distracts you from the tension you are feeling elsewhere.

Desensitization

Desensitization is a more specialized approach that normally involves a therapist. The principle is the same as that of allergy shots. The noxious stimulus is increased until the unwanted response no longer occurs in reaction to it. If, for example, you are afraid of dogs, you begin your conditioning by looking at pictures of very small, very friendly dogs. If you do not break out in hives or begin to perspire, you move on to pictures of larger, more ferocious dogs. Soon you will be able to walk down the opposite side of the street from a small, friendly dog with no negative results—and later, down the opposite side of the street from a large, ferocious dog. Finally, you will be able to pet a dog without experiencing any of the previous symptoms.

The same approach can work with tests. As a self-control approach, begin with an insignificant, "who cares" test, and keep taking practice tests until you are able to do it without panic and anxiety. When you are faced with a test that really matters, it will be just "another test." You may have already been through this process without really being aware of it. By this point, you have taken hundreds of tests, so you should be able to say, "So what? It's just another test. I've taken plenty of tests and passed them all or I wouldn't be here." Act out a positive script in your mind.

Breathing Exercises

Breathing awareness can reduce anxiety. Close your eyes and focus on your breathing. Say to yourself, "My chest is moving in and out." How fast is it moving? Track that speed. Is it shallow? Concentrate on that. Is it deep? This distracts your mind from the anxiety you are feeling and focuses it elsewhere. It is similar to the principles of Lamaze childbirth. For specific exercises, refer to *Little, Brown's NCLEX-RN Examination Review.*

Yoga or Zen meditation can also help focus your attention.

Guided Imagery

Guided imagery is an easy and enjoyable method of reducing anxiety. When you find yourself daydreaming during a stressful situation, "go" with that feeling. You are probably daydreaming because you want to escape from something unpleasant in your immediate environment. Fighting the feeling will increase your anxiety. Follow your thoughts and let yourself float. Let the pleasurable experience in. See examples also in the review book mentioned above.

We suggest that during the preexam period you think of peak experiences, the "highs," in your life. It might be the birth of your first child, the day you were accepted into nursing school, or the way you felt when you accomplished something you had worked hard for and very much wanted. Get in touch with that feeling and remember it. When your anxiety level begins to climb, blank out the pain and relive the peak experience. Pleasure and pain are usually mutually exclusive. If you are in great pain, substitute thoughts of pleasure.

To reiterate: These are self-help techniques. They must be practiced to be effective. Find out if they work for you by experimenting with them. Then when you need them, they will have become almost second nature.

5.   How to Take Tests

Before getting into the mechanics of answering questions, let us talk about the things you can do immediately before a test to simplify the process for yourself. We have already gone into some of these points in the chapter on being emotionally prepared, but they are important enough to be repeated here.

About a week before a nursing exam, ask where the exam will be given, how many questions will be on the test, what the time limits will be, and what you can bring with you. Knowing what to expect frees your mind to tackle the major concern—answering the test questions—so find out early.

You may need a watch during the exams. Plan to wear or carry a watch and expect to use it.

Be prepared for monitoring. Each school has a different system, but there are similarities. You may not be allowed to bring anything except pencils with you into the testing room—especially no books or notes. Many people feel insulted when asked to open their satchels before entering the testing room. Do not take it personally. Make it easy for yourself by removing all nonessentials before the exam. There is no point in complicating things for yourself by being asked to empty the contents of your backpack for a monitor to examine.

If you lean down to get a tissue from your coat during the exam, a monitor may come over to see what you are doing. Do not be offended. The monitors are simply doing their job, and they treat everyone in the same way.

Use the Question/Answer Booklet Correctly

Some schools use an electronically scored question/answer booklet for exams. You will need to use a special No. 2 pencil to mark your answers. If you do not use this pencil, your answers may not be clearly marked for the machine scoring. You will be instructed to fill in the circle that corresponds with the correct answer. Do not use check marks, vertical marks, or cross-outs. A shaded circle is the only acceptable mark.

Follow Directions

Perhaps the most important aspect of test taking is reading directions. "That's old stuff," you say. "Of course I read the directions." The point is to read the directions in front of you *for this exam,* and to read them thoroughly and carefully. Take nothing for granted. Assume nothing. It

may be a tired point, but it is the No. 1 source of trouble on all nursing exams.

Before you begin the test, the proctor will generally explain the directions and instruct you how the answer is to be marked. This is not the time to tune out.

If the proctor gives you a sample question to try out, take it seriously. It will offer an invaluable opportunity to figure out the test approach early in the test. Study the reasons given for the correct answer. It may be an obvious, easy answer, but try to pinpoint what makes it correct. What kind of thinking is behind the question? This may help you identify the "set" of the examination.

Ask for clarification of the directions if you have any doubts. Do not be concerned with appearing stupid. Find out what is expected of you.

Structure Your Time

You are now ready to begin your nursing test. Before you even look at the first question, examine the entire exam (unless, of course, it is administered by computer and does not allow for looking ahead). Note the number of questions and the type of questions. Are they long-winded, situational case descriptions, or are they short and to the point?

Take the number of questions and divide it into the length of time you will have to complete the exam. The answer will tell you about the speed you must maintain to complete the examination in the time allotted. Then set a schedule for yourself. For example, if you have 1½ hours for a 90-question test, plan to complete one question per minute to finish; note what questions you should be beginning at the end of 30 and 60 minutes.

These checkpoints help, particularly when you begin to tire. They offer reinforcement when you reach each one, and they jog you if you fall behind. If you miss a checkpoint, do not become discouraged. Recognize that you must speed up a bit. Work quickly, but not at the expense of accuracy.

Go Fast on the "Easy" Ones

It is time to begin question 1. Your strategy should be to accumulate the greatest number of points in the shortest amount of time. The goal is to get as many questions correct as possible.

Answer Each Question

Answer each question. If the question is long, break it down into its basic elements and tackle each part one by one. Do not give up immediately if you encounter an unfamiliar word. It may not be important.

Anticipate the Answer

Try this approach. After you have read the question, but before looking at the choices, anticipate the answer. Then check the choices to see if your answer is among them.

If you do not find the answer you anticipated, give that answer up. Do not cling tenaciously to your answer as if it were a prized possession. This should act as a warning beacon that something is wrong. Perhaps you did not read the question correctly. Do not alter words in an option to make it agree with your anticipated answer.

Relate Each Option to the Stem

Test each actual option against the actual question. Grammatically place the answer in the same sentence with the question. "Blank (insert choice

A, B, C, and D to see how each fits) is the most important element of treatment. . . ."

Do Not Get Stuck on the More Difficult Questions

Do not get bogged down with one question. If you have no idea what the answer is, give it a serious try, then *move on*.

Focus Your Reading

In reading the questions and the choices, skim past the "frills." Read the situation quickly, looking for the key words. Read the choices quickly, with the same focus. Drill yourself and put yourself in the mind-set to ignore unimportant words and to zero in on the key words.

Identify Key Words

What are these key words? Which one is *incorrect?* Which one is *correct?* Which one is most *important?* Which one is *least important?* Which *is?* Which *is not?* All of the following *except?* Are they asking for ounces or pounds? Hours or minutes?

Do Not Overthink

It is vital that you read the questions *as they are written*. Do not read anything into them. And do not misinterpret them. If the question is simple and obvious, pick a simple, obvious answer. Forget any subtleties you may know about the subject. Choose the answer intended. Recognize the level of sophistication of the question.

Answer the Condition in the Stem

Be sure you know what is being asked. If the question asks, "Which is true *only sometimes?*" the answer will be different than if the question asks, "Which is *never* true?"

Or perhaps the test writer is not looking for what is "true," but for what is "important." All the choices may be true, but *only one will be the most important*. Do not put down the first answer you read as correct, simply because it is true. It may not be true for the specific situation described, or it may be true but not of major importance to that situation.

If the question asks for a reason, the choice you make should be phrased in such a way that it *provides a reason*. If the question asks for an explanation, the answer should "explain." Ask yourself, "Is the answer to this question likely to be phrased in a positive or a negative manner?" Then look for a choice that meets that criterion.

Look first for a simple, straightforward, conservative, "garden variety" answer. It will save you time.

Give special attention to questions in which each word counts.

Patient-Focused Approach Is Important

A very important aid, and one too often overlooked, is to focus on the patient. What appears to be a correct answer can be wrong if it ignores the concerns and feelings of the patient.

"Feelings"-Oriented Answers May Be Emphasized

Keep the word *feelings* uppermost in your mind. Often the best choice may be the one that focuses on "feelings."

Watch for Negatives

Note that *incorrect* statements can be *correct* answers. If the question asks, "Which is an *incorrect* treatment?" the incorrect statement is the right answer.

All of the preceding suggestions assume that you know something about the questions being asked. What do you do about difficult or totally unfamiliar questions? Try inverted guessing.

Educated Guessing

General Versus Specific Answer

When you do not know the specific facts called for in a question, immediately turn to your powers of reasoning and search your related experience. Ask yourself, "Is the question asking for a general or a specific answer?" There will be a clue. Tests are generally not tricky. For example, "Which *one* is . . . ?" gives you the clue that the question requires a *specific*, not a general, answer.

Rephrase and Think of Clinical Examples

Paraphrasing the question may help. Change it into simple language. Reduce intellectual or abstract situations to the concrete. Substitute actual examples for concepts.

Use Process of Elimination: Wrong and Irrelevant

Narrow multiple-choice options (usually four per question) to two choices, if possible. First eliminate the answers you suspect are wrong. Then eliminate any choices you suspect are *true, but not relevant* to the question. If you know enough to eliminate two of the four choices, there is a 50-50 chance you will guess correctly.

Guess Between Two Choices

Should you guess? If the odds are 50-50, yes.

If all four answers seem to be correct, then go through a process of elimination.

What process can you use to help you make a guess or to check your guess to see if it might be correct? Read the following six suggestions.

Look for Patterns Among the Given Options

Compare the choices with each other. If several options look good, or if none of the options look good, try to examine patterns. If three of the four choices are similar, focus on the one that is different. It may well be the best choice.

Look for the Most Comprehensive Answer

Another method of elimination is to look for "telescoping" or "*umbrella*" answers. Eliminate the options that are contained in another option. If choice "C" takes in "A," "B," and "D," then it is likely that choice "C" is the best answer.

Avoid Global Terms: All, Always, Never, None, etc.

For example, if the words "it is always" are contained in an option, 9 times out of 10, it is the wrong answer. This is not a hard-and-fast rule, but if you are guessing, it is worth eliminating such an option.

Find the Plausible Distractors

Often the best answer may appear in the "B" or "C" position. This is certainly not true 100% of the time, but it is based on the premise that test writers are human. They are not infallible. The most attractive distractors are often placed first or last in the series of options. The best answer may then be more readily placed in the middle. Again, if you are guessing, take a second look at the middle options.

Choose an Answer That Makes Sense

And—all things being equal—if you are guessing, select the response you best understand. If you are leaning toward an answer you do not understand, discard it and choose one you do understand.

We have said this before, but *use all the time allotted.* You have already decided to be the last, rather than the first, to leave. Use any extra time to check your answers and to look again at questions you have left unanswered to see if now they make more sense.

Last Hunch May Be Better Than First Hunch

When checking your answers (before going on to the next question, or at the end of the test if that is possible), do not hesitate to change an answer if it seems appropriate to you on review. This is contrary to past popular belief, which was to keep your first-hunch answer. But current thinking on the subject says: If after you have answered a question, you think you have marked an incorrect choice, change it to what you now believe to be correct. If you are still unsure, then keep your original answer. Remember that you are not breaking an ironclad rule by changing your answers. What you do depends on information you may have remembered and its effect on your previous decisions.

Application of these test-taking guidelines and reminders can make taking nursing exams a less stressful process. It can also improve your scores. But there are no magic formulas. The responsibility for preparing yourself intellectually, emotionally, and physically for this challenge rests with you—the nursing student. And you *can* meet the challenge with confidence, skill, and success!

II. Case Management Scenarios with Critical-Thinking Exercises: Questions and Answers

Introduction

The National Council of State Boards of Nursing has identified measurable abilities that it feels a nurse must demonstrate in order to give competent care as a safe practitioner. These critical-thinking exercises are prepared to test for these abilities.

The practice questions and answers in this section are presented with detailed case management scenarios for practice and self-evaluation in applying the *nursing process*. The scenarios involve *integrated* health problems, and are organized within a *life-cycle* framework.

When you are ready to take the practice questions, set a time limit of 1 minute per question. The number of questions varies from case to case. You may choose to take a certain number of them at a time, check your answers, then take another segment. Use these questions as a "test run" for your nursing exams. *Do* set time limits for yourself.

Make full use of the answer and rationale sections provided. In addition to indicating the correct answer to each question, these sections give explanations of *why* the answer is *correct* and why the other choices are *incorrect*. Use these practice questions to identify areas in which you need more study or review—and then refer to your textbooks, classroom notes, or the references listed at the end of this book.

In Section III, you will find coding tables, summary grids, and the index to nursing problems. Use these study aids to identify areas with which you are having difficulty, and then focus your study on those areas.

6. Typical Test Designs for Critical-Thinking Exams in Clinical Nursing Areas

The following seven sections summarize the general structure and content focus of many exams that you may encounter in nursing school.

I. ***Exam questions are designed to test for safe, effective nursing practice.*** The following outline of categories of safe, effective practice is from a study by Angeline Jacobs and others, entitled *Critical Requirements for Safe/Effective Nursing Practice.* Nursing curricula are often designed to prepare students for nursing practice as identified in 2000 critical incidents collected and analyzed for this study, as well as 222 activities identified in the job analysis of what an entry-level nurse does in a study by Michael Kane and others. Nursing behaviors tested in school exams are often derived from these research studies, which identified behaviors relevant to current nursing practice (*critical requirements*) in each of the five clinical specialties.*

A. Exercises professional prerogatives based on clinical judgment.
   1. Adapts care to individual patient needs.
   2. Fulfills responsibility to patient and others despite difficulty.
   3. Challenges *inappropriate* orders and decisions by medical and other professional staff.
   4. Acts as patient advocate in obtaining appropriate medical, psychiatric, or other help.
   5. Recognizes own limitations and errors.
   6. Analyzes and adjusts own or *staff reactions* in order to maintain therapeutic relationship with patient.

B. Promotes patient's ability to cope with *immediate, long-range,* or *potential health-related change.*
   1. Provides *health care instruction* or information to patient, family, or significant others.
   2. Encourages patient or family to make decisions about accepting care or adhering to treatment regime.

*Sources: (1) Literature from National Council of State Boards of Nursing. (2) Adapted from *A Study of Nursing Practice and Role Delineation of Entry Level Performance of Registered Nurses* by the American College Testing Program for the National Council of State Boards of Nursing, Inc., Chicago (1986 and reviewed regularly since then).

     3. Helps patient recognize and deal with *psychological stress.*

     4. Avoids creating or increasing anxiety or stress.

     5. Conveys and invites *acceptance, respect,* and *trust.*

     6. Facilitates relationship of family, staff, or significant others with patient.

     7. Stimulates and remotivates patient, or enables patient to achieve *self-care* and *independence.*

C. Helps maintain patient comfort and *normal body functions.*

     1. Keeps patient clean and comfortable.

     2. Helps patient maintain or regain normal body functions.

D. Takes precautionary and *preventive* measures in giving patient care.

     1. Prevents infection.

     2. Protects skin and mucous membranes from injurious materials.

     3. Uses *positioning* or exercise to prevent injury or the *complications of immobility.*

     4. Avoids using injurious technique in administering and managing intrusive or other potentially traumatic treatments.

     5. Protects patient from falls or other contact injuries.

     6. Maintains surveillance of patient's activities.

     7. Reduces or removes *environmental hazards.*

E. Checks, compares, verifies, monitors, and follows up medication and treatment processes.

     1. Checks correctness, condition, and safety of medication being prepared.

     2. Ensures that correct medication or care is given to the right patient and that patient takes or receives it.

     3. Adheres to schedule in giving medication, treatment, or test.

     4. Administers medication by correct route, rate, or mode.

     5. Checks patient's readiness for medication, treatment, surgery, or other care.

     6. Checks to ensure that tests or measurements are done correctly.

     7. Monitors ongoing infusions and inhalations.

     8. Checks for and *interprets effect of medication, treatment,* or *care,* and *takes corrective action* if necessary.

F. Interprets symptom complex and intervenes appropriately.

     1. Checks patient's condition or status.

     2. Remains objective, further investigates, or verifies patient's complaint or problem.

     3. *Uses alarms and signals on automatic equipment as adjunct to personal assessment.*

     4. Observes and correctly *assesses signs of anxiety or behavioral stress.*

     5. Observes and correctly *assesses physical signs, symptoms, or findings,* and intervenes appropriately.

     6. Correctly *assesses* severity or *priority* of patient's condition, and gives or obtains necessary care.

G. Responds to emergencies.

     1. Anticipates need for *crisis care.*

     2. Takes instant, correct action in emergency situations.

     3. Maintains calm and efficient approach under pressure.

     4. Assumes leadership role in crisis situation when necessary.

H. Obtains, records, and exchanges information on behalf of the patient.
   1. Checks data sources for orders and other information about patient.
   2. Obtains information from patient and family.
   3. Transcribes or records information on chart, Kardex, or other information system.
   4. Exchanges information with nursing staff and other departments.
   5. Exchanges information with medical staff.
I. Utilizes patient *care planning*.
   1. Develops and modifies patient care plan.
   2. Implements patient care plan.
J. *Teaches* and supervises other staff.
   1. Teaches correct principles, procedures, and techniques of patient care.
   2. Supervises and checks the work of staff for whom she or he is responsible.

II. ***Nursing exam questions mainly reflect two levels of cognitive knowledge*** as described by Benjamin Bloom: *application* and *analysis. Recall* and *comprehension* levels (understanding) are usually *not* emphasized. About 80% of the test items tend to be:
A. *Application:* The use of abstractions in particular or concrete situations. They may be in the form of general ideas, *rules, procedures,* or general methods. The abstractions may also be technical *principles,* ideas, and *theories* that need to be remembered and applied.
B. *Analysis:* The breakdown of the whole into constituent parts or elements so that a rank *priority* of ideas can emerge and relationships between ideas can be made clear.

III. ***Categories of nursing knowledge*** included across clinical areas are:
A. Normal growth and development throughout the life cycle.
   1. *Recommendation:* Review theories of growth and development by Duvall, Sullivan, Piaget, Freud, and Erikson.
B. Basic human needs.
   1. *Recommendation:* Review Maslow and Havighurst.
C. Coping mechanisms used by individuals.
   1. *Recommendation:* Review most common adaptive behaviors, for example, blocking, compensation, denial, displacement, fixation, identification, introjection, projection, rationalization, reaction formation, regression, repression, sublimation, substitution, suppression, undoing.
D. Common health problems (actual or potential) in the major health areas and based on current morbidity studies.
   1. *Recommendation:* Review the 10 most common diseases, disorders, and causes of death.
E. Variations in health needs as affected by age, sex, culture, ethnicity, and religion.
   1. *Recommendation:* Be aware of *food* preferences and dietary restrictions; *belief systems* about causes of illness, methods of treatment, concept of death, concept of time; kinship structure and role of the male, female, and extended family; *ethnic variations in susceptibility* to certain diseases.

F. Nursing goals and interventions to assist individuals in maintaining life and health, coping with health problems, and recovering from the effects of injury or disease.
  1. *Recommendation:* Review nursing priorities for patients in *life-threatening* situations, *health teaching* and health maintenance situations, and *rehabilitation* situations.

IV. *Concepts relevant to general nursing practice are integrated throughout nursing exams:*

A. Management.
  1. *Recommendation:* Know the scope of RN functions, and what can be delegated to an LVN/LPN.

B. Accountability.
  1. *Recommendation:* Review major legal and ethical issues, areas of nursing responsibilities, and standards of nursing practice.

C. Life cycle.
  1. *Recommendation:* Review major health concerns, problems, and nursing care during birth, childhood, school age, adolescence, young adult and reproductive years, middle age, and older adult and geriatric years.

D. Patient environment.
  1. *Recommendation:* Review measures to protect from harm against airborne irritants, cold, and heat; review approaches to eliminate environmental discomforts, such as odors, noise, poor ventilation, dust; know *safety hazards*; review measures to maintain environmental order and cleanliness.

V. *Test items may emphasize four categories of patient needs* with 11 subcategories of activities designed to meet these needs (as on the RN licensure exam, NCLEX-RN).

A. You can expect that the most emphasis will be on meeting the patient's physical needs (*physiologic integrity*) in actual or potential life-threatening, chronic, recurring *physiologic* conditions and with patients who are at risk for complications or untoward effects of treatment. Subcategories include:
  1. *Physiologic adaptation:* meeting acute physical needs.
  2. *Reduction of risk potential:* monitoring patients at risk.
  3. *Provision of basic care:* performing routine nursing activities.

B. *Safe, effective care environment* subcategories include:
  1. *Coordinated care:* staff development, collaboration.
  2. *Environmental safety:* protecting the patient.
  3. *Safe and effective treatments and procedures:* ensuring safety during the procedures.

C. *Health promotion and maintenance* includes these subcategories:
  1. *Growth and development throughout the life span:* meeting patient needs related to parenting.
  2. *Self-care and integrity of support systems:* assisting patients with self-care and supporting patient's family.
  3. *Prevention and early treatment of disease:* immunizing/ screening.

D. *Psychosocial integrity,* with a focus on stress and crisis-related situations throughout the life cycle, includes these subcategories:

1. *Psychosocial adaptation:* meeting acute emotional and behavioral needs.
2. *Coping/adaptation:* helping patients to cope with stress.

VI. **A critical component of most nursing exams is nursing behaviors** (i.e., the five steps of the nursing process) applied to *patient situations,* in all stages of the *life cycle.*

A. The following are five specific behaviors in all of the patient situations listed in item **V** above.
1. Ability to assess (*assessment*).
2. Ability to analyze data and identify specific needs (*analysis*).
3. Ability to plan and set goals (*planning*).
4. Ability to implement specific actions (*implementation*).
5. Ability to evaluate outcome (*evaluation*).

B. Examples of questions focusing on specific nursing behaviors (i.e., aspects or steps of the nursing process) are given below.

1. *Assessment type.*

A multigravida in labor is admitted to the maternity unit. What information does the nurse need to evaluate the status of her labor?
A. Blood pressure.
B. FHR.
C. Contour of abdomen.
D. Duration, frequency, and intervals of contractions.

Comment: You need to know the relative importance of the signs and symptoms of labor to assess the status of labor when all the signs are correct. In this case, **D** is the most important. There is a need for interaction between the nurse and patient to assess the patient's progress in labor.

2. *Analysis type.*

A 30-year-old man with diabetes, who has previously been stabilized on insulin, calls the emergency room to relate that although his urine is negative for sugar and acetone, he feels as if he is about to pass out. He reports that he has been eating three regular meals a day and taking his insulin as directed. What would be a valid interpretation of his current condition?
A. He is not physically active enough.
B. He needs to change his insulin dose.
C. He has low blood sugar.
D. His insulin level is high.

Comment: In order to select **C** as the best answer, you need not only to understand diabetes, but also to assess the symptoms and to interpret them correctly.

3. *Planning type.*

A comatose patient is admitted to the hospital. Which nursing goal would be of prime importance at the outset?
A. Establish a flexible visiting schedule so that relatives can watch the patient closely.
B. Include the family in the immediate physical care.
C. Provide consistency of care by assigning the same nursing personnel to the patient.
D. Place the patient in a room where you can closely monitor the condition from the nursing station.

Comment: To choose the best response, **D,** you need to analyze the situation correctly and apply that knowledge in planning a strategy to achieve the goal of providing close observation of the patient's condition.

4. *Implementation type.*

An emaciated-looking patient with anorexia nervosa is admitted to the mental health unit. She refuses to eat lunch when it is served, just as she has refused most meals at home. What can the nurse say that would be most effective in encouraging her to eat?

A. "You will have to eat if you do not want us to tube-feed you."

B. "Here is a small sandwich. I will sit with you to keep you company while you eat."

C. "It is important that you eat your meals, as it is part of your therapy to help you gain weight."

D. "Aren't you hungry? It is a long time before you can eat dinner if you don't eat your lunch."

Comment: **B** is the best nursing action, to provide the basic need for nutrition when the client's psychological condition prevents her from taking care of her own physiologic need for food.

5. *Evaluation type.*

An obese man began his weight loss program after Christmas. The plan was for him to lose a minimum of 10 pounds a month. By February 1, he had lost 15 pounds. What action should the nurse take at this time?

A. Praise him for his results and ask him if he would like to set the next goal at 15 pounds.

B. Refer him to the dietitian to regulate his weight loss.

C. Encourage him to lose even more weight since he is doing so well.

D. Ask him why he lost more weight than planned.

Comment: The weight loss is controlled by the patient, with the nurse providing needed support and encouragement. Here, the best answer is **A** because it measures the results (15 pounds) against the goal (10 pounds) and works with the patient to change the goal appropriately.

VII. ***Exam questions focus on patient care situations and nursing measures that can be categorized into eight areas of human functioning:***

A. *Protective functions:* The patient's capacity or ability to maintain defenses and prevent physical and chemical trauma, injury, infection, and threats to health status. *Examples* include:

1. Communicable diseases (including sexually transmitted diseases).

2. Immunity.

3. Physical trauma and abuse.

4. Asepsis.

5. Safety hazards.

6. Poisoning.

7. Skin disorders.
8. Preoperative care and postoperative complications.
B. *Sensory-perceptual functions:* The patient's capacity or ability to perceive, interpret, and respond to sensory and cognitive stimuli (visual, auditory, tactile, taste, and smell). *Examples* include:
   1. Auditory, visual, and verbal impairments.
   2. Sensory deprivation and sensory overload.
   3. Aphasia.
   4. Brain tumors.
   5. Laryngectomy.
   6. Delirium, dementia, and other amnestic disorders.
   7. Body image.
   8. Reality orientation.
   9. Learning disabilities.
  10. Seizure disorders.
C. *Comfort, rest, activity, and mobility:* The patient's capacity or ability to maintain mobility; desired level of activity; and adequate sleep, rest, and comfort. *Examples* include:
   1. Joint impairment.
   2. Body alignment.
   3. Pain.
   4. Sleep disturbances.
   5. Activities of daily living.
   6. Neuromuscular and musculoskeletal impairment.
   7. Endocrine disorders that affect activity.
D. *Nutrition:* The patient's capacity or ability to maintain the intake and processing of the essential nutrients. *Examples* include:
   1. Normal nutrition.
   2. Diet in pregnancy and lactation.
   3. Obesity.
   4. Conditions such as diabetes, gastric disorders, and metabolic disorders that primarily affect the nutritional status.
E. *Growth and development:* The patient's capacity or ability to maintain maturational processes throughout the life span. *Examples* include:
   1. Childbearing and child rearing.
   2. Conditions that interfere with the maturation process.
   3. Maturational crises.
   4. Changes in aging.
   5. Psychosocial development.
   6. Sterility.
   7. Conditions of the reproductive system.
F. *Fluid-gas transport:* The patient's capacity or ability to maintain fluid-gas transport. *Examples* include:
   1. Fluid volume deficit and overload.
   2. Cardiopulmonary diseases.
   3. Acid-base balance.
   4. Cardiopulmonary resuscitation.
   5. Anemias.
   6. Hemorrhagic disorders.
   7. Leukemias.
   8. Infectious pulmonary diseases.

G. *Psychosocial-cultural functions:* The patient's capacity or ability to function in intrapersonal, interpersonal, intergroup, and sociocultural relationships. *Examples* include:
   1. Grieving; death and dying.
   2. Psychotic and anxiety behaviors.
   3. Self-concept.
   4. Therapeutic communication.
   5. Group dynamics.
   6. Ethical-legal aspects.
   7. Community resources.
   8. Spiritual needs.
   9. Situational crises.
   10. Substance abuse.

H. *Elimination:* The patient's capacity or ability to maintain functions related to relieving the body of waste products. *Examples* include:
   1. Conditions of the gastrointestinal system (vomiting, diarrhea, constipation, ulcers, neoplasms, colostomy, hernia).
   2. Conditions of the urinary system (kidney stones, transplants, renal failure, prostatic hypertrophy).

7.   Children and Adolescents

Case Management Scenarios and Critical-Thinking Exercises

Medical Diagnosis: Cleft lip and palate.

- ■ *Nursing Problems:*
  Risk for infection.
  Risk for injury.
  Knowledge deficit: feeding, medication administration.
  Altered nutrition requirements.
  Risk for altered parent–infant attachment.
  Parental role deficit.

- ■ *Chief Complaint:*
  Anne Jones is a newborn with unilateral cleft lip and palate.

- ■ *Past History:*
  Anne was born 3 days ago to Mr. and Mrs. Jones. She is their first child. Both are in their early 30s and are college educated. The paternal grandparents have told the mother that it is her fault: "There has never been a person with a hare lip in our family before." Mrs. Jones is very disturbed and very anxious about handling the infant; she has refused to see the child for the past day.

- ■ *Physical Exam:*
  Newborn infant with visible cleft of lip and cleft palate.
  Birth weight: 7 pounds, 6 ounces. No other defects noted.
  Heart: No murmurs. Pulse: 152 and regular.
  Lungs: Clear to auscultation.
  Skin: Pink, warm, and intact except for cleft; slightly jaundiced.
  Vitamin K given in the nursery on day 1. Child now taking formula
     via an Asepto syringe.

- ■ *Laboratory Data:*
  Hct: 56%. Hgb: 17 g. Bilirubin: 6.

■ *Plan:*

Child to go home with parents when parents can feed her well. She is to return at 3 months of age for cleft lip repair.

---

1. Immediately after Anne's birth, what should be included in planning nursing care of Mrs. Jones?
   A. Place her in a room with a woman who has had a set of twins.
   B. Restrict her visitors to allow her more rest.
   C. Have her watch the nurse feed the baby.
   D. Do not discuss the baby unless she brings the subject up herself.

2. In teaching Mrs. Jones to feed her daughter, Anne, which is inappropriate to include in nursing care?
   A. Place the tip of the Asepto at the front of Anne's mouth so she can suck.
   B. Rinse the mouth with sterile water after each feeding to minimize infections.
   C. Feed Anne in an upright position and bubble frequently to reduce air swallowing.
   D. Apply lanolin to the lips to reduce dryness associated with mouth breathing.

3. Part of the discharge teaching should be the signs and symptoms of a common complication for infants with cleft lip and palate. This is:
   A. Ear infections.
   B. Meningitis.
   C. Anemia.
   D. Seizures.

4. Mrs. Jones calls the clinic 2 months after taking Anne home. She tells the nurse that the baby has a temperature of 102°F, has been turning her head from side to side, and has been eating poorly. The nurse would advise her to:
   A. Cleanse Anne's ears with warm water.
   B. Give Anne acetaminophen, 0.3 mL of infant drops, and call back in 4 hours with her temperature.
   C. Bring Anne into the clinic for evaluation.
   D. Give Anne 4 oz. of water and retake her temperature in 1 hour.

5. Mrs. Jones brought Anne to the clinic, where otitis media on the left side was found on examination. Ear drops were prescribed. The nurse teaches Mrs. Jones to instill the ear drops by instructing her to:
   A. Pull the earlobe down and backward.
   B. Pull the earlobe up and backward.
   C. Pull the earlobe down and forward.
   D. Pull the earlobe up and forward.

6. At 3 months of age Anne returns to the hospital for repair of her cleft lip. On admission, the nurse assesses Anne's developmental status as appropriate. What would the nurse not expect to see?
   A. Smiling in response to her mother's face.
   B. Reaching for shiny objects but missing them.
   C. Holding her head erect and steady.
   D. Sitting with slight support.

7. Anne undergoes repair of her cleft lip. She returns to the unit with a Logan bow in place. She is awake and beginning to whimper. Her color is pink and pulse is 120 with respirations of 38. IV solution is infusing in her right hand at a rate of 15 mL/hour; the fluid is infusing sluggishly and the hand is edematous. The jacket restraint has loosened and one arm has partially come out. The nursing priority should be to:
   A. Recheck vital signs.
   B. Check the IV site for infiltration.
   C. Check to see if she has voided.
   D. Replace the restraints securely.

8. Following surgery, it is important to prevent Anne from crying excessively. To accomplish this, the nurse should:
   A. Give her a pacifier, to sooth her sucking needs.
   B. Place her on her abdomen, which is the position in which she normally sleeps.
   C. Ask Mrs. Jones to stay and hold her.
   D. Request a special nurse to hold Anne because the staff does not have the time to comfort her.

9. Anne is doing well and eating her normal formula via an Asepto syringe. The nurse wants to provide her with appropriate stimulation. The best toy for the nurse to give her is:
   A. A colorful rattle.
   B. A string of large beads.
   C. A mobile with a music box.
   D. A teddy bear with button eyes.

10. Anne is receiving ampicillin, 75 mg PO every 6 hours. Which is the appropriate method of administering the medication to her?
   A. Place the medication on a spoon and place it on the back of her tongue.
   B. Mix the medication with her formula to give it to her via Asepto syringe.
   C. Place the medication in a nipple and allow her to suck it down.
   D. Give via Asepto syringe to the side of the mouth.

11. While the nurse is instructing Mrs. Jones in Anne's care before the baby's release, which statement by Mrs. Jones should concern the nurse?
   A. "You know, I haven't taken the baby out much because people stare so."
   B. "My mother-in-law blamed me at first for the baby's problem but she's visiting more often now."
   C. "You know, she's so pretty now she will win a beauty contest just as I did."
   D. "My husband and I enjoy her so much and play with her a great deal."

12. In establishing long-term goals for Anne at discharge, the nurse would not include:
   A. Prevention of hearing loss.
   B. Promotion of adequate speech.
   C. Promotion of adequate parent-child relationship.
   D. Prevention of infection of surgical incision.

Medical Diagnosis: Hydrocephalus following myelomeningocele closure.

- **Nursing Problems:**
  Altered bowel elimination.
  Altered parenting.
  Impaired physical mobility.
  Seizures.
  Impaired skin integrity.
  Altered urinary elimination.

- **Chief Complaint:**
  David Stivic is a 2-month-old exhibiting increasing head size.

- **Past History:**
  David was born with a lumbosacral myelomeningocele, which was repaired at 1 week of age. He is the second child; the first child, a girl, is 3 years old and healthy. Mrs. Stivic had multiple problems with her pregnancy. David's birth weight was 6 pounds, 1 ounce.

- **Symptoms:**
  The child is irritable and difficult to feed; he experiences occasional vomiting.

- **Physical Exam:**
  Heart: No murmurs. Pulse: 126.
  Lungs: Clear to auscultation.
  Skin: Scar in lumbosacral area of back, approximately L1–3. Slightly erythematous area on back of head.
  Extremities: Bilateral clubfeet and dislocated hips.
  Bowels: Two stools per day, hard and dark.
  Bladder: Slightly distended; mother reports that the infant is constantly wet.
  No immunizations. History of frequent colds. Smiles in response to faces. Cannot hold head up.

- **Laboratory Data:**
  Chest x-ray: Clear.
  Hgb: 12 g. Hct: 40%.
  Urinalysis: Cloudy with many WBCs. Specific gravity: 1.015.
  Current weight: 9 pounds, 1 ounce.
  Temperature: 99.8°F (rectally).

13. The nurse should emphasize, to the families, that the most important goal of long-term care for the child with myelomeningocele is:
    A. Prevention of obesity.
    B. Promotion of skin integrity.
    C. Prevention of urinary tract infections.
    D. Promotion of sensory stimulation.

14. Following repair of the myelomeningocele, the nurse observes for signs of increased intracranial pressure based on knowledge that:
    A. The volume of circulating cerebrospinal fluid will increase with the infant's age.

B. Downward displacement of the medulla and posterior cerebral vermis into the vertebral canal can lead to noncommunicating hydrocephalus.

C. There is a decrease in the amount of surface area for absorption of the cerebrospinal fluid.

D. The space available for circulation of the cerebrospinal fluid has been decreased.

15. In the assessment of the infant for hydrocephalus, the first sign the nurse would note is:
A. Widening pulse pressure.
B. Headache.
C. Increasing irritability and poor feeding.
D. Increasing head circumference.

16. An individualized care plan for David would not include:
A. Daily measurement of head circumference.
B. Turn frequently and position with hips abducted.
C. Credé every 2–4 hours.
D. Feed 6 oz. every 3 hours.

17. While awaiting specific tests and studies in the hospital, David is alert and fairly active. An appropriate toy for him is:
A. A busy board.
B. Large, colorful blocks.
C. A rattle.
D. A music box.

18. Computed transaxial tomography (CAT) and other tests confirm the diagnosis of hydrocephalus. Surgery for shunt placement is scheduled. A ventriculoatrial shunt is to be placed. The parents ask the nurse to explain what this involves. The nurse's explanation will be based on knowledge that these shunts allow flow of cerebrospinal fluid from the lateral ventricle to the:
A. Right atrium by way of the internal jugular vein.
B. Peritoneum.
C. Ureter, where the fluid drains into the bladder.
D. Right atrium by way of the carotid artery.

19. David returns from the recovery room after placement of the shunt. The nurse's immediate plan of care would not include:
A. Position off the operative site.
B. Obtain baseline head circumference.
C. Elevate the head 60 degrees to promote cerebrospinal flow.
D. Monitor for signs of increased intracranial pressure.

20. Four days after surgery, the nurse determines that David is progressing well. However, on physical assessment, he or she would not expect him to be able to:
A. Hold head steady in a sitting position.
B. Hold a rattle for a brief time.
C. Smile in response to the human face.
D. Play with hands and fingers.

21. The revised nursing care plan when David leaves the recovery room and for 2 days afterward would emphasize the importance of keeping David flat in bed to:
A. Prevent ventricular collapse.
B. Avoid fluid overload.
C. Prevent destruction of the choroid plexus.
D. Prevent subdural hematoma.

22. In discharge teaching with David's parents, the nurse should include:
   A. How to pump the shunt.
   B. How to check the urine for glucose.
   C. The importance of daily weights.
   D. How to measure head circumference.

23. When should the nurse instruct the parents that it is not necessary to seek medical advice?
   A. When David develops a temperature and a reluctance to turn his head from side to side.
   B. When they note a foul odor to David's urine.
   C. When they note bulging of the anterior fontanelle when David is crying.
   D. When David has persistent, projectile vomiting.

24. David's parents should be aware that the placement of a new shunt may not be required when there is:
   A. Infection.
   B. Pulmonary embolism.
   C. Normal growth and development.
   D. Clogging or disconnection of the shunt.

25. Mrs. Stivic comes to the clinic 2 months after David's discharge. "I'm so tired. I spend all of my time with David. He cries a lot and I can't rest. I just don't know what to do." The nurse notes that David is alert, with normal vital signs and a flat fontanelle. The most helpful comment the nurse could make is:
   A. "Sometimes I bet you wish you hadn't had him."
   B. "Doesn't your husband help at all with the work?"
   C. "Has David been vomiting or eating less than usual?"
   D. "It must be very hard for you. Have you had any time away from the baby?"

26. During this visit the nurse also notes that David has had no immunizations yet. The nurse realizes that he has missed:
   A. 1 dose of DPT and 1 dose of OPV.
   B. 1 dose of DPT and 1 dose of MMR.
   C. 2 doses of DPT and 2 doses of OPV.
   D. 2 doses of DPT and his TB test.

Medical Diagnosis: Diarrhea and dehydration.

■ *Nursing Problems:*
Altered bowel elimination.
Diarrhea.
Fluid volume deficit.
Seizures.
Ineffective thermoregulation.

■ *Chief Complaint:*
Johnny Jones, 8 months old, has had diarrhea for 2 days.

■ *Past History:*
The child had been well until 3 days before admission, when he began having watery stools and his temperature increased. His family lives on a farm with the water supply coming from a well. His mother is well and has no health problems; the father has hypertension.

Johnny is one of three children; he has a 3-year-old brother and a 5-year-old sister, both of whom are healthy. Johnny weighed 8 pounds, 2 ounces at birth.

■ *Symptoms:*
Poor skin turgor, excoriated diaper area, depressed fontanelle, decreased weight and urine output.

■ *Physical Exam:*
Lethargic, pale infant.
Heart: No murmurs. Rate: 130.
Lungs: Clear to auscultation. Rate: 44.
Skin: Poor skin turgor. Dry mucous membranes. Fontanelle depressed. Lips dry and cracked. Buttocks erythematous and rash present.
No immunizations. No known allergies. Sits well and has begun to drink from a cup according to his mother. On junior foods and some table foods.

■ *Laboratory Data:*
Chest x-ray: Clear.
Hgb: 13 g. Hct: 45%.
K: 3.5 mEq/L.
pH: 7.28.
Temperature: 102°F (axillary).
Urinalysis: Cloudy and concentrated, with some WBCs. Specific gravity: 1.030.
Current weight: 17 pounds, 5 ounces. Normal weight: 18 pounds, 8 ounces.

---

27. Which is most useful in assessing the degree of Johnny's dehydration?
    A. Urinary output.
    B. Skin turgor.
    C. Mucous membranes.
    D. Weight.
28. Johnny is placed NPO and IV fluids are started. Which is the immediate goal of care?
    A. Restoration of intravascular volume.
    B. Prevention of further diarrhea.
    C. Promotion of skin integrity.
    D. Maintenance of normal growth and development.
29. 360 mL of D5W/0.2 is ordered to run over 8 hours intravenously. At what rate should the nurse run the fluids? (Microdrip = 60 gtt/mL)
    A. 32 gtt/min.
    B. 40 gtt/min.
    C. 45 gtt/min.
    D. 54 gtt/min.
30. Restraints have to be used with Johnny to maintain the IV line. Which statement best explains their use to the parents?
    A. They have to be used to keep Johnny still all the time.
    B. The restraints will prevent him from pushing the needle through his skin.

    C. He will get used to them if you leave him alone.

    D. As soon as he can take fluids again, the restraints will not be needed.

31. Johnny has had only two stools in the last 24 hours and has gained 5 ounces. He is more active and alert. His diet has been advanced to clear liquids. Which liquid is the most appropriate for the nurse to offer?

    A. 7-Up and ginger ale.

    B. Half-strength formula.

    C. Tea and clear broth.

    D. Pedialyte and glucose water.

32. After clear liquids are started, Johnny has four stools in 2 hours. The nurse's action is to:

    A. Continue oral feedings but increase the IV fluid rate.

    B. Take the pulse, temperature, and respirations.

    C. Stop feeding the child orally.

    D. Weigh the child.

33. Johnny is doing well and is alert and cheerful. Which diversion is appropriate for him?

    A. A colorful mobile.

    B. Large blocks to stack.

    C. A rattle and bell.

    D. A game of peek-a-boo.

34. The day before his discharge Johnny's temperature has risen to 103°F. The nurse comes into the room and finds Johnny having a tonic-clonic convulsion. The best action is to:

    A. Stay with the child and protect him from injury during the seizure.

    B. Inform the physician that seizure activity has occurred.

    C. Institute measures to reduce the child's temperature.

    D. Inform the parents that seizures often occur in children when they have elevated temperatures.

35. The seizure has ended and Johnny's vital signs are stable. His parents were in the room when the seizure occurred and are extremely upset. What is the best statement for the nurse to make to them?

    A. "Don't worry. This often occurs with young children and causes them no harm."

    B. "If we can keep his temperature down, this shouldn't happen again."

    C. "Have any of your other children ever had this happen?"

    D. "I know this must worry you. Has anything like this ever happened before?"

36. While assessing Johnny after the seizure, the nurse expects him to be able to:

    A. Stand erect while holding on to his mother's hand.

    B. Crawl and begin to say "Mama" and "Dada."

    C. Still have a Moro reflex.

    D. Sit down from a standing position without help.

37. The nurse's discharge plan with Johnny's mother included the need for immunizations and prevention of infection. Six weeks after discharge Mrs. Jones brings him to the clinic and he still has not been immunized. Which action is indicated?

    A. Repeat the instructions again, more clearly.

    B. Chastise her for not getting the immunizations.

C. Make an appointment at the clinic to have the immunizations done.
D. Assess why Mrs. Jones has not had the immunizations done.

Medical Diagnosis: Kawasaki disease (acute febrile syndrome of unknown etiology, mucocutaneous lymph node syndrome).

- **Nursing Problems:**
  Altered cardiac output.
  Risk for alteration in endocrine/metabolic processes.
  Altered fluid volume.
  Altered immune response.
  Ineffective thermoregulation.

- **Chief Complaint:**
  Jackie Shabata, 8 months old, is admitted to the pediatrics unit with a diagnosis of Kawasaki disease. This is her first hospitalization. She was a full-term infant who had been healthy prior to the onset of this illness. Jackie has received all her immunizations to date; an only child, she lives with both her natural parents. Her parents report she has had a fever and has refused to eat over the past 8 days, taking only sips from her bottle. She has lost almost 16 ounces during this time.

- **Physical Exam:**
  Jackie appears irritable and uncomfortable. On admission her vital signs are:
  Rectal temperature: 40°C.
  Apical pulse rate: 148.
  Respiratory rate: 36.
  Blood pressure: 78/50.

- **Plan:**
  Her physician orders:
  CBC, sedimentation rate, and platelet count.
  Daily weight.
  Acetaminophen (Tylenol) gr. $\overline{ss}$ PO q4h prn for temperature over 38°C or 101°F.
  IV 5% dextrose in 0.45% normal saline @ 55 mL/hour.

---

38. Jackie's parents want to know what Kawasaki disease is. The nurse should tell them that it is:
   A. An acute disease characterized by fever and having no known cause.
   B. A forerunner of rheumatic fever found in younger children.
   C. A newly discovered venereal disease originating in Japan.
   D. A virally induced cardiomyopathy following motorcycle accidents.

39. In reading through Jackie's chart, the nurse notices the term "mucocutaneous lymph node syndrome" and realizes this is the same as:
   A. Infantile polyarteritis nodosa.
   B. Kawasaki disease.
   C. Nephrotic syndrome.
   D. Still's disease.

40. Since Jackie has received all the immunizations appropriate for an 8-month-old, the nurse checks her record to confirm that she has been immunized against:
    A. Diphtheria, tetanus, pertussis, and polio.
    B. Diphtheria, tetanus, pertussis, and TB.
    C. Diphtheria, TB, polio, and German measles.
    D. Diphtheria, TB, polio, and measles.

41. During the initial nursing assessment, the nurse does not need to evaluate Jackie for:
    A. Bilateral conjunctivitis and strawberry tongue.
    B. Desquamation of fingers and toes.
    C. Petechiae and hematomas.
    D. Macular erythematous rash and lymphadenopathy.

42. Jackie's temperature is 40°C. Her parents ask the nurse what this "really means." The nurse should advise them that:
    A. It is too early to tell for sure.
    B. Jackie is quite sick, and this is most likely a complication.
    C. Jackie's temperature is within normal limits.
    D. Jackie's temperature is 104°F.

43. At 6 P.M., Jackie has a temperature of 102.7°F. The nurse must administer acetaminophen (Tylenol), gr. $\overline{ss}$, to Jackie per physician order. The label on the stock bottle reads: 60 mg = 0.6 mL. How many *minims* should the nurse give?
    A. 0.3.
    B. 4.5.
    C. 8.0.
    D. The nurse is unable to use the stock bottle, and must special order this dosage from the pharmacy.

44. The fever associated with Kawasaki disease is often erratic and high spiking. What action by the nurse is most appropriate, in terms of planning care for Jackie related to the fever?
    A. Ask the pediatrician about starting Jackie on antibiotics as soon as possible.
    B. Check Jackie's temperature every 1–2 hours.
    C. Administer antipyretics as ordered by the physician.
    D. Place Jackie on a hypothermia mattress if antipyretics alone prove ineffective.

45. Jackie's platelet count is 790,000/μL. The nurse should:
    A. Immediately notify the physician.
    B. Check for bleeding tendencies in Jackie's gums and rectal mucosa.
    C. Initiate reverse isolation and notify the physician.
    D. Place the lab report in Jackie's chart for the pediatrician to see.

46. The night nurse checks Jackie's IV line at midnight and notes that there is 225 mL left in the bottle that is hanging. At approximately what time does the nurse anticipate having to add a new bottle of IV fluid?
    A. Immediately, so there is enough fluid to last through the night.
    B. Just prior to change of shift, at 6:45 A.M.
    C. Whenever the bottle is nearly empty.
    D. At 3:30 A.M.

47. At 1 A.M., the nurse checks Jackie again and finds her temperature is 104.8°F. Jackie's IV bottle is almost empty. What action is most appropriate for the nurse to take *first*?

A. Administer acetaminophen (Tylenol), gr. $\overline{ss}$, as ordered by the physician.

B. Add another bottle of IV fluid.

C. Check the flow rate of Jackie's IV fluid.

D. Elevate the head of Jackie's bed and check for pulmonary edema.

48. Jackie has received four times the amount of IV fluid that was ordered for her between midnight and 1 A.M. What should the night nurse do in terms of IV fluids for Jackie during the remainder of the shift?

A. Check with the pediatrician for new IV orders.

B. Slow the flow rate enough so that Jackie will receive the amount ordered for the remainder of the shift.

C. Discontinue the IV and observe Jackie closely for fluid overload.

D. Adjust the flow rate to 55 mL/hour, per physician order.

49. During the night, Jackie's sedimentation rate comes back from the lab; it is 58 mm/hour. The nurse should know that:

A. This is a normal sedimentation rate for an 8-month-old.

B. Her sedimentation rate is lower than the norm, because of the extra IV fluid Jackie received during the night.

C. Her sedimentation rate is elevated, but has nothing to do with her diagnosis.

D. Her sedimentation rate is elevated and this is a typical lab finding in the child with Kawasaki disease.

50. When the day nurse weighs Jackie the next morning, she finds Jackie has gained 6 ounces since admission. Jackie still seems irritable and restless, with a cardiac rate of 180 bpm and regular, and a respiratory rate of 44 breaths/min. Her color is pale. The nurse should recognize that Jackie is:

A. Starting to gain weight due to better intake, but may be anemic.

B. Showing signs and symptoms of heart failure.

C. Stable at the moment.

D. Doing considerably better this morning.

51. Jackie is transferred to PICU for closer monitoring and her parents are very anxious about this sudden transfer. The nurse in PICU should:

A. Introduce them to the staff and do a thorough orientation to PICU.

B. Reassure them that the transfer was only routine procedure.

C. Allow them to discuss their concerns and fears.

D. Reinforce that Jackie has only a 1–2% chance for survival and then share their grief concerning her very grim prognosis.

52. Jackie is started on digoxin, 65 μg PO bid. The nurse administering medication to Jackie in PICU should "hold" the 7 P.M. dose of digoxin and notify the pediatrician if:

A. Jackie has finally fallen asleep for the first time all day.

B. Jackie's monitor is showing sinus tachycardia.

C. Jackie appears slightly jaundiced and her liver enzymes are elevated.

D. Jackie's apical rate is 98 and regular, with normal sinus rhythm on her monitor.

53. In checking Jackie for appropriate response to digoxin therapy, the nurse should look for:

    A. Nausea, vomiting, and diarrhea.

    B. Hypokalemia.

    C. Slowing of the heart rate and increased urinary output.

    D. Bigeminy or trigeminy.

54. While in PICU, nursing care planning for Jackie will not include:

    A. Allowing parents to visit frequently.

    B. Apnea and cardiac monitor.

    C. Play therapy appropriate to Jackie's age and condition.

    D. Restraining Jackie for IV therapy and confining her to her crib.

55. As Jackie's condition improves, she is sent back to the pediatric unit. Doctor's orders for Jackie include continuing digoxin and a sodium-restricted diet. In planning care for Jackie, the nurse must be sure to tell the play therapist that Jackie:

    A. Needs extra rest and cannot go to the playroom at this time.

    B. Is on a special low-sodium formula and should not receive milk.

    C. Has so many toys and dolls in her room that she does not need to go to the playroom at this time.

    D. Is receiving a medication for her heart and needs to remain on the cardiac monitor, although the therapist may play with her in her room.

56. At the discharge planning conference, the nurses are discussing when Jackie will go home. What precludes discharging Jackie?

    A. Rectal temperature = 37°C.

    B. Sedimentation rate = 22 mm/hour and platelet count = 319,000/μL.

    C. Stable weight.

    D. Poor appetite with minimal PO intake.

57. Jackie is going home with her parents; she will need to continue taking digoxin at home. Which is inappropriate health instruction to her parents?

    A. Do not mix digoxin with food or fluids; instead, offer digoxin alone.

    B. If one dose of digoxin is missed, you need not be concerned or attempt to make up the missed dose.

    C. It is best to give digoxin within 1 hour after feeding your child.

    D. If the child vomits within 15 minutes of receiving digoxin, you should repeat the dose once.

Medical Diagnosis: Congenital heart defect.

■ *Nursing Problems:*
Altered cardiac output.
Altered nutrition.
Altered parenting.

■ *Chief Complaint:*
Sam Peterson, 18 months old, was diagnosed shortly after birth as having transposition of the great vessels. He is showing increasing frequency of congestive heart failure and failure to thrive.

■ *Past History:*
There is maternal history of cardiac anomalies; Sam's mother has a sister with a cardiac problem. Sam is the youngest of three children,

with two older sisters 4 and 7 years old. Mother and father are both healthy. Father is employed as a plumber. Crippled Children's Services is helping with the bills. Sam was diagnosed at 2 days of age as having transposition of the great vessels and had a Roskind-Miller procedure done. He has had an increasing frequency of congestive heart failure, respiratory infections, and failure to thrive. (Height and weight are below the third percentile on the growth charts.) He is not achieving the normal motor tasks for his age. His parents express great concern over the child and are very anxious about his lack of progress.

■ *Symptoms:*
Failure to gain weight and increasing cyanosis.

■ *Physical Exam:*
Pale, thin, underdeveloped child with cyanotic lips and nailbeds and clubbing of the fingers and toes.
Heart: Harsh systolic murmur. Pulse: 115.
Lungs: Clear to auscultation. Rate: 36.
Skin: Cool, with clubbing of fingers and toes and cyanosis of lips, mucous membranes, and nailbeds. No known allergies.
Currently being treated with digoxin, chlorothiazide (Diuril), and KCl.

■ *Laboratory Data:*
Hgb: 17 g. Hct: 55%.
WBC: 5700/μL.
K: 3.7 mEq/L.
ECG: Right ventricular hypertrophy.
Chest x-ray: Cardiac enlargement and increased pulmonary vasculature.

Admission orders include continuation of his medication regime and a salt-restricted diet. Fluid restriction is also started.

---

58. Sam is admitted for a cardiac catheterization. The parents are present at all times in the hospital and do everything for the child—dress him, feed him, even play for him. You want to prepare the child and the parents for the procedure. It would be most appropriate for the nurse to:
   A. Realize that the child can understand simple explanations.
   B. Talk with the parents to assess their knowledge and how they can help with Sam's care.
   C. Take no specific action because the child and family have been through a cardiac catheterization previously.
   D. Ask the parents to stay away as much as possible because they upset the child.

59. In reviewing the procedure notes, the nurse determines that cardiac catheterization went well but will not yield information about:
   A. The status of the structural abnormalities.
   B. The pressure gradients.
   C. The degree of pulmonary vascularization.
   D. The oxygen levels in the chambers of the heart.

60. Sam is now 6 hours postcatheterization. Which concerns the nurse the most?

A. Sam complains of his head hurting.

B. His pulse is 110 and regular.

C. The skin temperature of his right leg has decreased.

D. His temperature has risen to 100.4° (rectally).

61. As the nurse administers antibiotics, Sam's parents ask why he needed to be placed on them prior to catheterization. The best answer is based on the nurse's knowledge that the antibiotics will prevent:

A. Urinary tract infection.

B. Pneumonia.

C. Otitis media.

D. Endocarditis.

62. Sam has had several episodes of congestive heart failure in the past few months. Which is the most useful to the nurse in assessing his current congestive heart failure?

A. The degree of clubbing of his fingers and toes.

B. Amount of fluid and food intake.

C. Recent fluctuations in weight.

D. The degree of sacral edema.

63. Sam's admission laboratory data reveal an elevated hematocrit. The nurse interprets this as being due to:

A. Chronic infection.

B. Recent dehydration.

C. Increased cardiac output.

D. Chronic oxygen deficiency.

64. Sam is given digoxin, 0.035 mg, at 8 A.M. and 8 P.M. each day. Before the 8 A.M. dose, the nurse takes his apical pulse and it is 85 and regular. Which interpretation is the most accurate?

A. He has just awakened and his heart action is slowest in the morning.

B. This is a normal rate for an 18-month-old child.

C. He may be going into heart block due to digoxin toxicity.

D. His potassium level needs to be evaluated.

65. Sam is eating poorly. While all of the following might play a part, which does the nurse consider most significant in planning Sam's care?

A. Delay in developmental milestones.

B. Decreased energy reserve.

C. Separation anxiety.

D. Frustrated autonomy needs.

66. In instructing the aide regarding what to feed Sam, the nurse emphasizes the importance of:

A. Cheese and ice cream.

B. Finger foods such as hot dogs.

C. Apricots and bananas.

D. Four glasses of whole milk per day.

67. Based on the results of the cardiac catheterization, open heart surgery to do a complete repair is planned. In answering the parents' questions, the nurse bases the explanations on:

A. The knowledge that the repair will include removal of the atrial septum and creation of an intracardiac baffle to redirect blood flow.

B. The knowledge that the repair will include patching of the ventricular septal defect and resection of the pulmonary stenosis.

C. The premise that it is the physician's responsibility to explain the surgical procedure.

D. The importance of directing the parents toward the postoperative recovery period.

68. Prior to surgery, Sam has a hypoxic episode. What should the nurse do first?

A. Administer oxygen and position him in the knee-chest position.

B. Position him on his side and give the ordered morphine.

C. Ask the parents to leave and start oxygen.

D. Give oxygen and notify the physician.

69. In planning postoperative care for Sam, the nurse expects him to be able to:

A. Button his shirt and tie his shoes.

B. Feed himself and drink from a cup.

C. Cut with scissors.

D. Walk up and down stairs.

70. The nurse discovers that Sam does none of the expected activities well or at all. His mother comments, "I worry constantly about him. I don't let his sisters tease him or play with him very much." Considering previous data, the nurse would interpret this to mean that:

A. Sam is physically incapable due to his cardiac defect.

B. His mother is overprotective and allows Sam few challenges to develop his skills.

C. Sam is just being stubborn because of his struggle for autonomy.

D. Sam is regressed due to the effects of hospitalizations.

71. It is now 10 days past Sam's surgery and he is very stable physically. He is alert and fairly active; he is playing well with his parents. Discharge is planned in the next few days. The nurse notes that Sam's parents are still very reluctant to allow him to do anything for himself. The nurse:

A. Reemphasizes the need for autonomy in toddlers.

B. Provides opportunities for autonomy when the parents are not present.

C. Reassesses the parents' needs and concerns.

D. Discusses the success of the surgery and how well Sam is doing.

Medical Diagnosis: Fracture of right arm.

- **Nursing Problems:**
  Abuse response patterns.
  Altered family processes.
  Altered parenting.

- **Chief Complaint:**
  Larry Walsh, 2½ years old, is admitted with a fracture of the right arm. Child abuse is suspected.

- **Past History:**
  Larry is one of five children under the age of 5; his father has been out of work for 6 months and is unable to find a job. His mother has

been trying to work part-time as a waitress. Larry has been seen in the emergency room four times in the past 8 months. His immunizations are not current. Larry's mother is being treated for anemia. The other four children are doing well, with no major health problems, although none of them has up-to-date immunizations.

- ■ *Symptoms:*
  Pain, edema, and decreased mobility of upper part of right arm.

- ■ *Physical Exam:*
  Child appears very quiet and seems afraid of adults.
  Clothes are dirty and torn and there is dirt under the nails.
  Heart: Normal with no murmurs. Rate: 88.
  Lungs: Clear to auscultation. Rate: 26.
  Musculoskeletal: Fracture of humerus. Arm placed in cast in the emergency room.
  Skin: Dirty. Several ecchymoses in various stages of resolution found over torso and lower legs.
  No known allergies. Received only first and second groups of immunizations. Not yet toilet trained but can feed self. No previous hospitalizations.

- ■ *Laboratory Data:*
  Chest x-ray: Clear.
  Hgb: 10 g. Hct: 32%.
  WBC: 7200/μL.
  Urinalysis: Clear.

---

72. On doing an admission history, the nurse notes that Larry's parents give contradictory explanations as to how Larry's fracture occurred. Because of the suspicion of child abuse noted on the ER notes, the nurse plans to prepare Larry for:
    A. Long bone x-rays.
    B. Brain scan.
    C. Repeat CBC.
    D. Placement in foster care.

73. The nurse plans her interactions with Larry's parents taking into account her knowledge that parents who abuse their children are often characterized as:
    A. Having reasonable expectations of their child's developmental skills.
    B. Having been abused or neglected themselves as children.
    C. Able to manage stressful situations appropriately.
    D. Having well-developed support systems among family and friends.

74. Larry's parents come to the unit to visit him the next day. The nurse expects:
    A. The parents to express concern about the extent of damage and Larry's current condition.
    B. The parents to appear guilty about the injury and bring Larry presents to compensate for it.
    C. The parents to play with, talk to, and hold Larry, which seems to make him happy.
    D. Excessive crying from Larry when his parents enter the room.

75. The LPN on the unit has an encounter with Larry's parents and angrily lectures them for hurting such a defenseless little child. The nurse's response to the LPN should emphasize:
  A. That no legal action has been taken and her comments make her vulnerable to slander charges.
  B. That being completely honest with the parents is the best policy.
  C. That she should not make premature judgments before all the information has been gathered.
  D. Recognition of the LPN's feelings.

76. The nurse's plan of care for Larry should include:
  A. Allowing the parents to visit 4 hours every day.
  B. Assigning one person on each shift to care for him.
  C. Encouraging the child to interact with many different people.
  D. Placing him in a room with several toddlers and encouraging peer interaction.

77. Child abuse is confirmed by the data and tests. The nurse's notes will assist in preparation of a report to the:
  A. Juvenile division of the police department.
  B. State attorney's office.
  C. Protective services department.
  D. Public health agency.

78. The daily nursing observations of Larry and his parents are important because:
  A. The nursing notes will be used in court to prove abuse.
  B. The notes will provide a database to plan family intervention.
  C. Accurate notes are required on all patients by hospital policy.
  D. The notes can prevent the parents from removing the child from the hospital.

79. In attempting to present a role model for appropriate parenting behavior for Mr. and Mrs. Walsh, the nurse should provide opportunities for Larry to:
  A. Pile seven or eight blocks on top of one another.
  B. Obey a set of complex commands.
  C. Brush his teeth and count three objects.
  D. Walk up and down stairs.

80. Larry is pale, easily fatigued, and has a hemoglobin of 10 g. He has been eating poorly in the hospital and is placed on liquid iron. It is not relevant for the nurse to teach the mother to:
  A. Prevent staining of teeth.
  B. Expect dark stools.
  C. Give the iron with orange juice.
  D. Look for the appearance of petechiae.

81. In planning the approach of care with Larry's parents, the nurse knows that the preferred treatment for abused children is to:
  A. Leave the child in the home and provide counseling for the parents.
  B. Remove the child from the home and place him in a foster home.
  C. Have the parents undergo psychiatric counseling while the child is hospitalized.
  D. Provide the parents with the number of a child abuse hotline so that they can call when they feel angry and are tempted to abuse the child.

82. Larry comes up to the pediatric unit 6 months after discharge home to his parents. He is very talkative, holds on to his mother's hand, and seems more lively than when you saw him last. His mother appears more relaxed and less critical of him. Which action is indicated?

A. Change the goals of the care plan since they are inappropriate.

B. Remark on how well the child and his parents are doing.

C. Say nothing, as things seem to be going very well.

D. Make an appointment for the family with the social worker.

Medical Diagnosis: Lead poisoning.

■ *Nursing Problems:*

Alteration in eating: unusual food ingestion.

Environmental conditions: poisoning.

Altered parenting.

Altered tissue perfusion.

■ *Chief Complaint:*

Steven Curtis, 3½ years old, is reported by his mother to have become more aggressive in the past month and to fight constantly with his siblings.

■ *Past History:*

Steven is the third of four children; his mother is separated from her husband. The family currently lives in an older apartment building in the poorer section of the city. One other child in the family—a boy, age 2—was seen recently for lead poisoning. The mother also states that Steven seems to eat anything with which he comes in contact—mud, newspaper, and so forth.

■ *Symptoms:*

Seems pale, mother noted constipation, and child seems more irritable in past few weeks.

■ *Physical Exam:*

Heart: Murmur present. Rate: 86 and regular.

Lungs: Clear to auscultation.

Skin: Intact with no lesions. Teeth: Lead lines visible.

Immunizations are current and there are no known allergies. Mother feels that Steven is developing more slowly than her other children did; he talked later than they did and she is having trouble with toilet training.

■ *Laboratory Data:*

Hct: 27%. Hgb: 9 g. WBC: 6000/μL.

Lead levels: 82 μg/dL and 86 μg/dL.

Urinalysis: Clear with specific gravity of 1.010.

Chest x-ray: Clear.

Bone x-rays: Long bones reveal lead deposition.

83. Steven is admitted to the pediatric unit for treatment of chronic lead poisoning. The nursing assessment should not focus on:

A. Mother-child interaction.

B. Level of consciousness and mental status.

C. Eating habits.

D. Peripheral circulation.

84. In building a database for eventual discharge planning, the nurse determines the most likely source of the lead to be:

A. Drinking from unglazed pottery mugs.

B. Eating paint chips.

C. Eating dirt.

D. Eating the newspaper.

85. The primary nursing goal is to:

A. Teach prevention of further lead ingestion.

B. Improve the child's nutritional status.

C. Prevent fatigue and discomfort.

D. Promote excretion of the lead.

86. Steven is pale, easily fatigued, and has a hemoglobin of 9 g. The nurse interprets this as being due to:

A. Pica.

B. Blood loss.

C. Decreased erythropoietin.

D. Hemoglobin S.

87. A 24-hour urine is ordered. Mrs. Curtis is not being helpful in the collection. The nurse bases the explanation of its importance on knowledge that this is being done to determine the presence of:

A. Corticosteroids.

B. Corproporphyrin.

C. Sodium and potassium excretion.

D. Hematuria.

88. In planning Steven's care, the nurse needs to be aware that the major complication of lead poisoning is:

A. Hematopoietic.

B. Renal.

C. Neurologic.

D. Cardiovascular.

89. When checking on Steven during morning rounds, the nurse enters the room and observes him having a tonic-clonic seizure. The nurse's initial action is to:

A. Prepare an injection of diazepam (Valium).

B. Suction his nasopharynx.

C. Maintain a patent airway.

D. Place a tongue blade between his teeth.

90. CaEDTA and BAL are ordered to be given by deep IM injection to Steven. He will be receiving eight injections per day. Which is inappropriate to include in the nursing care plan?

A. Allow the child to play with the needle after the injection.

B. Push fluids up to 200 mL/hour.

C. Record accurate intake and output.

D. Request an order for procaine to give with the injections to decrease pain.

91. What should the nurse assess daily during Steven's course of therapy?

A. Glucose and potassium levels.

B. BUN.

C. Urinary lead level.

D. CBC.

92. Which test is the most appropriate for assessing Steven's development?
    A. Stanford-Binet Intelligence Test.
    B. Denver Developmental Screening Test.
    C. Peabody Picture Vocabulary Test.
    D. Piers-Harris Self-Concept Scale.

93. Just before Steven is to be discharged, Mrs. Curtis says to the nurse, "Steven drives me crazy; he always wants to play out there in the street with his big brothers." Which reply is most helpful to Mrs. Curtis?
    A. "Don't worry about it. He will learn safety with experience."
    B. "You should never allow him outside unless an adult is with him."
    C. "You should keep him in the house with you. At age 3½ he must be protected from what he can't understand."
    D. "You should tell him to play in a certain area and enforce the rule with discipline."

94. At 3½ years, it is not expected that Steven is capable of:
    A. Night-time continence.
    B. Counting to 3 and dressing himself.
    C. Skipping on alternating feet.
    D. Having a vocabulary of 2000 words.

95. Which is an appropriate play activity for the nurse to plan for Steven in the hospital?
    A. Spinning tops.
    B. Finger painting.
    C. Dressing up and playing house.
    D. Cutting out pictures and pasting them on paper.

96. Which effect of lead poisoning is reversible and therefore does not need to be part of long-term goals for care?
    A. Learning disability.
    B. Hyperkinesis.
    C. Seizure activity.
    D. Persistent anemia.

97. Which value indicates that Steven is ready for discharge?
    A. Serum lead level of 40 μg.
    B. Glucose in the urine.
    C. Hemoglobin of 9 g.
    D. WBC of 5500/μL.

Medical Diagnosis: Tonsillectomy.

■ *Nursing Problems:*
Risk for altered body temperature.
Altered comfort (acute pain).
Risk for infection.
Altered nutrition requirements.

■ *Chief Complaint:*
Mary Hawkins, 4 years old, suffers frequent strep throats and shows decreasing hearing.

■ *Past History:*
Mary has had four strep throats in the past 7 months. She has also had several cases of otitis media, which have led to decreasing hear-

ing in the left ear. Mary is the youngest of three children; neither of the parents has any health problems. The children have had only minor illnesses and none has ever had surgery. Mrs. Hawkins is very concerned about the surgery and has told her daughter nothing about the planned admission.

■ *Symptoms:*
History of sore throats.

■ *Physical Exam:*
Heart: Normal with no murmurs. Rate: 90.
Lungs: Clear to auscultation.
Ears: Tympanic membranes clear.
Throat: Tonsils enlarged but not currently inflamed. Adenoids also enlarged. Child has nasal speech and is a mouth breather.
Mouth: Lips and mucous membranes slightly dry.
Immunizations current. Has had no previous surgeries. No known allergies.
Normal growth and development. Will start kindergarten in 3 months.

■ *Laboratory Data:*
Chest x-ray: Clear.
Hct: 42%. Hgb: 14 g.
Urinalysis: Clear with no RBCs or WBCs; pH of 7 and specific gravity of 1.015.
Clotting time: 6.5 minutes. Bleeding time: 3 minutes.

98. During examination of Mary's throat on admission, which is not indicative of an infection of the tonsils?
   A. Elevated temperature.
   B. Koplik spots on the buccal mucosa.
   C. Cervical lymph node enlargement.
   D. Tonsillar white patches.

99. Mary arrived at the hospital having been told she was going to "the doctor's house for a visit." The mother is emphatic that she does not want the child told about the surgery. Which is the best way for the nurse to manage this situation?
   A. Call the physician and have her or him talk to the mother.
   B. Prepare the child when the parents leave for the evening.
   C. Respect the mother's wishes and do not prepare the child.
   D. Assess the mother's reasons for refusing preparation.

100. Mary becomes very upset and agitated when her parents leave to go home for the night. She refuses to be comforted and is crying and calling out for them. What nursing action is most appropriate?
   A. Tell her that they will be back at 7 A.M.
   B. Bring another child her age to keep her company.
   C. Leave her alone so she will settle down on her own.
   D. Stay in the room and complete the nursing care of her roommate.

101. Prior to surgery, the nurse wants to gather certain data about Mary. It is irrelevant to:
   A. Check her teeth.
   B. Evaluate coagulation studies.

   C. Note the WBC count and urinalysis results.

   D. Check the SGOT and SGPT levels.

102. Just before Mary is to go to surgery, which assessment by the nurse necessitates a call to the surgeon?

   A. Temperature of 100.2°F (rectally).

   B. Mouth breathing.

   C. Urine specific gravity of 1.018.

   D. Bleeding time of 10 minutes.

103. Mary is to receive atropine IM prior to surgery. In explaining the medication to her, which statement is most appropriate?

   A. "All the other boys and girls have shots before surgery. They don't cry."

   B. "The needle will hurt a little bit, but if you don't fight me, I'll give you a balloon."

   C. "The doctor wants you to have this injection so you will go to sleep."

   D. "The needle will hurt like a prick. You choose the leg you want it in and then squeeze your mother's hand when I give it to you."

104. Which is *not* appropriate information about the side effects of atropine for Mary's mother to help alleviate her anxiety about Mary postoperatively? The medication:

   A. Causes a dry mouth.

   B. May produce a flushed face.

   C. Causes increased swallowing.

   D. May cause hallucinations.

105. In assessing Mary for hemorrhage postoperatively, the nurse does not expect to find:

   A. Frequent swallowing and drooling of serosanguinous mucus.

   B. Increased respiratory rate.

   C. Increased pulse and decreased blood pressure.

   D. Emesis of dark blood.

106. For Mary's postoperative care, which action is appropriate?

   A. Give warm tea and honey to decrease throat pain.

   B. Give aspirin for elevated temperature.

   C. Suction the oropharynx every 2 hours and prn.

   D. Place in semiprone position until responsive.

107. Which should the nurse plan to offer Mary?

   A. Orange juice.

   B. Lemonade.

   C. Vanilla ice cream.

   D. Green Jell-O.

108. Aspirin, 5 gr., PO or rectally, every 4 hours for pain or elevated temperature, has been ordered by the physician. The nurse should:

   A. Check the chart to see if Mary is allergic to aspirin.

   B. Note that this is an overdose and notify the physician that a smaller dose is needed.

   C. Go ahead and give the drug when Mary's temperature reaches 102°F.

   D. Question the order and not give the drug; notify the physician.

109. Mary is very angry with her mother after surgery. She tells her, "I hurt. I hurt. Why did you bring me here? I hate you." What is the best advice the nurse can give Mary's mother?

A. Try to regain Mary's trust by meeting all her needs and bringing extra presents.

B. Inform the mother that she should have let the nurse prepare the child before surgery, as was recommended.

C. Ignore Mary's words and allow her more freedom when she returns home, to distract her.

D. Allow Mary to talk about her surgery and play out her anger with her dolls.

110. Mary is to be discharged tomorrow. Which concern about effects of hospitalization can the nurse omit in discharge teaching with her mother?

A. Mary's likelihood of following the mother around the house as she does her work.

B. Nightmares.

C. Increased independence.

D. Temper tantrums and bedwetting.

111. The nurse has completed discharge teaching with Mrs. Hawkins. Which signs or symptoms should Mrs. Hawkins report to the physician to indicate that she understood the nurse's teaching?

A. Temperature elevation of 101°F (rectally) and dysphagia.

B. Minor bleeding between the fifth and tenth postop day, with scab loss.

C. White patches at the site of the tonsils.

D. Temperature elevation of 101°F (rectally) and pain in the right ear.

Medical Diagnosis: Acute lymphocytic leukemia.

- **Nursing Problems:**
  Altered feeling state: grief.
  Altered immune response.
  Altered nutrition.

- **Chief Complaint:**
  Amanda Curtis, 7 years old, experienced continued bleeding after having a tooth extracted.

- **Past History:**
  Amanda had been a normal, healthy child previously. She has had increased bruising of the extremities during the past 4 months. Mother reports that she also seems to tire more easily and has not been able to get rid of a cold she has had for 3 weeks. The parents are college educated and both work; the father is a lawyer and the mother works for an advertising firm. They have one other child—a boy, age 11—who is healthy. The maternal grandmother had breast cancer. No other known family health problems. Amanda developed normally and is now just starting second grade and doing well.

- **Symptoms:**
  Pallor, fatigue, and an increase in number and length of infections.

- **Physical Exam:**
  Pale, quiet, white female, 7 years old. Appears to be slightly underweight.

Heart: Questionable murmur heard. Rate: 86 and regular.

Lungs: Minor rhonchi on auscultation.

Skin: Ecchymoses on legs in various stages of resolution. Petechiae on throat and upper arms. Nailbeds pale.

Enlarged cervical and inguinal lymph nodes.

Enlarged liver.

Has lost 3 pounds in the past month. No known allergies. Has received all required immunizations.

■ *Laboratory Data:*

Chest x-ray: Clear.

Hct: 26%. Hgb: 8 g.

WBC: 12,000/μL, with several blast cells seen on peripheral smear.

Platelets: 100,000/μL.

Urinalysis: Clear.

Cerebrospinal fluid: Clear with no abnormal cells.

Vital signs: Blood pressure: 110/70. Pulse: 86 and regular. Respirations: 24. Temperature: 100.8°F (orally).

---

112. Amanda is suspected of having acute lymphocytic leukemia; she is admitted to the pediatric unit. The priority nursing assessment is:
    A. What Amanda knows about leukemia.
    B. Parents' coping abilities in previous grief situations.
    C. Amanda's previous experiences with illness and hospitalization.
    D. Amanda's cognitive and physical developmental abilities.

113. What is the priority nursing goal in selecting an appropriate roommate for Amanda?
    A. Promotion of peer contacts and stimulation.
    B. Prevention of infection.
    C. Provision of privacy.
    D. Promotion of sensory stimulation.

114. To confirm the diagnosis, Amanda is scheduled to have a bone marrow aspiration done. Which approach is *not* included in the nurse's preparation?
    A. Emphasize that there will be little pain.
    B. Tell Amanda that she can be up and about after she recovers from the premedication.
    C. Explain that the test will be done on her hip and a bandage placed over it afterward.
    D. Tell her that she will receive some medication to make her sleepy before the test is done.

115. In preparing Amanda for the procedure, the nurse explains that bone marrow aspirations in children are least likely to be done at the following site:
    A. Vertebral.
    B. Tibial.
    C. Sternal.
    D. Iliac crest.

116. Prior to her bone marrow aspiration, Amanda is given a DPT cocktail—Demerol (meperidine), Phenergan (promethazine), and Thorazine (chlorpromazine). Her aspiration was done on the left posterior iliac crest. Which causes the nurse the greatest concern 1 hour after the procedure?

A. Patch of blood about 2 inches in diameter on the pressure dressing.
B. Responds to her name slowly and pushes the nurse's hand away.
C. Respirations of 12 and decreased breath sounds.
D. Blood pressure of 110/72 and pulse of 90.

117. Based on the nurse's analysis of the previous situation, the best action is to:
A. Call the physician and report a suspicion of hemorrhage.
B. Elevate the head of the bed to aid respiratory effort.
C. Prepare an IV line to deliver fluids to expand the intravascular volume.
D. Provide more covers because she is hypothermic due to shock.

118. The bone marrow aspiration confirms the diagnosis of acute lymphocytic leukemia. Amanda is started on prednisone and vincristine. Which nursing behavior is *not* appropriate?
A. Increase daily fluid intake.
B. Provide a footboard for Amanda when she is in bed.
C. Assess for numbness and tingling in the extremities.
D. Assess for cardiac arrhythmias.

119. Which laboratory data should the nurse evaluate regularly in the child undergoing chemotherapy?
A. SGOT and SGPT.
B. LDH and WBC.
C. Potassium and pH levels.
D. CBC.

120. Amanda has nausea and vomiting and has developed stomatitis (mouth ulcers). Her nutritional intake is compromised. In selecting her lunch, the nurse recommends:
A. Pizza, coke, and cookies.
B. Tomato soup, toast, and ice cream.
C. Omelet, Jell-O, and milk shake.
D. Hot dog, potato chips, and popsicles.

121. Which activity will help a 7-year-old cope with the stress and fear engendered by her hospitalization and treatment?
A. Demonstrating procedures and treatments on another child her own age.
B. Explaining all procedures in great detail using a plastic model of the human body to do so.
C. Using puppets and dolls to explain procedures and permitting the child to handle the hospital equipment.
D. Diversional activities such as playing house, jigsaw puzzles, and books.

122. Amanda is doing well. It is 3:40 P.M. She has an IV running at 50 mL/hour, of which 25 mL remains, after which it will be discontinued. She has a small, sacral sore, which is healing, and the dressing was to have been changed at 3 P.M. She has IV penicillin due at 4 P.M. However, Amanda is looking forward to seeing her brother in the waiting room. In deciding which action to take first, the most important priority is to:
A. Change the sacral dressing.
B. Discontinue the IV immediately.
C. Allow her to visit her brother.
D. Administer the medication.

123. Amanda's parents are concerned about what they should tell her 11-year-old brother about her illness. The nurse should:
   A. Suggest that the topic be open conversation so that both information and feelings about leukemia can be discussed at home.
   B. Encourage the parents to realize that the brother is not really affected by this event since it is not a hereditary disease.
   C. Suggest that the parents share information about Amanda's condition.
   D. Encourage the parents to act as if everything were completely normal and avoid increasing family stress levels.

124. During her morning care one day, Amanda says to the nurse, "What is going to happen to me when I die?" The nurse's best response is:
   A. "You won't be alone. I'll be here with you."
   B. "I don't know."
   C. "Why do you think you're going to die?"
   D. "What do you think will happen?"

125. Mrs. Curtis is very critical of the care her daughter is receiving. Every day she has a new complaint. Today she says to you, "Amanda didn't get her bath until 10 o'clock today. I think it's disgraceful, considering the price of a room in this hospital." Before responding, the nurse is least likely to consider:
   A. That anger at the nursing staff is often seen in the grieving process.
   B. That Mrs. Curtis is looking for a reason to sue, particularly since her husband is a lawyer.
   C. That further data are needed about Mrs. Curtis and her current feelings and perceptions of Amanda's situation.
   D. Her own resentment at being criticized for giving poor care to a chronically ill child.

126. Amanda is discharged home in remission and on continuing chemotherapy. Her primary nurse phones Mrs. Curtis to discuss her progress 3 weeks after discharge. Her mother reports that she is keeping Amanda out of school to prevent infections. The nurse should:
   A. Explain that Amanda needs her peers and school work.
   B. Recommend further immunizations be done to protect her.
   C. Suggest an appointment with the psychiatrist to discuss why the parents are isolating Amanda.
   D. Agree that this is the best approach for the next year.

Medical Diagnosis: Sickle cell anemia.

- **Nursing Problems:**
  Altered comfort (acute pain).
  Altered tissue perfusion (gastrointestinal and peripheral).

- **Chief Complaint:**
  Ronny Simpson, 8 years old, has been experiencing increased pain in his legs and abdomen after playing a game of tag.

- **Past History:**
  Ronny is a black child who was diagnosed with sickle cell anemia at 2 years of age when he presented with hand-foot syndrome. He is

one of five children; none of his siblings has sickle cell anemia. His mother has hypertension; his father has sickle cell anemia and two cousins also have the problem. Ronny has had several crises over the past 6 years and has missed school frequently. Currently he is one grade behind where he should be in school.

- **Symptoms:**
Pain in legs and abdomen.

- **Physical Exam:**
Heart: Pulse: 100. Blood pressure: 100/66. Murmur present.
Lungs: Clear to auscultation; good air exchange; rate of 30.
Skin: Intact, with no lesions; slight diaphoresis.
Eyes: Icterus present.
Appears small and underweight for age. No known allergies, and immunizations are up-to-date. Has had three respiratory infections in the past year. No medications.

- **Laboratory Data:**
Hct: 20%. Hgb: 6 g.
Urinalysis: Clear, with some RBCs seen; specific gravity of 1.022.
Temperature: 100°F (orally).
Bilirubin: 25 mg/100 mL.

---

127. Of the following information recited by Mr. and Ms. Simpson in their explanations of sickle cell anemia, the one fact that the nurse needs to correct for them is that:
    A. Sickle cell anemia is due to heterozygous inheritance of a gene responsible for production of abnormal hemoglobin.
    B. Sickle cell anemia is seen in blacks and people of Mediterranean ancestry.
    C. Hemoglobin S sickles in conditions producing hypoxia.
    D. Sickle cell anemia is associated with a poor prognosis.

128. Ronny has severe abdominal and leg pain. His vital signs are blood pressure 100/68, pulse 94, and respirations 24. Which is inappropriate to include in the initial care plan?
    A. Oxygen by mask and cold applications to leg to relieve pain.
    B. Meperidine (Demerol) for pain and frequent vital signs.
    C. Forcing fluids and strict intake and output.
    D. Bedrest and diversionary activities as tolerated.

129. The best roommate for Ronny is:
    A. An 8-year-old boy 3 days after appendectomy with a temperature of 101°F.
    B. A 7-year-old boy with juvenile rheumatoid arthritis.
    C. An 8-year-old girl with a cardiac defect.
    D. A 9-year-old boy with pneumonia.

130. You want to assess Ronny's color. It is not helpful to inspect the:
    A. Mucous membranes.
    B. Nailbeds.
    C. Palms of hands.
    D. Earlobes.

131. Ronny normally has a hemoglobin of 7 and a hematocrit of 20. His anemia is due to:
    A. Poor dietary practices and decreased iron intake.

      B. Decreased erythropoietin activity.

      C. Blood loss due to the sickle cell crises.

      D. Increased fragility and decreased life span of RBCs.

132. Folic acid, 1 mg once a day, has just been ordered by the resident in charge of Ronny's case. The nurse should:

      A. Go ahead and give the medication with Ronny's favorite juice.

      B. Question the order because it is too high a dosage for a child.

      C. Give it to Ronny and explain its purpose and side effects in terms Ronny will understand.

      D. Question the order because the nurse knows vitamins are of questionable value in sickle cell anemia.

133. Ronny is doing better and is more comfortable. His IV has been discontinued and he wants to get out of bed. A few more days of bedrest are needed. Which activity is best for Ronny?

      A. Television programs most of the day.

      B. Phone calls from his friends at home and at school.

      C. Reading books about famous black people.

      D. Playing checkers and simple card games with his roommate.

134. Increased fluid intake and fluids that help produce an alkaline environment are recommended. The nurse needs to help Ronny and his parents understand that these fluids will:

      A. Increase his blood volume, decrease viscosity, and prevent sickling.

      B. Promote urinary output and decrease the chance of urinary tract infections.

      C. Promote caloric intake and good nutritional status.

      D. Prevent dehydration, which increases sickling of RBCs.

135. The basis of genetic counseling for the Simpson family stems from the knowledge that Mr. Simpson has sickle cell disease and Ms. Simpson:

      A. Must also have the disease.

      B. Must have the trait.

      C. Is disease free for the rest of her life.

      D. May or may not have the sickle cell trait.

136. There are other children in the family who have never had sickle cell disease. Therefore:

      A. Some of them may be free of the trait.

      B. None of them carries the trait.

      C. Each child carries the trait, but may not have the disease.

      D. Each child will pass the trait on to his or her children.

137. In establishing teaching goals for Ronny's discharge plan, the nurse may want to emphasize that Ronny's current sickle cell crisis was probably due to:

      A. Infection.

      B. Cold weather.

      C. Dehydration.

      D. Increased activity.

Medical Diagnosis: Rheumatic fever.

■ *Nursing Problems:*

Risk for alteration in circulation.

Altered comfort (pain).

Self-care deficit.

■ *Chief Complaint:*

Kathy Mendez, age 8, is brought to the public health clinic by her mother, who reports that lately the child has not felt like going to school.

■ *Past History:*

Kathy had rubella and chickenpox prior to entering school. In the past 2 years, she has had recurrent sore throats. The most recent complaint of sore throat, according to her mother, was 5 weeks ago. In the past 2 weeks, Kathy has had no energy, has complained of pain, has requested to stay at home, and appears to have lost weight. Other than immunizations required for school, Kathy has had no health care.

■ *Family History:*

No major illnesses were reported. Kathy has three brothers, ages 6, 9, and 11. Kathy's parents divorced 3 years ago and she has not seen her father since that time. Her mother works 50–55 hours per week away from home. Kathy and her family live in a three-room apartment in a government housing project.

■ *Physical Exam and Review of Symptoms:*

Weight: 5 pounds below normal for height and age.

Abdomen: Complains of generalized pain on palpation.

MS: Complains of pain in shoulders, elbows, and wrists during active range of motion.

Skin: Palpable subcutaneous nodules on knees, scapula, and thoracic spine. Macular rash on anterior trunk. Brown residue in skin creases and under nails.

Head: Mouth: Brown and yellow plaque lining gums and teeth. Hair: Oily and matted.

---

138. Kathy is admitted to the children's unit of the local hospital. Based on knowledge of rheumatic fever and related current adaptation needs and possible complications, the nurse evaluates which of the following as having *first* priority in Kathy's initial admission assessment?

   A. Reasons for Kathy's apparent lack of hygienic care.

   B. Apical pulse rate and coordination of voluntary muscles.

   C. Type, intensity, and duration of pain in subcutaneous nodules.

   D. Specific foods eaten within the past 24 hours.

139. During the assessment, Kathy's mother states that she thought Kathy's possible rheumatic joint pains were "growing pains." In health teaching regarding differentiation of the two pain types, the nurse replies to Kathy's mother:

   A. "During rest or sleep, pains from growth may lessen or disappear, but the joint pains of rheumatic fever do not lessen."

   B. "There are no characteristic differences between the two types of pain; therefore, neither one can be specified."

   C. "Rheumatic pain consistently occurs bilaterally in joints either above or below the diaphragm, whereas growth pains occur only in joints below the diaphragm."

   D. "The 8-year-old child is developmentally incapable of describing pain characteristics; therefore, it is not possible to differen-

tiate between rheumatic joint pain and 'growing pains' based on Kathy's complaints."

140. Based on Kathy's developmental stage, tentative diagnosis, and related health needs, the nurse plans to place Kathy in a room with:
   A. Eight-year-old Jolene, who has playroom privileges and was admitted with superficial bruises resulting from an automobile accident.
   B. Eight-year-old Alma, who is confined to bed with scarlet fever.
   C. Ten-year-old Marta, who is confined to bed with acute systemic lupus erythematosus.
   D. Nine-year-old Judy, who is up ad lib and admitted for glucose tolerance tests and insulin therapy regulation.

141. Kathy is diagnosed as having acute rheumatic fever. To destroy any remaining group A beta-hemolytic streptococci, penicillin is ordered for Kathy. Just prior to receiving an intramuscular injection, she cries and says to the nurse, "Will the shot hurt? I'm afraid." The nurse should reply:
   A. "You're scared? What would your friends think if they knew you were afraid?"
   B. "Yes, there will be a little pain, but it will soon be over."
   C. "You may go to the playroom as soon as you have your injection."
   D. "I will try not to hurt you. If you cry, it will only make it worse."

142. Polyarthritis migrates to Kathy's knees. Considering physical adaptation needs and appropriate nursing management related to this manifestation, the nurse should:
   A. Apply elastic bandages and leg splints to increase circulation while immobilizing Kathy's knees.
   B. Gently massage Kathy's legs and knees to increase circulation and promote healing.
   C. Perform passive range of motion and encourage Kathy to ambulate frequently to prevent contractures.
   D. Place a bed cradle over Kathy's legs to prevent physical or material contact.

143. In analyzing the school-age child's problems in adapting to hospitalization, the nurse acknowledges which aspect to be the *most* difficult for Kathy to accept?
   A. Separation from her mother and her family.
   B. Separation from her friends and possible alienation from her established group.
   C. Absence from school, possibly falling behind in homework, and failing in her grades.
   D. Restricted mobility, possible bodily injury, and lack of control over self-care.

144. The nurse recognizes Kathy's nervousness, weakness, and emotional instability as indications of Sydenham's chorea, a disorder associated with the rheumatic process. If this condition worsens and clinical manifestations increase and are intensified, nursing actions should include:
   A. Encouraging Kathy to build her strength, vent her emotions, and utilize stored excess energy by scheduling increased playroom time with school-age children.
   B. Preventing complications by providing absolute rest; padded bedding; spoon-feeding; and a diet high in calories, protein, and iron.

C. Alleviating Kathy's inferiority feelings by providing pencils, paper, and extra study time for keeping up with her schoolwork.

D. Preventing further psychological damage by removing all visiting restrictions, arranging psychological counseling, and instituting a self-care regimen for Kathy.

145. Kathy is prescribed large-dose acetylsalicylic acid therapy. In observing Kathy for indications of salicylate toxicity, the nurse does not expect to see:

A. Erythema marginatum, cardiac murmur.

B. Drowsiness, nausea, and vomiting.

C. Hyperpnea, purpuric manifestations.

D. Tinnitus, gastric bleeding.

146. In nursing evaluation of Kathy's progress, which manifestation indicates increasing severity in Kathy's health problem?

A. Decreasing erythrocyte sedimentation rate.

B. WBC count of 9000/μL.

C. Orthopnea and peripheral edema.

D. Pulse rate 88 and negative C-reactive protein.

147. Kathy is prescribed cardiac glycoside (digoxin) and steroid (cortisone) therapy. Which is not an important nursing action related to long-term steroid administration?

A. Prevent exposure to infection and check urine for glucose.

B. Administer with antacid or milk and assess for emotional lability.

C. Monitor blood pressure, daily weight, and skin for any changes.

D. Report possible infectious processes suggested by high temperature elevations.

148. Furosemide (Lasix) is added to Kathy's medication regimen. Which is the *most important* nursing action related to simultaneous administration of furosemide and digoxin:

A. Observe Kathy for headache, nausea, muscular weakness, disordered vision, and changes in apical heart rate and rhythm.

B. Monitor Kathy for changes in weight, skin turgor, texture of mucous membranes, and radial pulse rate.

C. Assess Kathy for absent or unequal peripheral pulses, ecchymoses, cyanosis, or numbness/tingling of the lower extremities.

D. Report elevated temperature, flushed face, anorexia, and bounding radial pulses.

149. Potassium chloride (KCl) is added to Kathy's IV solution. Which does not prevent possible life-threatening complications resulting from IV KCl administration?

A. Kathy's urinary output is greater than 30 mL/hour.

B. Kathy's cardiac status is monitored and recorded.

C. The solution is infused at a rapid rate.

D. The KCl is thoroughly mixed in the IV solution.

150. As Kathy's health status improves, she indicates she is bored and asks the nurse for "something entertaining to do." The nurse requests that Kathy decide on an activity, but omits which option from Kathy's selection?

A. Finger painting, stringing beads, and making puppets.

B. Knitting, leather crafts, and clay modeling.

C. Reading, playing jacks, and playing guessing games.

D. Playing checkers, Scrabble, or Monopoly with her roommate.

151. At discharge, Kathy's mother requests information about Kathy's home care. The nurse should reply:

A. "Kathy *must* have antibiotic therapy indefinitely and exactly as ordered."
B. "Kathy is now immune to further problems related to rheumatic fever."
C. "There are no restrictions on Kathy's activities at home or at school."
D. "Kathy should be provided a diet high in carbohydrates and fats and low in protein."

Medical Diagnosis: Asthma.

■ *Nursing Problems:*
Acid-base imbalance.
Impaired gas exchange.
Altered immune response.
Ineffective individual coping: mother.
Risk for infection.

■ *Chief Complaint:*
Tom Boyd, 11 years old, started wheezing during the night.

■ *Past History:*
Tom began having wheezing attacks 5 years ago at age 6 and was subsequently diagnosed as having asthma. He has multiple allergies to house dust; dog and cat dander; cigarette smoke; and some foods, such as chocolate. Father is also known to have respiratory problems and several allergies. His mother is healthy but seems very anxious about Tom and his care. She takes total responsibility for his care, including giving him all his medications; she has concentrated on Tom since his diagnosis. There are two other children, a 7-year-old boy and a 15-year-old girl. The younger boy has been having difficulty in school the past year. Tom has had several upper respiratory infections in the past few months and missed several days of school.

■ *Symptoms:*
Wheezing, increased pulse, and diaphoresis.

■ *Physical Exam:*
Heart: Pulse: 96 but regular. No murmurs heard.
Lungs: Audible wheezing. Breath sounds tight, with little air exchange. Coughing frequently. Mucus is thick and clear.
Skin: Intact, with no signs of breakdown; diaphoretic.
Immunizations up-to-date.
Current medications: Theophylline 3 times a day and cromolyn sodium 4 times a day.

■ *Laboratory Data:*
Chest x-ray: Hyperinflation of the lungs with beginning atelectasis in right middle lobe.
Hgb: 15 g. Hct: 40%.
Urinalysis: Clear, with no cells; specific gravity of 1.020.
Blood gases: Respiratory acidosis. pH 7.30. P_{CO_2}: 45. P_{O_2}: 89. HCO_3: 20.

152. Tom is admitted to the pediatric unit from the emergency room, where he received two doses of epinephrine without relief. His current vital signs are temperature of 100.8°F (orally), pulse of 110, and respirations of 36. The nurse wants to have which of the following readily available?
A. Tracheostomy set.
B. Oxygen.
C. Morphine sulfate.
D. ECG monitor.

153. Aminophylline is ordered to be given intravenously every 4 hours over a 20-minute interval. Which is the least essential to assess?
A. Cardiac rhythm.
B. Pulse rate.
C. Urinary output.
D. Increasing restlessness.

154. During his acute attack, Tom will be most comfortable in which position?
A. Head of bed elevated 15 degrees.
B. Turned on his side.
C. Sitting up.
D. Prone.

155. During morning report, two nurses hear that Tom has been admitted and say, "Oh, no. He's back in. It seems as though he has lived here the past few months." The head nurse's best response is:
A. "Yes, I agree it is really getting to be monotonous."
B. "His parents just don't seem to care enough, and they don't give him the care he needs."
C. "It's discouraging having him come back so often. It seems we aren't doing our discharge teaching right."
D. "I know it's difficult having him back so frequently. It is upsetting for us and for him and his family."

156. Two days later, Tom's secretions have become thicker and slightly yellow in color. His temperature is 102.4°F (orally) and breath sounds are decreased on the left side. The nurse should first:
A. Increase his fluid intake to mobilize secretions.
B. Administer acetaminophen (Tylenol) to decrease his temperature.
C. Increase the frequency of Tom's deep breathing and coughing.
D. Get a sputum culture.

157. The sputum culture is positive and a chest x-ray confirms pneumonia. Penicillin, 250 mg PO every 4 hours, is ordered. Before the penicillin is given, which nursing action is least relevant?
A. Assess his medical record for medication allergies.
B. Explain to him why he will be taking the medication.
C. Do a skin scratch test.
D. Note his urinary output and BUN.

158. Tom is trying very hard to become more independent and to manage his own care. But his mother says to the nurse, "I just get so scared when he wheezes. I'm afraid he will take too much of his medicine so I give it to him." The nurse interprets this as an example of:
A. Overprotection.
B. Regression.
C. Hostility.
D. Normal concern.

159. Tom is being discharged on theophylline 3 times a day and cromolyn sodium. What indicates that he has taken too much of either of these medications?
   A. Persistent cough.
   B. Prolonged rapid pulse.
   C. Diarrhea.
   D. Nausea and dizziness.

160. On a return visit to the clinic, which does *not* indicate Tom's increase in independence and responsibility for his own care?
   A. Cleans his own room and does his breathing exercises 3 times a day.
   B. Demonstrates his ability to take his own pulse and can name his medications and their effects.
   C. Has started taking swimming lessons with his friends.
   D. Has refused to join the Boy Scouts because being in large crowds of other kids might make him sick.

Medical Diagnosis: Scoliosis.

- **Nursing Problems:**
  Impaired physical mobility.
  Self-esteem disturbance.
  Impaired skin integrity.

- **Chief Complaint:**
  Susan Black, 13 years old, has one shoulder that appears higher than the other.

- **Past History:**
  Susan is a healthy, well-nourished female who has been developing normally. One month ago when she went to buy a new dress and could not find one to fit, she noted that one shoulder appeared to be higher than the other. Her mother has a history of multiple allergies; her father is healthy. One other child in the family—a brother, age 9—has allergies. There are no major illnesses; she had an appendectomy at age 10.

- **Symptoms:**
  None.

- **Physical Exam:**
  Well-developed adolescent female.
  Heart: Slight shift to the left; no murmurs.
  Lungs: Normal to auscultation.
  Skin: Intact. Beginning acne on face and shoulders.
  GU: Menstruating for 1 year; no pain, but still irregular.
  No known allergies. Taking no medications. Does not smoke or drink. Not sexually active. Attends junior high school, eighth grade, and is doing well.

- **X-ray:**
  Spinal x-ray reveals a 35-degree primary lateral curve with compensatory curve above and below the primary.

- **Laboratory Data:**
  Hct: 37%. Hgb: 12 g.
  Urinalysis: Clear with specific gravity of 1.015.

WBC: 6000/μL.

Chest x-ray: Lungs clear; heart displaced slightly to left.

---

161. Susan is being followed in the orthopedic clinic following the diagnosis of scoliosis. She is to be treated initially with a Milwaukee brace. Which assessment is least common?
    A. Thoracic asymmetry.
    B. Shoulder height asymmetry.
    C. One breast appears larger.
    D. Pain.

162. For evaluation of the extent of secondary problems that could occur, it is least important for the nurse to assess the status of the:
    A. Urinary system.
    B. Stability of the hip.
    C. Social and emotional equilibrium.
    D. Compensatory curvature of the spine.

163. Which point is inaccurate in teaching Susan and her parents about the brace?
    A. Eating raw fruits and vegetables and chewing sugarless gum.
    B. Use of cotton underclothes and loose-fitting garments.
    C. Daily skin check for redness and pressure points, to avoid breakdown.
    D. Permission to sleep without the brace at night.

164. Taking Erikson's stage of "identity versus role diffusion" into account while planning interventions for Susan, the nurse considers which to be inappropriate for her age group?
    A. Concerns about occupational or job identity.
    B. Psychosexual needs demanding fulfillment.
    C. Parents being the most significant interpersonal influences at this age.
    D. Reevaluation of bodily changes.

165. Susan returns to the clinic 6 weeks later. Her curvature is now 40 degrees. Which is an inappropriate action?
    A. Discussing why she is not wearing the brace as instructed.
    B. Asking her how she feels about having to wear the brace all day.
    C. Evaluating the fit of the brace.
    D. Assessing her cardiac and respiratory status.

166. What point does the nurse want to emphasize in discussing the need for treatment?
    A. Increase in the curvature will cause no pain.
    B. Increase in the degree of curvature will produce paralysis of the legs.
    C. Increase of the curvature will further impair cardiopulmonary function.
    D. The curvature will not increase further once Susan is past her growth period.

167. After the nurse has finished her teaching, Susan says, "I hate having to wear this. I feel so different from all my friends—like a freak." What is the nurse's best reply?
    A. "You have to wear the brace. The only other way to get better is surgery and you don't want that."

B. "Have you explained to your friends why you have to wear the brace? I'm sure they will understand."

C. "You only have to wear it for a few more months. You'll get used to it and then you won't mind it so much."

D. "I know it's very hard to be different from your friends. Have they said anything to you about the brace?"

168. When Susan visits the clinic a month later, the nurse notes that her acne has become worse. Her hair is long and rather greasy. The nurse should suggest that she:

A. Go to the doctor for birth control pills to help regulate her hormonal balance.

B. Reduce her intake of French fries, chocolate, and peanuts.

C. Cleanse her face several times a day and try a hairstyle off her face.

D. Evacuate comedones once a day.

169. Mrs. Black is upset with Susan and says that she will not do what she is told, wants to be with her friends all the time, and is even mentioning wanting to go steady with a boy. The nurse's most appropriate response is to:

A. Assure Mrs. Black that Susan will grow out of it and suggest that she not worry about it.

B. Agree with her that adolescents are a worry and very difficult to deal with.

C. Ignore the remarks. The nurse is concerned with Susan, not with her mother's problems.

D. Discuss what else might be worrying Mrs. Black at the moment.

170. Which behavior indicates that Susan is adapting well to her situation?

A. She is wearing her brace only at home.

B. She is attending school regularly.

C. She has dropped out of her after-school activities in order to stay home with her mother.

D. She is depending on her mother for all her hygiene needs.

Medical Diagnosis: Juvenile diabetes mellitus.

■ *Nursing Problems:*
Impaired gas exchange.
Management of therapeutic regime: noncompliance.
Altered nutrition.
Self-esteem disturbance.

■ *Chief Complaint:*
Theresa Fox, 16 years old, has been experiencing frequent episodes of ketoacidosis and poor control of diabetes.

■ *Past History:*
Theresa was diagnosed 3 years ago with juvenile diabetes mellitus. She presented with ketoacidosis. Her mother has diabetes and is on oral hypoglycemics and a controlled diet. Theresa has been hospitalized three times in the past 8 months with ketoacidosis. She has been treated for a urinary tract infection during the past month. She is also having difficulty with dietary restrictions.

■ *Symptoms:*
Nausea, polyuria, thirst.

■ *Physical Exam:*
Well-developed adolescent female.
Heart and lungs: Normal to auscultation. No murmurs.
Skin: Intact, with no signs of lipodystrophy. No lesions or signs of breakdown.
GU: Menstruating for 4 years; no problems.
No known allergies. Has been on NPH insulin once a day, with diet limited in carbohydrates. Attends high school and is doing well in 10th grade; has missed several days due to the flu in the past 6 months.

■ *Laboratory Data:*
Hct: 39%. Hgb: 13 g.
Glucose: 760 mg/100 mL.
BUN: 22 mg/100 mL.
Serum Cl: 100 mEq/L. Na: 130 mEq/L.
WBC: 18,500/μL.
Urinalysis: 4+ glucose. 4+ ketones. Many WBCs.
Blood gases: pH: 7.23. Pco_2: 28.

---

171. In working with adolescents with juvenile diabetes mellitus, which information is inaccurate?
    A. They need exogenous insulin.
    B. They have sudden onsets of the disease, frequently presenting with ketoacidosis.
    C. Oral hypoglycemics and diet are adequate for control when utilized precisely.
    D. Normal physical growth and development are not hindered with adequate control of the disease.

172. At home, Theresa's mother attempts to wake her daughter for school but finds her difficult to arouse, her skin feels warm, she has an axillary temperature of 101°F, and she is breathing deeply. Theresa has had the flu for several days. After Mrs. Fox has called the doctor and asked him to meet them in the emergency room, which action indicates that she has understood the nurse's teaching about diabetes?
    A. Giving Theresa an injection of glucagon to promote gluconeogenesis.
    B. Giving her several glasses of water for hydration.
    C. Giving her an injection of 10 units of regular insulin.
    D. Giving her some orange juice with sugar added to provide a rapid sugar source.

173. The physician decided to admit Theresa to the hospital. In response to Mrs. Fox's questions about what probably precipitated Theresa's current episode of ketoacidosis, the nurse doing the admission history suggests:
    A. Junk food binges.
    B. Infection.
    C. Increased physical activity.
    D. Adolescent growth spurt.

174. During the admitting physical assessment, the nurse least likely finds:
A. Increased blood glucose.
B. Doughy and clammy skin.
C. Decreased urine specific gravity.
D. Decreased potassium level.

175. Theresa has been treated for her ketoacidosis and is doing well. She is still on sliding-scale insulin according to her urinary glucose. She had 5 units of regular insulin at 10 A.M. The nurse enters her room at 11 A.M. and finds her pale, diaphoretic, and tremulous. The immediate action is to:
A. Give her 5 units of regular insulin.
B. Notify the physician of her condition.
C. Prepare a syringe with 50 mL of 50% dextrose.
D. Give her orange juice with extra sugar.

176. Theresa is on a "free diet." Which indicates that she has understood your teaching with respect to choosing an appropriate diet?
A. Frankfurter, ice cream bar, apple.
B. Hamburger, Coke, French fries.
C. Chicken, fresh banana, brownie.
D. Pizza slice, milk, Jell-O with fruit.

177. Theresa's glucose is now 150 mg and her urine glucose is +1 and negative for ketones. She is alert and up and about. However, she refuses to give her own insulin and check her own urine. She has been responsible for her own care for the past 3 years. She says to the nurse, "I just hate being different from my friends—always sticking needles in myself and not being able to do all the things they can." What is the best way for the nurse to handle this situation?
A. Recognize that it is typical of adolescent rebellion and ignore it.
B. Inform the physician of the situation and suggest that discharge plans be delayed until Theresa assumes responsibility for her own care.
C. Refer Theresa to the diabetic clinical specialist for additional teaching about diabetes.
D. Talk with Theresa about her feelings of exclusion and assess further how she is relating to her friends.

178. The nurse has found from assessment that Theresa has a decreased self-concept and feels less competent than and different from her peers. What would enhance Theresa's self-concept?
A. Every day mention how pretty and attractive she is.
B. Discuss the problem with her mother and father.
C. Encourage her to call her friends every day and keep in touch.
D. Have her help with tasks such as settling the younger children down at bedtime.

179. Theresa has become responsible for her own care again. Which is now the least important aspect of the nurse's teaching plan?
A. Eating a candy bar before playing several sets of tennis.
B. Meticulous foot care.
C. Staying away from friends with illnesses.
D. Increasing insulin as indicated with onset of illness.

180. Theresa comes back to the clinic 3 weeks later. She reports that she has been waking up in the middle of the night trembling, dia-

phoretic, and feeling very nervous and anxious. Which nursing suggestion is most helpful for her?

A. Increase in daily insulin.

B. Bedtime snack of cheese.

C. Bedtime snack of cookies and diet soda.

D. Increase in exercise level.

Answers and Rationale

1. **C. *Have her watch the nurse feed the baby.***

   C is the best answer in this situation. Seeing the child being fed successfully will help the mother gain confidence in herself. It will also help her begin to accept the child. ***A*** is an inappropriate choice because Mrs. Jones is grieving the loss of the perfect infant that she anticipated and is trying to accept her defective child. This roommate may not necessarily have any effect on helping or hindering that process. ***B*** is a poor choice because it isolates the mother and increases her concerns and anxieties. ***D*** is not appropriate because Mrs. Jones will need to start caring for Anne and verbalizing her concerns. She may feel reluctant to talk about the infant; the nurse needs to initiate discussion of the infant, emphasizing her good points.

2. **A. *Place the tip of Asepto at the front of Anne's mouth so she can suck.***

   A is the *inappropriate* instruction; the tip of the syringe should be directed to the side of the mouth, not into the cleft. Placing it at the front of the mouth will increase the chance of aspiration and nasal regurgitation. ***B*** *should* be done to help prevent formula accumulation in the nasopharynx, which can lead to infections. ***C*** is appropriate; the child will swallow a great deal of air and needs frequent bubbling to prevent vomiting and colic. ***D*** is also an *appropriate* action in that mouth care is essential for the child with cleft lip and palate.

3. **A. *Ear infections.***

   A is the correct answer; ear infections are common with cleft lip and palate because of ease of access of organisms to the middle ear. ***B, C,*** and ***D*** are all common problems in infancy but are not specifically associated with this defect.

4. **C. *Bring Anne into the clinic for evaluation.***

   C is the best answer. The child's symptoms are those of otitis media and she should be seen and treated immediately. ***A*** is inappropriate because the parent should never irrigate the ear or put objects into it; this should be done by a professional. ***B*** and ***D*** are incorrect because it is inappropriate to treat fevers at home in infants under 3 months of age. Fevers in young infants often indicate an infection, which needs professional treatment.

5. **A. *Pull the earlobe down and backward.***

   A is the correct answer for the child under the age of 3. For an older child or an adult, ***B*** is correct. ***C*** and ***D*** are incorrect for any age.

6. **D. *Sitting with slight support.***

   D is the correct answer because a 3-month-old cannot sit well with

minimal support. Behaviors *A, B,* and *C* are typical of a 3-month-old child.

7. ***D.   Replace the restraints securely.***

   D is the correct answer because protection of the suture line is the highest priority if there are no life-threatening problems. It is not a priority to take Anne's vital signs, *A,* because she is in no respiratory distress, her vital signs are normal, and her color pink. *B* is the second priority, and *C* should be done within an hour of return from the recovery room.

8. ***C.   Ask Mrs. Jones to stay and hold her.***

   C is the best choice. Anne's mother will be able to comfort her most easily and prevent the crying that will strain the suture line. *A* is incorrect in that sucking causes stress on the suture line. *B* is contraindicated after lip repair because the child can rub her face against the bed and damage the repair. *D* is unrealistic, expensive, and not as beneficial as having Anne's mother do it.

9. ***C.   A mobile with a music box.***

   C is the most appropriate toy in that it provides visual and auditory stimulation. *A* and *B* are inappropriate because they could be put in the mouth. *D,* a teddy bear with eyes that can come off, is a safety hazard because the eyes could be aspirated.

10. ***D.   Give via Asepto syringe to the side of the mouth.***

    D is the best answer in that the medication is given in the same way as a regular feeding, thus avoiding strain on the suture line. *A* is incorrect because the spoon could damage the suture line. *B* is incorrect for two reasons. First, it is a potentially inaccurate way to administer medication because the child may not drink all the formula; second, such a practice could cause the child to have difficulty eating due to the taste imparted to the food by the medication. *C* is incorrect because the child cannot use a nipple so soon after the repair.

11. ***C.   "You know, she's so pretty now she will win a beauty contest just as I did."***

    C is the best choice. This response should concern the nurse the most because it indicates potentially unrealistic goals on the part of the mother. *A* is a normal reaction; however, continued isolation of the child will be harmful. After surgery, the child's appearance will be improved and the mother should not isolate her so much. *B* is an indication that the rest of the family is beginning to accept the child. *D* is also a positive indication of the acceptance of the child by the parents.

12. ***D.   Prevention of infection of surgical incision.***

    D is the correct answer because it is a short-term goal and should no longer be a problem at discharge. *A* and *B* are long-term goals with any cleft lip and palate child because ear infections and palate deformity can lead to hearing loss and speech impediments. *C* is always a goal for any child with a congenital deformity where there is a high risk for impaired development of the parent-child relationship.

13. ***C.   Prevention of urinary tract infections.***

    C is the primary goal because chronic urinary tract infections lead to renal failure, which may cause death. *A, B,* and *D* are all important goals with any child who has problems with mobility but they are less important than the urinary problems.

14. **B. *Downward displacement of the medulla and posterior cerebral vermis into the vertebral canal can lead to noncommunicating hydrocephalus.***

    B is the correct answer in that many children with myelomeningocele have Arnold-Chiari syndrome, which is what this typifies. *A* is incorrect in that it is not a reason for hydrocephalus. With growth the amount of cerebrospinal fluid (CSF) produced increases but so do the spinal cord and absorptive surfaces. *C* and *D* are physiologically incorrect.

15. **D. *Increasing head circumference.***

    D is the first sign of hydrocephalus in infants in whom the anterior fontanelle and sutures have not closed. *A* and *C* are later signs. *B* is the first sign in the verbal child, from about 2 years of age and older; the infant may have a headache but cannot tell us.

16. **D. *Feed 6 oz. every 3 hours.***

    D is the inappropriate action. There is no indication of an increased need for calories and fluids; David's weight and hemoglobin are well within normal range. At this age he should be eating every 4–5 hours. In addition, he has been vomiting and this action would increase that likelihood. *A* is an appropriate action in assessing for increasing intracranial pressure. *B* prevents skin breakdown and is the correct position for the child with dislocated hips. *C* is an appropriate action due to the bladder involvement; it helps empty the bladder and prevent urinary stasis and infection.

17. **C. *A rattle.***

    C is the best choice. At 2 months he can grasp and hold a rattle or object. *A,* a busy board, is too advanced and would be appropriate for an older infant. *B,* large blocks, would also be better for the older child. *D* is not a good choice; while a music box provides auditory stimulation, it does not encourage motor development as the rattle does.

18. **A. *Right atrium by way of the internal jugular vein.***

    A is the correct definition of what a ventriculoatrial shunt does. *B* is the explanation of a ventriculoperitoneal shunt and *C,* of the ventriculoureteral shunt. *D* is incorrect in that the venous system with its lesser pressures, not the arterial system, is used.

19. **C. *Elevate the head 60 degrees to promote cerebrospinal flow.***

    C would not be included in the plan of care. This is the first shunt and too rapid decompression could lead to a subdural hematoma. The child should be kept flat for a few days after surgery. *A* is appropriate as it will promote functioning of the shunt and prevent skin breakdown. *B* is a correct procedure in that it helps assess for increased intracranial pressure. *D* is appropriate for the same reason, because the shunt occasionally malfunctions.

20. **A. *Hold head steady in a sitting position.***

    A is probably the task David cannot do because of the increased weight of the head with hydrocephalus. *B, C,* and *D* are three tasks he should be able to do as any normal child can.

21. **D. *Prevent subdural hematoma.***

    D is the correct answer because too rapid decompression could lead to subdural hematoma. *A* and *C* will not occur, because the shunt allows a slow, constant outflow of fluid. *B* will not occur; the amount of fluid shunted into the circulatory system usually does not cause a problem.

22. **D. *How to measure head circumference.***

    D is the correct answer because it is one method of evaluating for increasing intracranial pressure. *A* is incorrect because the shunt should work properly without any assistance. *B* is incorrect because glucose in the urine is not expected in relation to any of David's problems; it is therefore inappropriate teaching. *C* is unnecessary; David is not in congestive heart failure, nor is he having other fluid problems.

23. **C. *When they note bulging of the anterior fontanelle when David is crying.***

    C is the one instance in which the parents should not call the physician, as the fontanelle normally bulges with crying. *A* may indicate signs of meningitis and medical help must be sought. *B* is indicative of a urinary tract infection and *D,* of increasing intracranial pressure.

24. **B. *Pulmonary embolism.***

    B does not require placement of a new shunt and it does not interfere with the flow of cerebrospinal fluid. *A, C,* and *D* are situations that eventually will require replacement of the shunt.

25. **D. *"It must be very hard for you. Have you had any time away from the baby?"***

    D is the best choice in that it reflects the mother's concerns and asks for further data. With the mother receiving no relief from the constant stress of caring for this infant, one aspect the nurse should be concerned about is the potential for abuse. *A* is not therapeutic and may make Mrs. Stivic feel even more guilty. *B* is not the best answer in that it does not reflect the mother's concerns and also makes an assumption without sufficient data. *C* has nothing to do with the mother's concerns.

26. **C. *2 doses of DPT and 2 doses of OPV.***

    C is the correct answer. David should have received 2 doses of both DPT and OPV. Recall that he is now about 5 months old. (He was 2 months on admission, plus the hospital stay and this clinic visit 2 months later.) *A* is incorrect in that the first round of immunizations is given at about 2 months of age and the second, at about 4 months of age. *B* is incorrect in that the MMR is given at 15 months of age and *D* is incorrect in that TB is done at about 10 months of age.

27. **D. *Weight.***

    D is the correct answer; weight is the most accurate way to determine degree of dehydration. Comparison of his normal weight and his current weight (in the case situation) shows about a 10% weight loss, which is moderate dehydration. *A, B,* and *C* are all signs of dehydration but are of little use in determining degree.

28. **A. *Restoration of intravascular volume.***

    A is the correct answer. This type of dehydration in infants leads to shock and must be prevented. The goal of care is to restore the lost fluids. *B* is not the primary goal but a secondary one; the diarrhea should decrease with the child NPO. *C* is another goal that can be met later. *D* is also not primary in that the physiologic stability is of primary importance initially.

29. **C. *45 gtt/min.***

    C is the correct answer. The calculation is as follows:

    $360 \div 8 = 45$ mL/hour.

    $(45 \div 60) \times 60 = 45$ gtt/min.

    A, B, and *D* are incorrect calculations.

30. **D. As soon as he can take fluids again, the restraints will not be needed.**

    D is the best way to initiate discussion of the restraints. It tells the parents that they are only temporary. *A* is an incorrect statement; the restraints are removed periodically to check for skin breakdown and to move the joints. Also the child can be held with the IV line in place and then the restraints are not needed. *B* is incorrect because explaining the procedure to the parents in this way is frightening and anxiety producing. *C* is false because by limiting the infant's mobility and his way of expressing anxiety, there is little chance he will grow used to the restraints. The answer is also incorrect in that it indicates that the child should be left totally alone in bed, which is not beneficial for either the child or the parents.

31. **D. Pedialyte and glucose water.**

    D is the best answer. It provides electrolytes and calories in an easily digested form. *A* is incorrect in that the carbonation is irritating to the bowel and the high sugar content will restart the diarrhea. *B* is incorrect because even half-strength formula is not a clear liquid and should not be offered until all the diarrhea has stopped. *C* is incorrect in that few infants will drink the salty broth and tea is a poor choice because the xanthine in it is a mild diuretic, which could enhance the dehydration.

32. **C. Stop feeding the child orally.**

    C is the correct answer. The diarrhea has restarted and placing the child NPO again will aid in stopping GI peristalsis. *A* is incorrect in that the feedings continue and irritate the GI tract. *B* is incorrect because it is not the first action to take. However, the data would be useful in evaluating an increase in degree of dehydration. *D* is not the immediate priority but again would help determine the amount of water loss.

33. **D. A game of peek-a-boo.**

    D is the best choice. Peek-a-boo is an age-appropriate game for an 8-month-old and also involves personal interaction, which will help social development. *A* is too young for him; he needs something or someone to interact with, not just look at. *B* is too advanced; large blocks are appropriate for a child over 1 year of age. *C* is again rather young for this child and more appropriate to a 4-month-old.

34. **A. Stay with the child and protect him from injury during the seizure.**

    A is the best answer in that it provides for the safety of the child and permits observation of the progression and duration of the seizure. *B* would be an appropriate action after *A;* in fact, the best action would be to have another nurse inform the physician while you stay with the infant. *C* is not correct as the first action; it would be indicated later, as this is probably a febrile convulsion. *D* is not the best answer for the child's safety. Again this is an action to take after the child's physical safety has been ensured.

35. **D. "I know this must worry you. Has anything like this ever happened before?"**

    D is the best answer; it acknowledges the parents' concerns and asks for further data. *A* is incorrect because seizure activity in children can lead to brain cell hypoxia and cell death. It is true that febrile seizures often stop by age 6; however, there is nothing

to indicate that this is the situation here. It is also poor reassur-ance—any parent seeing a seizure is going to worry. *B* is incorrect in that it does not acknowledge the parents' anxieties and gives false information; there is no guarantee that the seizure will not occur again. *C* is incorrect in that it does not address the issue at hand—this infant's seizure.

36. *B.*   *Crawl and begin to say "Mama" and "Dada."*

B is the correct answer. *A* and *D* are abilities of the infant of about 11–12 months. *C* is incorrect in that the Moro reflex should be gone by now; if it is still present, it indicates neurologic deficit.

37. *D.*   *Assess why Mrs. Jones has not had the immunizations done.*

D, collecting data as to why the immunizations have not been started, is the best answer. *A* is incorrect in that it assumes that Mrs. Jones did not understand the instructions, an assumption that has no basis in fact. *B* is a poor choice in that it could make the mother angry, defensive, and unlikely to cooperate. *C* is a poor choice in that it takes the locus of care away from the parent, who has primary responsibility for the child's care.

38. *A.*   *An acute disease characterized by fever and having no known cause.*

A is the correct answer. Kawasaki disease is an acute febrile syn-drome of unknown etiology. The disease occurs worldwide and does not seem to be prevalent in any particular geographic, seasonal, socioeconomic, or environmental pattern. Children from 3 months through 14 years old can be affected. *B* is incorrect because there is no known forerunner of rheumatic fever. *C* and *D* are both in-correct because they have no basis in fact.

39. *B.*   *Kawasaki disease.*

B is the correct answer. Dr. Kawasaki first identified this syn-drome as mucocutaneous lymph node syndrome (MLNS) in 1967, based on the signs and symptoms he identified in children having this disease. *A,* infantile polyarteritis nodosa, is a similar disorder, although characterized by a 100% fatality rate. *C,* nephrotic syn-drome, is a kidney disorder found primarily in toddlers and pre-schoolers. *D,* Still's disease, is also known as juvenile rheumatoid arthritis and is generally a disease of school-age children.

40. *A.*   *Diphtheria, tetanus, pertussis, and polio.*

A is the correct answer. According to the guidelines of the Ameri-can Academy of Pediatrics, routine immunizations for an 8-month-old include a primary series of 3 DTP shots for diphtheria, teta-nus, and pertussis and 2 or 3 trivalent oral polio vaccines. These are given most commonly at 2, 4, and 6 months of age. Additional immunizations such as HB (hepatitis) and Hib (*Haemophilus in-fluenzae* type B) are also recommended. *B* is incorrect because while children are tested for TB, a vaccine is not given. *C* and *D* are incorrect because the vaccines against German measles and "regular" measles are not given until 15 months; they can be com-bined with the mumps vaccine in one injection for MMR (measles, mumps, rubella).

41. *C.*   *Petechiae and hematomas.*

C is the correct answer. Kawasaki disease is *not* evidenced by bleeding tendencies; therefore, the nurse will not find petechiae or hematomas on Jackie. *A,* bilateral conjunctivitis, *B,* desquamation

of fingers and toes, and *D,* rash and lymphadenopathy, are all usually found in the child with Kawasaki disease.

42. **D.   *Jackie's temperature is 104°F.***

    D is the correct answer: 40°C = 104°F. To convert from centigrade to Fahrenheit, use this formula:

    (9/5 × °C) + 32 = °F.
    (9/5 × 40) + 32 = 104°F.

    Response *A* evades the parents' question and is not true. *B* is incorrect because high-spiking fevers are common in Kawasaki disease. *C* is incorrect; a normal temperature on the centigrade scale is 37°C.

43. **B.   *4.5.***

    B is the correct answer; the nurse should administer 4.5 minims. Knowing that there are 15 or 16 minims per milliliter, the nurse must first convert 0.6 mL to minims: 9 minims. Then, the nurse must convert gr. ss to mg: 1 gr. = 60 mg; thus, gr. ss, or ½ gr., is equal to 30 mg. Finally, the nurse can calculate how much to give Jackie:

dose desired	:	dose on hand	=	x : amount on hand
30 mg	:	60 mg	=	x : 9 minims
		60 (x)	=	30 (9)
		x	=	270/60
		x	=	4.5 minims

    This amount should be measured in a TB syringe for accuracy. *A* and *C* are incorrect calculations. *D* is incorrect because the nurse, with careful calculations, should be able to use the stock bottle of acetaminophen (Tylenol).

44. **B.   *Check Jackie's temperature every 1–2 hours.***

    B is the correct answer. The conscientious nurse would check Jackie's temperature and document it in the nurse's notes at least every 1–2 hours, knowing that a sudden, sharp rise in temperature might cause a febrile convulsion. *A,* antibiotics, are ineffective as a treatment for Kawasaki disease. *C,* giving antipyretics alone, might not control Jackie's fever, and the pediatrician will have to be notified. *D* is incorrect because an 8-month-old infant should never be placed on a hypothermia mattress without a physician's order.

45. **A.   *Immediately notify the physician.***

    A is the correct answer. The normal platelet count for an infant is 200,000–470,000/μL. Since Jackie's count is quite elevated, Jackie is prone to develop the complication of multiple thromboses and her physician should be promptly notified. Aspirin may be used to help prevent clot formation secondary to the increased platelet count. *B* is incorrect because bleeding tendencies are found with a decreased platelet count. *C* is incorrect because the child with Kawasaki disease does not need reverse isolation. *D* is incorrect because this lab report deviates too far from the norm to simply be placed in the chart.

46. **D.   *At 3:30 A.M.***

    D is the correct answer; the IV will need changing around 3:30 A.M. If Jackie's IV is adjusted properly, to run at 55 mL/hour as the physician ordered, by 3:30 A.M. there would be about 30–35 mL left in her bottle and this is the best time to hang a new bottle. *A,* hanging a new bottle immediately, not only would waste a mod-

est amount of expensive IV fluid, but also might not encourage the nurse to keep going back hourly to check Jackie's flow rate and IV site. If the bottle were left hanging until 6:45 A.M., *B,* it would probably be empty and Jackie would not have received the amount of IV fluid ordered. *C,* waiting until "whenever" the bottle empties, shows a lack of careful planning by the nurse!

47. **C.**   ***Check the flow rate of Jackie's IV fluid.***

     C is the correct answer. Whenever IV fluid appears to have run in much more quickly than was ordered by the physician, the nurse's *first* action should be to check the flow rate and adjust it either to the correct rate of flow or to a KVO rate, depending on how much extra fluid has been received by the patient. Following this, the physician should be notified, and the patient should be checked for signs of fluid overload and placed in semi-Fowler's or high Fowler's position to facilitate breathing, choice *D. A* is inappropriate as the *first* nursing measure to be taken at this time; Jackie has been running high-spiking fevers so there is relatively little danger of an immediate febrile convulsion. Valuable time will be lost while the nurse calculates, pours, and administers acetaminophen (Tylenol). *B* is also inappropriate as a *first* nursing action: Slowing the flow rate will give the nurse adequate time before having to add another bottle of fluid.

48. **A.**   ***Check with the pediatrician for new IV orders.***

     A is the correct answer. The physician should be notified and will need to write IV orders for the remainder of the shift. *B* is incorrect since a nurse may not change the prescribed flow rate of IV fluid without a new doctor's order. *C* is incorrect; the IV should never arbitrarily be stopped since Jackie may need to receive emergency drugs through this line. *D* reflects poor nursing judgment because additional fluid may overload Jackie's circulatory system.

49. **D.**   ***Her sedimentation rate is elevated and this is a typical lab finding in the child with Kawasaki disease.***

     D is the correct answer. The normal sedimentation rate for an infant is 0–10 mm/hour. Jackie's sedimentation rate is well above normal. One lab finding typical of Kawasaki disease is an elevated sedimentation rate. Therefore, *A, B,* and *C* are incorrect.

50. **B.**   ***Showing signs and symptoms of heart failure.***

     B is the correct answer. Early signs and symptoms of heart failure in infants include tachycardia, tachypnea, and a sudden weight gain. Knowing that infants and children with Kawasaki disease are at high risk for cardiac involvement, and given Jackie's signs and symptoms, the nurse is correct in suspecting heart failure. *A* reflects inappropriate nursing judgment related to signs and symptoms. *C* and *D* show lack of nursing knowledge related to heart failure in infants.

51. **C.**   ***Allow them to discuss their concerns and fears.***

     C is the correct answer. As with any patient demonstrating signs of anxiety, the best nursing intervention is to allow these parents to discuss their fears. This helps the nurse establish herself or himself as a concerned, trustworthy person, which is vital to a therapeutic relationship. *A* is incorrect; when clients have a high anxiety level, "thorough" orientation is inappropriate at that time. *B* is not a true statement, as this was not a routine transfer. *D* is also incorrect; Jackie has a 98–99% chance for survival.

52. *D.* *Jackie's apical rate is 98 and regular, with normal sinus rhythm on her monitor.*

    D is the correct answer. When giving digoxin to an infant, the apical rate must be checked for 1 full minute. An apical rate of less than 120 may be an early sign of digoxin toxicity. *A* is incorrect because digoxin is an important part of Jackie's treatment and the dose cannot be delayed, even if she is asleep. *B* is incorrect because tachycardia is one sign of congestive heart failure (CHF), for which Jackie needs her digoxin. *C* is also incorrect; in some children with Kawasaki disease, jaundice, elevated bilirubin levels, and elevated liver enzyme levels may be present; however, this will not affect the administration of digoxin.

53. *C.* *Slowing of the heart rate and increased urinary output.*

    C is the correct answer. Digoxin is a cardiotonic. It should slow the heart rate to normal and it also improves renal perfusion—increasing urinary output. *A, B,* and *D* all reflect adverse effects of digoxin therapy and are therefore incorrect choices.

54. *D.* *Restraining Jackie for IV therapy and confining her to her crib.*

    D is the correct answer because although Jackie will probably need some restraints during IV therapy, she need *not* be confined to her crib. Her parents and nurses should certainly hold and rock her. *A, B,* and *C* are appropriate nursing measures for Jackie.

55. *B.* *Is on a special low-sodium formula and should not receive milk.*

    B is the correct answer. Low-salt formulas such as Lonalac and PM 60/40 are often used for infants with congestive heart failure to prevent increased accumulation of fluids. All parties caring for Jackie should be advised of her special dietary needs. *A* and *C* are incorrect; since Jackie's condition has improved, there is no reason she should not go to the playroom. *D* is incorrect; even though she is still on digoxin, there is no need for her to remain on a cardiac monitor unless ordered by the physician.

56. *D.* *Poor appetite with minimal PO intake.*

    D is the correct answer. If Jackie were eating and drinking poorly, she would not be ready for discharge. In fact, she might even be showing signs of digoxin toxicity and the dose of this drug would need adjusting. *A,* 37°C, is a normal 98.6°F temperature and indicates readiness for discharge, as do *B* and *C*.

57. *C.* *It is best to give digoxin within 1 hour after feeding your child.*

    C is the best answer. Digoxin should *not* be given within an hour after eating since this decreases its absorption. Digoxin is best given 1 hour before or 2 hours after meals for best absorption from the stomach. The nurse would be correct in teaching choices *A, B,* and *D*.

58. *B.* *Talk with the parents to assess their knowledge and how they can help with Sam's care.*

    B is the best answer. It acknowledges the parents' concern (from the history) and gathers a database to start preparation. *A* is incorrect because the child is too young to understand explanations. *C* is not correct because the parents need to know what will happen *this* time; the previous catheterization was probably performed at a few weeks of age. *D* is inappropriate because the child

is vulnerable to separation anxiety and the parents have also expressed their need to be present.

59. C.   *The degree of pulmonary vascularization.*

C is the information not revealed by the catheterization. Pulmonary vasculature is usually seen either with a chest x-ray or an angiogram. Cardiac catheterization yields information about *A, B,* and *D.*

60. A.   *Sam complains of his head hurting.*

A would cause the greatest concern since it might indicate a cerebral infarct, for which the child would be at risk considering that this child is NPO prior to a catheterization. This means a degree of dehydration plus the possibility of an embolus due to the invasive procedure. *B* is not a concern as it is within the range of his normal pulse. *C* should not concern the nurse because a decreased temperature of the leg in which the catheterization is done is to be expected; it would be of concern if it persisted for an extended time and was accompanied by absent pulses. *D* is not too much of a concern considering the normal temperature rectally and the degree of dehydration, as noted above.

61. D.   *Endocarditis.*

D is the correct answer; all children with cardiac defects are prone to infection with invasive procedures. In this case a cardiac infection is most likely. *A, B,* and *C* are not complications commonly seen after cardiac catheterization.

62. C.   *Recent fluctuations in weight.*

C is the correct answer. Systemic edema is best assessed through weight gain, particularly in the child under the age of 2. *A* is a long-term complication of cyanotic heart disease and chronic hypoxemia. *B* is incorrect in that it does not indicate congestive heart failure directly but may add to it indirectly. *D* is incorrect in that peripheral edema is not the best way to assess congestive failure in the toddler.

63. D.   *Chronic oxygen deficiency.*

D is the correct answer, reflecting chronic hypoxia. In children with cardiac anomalies, the bone marrow produces more RBCs, leading to polycythemia and increased problems with clots forming (as noted in question 60). *A* is incorrect in that infection usually is seen with a WBC elevation. *B* is incorrect in that the child did not enter the hospital with symptoms of dehydration and a cyanotic defect is associated with polycythemia. *C* is incorrect in that cardiac output indirectly affects the hematocrit.

64. C.   *He may be going into heart block due to digoxin toxicity.*

C is the correct answer because Sam's pulse is much slower than his admission pulse and bradycardia is a sign of digoxin toxicity. *A* is incorrect because even though the pulse does slow with decreased activity, this pulse rate is more than 20 beats slower than his normal rate. *B* does not apply, since Sam's normal heart rate is different from that of a normal 18-month-old. When digoxin is administered, evaluation of its effects must be individualized. *D* is incorrect in that potassium is more likely to affect rhythmicity than rate.

65. B.   *Decreased energy reserve.*

B, the decreased energy reserve associated with congestive heart failure, is the *most likely* reason Sam is eating poorly. Therefore, the nurse should plan rest periods before meals and smaller and

more frequent feedings. Both will provide the necessary oxygen and energy, which must be the nurse's first priority for Sam. *A, C,* and *D* are secondary considerations.

66. *C.* ***Apricots and bananas.***

    C is the correct answer in that these foods provide potassium. Note from the case situation that Sam's potassium is low; since he is on diuretics, he needs potassium in his diet. *A* is incorrect in that milk products are high in sodium and orders indicate salt and fluid restriction. *B* is also incorrect in that preserved meats, such as hot dogs, are high in sodium. *D* is incorrect because both high sodium content and excess fluids are not appropriate for an 18-month-old on fluid restrictions.

67. *A.* ***The knowledge that the repair will include removal of the atrial septum and creation of an intracardiac baffle to redirect blood flow.***

    A is the correct answer and describes the Mustard procedure done to repair a transposition. *B* is the repair for tetralogy of Fallot; *C* is incorrect because parents often find explanations by the physician confusing and many times hesitate to "take up the doctor's time" with questions. *D* is incorrect because such an approach may ignore real concerns that the parents have for the preoperative and intraoperative stages of Sam's care.

68. *A.* ***Administer oxygen and position him in the knee-chest position.***

    A is the preferred treatment for hypoxic episodes. These episodes are particularly prevalent in children with tetralogy but can occur in any child with a cyanotic cardiac defect. Oxygen and the knee-chest position are most helpful. The position in *B* is incorrect; morphine is given to relieve pain and anxiety while not affecting the blood pressure. *C* is incorrect in that asking the parents to leave will probably increase their anxiety; however, the main reason this is a poor choice is that it does not deal with the child exclusively. *D* is not as good a choice as *A* in that positioning is omitted. The physician should be notified after the other appropriate actions have been taken.

69. *B.* ***Feed himself and drink from a cup.***

    A, C, and *D* are all skills of older children.

70. *B.* ***His mother is overprotective and allows Sam few challenges to develop his skills.***

    B is the best answer. Data from the history and from her comments indicate that the parents have given the child few opportunities to develop normal skills. *A* is incorrect because while the defect may slow down acquisition of skills, Sam will be able to achieve them eventually. *C* is incorrect in that there are no data supporting the idea of a struggle for autonomy. *D* is incorrect in that Sam's behavior is no different from when he first entered the hospital—note that the case situation indicated developmental lags.

71. *C.* ***Reassess the parents' needs and concerns.***

    C is the best answer. The behavior of the parents indicates they are still overprotecting Sam so the nurse needs to reassess and to plan other goals and interventions. *A* is not the best choice in that it is intervention without a basis—there is no indication that the parents do not understand the needs of children. *B* is inappropriate because the child is going home and the parents will have to

present the activities and stimulation if Sam is to continue to develop. *D* is also not the best answer in that it assumes that the parents do not realize the results of the surgery—this action is included in the correct answer *C*.

72. A.   Long bone x-rays.

A is the test done to confirm the other data. Long bone x-rays will reveal old or partially healed bone fractures, which could indicate abuse. *B* is appropriate if the child had neurologic problems, but none are indicated in the history. *C,* a repeat CBC, will not add to the database of suspected child abuse. *D* is incorrect because there is not a diagnosis yet. Removing the child from the home, even for a short period, is not the treatment of choice.

73. B.   Having been abused or neglected themselves as children.

B is the correct answer. Many abusive parents were abused themselves and this is the only way they know to parent children. *A* is not true; in fact, parents who abuse their children generally have unrealistic development expectations, which leads to abuse when the children cannot meet parental expectations. *C* is incorrect because abuse is associated with multiple stresses, as Larry's history indicates, and abusive parents have an inability to deal with stress. *D* is a poor choice because abusive parents are frequently isolated from sources of support that will help them cope with stresses.

74. A.   The parents to express concern about the extent of damage and Larry's current condition.

A is the correct answer. Once an abused child has been hospitalized and the immediate stresses have been reduced, abusive parents may feel quite concerned about the child's welfare. *B* is incorrect because abusive parents may not link their actions and the need to compensate for them with a sense of guilt about inflicting injury. *C* is incorrect because abused children are often fearful of their parents. *D* is incorrect because abused children learn very early that crying often brings punishment or injury from their parents.

75. D.   Recognition of the LPN's feelings.

D is the best choice. Recognizing the LPN's feelings and dealing with them first will aid the nurse in going on to other concerns. *A* is a poor choice in that slander is not really a concern, but it is correct in that legal action has not been initiated. *B* is incorrect because accusing and lecturing the parents will make them angry and defensive and impair further relations between the parents and the nursing staff. *C* is not the best answer in that it does not deal with the LPN's primary feelings about the parents. This should be noted prior to any further interaction.

76. B.   Assigning one person on each shift to care for him.

B is the best answer. This approach provides a consistent set of caretakers and allows trust to develop between the child and the nurse. *A* is inappropriate since it restricts the parents, who have the right to see their child, and also restricts opportunities for the nurse to work with them. *C* is incorrect in that multiple caretakers will increase the child's loneliness and will not meet the need for a consistent person with whom he can identify. *D* is not useful because at this age peer interaction is minimal. Also, the child needs an adult to trust; children would not meet this need.

77. **C. *Protective services department.***

    C is the correct answer. All 50 states have child abuse laws that provide for reporting such cases, usually to some form of state protective agency. The police, *A,* are not notified routinely unless the child's life is endangered. *B* is incorrect because legal action is not decided on until after the case has been reported. *D* is incorrect in that initial reporting is not done to such agencies although they may be involved in follow-up care.

78. **B. *The notes will provide a database to plan family intervention.***

    B is the best answer. The interactions will help determine what type of therapy will be effective. *A* would be important if the case was going to court, but there are no data about this so it is an unwarranted assumption. *C* is incorrect because it applies to any patient; the question asks about the abused child and his parents. *D* is not the best answer because the intervention is the primary reason for observation; this choice might be a secondary reason.

79. **D. *Walk up and down stairs.***

    D is the correct answer because it reflects the developmental capabilities of a 2½ year old. *A, B,* and *C* are abilities of the 4- or 5-year-old.

80. **D. *Look for the appearance of petechiae.***

    D is the answer. Iron is given to build hemoglobin and does not affect coagulation of the blood. *A,* staining of the teeth, and *B,* dark stools, are frequent side effects of iron. The vitamin C in the orange juice, *C,* will improve absorption of the iron.

81. **A. *Leave the child in the home and provide counseling for the parents.***

    A is the primary treatment choice since the child usually does better with his natural parents. The therapy is aimed at helping the parents change their behaviors and learn to cope with their stresses. *B* is not the treatment of choice unless the child is in danger of further physical abuse. *C* may be done, but the child cannot be hospitalized for the entire course of these lengthy treatments. *D* will be included in the treatment plan but it is not the entire plan. *A* is really inclusive of this choice and therefore is the more complete answer.

82. **B. *Remark on how well the child and his parents are doing.***

    B is the correct answer. The behaviors are indicative that the treatment plan is working, and reinforcement of the improved behaviors is indicated. Since these behaviors are indicative of an improving relationship, there is no need to change the goals of care and *A* is incorrect. *C* will not be helpful since the family needs all the support it can get; saying nothing is not supportive of any of the observed changes. *D* is incorrect because it is not indicated by the data provided.

83. **D. *Peripheral circulation.***

    D is the one area that does not require assessment since there is nothing in the history indicating decreased circulation, and lead poisoning causes anemia, not decreased circulation. *A* is an important area to assess because impaired parent-child relationships are often associated with lead poisoning. *B* is important for two reasons—the history reveals changes in mental status and the primary effect of lead on the body is neurologic. *C* is important be-

cause the history reveals pica and data about eating habits will be primary to planning the nursing care.

84. **B. *Eating paint chips.***

B is the correct answer. Most lead poisoning in children is due to old paint from buildings that are old and poorly maintained, and the child's case situation mentions that he lives in such a building. *A* is a very unlikely source of lead poisoning in a child although it may be seen frequently in cases of adult lead poisoning. *C* is incorrect because dirt usually does not contain lead. *D* is incorrect as there is no lead in the ink or the paper used to print most newspapers. The fact that this is in the case situation is indicative of pica.

85. **D. *Promote excretion of the lead.***

D is the most important nursing goal to help prevent permanent damage. *A* is an important goal for discharge planning and follow-up to avoid *continued* poisoning. *B* and *C* are important nursing goals but can be dealt with later in the care.

86. **A. *Pica.***

A is the correct answer because ingestion of nonnutritive substances—and therefore a poor diet and lack of iron—lead to iron deficiency anemia. *B* is incorrect because there is no history of blood loss. *C* is not the usual cause of the anemia seen with chronic lead poisoning. *D* is incorrect in that the child is not noted to be black or diagnosed with sickle cell anemia.

87. **B. *Corproporphyrin.***

B is the correct answer and is part of the diagnostic studies performed to identify lead poisoning. *A, C,* and *D* have nothing to do with the diagnosis of lead poisoning.

88. **C. *Neurologic.***

C is the correct answer because neurologic damage is the side effect that produces the major, serious, long-term complications such as seizure disorders, learning disabilities, hyperkinesis, and other problems. *A* and *B* are also found with lead poisoning but usually do not lead to death or to other long-term side effects. *D,* the cardiovascular system, is not affected.

89. **C. *Maintain a patent airway.***

C is the correct answer because this will promote respirations. *A* is an action to take once the safety of the child is ensured. *B* also is done once the child is assessed not to be swallowing or to be having difficulty with secretions. *D* is inappropriate in that the child is already convulsing and forcing an object between his teeth could break a tooth and cause other problems.

90. **A. *Allow the child to play with the needle after the injection.***

A is the best answer because it is the action that should *not* be included in the care plan. This child is 3½ years old and playing with the needle could be dangerous. The syringe might be appropriate but not the needle. *B* is an appropriate action in that increased fluids will flush the lead through the kidneys and help prevent kidney damage. *C* is an associated correct action, for the same reason. *D* is a correct action in that this medication will help decrease the pain of the many painful injections.

91. **B. *BUN.***

B is the most important piece of data to assess on a daily basis due to the toxicity of lead to the kidney. *A* has nothing to do with the lead or the treatment. *C* does not really tell the nurse about

ongoing lead levels—a blood lead level would be better. *D* is not appropriate as the anemia will not resolve itself due to the chelation therapy—nutritional and iron intake must be improved.

92. *B.   **Denver Developmental Screening Test.***

B is the correct answer. This test is appropriate for children through the age of 6 and can be easily and quickly given by the nurse on a regular basis. *A* is an IQ test only and does not give the needed developmental information; also, the nurse has neither access to it nor the time to administer it. *C* and *D* are tests that are appropriate for the older, school-age child. Also, the Piers-Harris reflects self-concept only, not the developmental progress that the nurse wants to assess.

93. *D.   **"You should tell him to play in a certain area and enforce the rule with discipline."***

D is the best answer because it stresses setting limits and enforcing them consistently to prevent accidents. *A* is inappropriate because the child does not have the ability to learn and formulate rules by himself and he will probably get hurt before he develops that ability. *B* is unrealistic and includes unnecessary and harmful restrictions. *C,* keeping a child in the house at all times, is also unrealistic.

94. *A.   **Night-time continence.***

A is the correct answer. Complete night-time dryness is not normally achieved until the child is older. Also recall from the case situation that the mother spoke of her difficulty in training the child. *B, C,* and *D* are all normal achievements for a 3-year-old.

95. *B.   **Finger painting.***

B is the best choice. It allows Steven to develop motor coordination and is not dangerous. *A* is an activity that is enjoyed more by school-age children. *C* is seen more with the older preschool child of about 5, and *D* is inappropriate because of Steven's neurologic problems and the danger that he might cut himself.

96. *D.   **Persistent anemia.***

D is the correct answer because anemia can be reversed. *A, B,* and *C* reflect neurologic—and therefore permanent—damage, which requires long-term follow-up.

97. *A.   **Serum lead level of 40 μg.***

A is the correct answer because it reflects the downward trend of the lead level. *B* might cause concern but would be followed through the outpatient clinic. *C,* the hemoglobin, will increase with improved nutrition and can also be followed on an outpatient basis. The WBC count in *D* reflects a normal value and shows little change from the admission lab data.

98. *B.   **Koplik spots on the buccal mucosa.***

B is the correct answer. Koplik spots are present briefly at the beginning of a case of *measles. A, C,* and *D* may all be indicative of infection of the tonsils.

99. *D.   **Assess the mother's reasons for refusing preparation.***

D is the correct approach. Despite the mother's adamant position, the nurse needs to determine the reasons for such a potentially harmful position. *A* is inappropriate because the physician is not in any better position to influence the mother over the phone than the nurse is. Such an approach could lead to tension between the nursing staff and the mother. It will also force the physician to take sides in the situation. *B* is inappropriate because Mary may

feel frightened and abandoned by her parents if she is told something different once they leave. *C* is inappropriate without first assessing the reasons for refusal and attempting to support the parents' concerns first.

100. *D. Stay in the room and complete the nursing care of her roommate.*

D is the best answer. The child is in the phase of protest of separation anxiety and refuses to be comforted; the best approach is to be within the child's sight. *A* is incorrect since the 4-year-old has little concept of time. *B* is not appropriate because she is missing her parents, not children of her own age. *C* is incorrect because the child is anxious and the presence of someone familiar, such as the nurse, will be helpful.

101. *D. Check the SGOT and SGPT levels.*

D is the data the nurse does *not* evaluate since there is no indication of liver dysfunction in the child's history. *A* is evaluated since the 5-year-old child is beginning to lose teeth about this time and a loose tooth could be knocked out during anesthesia and aspirated. *B* is essential in that this is Mary's first surgery and hemorrhage is the main complication postoperatively. *C* is evaluated in that they are indicators of infection and surgery would be postponed if infection were indicated.

102. *D. Bleeding time of 10 minutes.*

D necessitates a call to the physician in that normal bleeding time is 3–9 minutes. *A* is expected in that the child has been NPO prior to surgery and will be slightly dehydrated. *B* is an expected behavior with enlarged adenoids. *C* is again to be expected with decreased fluids prior to surgery.

103. *D. "The needle will hurt like a prick. You choose the leg you want it in and then squeeze your mother's hand when I give it to you."*

D is the best choice. It informs the child that the injection will hurt and offers her a choice and some control over the situation. *A* is a poor choice in that it belittles the child and tells her to suppress her normal reaction to pain. *B* reflects an attempt to bribe the child and is inappropriate. *C* is incorrect in that it does not acknowledge the frightening aspect of the injection and also tells the child it will put her to sleep. Remember that at this age sleep is often confused with death and therefore may make the whole process more frightening.

104. *D. May cause hallucinations.*

D will certainly increase the mother's anxiety; hallucinations are an infrequent side effect of atropine. *A* and *B* are frequent normal side effects of atropine. *C* will also be a result of the atropine due to decreased secretions.

105. *D. Emesis of dark blood.*

D does not indicate current hemorrhage; an emesis of old (dark) blood simply reflects a past event. *A* is a classic sign of bleeding after throat surgery. *B* and *C* indicate shock with the typical signs of increased pulse and respirations and decreased blood pressure.

106. *D. Place in semiprone position until responsive.*

D is the appropriate action because it allows the drainage of pooled secretions caused by decreased swallowing due to pain. *A* is incorrect in that hot fluids will increase, rather than decrease, bleeding and pain. *B* is inappropriate since aspirin increases the

clotting time, which prolongs any bleeding. *C* is incorrect because suctioning, unless absolutely necessary, is contraindicated in that it will traumatize the surgical site and cause further bleeding.

107. *D.　Green Jell-O.*

D is the best answer because gelatin (Jell-O) is chilled, which will help promote clotting and decrease pain. *Green* Jell-O is especially good because if the child vomits, there will be no confusion as to possible bleeding. *A* and *B* are acidic and will increase the child's pain. *C* is not recommended since milk and milk products increase mucus production and lead to more swallowing and pain.

108. *D.　Question the order and not give the drug; notify the physician.*

D is the correct answer. Aspirin decreases the ability of the platelets to clot and could lead to bleeding difficulties. Acetaminophen (Tylenol) is preferred for pain and temperature control. *A* is appropriate with any drug and is not the best answer, considering the child's surgery. *B* is not the best choice because 5 gr. is an appropriate dosage—the usual dosage of ASA is 1 gr. per year of age. *C* is normally an appropriate use of the drug, but not considering the child and her surgery.

109. *D.　Allow Mary to talk about her surgery and play out her anger with dolls.*

D is the correct answer because it allows for ventilation of feelings and anxiety. *A* is inappropriate in that the nurse is advising the mother to bribe the child to forget and because it does not deal with the primary problem. *B* is a poor choice in that it will increase the mother's anxiety and guilt and do nothing for the child. *C* is a poor choice in that it involves ignoring the problem, which is not an appropriate response.

110. *C.　Increased independence.*

C is the correct answer since this is the behavior that the mother should *not* expect. The child has been through a traumatic experience and may regress, showing increased dependency. *A, B,* and *D* are all behaviors that may be seen with young children after hospitalizations. The mother should be warned to expect them but told that they are usually self-limiting and will decrease as the child begins to feel safe again.

111. *D.　Temperature elevation of 101°F (rectally) and pain in the right ear.*

D necessitates a call to the physician because the symptoms indicate an ear infection. *A* is a low-grade temperature and dysphagia is to be expected for several days after a tonsillectomy and adenoidectomy. *B* and *C* are also normally seen after surgery as healing progresses and do not necessitate bringing the child to the clinic or to a physician.

112. *C.　Amanda's previous experiences with illness and hospitalization.*

C is the correct answer. Assessment of Amanda's situation and her previous experiences is essential; this will be a hospitalization involving several painful diagnostic procedures and treatments. The nurse will have to have data on which to base her approaches. *A* is incorrect in that the child is just being admitted; it will be appropriate for later on in her hospitalization. *B* is incorrect in that initially the nurse is concerned with Amanda, but this is the next

priority after assessing Amanda. **D** is incorrect as the immediate nursing priority, but again can be assessed at a later date.

113. B. *Prevention of infection.*

B is the correct answer. The data reveal a WBC of 12,000/μL but there is no way to tell how many of these are effective WBCs in the immune process, so prevention of infection is the most important goal. **A, C,** and **D** are important to consider, but only after the nurse has taken steps to prevent infection.

114. A. *Emphasize that there will be little pain.*

A is the correct answer in that pain *will* occur with the procedure and the nurse should always be honest in preparing a child for a procedure. **B, C,** and **D** reflect true facts about bone marrow aspiration. Some children do not require premedication as they become more accustomed to the tests, but for the first such test medication is used.

115. C. *Sternal.*

C is the correct answer. With children, the sternal site is not frequently used, although it can provide the needed tissue. It is not used because there is too great a danger of missing the site and causing a pneumothorax. In addition, it is a very frightening site for the child. **D,** the iliac crest, is the preferred site with **A,** vertebral, and **B,** tibial, used less frequently.

116. C. *Respirations of 12 and decreased breath sounds.*

C are the manifestations that should cause the greatest concern. Note that the child received three drugs prior to the test, all of which affect respirations. Their depressant effect, plus the fact that the child's normal respirations were 24 on admission, should cause concern. **A** indicates bleeding, but not too great an amount; it should be watched, particularly since the platelet count is 100,000/μL. **B** indicates the child is doing well and shows she is beginning to wake up and respond. **D** indicates normal vital signs for the child based on the admission vitals.

117. B. *Elevate the head of the bed to aid respiratory effort.*

B is the correct answer. Elevation of the head of the bed aids respiratory effort by providing for maximum expansion. **A** is incorrect because there is presently no indication of hemorrhage. **C** is incorrect for the same reason; there is no evidence of shock. **D** is incorrect for two reasons. There are no data to support shock; second, providing more covers is inappropriate in that it will lead to peripheral vasodilation, causing shifting of blood from vital organs to the periphery.

118. D. *Assess for cardiac arrhythmias.*

D is the incorrect intervention and, therefore, the best answer. Arrhythmias are not common with these two drugs. **A** is appropriate due to the constipation that vincristine causes. **B** and **C** are also seen with vincristine due to peripheral neuropathies seen with the drug.

119. D. *CBC.*

D is the correct answer due to the fact that chemotherapeutic cancer drugs interfere with rapidly reproducing normal cell activity; the bone marrow is one of the most rapidly reproducing tissues. **A** is incorrect in that it reflects primarily liver activity and the liver will not be affected quickly. **B** is incorrect due to the LDH, which is an enzyme that is usually assessed with cardiac function—that is, after a myocardial infarction. **C** is incorrect in that these levels are

dependent on many other factors, such as hydration and excretion; the chemotherapy will have an indirect, rather than a direct, effect on them.

120. **C.   Omelet, Jell-O, and milk shake.**

  C is the best answer. The eggs and milk shake provide protein and calories, which Amanda needs, and the Jell-O is bland and cold and will not cause increased ulcer pain. **A** is incorrect in that the spices in pizza and the chewing of cookies will increase pain due to the ulcers and therefore lead to less intake and further nutritional deficit. Tomato soup in **B** is hot, and cold liquids are better with stomatitis. In **D,** although the foods would be attractive to a 7-year-old, the hot dog and potato chips would cause increased pain.

121. **C.   Using puppets and dolls to explain procedures and permitting the child to handle the hospital equipment.**

  C is the best answer. The objects used in explaining the procedures are nonthreatening and appropriate for her age. **A** is incorrect in that it is an unethical action, as well as being very threatening and frightening for the child. **B** is inappropriate in that the child of 7 has very little knowledge of the body and is unable to think abstractly and to transfer knowledge from the model to herself. **D** is incorrect in that the question asked for activities to explain procedures; these activities are age appropriate but will do little to respond to specific illness stresses.

122. **D.   Administer the medication.**

  D is the correct answer. It is important to get the IV antibiotic in before the IV is finished. **A, B,** and **C** are all done later. Probably the best order is: **D,** give the medication; **B,** discontinue the IV; **C,** let her visit with her brother; and then **A,** change the dressing later when there is time, since the sore is healing well.

123. **A.   Suggest that the topic be open conversation so that both information and feelings about leukemia can be discussed at home.**

  A is the best answer since it allows exchange of information and feelings among the family; at 11 years of age, the brother will need information and support and these actions will help prevent fantasizing. **B** is a poor choice in that it says that the brother is only concerned with whether or not he will be affected, which is a very limited and untrue view of the situation. **C** is a poor choice in that only information is being shared; **A** is the more inclusive answer. **D** is simply avoiding the situation and does not prepare the brother for illness and the possible death of his sister, which will lead to further problems later.

124. **D.   "What do you think will happen?"**

  D is the best choice as it allows Amanda to express other feelings and allows the nurse to gather further data. **A** makes an unwarranted assumption that being lonely is what Amanda is concerned with; also, the nurse is guaranteeing something she may be unable to follow through with. **B** completely shuts off the conversation and allows no further discussion. **C** is incorrect because "why" questions are difficult for the young school-age child. It asks her for a higher level of abstraction than **D.**

125. **B.   That Mrs. Curtis is looking for a reason to sue, particularly since her husband is a lawyer.**

  B is the exception and therefore the best answer. Few parents are looking to sue for malpractice. **A** is seen in learning of the diagno-

sis and not being able to do anything about it; in cases of leukemia, the parents often feel guilty and angry at themselves for not taking the child to the doctor earlier and will project this anger onto the professional staff. **C** is an appropriate consideration because there may be other reasons for her anger. **D** is a normal reaction on the part of staff on being criticized and this recognition of her own responses should be part of the nurse's considerations.

126. **A.** ***Explain that Amanda needs her peers and school work.***
A is the best choice; with remission, the child can usually return to school and needs peer contact and normal developmental tasks to continue. **B** is inappropriate because the case situation reveals that all immunizations are up-to-date. In addition, the child is on immunotherapy, which decreases resistance, and immunizations must be done very carefully to prevent complications. **C** reflects jumping to conclusions without data to support them. **D** is incorrect because it is untrue, as noted in the explanation for **A.**

127. **A.** ***Sickle cell anemia is due to heterozygous inheritance of a gene responsible for production of abnormal hemoglobin.***
A is the statement that is *not* true and is therefore the best answer. Sickle cell anemia is due to homozygous, not heterozygous, inheritance. **B** is true; sickle cell anemia is primarily seen in blacks and people of Mediterranean ancestry; **C** is true in that conditions of hypoxia, such as increased exercise and high altitudes, cause sickling. **D** is true in that the chronic anemia and multiple infections associated with sickle cell anemia often lead to an early death.

128. **A.** ***Oxygen by mask and cold applications to leg to relieve pain.***
A is the best answer because cold causes further sickling and increases acidosis, which also increases the sickling process. **B** is an appropriate action in that powerful analgesics are often required with crisis pain. **C** and **D** are appropriate actions because the main treatment of crisis is pain control and fluids to decrease blood viscosity. Bedrest is to prevent further energy expenditure and more sickling.

129. **B.** ***A 7-year-old boy with juvenile rheumatoid arthritis.***
B is the best choice because it provides a child of like age and sex who also has decreased mobility. **A** is incorrect because the child seems to be developing an infection and Ronny is very susceptible to infections. **C** is incorrect because the patient is a girl and modesty is important to the school-age child. School-age children are usually placed in a room with children of the same sex. **D** is incorrect because the child has an active respiratory infection, to which children with sickle cell anemia are particularly susceptible.

130. **D.** ***Earlobes.***
D is the evaluator that is incorrect. It is impossible to assess a black child's color well by checking the earlobes. **A, B,** and **C** reflect areas of the body with decreased pigment, which makes it possible to assess Ronny's color. Another place would be the conjunctiva.

131. **D.** ***Increased fragility and decreased life span of RBCs.***
D is the correct answer. The sickle hemoglobin leads to a shortened life span of about 15–20 days for RBCs; this contrasts with the normal RBC life span of 120 days. **A** is an incorrect choice because diet is the primary cause of iron deficiency anemia and there

are no data indicating that Ronny's diet is poor. *B* will lead to decreased RBC production and is not related to RBC life span. *C* is incorrect because there is no blood loss with crises.

132. *D.* ***Question the order because the nurse knows vitamins are of questionable value in sickle cell anemia.***

D is the correct answer. Vitamins are of use only in circumstances of normal dietary deficiency and there is no indication of this in the case situation. Folic acid in no way halts or alters sickling. *A, B,* and *C* indicate poor analysis of the situation and the treatment rationale and are therefore incorrect.

133. *D.* ***Playing checkers and simple card games with his roommate.***

D is the best choice because the activities are appropriate developmentally and also involve him in some peer contact. *A* is a very isolating activity, presents the problem of finding appropriate shows, and does not involve peer contact. *B* is incorrect because phone calls are a very transitory activity and do not solve the problem of continued diversion. *C* could be appropriate, but since the case situation indicates that Ronny has been missing school and is a grade behind, reading might be a frustrating activity. Also, for 8-year-olds games hold more attraction than books.

134. *D.* ***Prevent dehydration, which increases sickling of RBCs.***

D is the correct answer because the fluids help prevent dehydration and the acidosis seen with it, which will increase sickling. *A* is incorrect in that increasing blood volume will not deal with the problem. *B* is incorrect because it is not the primary reason for increasing fluids (the primary reason is the relationship between hydration and sickling), although a side effect of the fluids will be prevention of urinary tract infections. *C* is not the best choice in that it does not address the hydration goal.

135. *B.* ***Must have the trait.***

B is the correct answer. The genetics of the disease require that the child receive one trait from each parent to express the disease. *A* is incorrect by history and *C* is incorrect physiologically because both parents need not show the disease, although both carry the trait. *D* is incorrect because Ms. Simpson *must* at least have the trait for her child to have the disease.

136. *C.* ***Each child carries the trait, but may not have the disease.***

C is correct because if Ronny's father has the disease, he will pass the gene to all of his children; they will either carry the trait (if they do not also get the gene from their mother, a carrier) or have the disease (if they do get the gene from their mother). If these children do not have the disease, they at least carry the trait. *A* and *B* are incorrect because *all* the children will at least carry the trait. *D* is incorrect because the children's choices in marital partners will determine whether their children will carry the trait.

137. *D.* ***Increased activity.***

D is the correct answer. The symptoms were reported after an episode of playing. *A* is incorrect; there is a history of infections but no current data indicating that one is present. *B* and *C* are also incorrect because there is no indication that either of these factors were present.

138. *B.* ***Apical pulse rate; coordination of voluntary muscles.***

B is correct because tachycardia may indicate carditis, a major manifestation of rheumatic fever. Other signs include heart mur-

mur and cardiomegaly, which result from inflammation, necrosis, scarring, valvular stenosis, and damage to the myocardium. If untreated, congestive heart failure may occur. Chorea, a major CNS manifestation of rheumatic fever, affects prepubescent girls more often than boys. Signs include incoordination of small or fine muscles, twitching, weakness, inattention, and emotional lability. The nurse must assess these factors to establish a baseline from which to evaluate progress. Although *A* indicates a teaching need, it is incorrect because this answer relates to a *causal* factor, not current needs or complications of rheumatic fever. Thus, *A* is post hoc in relation to the question. Also, *A* is threatening to the patient and the mother. *C* is incorrect because the nodules are painless. *D,* which may indicate a teaching need, is incorrect because a 24-hour diet recall is not specifically related to current manifestations of rheumatic fever. A balanced diet is necessary for healing. Thus, the nurse should include this, as well as hygiene, in planning and intervening.

139. **A.** *"During rest or sleep, pains from growth may lessen or disappear, but the joint pains of rheumatic fever do not lessen."*
   A is correct since the two pain types may be differentiated by these characteristics. *B* is incorrect for reasons stated in *A*. *C* is incorrect, as neither of the pain types is restricted anatomically. Rheumatic fever joint pains migrate from knees and ankles to the wrists, elbows, and hips. *D* is not correct because the 8-year-old can define intensity of pain on a scale of 1–10; can point to the region of pain; and can, with a limited temporal concept, generally describe when pain does and does not occur.

140. **C.** *Ten-year-old Marta, who is confined to bed with acute systemic lupus erythematosus.*
   C is the correct answer, as the school-age child (6–12) who has rheumatic fever should be allowed interaction with other school-age children who are the same sex, have similar therapeutic regimens, and have no infectious or communicable conditions. Systemic lupus erythematosus, a long-term noncommunicable inflammatory disease, requires strict bedrest in acute stages and manifests problems and other management needs similar to rheumatic fever. *A* is incorrect because the child has playroom privileges. The school-age child with rheumatic fever may become depressed because he or she is confined to bed and not allowed activity, which is extremely important at this industrious age. *B* is incorrect. The rheumatic fever child *must* be kept from persons with upper respiratory tract infections (e.g., scarlet fever) as reinfection with streptococci increases the threat of additional cardiac damage. *D* is incorrect for the same reason as *A.*

141. **B.** *"Yes, there will be a little pain, but it will soon be over."*
   B is the correct answer. The school-age child fears body injury, needs and can accept honest and specific explanations of procedures and accompanying sensations, and understands the concept of time. Explanation helps the child to maintain control and to cooperate during the procedure. *A* is incorrect. This answer promotes shame and distrust, does not acknowledge the child's needs, is threatening, and intensifies fears. A child of this age is very concerned about peer attitudes and embarrassed by loss of self-control. *C* is incorrect because this answer does not acknowledge the

child's question and is untrue. Acute rheumatic fever patients are confined to bed for prevention of serious cardiac damage and promotion of comfort. *D* does not answer the child's question, threatens the child, intensifies fear, and inhibits the expression of emotions regarding a natural fear. The child should be offered alternative outlets for pain or fear, such as squeezing a hand or counting. If the child does scream or cry, he or she should not be shamed, but reassured that such expression is normal and accepted.

142. *D.* ***Place a bed cradle over Kathy's legs to prevent physical or material contact.***

   D is correct. The child with rheumatic fever and polyarthritis often self-immobilizes and cannot tolerate movement, manipulation, or anything touching the affected areas. A bed cradle prevents the weight of bedclothes on the extremely painful joints. Nursing care is organized to include only absolutely necessary activities, and pain management includes proper body alignment and support of affected joints with pillows. *A, B,* and *C* are incorrect because these actions enhance, not relieve, pain.

143. *D.* ***Restricted mobility, possible bodily injury, and lack of control over self-care.***

   D is the correct answer. The school-age child is industrious in nature and desires to be active. Illness and forced immobility may be regarded as a sign of inferiority, failure, or punishment for a misdeed. The nurse must explain the reasons for immobility; reassure of blamelessness; and allow appropriate outlets for anger, anxieties, and energy in order to prevent depression. A child of this age understands the reasons for immobility and should be given the opportunity to express related emotions. The nurse allows maximum possible self-care independence, as this aids in preventing self-image destruction. Honest explanations and involving the child in planning and implementing care will reduce fear of bodily injury. Although an 8-year-old child is not totally free of separation anxiety, he or she can tolerate parental separation because of progressing reality orientation and desires to form relationships outside the family and home. Therefore, *A* is not the *most* difficult to accept and is incorrect. *B* is incorrect because it is not the *most* difficult problem; however, a child of this age does worry about peer group rejection and peers' attitudes and conversations in his or her absence. *C* is incorrect; it is a concern, but not the *most* difficult aspect of hospitalization.

144. *B.* ***Preventing complications by providing absolute rest; padded bedding; spoon-feeding; and a diet high in calories, protein, and iron.***

   B is correct. CNS inflammation results in choreiform movements involving the voluntary muscles. The child may become incontinent, clumsy, and spastic and may demonstrate twitching, facial grimaces, and other involuntary, purposeless, and irregular movements. Constant movement utilizes large amounts of energy (requiring extra calories, protein, vitamins, and minerals) and is abrasive to bony prominences (requiring frequent skin care). Rest is essential to prevent exhaustion or death from cardiac disease. Injury may result from bed frames and sharp eating utensils; therefore, soft toys, padded beds, and spoon-feeding are provided. *A* is incorrect since bedrest is enforced. *C* is incorrect because the lack

of voluntary small muscle control, inattention, and activity restrictions prohibit these actions. Writing requires coordination of small, complex muscle structures. Deterioration in penmanship promotes anxiety, frustration, inferiority feelings, and tension—which, in turn, exacerbate choreiform movements. As improvement in condition occurs, short study periods are allowed and utilization of muscle structures gradually progresses from large to small. *D* is incorrect because the child with this manifestation: (1) is incapable of self-care; (2) must have a planned, balanced diet and restricted activity; and (3) needs emotional support for a CNS disorder rather than psychological counseling for an emotional disorder.

145. **A.** *Erythema marginatum and cardiac murmur.*
    A is the only *incorrect* indication, so it is the best answer. Erythema marginatum, a major manifestation of rheumatic fever, is a nonpruritic, pink, macular rash appearing on warm body areas. Cardiac murmur signifies carditis, another major manifestation resulting from Aschoff's bodies (inflammatory lesions) on the endocardium, which scar and cause stenosis of heart valves. *B, C,* and *D* are common signs of salicylate toxicity. Aspirin should be given with meals to avoid gastric irritation and bleeding. CNS changes indicate neurologic toxicity. Hyperventilation is a respiratory compensation for metabolic acidosis. Bleeding tendencies result from increased prothrombin time.

146. **C.** *Orthopnea and peripheral edema.*
    C is correct because these are signs of congestive heart failure (CHF), which may occur from inflammation, scarring, and necrosis of the myocardium and consequent inability of the heart to function as a pump. Other signs of CHF include neck vein distention, tachycardia, generalized edema, weight gain, hepatomegaly, dyspnea, and rales. *A* is incorrect, as *increased* ESR indicates increased inflammation and increased severity of rheumatic fever. *B* is incorrect because 9000/μL is within the normal range for the WBC count. *D* is incorrect because the *presence* of C-reactive protein in plasma indicates inflammation. *Increased,* not normal, heart rate indicates increased severity of rheumatic fever.

147. **D.** *Report possible infectious processes suggested by high temperature elevations.*
    D is the only *incorrect* action, so it is the best answer. Fever does not usually accompany infection during long-term steroid therapy because the drugs are immunosuppressive, inhibit the inflammatory response, and mask signs and symptoms of infection. The nurse monitors for other indicators, such as lethargy and anorexia. *A* includes important nursing actions. Steroids increase susceptibility to infection by suppressing the immune system and causing a negative nitrogen balance. The urine is checked for glucose because steroids enhance gluconeogenesis and produce insulin antagonism, with subsequent hyperglycemia and spilling of sugar in the urine. *B* is also relevant. Steroids increase secretion of gastric hydrochloric acid, which may aggravate or create an ulcer. The nurse monitors for melena and hematemesis. Extreme mood swings (euphoria, severe depression) may result from CNS effects. *C* is important, as steroids promote sodium and water retention, edema, and weight gain. The nurse must also assess for indications of hypokalemia, as potassium is replaced in the cell by so-

dium and the K⁺ moves into the vascular space and is excreted in the urine. Steroid therapy may also cause Cushing's syndrome, with manifestations including skin striae, acne, hirsutism, "moon face," and "buffalo hump." It is important for the nurse to explain to the patient and family that these problems disappear when therapy is discontinued.

148. **A.   Observe Kathy for headache, nausea, muscular weakness, disordered vision, and changes in apical heart rate and rhythm.**

   A is the correct answer. The signs and symptoms listed may indicate life-threatening digitalis toxicity, which may be induced by decreased serum potassium as a result of the potassium-depleting effects of furosemide (Lasix) administration. The nurse assesses for arrhythmias, bradycardia, tachycardia, or a rapid-weak pulse on arising, as well as for colored or blurred vision, dilated or contracted pupils, and confusion. A resting apical rate (assessed for 1 full minute) less than 60 or greater than 100 bpm, or indications of any other signs or symptoms, warrants immediate notification of the appropriate health team member (prescribing doctor). **B** includes important observations related to furosemide therapy (blood volume and circulation), but is not specific to digitalis toxicity. (Comparison of apical-radial pulses *would* be indicated, but not radial pulse rate alone. An apical pulse rate greater than the radial rate may indicate ineffectual heartbeats and decreased cardiac output.) **C** is incorrect because these signs may indicate peripheral circulatory or neurologic problems, but are not specific to digitalis toxicity. **D** is not correct; although anorexia may accompany signs and symptoms of digitalis toxicity, the other listed signs do not. (In digitalis toxicity the temperature may be decreased, the face pale, and the pulse weak.)

149. **C.   The solution is infused at a rapid rate.**

   C is the only *incorrect* activity and is therefore the best answer. Rapid IV administration of potassium may cause cardiac arrhythmias and pain or burning at the infusion site. KCl is infused at a slow to moderate rate in concentrations of not more than 40–60 mEq/L. **A** is a correct action. The nurse ensures adequate urine output prior to IV KCl infusion in order to prevent toxicity. If kidney function is impaired and daily urinary volume is low (less than 500–600 mL), potassium should not be administered. **B** is a correct action. Cardiac arrhythmias may occur with IV KCl infusions; therefore, ECG monitoring and repeated serum potassium determinations must be ensured. **D** is also a correct action. KCl should not be administered undiluted and must be *thoroughly* mixed in solution to prevent toxic effects of unmixed (or bolus) infusion. Thorough mixing is accomplished by complete inversion (turning upside down) of the KCl in solution at *least* eight rotations.

150. **C.   Reading, playing jacks, and playing guessing games.**

   C is the answer because it includes an inappropriate action, playing jacks, which requires physical exertion and the use of small muscles. Reading and playing guessing games are appropriate actions. **A, B,** and **D** are appropriate options, since all of these activities require minimal physical activity. To prevent cardiac complications, children recovering from rheumatic fever are permitted only gradual activity increases as inflammation subsides.

151. **A.** *"Kathy must have antibiotic therapy indefinitely and exactly as ordered."*

    A is the correct response. Antibiotic administration, either monthly (IM) or daily (PO) is essential for several years (or for a lifetime) after the initial attack to prevent recurrent rheumatic fever attacks and consequent serious cardiac damage or rheumatic heart disease. *B* is incorrect. Recurrent attacks may occur after exposure to group A beta-hemolytic streptococci. The child must be kept from persons having upper respiratory infections, must maintain personal hygiene, and should live in a warm, uncrowded environment, if at all possible. *C* is incorrect. The child recovering from rheumatic fever is restricted in both physical and mental activities. These children usually have short study periods in special schools or at home. No physically competitive sports or individually exertive activities are allowed. *D* is incorrect. To prevent complications and promote healing, a balanced diet high in protein, calories, and iron is essential.

152. **B.** *Oxygen.*

    B is the correct answer. The blood gases reveal respiratory acidosis and oxygen will be needed as part of the therapy. Sodium bicarbonate will probably also be needed to correct the acid-base problem. *A* is incorrect since asthma is a lower respiratory problem, not an upper airway obstruction. Epiglottitis is an example of an upper airway problem where a tracheostomy set should be available. *C* is incorrect in that morphine causes respiratory depression and is harmful in this situation. *D* is incorrect because there is no indication of cardiac irregularity.

153. **C.** *Urinary output.*

    C is the answer. Although aminophylline will increase the urinary output, the child is not dehydrated. *A, B,* and *D*—irregularity in cardiac rhythm, rapid pulse, and restlessness—are all signs of beginning toxicity with aminophylline.

154. **C.** *Sitting up.*

    C is correct in that it is the preferred position of asthmatics during severe attacks and it allows for maximum expansion of the lungs. *A* is incorrect because the elevation is too little; this position still allows the diaphragm to minimize expansion of the lungs. *B* and *D* are incorrect for the same reasons; they do not allow for maximum expansion of the lungs.

155. **D.** *"I know it's difficult having him back so frequently. It is upsetting for us and for him and his family."*

    D is the best answer in that the head nurse acknowledges the feelings of the staff nurses and also points out the stresses on the family. *A* is incorrect in that it agrees with the staff and reinforces their attitudes rather than emphasizing the child's difficulties. *B* is poor in that it is an unwarranted assumption and there is nothing in the database to support it. *C* is incorrect for the same reason; it assumes that the teaching has been poor and there is nothing in the database to suggest this.

156. **D.** *Get a sputum culture.*

    D is the best answer. A respiratory infection appears to have initiated this attack and the data provided in the question indicate that it has become worse. The mucus has changed in consistency and color, which along with the increased temperature suggests a bacterial infection. The best choice is to get a culture while the

temperature is elevated. **A** is incorrect as a *first* action; it is done after obtaining the culture. **B** is also an action for later. Because of his age, Tom is not in danger of a febrile convulsion, so antipyretics could wait the few minutes it would take to get the culture. **C** is also appropriate after obtaining the culture.

157. **D.   *Note his urinary output and BUN.***

D is the incorrect action because there is no indication of renal problems in the database. Also, the dosage is not so high that excretion should be a problem. Therefore, it is unnecessary to note urinary output and BUN. **A** and **C** are correct actions in that his history and that of his family reveal multiple allergies and make him at high risk for an allergy to penicillin. **B** is a correct action since Tom is 11 years old and of normal mental development; part of his care will always be explaining his therapy and why it is being done.

158. **A.   *Overprotection.***

A is the correct answer. Tom is 11 and has had asthma for 5 years. He should be more responsible for his care, and there is nothing in his history to indicate that he could not be. **B** is incorrect in that the statement does not indicate regression on either the mother's or Tom's part. **C** is incorrect in that the mother is not expressing hostility directly—that is, she has not said she is angry. **D** is incorrect in that although the mother is expressing normal concern, the fact that Tom is 11 means that he should have some independence and be involved in the management of his own care.

159. **B.   *Prolonged rapid pulse.***

B is the correct answer. Theophylline is a bronchodilator, and a sign of toxicity is prolonged increased pulse. A transient increase in heart rate is expected. **A** is incorrect in that although it is a side effect of cromolyn sodium, it is not indicative of toxicity. **C** and **D** are not signs of toxicity with these drugs, although nausea may be seen with any drug taken orally.

160. **D.   *Has refused to join the Boy Scouts because being in large crowds of other kids might make him sick.***

D is the correct answer in that it shows that Tom is isolating himself from his peers. There is no greater risk of infection in Scout groups than in school. The same precautions should be used in both to minimize respiratory infections. **A, B,** and **C** are all positive actions that indicate his involvement in environmental control of allergens, knowledge of medications, and improvement of respiratory capacity.

161. **D.   *Pain.***

D is the answer; pain is *not* a common symptom of scoliosis. **A, B,** and **C** are all reported with lateral curvature of the spine.

162. **A.   *Urinary system.***

A is the answer. Susan is going to use a brace, so she will not be totally immobilized. There is nothing in her case situation to indicate that she is at high risk for urinary infections. **B** must be assessed because the hip is often affected by the curvature. **C** is seen with anyone undergoing extended treatment for a problem—particularly an adolescent who is undergoing multiple other bodily and psychological changes. **D** is found with scoliosis and is reported in the physical findings.

163. **D.   *Permission to sleep without the brace at night.***

D is the answer. The brace is to be worn at *all* times. **A** is included in the teaching because the brace limits mouth movement, and

chewing raw vegetables and fruits will help prevent jaw mobility difficulties; also, dental hygiene may be impaired to some degree so *sugarless* gum is important to protect teeth while providing jaw action. Cotton clothes, *B,* are appropriate because they allow evaporation of moisture. *C* is an important teaching point in that any brace can lead to skin breakdown.

164. *C.   Parents being the most significant interpersonal influences at this age.*

   C is the answer; in adolescence the peer group becomes more important to the adolescent than her parents. *A, B,* and *D* are all characteristic of identity formation.

165. *A.   Discussing why she is not wearing the brace as instructed.*

   A is the inappropriate response. The nurse is making an assumption without a database; Susan may not have been wearing the brace but other factors could be operating to increase the curvature. *B* will elicit appropriate data, as it will influence her compliance with the regime. *C* is a correct nursing action in that an improperly fitting brace that is not functioning properly could lead to increased curvature. *D* is a necessary action since a main problem with scoliosis is that it causes cardiac shift and compresses the lungs.

166. *C.   Increase of the curvature will further impair cardiopulmonary function.*

   C is a major point; further curvature may lead to Susan being short of breath and fatigued. *A* is incorrect because increasing curvature leads to crowding and nerve impingement on occasion. It is important to note that pain is not an *early* symptom of scoliosis. *B* is incorrect in that paralysis is not seen with scoliosis. *D* is incorrect because at 13 Susan is not past her growth period and the curvature can increase even after cessation of the growth spurt.

167. *D.   "I know it's very hard to be different from your friends. Have they said anything to you about the brace?"*

   D is the most therapeutic answer in that it acknowledges Susan's feelings and also seeks further data before planning any further interventions. *A* is poor in that it contains a threat and scare tactics, which never work for long. *B* is poor in that it does not allow Susan to express her feelings further and also assumes that her friends will understand, which they may not do. *C* is false reassurance; there is no guarantee that the treatment will last only a few months. In addition, Susan is an adolescent with many body changes occurring and this is one more change to which she must adapt. Telling her she will get used to it is a very superficial response.

168. *C.   Cleanse her face several times a day and try a hairstyle off her face.*

   C is the correct answer. Cleansing of the face and hair will reduce the oil. *A* is incorrect in that birth control pills are not used for acne. *B* is not correct because there is no proven association between a diet of fatty foods and acne. *D* is incorrect in that it can lead to permanent scarring.

169. *D.   Discuss what else might be worrying Mrs. Black at the moment.*

   D is the best approach—assessing what the mother's feelings and concerns are at present may help identify issues that the nurse

can help Mrs. Black resolve, along with understanding more about normal adolescent behaviors. *A* is a poor response in that it is a superficial statement that cuts off further discussion and is of no help to the mother in dealing with her daughter. *B* is also a poor response since it leaves no opening for discussion and is of no help to the mother, who is asking for help. *C* is incorrect because the nurse must be concerned with the whole family; Mrs. Black's problems will affect her daughter.

170. *B. She is attending school regularly.*

B is a good indication that Susan is carrying on with her normal activities and is not isolating herself from her friends. *A* does not indicate follow-through in her treatment. *C* indicates poor adaptation in that she is isolating herself from her peers and activities; this will be harmful to her psychological health and self-concept. *D* indicates regression, dependence, and poor adaptation. Susan will have to have some help with the brace, but total dependence is unhealthy.

171. *C. Oral hypoglycemics and diet for control when utilized precisely.*

The answer is *C*. Juvenile diabetes mellitus is characterized by a relative lack of insulin, and oral hypoglycemics are of *no* use and may even be harmful. A diet control will be necessary, as with all diabetics. *A* is true; the juvenile diabetic has to rely on injectable insulin. *B* is also true; in contrast to mature-onset diabetes, juvenile diabetes is very sudden and often occurs with ketoacidosis as the first symptom. *D* is a correct approach in that good control should not hinder the child from growing and developing normally. If the child is in very poor control, he may be smaller and thinner than his peers.

172. *B. Giving her several glasses of water for hydration.*

B is the correct answer. Theresa is exhibiting signs of ketoacidosis due to hyperglycemia, which has led to dehydration. The best action after notifying the physician is to begin rehydrating her; note that she is difficult, but not impossible, to arouse and should be able to swallow. *A* and *D* will add further to her hyperglycemia and be of no use; she does not need glucose. *C* is inappropriate until further assessment is done, probably in the hospital where blood glucose levels can be monitored while insulin is being given. In this home situation, hydration is the most appropriate action.

173. *B. Infection.*

B is the correct answer since infection can precipitate hyperglycemia. Theresa's case situation indicates a recent urinary tract infection plus the history of the flu. *A* could be correct but there are no data to support this inference. *C* is incorrect because increased physical activity leads to hypoglycemia, not hyperglycemia. *D* is a possibility since insulin demands are increased with puberty; however, the data most directly support infection as the cause of Theresa's ketoacidosis.

174. *C. Decreased urine specific gravity.*

C will not be expected; with dehydration the specific gravity will be increased, not decreased. *A* will be expected since the cause of ketoacidosis is hyperglycemia. *B* is seen with hypertonic dehydration, in contrast to the tenting seen with isotonic dehydration. *D* is seen in that the nausea, vomiting, and decreased fluid intake lead to potassium depletion.

175. D. *Give her orange juice with extra sugar.*

D is the appropriate action since Theresa is exhibiting the classic signs of hypoglycemia. She also had regular insulin an hour previously and onset of action is 30–60 minutes after administration. Glucose is needed. *A* is incorrect; the data indicate hypoglycemia; insulin will exacerbate the condition. *C* provides glucose, but glucose is needed immediately—preferably by the oral route. *B* should be done after giving the glucose and noting its effect so the physician is aware of the patient's status.

176. D. *Pizza slice, milk, Jell-O with fruit.*

D is the best answer because it provides more protein and less carbohydrate than the other choices. *A*, *B*, and *C* have more carbohydrate than *D* does.

177. D. *Talk with Theresa about her feelings of exclusion and assess further how she is relating to her friends.*

D is correct in that Theresa needs to ventilate her feelings; the nurse also needs to gather more data to aid in planning how to manage her noncompliance. Rebellion and noncompliance are common in adolescents with chronic diseases, particularly since these diseases make them different from their peers at a time in development when similarity to peers is important. *A* is incorrect in that the patient is expressing her feelings and these need to be explored, not ignored. *B* is indicated, but at a later time when further data have been collected and appropriate interventions planned. *C* is not appropriate at this time; no data presented indicate that a lack of knowledge about diabetes is causing the noncompliance.

178. D. *Have her help with tasks such as settling the younger children down at bedtime.*

D is the correct answer. In this way the nurse is including her in the activities of the unit and entrusting her with responsibilities, which will enhance her feelings of competence. *A* is helpful but does not address the issue of competence. *B* will be appropriate later, in helping her parents understand how she feels about her disease, but is not immediately helpful in promoting her self-concept. *C* is incorrect because although peer contact is important, it is not immediately useful in dealing with the feelings of incompetence.

179. B. *Meticulous foot care.*

The answer that reflects the *least* important aspect of the teaching plan is *B.* Foot care and vascular changes are less important in the juvenile diabetic than in the adult. The changes will not be seen for several years; remember from the history that Theresa has been diagnosed for only 3 years. *A* is appropriate to include in the teaching plan because sudden increases in physical activity cause hypoglycemia and increasing glucose intake beforehand is appropriate. *C* is useful in preventing hyperglycemia and ketoacidosis, while *D* is also taught to patients to help prevent ketoacidosis. This teaching is done with strict guidelines, but the adolescent can learn to manipulate his or her insulin to some extent.

180. B. *Bedtime snack of cheese.*

B is the best suggestion. Theresa is experiencing nighttime hypoglycemia due to her peak action time of NPH insulin. A bedtime snack of cheese, a slowly digesting protein, will counterbalance the peak action of the insulin. *A* is incorrect because the symptoms are

associated with hypoglycemia. More insulin will compound the problem. *C* is incorrect because the cookies are high in carbohydrate and will not be available at peak action time for insulin. The diet soda has no sugar to combat the hypoglycemia. *D* is incorrect because exercise enhances insulin-induced hypoglycemia.

8. Young Adult and Reproductive Years

Case Management Scenarios and Critical-Thinking Exercises

Medical Situation: Reproductive health maintenance for adolescents.

■ *Nursing Problems:*
Altered health maintenance.
Altered immune response.
Impaired skin integrity.

The nurse is asked to plan a course about reproductive health maintenance for a group of 14–16-year-old females. General concepts of anatomy and physiology and the adolescents' level of growth and development help structure the course outline and method of presentation.

1. The nurse can best encourage adolescents' interest and participation in the class by:
   A. Tuning in to her or his own feelings and values about the topic.
   B. Emulating the adolescents' dress and language.
   C. Sharing her or his own experiences as an adolescent.
   D. Explaining how this knowledge will help them later throughout their life span.
2. The nurse instructs the group that during the menstrual cycle, the fertile time occurs:
   A. On day 14 of the cycle.
   B. Just before menstruation.
   C. When the ovum is present.
   D. On day 1 of the menstrual cycle.

3. The cervical mucus of one of the students is clear, copious, like egg white, stretchable to 5 cm, and slippery. The student knows that at this time she:
   A. Is pregnant.
   B. Has a mild leukorrhea.
   C. Is about to menstruate.
   D. Is about to ovulate.

4. Based on the calendar method, the next probable fertile period of a woman whose longest cycle is 35 days and whose shortest cycle is 26 days includes days:
   A. 8 through 24.
   B. 15 through 17.
   C. 12 through 21.
   D. 13 through 17 or 18.

5. The nurse instructs the group that which skin change occurs both during pregnancy and with the use of oral contraceptives, but may not fade after the pill is discontinued?
   A. Chloasma.
   B. Palmar erythema.
   C. Telangiectasia.
   D. Pruritus.

6. Women taking oral contraceptives are cautioned that a hazardous side effect of these contraceptive is:
   A. Ectopic pregnancy.
   B. Cholelithiasis.
   C. Gastroenteritis.
   D. Thromboembolic disease.

7. Jan, one of the students in the group, confides to the nurse that she is pregnant. She is wearing loose clothing and is dieting in an attempt to keep the pregnancy a secret. Jan and her boyfriend, Mike, have decided to "go through with" the pregnancy and then place the baby for adoption. They think they have "lues" and ask the nurse where they can go for treatment. They decide to attend a Teenage Parent clinic. If they have just acquired lues, the clinic nurse expects to obtain the following history:
   A. Appearance of a chancre.
   B. Infection 1 week ago.
   C. A positive test for serology.
   D. Pain in the joints and loss of position sense.

8. Mike and Jan's histories and physical exams reveal that they both have secondary syphilis and that she is 18 weeks pregnant. The nurse knows that:
   A. Lesions of secondary syphilis are not infectious.
   B. Obliterative endarteritis is a possible complication.
   C. Gummas will appear now throughout all body tissues.
   D. The spirochete crosses the placenta after the 16th week of gestation.

9. Which is true about gonorrhea?
   A. It can result in pharyngitis, tonsillitis, perianal pruritus, and burning.
   B. Ophthalmia neonatorum is the only effect of congenital gonorrhea.
   C. Small lesions or blisters occur in the vagina and on the glans penis.
   D. It can be transmitted only by sexual contact.

Medical Situation: Normal pregnancy: health supervision.

■ *Nursing Problems:*
Anxiety.
Altered comfort pattern (discomfort).
Impaired digestion.
Impaired family coping.
Risk for infection.
Altered nutrition: more than body requirements.
Altered parenting.
Impaired physical mobility.
Sexual dysfunction.
Ineffective thermoregulation.

■ *Chief Complaint:*
Wendy Wilkes is a 22-year-old married black woman who states she is "8 weeks pregnant by dates."

■ *Present Status:*
Wendy has been amenorrheic times two periods. She has had "mild nausea" with no vomiting, mastalgia, urinary frequency, and fatigue for about 5 weeks. This first pregnancy was planned.

■ *Past Health History:*
Wendy states she has always been healthy. She experienced menarche at age 11. Her menstrual cycles have always been regular, occurring every 28–30 days. She disclaims dysmenorrhea. She has never used contraception. She has kept her immunizations current, seldom has colds, and has never been hospitalized.

■ *Family History:*
Hypertension has been diagnosed in both sets of grandparents and in Wendy's mother. One grandparent died of a cerebrovascular accident at age 56. Wendy's youngest brother died in a sickle cell crisis at age 7. Her husband's family denies sickle cell disease. Neither she nor her husband has been tested for the presence of sickle cell trait. Her husband, Ed, is an auto mechanic. The couple plan to attend childbirth classes.

■ *Physical Exam:*
Pregnancy test: Positive. Rubella titer: 1:6.
Vital signs: Temperature: 37.1°C. P: 76. R: 18. BP: 134/86.
Weight: 140 pounds. Height: 5 feet, 6 inches.
Findings within normal limits for other organ systems.
Pelvic examination reveals uterine enlargement coincident with 8 weeks' gestation; Hegar's and Chadwick's signs positive; gynecoid pelvis; pelvic measurements adequate.

■ *Psychological Status:*
Wendy states that she and her husband are pleased and excited about the pregnancy. This will be the first grandchild for both families.

10. Wendy's rubella titer is 1:6. In planning Wendy's care, the nurse needs to know that Wendy:

    A. Has had rubella.

    B. Is presently infected with rubella.

    C. Is a candidate for rubella vaccination shortly after birth.

    D. Must wait until she weans the baby before she can get the vaccine.

11. While being prepared for a pelvic examination, Wendy confides in the nurse, "I can't stand this part of the examination. It's so hard to relax." Which nursing action *best* assists Wendy to relax her perineal structures?

    A. Coach her to keep eyes, mouth, and hands open.

    B. Give Wendy a hand or an exam table handle to squeeze.

    C. Provide ongoing explanation during the examination.

    D. Ensure privacy and drape her comfortably.

12. Wendy works as a general surgical scrub technician in the local hospital. The nurse will most likely recommend that for the duration of her pregnancy Wendy take:

    A. Only short surgical cases to avoid long periods of standing.

    B. Frequent walks or rest with legs up between cases.

    C. Only "clean" cases.

    D. An assignment on another unit.

13. Wendy states that she and her mother, age 48, are "driving the rest of the family wild." Wendy says that both of them need to change clothes more than once a day because of bouts of perspiration and prefer to keep the home thermostat set on low. Which nurse demonstrates the *best* understanding of the physiologic phenomenon described?

    A. Nurse L: "Mothers tend to identify with their daughters during pregnancy."

    B. Nurse M: "It could be the flu bug. Does anyone else in your household show these symptoms?"

    C. Nurse N: "This is often seen when one is anxious. Let's talk about what is going on with you two."

    D. Nurse O: "Changes in female hormone levels occur in both pregnancy and menopause."

14. The nurse knows that Wendy has understood anticipatory guidance regarding heartburn when Wendy states that she:

    A. Avoids bending at the waist and does "flying" exercises.

    B. Eats some unsalted, butterless popcorn before getting out of bed in the morning.

    C. Has stopped eating starch or red clay.

    D. Gets plenty of fluids, exercise, and foods containing roughage.

15. Which statement to Wendy *best* demonstrates the nurse's knowledge of placental function?

    A. "Lie on your side to rest or sleep."

    B. "Your baby will obtain all the nutrients he or she needs from your stores."

    C. "Avoid taking any drug unless it is prescribed by your physician."

    D. "Don't worry about your bladder infection. The urinary and reproductive tracts are not connected."

16. When Wendy's husband asks about the best remedy for leg cramps, the nurse instructs him to:

    A. Massage Wendy's leg to relax the calf muscle.

    B. Report the incident to Wendy's physician.

    C. Help Wendy to place her full weight on the affected leg and lean forward on it.

    D. Direct Wendy to point the toes of the affected leg, to stretch the cramped muscle.

17. Wendy is now in her third month of pregnancy. Anticipatory guidance, coincident with the patient's readiness for learning, is a significant intervention in the plan of care for the pregnant woman. The choice of which topic demonstrates that the nurse knows the needs of women during the first trimester?

    A. Feelings and attitudes toward pregnancy.

    B. Preparation for labor.

    C. Fetal growth and development.

    D. New role as mother, and the new baby.

18. Wendy is given written instructions to notify her physician or midwife immediately if she or her family notices any signs and symptoms of complication. Which is not included?

    A. Severe, persistent headache.

    B. Tight finger rings and puffy eyes.

    C. Ptyalism and gums that bleed when teeth are brushed.

    D. Dull facial expression or affect.

19. Vitamin C absorption in pregnancy is hindered by the decreased hydrochloric acid in the stomach. The nurse counsels Wendy that the breakfast highest in vitamin C and protein is:

    A. Orange juice, Cream of Wheat with milk, toast and margarine, coffee.

    B. Grapefruit juice, beef liver with scrambled egg, toast, coffee.

    C. Eggs with bacon and fried tomato, toast, coffee.

    D. Banana, French toast with syrup, milk.

20. One pregnant friend of Wendy's does not eat meat. The nurse recommends which diet to meet Wendy's friend's protein needs?

    A. Cottage cheese and tomatoes, peanut butter sandwich, milk.

    B. Macaroni and cheddar cheese, an orange, coffee.

    C. Dried peas, egg sandwich, skim milk.

    D. Dried beans, lettuce and tomato sandwich, milk.

21. During pregnancy, 30–60 mg of supplemental iron is recommended. Wendy, like many mothers, will not eat liver, which is high in iron. The best menu to supply iron is:

    A. Pork chops, sweet potatoes, cooked dried beans, raisin pudding.

    B. Beef sandwich, spinach salad, dried peaches, milk.

    C. Bacon and eggs, toast with butter, milk.

    D. Chicken, white potatoes with gravy, green peas, an orange, coffee.

22. In addition to the normal diet, the daily recommended calorie increase for a pregnant woman is:

    A. 200.

    B. 300.

    C. 800.

    D. 1000.

23. Wendy understands the relationship between good nutrition and its effects on the maternal-fetal unit, so she follows nutritional counseling conscientiously. It is often more difficult for an adolescent to understand and comply. In counseling the teenage pregnant mother about nutrition, the nurse does not need to consider:

    A. Maintenance of a trim figure to promote good body image.

    B. Cultural, economic, and educational background.

    C. Growth needs of the mother and infant.

    D. Individual dietary habits of teenagers.

24. When a pregnant woman prefers nonfat to whole milk, the nurse need not worry about replacing the deficient nutrient if the woman's diet contains sufficient quantities of:

    A. Egg yolk.

    B. Wheat germ oil.

    C. Deep-yellow or green vegetables.

    D. Egg yolk and deep-yellow or green vegetables.

25. Wendy has a dislike for milk. The nurse explains to Wendy that to meet her calcium needs during pregnancy, some items can be substituted for 1 cup of milk. Which one is not a good substitute?

    A. 1 cup of yogurt.

    B. 1½ oz. of hard cheese.

    C. ¼ cup of cottage cheese.

    D. 3 cups of vanilla ice cream.

26. On Wendy's next visit to the clinic, the nurse determines that her nutritional needs are being met because Wendy:

    A. Is at her prepregnant weight.

    B. Has increased the amount of meat in her diet.

    C. Is gaining the desired amount of weight.

    D. Does not eat between-meal snacks.

When Wendy is 28 weeks pregnant, she complains of easy fatigability and palpitations. Hematologic evaluation reveals:

    Hgb: 4 g/dL.

    RBC: 2.5 million/μL; microcytic, hypochromic.

    WBC: Normal.

    Serum iron: 30 μg/dL (N = 90–150 μg/dL).

    Fe-binding capacity: 350–500 μg/dL (N = 250–350 μg/dL).

    Retic/platelets: Normal to high.

Iron deficiency anemia is diagnosed and oral supplementation of iron begun.

---

27. Which foods best ensure Wendy's iron dietary intake?

    A. Oranges, bananas, spinach.

    B. Raw cabbage, brown sugar.

    C. Dried apricots, prune juice, chili con carne.

    D. Enriched milk and milk products.

28. Which direction is most beneficial to Wendy in regard to her iron supplements?

    A. Take daily in 1 dose before a meal, with milk.

    B. Take divided doses with meals, with water.

    C. Take divided doses after meals, with a glass of fruit juice.

    D. Take daily in 1 dose after a meal, with fruit juice.

29. When discussing prenatal classes with Wendy and Ed, the nurse informs them that the *primary* purpose of these classes is to assist the woman (couple) to:

    A. Minimize the amount of analgesia and/or anesthesia needed.

    B. Experience childbirth without fear or pain.

C. Prepare for childbirth.

D. Minimize the possibility of dystocia and medical intervention.

30. The nurse encourages Wendy and Ed to write down any questions they have. Ed says that the Lamaze instructor spoke of the "gate control" theory about pain control. The nurse explains that the "gating" theory refers to the mechanism whereby:

A. Nonpain signals from other parts of the nervous system can greatly alter the degree of transmission of pain signals.

B. In the presence of pain signals, learned responses can override and eliminate the perception of pain, that is, shut the gate.

C. Relaxation of voluntary muscles dulls pain receptors.

D. Pain control is with analgesics.

31. During her seventh month of pregnancy, Wendy tells the nurse about her negative feelings toward motherhood and sex, and her doubts about her husband's qualifications as a father. Wendy is:

A. In need of a psychiatric referral.

B. A candidate to become an abusing parent.

C. Trying to resolve previously repressed conflicts.

D. Attempting to shock the nurse.

32. During her eighth month, Wendy says that her "panties are wet." She is concerned that her "water bag broke." To assess for ruptured membranes, the nurse uses the phenaphthazine (Nitrazine) test. To perform the test, the nurse:

A. Puts on a sterile glove and inserts litmus paper into the vagina, placing it near the cervix.

B. Does a finger stick and draws blood into a capillary tube to be centrifuged.

C. Places a drop of urine on a slide and mixes it with antiserum for 30 seconds.

D. Collects urine for 24 hours, storing it in ice, and sends it to the laboratory.

33. Nursing care for a mother with premature ruptured membranes (PROMs) includes frequent assessment for:

A. Fever.

B. Cervical changes.

C. Coagulopathy.

D. Fetal lung maturity.

34. Wendy arrives on the labor unit. As the nurse is taking Wendy's history and performing an admission examination, Wendy tells the nurse that she is "afraid of the doctor's examinations." Which is the *best* response the nurse could make?

A. "Our patients consider this physician to be one of the best."

B. "Would you like to talk about what you mean when you say you are afraid?"

C. "Don't be afraid. I'll be here with you during the examination."

D. "It's not unusual for new mothers-to-be to be afraid."

35. While Wendy is in labor, the nurse performs Leopold's maneuvers. Leopold's maneuvers refer to palpation of the uterus through the abdominal wall in order to:

A. Determine fetal presentation, position, lie, and engagement.

B. Assess frequency, duration, and quality of contractions.

C. Measure the height of the uterine fundus.

D. Estimate the fetal weight.

36. The nurse who is responsible for determining priorities for Wendy's care needs to know that the woman who labors unattended or

the woman who is discharged from the hospital 12–24 hours after birth without verbalizing understanding of discharge teaching, and then develops a postpartum complication at home, can initiate legal action for:

A. Abandonment.

B. Battery.

C. Ethical dereliction of duty.

D. Prudent behavior.

37. Wendy gave birth 1½ hours ago. She received no analgesia or anesthesia. She is alert and physically active in bed. She says she needs to urinate. The nurse's *most* therapeutic response is:

A. "I'll walk you to the bathroom and stay with you."

B. "You can get up any time you want to now."

C. "Make sure you wash your hands before and after, and wipe yourself once with each tissue from front to back."

D. "Lean forward a little as you void. This will keep the urine off your stitches and make you more comfortable."

38. To provide accurate anticipatory guidance, the nurse needs to know that primary breast engorgement results from:

A. A stasis of milk in the breast.

B. Increased circulation of blood and lymph in the mammary glands.

C. Increased fluid intake (oral and parenteral) during the perinatal period.

D. Increased milk production with each successive pregnancy.

39. The following behavior *best* describes the postpartum mother following a normal birth:

A. Assertive and independent.

B. Difficulty with hearing, retaining, and making connections.

C. Eager to make decisions.

D. Preoccupied with her own needs.

40. The postpartum mother is considered to be recovering for the first 2 hours after birth. Her fundus should be:

A. Firm, three fingerwidths above the umbilicus.

B. Firm, below the umbilicus in the midline.

C. Slightly boggy, two fingerwidths above the umbilicus.

D. Slightly boggy, below the umbilicus in the midline.

41. In the taking-in phase, the nurse expects that the mother is very concerned about:

A. The normalcy of her infant.

B. Herself and her needs.

C. The baby's need for nourishment and touch.

D. Her husband's response to the baby.

42. What is least relevant in decreasing discomfort of an episiotomy?

A. Application of an ice pack soon after repair reduces the amount of discomfort experienced later.

B. Sitting down on and getting up from a chair with the posture straight and the buttocks together decreases discomfort.

C. Massaging the perineum promotes absorption of the hematoma.

D. Warm sitz baths or heat lamp treatments ease episiotomy discomfort.

43. In doing health teaching, the nurse usually recommends that parents refrain from sexual intercourse for 4–6 weeks following birth, or until lochia has ceased, because:

A. Of the possibility of discomfort and infection in the mother.
B. This is an unsafe time to become pregnant.
C. It could tear the episiotomy site.
D. It could delay involution of the uterus.

Medical Situation: Normal pregnancy.

■ *Nursing Problems:*
Anxiety: related to loss.
Altered bowel and urinary elimination.
Altered cardiac output.
Altered comfort pattern (discomfort, fear).
Impaired digestion.
Impaired skin integrity.

■ *Chief Complaint:*
Debby Primings is a 26-year-old white woman who is 4 months pregnant.

■ *Present Status:*
Ms. Primings has been experiencing morning sickness at least 4 times a week since the sixth week of pregnancy. She usually feels nauseated and vomits about 30–40 mL of clear mucus. Remaining in bed and eating a cracker very slowly has helped to alleviate some of her feelings of sickness. She has gained about 5 pounds. Within the last month she has experienced less urinary frequency.

■ *Past Health History:*
Ms. Primings has been fairly healthy throughout her life. She started her menstrual periods at age 15. She usually has a period every 30–40 days but has never been on a regular schedule. She was on birth control pills for about 6 months, but she stopped taking them because they made her "feel sick" in the morning. Ms. Primings states that she usually has one or two colds per year. Every once in a while she gets the flu. Other than a slight cold or the flu, she "doesn't remember being really sick." Ms. Primings's husband is "pretty healthy." However, he did have a positive tine test 2 years ago; his chest x-ray was negative. There are no other significant findings regarding his health.

■ *Family History:*
Mr. and Ms. Primings have been married 5 years. They lost one baby at 3½ months' gestation 2 years ago. Mr. Primings is employed as a plumber and works 8–10 hours a day, 6 days a week. Ms. Primings is a secretary for a family practice physician. There is no significant history regarding the extended family. However, this will be the first grandchild for both Mr. and Ms. Primings's parents, who are all living.

■ *Physical Exam:*
Head: No significant findings.
Neck: No lymph node enlargement, tenderness, or pain. ROM appropriate.
Heart: No murmur noted; pulse strong and steady = 80 bpm.
Lungs: Clear to auscultation; respirations strong and rhythmic. Rate: 22/min.

Vital signs: Temperature: 98.8°F (37.1°C). BP: 122/80. Pulse: 74.

Extremities: ROM appropriate; no pain; tenderness; slight cramping in legs.

Abdomen: Liver and spleen nonpalpable; fundus: slight rise noted; no pain or tenderness; Ms. Primings complains of anorexia, nausea, and constipation.

■ *Elimination System:*

Urinates 5–6 times daily. During the first 8 weeks of pregnancy, she urinated 8–10 times per day.

Usually has at least one stool each day. Within the past month has been having only one stool every 2–3 days with slight gas.

Urine: SG: 1.015. Protein: Negative. Sugar: Negative.

■ *Overall Assessment:*

No spotting or bleeding noted. Can eat very little at one meal without feeling "stuffed and nauseated." There has been a decrease in stool production. Appears somewhat concerned about being able to carry this baby to term due to the loss of her first pregnancy. Husband attends most of the prenatal sessions.

---

44. A late objective sign that may cause Ms. Primings to suspect she might be pregnant is:
    A. Mastalgia.
    B. Amenorrhea.
    C. Abdominal enlargement.
    D. Lassitude and easy fatigability.

45. Ms. Primings wonders what she can expect at her initial prenatal visit. Which is not indicated at the first visit and therefore will not be included in the nurse's answer?
    A. Blood tests for Rh, type, hemoglobin, syphilis.
    B. Urine test for pregnancy, protein, sugar, estriol levels.
    C. Blood pressure, pulse, weight, height.
    D. FHR and fundal height.

46. The nurse knows that routine prenatal care for Ms. Primings will not include:
    A. Personal and family history.
    B. Diet counseling to promote appropriate weight gain.
    C. X-ray pelvimetry prior to the onset of labor.
    D. VDRL, gonorrhea cultures, blood type, and Rh factor.

47. Ms. Primings appears quite apprehensive about her pregnancy. She states that this will be her first baby; although she was pregnant before, she lost that baby at about 3½ months' gestation. During this pregnancy, Ms. Primings is:
    A. Gravida 1, para 0.
    B. Gravida 1, para 1.
    C. Gravida 2, para 0.
    D. Gravida 2, para 1.

48. In planning health teaching to help allay Ms. Primings's anxiety, the nurse needs to explain that in addition to producing hormones, the placenta:
    A. Acts as a barrier to infection and as a shock absorber; it weighs about one-sixth of the fetus's weight at term.
    B. Secretes nutrients for the fetus and removes waste.

    C. Acts as the fetus's organ of respiration, nutrition, and elimination.

    D. Produces the hemoglobin necessary for the fetus during the first trimester.

49. While she is having a pelvic examination, Ms. Primings can best be helped to relax if the assisting nurse:

    A. Coaches her on breathing normally and keeping her hands relaxed.

    B. Assures her that there is really nothing to the pelvic examination.

    C. Distracts her attention from the procedure.

    D. Offers a hand for Ms. Primings to squeeze.

50. In conducting health teaching, the nurse needs to know that increased absorption of phosphorus is thought to be responsible for which pregnancy-associated discomfort?

    A. Leg cramps.

    B. Backache.

    C. Heartburn.

    D. Constipation and hemorrhoids.

51. Ms. Primings asks the nurse how the baby breathes in utero. To explain, the nurse draws a picture showing the cardiovascular structure in fetal circulation that contains the highest concentration of oxygen. The nurse draws the:

    A. Umbilical artery.

    B. Umbilical vein.

    C. Left ventricle.

    D. Arch of the aorta.

52. The nurse can anticipate that Ms. Primings is most likely to notice and bring to the nurse's attention which normal physiologic change?

    A. Physiologic anemia.

    B. Palpitation.

    C. Increased blood pressure.

    D. Increase in the anterior-posterior thoracic diameter.

53. Ms. Primings is advised to avoid lying down soon after eating, avoid bending at the waist, to do "flying" exercises (rotating the extended arms at the shoulder in a wide cycle), and to eat frequent, small meals to avoid which discomfort during pregnancy?

    A. Morning sickness.

    B. Heartburn.

    C. Pica.

    D. Backache.

54. The nurse can advise Ms. Primings that she will once again experience frequency and an increased tendency to varicosities, but will be able to breathe easier when what occurs?

    A. Transition.

    B. Quickening.

    C. Effacement.

    D. Lightening.

55. Which emotional change is Ms. Primings likely to experience in the last trimester of pregnancy?

    A. Irritability.

    B. Anxiety about coming labor.

    C. Introspection.

    D. Ambivalence.

Medical Situation: Pregnancy and nutrition.

■ *Nursing Problems:*
Altered bowel elimination (constipation).
Altered comfort pattern (discomfort).
Altered nutrition.

■ *Chief Complaint:*
Bertha London is a 25-year-old white woman who is approximately 20 weeks pregnant.

■ *Present Status:*
Bertha suspected that she might be pregnant when she missed two menstrual periods. She became more certain when her breasts became tender and swollen and she began vomiting on arising early each morning. Her physician confirmed that she was 10 weeks pregnant. This pregnancy was unplanned.

■ *Past Health History:*
Bertha has been essentially healthy throughout her life. She had measles as a child, but she is not certain what type she had. She also thinks she had chickenpox when she was about 8 years old. Bertha began menstruating when she was 11 years old. Her cycle has been consistent every 28–30 days. She has never before been pregnant. She usually experiences one or two colds a year. She has had the flu at least once in the past 3 years.

■ *Family History:*
Bertha has been married for 3 years to Chet, a 26-year-old white man. Bertha is employed as a high school science teacher and teaches 10th graders. Chet is a social worker for the local department of social services. Bertha is an only child. Both of her parents are living and essentially healthy. Her mother is 46 and her father is 50. They live about 20 minutes from Bertha and Chet. Bertha's parents were unable to conceive any other children. No reasons for this inability were identified because both parents decided not to undergo fertility testing.

Chet's parents live about 4 hours from the couple. Chet is the youngest of three children. He has a brother who is 30 and a sister, age 28. Both of these siblings are healthy. Chet's father is 54 and his mother is 52. Chet's father has been hypertensive for the past 7 years; he is currently taking prescribed medications and is adhering to a low-sodium diet. Chet's mother suffered one miscarriage prior to her first successful pregnancy with Chet's brother. The miscarriage occurred at 8 weeks' gestation, and no reason for its occurrence was ever discussed with Chet's parents.

■ *Physical Exam:*
Head and neck: No swelling of lymph nodes; no pain or edema noted; normal ROM. Eyes, nose, and ears clear; no inflammation, edema, or drainage.
Chest: Heart sounds audible; no murmur, splitting, etc. Lungs bilaterally clear to auscultation.
Breasts: Moderately enlarged, tender to touch, and veins engorged. No inflammation or drainage observed.

Abdomen: Spleen and liver nonpalpable; slight rise in fundus. No significant scars noted.

Extremities: ROM within normal limits; no pain or distortions; no edema observed.

Vital signs: Temperature: 98.8°F (37.1°C). P: 78. R: 10. BP = 112/80.

Weight: 115 pounds. Height: 5 feet, 1 inch.

Urine tests: Pregnancy: Positive. Sugar: Negative. Protein: Negative.

Acetone: Negative.

Hematocrit: 32%.

■ *Significant Complaints:*

Bertha states that she vomits every morning and feels somewhat fatigued throughout the day, especially around 3 P.M. She has been experiencing a need to urinate once or twice per hour. She also has very little, if any, appetite, especially early in the morning. She has tried eating saltines to prevent "this feeling of sickness"; however, nothing seems to work.

---

56. Before assuming Bertha may be anemic, what does the nurse need to consider as another possible explanation?

  A. Mothers are supposed to be a little anemic during pregnancy.

  B. The mother's blood volume is decreased during pregnancy, so this is a true anemia.

  C. The mother's blood volume is decreased during pregnancy, so her hematocrit should be higher.

  D. The mother's blood volume is increased during pregnancy, so there might normally be a decrease in hematocrit.

57. It is difficult to maintain the recommended intake of dietary iron during pregnancy. Bertha is given iron pills to meet her pregnancy needs. Which is inaccurate information for the nurse to impart to the patient?

  A. Stools will be loose.

  B. Take the pills after food to prevent GI distress.

  C. Stools will be black in color.

  D. Drink fluids and take moderate exercise.

58. Because of reduced motility of the GI tract during pregnancy, the nurse can expect Bertha to complain of constipation. The nurse recommends to her that she:

  A. Take glycerine suppositories at bedtime.

  B. Increase roughage in the diet.

  C. Drink warm milk at bedtime and on rising.

  D. Decrease walking up stairs.

59. Bertha is complaining of leg cramps during her pregnancy. In discussing her diet, she states that she loves milk shakes, and sometimes drinks 2 quarts of milk a day. The nurse recommends that she:

  A. Decrease her milk intake to 1 quart a day.

  B. Increase her milk intake to include yogurt.

  C. Decrease her ambulations.

  D. Use skim milk for milk shakes.

In her fifth month of pregnancy, Bertha presents with the following hematologic laboratory values:

Hgb: 6 g/dL.
RBC: 2.5 million/μL.
MCV (mean corpuscular volume): Normal.
WBC: Hypersegmented with leukocytopenia.
Serum Fe: High normal 150 μg/dL (N = 90–150 μg/dL).
Serum vitamin B$_{12}$: Normal.

These values, her age, pregnancy, and complaints of lassitude, anorexia, and mental depression suggest folic acid deficiency anemia. She has no history of alcoholism or protracted vomiting.

---

60. In addition to oral or parenteral folic acid, 5–10 mg/day, the nurse suggests that Bertha eat folacin-containing foods. Which foods does the nurse suggest, as they have the most folic acid?
   A. Asparagus and kidney beans.
   B. Chicken and beef.
   C. Bread and bread products.
   D. Milk and milk products.

61. Since Bertha is of normal weight for her age and height, during pregnancy the nurse advises her to gain:
   A. No more than 20 pounds.
   B. 20–24 pounds.
   C. 25–30 pounds.
   D. As little as possible.

Medical Situation: Pregnancy: complications related to diabetes.

- ***Nursing Problems:***
  Anxiety.
  Risk for alteration in endocrine/metabolic processes.
  Altered urinary elimination.

- ***Chief Complaint:***
  Leslie Osborne is a 23-year-old white woman who is 3 months pregnant with her first child.

- ***Present Status:***
  Ms. Osborne is a known diabetic. She has had diabetes "all of her life." She has come to the doctor's office for her third prenatal examination. She has been feeling very restless and tired, and continues to complain of nausea and vomiting each morning. In fact, she has been "vomiting sometimes twice a day."

- ***Past Health History:***
  Ms. Osborne has been taking insulin since childhood. She states that she cannot remember ever not taking insulin. Her medical records show that she was diagnosed as a diabetic at age 4. Ms. Osborne has never before been pregnant and this is an unplanned pregnancy. However, she has decided to follow through with the pregnancy.

- ***Family History:***
  Ms. Osborne is not and has never been married. She has no immediate plans to marry. She is contemplating giving up her baby for

adoption, but at this time she is not sure what she will do. Ms. Osborne is employed as a librarian in the university library. She intends to work as long as possible. The father of the baby is unaware that Ms. Osborne is pregnant, and Ms. Osborne has not decided whether she will tell him about the pregnancy.

■ *Physical Exam:*

Head: No complaints of headaches, blurred vision, etc.

Neck: No swelling of lymph nodes; no pain on palpation.

Lungs: Clear to auscultation.

Heart: Heart sounds present; no splitting or murmurs noted.

Abdomen: Liver and spleen nonpalpable; uterine size consistent with 3 months' gestation.

Extremities: ROM appropriate; no edema, pain, or tenderness.

Urinary system: Complains of urinary frequency. She voids at least 2 or 3 times per hour. SG: 1.006. Hematest: Negative. Protein: Negative.

Vital signs: Temperature: 98.2°F (36.8°C). P: 72. R: 20. BP: 122/74.

---

62. In assessing Ms. Osborne's overall level of functioning, the nurse is concerned primarily that Ms. Osborne:
    A. Is not married.
    B. Is 23 years old.
    C. Feels tired and restless.
    D. Is experiencing nausea and vomiting.

63. Ms. Osborne asks the nurse what she thinks Ms. Osborne should do in terms of this pregnancy. The *most* appropriate response is to:
    A. Tell Ms. Osborne that she is still within the time limits for obtaining an abortion.
    B. Refer Ms. Osborne for prenatal counseling.
    C. Assess Ms. Osborne's feelings about this pregnancy.
    D. Inform Ms. Osborne that this is essentially her own decision.

64. Ms. Osborne tells the nurse that as a diabetic pregnant mother she is concerned about having a healthy baby. She wants to cooperate with the physician in every way, but needs to know what to expect during pregnancy. The nurse informs her that the medical and nursing management will not include the need to:
    A. Monitor her weight closely to ensure appropriate weight gain.
    B. Adjust her insulin dosage as needed to prevent a glucose imbalance.
    C. Attempt to carry the baby to term to ensure adequate fetal development.
    D. Assess for and treat all infections promptly.

65. Ms. Osborne complains of nausea and vomiting that occur periodically throughout the day. The *most* appropriate nursing intervention is to:
    A. Assess the frequency and amount she vomits each day.
    B. Tell her to relax and try lying down when she feels nauseated.
    C. Suggest that she ask the doctor for an antiemetic medication.
    D. Inform her that the nausea will probably decrease after the first trimester.

66. In developing a plan of care for Ms. Osborne, the nurse incorporates knowledge of the effects of diabetes mellitus on childbearing and the effects of childbearing on diabetes. These effects include:

A. Insulin requirements are highest during the first trimester and early puerperium.

B. Fetal macrosomia is due to excessive water retention.

C. Vaginal birth before 35 weeks is indicated.

D. Hospitalization may be required for mental and physical rest, to treat infection, to adjust insulin requirements, or to change therapy from tolbutamide to regular insulin.

67. At her third clinic visit, Ms. Osborne's urinalysis reveals glucosuria. Ms. Osborne can expect:

A. An increase in insulin dosage.

B. A decrease in caloric intake.

C. Frequent glucose monitoring.

D. Liver function studies.

68. Ms. Osborne is hospitalized for nausea and vomiting during her third month of pregnancy. The most serious nursing concern is:

A. That the baby will be deprived of nourishment.

B. That the mother is rejecting her pregnancy.

C. That the mother may go into ketoacidosis.

D. Starting an intravenous feeding and preventing it from infiltrating.

69. The nurse knows that Ms. Osborne's expected date of birth is difficult to determine because a mother with uncontrolled diabetes may:

A. Have babies with congenital problems.

B. Have irregular menstrual cycles.

C. Tend to gain very few pounds.

D. Have small babies.

70. In planning health teaching for Ms. Osborne, the nurse needs to know that although babies of diabetic mothers often are large (more than 9 pounds), some babies are small because of:

A. Atherosclerosis of the placenta.

B. Increased blood sugar.

C. Size of the mother.

D. Hypoglycemia in utero.

Medical Situation: High-risk pregnancy: nutrition.

■ *Nursing Problems:*

Altered cardiac output.

Fluid volume excess.

Altered nutrition requirements.

Altered parenting.

Self-care deficit: medication administration.

Self-esteem disturbance.

Altered urinary elimination.

■ *Present Status:*

Doris Garvey is a 29-year-old married woman, $G_8P_3AB_4$, with two living children. Her expected date of birth is 6/30/96. The father is ambivalent about this pregnancy. He verbalizes that he "doesn't care what happens as long as my wife is OK." He comments that they do not want any more pregnancies. Ms. Garvey's skin is dry, her hair is dull and unkempt, she walks slowly with her head tilted somewhat downward, and she looks as if she has not been sleeping well lately.

■ *Past Pregnancy History:*

Year	Weeks Gestation	Weight Gain	Length of Labor	Complications	Sex/Weight
1982	7	—	SAB	None	—
1983	36	40 lb	13 hr	None	Female 4 lb, 12 oz.
1985	6	—	SAB	None	—
1986	34	45 lb	7 hr	Postpartum hemorrhage	Female 3 lb, 10 oz.
1990	8	—	SAB	None	—
1992*	32	42 lb	4 hr	PROMs; neonatal deaths; postpartum infection	Male 1 lb, 14 oz.; Male 2 lb, 2 oz.
1994	6	—	SAB	Hemorrhage	—
1995	Current pregnancy				

*Note that birth of twins in 1992 is counted as *one* birth in her *para* number.
SAB, spontaneous abortion; PROMs, premature rupture of membranes.

■ *Family Health History:*
Maternal and paternal parents are living and essentially well. Ms. Garvey has six siblings, all living and well. Her older children, 12 and 9 years old, are essentially well. They go to a special school because "the teachers say that they are a little slow."

■ *Present Pregnancy Prenatal Record:*
Pregravid weight: 170 pounds. Height: 5 feet, 4 inches.

Date	Weeks Gestation	Weight	BP	Urine	Hct	Fundal Height	FHR
11/5/95	10	168½	138/82	Neg	35%	—	—
12/3/95	14	166½	138/80	Neg	35%	—	—
12/17/95	16	167	140/86	Neg	34%	—	—
1/2/96	18	169	138/82	Neg	33%	16 cm	156
1/16/96	20	169	138/84	Neg	30%	16 cm	154
1/30/96	22	172	138/84	P: + G: +	29%	17 cm	154

P, protein; G, glucose.

---

71. After plotting Ms. Garvey's weight gain on the graph for desirable prenatal gain in weight, the nurse's assessment of Ms. Garvey's weight gain is that it is:
A. Consistent with average growth curves.
B. Adequate in light of her obesity.
C. Inadequate.
D. Excessive.

72. Several factors put Ms. Garvey at nutritional risk and therefore increase her risk during this pregnancy. Which does not put Ms. Garvey at nutritional risk?
A. Weight gain inconsistent with desirable prenatal gain.
B. Poor reproductive history.
C. Falling hematocrit (Hct) values.
D. Age at 29 years.

73. In counseling a woman about nutrition, the nurse has several strategies from which to choose. Which one is the *least* helpful for Ms. Garvey?

A. Discuss with her how nutrition affected her pregnancies in the past, because she wants a good pregnancy now.

B. Ask her which foods (containing nutrients necessary for forming blood) she would be able to or want to add to the family's diet.

C. Discuss how her physical appearance and feelings now relate to her physiologic and nutritional state.

D. Mutually identify behavioral objectives with emphasis on short-term goals.

74. The simplest and least expensive test that can be employed to identify a person's nutritional status and to evaluate the effectiveness of therapy is the laboratory assessment for serum levels of:

A. Folate.

B. Protein.

C. Mean corpuscular hemoglobin concentration (MCHC).

D. Hematocrit.

75. If tests indicate that Ms. Garvey's nutritional status is compromised by a lack of blood-forming nutrients, the nurse needs to help Ms. Garvey pick out and add to her diet foods that contain blood-forming nutrients. The nurse therefore omits foods high in:

A. Vitamin D.

B. Iron.

C. Folacin.

D. Vitamin B$_6$.

76. Many nutrients are needed to form blood. However, there is one food group that can be emphasized in Ms. Garvey's nutritional plan. This food group is:

A. Leafy green and yellow vegetables.

B. Fortified or natural breads and cereals.

C. Yellow fruits.

D. Animal proteins such as meat and organs.

77. Ms. Garvey's anemia most likely has been long-standing because of the following facts from her database. Which finding is not a sequela to anemia?

A. Small-for-gestational-age (SGA) neonates.

B. Hemorrhage.

C. Hypertension.

D. Perinatal infection.

78. Ms. Garvey tells you that she would like to lose some weight because "I'm so fat. I'm afraid my husband will start looking at other women." Which statement *best* demonstrates application of knowledge of pregnancy and the woman's needs during pregnancy?

A. "Let's plan a diet that will provide the baby with the food he or she needs and help you trim off some pounds."

B. "It is important to gain weight with good food to feed the baby and fix your anemia first."

C. "You sound like you don't like the way you look right now. Do you want to talk about it some more?"

D. "Would you like me to refer you to the social worker to get food stamps and to talk with you and your husband?"

79. Ms. Garvey tells you that she too wants to have her "tubes tied." When providing counseling in matters such as this, the nurse and physician must keep in mind the patient's rights. Which is *not* one of the three elements of a legally effective "informed consent"?

A. It must be given voluntarily.

   B. Persons must be given information about the procedure and its consequences and any alternative procedures and their consequences.

   C. The person giving consent must be capable of comprehending the information.

   D. The consent form must be written in English, but an interpreter should be used when needed for the person's comprehension.

80. If a procedure that was not clearly specified in the original informed consent is performed, the nurse needs to consider that which of the following legal actions may be initiated?

   A. Abandonment.

   B. Battery.

   C. Ethical dereliction of duty.

   D. Invasion of privacy.

81. Among the data provided for Ms. Garvey's present pregnancy at 20 weeks, which is not associated with increased fetal mortality?

   A. Weight: 169 pounds.

   B. Blood pressure: 138/84.

   C. Fundal height: 16 cm.

   D. FHR: 154.

82. At 36 weeks Ms. Garvey comes to the clinic for her checkup. The nurse suspects she may be developing preeclampsia when the nurse notes:

   A. Decreased systolic blood pressure, ankle edema, 1+ proteinuria.

   B. Swollen fingers, increased diastolic blood pressure, 1+ proteinuria.

   C. Proteinuria, 1+ sugar in the urine, hypotension, weight loss.

   D. Ankle edema, 2+ glycosuria, increased systolic blood pressure.

83. Ms. Garvey is diagnosed as preeclamptic and admitted to the hospital. What is an inappropriate nursing intervention?

   A. Place in a room with mothers to keep her company.

   B. Put on bedrest, preferably on left side.

   C. Check urine for protein and BP every 4 hours.

   D. Place in quiet room and limit visitors.

84. Which equipment should be placed in the room in preparation for an anticipated emergency admission of a woman with severe preeclampsia?

   A. Suction apparatus.

   B. TV set for distraction.

   C. Telephone to maintain contact with her family.

   D. Urine collection bottle set in ice bucket for determination of 17-hydroxysteroids.

85. Preeclamptic mothers often receive an anticonvulsant, magnesium sulfate, which controls cerebral irritability and secondarily causes a lowering of blood pressure. Which sign indicates that the mother should be carefully assessed for magnesium toxicity?

   A. Respirations at 20/min.

   B. Decreasing urinary output.

   C. Blood pressure of 150/84.

   D. Hyperreflexia.

86. The nurse is responsible for planning to have the following antidote for magnesium sulfate close at hand:

   A. Calcium gluconate.

    B. Naloxone (Narcan).

    C. Potassium chloride.

    D. Terbutaline.

87. Ms. Garvey has an order for 1000 mL of D5 lactate Ringer's with 10 g of magnesium sulfate to run in over 10 hours. With a macro drop factor of 15 gtt/mL, Ms. Garvey should receive which number of drops of fluid per minute?

    A. 14–16.

    B. 24–26.

    C. 34–36.

    D. 44–46.

88. A blood pressure reading of 120/80 may be indicative of pre-eclampsia if:

    A. There has been a weight gain of 2 pounds for each of the previous 2 weeks.

    B. The woman is carrying a hydatidiform mole.

    C. The woman has had ankle swelling each evening for the previous 2 weeks.

    D. Her systolic pressure has increased by 30 mm Hg; the diastolic, by 15 mm Hg.

A cesarean birth is planned for Ms. Garvey. A healthy 6-pound, 6-ounce boy is born. Ms. and Mr. Garvey are happy, but Mr. Garvey states that this is definitely the last pregnancy his wife will have. Questions 89 through 95 deal with nursing care of a woman following cesarean birth.

---

89. A check for Homans' sign is performed postpartum, especially on Ms. Garvey, to assess for:

    A. Infection.

    B. Edema.

    C. Phlebitis.

    D. Hemorrhage.

90. Methylergonovine maleate (Methergine), an oxytoxic, has a side effect of raising the blood pressure. It is sometimes given to mothers after cesarean birth. The nurse should question giving this medication to Ms. Garvey if she had a history of:

    A. Postpartum hemorrhage.

    B. Toxemia.

    C. Diabetes.

    D. Hyperemesis gravidarum.

91. In spite of the recovery process, cesarean birth mothers are often able to have their newborns with them as early and as much as vaginally delivered mothers. The most important reason for mothers to have their newborns with them at frequent intervals for short periods is:

    A. Mother-infant bonding.

    B. Breastfeeding to prevent engorgement.

    C. Breastfeeding for oxytocin stimulation of the uterus.

    D. Active involvement in their baby's care.

92. A mother who has had a cesarean birth is given medication for pain. She wishes to start breastfeeding. In analyzing data at hand, the nurse needs to know that:

A. Cesarean birth mothers should not breastfeed.

B. Medication should not be given to breastfeeding mothers.

C. Medication should only be given an hour prior to breastfeeding.

D. Medication should be given at such a time that it will not be in the breast milk when nursing.

93. Due to the incision, cesarean birth mothers often find it difficult to breastfeed. The nurse encourages:

A. Bottle feeding until the mother is free of pain and her incision is healed.

B. The mother to try various breastfeeding positions to find the one that is most comfortable for her.

C. Pumping the breasts and giving the breast milk to the baby by bottle.

D. The mother not to bother with breastfeeding since she needs her energy to heal her own body.

94. It is the third postpartum day for a cesarean birth mother. She is crying and states, "I feel so abnormal not being able to have my baby the normal way." The nurse would respond:

A. "I will ask the doctor to explain to you why the cesarean birth was necessary."

B. "It's normal to feel guilty. You will feel better in a few days."

C. "You are upset because you did not have a vaginal birth. It must have been hard for you to accept."

D. "Many of our mothers have cesarean births. I will ask one of them to come and talk to you."

95. The nurse arranges with Ms. Garvey to make a home visit 3 days after her discharge from the hospital. The nurse determines that she has not accomplished her discharge planning goals for Ms. Garvey when the mother states:

A. "I'm so tired; the baby breastfeeds every hour."

B. "I'm trying to rest when the baby sleeps."

C. "I'm enjoying breastfeeding and the baby seems content."

D. "My lochia is brownish in color and decreasing each day."

96. Ms. and Mr. Garvey decide that he will have a vasectomy. In the plan of care, the nurse prepares Mr. Garvey for vasectomy based on analysis of data that indicate that:

A. He understands that no change in testosterone levels will result.

B. Fertility can be restored by reanastomosis.

C. Recovery may be complicated by balanitis or phimosis.

D. He will be sterile as soon as the surgery is performed.

Medical Situation: Conception/contraception/hormones.

■ *Nursing Problems:*
Anxiety.
Altered health maintenance.
Health-seeking behaviors: function of estrogen, oral contraceptives, infertility.

■ *Chief Complaint:*
John Jackson, 25, and Estelle Jackson, 24, have been trying to conceive a child for the past year without success.

■ *Present and Past Health History:*

Mr. Jackson has been relatively healthy throughout his life. His BP normally is 130/78. His vital signs are within normal limits: 98.8°F temperature, pulse 78, respirations 20. He has had no major illnesses or surgery during his life. He is unsure whether he is capable of fathering children and has never been given any tests to screen his fertility. There are no other significant findings at this time.

Ms. Jackson does not remember being seriously ill as a child. She did have chickenpox and mumps as a child. She began menstruating at age 13. She has been regular since her second period, experiencing menstruation every 28–30 days. She has never been pregnant and began taking oral contraceptives when she was 17 years old. She discontinued the oral contraceptives 1 year ago, when she and her husband decided to have a child. Her vital signs are: Temperature: 99°F; P: 72; R: 20; BP: 110/70. Urine analysis was negative for pregnancy. There are no other significant findings.

■ *Family History:*

Mr. Jackson is one of two children. He has an identical twin brother, Jason. His mother miscarried several times before she had these boys. Mr. Jackson's family history is essentially unremarkable. Both his parents and grandparents are living and healthy. Mr. Jackson is a high school coach.

Ms. Jackson is an only child; her mother died during her birth. The cause of death was preeclampsia and obesity. Her father remarried and has had two other daughters. Ms. Jackson's father has been diagnosed as being hypertensive. His BP is about 180/90, for which he takes medication daily. Ms. Jackson is employed as a computer programmer.

■ *Psychological Evaluation:*

In evaluating Mr. and Ms. Jackson, the infertility clinic nurse found them to be moderately anxious about what to expect. They were concerned about the outcome, the expense of the testing procedures, and the future of their relationship as a couple.

---

97. In planning health teaching with Mr. and Ms. Jackson about the physiologic effects of progesterone, the nurse knows that one of the following is not an effect of progesterone:
   A. Slightly elevated basal body temperature.
   B. Movement of fertilized ovum through the oviduct.
   C. Formation of the corpus luteum.
   D. Inhibition of uterine contractility during pregnancy.

98. Ms. Jackson tells the nurse that she has heard that estrogen is important in facilitating reproduction. The nurse should emphasize that a prominent function of estrogen is not:
   A. Growth of spiral arteries in the endometrium.
   B. Development of the mammary duct system.
   C. Inhibition of uterine contractility during pregnancy.
   D. Augmentation of the quantity of cervical mucus and its receptivity to sperm.

99. In teaching Mr. and Ms. Jackson about conception, the nurse states that conception usually occurs in the presence of high levels of:

A. Estrogen and luteinizing hormone (LH).
B. Follicle-stimulating hormone (FSH).
C. Progesterone.
D. Human chorionic gonadotropin (HCG).

100. In providing care for Mr. and Ms. Jackson, the clinic nurse realizes that the couple will *primarily* need:
A. Health teaching regarding conception.
B. Information about various positions to use during intercourse.
C. Further physiologic assessment.
D. Psychotherapy for emotional support.

101. Which factor is the *most* significant in assessing the present inability of the Jacksons to conceive?
A. The couple's ages.
B. Ms. Jackson's history of taking oral contraceptives.
C. Ms. Jackson's father's being hypertensive.
D. Ms. Jackson's being an only child.

102. In planning the infertility tests to be performed, the nurse plans for which test *first?*
A. Sperm analysis.
B. Sonogram.
C. Dilation and curettage.
D. Basal body temperature evaluation.

103. Ms. Jackson asks the nurse "whether taking oral contraceptives could have made her sterile." The *most* appropriate response includes:
A. "Yes, many women become infertile after taking birth control pills."
B. "No, this is unlikely."
C. "There is no indication at this point that you are sterile."
D. "There is a 30% chance that oral contraceptives decreased your ability to conceive."

104. Mr. Jackson, in confidence, expresses concern over the couple's inability to conceive. The nurse needs to be aware that he is *most* likely primarily concerned about:
A. The possibility of not being able to have children.
B. His inability to father children.
C. The couple's overall sex life.
D. His wife's inability to become pregnant.

Medical Diagnosis: Infertility/spontaneous abortion.

■ *Nursing Problems:*
Risk for alteration in endocrine/metabolic processes.
Altered health maintenance.
Altered parenting.
Sexual dysfunction.

■ *Chief Complaint:*
Joan and Mike Hamlin have been unable to conceive after attempting for 4 years.

■ *Present Health History:*
Joan is negative for chronic or concurrent disorders. Joan (27 years old; 5 feet, 6½ inches; 130 pounds) and Mike (27 years old; 5 feet, 11

inches; 165 pounds) have been married for 4 years. They had intended to have children as soon as possible after they married. As time went on without conception, they became concerned.

■ *Past Health History:*

Joan had all the usual childhood infectious diseases. She has had no hospitalizations. She gets a cold about once every 3–4 years. Seven years ago she had an IUD inserted but then had it removed within 6 months. At that time, she started oral hormonal contraception, which she discontinued 1 month prior to marriage. Her normal menstrual cycle recurred within 8 months.

■ *Family History:*

There is no history of infertility in maternal or paternal families or among Joan's siblings. Mike's family history is negative, also. Joan's family history is negative for hypertension and diabetes mellitus. Her maternal grandmother died of breast cancer at age 76; her maternal grandfather is alive and well at 80 years. Her paternal grandmother is alive and well at the age of 79; her paternal grandfather died from a tractor accident on his farm at the age of 62.

■ *Review of Systems:*

Nutrition assessment: Excellent.

Menarche at age 11. Periods regular at about 28 days; flow: ×4 days, heaviest for first 2 days. Denies dysmenorrhea. Couple has intercourse "regularly."

Findings negative for all organs and systems, except for the following: After IUD insertion, experienced intermittent heavy vaginal bleeding and uterine cramping. Within 2 months, experienced lower abdominal pain, which was treated by removal of the IUD and oral antibiotic therapy. Has had no recurrence of this symptomology.

Has had two episodes of "yeast" infection, both successfully treated with nystatin (Mycostatin) vaginal cream. Does not douche.

■ *Physical Exam:*

Within normal limits: Vital signs, height, weight. Development of female anatomy. Indices for good nutritional status: hair, skin, eyes, teeth, hemoglobin, hematocrit.

Negative findings: All organs and systems.

■ *Psychological Assessment:*

Alert, appropriate affect. Well-groomed, good posture, articulate. Couple sit close to each other, touch each other, and look at each other frequently. Both are "certain" that they want children.

---

105. The nurse is adding to the database for this couple. Which question provides the *least* relevant data?
    A. "How often do you have sexual intercourse?"
    B. "Do you experience orgasm each time you have intercourse?"
    C. "What is your husband's occupation?"
    D. "Was your pelvic inflammatory disease due to gonorrhea?"
106. When taking Joan's history, the nurse in the fertility clinic knows that data relevant to infertility include:
    A. A 28–30-day menstrual cycle.

B. Pubic hair and breast development at age 11.

C. Epidemic parotitis at age 14.

D. Gonorrhea treated 2 years ago.

107. Which finding during Joan's physical examination requires additional assessment when seeking the etiology of this couple's infertility?

A. Uterus is anteverted.

B. Abdominal scars noted in the right lower quadrant.

C. Clitoral size is 0.5 cm.

D. Pubic hair is thick and covers the mons and vulva.

108. Which statement is the *most* therapeutic nursing action to assist Joan and Mike to cope with infertility?

A. "Your sexual relationship is considered normal."

B. "Tell your friends and family that you have chosen not to have children."

C. "Your angry feelings are understandable and an expected response."

D. "When you are over your initial response, you may want to consider an alternative such as adoption."

109. Prior to instructing Joan and Mike in the procedure for assessing Joan's cervical mucus, the nurse's *first* action is to:

A. Show her pictures of cervical mucus common to each phase of the menstrual cycle.

B. Explain the effects of the ovarian hormones, estrogen and progesterone, on cervical mucus.

C. Demonstrate, using a pelvic model, how a specimen is obtained.

D. Assess her readiness to touch her external genitalia and her cervix.

110. Mike wants to know how to go about collecting semen for analysis. Using pelvic models, the nurse demonstrates and instructs him to:

A. Apply a regular condom on the penis to collect ejaculate.

B. Ejaculate directly into a clean glass jar.

C. Soak up ejaculate from Joan's vagina using sterile cotton balls.

D. Remove ejaculated sperm from his wife's vagina with a vaginal spoon and place it directly into a clean test tube, which is then packed in ice.

111. The nurse knows that the couple understood which fertility test when Mike and Joan arrive at the clinic at the expected time of ovulation and within 6–8 hours after coitus?

A. Complete semen analysis.

B. Sims-Huhner test.

C. Buccal smear.

D. Rubin's test.

112. Joan's last normal menstrual period was March 18, but she spotted around April 15. Her estimated date of birth is:

A. December 25.

B. December 11.

C. January 22.

D. January 18.

113. Joan has been amenorrheic for two successive cycles, and has been experiencing some nausea, breast tenderness, and fatigue. Now she is admitted to the hospital for some spotting and cramping sensations. Vaginal examination reveals no cervical dilation. The

nurse prepares Joan for diagnostic and therapeutic measures related to:

A. Threatened abortion.
B. Inevitable abortion.
C. Tubal pregnancy.
D. Incompetent cervical os.

Medical Diagnosis: Hemorrhage/placenta previa.

■ *Nursing Problems:*

Risk for alteration in circulation.
Altered feeling state: grief.
Altered fluid volume: deficit.
Impaired gas exchange.
Knowledge deficit.
Impaired physical mobility.
Self-care deficit: medication administration.

■ *Chief Complaint:*

Judy Moss, an expectant mother, is admitted to the labor unit via gurney. She is a 31-year-old white primigravida at 36 weeks' gestation. Her husband, Donald, is with her. She reports vaginal bleeding that started 1½ hours ago.

■ *Present Health History:*

"Immediately after getting out of bed this morning, blood started to run down my legs and into my slippers."

■ *Past Health History:*

Negative for chronic or concurrent medical disorders.

■ *Obstetric History:*

Course of pregnancy: Unremarkable.
Blood type: A, Rh (D) negative.
Coombs: Negative.
Husband: B, Rh positive.
BP: 132/84.
Weight gain: 26 pounds over baseline weight of 130.
Height: 5 feet, 5 inches.
Pelvic measurements: Gynecoid, adequate.
VDRL: Negative. Gonorrhea: Negative.
Plans to have a nonmedicated vaginal birth with husband acting as coach.

■ *Review of Systems:*

Increasing fatigue.
Return of urinary frequency, mild constipation, itching and uncomfortable hemorrhoids.
Inability to sleep because of large abdominal size and fetal movements.
Difficulty sitting and walking because of sensation of pelvic looseness.
Good appetite. Last ate dinner 12½ hours ago; nothing by mouth since then.
No respiratory problems now, but had a cold 3 weeks ago.

■ *Physical Exam:*

Vaginal discharge: Perineal pad contains one small blood spot.

Abdominal examination: Uterus relaxed. One fleeting mild contraction, 30–40 seconds in 10 minutes. Fetus active. Possible breech presentation. FHR in RUQ at 144 bpm.

Vaginal examination: Deferred.

Hct: 35%. Hgb: 11.5 g.

Sonogram: Placenta located in lower uterine segment over internal os.

Diagnosis: Complete (total) placenta previa with fetal breech presenting.

Plan: Cesarean birth after fetal lung maturity is ensured.

■ *Psychological Status:*

Anxious and concerned for the fetus.

---

114. While transferring Judy Moss from the gurney to the bed, the most appropriate question the nurse can ask regarding the vaginal bleeding is:
    A. "Was the vaginal bleeding accompanied by pain?"
    B. "Had you had intercourse just before the bleeding started?"
    C. "Has this happened before during your pregnancy?"
    D. "Do you usually bleed easily, like a bloody nose?"

115. The nurse settles Ms. Moss in bed. Which appropriate nursing action is taken *first?*
    A. Auscultate and record fetal heart rate and rhythm.
    B. Palpate tone of uterine muscle.
    C. Assess amount of bleeding presently on the perineal pad.
    D. Complete the intake history and physical examination.

116. The intern arrives on the unit and asks for Ms. Moss's room number. He asks the nurse to accompany him while he assesses Ms. Moss for cervical dilation and effacement and fetal presentation. The nurse's *best* response is:
    A. "Ms. Moss has just been given a sedative. Can you come back later?"
    B. "Ms. Moss has had vaginal bleeding. Vaginal examinations are contraindicated."
    C. "You will need to check with the resident first."
    D. "I will be glad to go with you. She is very anxious."

117. To prepare Judy for a vaginal examination under double setup, the nurse prepares a sterile field:
    A. In the labor room.
    B. In the birthing room.
    C. In the operating room.
    D. Nowhere. Double setup refers to an intravenous infusion line suitable for regular fluids and for blood transfusion.

118. Vaginal bleeding has stopped. The tracing of the FHR indicates a rate between 136 and 144; beat-to-beat variability, between 5 and 20; oscillatory cycles, 2–6/min; no periodic variability (e.g., deceleration). In the presence of these findings, the nurse:
    A. Continues to monitor maternal and fetal status with no change in care plan.
    B. Starts oxygen by face mask at 10–12 L/min.

C. Notifies the physician stat and prepares for cesarean birth.

D. Reassures the family that the crisis is over.

119. When providing nursing care to a woman with a fetal monitor in place, the nurse keeps in mind the following case:

> The physician ordered a fetal monitor for a woman in labor and then proceeded to meet his surgery schedule. The nurse applied the monitor. The physician returned to find a late deceleration pattern that had persisted for 2 hours and went unrecognized by the nurse. Within 30 minutes, a stillborn fetus was born.

In the situation described, which statement is true?

A. The physician is solely responsible/accountable.

B. The nurse accepted total responsibility/accountability for recognizing ominous FHR patterns.

C. The nurse was not expected to take independent action to prevent harm to the fetus.

D. The late deceleration pattern was unrelated to the intrauterine fetal demise.

120. Ms. Moss is not bleeding now. She and her husband appear to be—and state that they are—less anxious and more hopeful about continuing the pregnancy at this time. Mr. Moss reveals he does not understand "about the Rh thing." He continues, "I heard the Rh can hurt the baby. How would we know if the baby is hurt?" Which is the *most* appropriately stated nursing problem to enter into Ms. Moss's Kardex?

A. Mr. Moss would like an explanation of Rh incompatibility.

B. The Mosses want to know how the Rh factor affects the baby.

C. The Mosses need to learn more about the Rh factor.

D. Mr. Moss's concern is increased because he does not understand isoimmunization—its cause or diagnosis.

121. Ms. Moss is to remain on bedrest for at least a week. Which nursing action is inappropriate?

A. Ask her to cough and deep breathe at least 4 times per day.

B. Encourage her to wiggle her toes and flex and extend her feet against the foot board whenever she thinks of it.

C. Prepare her for L/S ratio determination.

D. Assist her in choosing a diet high in roughage and fluids.

Two weeks later, Ms. Moss begins to bleed slightly. A cesarean birth is accomplished. Mr. Moss stayed with Judy during the birth. Both saw and held their prematurely born son before he was admitted to the premature nursery. Baby Boy Moss is assessed to be large for gestational age (LGA). Gestational age is estimated at 37 weeks.

Judy's glucose tolerance test is positive for (gestational) diabetes. Her hemoglobin A_{1c}, at 10% or more, implies abnormal glucose metabolism over the past several weeks.

---

122. Based on these new findings, what is least relevant in the neonate's care?

A. Observation for signs of respiratory distress syndrome (RDS).

B. Evaluation for blood glucose below 30 mg/dL.

C. Laboratory assessment for serum bilirubin above 15 mg/dL.

D. Examination of diapers for pinkish staining.

123. At 1 minute after birth, Baby Boy Moss's heart rate is 92; he has made no respiratory effort; his muscle tone shows some flexion; he grimaces in response to a slap on the soles of his feet; and his color is pale. Based on his Apgar score, the nurse:

A. Does nothing.

B. Takes him to the nursery for further observation.

C. Administers oxygen by mask until he is pink.

D. Begins resuscitative actions immediately.

124. At cesarean delivery, the diagnosis of complete placenta previa is confirmed. In the first 24 hours after birth, the nurse must know that the woman who had placenta previa is at greater risk for hemorrhage than other new mothers due to:

A. Uterine atony.

B. Retained placental fragments.

C. Pelvic hematoma.

D. Location of decidua basalis.

125. To prevent or treat postpartal hemorrhage, defined as the loss of 500 mL or more of blood within the first 24 hours after birth, the nurse administers which drug?

A. Metronidazole.

B. Methylergonovine maleate.

C. Estradiol valerate.

D. Naloxone (Narcan).

126. Judy Moss and her husband feel that breast milk is better for their baby. She wants to supply milk for their son during his hospitalization and to have sufficient milk for him when he comes home. What is the most appropriate action to help her establish and maintain lactation when the baby cannot nurse?

A. Increase fluid and caloric intake.

B. Empty the breasts every 3–4 hours.

C. Receive an injection of oxytocin to stimulate the letdown reflex.

D. Take oral estradiol valerate as prescribed.

127. The specimen of cord blood obtained from Baby Boy Moss at delivery reveals he is Rh positive, Coomb's negative. Judy is a candidate to receive Rho (D) immune globulin (i.e., RhoGAM). Rho (D) immune globulin protects a future Rh-positive fetus of Ms. Moss by:

A. Beginning to build maternal immunity to the Rh factor.

B. Destroying fetal Rh-positive RBCs in the maternal blood system.

C. Inactivating maternal anti-Rh factor antibodies.

D. Supporting the maternal immunologic rejection of fetal Rh-positive RBCs.

128. Prematurity places the neonate at high risk. The nurse knows that prematurity is associated with:

A. Moderate obesity, maternal age between 28 and 32 years, exercise such as swimming, or too much intercourse.

B. High hemoglobin and hematocrit, increased water conservation by the kidneys, and constipation.

C. Lack of surfactant and immaturity of retinal and other blood vessels.

D. Parental preference for this child over the other normal-term siblings because of the emotional and financial investment.

Baby Boy Moss dies.

---

129. When a neonate dies, the *initial* response of the parent(s) is often:
A. Denial and shock.
B. Anger.
C. Depression and acute grief.
D. Resolution.

130. The nurse can *best* comfort/reassure Ms. Moss by saying:
A. "It is God's will. He works in ways we sometimes do not understand."
B. "It is better he died now before he suffered through any surgeries and before you got too attached to him."
C. "You can be thankful that you are still young and can have another baby."
D. "This must be a very difficult and sad time for you."

Medical Situation: Complications in pregnancy and labor: abruptio placentae.

■ *Nursing Problem:*
Fluid volume deficit.

■ *Chief Complaint:*
Mavis Scott, a 31-year-old black woman, arrives in the emergency room complaining of severe abdominal cramping. She is 30 weeks pregnant and is experiencing intense pain during contractions.

■ *History of Present Illness:*
Ms. Scott's husband states that shortly after dinner his wife began complaining of cramps: "She began to hurt so bad that she couldn't walk." Mr. Scott called the physician, who told him to bring his wife to the hospital immediately. On arrival at the hospital, the nurse identified that the contractions were irregular, the uterus was not relaxing between contractions, and vaginal bleeding was apparent.

■ *Past History and Family History:*
This is Mr. and Ms. Scott's third pregnancy. They have a son, age 3, and a daughter, age 2. This pregnancy was unplanned, but the family wants this baby. Ms. Scott did not have problems with her other two pregnancies. Throughout this present pregnancy, she has "felt pretty good." Ms. Scott is healthy. Except for her two earlier births, the only time she was hospitalized was for a broken arm when she was 10 years old. Mr. Scott is very supportive of his wife and is concerned about both his wife and child. Mr. Scott is employed as a cab driver and Ms. Scott is a homemaker. Prior to having the children, Ms. Scott worked as a nursing aide in a nursing home.

■ *Physical Exam:*
Vital signs: Temperature: 99°F. P: 88. R: 28. BP: 100/70.
Lab values: Hct: 40%. Decrease in clotting time. Urine: Negative for sugar, protein, and blood.
Lungs: Clear to auscultation.
Heart: Pulse strong and regular; no murmurs or splitting noted.
Abdomen: Spleen and liver nonpalpable; fundal height halfway between umbilicus and xiphoid process. Contractions hard and irregular, lasting 30–60 seconds; membranes intact.
Extremities: No edema, pain, or tenderness; ROM intact.
Head: All neurologic signs within normal limits.
Eyes: PERL.
Reflexes: Slightly hyperactive.
Fetus: FHR: 140 bpm (steady rhythm noted); active.

■ *Psychological Status:*
Ms. Scott is very anxious; she is crying and shaking intermittently. Mr. Scott remains with his wife and refuses to leave her bedside. He appears as if he is "being strong for her."

---

131. A vaginal examination is contraindicated for Ms. Scott because she is bleeding during the last trimester of pregnancy. The nurse knows that an examination could:
    A. Cause hemorrhage from a low-lying placenta.
    B. Be painful and stimulate labor.
    C. Introduce infection.
    D. Rupture the membranes.
132. Ms. Scott is diagnosed as possibly having abruptio placentae. A major diagnostic sign to look for as part of the nursing assessment is:
    A. Painless bleeding.
    B. Hypotension.
    C. Painful bleeding.
    D. Nausea and vomiting.
133. A diagnosis of abruptio placentae is made. Abdominal palpation will reveal:
    A. A soft, boggy uterus with a distended bladder.
    B. A hard, tender, boardlike uterus.
    C. A distended bladder with hypertension.
    D. Mild contractions every 2 minutes, lasting 30 seconds.
134. What assessment does the nurse need to make because it is associated with abruptio placentae as a serious complication?
    A. Tachycardia.
    B. Decreased urinary output.
    C. Hypertension.
    D. Coagulopathy.
135. In providing care for Ms. Scott, it is essential to maintain her in which position?
    A. Semi-Fowler's.
    B. Prone.
    C. Side-lying.
    D. Supine.

Medical Situation: Normal labor and birth.

- ■ *Nursing Problem:*
  Altered health maintenance.

- ■ *Chief Complaint:*
  Allene Nelson is a 25-year-old white woman who arrives at the hospital complaining of contractions occurring every 5–18 minutes and lasting for about 30 seconds.

- ■ *Present Status:*
  Allene is 40 weeks pregnant with her first child. Her pregnancy has been essentially unremarkable. She has gained 30 pounds and has "felt good" throughout her pregnancy. She experienced a slight amount of nausea during the first 2 months of pregnancy; however, she never vomited and has had a good appetite throughout the pregnancy. Her last prenatal visit was 2 days ago. The cervix was dilated 1 cm at that time, and she was experiencing a backache. Her husband, Tim, has accompanied her to every prenatal visit and has been very supportive.

- ■ *Past Health History:*
  Allene began menstruating at age 13. Her periods usually occur every 29 days and she experiences moderate abdominal cramps, which are relieved by taking aspirin every 4 hours for the first 24 hours. Allene has had all of her immunizations. She had chickenpox when she was 5 and does not remember being sick otherwise.

- ■ *Family History:*
  This pregnancy was planned. Both Allene and Tim are looking forward to the birth of their first baby. Allene and Tim have been married 4 years. Tim is a lab technician and Allene is a secretary at the same medical laboratory. Allene worked until last week and intends to return to work when the baby is 6 months old. The couple attended childbirth classes, and Tim is planning to be present throughout the birth.

- ■ *Physical Exam:*
  Vital signs: Temperature: 99.2°F. P:84. R:24. BP: 110/80.
  FHR: 160 bpm.
  Dilation: 4 cm; effacement: 100%; station: −1 cm.
  Contractions: Approximately every 8 minutes, lasting about 30 seconds; mild.
  Membranes: Intact.
  Bloody show: Present.
  Overall physiologic status good. No apparent heart, lung, or other internal disruptions.

- ■ *Psychological Status:*
  Allene is very excited and anxious for the baby to arrive. She is very talkative and appears to be enjoying her husband's attention. Tim appears somewhat anxious. He is providing support for Allene by massaging her back, getting the nurse when needed, and sponging her forehead with cool water.

136. On admission to the labor unit, Allene presents with the following signs: 100% effaced, 4 cm dilated, vertex at −1 station, membranes intact. The nurse's *next* appropriate action is to:

A. Start an intravenous infusion.

B. Check the FHR.

C. Set up for imminent birth.

D. Prepare her for amniotomy.

137. While listening to the FHR, the nurse is least likely to hear:

A. Funic souffle.

B. Fetal respiratory movements.

C. Uterine souffle.

D. Maternal intestinal noises.

138. The nurse continues the admission assessment with palpation of Allene's abdomen. Palpation to assess for fetal presentation, position, lie, and engagement is known as the maneuver of:

A. Leopold.

B. Scanzoni.

C. Ritgen.

D. Shirodkar.

139. After entering the findings from the admission assessment, the nurse reviews Allene's prenatal record. The nurse notes that Allene's pelvis is described as gynecoid. When the patient asks the nurse what "gynecoid" means, since she heard the doctor use it to describe her pelvis, the nurse draws the following picture:

A. Oval, widest from front to back.

B. A transverse oval.

C. Somewhat heart shaped, widest from side to side.

D. Somewhat heart shaped, widest from front to back.

140. The nurse knows that the part(s) of the bony pelvis that serve(s) as a landmark for determining the station of the presenting part is (are) the:

A. Ischial spines.

B. Ischial tuberosities.

C. Sacral promontory.

D. Lower border of symphysis pubis.

141. The Nelsons inform the nurse that they are planning to use the Lamaze method for labor. To assist this couple, the nurse needs to know that the Lamaze method suggests the following breathing pattern during early labor:

A. Deep chest.

B. Shallow chest.

C. Shallow abdominal.

D. Deep chest at beginning and end of contraction; rapid and shallow chest during peak of contraction.

142. In planning care for Allene, the nurse need not be concerned with:

A. Monitoring the FHR.

B. Examining the vagina.

C. Providing support for her husband.

D. Rupturing the membranes.

143. Allene is becoming fatigued and more uncomfortable. Tim asks the nurse what he can do to help. The *most* appropriate response includes:

A. "Relax and remain calm."

B. "Call the physician whenever your wife needs her."

C. "Reassure your wife."

D. "Keep your eye on the fetal monitor."

144. Tim asks the nurse when he will know that Allene has entered the transitional stage of labor. The *most* appropriate response includes the information that his wife will become:
   A. More irritable and complain of nausea.
   B. Less talkative and more serene.
   C. More talkative and excited.
   D. More serious and less energetic.

145. Allene notices a leaking of clear fluid from her vagina. During the vaginal examination, the nurse cannot tell if the membranes have broken. To test whether the "bag of waters" has broken, the nurse may employ which test?
   A. Phenaphthazine (Nitrazine).
   B. Guthrie assay.
   C. Gravindex.
   D. Pap smear.

146. The nurse applies the fetal monitor ordered by the physician, and the FHR and contractions are recorded continuously on a strip. Fetal monitor readings indicate a uniform shape of FHR deceleration pattern, which *coincides with* the uterine contraction; FHR returns to baseline as the contraction ends. The nurse:
   A. Reports this pattern immediately to the physician.
   B. Turns the woman onto her side, administers oxygen, and stops the oxytocin induction if induction is in progress.
   C. Alerts surgery and anesthesiology departments that an emergency cesarean birth is pending.
   D. Does nothing; there is no clinical significance to this pattern.

147. Allene has had no regional anesthesia. Which symptom indicates that she is approaching the second stage?
   A. Elevation of temperature and pulse.
   B. Nausea, shaking legs, increased irritability.
   C. Malar flush, increasing seriousness.
   D. Tingling or numbness of the fingers and/or toes.

148. Allene begins to complain of nausea, belches, and begins to perspire over her upper lip, and her legs begin to shake. The nurse's *first* action is to:
   A. Notify the physician because she is starting to convulse.
   B. Check to see if the cervix is dilated 10 cm.
   C. Check for hypotension.
   D. Cover her with extra blankets and keep her quiet.

Medical Situation: Complications during labor and birth: placenta previa; cesarean surgery.

■ *Nursing Problems:*
   Altered family processes.
   Altered feeling processes (anxiety).
   Fluid volume deficit.
   Impaired gas exchange.
   Risk for injury.
   Altered tissue perfusion.

■ *Chief Complaint:*

Dorothy Toth, a 31-year-old white woman, is 38 weeks pregnant with her third baby. She woke up this morning and discovered that she had vaginal bleeding.

■ *Present Status:*

Ms. Toth called her obstetrician, who suggested that she come into the hospital to be examined. Ms. Toth was not experiencing any cramping or pain; however, the bleeding was continuous and Ms. Toth was very apprehensive.

■ *Past Health History:*

Ms. Toth had no complications during her other two pregnancies. During this pregnancy, she felt unusually tired. However, she felt this was related to having two small boys, ages 6 and 4, to care for while being pregnant.

■ *Family History:*

Mr. and Ms. Toth have been married for 8 years. They did not plan the three pregnancies; however, "they only used the diaphragm whenever they felt like it." They enjoy children but do not feel they can afford more than three. Mr. and Ms. Toth attended childbirth classes with all three pregnancies, and Mr. Toth has attended both previous births and intends to be present for this birth. Mr. and Ms. Toth have decided that Ms. Toth will have her tubes ligated following the birth of this baby.

■ *Physical Exam:*

Vital signs: Temperature: 99.4°F. P: 98. R: 30. BP: 110/70.
Hct: 33%.
FHR: 130 bpm.
Bleeding: Moderate and profuse.
Heart: Sounds present, strong and regular pulse; no murmur or split-ting.
Lungs: No congestion, wheezes, rales, or rhonchi noted.
Elimination: Urine: Negative for protein, sugar, acetone, and blood; bowel sounds present.
Abdomen: Liver and spleen nonpalpable. Fundus: No contractions; membranes intact.
Neurologic signs: Eyes: PERL. Reflex: Appropriate.

■ *Psychological Status:*

Ms. Toth is very anxious. She appears afraid and has verbalized fear of possibly losing this baby. Mr. Toth paces constantly and questions everybody about the care his wife is receiving. He appears very con-cerned about his wife and baby.

---

149. In assessing Mr. Toth's constant questioning of the care his wife is receiving, the nurse needs to realize that Mr. Toth is:
A. Anxious about the hospital procedures being utilized.
B. Concerned about the medical staff's level of competence.
C. Anxious about his wife's and baby's condition.
D. Concerned about not being able to remain with his wife.

150. In assessing Ms. Toth's condition, the nurse needs to realize that the signs and symptoms of placenta previa are:

A. Contractions occurring every 5–6 minutes.

B. Low abdominal pain.

C. Painless vaginal bleeding.

D. Increased uterine muscle tonus.

151. In implementing care for Ms. Toth, the nurse should avoid:

A. Positioning Ms. Toth in a high Fowler's position.

B. Monitoring maternal and fetal vital signs.

C. Permitting vaginal examinations.

D. Providing a quiet environment.

152. Ms. Toth begins crying uncontrollably. The *most* appropriate nursing intervention includes:

A. Telling Mr. Toth to ask his wife what is wrong.

B. Sitting quietly with the couple.

C. Assessing Ms. Toth's reasons for crying.

D. Asking the physician to order her a sedative.

153. Electronic fetal monitoring is begun. Fetal monitor readings indicate a uniform shape of FHR deceleration pattern, which *begins after* the contraction is established and returns to baseline after the contraction ends. The nurse:

A. Does nothing; there is no clinical significance to this pattern.

B. Turns Ms. Toth onto her side, administers oxygen, and stops oxytocin induction if induction is in progress.

C. Alerts the surgery and anesthesiology departments that an emergency cesarean birth is pending.

D. Prepares Ms. Toth for immediate birth, either vaginal or abdominal.

154. Ms. Toth is scheduled for cesarean birth. In preparing her for the cesarean birth, the nurse would question which order?

A. Observe for contractions and ruptured membranes.

B. Continue fetal monitoring prior to surgery.

C. Insert an indwelling catheter.

D. Give meperidine hydrochloride, 100 mg IM 1 hour prior to surgery.

155. The nurse checks Ms. Toth's vital signs; 8 hours ago she gave birth by cesarean surgery. She is restless, but has no specific complaints. Her pulse is 94 and her BP is 108/70. The nurse should be suspicious of:

A. Anxiety over the birth.

B. Beginning shock.

C. Preeclampsia.

D. "Postpartum blues."

Medical Situation: Postpartum.

■ *Nursing Problems:*

Risk for altered feeling processes (anxiety).

Risk for fluid volume excess.

Health-seeking behavior (contraception).

Altered role performance.

■ *Present Status:*

Ann Bayless is a 24-year-old gravida 2, para 2. She has just given birth to a 7½-pound boy and is being moved into the recovery room.

■ *Past History:*

Normal prenatal course. Husband and wife attended Lamaze classes and looked forward to their roles during labor and birth. They did not have prenatal classes for their first child.

■ *Family History:*

Both parents are only children and are college graduates. Joe Bayless teaches English in a local high school and Ann is a registered nurse, although she has never worked. They have no religious preference.

■ *Physical Exam:*

Breasts: Erect nipples, free of cracks or fissures.
Fundus: Firm at umbilicus.
Lochia: Moderate rubra.
Perineum: Episiotomy (midline); 1-degree laceration, moderate pain.
Legs: Negative Homans' sign.
Reflexes: Negative.
Vital signs: Within normal limits.
Urinalysis: Free of sugar, protein, and WBCs.
Hematocrit: 37%.

---

156. In assessing Ann, the nurse should determine that which behavior demonstrates postpartum response to a *normal* birth?
    A. Anxiousness to discuss her labor and birth.
    B. Difficulty in relating to her labor experience.
    C. Seriousness and weariness.
    D. Concern about the postpartum environment.

157. The nurse finds Ann's fundus to be "atonic and boggy." The *first* nursing intervention is to:
    A. Check Ann's blood pressure and pulse.
    B. Catheterize the bladder for urinary retention.
    C. Administer an oxytocic drug.
    D. Massage the uterine fundus.

158. Fifteen minutes later the nurse checks Ann and finds her fundus firm, 2 fingerbreadths above the umbilicus, and shifted to the right. The nursing intervention is to:
    A. Check for a distended bladder and catheterize Ann if it is distended and she cannot void adequately.
    B. Take her vital signs.
    C. Get an order for methylergonovine maleate (Methergine).
    D. Chart that her fundus has failed to involute.

159. Due to religious convictions, the nurse is not in favor of artificial means of family planning. Ann asks the nurse, "What contraceptive would you recommend?" The nurse's most appropriate response is:
    A. "I don't believe in contraceptives, so you will have to ask somebody else."
    B. "Contraceptives have bad side effects; I don't recommend them."
    C. "There are several contraceptives. I will give you the pros and cons so that you can make the decision. Also, consult your doctor about what is best for you."

D. "The pill is the most effective and I recommend you start taking it right away so that you will be protected when you begin having intercourse."

160. Ann confides to the nurse that she does not know how she will be able to give enough attention to her husband, her 2-year-old, and her newborn. The nurse's most appropriate response is:
A. "This is the time your newborn needs you most; your husband and child will understand."
B. "Plan times when you can be with each one alone for a certain period."
C. "Ask your husband to assume the care of your 2-year-old."
D. "Suggest to your husband that this is a time when he can be free to go out with the boys and not worry about you."

Medical Situation: Postpartum.

- *Nursing Problems:*
  Ineffective breastfeeding.
  Altered comfort (pain related to engorged breasts).
  Health-seeking behaviors (exercises, breastfeeding).
  Psychomotor retardation.
  Management of therapeutic regime: effective individual.

- *Present Status:*
  Audrey Johnson, 31 years old, is gravida 2, para 2. She had a normal birth 8 hours ago. She is anxious to start caring for herself and her infant, and is excited about breastfeeding.

- *Past and Family History:*
  Ms. Johnson had a normal prenatal course. She has a 5-year-old son. Her husband works as a plumber and she works as a bookkeeper part-time. There are no health problems on either side of the family.

- *Physical Exam:*
  Breasts: Soft, erect nipples; no fissures or bleeding.
  Fundus: Firm at umbilicus.
  Lochia: Moderate rubra.
  Extremities: Negative Homans' sign; no edema.
  Reflexes: Normal.
  Perineum: Episiotomy (mediolateral) healing; minimal pain, ice to area.
  Vital signs: Within normal limits.
  Hematocrit: 36.5%.
  Urine: Free of protein and sugar.

161. Ms. Johnson is anxious to learn postpartum exercises. Which exercise is inappropriate for the nurse to instruct?
A. Kegels.
B. Pelvic tilt.
C. Sit-ups.
D. Head raising.

162. Ms. Johnson wants to breastfeed and assume care of her newborn. The nurse determines that she is in the "taking-hold" phase due to her:
    A. Self-centeredness.
    B. Readiness to concern herself with her infant.
    C. Concern about personal weight gain and diet.
    D. Reluctance to assume mothering role.

163. The nurse instructs Ms. Johnson that during the postpartum period a brassiere should:
    A. Not be worn.
    B. Be worn to support and lift the breasts.
    C. Be worn to press the breasts flat against the chest wall.
    D. Be worn to apply constant, firm pressure toward the midline.

164. Ms. Johnson develops engorged breasts. The nurse teaches her to:
    A. Apply heat to the breasts prior to nursing and express some milk.
    B. Apply heat to the breasts after nursing.
    C. Apply cold packs to the breasts prior to nursing.
    D. Take pain medication 1 hour prior to nursing.

165. Ms. Johnson experiences "postpartum blues" due to feelings of inadequacy about breastfeeding. The nurse's most appropriate response is to:
    A. Make inquiries about a psychiatric consult.
    B. Explain that these feelings are normal and why they occur.
    C. Separate mother and baby until the depression ceases.
    D. Utilize suicidal precautions.

166. In preparation for Ms. Johnson's discharge, which instruction by the nurse is inappropriate?
    A. Resuming her prepregnancy life-style within 1 week.
    B. Strategies for dealing with new responsibilities.
    C. Plans for her own rest and relaxation.
    D. Resources to call in an emergency.

167. One week later Ms. Johnson phones the clinic and says, "I decided not to continue breastfeeding because the baby liked the bottle better." The nurse's most appropriate response would be:
    A. "You should feed the baby the way that works best for both of you."
    B. "You should not have given up breastfeeding so soon."
    C. "Breastfeeding is better for the baby, but you may as well stay with bottlefeeding now."
    D. "Your baby won't gain weight and may develop milk allergies."

Medical Situation: Postpartum complications/neonatal complications.

■ *Nursing Problems:*
Altered family processes.
Altered feeling state: grief.
Altered parenting.

■ *Chief Complaint:*
Henrietta Young is a 36-year-old white woman who has just given birth via cesarean surgery to a stillborn infant at 28 weeks' gestation.

- *Present Health History:*

  Ms. Young began experiencing contractions early this morning around 4 A.M. She started bleeding profusely and her husband took her to the hospital immediately. Upon admission, her vital signs were: temperature: 97.6°F, P: 88, and R: 28; BP was 100/60; her pulse was very weak and thready. She was taken to the birthing room immediately because the FHR was not audible. She gave birth to a stillborn baby boy at 5:15 A.M.

- *Past Health History:*

  Ms. Young has one child, Bryant, age 6. She has experienced two miscarriages; one occurred about 8 years ago and another occurred approximately 2 years ago. Both of these miscarriages occurred at about 14 weeks' gestation. Her physician told her that the miscarriages were due to a hormone imbalance. However, her pregnancy with Bryant was essentially unremarkable, with the exception of 2 days of slight spotting at 8 weeks' gestation. Ms. Young had planned to have a tubal ligation immediately following this birth because she felt that she did not want to subject herself and family to other pregnancies.

- *Family History:*

  Mr. and Ms. Young have been married 10 years. Both of their parents are living and have no major health problems. Bryant attends a private school and is doing very well; that is, he is in the first grade doing second-grade reading and math. Mr. Young is employed as a graphic designer and Ms. Young is employed as a home economics teacher at a junior college.

- *Physical Exam:*

  First hour postpartum:

  Vital signs: Temperature: 97.8°F. P: 84. R: 26. BP = 100/70. Pulse is stronger and steadier than it was on admission.

  Lochia: Minimal.

  Incision site: Inflamed, edematous, and draining small amounts of sanguineous liquid; area is tender to touch.

  Bladder: Catheter draining well. SG: 1.015.

  Analgesia-anesthesia: Spinal.

  IV therapy: IV fluids infusing well; IV site not tender, no edema or inflammation.

  Level of consciousness: Ms. Young was given an analgesic immediately following surgery. She is responsive when stimulated. However, she is generally nonverbal. Her husband is visiting. He is visibly upset and has been quietly crying intermittently.

---

168. The nurse observes Mr. Young quietly crying. The *most* appropriate nursing action is to:
   A. Give him privacy.
   B. Talk with him about his wife's condition.
   C. Assess his plans for the baby.
   D. Ask him to leave the room.

169. Mr. Young asks the nurse what she thinks his wife's reaction will be to the loss of their baby. The *most* therapeutic response is to:
   A. Ask Mr. Young how he thinks she will react.

B. Tell him you are not sure.

C. Inform him that there is no uniform pattern of reaction.

D. Validate with him how his wife usually copes with losses.

170. Mr. Young asks the nurse what he should tell his son, Bryant. The *most* appropriate response the nurse could suggest is to:

A. Tell Bryant what has occurred at a level he can comprehend.

B. Ask Bryant what he thinks death involves.

C. Wait until Bryant's mother comes home so that both parents can tell him about his brother's death.

D. Inform Bryant that his brother is now in heaven.

Medical Situation: Normal-term neonate.

■ *Nursing Problems:*
Altered comfort (discomfort related to inverted nipples).
Health-seeking behavior: infant nutrition.

■ *Present Status:*
Michael, a black newborn, is 2 hours old, and weighs 7½ pounds (3405 g). Normal newborn; 39 weeks' gestation; uncomplicated labor and birth. Apgar scores of 9-9; second child of 34-year-old mother.

■ *Past History:*
Mother stopped working at the 28th week of this gestation; she does not plan to return to work. Father has a position in a law firm. Mother is anxious to care for her newborn.

■ *Physical Exam:*
Fontanelles: Within normal limits.
Reflexes: Appropriate and intact.
Muscles: Good tone; attitude of general flexion.
Color: Good oxygenation; hands and feet slightly blue when at rest.
Activity: Cries vigorously.
Stool: One meconium passed at 1 hour of age.
Urine: Once.
Chest: Clear.
Temperature: 36.8°C (axillary).
Respirations: 35 and irregular; without grunting, flaring, or retractions.
Bowel sounds: Present.
Extremities: No hip clicks.
Genitalia: Descended testicles; scrotum well creased.
Buttocks: Nickel-size mongolian spot.

171. The nurse instructs Michael's mother that in comparison with commercial formulas, breast milk:

A. Has a higher protein content.

B. Has a lower protein content.

C. Has a lower carbohydrate content.

D. Can cause milk allergy.

172. The nurse continues her health teaching and states that breastfed infants normally do not:

A. Have less chance of GI infection than formula-fed infants.

   B. Gain more weight than formula-fed infants.

   C. Have an increased chance of loose stools.

   D. Have less chance of milk allergy.

173. Because of severely inverted nipples, Michael's mother wishes to decrease the milk in her breasts and bottlefeed. The nurse would teach her to:

   A. Decrease stimulation to the breasts.

   B. Stimulate the nipples every 4 hours.

   C. Place heat on the breasts.

   D. Put the baby to the breast once a day.

174. The nurse bases her instruction to Michael's mother on knowledge that a 7½-pound (3.4-kg) infant from birth to 5 months of age needs approximately the following calories per kilogram per day (FDA guidelines):

   A. 100–110.

   B. 115–120.

   C. 150–160.

   D. 180–185.

175. It is determined that Michael needs 120 cal/kg/day. Commercial formulas are prepared to supply 20 cal/oz. How many ounces should Michael receive per feeding if he is fed every 4 hours?

   A. 2–3.

   B. 3–4.

   C. 4–5.

   D. 5–6.

Medical Situation: Normal-term neonates.

■ *Nursing Problems:*

Ineffective airway clearance.

Risk for disorganized infant behavior.

Risk for altered endocrine/metabolic processes.

Health-seeking behaviors related to effective breastfeeding, nutritional needs of infants.

Impaired home maintenance management.

Risk for infection.

Risk for injury.

Knowledge deficit (infant feeding and care).

Risk for altered neurologic/sensory processes.

Altered nutrition.

Risk for impaired skin integrity.

■ *Chief Concern:*

Baby Boy Zale is one baby in a neonate nursery. Baby Boys Allen, Bohack, and Chelain are also in the neonate nursery at this time. The chief concern is establishing health supervision for these normal neonates. The history below applies to Baby Boy Zale.

■ *Present Health History:*

Baby Boy Zale was born 2 hours ago with an Apgar score of 9 at 1 minute and 9 at 5 minutes. The vertex birth was spontaneous. He nursed at the breast soon after the cord was cut. No gross abnormalities were noted at birth. Prophylactic eye care was completed and vitamin K was administered at 2 hours of age.

■ *Past Health History:*

Baby Boy Zale is the product of the second pregnancy of a 23-year-old married white woman. The mother was not anemic, took no drugs, does not smoke, had no concurrent medical problems, and availed herself of early and continuous prenatal care. Labor was spontaneous and unmedicated.

■ *Family History:*

The family history is essentially negative. The maternal grandmother died of endometrial carcinoma at age 75. The father of the baby is 25 years old, blood type A, Rh positive, and works as a certified public accountant. The older brother is 3 years old, has had all immunizations, is followed routinely at the well baby clinic, and is apparently healthy.

■ *Physical Exam:*

Gestational age: 40 weeks.

Apical pulse: 132–144 in left midclavicular line, fifth intercostal space.

Posture: General flexion.

Respirations: 56/min (taken when active).

Weight: 3515 g (7 pounds, 12 ounces); length: 50 cm (20 inches).

Head circumference: 34 cm (13½ inches).

Chest circumference: 32 cm (12½ inches).

Axillary temperature: 36.5°C (97.6°F).

Findings of physical assessment: Reflexes within normal limits. Genitals: male, testes descended. Passed meconium and voided during physical exam.

---

176. Which statement related to administration of prophylactic eye care to the newborn is not accurate?
    A. Eye prophylaxis is a legal requirement.
    B. Ophthalmia neonatorum may be caused by gonococcal or pneumococcal infection.
    C. Eye prophylaxis may be delayed up to 1 hour after birth to facilitate parent-child attachment.
    D. The medication is placed into the conjunctival sac of the lower lid.

177. When providing eye prophylaxis, the nurse keeps in mind the following case.

    The nurse handed the physician an incorrect strength of eye medication drops for instillation into the neonate's eyes. The physician instilled the medication, and the infant's vision was impaired.

    The possible outcome of this situation may be that a charge of negligence can legally be brought against:
    A. The nurse.
    B. The physician.
    C. Both the nurse and the physician.
    D. The pharmacist for supplying an incorrect strength.

178. Vitamin K_1 (Aquamephyton) is injected intramuscularly to Baby Boy Zale because his GI tract is incapable of producing it for several days. Vitamin K_1 acts to catalyze the synthesis of prothrombin

in the liver. The nurse giving vitamin K₁ to the neonate needs to know that which of the following may be unsafe?

    A. Using the greater trochanter and posterior superior iliac spine as landmarks.

    B. Using the vastus lateralis muscle as the injection site.

    C. Injecting the needle at a 45-degree angle.

    D. Injecting the needle in the direction of the knee.

179. Baby Boy Zale has been sleeping quietly since he fell asleep at 1 hour after birth. Now, 4½ hours after birth, he awakens, regurgitates mucus, and has a gagging episode. His mother calls out for the nurse. The nurse:

    A. Holds the baby upright and pats him on his back.

    B. Immediately takes him to the nursery for suctioning.

    C. Helps the mother hold him face down with the head lower than the buttocks.

    D. Reassures the mother that this is normal and not to worry.

180. To measure Baby Boy Zale's head circumference, the nurse places the measuring tape:

    A. From the bregma, around the lower portion of the occiput, and back to the bregma.

    B. On the forehead just above the eyebrows, over the junction of the ear on the head, and back to the forehead.

    C. Over the eyebrows, over the top of the earlobes, and back over the eyebrows.

    D. From the mentum to the vertex and back to the mentum.

181. Baby Boy Zale's head circumference is 34 cm; his chest circumference, 32 cm. Based on the findings, the nurse:

    A. Refers him to be appraised for psychomotor retardation.

    B. Knows that the physician will want to transilluminate his cranial vault.

    C. Measures his occipitofrontal circumference daily.

    D. Records the findings and does nothing further.

182. The nurse assesses each neonate to identify existing or potential problems. If an examination of Baby Boy Zale revealed the following findings, which requires further assessment and/or therapy?

    A. Flat hemangiomas, head circumference between 31 and 36 cm, hemoglobin between 14 and 19 g.

    B. Acute hearing, erythema toxicum, wrinkles covering the soles of the feet.

    C. Single palmar crease, snuffles, apical pulse between 120 and 140/min at rest.

    D. Molding, smegma, acrocyanosis, respiratory rate of 40/min at rest.

183. As the charge nurse, your responsibilities include checking all orders for feeding the neonates in your nursery. Which information is the *most* important in evaluating the nutrient needs of a normal-term neonate such as Baby Boy Zale?

    A. Kilograms of body weight.

    B. Percentage of body fat.

    C. Route of feedings, for example, oral versus gavage.

    D. Crown-to-heel length in centimeters.

184. In the plan of care for a normal newborn like Baby Boy Zale, the nurse incorporates knowledge that:

    A. The neonate may vomit if his bottle is propped.

B. He may lose up to 4 ounces within the first days of life.

C. His stomach holds about 120 mL at any one time.

D. He needs about 90 cal/kg/24 hours or 45 cal/pound/24 hours.

185. When an object touches the newborn's cheek and/or lips, he turns toward the stimulus and opens his mouth. This reflex is known as:

A. Babinski.

B. Suck and swallow.

C. Rooting.

D. Moro.

186. Baby Boy Allen, a full-term neonate, was born 16 hours ago. Birth weight: 7 pounds, 6 ounces. Length: 20 cm. He is very active when awake and is on oral feedings. The fluid requirement for Baby Boy Allen is:

A. 32 oz. (960 mL) per day.

B. 24 oz. (720 mL) per day.

C. 16 oz. (480 mL) per day.

D. 12 oz. (360 mL) per day.

187. Ms. Bohack, 2 days postpartum, seems preoccupied. Her baby has just eaten and is now back in the nursery. She tells you the baby weighed 8 pounds, 2 ounces (3685 g) at birth; today he weighs 7 pounds, 10 ounces (3459 g). Ms. Bohack is worried because the baby "lost so much." The nurse's reply is based on the knowledge that Baby Boy Bohack can lose up to how many ounces and still be within normal range?

A. 7 ounces.

B. 9 ounces.

C. 11 ounces.

D. 13 ounces.

188. Baby Boy Chelain, birth weight 9 pounds, 10 ounces, was born 3 hours ago. Apgar scores are 8 at 1 minute and 9 at 5 minutes. At this time, the nurse notes athetoid posturing and movements, with slight tremors of the arms. The nurse's *first* response is:

A. Feed him glucose water because these behaviors could be signs of hypoglycemia.

B. Swaddle him because these are signs of drug (heroin or alcohol) dependence.

C. Notify the physician because these behaviors are signs of increased intracranial pressure.

D. Obtain and assess a blood sample by heelstick because these behaviors could be signs of hypoglycemia.

189. Ms. Allen is concerned because she thinks her son has lost too much weight. She adds, "Maybe I should just give him a bottle and not try to breastfeed. That way I know what he gets." Baby Boy Allen's weight was 7 pounds, 6 ounces (3345 g). Today, 2 days later, he weighs 6 pounds, 14 ounces (3118 g). Which is the *best* way for the nurse to deal with this situation?

A. Explain to Ms. Allen that he can lose 3 more ounces and still be within normal limits for weight loss.

B. Bring a bottle of glucose water for Ms. Allen to give after breastfeeding, until her milk comes in.

C. Tell her he is OK and that you will ask her pediatrician to come by to see her.

D. Support Ms. Allen's statement about the fact that an accurate record is possible when feeding is by bottle.

190. Ms. Zale is sitting up in bed, looking comfortable. She is holding her son and feeding him a formula. Ms. Zale says he seems to get full fast and he has regurgitated some of the formula already. She adds, "Maybe I just don't know how to feed him." As the baby sucks on the nipple, large bubbles are seen in the bottle. The nurse tells the mother:
   A. "Tighten the bottle cap a little."
   B. "Hold the baby in a more upright position."
   C. "Bubble (burp) him after each ½ oz. of formula."
   D. "Let me show you how. Then you can try it again."

191. Which *best* demonstrates that Ms. Allen has incorporated knowledge about breastfeeding?
   A. She weighs her baby and records his weight before and after feeding.
   B. She touches her nipple to the baby's cheek at the start of the feeding.
   C. She bubbles (burps) her baby after his feeding.
   D. She protects her clothing by placing small squares of cellophane wrap in her nursing bra.

192. A heelstick is performed before discharge of Baby Boy Zale to obtain a blood specimen to rule out:
   A. Kernicterus.
   B. Hypoglycemia.
   C. Thrush.
   D. Phenylketonuria (PKU).

193. The nurse determines that Ms. Allen needs further health instruction in appropriate care for her son when the nurse observes her:
   A. Cleanse each eye with a different cotton swab or corner of a washcloth from the inner canthus outward.
   B. Retract his prepuce completely to cleanse the glans penis of smegma and other debris.
   C. Rinse shampooed hair thoroughly, then use a fine-toothed comb or brush.
   D. Cleanse his nares and ears with twists of moistened cotton.

194. American nurses often work with people from a wide variety of cultures, both within the United States and in other countries. Nurses are asked questions about the preparation of formulas to meet the nutritional needs of infants. Which information about the infant's digestive capabilities and about formula preparation is inaccurate?
   A. Fat is not easily digested by the gastric fluids.
   B. Milk sugar (lactose) is not digested in the stomach.
   C. Corn and coconut oil mixtures are not absorbed to the same extent as the fat in human milk.
   D. The infant is developmentally ready for solid foods at about 4–6 months. There are no nutritional advantages to giving him solids before that time.

195. The postpartum nurse is asked to evaluate a pamphlet to be used to provide anticipatory guidance regarding breastfeeding. Which is the only accurate statement?
   A. Breastfeeding provides immune bodies to the neonate and contraception for the mother for 8 months.
   B. Breast milk jaundice is a possible complication that necessitates weaning.

    C. Breast milk contains more protein and less carbohydrate than cow's milk.

    D. Women who have breastfed are less prone to develop breast cancer later in life.

196. Which commonly occurring neonatal problem is due to overdressing or excessive warmth?

    A. Diaper rash.

    B. Miliaria.

    C. Seborrheic dermatitis.

    D. Diarrhea.

197. When working in the newborn nursery, the nurse keeps the following case in mind:

> The newborn nursery is desperately short of staff. The nurse agrees to work an extra shift. During the 14th hour on duty, the nurse makes a medication error.

Since the nurse was filling a "desperate" staffing need and therefore was fatigued, the outcome may be that:

    A. The hospital accepts the responsibility for the negligent act.

    B. The nurse is held to the same standard of care as always.

    C. No problem exists if the patient does not bring legal action.

    D. Legal responsibilities are met if an incident report is filed.

Answers and Rationale

  1. *A.* ***Tuning in to her or his own feelings and values about the topic.***

A is correct; before beginning a class for adolescents, the nurse needs to assess her or his own feelings. If these feelings are negative, this attitude will be conveyed to the adolescents. *B* is incorrect; adolescents assume their own style of dress and language to assert their difference and independence from adults. In addition, adolescents still want to have adults there to provide some limits, to give direction, and so forth. Adolescents prefer to have adults behave like adults. *C* is incorrect because the adult focusing on her or his own experience as an adolescent does not help the adolescent focus on her own reality now. *D* is incorrect because the younger adolescent is still operating on the concrete level of cognition (Piaget); that is, the adolescent has to have an experience first before she can understand it or its implications for the future.

  2. *C.* ***When the ovum is present.***

C is correct; a woman is fertile only during the time that an egg is present. The egg lives for 24 hours but is thought to be most fertilizable for 12 of those hours. The precise moment of ovulation is difficult to determine. In general, ovulation occurs 14 days before the next menstrual flow begins. *A* is incorrect because ovulation occurs on day 14 of a *28-day cycle only*. *B* is incorrect because ovulation occurs 14 days before the next cycle begins. Just before menstruation, the endometrium is beginning to die. *D* is incorrect; day 1 of the cycle is the first day of menstrual flow, that is, when the endometrium is being sloughed off.

  3. *D.* ***Is about to ovulate.***

D is correct. Just before ovulation, the cervical mucus is clear, copious, stretchable (good spinnbarkeit), and slippery. This is the

characteristic assessment in the presence of high levels of estrogen. *A* is incorrect. During pregnancy, high levels of estrogen and progesterone are present; therefore, the cervical mucus does not have the consistency of egg white and is not stretchable. *B* is incorrect; leukorrhea is a white or yellowish discharge that does not have all the characteristics listed. *C* is incorrect; just before menstruation, when the levels of progesterone and estrogen fall, the mucus is sticky and less copious.

4. **A. *8 through 24.***

   A is correct. To calculate the fertile period, subtract 18 from the shortest cycle in the previous 12 months, and 11 from the longest. $26 - 18 = 8$; $35 - 11 = 24$. This formula works in the following manner: In the 26-day cycle, ovulation occurs on day 12; sperm live for 3 days, so there should be no sperm in the vaginal vault for 3 days before ovulation. For added protection, subtract 1 more day. Therefore, the fertile period begins on day 8. In the 35-day cycle, ovulation occurs on day 21; the ovum lives for 1 day (or so); add another 2 days for added protection. The fertile period then ends on day 24. *B, C,* and *D* are incorrect because of the explanation for *A.*

5. **A. *Chloasma.***

   A is correct. Chloasma, or mask of pregnancy, consists of brownish blotches, irregular in shape and size, caused by the high levels of estrogen. Estrogen stimulates production of melanin-stimulating hormone (MSH), which increases the level of anterior pituitary melanotropin. *B* and *C* are incorrect. Although both palmar erythema and spider nevi (telangiectasia) result from high levels of estrogen, they *do* fade after birth and when the oral contraceptive estrogen dosage is changed or discontinued. *D* is incorrect; pruritus is due to excretion of bile salts through the skin in some women during pregnancy. Treatment for mild cases is bathing. For more severe cases, phenobarbital is prescribed. Phenobarbital works in the liver to stimulate the production of an enzyme that relieves this condition.

6. **D. *Thromboembolic disease.***

   D is correct; thromboembolic disease is the most serious of the side effects and carries an increased risk of morbidity and mortality. Pulmonary embolism and cerebral thrombosis may occur, especially in women over 35 years old. However, there is a much higher risk—6–10% higher—of thromboembolic disease during pregnancy than when one is taking the pill. *A, B,* and *C* are incorrect because ectopic pregnancy, cholelithiasis, and gastroenteritis are not associated with the ingestion of oral contraceptive hormones. *A,* ectopic pregnancy, is sometimes related to the presence of an IUD. Ovulation continues to occur with this contraceptive method and there is an increased incidence of pelvic inflammatory disease in IUD wearers. When inflammation heals, scar tissue forms. This scar tissue may partially occlude the fallopian tubes, preventing a fertilized egg from moving into the uterine cavity.

7. **A. *Appearance of a chancre.***

   A is correct; appearance of a chancre brings most people in for treatment. This primary lesion usually forms at the initial site of infection. The chancre is a dull red, firm, and painless papule that will heal spontaneously. *B* is incorrect; within a few hours, organisms multiply and spread to all tissues, but the chancre does not

appear until 3–4 weeks after infection. *C* is incorrect because the tests for serology are not positive until at least 1 week after the chancre appears. The test is for antibodies to *Treponema pallidum.* *D* is incorrect because pain in the joints and loss of position sense are two characteristics of tertiary syphilis, known as tabes dorsalis.

8. **D.** **The spirochete crosses the placenta after the 16th week of gestation.**

    D is correct. The spirochete can cross the placenta after the 16th–18th week; however, treatment with high doses of penicillin now should cure the disease for the parents and the fetus. If treated now, there will be no residual fetal damage. *A* is incorrect; secondary lesions, called condylomata lata, appear about 4–6 weeks after the chancre heals and are extremely infectious. Skin lesions can occur anywhere over the body, but are characteristic over the palms of the hands and soles of the feet. *B* is incorrect; obliterative endarteritis, destruction of the arteries and arterioles, causes cell damage to the eyes, cardiovascular system, brain, and spinal cord, and is characteristic of tertiary syphilis. *C* is incorrect; gummas, hard nodules of dead tissue throughout the body, are characteristic of tertiary syphilis.

9. **A.** **It can result in pharyngitis, tonsillitis, perianal pruritus, and burning.**

    A is correct. In addition to pharyngitis, tonsillitis, and perianal pruritus, gonorrhea may result in pelvic inflammatory disease and sterility; thick, purulent vaginal discharge; infection of Skene's and Bartholin's glands; and infection of the urinary tract. Arthritis, meningitis, and endocarditis may be systemic infections. *B* is incorrect; systemic congenital gonorrhea results in purulent discharge from the newborn's eyes, nose, and vagina or anus. The infant may have a cough, be lethargic, and be a poor feeder. He or she may also be born prematurely from premature rupture of the membranes. *C* is incorrect; symptoms of vaginal infection are profuse, purulent drainage and bartholinitis. The male may experience dysurea, a discharge from the urinary meatus, penile pain, and frequency. Blisterlike lesions are characteristic of herpes. *D* is incorrect; if infected drainage soils bed linens and clothing, one can become infected through contact with these soiled articles. Also, the newborn may receive a systemic infection via the placenta or by direct contact with the infected birth canal.

10. **C.** **Is a candidate for rubella vaccination shortly after birth.**

    C is correct; Wendy has not had rubella or is serologically negative because her rubella titer is less than 1:8. She is scheduled for rubella vaccine shortly after birth. Because the live, attenuated rubella virus is not communicable, she may receive the vaccine even if she is lactating. However, the virus can be teratogenic, so contraception is mandatory for 2 months following vaccination. *A, B,* and *D* are incorrect because of the rationale given for *C.*

11. **A.** **Coach her to keep eyes, mouth, and hands open.**

    A is correct because while keeping one's eyes, mouth, and hands open, it is almost impossible to contract the perineal musculature. *B* is incorrect because squeezing anything facilitates tensing all other parts of the body. *C* and *D* are both therapeutic nursing actions, but *A* *best* assists Wendy to relax her perineal muscles.

12. ***D.   An assignment on another unit.***
    D is correct because Wendy needs to avoid inhalant anesthetics that leak out into the surgical suites. An increased incidence of spontaneous abortion and congenital disorders has been noted among women who are in contact with air polluted with anesthetic gases, or who are married to anesthesiologists. ***A, B,*** and ***C*** are incorrect because the seriousness of exposure to anesthetic gases is the most important consideration in this instance.

13. ***D.   Nurse O: "Changes in female hormone levels occur in both pregnancy and menopause."***
    D is correct because the changes in female hormone levels that occur in both pregnancy and menopause cause vasomotor instability. ***A*** is incorrect; although identification between mother and daughter may occur, shared feelings of increased body warmth have not been noted. ***B*** is incorrect because the data as stated do not support an infectious process as the etiologic factor. ***C*** is incorrect; although feelings of increased perspiration and warmth do accompany anxiety, the data given here do not support that conclusion.

14. ***A.   Avoids bending at the waist and does "flying" exercises.***
    A is correct. Heartburn occurs when food is regurgitated from the stomach into the esophagus through the relaxed cardiac sphincter. The "flying" exercise promotes good posture, thus permitting food to reenter the stomach; avoidance of bending at the waist prevents food from spilling out of the stomach into the esophagus. ***B*** is incorrect because eating some type of dry (not fatty) carbohydrate before getting out of bed is a method for relieving *nausea,* not heartburn. Excessive salt is avoided, especially during pregnancy. ***C*** is incorrect because it has nothing to do with heartburn; the unusual (usually culturally influenced) practice of eating starch or red clay is termed pica. ***D*** is incorrect because health habits that include getting plenty of fluids, exercise, and food with roughage help to prevent or minimize *constipation,* not heartburn.

15. ***C.   "Avoid taking any drug unless it is prescribed by your physician."***
    C is correct; drugs ingested by the mother cross the placenta and may adversely affect fetal development or function. ***A*** is incorrect. Lying on one's side does take the weight of the uterus off the ascending vena cava, but the positive physiologic effect of the left lateral position is not due to placental function. ***B*** is incorrect because not all vitamins and minerals can be stored. The baby's nutritional needs require the mother's daily intake of certain nutrients—for example, vitamin C. ***D*** is incorrect because infectious agents of small molecular size do cross the placenta. In addition, maternal fever has an adverse effect on the fetus. Because of this, all maternal infections must be diagnosed and treated appropriately.

16. ***C.   Help Wendy to place her full weight on the affected leg and lean forward on it.***
    C is correct. Stretching the affected muscles—by standing on the affected leg and leaning forward, by extending the knee and flexing the toes toward the tibia, or by standing on a cold floor—forces the muscle to relax. ***A*** is incorrect because the pain may be from thrombophlebitis or a thrombus, and massage could dislodge the clot into the bloodstream. ***B*** would be correct *only* if pain were felt when the muscle was stretched; in that case the physician should

be notified immediately. *D* is incorrect because pointing one's toes may cause cramping of the gastrocnemius (Charley horse). One should remember to point the heel, not the toes.

17. **A. *Feelings and attitudes toward pregnancy.***
 A is correct; feelings and attitudes toward the pregnancy are important during the first trimester. Other concerns and questions at this time include the discomforts being experienced (such as nausea, frequency of urination, fatigue, and increase in vaginal mucus) and content pertaining to general hygiene (such as rest, sleep, douching, and marital relations). Financial problems are also a major concern for many now. The Bradley method of preparation for childbirth features an "early bird" class to answer these questions. *B* is incorrect because preparation for labor and child care is begun in the eighth month by Bradley, as well as by Lamaze and other expectant-parent teachers. *C* is incorrect because it is more appropriate to the middle trimester. At that time the mother is very interested in fetal growth and development and in learning what discomforts are normal, how long they will last, and what can be done to relieve them. The nurse instructs the mother regarding danger signals at that time. *D* is incorrect because most discussions regarding questions and concerns about the new baby and the new role as parents are most appropriate during the last trimester. Most preparation for labor is done during the last trimester, as well.

18. **C. *Ptyalism and gums that bleed when teeth are brushed.***
 C is the answer; ptyalism and gums that bleed easily result from hyperemia due to the increased vascularization during pregnancy. It is *not* necessary to notify the physician about them; gentle brushing and good oral hygiene are the only treatments recommended. Excessive salivation may increase the discomforts of nausea. If bleeding of the gums becomes severe, however, the physician should be notified immediately. *A* is incorrect because severe, persistent headache is one of the symptoms of preeclampsia—probably due to generalized arteriospasm and edema of the CNS. *B* is incorrect because tight finger rings and puffy eyes are symptomatic of preeclampsia. Edema with weight gain in excess of 1 pound per week is one of the three cardinal symptoms/signs. The other two are hypertension and proteinuria. *D* is incorrect because a change in the woman's facial expression or affect, often accompanied by a puffy, edematous appearance, is seen with severe preeclampsia.

19. **B. *Grapefruit juice, beef liver with scrambled egg, toast, coffee.***
 B is correct because the breakfast described is high in both vitamin C and protein. The *A* breakfast is high in vitamin C, but the Cream of Wheat is lower in protein than the liver in the *B* breakfast. *C* is incorrect because the fried tomato has less vitamin C than orange or grapefruit juice. *D* is incorrect because this breakfast has less vitamin C and less protein than the other three choices.

20. **A. *Cottage cheese and tomatoes, peanut butter sandwich, milk.***
 A is correct because this combination contains the most protein (peanut butter), vitamin C (tomato), and calcium (milk). *B, C,* and *D* all have less protein than choice *A*.

21. **A. *Pork chops, sweet potatoes, cooked dried beans, raisin pudding.***

    A is correct because it has the most iron. *B, C,* and *D* are incorrect because all have less iron than choice *A.*

22. **B. *300.***

    B is correct; a calorie increase of 300 is recommended for (1) growth of fetus, placenta, and maternal tissues; (2) sparing of protein for tissue synthesis; and (3) weight gain needed as a buffer against food deprivation (also prevents catabolism of maternal tissues). *A* is incorrect because 200 calories is too small an increase. *C* and *D* are incorrect because each is much too large an increase.

23. **A. *Maintenance of a trim figure to promote good body image.***

    A is the answer; the nurse should *not* counsel any pregnant woman on maintaining a trim figure during pregnancy; this is an impossibility. The factors in *B* (cultural, economic, and educational background) *will* affect each teenager differently and *must* be considered to provide a healthy prenatal course. *C,* growth needs of the mother and the infant, *will* vary with each mother and should be considered. *D,* dietary habits, *will* vary and *must* be considered for each individual.

24. **D. *Egg yolk and deep-yellow or green vegetables.***

    D is correct. In the American diet fat-soluble vitamin A is easily acquired through sources other than whole milk—egg yolk, deep-yellow or green fruits and vegetables, milk products, and liver. *A* and *C* are incorrect because they are not as complete as *D. B* is incorrect because wheat germ oil is a source of vitamin B, not vitamin A.

25. **D. *3 cups of vanilla ice cream.***

    D is the answer; substitution of ice cream would add extra calories and too many carbohydrates, leading to obesity, and could leave little appetite for other essential food groups, perhaps leading to poor nourishment. *A, B,* and *C* are not the answers because each *is* acceptable as a milk substitute to meet nutritional requirements.

26. **C. *Is gaining the desired amount of weight.***

    C is correct; Wendy is meeting her nutritional needs if her weight gain is appropriate. *A* is incorrect; if she is at her prepregnant weight, she is not maintaining a nutritional state that is meeting her pregnancy needs or the growth needs of her infant. *B* is incorrect; increasing the amount of meat in her diet may not meet her RDA requirements. *D* is incorrect; she may need nutritional between-meal snacks to meet both her and the baby's needs.

27. **C. *Dried apricots, prune juice, chili con carne.***

    C is correct because all these foods are high in iron. Egg yolk, liver, and Fe-enriched bread and bread products are also good food sources. Use of iron frying pans and pots adds some iron to one's intake. Steaming vegetables, rather than boiling them, and using all the fluid from cooking conserve the iron content. Spinach, *A,* is not as good a source of iron because its high bulk content decreases absorption. *B* and *D* are incorrect because *C* supplies more iron.

    Note: Some people with nutritional iron deficiency anemia have normocytic, normochromic RBCs.

28. **C.** *Take divided doses after meals, with a glass of fruit juice.*

    C is correct because vitamin C seems to enhance iron absorption, and because ingestion after meals seems to lessen GI irritation. Also, since bulk in food decreases absorption, iron medications are taken *after* the meal. *A* is incorrect because milk and phytic acid (found in some whole-grain cereals like oatmeal) prevent absorption. *B* and *D* are incorrect because of the rationale given for *C*.

    A normal diet contains about 6 mg of iron per 1000 cal, so 3000 cal (well above the intake of most women) is needed to meet the RDA of 18 mg of iron. Pregnancy increases the demand for iron; therefore, supplementation is still being recommended even though there is some concern about possible effects on embryonic/fetal development.

29. **C.** *Prepare for childbirth.*

    C is correct; preparation for childbirth is the primary purpose of the classes, although this preparation may be instrumental in helping the woman (couple) achieve some success with *A* and *B. D* is incorrect because some forms of dystocia are unavoidable. There are no guarantees regarding childbirth or its outcome. Childbirth preparation classes counter ignorance and misconceptions, and offer alternatives. This is ego building for the woman (couple) and increases her (their) ability to solve problems and to cope.

30. **A.** *Nonpain signals from other parts of the nervous system can greatly alter the degree of transmission of pain signals.*

    A is the correct definition of the "gating" mechanism. *B, C,* and *D* are incorrect; the specific details of the "gating" mechanism are as yet unknown.

31. **C.** *Trying to resolve previously repressed conflicts.*

    C is correct; previously unresolved (repressed) issues can be revived during pregnancy. During this time, the woman is vulnerable, but she is also open to assistance in resolving unfinished developmental tasks or emotional problems, so that the experience can be one of maturation and growth. The woman may be concerned that her thoughts are bizarre. She can, however, distinguish fantasy from reality, which clearly distinguishes her from the psychotic. Therefore, *A, B,* and *D* are incorrect.

32. **A.** *Puts on a sterile glove and inserts litmus paper into the vagina, placing it near the cervix.*

    A is correct; phenaphthazine (Nitrazine) paper is litmus paper used to determine pH values. Therefore, the paper turns blue-green if the membranes have ruptured. The litmus paper must be placed near the cervix so that the color will change in response to the presence of amniotic fluid and to avoid false readings from urine or mucus. *B, C,* and *D* are all incorrect because they do not pertain to testing for amniotic fluid.

33. **A.** *Fever.*

    A is correct because fever is an indication of a possible complication of PROM—namely, chorioamnionitis. Intrauterine infection can cause intrauterine fetal death or neonatal pneumonia. *B* is incorrect. Vaginal examinations are avoided, if possible, for two reasons: (1) There is an increased risk of infection and (2) touching the cervix may elicit Ferguson's reflex, for example, uterine contractions. *C* is incorrect because coagulopathy (i.e., DIC) is not associated with PROM. *D* is incorrect. Only when the physician is

considering effecting birth, perhaps after treatment with beta-methasone, is the risk of amniocentesis considered, to estimate fetal lung maturity (L/S ratio).

34. **B. *"Would you like to talk about what you mean when you say you are afraid?"***

   B is correct because it encourages the mother-to-be to communicate her feelings and concerns. *A, C,* and *D* ignore Wendy's concerns, deny her right to have concerns, and try to offer reassurance for as-yet-unidentified fears—fears that may have nothing to do with the "doctor's examinations."

35. **A. *Determine fetal presentation, position, lie, and engagement.***

   A is correct. Abdominal palpation, the four maneuvers of Leopold, provides data regarding fetal presentation, position, lie, and engagement. After determining fetal position, the nurse then also knows approximately where the FHR may be heard the loudest. *B, C,* and *D* are incorrect descriptions of Leopold's maneuvers.

36. **A. *Abandonment.***

   A is correct; the woman who chooses to return home soon after birth needs to have signed an informed consent and to have been prepared for her physical recovery and care of the neonate through some educational program during her pregnancy. Many hospitals arrange for RNs to make home visits during the early days after birth. A woman in labor must be attended at all times. *B, C,* and *D* are incorrect statements of the woman's legal rights.

37. **A. *"I'll walk you to the bathroom and stay with you."***

   A is correct. Even if the woman has had an unmedicated labor, she is subject to orthostatic hypotension, which can develop during the first 48 hours after birth. The rapid change in intraabdominal pressure at birth results in splanchnic engorgement. The woman must be forewarned about the possibility of light-headedness and should be instructed to sit down immediately anywhere if she feels faint. *B* is incorrect because of the danger of orthostatic hypotension after birth. *C* and *D* are incorrect because orthostatic hypotension poses a safety hazard for the new mother. However, the instructions to wash hands, to wipe from front to back, and to lean forward when voiding are all appropriate *after* the woman's safety is ensured.

38. **B. *Increased circulation of blood and lymph in the mammary glands.***

   B is correct because primary breast engorgement represents an increase in the normal venous and lymph system of the breasts. It is *not* due to an overaccumulation of milk, to increased fluid intake, or to production of more milk with successive pregnancies. Therefore, *A, C,* and *D* are incorrect.

39. **D. *Preoccupied with her own needs.***

   D is correct; the mother generally has a great need to talk about her perceptions of labor and birth. *A* is incorrect because the mother tends to be passive and somewhat dependent after a normal birth. *B* is incorrect because a new mother's hearing is very acute; she thinks everything she hears refers to her. *C* is incorrect because the new mother follows suggestions and is hesitant about making decisions.

40. **B. *Firm, below the umbilicus in the midline.***

   B is correct because vasoconstriction is causing firmness and descent of the uterus into the pelvis. *A* indicates a possible full blad-

der, which could eventually affect involution by preventing descent of the uterus into the pelvis. *C,* a slightly boggy fundus two finger-widths above the umbilicus, indicates vasodilation, with the possibility of hemorrhage; it would also mean a full bladder with urinary stasis. *D,* a boggy fundus below the umbilicus in the midline, indicates vasodilation, with the possibility of hemorrhage.

41. **B. *Herself and her needs.***

    B is correct because the mother needs mothering before she can nurture. *A* is incorrect because in the taking-in phase the mother is concerned about herself first, and her infant only *after* her own needs are met. *C* and *D* occur, not in the taking-in period, but in the taking-hold period.

42. **C. *Massaging the perineum promotes absorption of the hematoma.***

    C is the least relevant and therefore the answer of choice. *A, B,* and *D* are not the answers because they *are* all good nursing actions to increase the comfort of episiotomy repair.

43. **A. *Of the possibility of discomfort and infection in the mother.***

    A is correct. Until small tears are healed, there is the possibility of infection and discomfort. Discomfort may also occur because the vagina is hormone poor at this time. Healing is usually complete by 3 weeks, which corresponds with the cessation of lochial flow. *B* is incorrect; whether this is an unsafe time to become pregnant would depend on the person and the circumstances. *C* is incorrect because if the episiotomy is correctly repaired, tearing will not occur. Some episiotomies or tears are purposely not repaired; intercourse is dependent on the comfort of the couple. *D* is incorrect because intercourse actually aids in involution due to the uterine contractions that occur during intercourse.

44. **C. *Abdominal enlargement.***

    Abdominal enlargement does not generally occur *until the end* of the first trimester; it could also be due to rapid weight gain or "gas." *A, B,* and *D* are incorrect because they are all presumptive symptoms of pregnancy; *A* and *D* are *subjective* symptoms, and *B* is an *early* objective sign.

45. **B. *Urine test for pregnancy, protein, sugar, estriol levels.***

    A urine test for *estriol* levels was often used in the past to assess for fetal-placental-maternal well-being, but it is not used currently, and would *not* be indicated at the *first* prenatal visit. *A* and *C* are not the answers because they *are* appropriate tests to perform. *D* is less correct than *B.* If the pregnancy is under 10 weeks, fetal heart rate and fundal height may not be appropriate; however, some women wait much longer than the second missed menstrual cycle before "suspecting" pregnancy.

46. **C. *X-ray pelvimetry prior to the onset of labor.***

    C is correct because x-ray pelvimetry is not ordered prior to the onset of labor; it might be ordered *if* the lack of progress of labor indicates. *A, B,* and *D* are all appropriate procedures performed during routine prenatal care.

47. **C. *Gravida 2, para 0.***

    "Gravida" refers to a pregnancy regardless of its duration. "Para" refers to past pregnancies carried to the point of viability (24 weeks). Since this is the second pregnancy, the woman is gravida 2. The first pregnancy was approximately 14 weeks' gestation, and

therefore was not carried to the point of viability. Ms. Primings is classified as gravida 2, para 0. *A, B,* and *D* are all incorrect based on the explanation for *C.*

48. *C.   Acts as the fetus's organ of respiration, nutrition, and elimination.*

C is correct because movement of nutrients, gases, and wastes is by diffusion and osmosis. This exchange occurs through the chorionic portion. The hormones secreted include human chorionic gonadotropin (HCG), human chorionic somatomammotropin (HCS), estrogen, and progesterone, which is called pregnanediol during pregnancy. The placenta is a barrier against larger molecules of infectious agents and many drugs, preventing their passage through to the fetus. Generally, maternal and fetal blood do not mix. At term, the placenta weighs about 1 pound (or one-sixth of the newborn's weight) and is 7–8 inches in diameter. The word *placenta* is a Greek word meaning "flat cake." *A* is incorrect because the amniotic fluid acts as the shock absorber. *B* is incorrect because nutrients are not secreted. They are passively received from the mother and actively transported to the fetus. *D* is incorrect because all blood elements are produced within the fetal system.

49. *A.   Coaches her on breathing normally and keeping her hands relaxed.*

A is correct because it is almost impossible to tense the abdominal and perineal muscles if the mouth and hands are relaxed and when one is concentrating on breathing. *B, C,* and *D* are incorrect because they contribute to increased tension without reassurance or therapeutic value.

50. *A.   Leg cramps.*

A is correct; increased absorption of phosphorus upsets the calcium-phosphorus ratio. Treatment consists of discouraging ingestion of more than 1 quart of milk and prescribing calcium pills. *B, C,* and *D* are unrelated to the calcium-phosphorus ratio.

51. *B.   Umbilical vein.*

B is correct; the umbilical vein conveys oxygenated blood from the placenta to the fetus. The two umbilical arteries, *A,* return depleted blood from the fetus to the placenta for replenishment. Both the left ventricle, *C,* and the arch of the aorta, *D,* contain blood from the umbilical vein, which has been diluted with deoxygenated fetal blood.

52. *B.   Palpitation.*

B is correct; palpitation, a subjective symptom, may occur occasionally during a normal pregnancy; the pulse rate may increase by as much as 10 bpm. *A, C,* and *D* are incorrect because physiologic anemia, increased blood pressure, and increase in the anterior-posterior chest diameter (all objective signs) are not changes that the woman will necessarily notice. She may become aware that she needs a larger bra, but probably attributes this to the increased size of the breasts.

53. *B.   Heartburn.*

Heartburn is caused by the compression and displacement of the stomach upward with the increasing size of the uterus and from progesterone-provoked reverse peristalsis, which results in reflux of stomach contents into the esophagus. It has nothing to do with the heart. *A* is incorrect; morning sickness occurs in about 50% of

women and is thought to be caused by both hormonal and psychological influences. Eating small amounts of dry carbohydrate before getting out of bed or during the day may suffice to relieve this distressing symptom. *C* is incorrect; pica is a craving for unusual nonfood substances such as red clay, plaster, or cornstarch. It is sometimes associated with anemia. Pica is a culturally controlled phenomenon. *D* is incorrect; backache is relieved by rest, good posture, sturdy shoes, and pelvic rock exercises.

54. **D. *Lightening.***

   D is correct; lightening is the descent or settling of the presenting part into the pelvis, which then results in the pressure symptoms given. Lightening is expected to occur about 2 weeks before labor in the nullipara and to occur at the time of labor in the multipara. *A* is incorrect; transition is the period between 7 and 10 cm; some refer to this as a rim of cervix. This period is usually described as the most difficult time during labor. The powerful, overwhelming contractions are closer together and longer; the mother may have the urge to push and is instructed not to do so until the cervix is completely (10 cm) dilated. *B* is incorrect; quickening is the feeling of life, a stirring that is felt by the nullipara about the 18th–20th week, but may be felt by the multipara earlier. Some women describe these early movements as similar to the fluttering of butterfly wings. *C* is incorrect; effacement is the process by which the cervix becomes incorporated into the lower uterine segment. Effacement is caused by contractions of the muscle layer, the middle layer of the uterus, and pressure from the amniotic sac and presenting fetal part.

55. **C. *Introspection.***

   Introspection is characteristic of the seventh month of pregnancy for the woman (and for most prospective fathers, also). Introspection allows the prospective parent to reflect on and experiment with different philosophies regarding child rearing, religion, sexuality and sexual expression, and relations with others, including one's parents, spouse, etc. *A* is incorrect because irritability is more characteristic of early pregnancy. *B* is incorrect because anxiety regarding labor is generally not seen at this time; labor is still far in the future. *D* is incorrect because ambivalence is usually more characteristic of early pregnancy.

56. **D. *The mother's blood volume is increased during pregnancy, so there might normally be a decrease in hematocrit.***

   D is correct; the nurse must determine if anemia is actually due to low hematocrit or to increased volume of fluid in pregnancy. *A* is incorrect; mothers are *not* supposed to be anemic during pregnancy but may appear anemic as determined by hematocrit, due to hypervolemia. *B* is incorrect; the mother's blood volume is not decreased, but increased, so this is not necessarily a true anemia. *C* is incorrect because the mother's blood volume is not decreased, but increased, during pregnancy.

57. **A. *Stools will be loose.***

   A is the best answer; the nurse would *not* inform Bertha that her stools will be loose. If Bertha is taking iron, her stools will be hard and constipating. *B, C,* and *D* are not the answers because these three statements *are true*. Iron can be irritating to the GI tract, and food will provide a buffer against irritation. Iron will also

cause the stools to be black. Fluids and exercise will aid in preventing constipation, which often occurs with iron ingestion.

58. **B.   *Increase roughage in the diet.***

  B is correct because increasing roughage in the diet will help prevent constipation. *A* is incorrect because only a physician can prescribe drugs. *C* is incorrect because milk can cause further constipation. *D* is incorrect because activity will increase circulation throughout the body and will have an effect on the overall muscle tone of the body.

59. **A.   *Decrease her milk intake to 1 quart a day.***

  A is correct because too much milk can produce a calcium-phosphorus imbalance, causing leg cramps. *B* is incorrect because increased milk intake will aggravate the problem. *C,* decreasing ambulation, will increase the pain and decrease circulation. *D,* use of skim milk, will not alleviate the calcium-phosphorus imbalance.

60. **A.   *Asparagus and kidney beans.***

  A is correct because the best sources of folic acid are organ meats (liver, kidney), yeast, and mushrooms; asparagus, broccoli, lima beans, and spinach; and lemons, bananas, strawberries, and cantaloupes. *B, C,* and *D* are poorer sources of folacin, with milk lowest in folacin content.
  Note: Absorption of folacin is decreased in pregnancy, in women taking the pill, and with alcoholism.

61. **C.   *25–30 pounds.***

  C is the correct gain for a woman of normal weight. *A* and *B* are incorrect because low weight gain tends to be associated with babies that are small for their gestational age. *D* is incorrect; dieting during pregnancy is contra-indicated because of the danger of insufficient nutrients to meet the demands of pregnancy and because of the danger of ketoacidosis. Underweight women need to gain *more* than 25–30 pounds.

62. **D.   *Is experiencing nausea and vomiting.***

  D is correct because nausea and vomiting may prevent adequate intake and therefore predispose the diabetic to ketosis and acidosis. *A* is incorrect because it is a value judgment that may be irrelevant unless the woman perceives this as a problem. *B* is incorrect; as the diabetic mother becomes older, there is an increased risk of complications. *C,* although significant, is incorrect because fatigue and restlessness during the first trimester are normal.

63. **C.   *Assess Ms. Osborne's feelings about this pregnancy.***

  C is correct because asking Ms. Osborne about her feelings and needs provides an opportunity for her to be involved in decisions about her life. The nurse functions as a facilitator of decision making by offering support and guidance. *A, B,* and *D* are incorrect because the nurse is making assumptions about Ms. Osborne's needs and feelings. Ms. Osborne needs to be encouraged to participate in decision making so that she will arrive at a decision that is "right" for her and to increase her own coping ability.

64. **C.   *Attempt to carry the baby to term to ensure adequate fetal development.***

  C is the answer because diabetic mothers give birth *by induction or cesarean surgery* when fetal lung maturity has been determined, to prevent fetal death if the diabetic condition is poorly controlled. *A, B,* and *D are all appropriate* to tell the mother to ex-

pect. The goal of all intervention is to control her diabetes for her own sake and for the well-being of her fetus.

65. **A. *Assess the frequency and amount she vomits each day.***
    A is essential to determine if any medical or nursing intervention is needed such as for starvation and dehydration (hyperemesis gravidarum). Usually, no intervention is warranted. Health teaching includes instructing her to avoid an empty stomach; eat small, frequent meals; and eat dry carbohydrate prior to rising. *B* is incorrect. Teaching her active relaxation may be appropriate after validating the frequency, type, and amount of vomitus and assessing her pattern of weight gain. *C* is incorrect because instead of assessing, the nurse merely encourages the woman to share her complaints with the physician. *D* is inappropriate because it does not assist in alleviating her symptoms and anxiety now. This statement may be false reassurance if Ms. Osborne continues to experience nausea and vomiting.

66. **D. *Hospitalization may be required for mental and physical rest, to treat infection, to adjust insulin requirements, or to change therapy from tolbutamide to regular insulin.***
    D is correct because any stress can lead to acidosis, which is hazardous to fetal well-being. Tolbutamide does not protect against acidosis-ketosis. Oral hypoglycemics are not used during pregnancy because they may be teratogenic. *A* is incorrect because insulin requirements remain the *same or decrease* during the first trimester, increase rapidly during the second and third trimesters, and are variable during the early puerperium. *B* is incorrect because fetal macrosomia is due to excessive growth of all tissues except the brain. Maternal hyperglycemia results in fetal hyperinsulinism and excessive fetal growth. *C* is incorrect because if the intrauterine fetus is not at risk and lung maturity has not been achieved, birth is *not* indicated.

67. **C. *Frequent glucose monitoring.***
    C is correct because frequent blood sugar monitoring is indicated if glucosuria is found. *A* and *B* are incorrect because alterations in the management of diabetes during pregnancy are accurate only if based on serum values of glucose. *D* is incorrect because creatinine clearance and 24-hour total urine protein level are obtained to assess *renal* status.

68. **C. *That the mother may go into ketoacidosis.***
    C is correct; ketoacidosis is detrimental to the mother and the infant. *A,* deprivation of nourishment to the baby, might eventually be a concern, but it is not a serious problem at this point unless the mother goes into acidosis due to an inadequate carbohydrate intake. *B* is incorrect because it is merely conjecture at this point, although some research indicates this to be true. *D* is incorrect because it is not the *most* serious concern, although it is an important nursing action.

69. **B. *Have irregular menstrual cycles.***
    B is correct because menstrual cycle irregularity makes it difficult to estimate the date of birth. *A* is incorrect because it has nothing to do with the date of birth. *C* is not true; diabetic mothers are usually overweight in pregnancy. (Also, this has nothing to do with date of birth.) *D* is not necessarily true, unless severe atherosclerosis is present; diabetic mothers usually give birth to babies

weighing more than 9 pounds. Also, this has nothing to do with date of birth.

70. A.   Atherosclerosis of the placenta.

A is the correct cause of small babies in diabetic mothers. *B* is incorrect; increased blood sugar will cause the baby to be large. *C* is incorrect because the size of the mother has no bearing on the size of the infant. *D* is incorrect; hypoglycemia in utero does not necessarily follow; the baby will tend to be large.

71. C.   Inadequate.

C is correct; her inadequate weight gain puts this pregnancy at risk. An inadequate prenatal weight gain is 0.9 kg (2 pounds) or less per month during the second and third trimesters. *A, B,* and *D* are all incorrect because of the rationale given for *C.*

72. D.   Age at 29 years.

D is the answer. Her age places her in the low-risk category under usual conditions. *A, B,* and *C* are not the answers because all of these statements are *true.* Inadequate weight gain, poor reproductive history with several pregnancies in rapid succession, and a dropping hematocrit indicate inadequate nutrition and depleted maternal stores, both of which are associated with lowered birth weight, intrauterine growth retardation (IUGR), and pregnancy wastage.

73. A.   Discuss with her how nutrition affected her pregnancies in the past, because she wants a good pregnancy now.

A is correct. A discussion about how her poor nutrition most likely led to her poor reproductive history indicates to Ms. Garvey that she has been "bad," and leaves her with a sense of lowered self-esteem and perhaps guilt. The negative feelings can alienate her from medical-nursing supervision and counseling. *B, C,* and *D* are not the answers because all of these statements *are* helpful. In addition, the nurse needs to identify her learning needs and, ultimately, to evaluate whether Ms. Garvey has actually gained the knowledge she needs.

74. D.   Hematocrit.

D is correct because hematocrit levels provide data for nutritional assessment and evaluation of therapy simply and for the least expense. *A, B,* and *C* are incorrect because although these values add important data, they are expensive to obtain and require laboratory examination.

75. A.   Vitamin D.

A is the best choice because vitamin D promotes a positive *calcium* balance during pregnancy. *B, C,* and *D* are not the answers because iron, folacin, and vitamin B_6 *are* all *blood-forming* nutrients.

76. D.   Animal proteins such as meat and organs.

D is correct; animal proteins such as meat and organs provide protein of high biologic value, as well as iron, folacin, and vitamin B_6. *A* is incorrect; although leafy green vegetables are an exchange group containing folacin, meat provides a better supply and balance of all the nutrients needed to form blood. These vegetables do supply iron and magnesium and several vitamins, for example, A, E, B_6, and riboflavin. *B* is incorrect because of the rationale for *D.* Grain products do provide vitamins (niacin, riboflavin, and thiamine) and minerals (iron, phosphorus, and zinc), however. *C* is incorrect; yellow fruits do not have the necessary protein, iron, and B vitamins, but they do provide significant amounts of vitamin A.

77. **C. *Hypertension.***

C is the answer because hypertension is *not a consequence of anemia.* *A, B,* and *D* are not good because maternal anemia *can* result in SGA neonates, hemorrhage, and perinatal infection.

78. **B. *"It is important to gain weight with good food to feed the baby and fix your anemia first."***

B is correct because the immediate need is to feed this fetus and to resolve maternal anemia for both the mother and the fetus. A poor reproductive experience now, added to previous ones, further decreases her self-esteem as a woman and her feelings of self-worth. *A* is incorrect; Ms. Garvey needs to take in sufficient calories to prevent ketonemia—a condition incompatible with fetal enzyme function. *C* is incorrect because the first priority is to meet maternal and fetal nutritional needs. The nurse may plan to spend some time later allowing Ms. Garvey to ventilate and to identify her real concerns and probable ways of solving them. Poor nutritional status can result in depression. *D* is incorrect because the nurse is not meeting her role expectations by referring this type of problem to another professional person.

79. **D. *The consent form must be written in English, but an interpreter should be used when needed for the person's comprehension.***

D is correct because the consent form can be written in any language. The important issue is the assurance that the involved person understands it thoroughly. *A, B,* and *C* are incorrect because all of these conditions must be met for an informed consent.

80. **B. *Battery.***

B is correct; the person (physician) is liable for battery, unless the consent form provides for reasonable and necessary extensions. *A, C,* and *D* are incorrect because they are not the legal actions that can be initiated in this situation.

81. **D. *FHR: 154.***

D is correct because an FHR of 154 is common at this gestational age. The parasympathetic system takes longer to mature—that is, the faster-maturing sympathetic system is not controlled by the heart rate, lowering effects of the parasympathetic system until later in gestational life. *A* is incorrect; by itself, a weight of 169 pounds does not place the pregnancy at risk. However, loss of weight since the onset of pregnancy indicates that she has not taken in the number of calories needed for optimum protein utilization (3000 cal/kg/24 hours). The result is catabolism of fat stores, which in turn produces ketonemia. Fetal enzyme systems cease in an acidotic environment. *B* is incorrect because by midpregnancy, the systolic and diastolic values are expected to drop by 10–15 mm Hg. If the diastolic value is over 75 mm Hg during the second trimester, there is a statistically significant rise in fetal mortality. *C* is incorrect. Fundal height at 20 weeks, according to McDonald, is expected to be 18 cm, and according to Sandberg, should be 20 cm.

82. **B. *Swollen fingers, increased diastolic blood pressure, 1+ proteinuria.***

B is correct because swollen fingers, increased diastolic blood pressure, and 1+ proteinuria are all signs of preeclampsia. *A* is incorrect because not all the items listed are signs of preeclampsia; the blood pressure should be elevated, rather than decreased. *C* is in-

correct because sugar in the urine is not indicative of preeclampsia. *D* is incorrect because not all signs listed are indicative of preeclampsia.

83. *A. **Place in a room with mothers to keep her company.***

A is the best answer because the nurse should *not* take this action. The mother needs quiet, decreased stimulation, and privacy. *B, C,* and *D* are not the answers to the question because they *are* good nursing measures.

84. *A. **Suction apparatus.***

A is correct because suction may be needed to maintain a clear airway if the woman convulses. Other essential equipment includes oxygen, tubing, and mask; intravenous fluids and infusion sets readied for use; emergency delivery pack; padded tongue blade or plastic airway; drugs such as magnesium sulfate, phenobarbital, calcium gluconate, morphine, and antihypertensives; a reflex hammer; indwelling catheter and collection bag; ophthalmoscope (to evaluate the eye grounds); and an electronic FHR monitor. *B* and *C* are incorrect because sudden noise may precipitate a convulsion. *D* is incorrect because this test is unrelated. However, the amount of urine she produces is crucial. An increased urinary output is indicative of decreasing severity of the disease. Urine must be checked for protein. Urine estriol levels are calculated serially—that is, daily—to assess the functioning of the maternal-placental-fetal unit.

85. *B. **Decreasing urinary output.***

B is correct because magnesium sulfate is eliminated through the kidneys; if it is not eliminated, it may increase CNS problems and cause respiratory depression. *A, C,* and *D* are incorrect because these signs would indicate that the drug should be continued; the drug is intended to decrease these signs.

86. *A. **Calcium gluconate.***

A is the correct antidote for magnesium sulfate. *B* is an antidote for meperidine (Demerol) or morphine. *C* is an electrolyte, not an antidote for magnesium sulfate. *D* is a drug that can be used to stop labor contractions.

87. *B. **24–26.***

B is correct. *A* is too slow a rate and would not correct the problem. *C* and *D* are too fast; these rates will give too much and could cause too severe depressive effects.

88. *D. **Her systolic pressure has increased by 30 mm Hg; the diastolic, by 15 mm Hg.***

D is correct because the significance of the blood pressure can be determined only by assessing any reading in relation to previous readings and to other findings. *A* is incorrect; a rapid weight gain can be caused by overeating, especially over holidays. *B* is incorrect because although hydatidiform mole does present with signs of preeclampsia, these and other signs generally appear by weeks 12–14. Preeclamptic changes in blood pressure are still assessed on the basis of *D.* The signs of true preeclampsia of pregnancy appear after week 24. *C* is incorrect because ankle swelling, without pretibial pitting, may occur during hot weather or in the woman who is on her feet for many hours each day.

89. *C. **Phlebitis.***

C is correct because pain in the calf can reflect phlebitis. *A, B,* and *D* are incorrect because pain in the calf does not necessarily reflect infection, edema, or hemorrhage.

90. **B.   Toxemia.**

    B is correct; methylergonovine maleate (which increases the blood pressure) is not given to a woman with a history of toxemia because one indication of toxemia is high blood pressure. Administration of the drug could obscure the symptoms of toxemia. **A, C,** and **D** are incorrect because none of these conditions involves high blood pressure.

91. **A.   Mother-infant bonding.**

    A is correct; it is important to foster a strong mother-child relationship to prevent or at least minimize problems as the child gets older. **B,** breastfeeding to prevent engorgement, is important, but secondary to **A;** also, not all mothers breastfeed. **C** is important; however, not all mothers breastfeed and oxytocin can be given orally. **D** is incorrect because the mother cannot be active in the early phase of recovery due to her physical condition.

92. **D.   Medication should be given at such a time that it will not be in the breast milk when nursing.**

    D is the correct choice. **A** is incorrect because it is not an accurate statement. **B** is incorrect because medication can be given to nursing mothers, but not at times that would coincide with the medication being in the breast milk at the time of feeding. **C** is incorrect because giving medication an hour before breastfeeding could cause CNS depression in the newborn.

93. **B.   The mother to try various breastfeeding positions to find the one that is most comfortable for her.**

    B is correct; a cesarean birth mother can breastfeed with comfort if given assistance. The "football hold" is used by many mothers. **A** is incorrect because it would lead to the mother's milk supply drying up. **C** is not the best answer, because the infant is deprived of bonding. **D** is incorrect because the healing process is enhanced by breastfeeding due to oxytocin stimulation to the uterus.

94. **C.   "You are upset because you did not have a vaginal birth. It must have been hard for you to accept."**

    C is the correct answer because it gives the mother an opportunity to respond. **A** is incorrect; it is the nurse's responsibility to provide an opportunity for the mother to ventilate her feelings and not put the mother's statement off for the doctor to answer. **B** is incorrect because the nurse is inferring that the mother feels guilty. In **D,** also incorrect, the nurse is again not giving the mother an opportunity to vent her feelings, but rather is hoping another mother will give assistance.

95. **A.   "I'm so tired; the baby breastfeeds every hour."**

    A is correct; this comment indicates that the mother is not getting enough rest, and that the baby is feeding too often and may not be getting enough milk at each feeding. **B, C,** and **D** are not correct; these comments indicate that the nurse's discharge planning *was* effective. **B** is a good way for the mother to recover from surgery. **C** indicates that the mother is rested and happy in her postpartum period. **D** indicates normal status for lochia at this time.

96. **A.   He understands that no change in testosterone levels will result.**

    A is correct; Mr. Garvey should have no change in levels of testosterone, follicle-stimulating hormone, or luteinizing hormone after vasectomy. **B** is incorrect; reinstating *tubal patency* through a reanastomosis is successful in many vasectomized males, but if he

develops autoimmunity to his own sperm, he would still remain sterile. *C* is incorrect because both conditions are unrelated to vasectomy. Balanitis is an inflammation or infection of the glans penis. Phimosis is a narrowed foreskin or prepuce, a condition that prevents its easy movement over the glans. Vasectomy is performed through a small incision in each scrotal sac. *D* is incorrect. Following vasectomy, the man is cautioned that because of the sperm already in the upper part of the sperm duct, he is not considered sterile until after 10 ejaculations or after 3–6 months.

97. **C. *Formation of the corpus luteum.***

C is the answer because progesterone is produced by the corpus luteum. Stimulus for the formation of the corpus luteum comes from the anterior pituitary and is known as the luteinizing hormone. *A, B,* and *D are* physiologic effects of progesterone.

98. **C. *Inhibition of uterine contractility during pregnancy.***

C is correct because uterine contractility is inhibited by *progesterone* during pregnancy. *A* is incorrect because estrogen begins the process of developing the spiral arteries in the innermost lining of the uterus, the endometrium; progesterone assists estrogen in this function during the second half of the cycle. *B* and *D* are incorrect because they are physiologic effects of *estrogen.*

99. **A. *Estrogen and luteinizing hormone (LH).***

A is correct; after ovulation, luteinizing hormone is responsible for the development of the corpus luteum from the now-empty graafian follicle. The corpus luteum continues to produce progesterone until about the end of the third month, when the placenta takes over this function. *B* is incorrect because the follicle-stimulating hormone (FSH) is produced by the anterior pituitary in response to a low circulating level of estrogen and progesterone toward the end of a cycle if conception has not occurred. Its target organ is the ovary, where it stimulates the maturation of the graafian follicle in which an egg is beginning to be prepared for ovulation. *C* is incorrect because progesterone is the main hormone during the last half of the cycle, known as the secretory phase, when the endometrium is readied for implantation if an egg is fertilized. *D* is incorrect because human chorionic gonadotropin (HCG) is produced as soon as nidation (also called implantation) occurs; chorionic villi continue to produce this throughout the pregnancy. This hormone stimulates the production of pituitary growth hormone. The presence of HCG in the mother's urine from the third week after conception is the basis for pregnancy tests.

100. **C. *Further physiologic assessment.***

C is correct; a complete physical assessment of both the male and the female needs to be performed to rule out any physiologic abnormalities that may prevent conception. *A* and *B* are incorrect because they are not the *primary* tests required. Both may need to be performed after completing the physical exams to ensure physical capability. *D* is less correct than *C.* Many clinics will perform preliminary health teaching in the area of conception during the initial screening process; psychotherapy may be appropriate if the couple perceives this as a need and the health team identifies this as an appropriate health action.

101. **B. *Ms. Jackson's history of taking oral contraceptives.***

B is of primary importance because many couples experience some degree of difficulty conceiving for several months when the wife

has taken oral contraceptives. With a little more time for hormonal readjustment, this couple may find that they will have little difficulty conceiving. *A* is insignificant because the couple are at the optimum fertility age. The fact that Ms. Jackson started taking oral contraceptives about 7 years ago is more significant. *C* probably has little to do with the possibility of Ms. Jackson's being sterile or infertile. *D,* the fact that Ms. Jackson's mother died during childbirth, is significant, but not the *most* significant factor at this time. A family history of problems with pregnancy may be of more significance if Ms. Jackson is unable to conceive within another year, provided that her husband's sperm analysis is within normal limits. In assessing the reason for death, it appears that Ms. Jackson's mother died due to preeclampsia. The cause of death would be highly relevant if Ms. Jackson becomes pregnant.

102. *A.   Sperm analysis.*

   A is correct because it is generally a less complicated procedure and less expensive than testing the female. *B* and *C* are incorrect because these tests are more expensive and time-consuming. Also, these tests may not be necessary if the sperm count is found to be too low for conception or the percentage of abnormal sperm is not within normal limits and the female is experiencing no other problems. *D* is incorrect because although not expensive, this test is usually performed by the patient over several months. Basal body temperature monitoring can be a very frustrating procedure and can be highly inaccurate.

103. *C.   "There is no indication at this point that you are sterile."*

   C is the correct response; more data need to be collected and analyzed before Ms. Jackson is told (by the physician) that she may be infertile or sterile. *A* is untrue; there are no definitive studies to support this assumption. There is, however, a small chance that permanent suppression of the anterior pituitary may occur. *B* is incorrect because there are no data to support this statement. *D* is incorrect because there are no concise statistical data to support what percentage of women taking oral contraceptives may have miscarriages, have problems becoming pregnant, or become infertile.

104. *B.   His inability to father children.*

   B is correct; many men fear being a failure as a male if they are unable to produce offspring. This is probably why Mr. Jackson wanted to see the nurse confidentially. *A* is probably a concern but is secondary to his self-image and perception of his role as a male. *C* may be a problem if the couple is experiencing frustration. However, some of the frustration may be enhanced by Mr. Jackson's feelings regarding his own self-esteem and "manly" image. *D* is important but secondary to the male's feelings of uncertainty about himself. In fact, the male will probably blame the female for the couple's inability to conceive if he is unable to face the possibility of his own deficits.

105. *B.   "Do you experience orgasm each time you have intercourse?"*

   B is correct because this question is *irrelevant*. Female orgasm is unnecessary for fertilization; male ejaculation is essential. *A* is incorrect because both frequency and timing of intercourse affect fertility. Intercourse with ejaculation must occur during the 12–24

hours after ovulation. *C* is incorrect; the husband's occupation *is* relevant. The male's sperm count decreases if he wears tight-fitting or thermal clothing or if he sits for long hours, such as while driving a truck long distances. Excessive heat to the testicles is the etiologic factor. *D* is incorrect. Pelvic inflammatory disease (PID) due to gonorrhea (or any other cause) heals by scar tissue formation. Scar tissue may obliterate or distort the fallopian tubes, thus preventing the passage of sex cells (ova and sperm) and fertilization.

106. **D.**   ***Gonorrhea treated 2 years ago.***

     D is correct because if the upper reproductive tract (i.e., the uterus, tubes, and ovaries) is infected, scar tissue forms as the inflammation subsides. Scar tissue in the fallopian tubes prevents normal passage of sperm, ova, or a fertilized egg. *A* is incorrect; a 28–30-day menstrual cycle is normal and unless the cycles are anovulatory, the time of her ovulation or her period of fertility is easily determined. *B* is incorrect; the appearance of pubic hair and breast development are pubescent changes that appeared at the expected time, that is, between the ages of 11 and 14 years. *C* is incorrect; epidemic parotitis, or mumps, in the *male,* especially during adolescence, leads to sterility if both testes are affected. There is no analogous occurrence in the female.

107. **B.**   ***Abdominal scars noted in the right lower quadrant.***

     B is correct because the presence of abdominal scars indicates a possibility of scar formation or adhesions, which may interfere with reproductive capacity. *A* is incorrect because an anteverted uterus is a normal finding. *C* is incorrect because clitoral size of 0.5 cm is within normal limits. *D* is incorrect because the hair pattern described is within normal limits for the woman of childbearing age.

108. **C.**   ***"Your angry feelings are understandable and an expected response."***

     C is correct because it is the only response that acknowledges this couple's feelings and their right to have these feelings. *A, B,* and *D* are *not* therapeutic.

109. **D.**   ***Assess her readiness to touch her external genitalia and her cervix.***

     D is correct because until Joan is able to come to terms with her feelings about touching herself, she will be unable to hear how to obtain and assess a specimen of cervical mucus. *A, B,* and *C* would all be appropriate nursing actions *after* the nurse has determined whether Joan is able to obtain a specimen.

110. **B.**   ***Ejaculate directly into a clean glass jar.***

     B is the correct way to collect semen for analysis. The man ejaculates directly into a glass jar, seals it, and takes it to the laboratory without either chilling it or warming it. *A* is incorrect because the residual rubber solvents and sulfur in condoms adversely affect sperm; however, the Milex Corporation does make a condom especially for collecting semen for analysis. *C* is incorrect because the total volume of semen and number of sperm could not be retrieved from the cotton balls. *D* is incorrect; although retrieval of sperm from the vaginal vault with a special spoon (Doyle) is acceptable, the semen is to be neither chilled nor warmed.

111. **B.**   ***Sims-Huhner test.***

     B is correct. The Sims-Huhner test is a postcoital test to determine sperm movement and survival in the vaginal environment. *A*

is incorrect; to prepare for a semen analysis, a specimen must be obtained from the male either by masturbation or by coitus (with special precautions) and examined for volume, pH, sperm density and motility, and percentage of sperm with normal morphology. *C* is incorrect; a scraping of the buccal mucosal cells is examined for chromosomal aberrations that cause infertility (e.g., Klinefelter's syndrome). *D* is incorrect; a Rubin's test is performed to ascertain tubal patency. With the woman in the lithotomy position, carbon dioxide gas is administered under pressure through the cervix. The woman prepares for this test by voiding.

112. *A. December 25.*

A is the correct calculation. Nagele's rule states, "From the last menstrual period, count back 3 months, then add 7 days." *B* is incorrect; if you calculated December 11, you have subtracted 7 days instead of adding them. *C* and *D* are incorrect; if you calculated January 22 or 18, you assumed that the *spotting* Joan experienced was a menstrual period. It is not uncommon for the expectant mother to have some small amount of spotting during the time she would have experienced menstruation if she had not conceived.

113. *A. Threatened abortion.*

A is correct; with threatened abortion, some vaginal bleeding and cramping are noted but no cervical dilation is seen. Treatment would probably be bedrest and the avoidance of orgasm, sexual intercourse, and vaginal examinations. *B* is incorrect because with inevitable abortion, considerable bleeding and cramping are noted. The cervix is dilating and the membranes may have ruptured. *C* is incorrect because the symptoms of a tubal pregnancy usually occur after the first missed period. Rupture of the tube usually results in a sudden, sharp pain in the lower abdominal quadrant, with symptoms of acute shock out of proportion to the amount of vaginal bleeding observed. *D* is incorrect; incompetent cervical os is the most common cause of second-trimester abortion. Treatment consists of the application of a purse-string suture, using the technique of Shirodkar or MacDonald, applied prior to cervical dilation. This is intended to prevent premature delivery, which would probably occur in the sixth month, when the fetus weighs between 1¼ and 1½ pounds. An incompetent cervix tends to occur more frequently when the cervix has been forcibly dilated in the past, such as for elective abortion.

114. *A. "Was the vaginal bleeding accompanied by pain?"*

A is correct. The bleeding of placenta previa is usually painless; that of premature separation of the placenta is usually accompanied by pain. The answer to this question adds to the database. *B* is incorrect; although coitus may result in a few drops of dark blood because of increased cervical friability, frank bleeding is uncommon. Also, this question sounds intimidating and accusatory. *C* is incorrect because it is not as appropriate a question as *A.* Episodes of spotting (a few drops of dark bleeding) may have occurred during the pregnancy, but it is unlikely that a frank hemorrhage had occurred previously. It is *most* important to determine the presence or absence of pain. *D* is incorrect because the question does not address the most commonly seen causes of hemorrhage during late pregnancy—placenta previa and premature separation of the placenta.

115. **C. *Assess amount of bleeding presently on the perineal pad.***
 C is the nurse's *first* action. With severe hemorrhage, oxygen is delivered by face mask at 10–12 L/min, and the woman is prepared quickly for examination under double setup. Blood work is needed stat. An intravenous infusion is begun, to facilitate stabilization of fluid and hematologic status. The woman is prepared for a possible emergency cesarean birth. *A* is incorrect because time should not be wasted in trying to find the FHR if hemorrhage is brisk. Both mother and fetus are best served by the rationale for *C*. As soon as possible, institute electronic fetal monitoring. *B* is incorrect because tone of uterine muscle is irrelevant if bleeding is brisk. (See rationale for *C*.) If hemorrhage has stopped or is minimal, the FHR is obtained first, then uterine tone is assessed. Increased uterine tone or rigidity is indicative of premature separation of the placenta; a uterus that relaxes well is found with placenta previa. *D* is incorrect because the history and physical examination are the last on this priority list.

116. **B. *"Ms. Moss has had vaginal bleeding. Vaginal examinations are contra-indicated."***
 B is correct; that vaginal examination, except under double setup, is contra-indicated is expected knowledge of the "reasonably prudent" nurse. To allow the intern to proceed could result in a massive hemorrhage from Ms. Moss and a lawsuit against the nurse. *A* and *C* are incorrect because they do not communicate clearly the rationale for *B*. Answer *D* is incorrect because it implies the nurse lacks knowledge of Ms. Moss's condition.

117. **C. *In the operating room.***
 C is correct. Double setup refers to a sterile vaginal examination in an operating room. Staff and equipment are ready for either a vaginal or an abdominal birth if profound hemorrhage occurs. *A, B,* and *D* are incorrect because double setup refers to readiness for immediate birth should profound hemorrhage occur.

118. **A. *Continues to monitor maternal and fetal status with no change in care plan.***
 A is correct; FHR tracings indicate adequate fetal oxygenation, so no change in the care plan is indicated. *B* is incorrect because the maternal-placental-fetal unit is adequately oxygenated at the moment. *C* is incorrect because no crisis exists at the moment. *D* is incorrect because no one can predict accurately the outcome of a "normal" labor, let alone the outcome of a possibly compromised pregnancy.

119. **B. *The nurse accepted total responsibility/accountability for recognizing ominous FHR patterns.***
 B is correct. Nurses who perform in highly specialized areas of care are expected to perform as a comparably trained and reasonably prudent nurse would perform in that area, and the nurse must take appropriate action to prevent harm to the infant. *A* and *C* are incorrect because the nurse is responsible for recognizing an ominous FHR pattern and taking appropriate action. *D* is incorrect because a late deceleration pattern is associated with fetal jeopardy, which requires medical-nursing intervention.

120. **D. *Mr. Moss's concern is increased because he does not understand isoimmunization—its cause or diagnosis.***
 D is correct. An appropriately stated nursing problem has three components: the person involved, the condition or problem that is

increased or decreased, and an intervention that can be initiated by a nurse without a physician's order. *A, B,* and *C* are incorrect because they lack the three components of a nursing problem.

121. *C. Prepare her for L/S ratio determination.*

C is correct. Amniocentesis for an L/S ratio for this 36–37-week pregnancy is *not* warranted. Even if the ratio is below 2:1, the fetus would be delivered if hemorrhage were to occur. *A, B,* and *D* are incorrect because anyone on bedrest *is* prone to respiratory, cardiovascular, and GI problems associated with immobility. Coughing and deep breathing, *A,* are interventions to prevent hypostatic pneumonia; *B,* wiggling toes, prevents thrombus formation; and *D,* diet with roughage and fluids, prevents constipation and stimulates appetite.

122. *D. Examination of diapers for pinkish staining.*

D is the answer because pinkish staining is due to the presence of urates, a common variation in *normal* newborn characteristics. *A* is not the answer because an increased incidence of RDS *is* associated with maternal diabetic condition regardless of gestational age. *B* is not the answer because hypoglycemia *is* an expected response. During fetal life, persistent maternal hyperglycemia results in fetal hyperinsulinism. At birth, the maternal supply of glucose stops abruptly and hypoglycemia occurs. *C* is not the answer because hyperbilirubinemia *is* associated with maternal diabetes. The cause is not fully understood.

123. *D. Begins resuscitative actions immediately.*

D is correct. This neonate, with an Apgar score of 3 at 1 minute, is severely depressed and requires immediate resuscitation. In addition, he is suffering from asphyxia neonatorum, which is defined as the absence of respiratory effort for 30–60 seconds after birth. Asphyxia neonatorum may be due to anoxia from prolapsed cord or placenta previa or abruptio placentae; from cerebral injury, such as intracranial hemorrhage; or from narcosis from the analgesics or anesthetics given to the mother. If respiratory depression is due to narcotics, nalorphine (Nalline), levallorphan tartrate (Lorfan), or naloxone (Narcan) may be administered. As resuscitation procedures are being done, the neonate must be kept warm or the rapidly developing acidosis will be increased because of the effects of chilling. If the infant is born bathed in meconium, an endotracheal tube may need to be inserted to suction out the tenacious meconium and other debris from the upper respiratory tract prior to the administration of oxygen. *A* and *B* are incorrect because the baby is severely depressed and requires immediate intervention. *C* is incorrect because the baby is severely depressed and requires active resuscitation, starting with tracheal suctioning.

124. *D. Location of decidua basalis.*

D is correct. Because the placenta has been implanted in the lower uterine segment, there is a greater possibility of hemorrhage. The "living ligature" uterine muscles are located mostly in the body (corpus) of the uterus. The muscle fibers in the lower uterine segment do not contract as strongly. Hemorrhage may occur even when the fundus feels firmly contracted. *A* is incorrect; Ms. Moss has just as great a chance of uterine atony followed by hemorrhage as any other new mother, especially those mothers who had spinal anesthesia for delivery. *B* is incorrect; the possibility of retained placental fragments is almost nonexistent when birth is by

cesarean surgery. *C* is incorrect; pelvic hematoma may occur in any new mother, especially when birth is surgical or if a vaginal birth was very difficult.

125. **B. Methylergonovine maleate.**

   B is correct. Methylergonovine maleate (Methergine) produces progressive contractions of uterine muscle. This derivative of ergonovine stimulates stronger and lengthier contractions and has less tendency to raise the blood pressure. However, do not give to the woman if she is hypertensive. Notify the physician if she needs an alternative oxytocic. *A* is incorrect; metronidazole (Flagyl), is specific for the treatment of *Trichomonas vaginalis*. *C* is incorrect; estradiole valerate is an estrogen-androgen combination that exerts an antilactogenic effect. There is only a negligible incidence of engorgement if this drug is injected just before or just as the placenta separates, before the release of lactogenic hormone from the posterior pituitary. It is contraindicated in women with thrombophlebitis, epilepsy, migraines, cardiac or renal disease, and asthma. Because exogenous estrogen has been implicated in endometrial carcinoma, the woman signs an informed consent prior to receiving it. *D* is incorrect. Naloxone (Narcan) is a narcotic antagonist that can be given to the mother along with the narcotic during labor without affecting pain relief; or it may be given to an infant whose respiratory depression is due to narcotics. Nalorphine and levallorphan titrate (Lorfan) are not the narcotic antagonists of choice because if the neonate's depression is not due to narcosis, the neonate will become more depressed. Naloxone is a narcotic antagonist that if given to a neonate whose depression is not due to narcosis, does not deepen the depression.

126. **B. Empty the breasts every 3–4 hours.**

   B is correct. The empty breast is the appropriate stimulus to establish and maintain lactation. *A* is incorrect; additional added fluids and calories and good general nutrition do facilitate lactation, but emptying the breast is the specific stimulus needed. *C* is incorrect. On rare occasions, a woman does not lactate spontaneously; for the affected woman, one injection of oxytocin may be needed to initiate lactation and to stimulate the let-down reflex. *D* is incorrect; estradiol valerate (Delestrogen) is a depoestrogen, which is an antilactogenic drug.

127. **B. Destroys fetal Rh-positive RBCs in the maternal blood system.**

   B is correct. Rho (D) immune globulin promotes lysis of fetal Rh-positive RBCs circulating in the maternal bloodstream before the mother can produce antibodies. RhoGAM is administered, within 72 hours of birth, to the Rh-negative woman who gives birth to an Rh-positive fetus if she is Coombs negative, that is, does not have antibodies against the Rh factor. *A, C,* and *D* are incorrect. Rho (D) immune globulin provides passive immunity by acting like antibodies. Following any future pregnancy, whether it terminates in abortion or the birth at term of an Rh-positive baby, Ms. Moss will receive another injection of Rho (D) immune globulin.

128. **C. Lack of surfactant and immaturity of retinal and other blood vessels.**

   C is correct. Lack of surfactant predisposes the infant born at 37 weeks or less to respiratory distress syndrome (hyaline membrane disease). Immature retinal blood vessels in the presence of high

arterial oxygen tension may result in visual impairment due to retrolental fibroplasia. Immature vessels elsewhere may rupture and cause thrombosis and hemorrhage. *A* is incorrect because none of these factors is related to the incidence of prematurity. Related factors include maternal age 17 years or less, multiple gestation, high parity, poor nutrition, severe physical or emotional stress, premature rupture of membranes due to infections such as gonorrhea, incompetent cervical os, fetal congenital anomalies, toxemia, and diabetes. *B* is incorrect because premature infants tend to be anemic due primarily to fragile capillary walls and hemorrhage and to poorly developed ability to form new erythrocytes. The premature kidneys do not conserve water well. Water retention and edema may occur. The premature infant does not absorb nutrients from the intestine efficiently. *D* is incorrect because children who are ill or premature at birth and require early and prolonged separation from the parents are more often victims of child abuse and neglect. Early and consistent contact to facilitate bonding, involvement in the care of the infant, and assistance with finances are necessary to prevent this unhealthy result.

129. *A. Denial and shock.*

    A is correct; denial and shock are often the *first* responses to loss. *B, C,* and *D* are often the *next* steps in the grieving process, in that order. The nurse's role is to evaluate the grieving process so that the involved people can move toward resolving the loss. The acute phase lasts about 6 weeks; the total process, about 1 year. Parents need to be reassured that what they are or will be experiencing is symptomatic of this process.

130. *D. "This must be a very difficult and sad time for you."*

    D is correct. The nurse is serving as a role-model in facing grief, open communication, and feeling safe and comfortable in dealing with unpleasant situations. *This* baby is important *now*—the woman does not want or need to focus on other children. Occasionally a mother must withdraw, as if to take the experience a small piece at a time. *A, B,* and *C* are incorrect because of the rationale given for *D.*

131. *A. Cause hemorrhage from a low-lying placenta.*

    A is the correct reason why vaginal examinations are contraindicated when bleeding is present in the last trimester. *B* is untrue; vaginal examinations are not more painful, and an exam does not stimulate labor. *C* is incorrect; an examination using sterile techniques does not cause infection and is not contraindicated. *D* is not true; correctly administered vaginal examinations do not rupture membranes.

132. *C. Painful bleeding.*

    C is correct; painful bleeding—and particularly, the element of abdominal pain—is a major sign of abruptio placentae. *A, B,* and *D* are incorrect because they are not specific to this condition. *A* is a symptom of previa. *B* could be a sign of *any* hemorrhage, vena cava syndrome, or bacteremic shock. *D* could be a sign of various conditions.

133. *B. A hard, tender, boardlike uterus.*

    B is the correct answer because the uterus reacts to placental abruption with a sustained tetanic contraction. *A* is incorrect because it is not related to abruption. *C* could be present in other conditions; one would not necessarily find this with abruptio pla-

centae. Also, abruptio tends to lower blood pressure. *D,* mild contractions, is highly unlikely; the woman would have severe pain.

134. D. *Coagulopathy.*

D is the correct answer because clotting factors, especially platelets, are used up and disseminated intravascular coagulation (DIC) can occur. *A* is not a complication, but a sign, of abruptio placentae; thus, it is incorrect. *B* is incorrect because decreased urinary output might be a compensatory mechanism in early shock. *C* is incorrect because hypertension is one etiologic factor for abruption, not a result of abruption.

135. C. *Side-lying.*

C is correct because this position prevents pressure on the vena cava, which contributes to a diminished blood supply to the heart, thereby decreasing cardiac output and causing hypotension and decreased perfusion of the uterus and fetus. *A* is incorrect because it could increase discomfort and pain. *B* is incorrect because lying on one's stomach is highly uncomfortable during the 30th week of pregnancy and could depress the vena cava, thereby contributing to shock. *D* is incorrect because it is uncomfortable, may cause difficulties in breathing, and could result in the vena caval syndrome described in the rationale for *C.*

136. B. *Check the FHR.*

B is the correct choice at this time. *A, C,* and *D* are incorrect because Allene is still early enough in labor that these actions are not indicated.

137. B. *Fetal respiratory movements.*

B is best because although the fetus does inhale and exhale amniotic fluid within the upper respiratory tree and even makes crying motions, no sound is heard since there is no air available to force through the larynx. *A, C,* and *D* are not the answers because these sounds *can* be heard. Both the funic and the uterine souffle may be heard. The funic souffle is identical to the FHR and represents the sound of blood rushing through the umbilical vessels. The uterine souffle is identical to the mother's heart rate and represents her pumping blood to the placenta. To differentiate between the two, listen to the fetal heart tones and palpate the maternal pulse simultaneously. This is especially important if the FHR has decelerated to about the level of the maternal pulse, or vice versa. Maternal intestinal noises can also be heard.

138. A. *Leopold.*

A is correct; Leopold's maneuver refers to the palpation of the uterine contents through the abdominal wall to assess fetal presentation, position, lie, and ballottement. *B* is incorrect; the Scanzoni maneuver involves the rotation of the fetal head to the occiput anterior position with forceps such as Luikart's. *C* is incorrect; during normal spontaneous vertex birth, the Ritgen maneuver is used to control the birth of the infant's head between contractions. Pressure is applied to the infant's chin through the mother's perineum with one hand while applying pressure to the occiput or crown with the other. *D* is incorrect; Shirodkar refers to the application of a purse-string suture around the internal cervical os to support an incompetent cervix until the time of birth.

139. C. *Somewhat heart shaped, widest from side to side.*

C is correct; the fetal head enters the true pelvis in the transverse position. The configuration of the mid-pelvis and pelvic outlet di-

rect the internal rotation of the presenting part during descent. *A* describes the anthropoid pelvis. *B*, a transverse oval, describes the platypelloid pelvis. *D* describes the mixed pelvis (i.e., gynecoid-android).

140. *A. Ischial spines.*

A is correct; the ischial spines serve as a landmark. *B, C,* and *D* are incorrect. The tuberosities and symphysis pubis form part of the outlet of the pelvis and the sacral promontory cannot be felt during vaginal examination in the normal gynecoid pelvis.

141. *B. Shallow chest.*

B is correct; the Lamaze method suggests this breathing pattern during early labor. *A, C,* and *D* do not facilitate optimal relaxation and comfort, the goal for early labor.

142. *D. Rupturing the membranes.*

D is best because this is the *physician's* responsibility. *A, B,* and *C* *are* all appropriate *nursing* actions based on the nurse's responsibility and obligations.

143. *C. "Reassure your wife."*

C is correct because this is the most important function of the husband throughout labor and birth. *A* is an evasive and ineffective response and does not allow the husband to feel needed. *B* is incorrect because this is the nurse's responsibility, and the nurse should be readily available to the woman in labor. *D* is incorrect because it is a nursing function that should not be delegated to the husband. However, if the husband is interested, he can be informed about how the monitor works. In childbirth classes, both parents are taught the monitor's purpose during childbirth. The nurse in the labor room needs to assess and reinforce this information.

144. *A. More irritable and complain of nausea.*

A is correct; during the transitional phase (8–10-cm dilation), most women begin feeling as if they cannot make it; become more irritable; experience nausea, vomiting, and belching; and begin to perspire more over their upper lip and between their breasts. *B* and *D* are incorrect because they are seen during the midactive phase (4–8 cm). *C* is incorrect because it is seen during the latent and early phases (0–4 cm).

145. *A. Phenaphthazine (Nitrazine).*

A is correct; phenaphthazine (Nitrazine) paper is litmus paper used to determine pH values. Vaginal secretions are acidic and amniotic fluid is basic; therefore the paper turns a blue-green to deep blue. To do the test, the nurse puts on a sterile glove and inserts the paper into the vagina near the cervix. *B* is incorrect because the Guthrie method tests for the phenylalanine concentration in blood drawn from a heelstick. It is usually done on the third or fourth day (after the child has ingested some milk), and should be repeated in 4 weeks' time. *C* is incorrect because Gravindex is a simple slide test for pregnancy whereby a drop of urine is placed on a slide and mixed with antiserum for 30 seconds. More antiserum is added and the slide is rocked gently for 2 minutes. If no agglutination occurs, the test is positive for pregnancy; it is accurate 90% of the time. *D* is incorrect because the Pap smear is the examination of cells from the squamocolumnar junction of the cervix to identify cytologic changes that may be indicative of malignancy.

146. D. *Does nothing; there is no clinical significance to this pattern.*

D is correct because this pattern represents early deceleration caused by head compression. In general, perfusion of the placenta and fetus and normotension are enhanced by the side-lying (or semi-Fowler's) position during labor. *A, B,* and *C* are not relevant for this FHR pattern.

147. B. *Nausea, shaking legs, increased irritability.*

B is correct; these are a few of the symptoms of second stage, when the cervix is drawn up to become part of the lower uterine segment. Other symptoms may include feelings that she cannot go on, perspiration, belching, and the urge to defecate. *A* is incorrect because elevation of temperature and pulse are indicative of increasing exhaustion due to prolonged labor and to a lack of fluid intake, as well as of infection. *C* is incorrect because malar flush, increasing introspection, and seriousness are characteristics of about 5-cm dilation. *D* is incorrect because tingling or numbness of the fingers and/or toes and carpopedal spasms are indicative of a decreased P_{CO_2} due to hyperventilation. Treatment is to change the breathing pattern through coaching and role modeling, and to ask her to breathe into a paper bag.

148. B. *Check to see if the cervix is dilated 10 cm.*

B is correct because the symptoms described are signs of complete dilation and Allene will need to be coached for pushing. *A, C,* and *D* are incorrect since the signs herald the beginning of the second stage of labor and are not pathologic.

149. C. *Anxious about his wife's and baby's condition.*

C is correct; Mr. Toth is concerned about his wife's and infant's condition. *A* is incorrect; although this may be an underlying factor, he is *primarily* concerned about his wife and the baby's condition. *B* may be a contributing factor, but he has not questioned the staff's level of competency. He is concerned about the baby and his wife's condition and treatment of those conditions. *D* is incorrect because there is nothing to indicate that he cannot be with his wife. In fact, he may be able to provide support for her and help lower her anxiety level.

150. C. *Painless vaginal bleeding.*

C is correct because it is one of the primary symptoms of placenta previa. *A* and *B* are incorrect because there are usually no contractions and the bleeding is usually painless. *D* is incorrect because it is a symptom of abruptio placentae.

151. C. *Permitting vaginal examinations.*

C is correct because vaginal examinations may initiate hemorrhage and/or labor. *A* is incorrect because it *is* an *appropriate* action if the placenta is marginal. *B* is incorrect because this action *is* absolutely essential to provide optimum nursing management. *D* is incorrect because a quiet environment *is* important to maintain a low level of anxiety and prevent an increase in bleeding.

152. C. *Assessing Ms. Toth's reasons for crying.*

C is correct because it is important to ascertain her reasons for crying so that the nurse can provide the appropriate intervention. *A* is incorrect; the burden of assessing Ms. Toth's condition should not be transferred to her husband. He may assist in comforting his wife and identifying her feelings, but the nurse is the primary caretaker. *B* may be appropriate, but only after the nurse has as-

sessed Ms. Toth's condition. *D* is inappropriate because further assessment is required before any intervention is initiated.

153. **B. *Turns Ms. Toth onto her side, administers oxygen, and stops oxytocin induction if induction is in progress.***

B is correct because the FHR pattern describes late deceleration, an indication of uteroplacental insufficiency. To increase perfusion of the uteroplacental unit, keep the uterus off the vena cava, oxygenate maternal blood, and hydrate. *A* is incorrect; the nurse who "does nothing" or does not recognize the implications of this tracing is negligent. *C* is not as correct as *B;* although there may be occasions when the nurse initiates contact with surgery, this is normally within the physician's role. *D* is incorrect because the *immediate* need is to deliver oxygen to the fetus. Preparation for birth may follow immediately in some cases.

154. **D. *Give meperidine hydrochloride, 100 mg intramuscularly 1 hour prior to surgery.***

D is correct; a nurse should *question* this order because the drug has an adverse effect on the newborn, causing respiratory depression. *A, B,* and *C* are incorrect choices because *A* is a *correct* assessment for the safety of the fetus; *B* is an *appropriate* nursing action for the safety of the fetus; and *C* is an *appropriate* nursing action following a doctor's order.

155. **B. *Beginning shock.***

B is correct; the restlessness and tachycardia alert the nurse to shock. *A* is incorrect because the blood pressure is not appropriate to a person who is anxious; she may be anxious, but the data presented do not relate to anxiety. *C* is incorrect because these are not the correct vital signs for preeclampsia (the blood pressure is too low). *D* is incorrect because postpartum blues do not necessarily change vital signs and usually do not occur so soon after birth.

156. **A. *Anxiousness to discuss her labor and birth.***

A is correct; it is important for mothers to go over the details of the labor and birth, to realize and integrate their accomplishment. *B* is incorrect; mothers usually do not have difficulty remembering their labor and are anxious to talk about it, even if they feel they did not do what they wanted or it did not go as planned—it is an important time for catharsis. *C* is incorrect; after a *normal* birth mothers usually gain new energy and are less weary and tired than during labor; a mother might be weary and serious after a *problem* birth. *D* is incorrect; as long as she is satisfied with her performance, the mother is not too concerned about her environment.

157. **D. *Massage the uterine fundus.***

D is correct because massage might solve the problem without the need for drugs or other measures. However, the nurse should avoid overmassage, which may fatigue the muscle. *A* is not the *first* nursing intervention; while the nurse is making these checks, the woman could be hemorrhaging; this would be a later intervention. *B* is incorrect; catheterization is not indicated unless there is evidence of a distended bladder. *C* is incorrect; this would not be the *first* intervention, but might be ordered by the doctor.

158. **A. *Check for a distended bladder and catheterize Ann if it is distended and she cannot void adequately.***

A is correct; a full bladder elevates the uterus, but because the uterus is firm, the nurse should not worry about hemorrhage. *B* is

incorrect; these data do not give information about the fundus. *C* is incorrect; Ann does not need methylergonovine maleate (Methergine) because there is no evidence of uterine relaxation. *D* is incorrect; the uterus is not failing to involute; it is high simply because of a full bladder.

159. *C.* **"There are several contraceptives. I will give you the pros and cons so that you can make the decision. Also, consult your doctor about what is best for you."**

C is correct because the nurse gives the facts and lets the woman, in consultation with her physician, make the decision. *A* is incorrect; the woman's question is ignored due to the nurse's bias. *B* is incorrect; the statement is not completely true and should be explained. *D* is incorrect; the nurse cannot decide what is best for the woman.

160. *B.* **"Plan times when you can be with each one alone for a certain period."**

B is correct because this approach includes each member of the family and all will feel that they are accepted and cared about. *A* is incorrect; the entire family must be included in caring for and accepting the newborn to avoid the husband's and child's feeling left out. *C* is incorrect; the husband and child will feel replaced by the newborn, and they may have problems accepting the newborn. *D* is incorrect; the husband will feel rejected and may go outside of the home for support and not return; he may feel that the relationship has changed and cannot be healed in the future.

161. *C.* **Sit-ups**

C is best because sit-ups are too strenuous immediately postpartum. The exercises in choices *A, B,* and *D* are all *appropriate* at this time and will not harm the mother. Kegels strengthen perineal muscles and enhance sensation and muscle tone. Pelvic tilt will help her regain muscle tone and good posture. Head raising will strengthen abdominal muscles.

162. *B.* **Readiness to concern herself with her infant.**

B is correct because having met her own needs during the "taking-in" phase, the mother is ready to concern herself with her infant. *A, C,* and *D* are incorrect; these behaviors are characteristic of the "taking-in" phase.

163. *B.* **Be worn to support and lift the breasts.**

B is the correct answer; a brassiere gives support and comfort. *A* is incorrect; not wearing a brassiere can lead to sagging breasts and discomfort in the shoulders and back from heavy breasts. *C* and *D* are incorrect because they will not support the breasts in the breastfeeding state and could cause pain and possible injury to the breasts.

164. *A.* **Apply heat to the breasts prior to nursing and express some milk.**

A is correct; this will soften the breasts, and manual removal of some milk will allow the baby to take hold of the nipple more easily and remove more milk. *B* is incorrect; this will fill the breasts with more milk, which will not be removed until the next feeding, causing more discomfort and possibly increasing the problem. *C* is incorrect because it will cause vasoconstriction, inhibiting the release of milk and adding to the problem. *D* is incorrect because the mother may not be in pain, and taking medication at this time

could mean that the medication will be in the breast milk at the
time of feeding.

165. B.   *Explain that these feelings are normal and why they occur.*

B is correct; the nurse should offer an explanation for the mother's
feelings and present her with options to correct the problem. *A* is
incorrect; this condition is normal and can be helped by consideration, support, and health teaching. *C* is incorrect because bonding
is essential at this time and can help to solve the problem, although the mother may need additional rest. *D* is inappropriate; it
is an extreme solution to a normal situation.

166. A.   *Resuming her prepregnancy life-style within 1 week.*

A is the incorrect instruction because the nurse has no way of
knowing whether the mother's prepregnancy life-style is appropriate; even if this information were known to the nurse, resumption
of the prepregnancy life-style after 1 week most likely is inappropriate. *B, C,* and *D* are appropriate instructions. The mother
should be given health teaching regarding herself, the baby, and
family responsibilities. Planning for rest and relaxation helps the
mother avoid exhaustion and frustration. And knowing that she
has planned ahead for emergencies is reassuring and could be lifesaving.

167. A.   *"You should feed the baby the way that works best for
both of you."*

A is correct; since the mother has changed the method of feeding,
she should be supported in her choice, since it probably is best for
her situation. *B* is judgmental and could cause the mother to feel
guilty. *C* is untrue; whether breastfeeding or bottle feeding is better for a baby depends on the circumstances. *D* is an inappropriate
response; it could make the mother feel guilty and it is not necessarily true; the baby may or may not gain more weight and may or
may not be allergic.

168. A.   *Give him privacy.*

A is the most appropriate response of the four choices given. However, the nurse still needs to be readily available so that the
woman does not feel isolated. *B* is inappropriate because Mr.
Young also needs time to grieve for his loss. *C* is incorrect; it is too
premature, will serve to stimulate his anxiety, and may lead to
feelings of hostility or depression. *D* is inappropriate because
these parents may feel the need to be close to each other at this
time.

169. D.   *Validate with him how his wife usually copes with
losses.*

D is correct because each individual reacts differently to various
stressors. It is important to identify past reactions in order to begin to identify how an individual *may* react or attempt to cope
with a new crisis. *A* is incorrect because it places pressure on Mr.
Young to answer his own question, without providing adequate
emotional support. *B* may be true, but is not an appropriate nursing response. The nurse needs to attempt to identify how the
woman may react to this stressor in order to facilitate planning for
nursing intervention. *C* is also a true statement, but this is not an
individualized response and tends to be perceived as nonempa-
thetic.

170. **A.** *Tell Bryant what has occurred at a level he can comprehend.*

   A is correct; it is important to provide Bryant with an accurate explanation that he can understand, to alleviate any fears and assist him in coping with his feelings. *B* is incorrect because it is too abstract and evasive an approach for a 6-year-old child. *C* is incorrect because while Bryant is waiting for his mother to return home, all types of anxieties and fears will build up; he needs to be informed now in order to prevent this. Also, it may be too stressful for Bryant's mother to have to come home and tell him of this loss. *D* is incorrect because it is too abstract and assumes that this family has this belief.

171. **B.** *Has a lower protein content.*

   B is correct; breast milk has a lower protein content, but its protein is more easily digested and more fully utilized than that in formula. *A* is incorrect because breast milk has a lower, not a higher, protein content. *C* is incorrect; breast milk has a higher carbohydrate content; formulas must have carbohydrates added. *D* is incorrect; ingestion of commercial formulas may lead to milk allergy.

172. **B.** *Gain more weight than formula-fed infants.*

   B is the incorrect instruction; breastfed babies gain weight *less* rapidly than do formula-fed babies. *A, C,* and *D* are true of breastfed babies. *A* is due to the immunoglobins of breast milk. *C* is due to the consistency of breast milk and more frequent feeding. *D* is true because breast milk is more compatible with the infant's metabolism and digestive system.

173. **A.** *Decrease stimulation to the breasts.*

   A is correct; lack of stimulation will aid in decreasing milk letdown and negative lactogenic hormone feedback. *B, C,* and *D* are all incorrect because these activities support lactation. *B* increases milk production. *C* increases vasodilation and stimulates the letdown reflex. *D,* emptying the breasts, stimulates milk production.

174. **B.** *115–120.*

   The correct amount is 117 cal/kg/day. *A* is an inadequate number of calories for growth. *C* and *D* are too many calories; the infant will gain too much weight, which can lead to adult obesity.

175. **B.** *3–4.*

   Michael is fed 6 times a day. He requires 20 oz./day to provide 120 cal/kg/day (400 cal total). 20 oz. divided by 6 = $3\frac{1}{3}$ oz. per feeding. *A* is incorrect because 2–3 oz. $\times$ 6 = 12–18 oz.—less than Michael's needs. *C* and *D* are incorrect because both would exceed Michael's needs and lead to diarrhea or to obesity from overfeeding.

176. **C.** *Eye prophylaxis may be delayed up to 1 hour after birth to facilitate parent-child attachment.*

   C is the correct answer because it is *not* true. Eye prophylaxis may be delayed up to *2 hours* after birth to facilitate parent-child attachment through eye-to-eye contact. *A, B,* and *D* are not the answers because each of these statements is *true* about prophylactic eye care of the newborn.

177. **C.** *Both the nurse and the physician.*

   C is correct because both the nurse and the physician are responsible and accountable for checking the label prior to using a medication. *A, B,* and *D* are therefore incorrect.

178. **A. *Using the greater trochanter and posterior superior il-iac spine as landmarks.***

A is the best choice; the landmarks for IM injections are *not* the greater trochanter and the posterior superior iliac spine. That would place the needle in the posterior gluteal muscle, which, in neonates, is small, poorly developed, and close to the sciatic nerve. The landmarks for IM injection in neonates are the greater trochanter and the knee. *B, C,* and *D* are not the answers because all of these *are* true about giving a neonate an IM injection.

179. **C. *Helps the mother hold him face down with the head lower than the buttocks.***

C is correct. Between the fourth and eighth hour after birth, the baby experiences the second period of reactivity. This period is characterized by alertness, gagging with regurgitation of mucus, and passage of meconium. Although the baby's response is normal, to prevent aspiration, the baby is held face down with the head lower than the buttocks to assist gravity drainage. At this time, the nurse increases the mother's confidence by assisting the mother to take the appropriate action. *A* is incorrect; holding the baby upright and patting his back encourages inhalation and aspiration. *B* is incorrect; the baby's mucus must be drained stat and the nurse's racing the baby to the nursery would be very frightening for the mother. The mother may then feel she is not capable of appropriate action. *D* is incorrect because the baby's mucus must be drained stat. Also, the mother is rightly concerned; telling her not to worry is false reassurance.

180. **B. *On the forehead just above the eyebrows, over the junction of the ear on the head, and back to the forehead.***

B is correct; head circumference is measured by placing the tape on the forehead just over the eyebrows and then around the head over the junction of the ear with the head. The head circumference may need to be remeasured after molding and caput succedaneum have resolved, usually after 2–3 days. *A, C,* and *D* are incorrect because of the rationale given for *B.*

181. **D. *Records the findings and does nothing further.***

D is correct. The head (occipitofrontal) circumference is about 2 cm (¾ inch) more than the chest circumference at the nipple line. If the head measures under 32 cm, microcephaly must be ruled out. If the head is 4 cm or more larger than the chest, hydrocephaly is suspected. *A, B,* and *C* are all inappropriate actions because the head and chest circumferences are within normal limits.

182. **C. *Single palmar crease, snuffles, apical pulse between 120 and 140/min at rest.***

C is correct because single palmar creases are associated with genetic defect, Down syndrome, or trisomy 21. Assess for incurvature of the fifth fingers, thick tongue, extra epicanthal folds, excessive joint motility, and poor feeding. Snuffles refers to the irritating rhinitis of syphilis, which cause excoriation and formation of scar tissue called rhagades. Assess for other symptoms of syphilis, and wear gloves when working with this infant. The pulse is within the normal limits. *A* is incorrect. Flat hemangiomas are also known as stork bites, and are the small, irregular red blotches over the bridge of the nose, upper eyelids, and the nape of the neck caused by rupture of capillaries from the stress of vaginal birth. Head circumference and hemoglobin levels are within normal lev-

els. *B* is incorrect because all findings are within normal limits. Erythema toxicum is not fully understood and seems to have no clinical significance. *D* is incorrect. Molding is the overlapping of cranial bones to accommodate the head to the birth canal. Smegma is the white mucus formed between the labia majora and under the prepuce. Acrocyanosis is blueness of the hands and feet, primarily within the first week of life, especially when the neonate is chilled and at rest. Acrocyanosis is due to immaturity of the peripheral vascular system. Respirations are within normal limits for a neonate at rest.

183. **A. *Kilograms of body weight.***

A is correct; nutrient needs for the neonate are calculated as kilocalories per kilogram of body weight. *B, C,* and *D* are incorrect because nutrient needs are calculated per kilogram of body weight.

184. **A. *The neonate may vomit if his bottle is propped.***

A is correct. In addition to the increased hazard of aspiration, if the bottle is propped, formula may enter the neonate's eustachian tube and predispose the baby to ear infections. *B* is incorrect. The neonate may lose 5–10% of his birth weight normally. To determine the percentage of weight loss, multiply the number of pounds by 16 ounces and divide by 10. Example: Birth weight is 8 pounds. $8 \times 16 = 128$ ounces. 10% of 128 ounces $= 12.8$ ounces. *C* is incorrect; the neonate's stomach capacity is between 50 and 60 mL, or about 2 oz. *D* is incorrect; the neonate's caloric needs are high—about 115 kcal/kg/day for the first 6 months, or about 52 cal/pound/day. The gastrocolic reflex is stimulated during feeding, and he may have a bowel movement immediately following each feeding.

185. **C. *Rooting.***

C is correct; the rooting reflex is activated whenever the cheek and lips are touched, and the infant is awake and hungry. The mother is advised to use the reflex for introducing the nipple into the infant's mouth. *A* is incorrect; the Babinski reflex is normally present in early neonatal life. When the plantar surface of the neonate's foot is stroked firmly, the toes fan out. *B* is incorrect; suck and swallow reflexes are present and complete in the full-term infant. However, even in the full-term infant, suck and swallow may not be synchronized at birth, and may result in gagging and coughing. *D* is incorrect; the Moro reflex is a response to a falling sensation, a disorientation in space, or a noise. The arms and legs are extended, then returned to the midline in an embracing motion, and the fingers form the letter "C." The newborn usually cries. Adequate reflex responses indicate an intact musculoskeletal and neurologic system from the brain stem downward. Reflexes disappear as myelinization is completed.

186. **D. *12 oz. (360) mL per day.***

D is correct. 7 pounds, 6 ounces $= 3.345$ kg $\times 105$ mL per day $= 12$ oz. (360 mL) per day. *A, B,* and *C* are incorrect because the full-term neonate's fluid requirement per day is 105 mL per kilogram of body weight.

187. **D. *13 ounces.***

D is correct. The baby can lose up to 10% of his birth weight, or 13 ounces. 8 pounds, 2 ounces $= 130$ oz. 10% of 130 $= 13$ ounces. *A, B,* and *C* are incorrect calculations.

188. *D.* *Obtain and assess a blood sample by heelstick because*
these behaviors could be signs of hypoglycemia.

D is correct; slight, persistent tremors could be signs of hypoglycemia. This baby is larger but still within normal limits and the
movements described *are within normal limits* for newborns. If
any doubt exists, the nurse gathers more data (assesses for blood
glucose level according to protocol, e.g., heelsticks at 1, 2, 4, and 6
hours) prior to initiating an intervention. *A* is incorrect because
the nurse acts *prior to completing an assessment*. *B* is incorrect because the data do not correlate with drug dependence. Signs of
drug dependence, such as hypertonicity, high-pitched cry, frantic
fist sucking, and poor feeding, likely will have been evident prior
to this episode of tremors. *C* is incorrect because signs of increased
intracranial pressure are a full, bulging fontanelle, abnormal respirations with cyanosis, and reduced responsiveness.

189. *A.* *Explain to Ms. Allen that he can lose 3 more ounces and*
still be within normal limits for weight loss.

A is correct; the full-term neonate of average weight for gestational age can lose up to 10% of body weight after birth. A matter-
of-fact answer such as this can be very reassuring to the mother. *B*
is incorrect; the neonate is born with enough fluids and energy
(usually) to sustain him until his mother's milk comes in. By
bringing a bottle of glucose to the mother, the nurse implies that
Ms. Allen is inadequate as a breastfeeding mother. In addition, the
baby may be too satisfied to empty his mother's breast completely,
thus preventing adequate lactation; and some are questioning
whether giving glucose at this early age sets a pattern for the ingestion of refined sugar. *C* is incorrect; the nurse is giving false reassurance and is passing the responsibility for explaining the situation to the physician. *D* is incorrect; an accurate record is *not* the
answer to the mother's concern about the well-being of her newborn.

190. *A.* *"Tighten the bottle cap a little."*

A is correct; a loose bottle cap pulls in extra air. The baby swallows the air, filling his stomach rapidly and causing regurgitation
of formula and gagging. If the cap is too tight, the bubbles are tiny
and the baby has to work hard to get any milk. *B* is incorrect; an
upright position does facilitate burping, but the amount of air the
baby is taking in must be decreased first. *C* is incorrect; the baby
is burped often during feeding, but usually once after each ounce
is sufficient. *D* is incorrect; by this behavior, the nurse is implying
that the mother cannot feed her child correctly. The nurse is going
to show the mother how well she can do it.

191. *B.* *She touches her nipple to the baby's cheek at the start of*
the feeding.

B is correct; Ms. Allen elicits the rooting reflex by allowing her
nipple to touch the baby's cheek or mouth. The baby turns his
head toward the nipple and opens his mouth to receive it. *A* is incorrect; ritualistic weighing of the baby to note "how much the
baby took" detracts from the naturalness of breastfeeding, and can
increase the mother's anxiety. Anxiety has an antilactogenic effect.
C is incorrect. Even breastfed babies take in air during feeding,
even though it is less than the amount of air taken in during bottle feeding. Breastfed babies are burped *at least* when changing
them from one breast to the other and also at the end of the feed-

ing. *D* is incorrect; placing any nonabsorbent material over the nipple does not let the nipples dry well. This causes maceration and possibly mastitis.

192. **D. *Phenylketonuria (PKU).***

D is correct; PKU is an inherited recessive error of metabolism. The enzyme that is missing is phenylalanine hydroxylase, which is needed to convert phenylalanine, one of the 26 essential amino acids, into tyrosine. Tyrosine is needed for the production of melanin; therefore, those with this affliction tend to have light skin and hair. Lofenelac is a phenylalanine-poor formula for the neonate. *A* is incorrect; kernicterus results when the breakdown products of bilirubin are deposited in the nuclear masses and basal ganglia of the medulla. This area is responsible for muscle coordination. Kernicterus is usually associated with Rh incompatibility, but may result from jaundice from any cause. However, the jaundice associated with breastfeeding has never been known to result in kernicterus. *B* is incorrect; hypoglycemia is a condition characterized by an excessive lowering of blood sugar due to high levels of circulating insulin in the infant of the diabetic mother. It may also be due to the rapid utilization of glucose, to sepsis, or to exertion from respiratory distress and the like. *C* is incorrect; thrush is a fungal disease caused by the organism *Candida albicans.* It is usually found only in the neonate's mouth but may spread throughout the GI tract or become generalized with septicemia. Septicemia usually occurs only in a very sick, debilitated neonate.

193. **B. *Retract his prepuce completely to cleanse the glans penis of smegma and other debris.***

B is the answer. Retraction of the foreskin (prepuce) of the uncircumcised male must be done gently and not forced. It is common to find the foreskin adherent to the glans penis for at least 3 months to several years. The foreskin is replaced after the glans penis is washed with mild soap and warm water. *A, C,* and *D* are not the answers because they are all *appropriate* care activities. Each eye is washed with a clean material from the inner canthus outward to prevent spread of any infection from one eye to the other. The hair must be rinsed of soap thoroughly and then combed with either a fine-toothed comb or a brush to prevent the development of cradle cap (seborrheic dermatitis). The nares and ears are washed with a soft, pliable material to avoid trauma to delicate tissues.

194. **C. *Corn and coconut oil mixtures are not absorbed to the same extent as the fat in human milk.***

C is the answer because corn and coconut oil mixtures *are* absorbed to the same extent as the fat in human milk; both oils are therefore used in the preparation of formulas. *A, B,* and *D* are not the answers because each of these statements is *true* about the neonate's digestive capabilities and formula preparation.

195. **D. *Women who have breastfed are less prone to develop breast cancer later in life.***

D is correct. Studies have shown that women who have breastfed are less prone to develop breast cancer later in life than are women who have not lactated. *A* is incorrect; immune bodies *do* cross the placenta to the fetus during the last trimester to protect the neonate against diseases that the mother herself has had. These immune bodies last only about 6–8 months. Breastfeeding

is a most unreliable contraceptive. Ovulation may occur while lactating as early as the 39th day or as late as 12 months. The average time of the first ovulation is about 8 months. The woman who assesses her cervical mucus can identify impending ovulation. *B* is incorrect. Breastfeeding jaundice results from the presence of pregnanediol or progesterone in the mother's milk. Pregnanediol inhibits the enzyme glucuronyl transferase and is needed to convert bilirubin to its soluble excretable form. Even severe breast milk jaundice is not known to cause kernicterus. However, most physicians suggest that the mother wean the infant. *C* is incorrect; breast milk contains less (not more) protein and more carbohydrate than cow's milk. The protein and carbohydrate in breast milk, which is bluish-white in color, are utilized much more efficiently and are better tolerated by the neonate than those from cow's milk. Prepared formulas, such as Similac and Enfamil, approximate breast milk more closely than does cow's milk.

196. **B. *Miliaria.***

B is correct. Miliaria, more commonly known as prickly heat, is due either to overdressing the neonate or to a hot, humid environment. Treatment includes removing excess clothing and giving tepid baths, being especially careful to rinse and dry thoroughly all skinfolds. *A* is incorrect; diaper rash is a chemical dermatitis due to ammonia and nitrogen products in the urine or stool and/or due to cloth diapers from which the soap was not completely removed. Changing diapers often, rinsing them well, drying diapers in a hot dryer or in direct sunlight, and exposing the infant's buttocks to air for periods of time cure the rash. *C* is incorrect; seborrheic dermatitis is more commonly known as cradle cap. Although it starts on the scalp above the forehead, it spreads downward over the cheek, neck, and then chest. It often occurs because mothers shy away from washing over the anterior fontanelle. Treatment is daily washing with mild soap and water. *D* is incorrect; diarrhea may be due to overfeeding, high sugar content in formula, or GI or other infections. This stool is usually forcefully expelled, may be greenish, and leaves a watery ring on the diaper around the loose stool. This is a potentially fatal condition, since the neonates tolerate this water loss poorly. The parent is advised to notify the physician.

197. **B. *The nurse is held to the same standard of care as always.***

B is correct; regardless of when or for how long nursing care is given, the nurse is held to the same standards that apply during normal shifts. Fatigue does not excuse poor performance. *A, C,* and *D* are incorrect assessments of the nurse's responsibility.

9. Adult Health Problems

Case Management Scenarios and Critical-Thinking Exercises

Medical Diagnosis: Diabetic ketoacidosis.

- **Nursing Problems:**
  Fluid volume deficit.
  Altered nutrition.

- **Chief Complaint:**
  Dennis Rose, an 18-year-old high school basketball star, was admitted to the hospital for treatment of polyuria, polydipsia, and dry cough.

- **History of Present Illness:**
  The patient was well until 10 days prior to admission, when he developed gastroenteritis followed by bronchitis. He had diarrhea, abdominal pain, intermittent vomiting, and dry cough.

- **Past History:**
  Measles at age 7. No allergies.

- **Family History:**
  Hyperthyroidism in maternal grandmother. Adult-onset diabetes mellitus in maternal uncle.

- **Review of Symptoms:**
  20 pound weight loss over the past 6 days.

- **Physical Exam:**
  Head and neck: Eyes sunken. Tongue dry. Fruity breath odor.
  Lungs: Clear on auscultation.
  Abdomen: Benign.
  Extremities: Skin dry with decreased turgor.
  Vital signs: BP: 110/70. Temperature: 98.6°F. P: 110. R: 40.

- **Laboratory Data:**
  Na: 138.
  K: 3.2.

CO_2: 6.
Cl: 102.
Glucose: 860.
WBC: 17,000/μL.
Hgb: 18.1g.
Hct: 54%.
Urinalysis: 4+ glucose; ketones—large.

---

1. The nurse knows that the condition that does not *precipitate* diabetic ketoacidosis in the patient is:
   A. Diarrhea.
   B. Intermittent vomiting.
   C. Bronchitis.
   D. Dehydration.
2. The nurse knows that Dennis will require insulin injections. Which statement best explains the rationale for insulin therapy?
   A. The juvenile diabetic requires insulin to supplement circulating insulin.
   B. The juvenile diabetic is insulin dependent.
   C. The injections are temporary until the pancreas increases insulin production.
   D. All diabetics must receive insulin.
3. Which admitting symptom indicates the incomplete lipid metabolism that occurs in diabetes?
   A. Polyuria.
   B. Polydipsia.
   C. Fruity breath.
   D. Weight loss.
4. The nurse knows that the presence of hypokalemia increases the risk of the patient's developing:
   A. Cardiac irritability.
   B. Renal failure.
   C. Cardiac depression.
   D. Respiratory failure.
5. Dennis has lost approximately 20 pounds in 6 days. He has been started on a 2800-calorie ADA diet. Which food is included in his diet?
   A. Whole-grain cereal with milk.
   B. Seafood casserole.
   C. Banana cream pie.
   D. Fried chicken.
6. Dennis has been asked to save his urine for testing. Which instruction should the nurse give him to ensure an accurate test?
   A. Supply only a single specimen of urine.
   B. Ingest at least 2 glasses of water before voiding specimen.
   C. Empty bladder ½ hour before testing and also supply a second voided specimen for testing.
   D. Test urine that has collected in bladder for 2–3 hours.
7. Blood glucose fingersticks are usually done for daily diabetic testing. Which test is done to assess a patient's compliance for the previous 3 months?
   A. Glycosylated hemoglobin measurement.
   B. Glucose tolerance test.

    C. Postprandial blood sugar measurement.

    D. Connecting peptide test.

8. Dennis is concerned about returning to basketball. The nurse needs to tell him that he will be able to continue the sport, but the exercise will require him to make the following adjustments:

    A. Increase his caloric intake, but continue the same insulin dosage.

    B. Reduce both his caloric intake and the insulin dosage.

    C. Continue the same diet, but increase the insulin dosage.

    D. Reduce his caloric intake and continue the same insulin dosage.

9. The nurse anticipates that the major problem for a diabetic of this age following discharge will be:

    A. Learning self-injection of insulin.

    B. Remembering to test his urine.

    C. Compliance with the dietary restrictions.

    D. Limiting his activities.

10. Following instructions on self-administration of insulin, the patient should be able to demonstrate his knowledge that tissue hypertrophy or atrophy (lipodystrophies) is prevented by:

    A. Using the same injection site.

    B. Administering room-temperature insulin.

    C. Injecting the insulin into the adipose tissue.

    D. Chilling the insulin before injection.

11. Dennis is being maintained on 12 units of NPH before breakfast, and 6 units before dinner. Which sign of an insulin reaction should the nurse watch for?

    A. Diaphoresis during the night.

    B. Kussmaul breathing in the late afternoon.

    C. Flushed, dry skin 6 hours after administration.

    D. Excessive hunger at meals and between meals.

Medical Diagnosis: Fracture of C-7 vertebra and multiple trauma.

■ *Surgical Treatment:*
Immobilization of fractured vertebra.

■ *Nursing Problems:*
Impaired physical mobility.
Knowledge deficit: about treatment.

■ *Chief Complaint:*
Jack Jenkins, a 19-year-old student, is seen in the emergency room after a motorcycle accident. He has suffered multiple traumatic injuries.

■ *History of Present Illness:*
Loss of consciousness at time of accident. Skin lacerations on face and forearm. No obvious long bone injuries. When alert, complains of loss of feeling in lower legs.

■ *Past History:*
Usual childhood illnesses.

- *Family History:*
  No history of diabetes, cancer, or mental illnesses.

- *Review of Systems:*
  Unable to answer questions at time of exam.

- *Physical Exam:*
  Vital signs: BP: 90/60. P: 88. R: 24.
  Head laceration 5 cm.
  Pupils equal and reactive to light.
  No sensation in legs and feet.
  No respiratory distress.
  Vomited undigested food; no abdominal distention or tenderness.
  Urine clear.

- *Laboratory and X-ray Data:*
  X-ray: Fracture of C-7 vertebra.
  Hgb: 13 g.
  Hct: 40%.
  Urinalysis: No RBCs.

---

12. In setting initial priorities of care for Jack, the nurse should spend the least amount of time:
    A. Maintaining proper body alignment.
    B. Immobilizing fractures of the long bones.
    C. Establishing and maintaining an airway.
    D. Scrubbing superficial lacerations.
13. It is determined that Jack has a head and back injury. A Stryker frame has been ordered for him. The nurse explains to Jack that the advantage of the frame is that:
    A. He can be turned while his spine is kept immobilized.
    B. It has automatic support in a position of good anatomic alignment.
    C. It provides even weight distribution.
    D. He will be able to turn himself.
14. Jack's area of permanent complete spinal cord damage is at the lowest end of the cervical spine. The nurse knows that it is unrealistic for Jack to expect to achieve the rehabilitative goal of:
    A. Walking.
    B. Writing.
    C. Feeding himself.
    D. Driving a car.
15. When testing the motor strength of a patient with a head injury, it is best for the nurse to:
    A. Observe the patient feeding himself or herself.
    B. Ask the patient to cross the legs.
    C. Ask the patient to squeeze the nurse's hands.
    D. Ask the patient to identify a pinprick on the foot.

Medical Diagnosis: Fracture of the right radius and ulna.

- *Nursing Problems:*
  Altered health maintenance.
  Risk of injury.

Knowledge deficit: cast management.
Altered nutrition: more than body requirements.

■ *Chief Complaint:*

Sally Peters, a 20-year-old student, received an injury to her right arm when she fell off a chair while hanging curtains.

■ *History of Present Illness:*

Pain, swelling, tenderness, and loss of motor function in the right arm. No other reported injuries.

■ *Past History:*

No diabetes or cancer. No other siblings in family.

■ *Review of Systems:*

Healthy until this injury. Attending school. No unusual childhood illnesses.

■ *Physical Exam:*

Vital signs: BP: 120/66. P: 72. R: 16.
Obvious deformity of right forearm; able to feel sensations when touched.
No cyanosis noted.
Right radial pulse: 72.

■ *Laboratory Data:*

Fracture, right radius and ulna.

---

16. X-ray reports document that a bone is broken into several fragments. The nurse knows that this type of fracture is known as:
    A. Impacted.
    B. Greenstick.
    C. Transverse.
    D. Comminuted.

17. While Sally is being examined, another patient is brought in to the emergency room by a friend. His left arm has an obvious deformity. There is some bleeding at the site, and when the nurse moves the patient's sleeve, the bone ends are penetrating the skin surface. The nurse should first:
    A. Reduce the fracture if possible.
    B. Leave the wound open to allow drainage and prevent hematoma.
    C. Immobilize the left arm.
    D. Thoroughly cleanse the wound.

18. The patient in the bed next to Sally's is in traction. The nurse caring for the patient in traction:
    A. Allows the weights of her traction to hang freely on the pulleys.
    B. Decreases the amount of weight on her traction each day.
    C. Removes the weight to inspect the site daily.
    D. Lifts the weights to help her move in bed.

19. The nurse is planning Sally's discharge teaching. Which statement should the nurse omit from Sally's health care instructions?
    A. Careful attention must be paid to circulation monitoring.
    B. There should be no exercise of the affected extremity.
    C. The cast should be kept dry at all times.

>   D. Pain will be lessened if the arm remains elevated.
20. Sally's roommates bring her back to the emergency room because of a possible cast complication. Their action was most likely based on Sally's complaint that:
>   A. The cast was too heavy.
>   B. She felt irritable and depressed.
>   C. The cast was still damp 6 hours after application.
>   D. Her fingers were blue and felt cold.
21. The nurse and Sally discuss the importance of a good diet to help bone healing. The nurse knows that the food substance determined to be least helpful in promoting healing is:
>   A. Iron.
>   B. Sodium.
>   C. Calcium.
>   D. Vitamin D.

Medical Diagnosis: Partial-thickness and full-thickness burns.

- ***Nursing Problems:***
  Risk of altered fluid volume.
  Risk of injury.
  Altered nutrition: more than body requirements.
  Self-care deficit: medication administration.

- ***Chief Complaint:***
  Jim Harte, a 24-year-old construction worker, enters the emergency room with burns resulting from a fire in the auto engine he was repairing.

- ***History of Present Illness:***
  Burns 30 minutes ago. Partial-thickness burns of face and chest; full-thickness burns of left leg.

- ***Past Health History:***
  Usual childhood illnesses. Weight 180 pounds. No history of drug or alcohol abuse.

- ***Family History:***
  No significant findings.

- ***Review of Systems:***
  No respiratory or cardiac problems. No major illnesses or infections.

- ***Physical Exam:***
  Head, neck, and chest: Partial-thickness (second-degree) burns.
  Left leg: Full-thickness (third-degree) burns over entire leg.
  No other burns noted. No fractures. Breathing patterns normal; no crackles or wheezes.

22. The nurse is notified that a Jim Harte, who has received burns, is on his way to the emergency room. The nurse remembers that characteristic signs of a partial-thickness (or second-degree) burn include:
>   A. Blisters covering the entire burn site.

B. Mild erythema.

C. Absence of pain when touched.

D. Color of the skin is white or black.

23. Jim's brother was with him at the time of the burn. He demonstrated incorrect burn first aid by:

A. Sending for the ambulance quickly.

B. Applying cold water to the area.

C. Applying first aid cream to the burn.

D. Preventing Jim from being chilled by getting him a blanket.

24. Besides the burns on his torso, Jim also has a full-thickness (or third-degree) circumferential burn on his left leg. The nurse should anticipate that he may need an escharotomy, which can be described as:

A. Debridement of the burn area.

B. A skin graft to the burn site.

C. Tube insertion for irrigation of the wound.

D. A lengthwise incision through the burn area.

25. The nurse explains to Jim that the disadvantage of silver sulfadiazine (Silvadene) ointment in burn therapy is due to the fact that:

A. Application is fast.

B. The ointment does not affect electrolyte imbalance.

C. The ointment depresses granulocyte formation.

D. Application is painless.

26. During the first 48 hours after a major burn, the nurse should see evidence of:

A. Plasma to interstitial fluid shift.

B. Interstitial tissue to plasma shift.

C. Potassium deficit.

D. Metabolic alkalosis.

27. During the recovery stage, the nurse plans the ideal diet with Jim. The nurse will be able to determine that Jim's caloric intake is adequate if he tells her that he takes in:

A. 3500–5000 calories.

B. 2000–2500 calories.

C. 1800–2500 calories.

D. 2000–3000 calories.

28. The nurse plans a high-protein diet with Jim to counteract his negative nitrogen balance. The nurse would need to review baseline theory when he or she states that this condition is caused by:

A. Tissue destruction and protein loss.

B. Stress response.

C. Decreased protein intake during initial burn phase.

D. Tissue anabolism resulting from immobility.

Medical Diagnosis: Acute viral hepatitis.

■ *Nursing Problems:*

Risk of injury related to biochemical regulatory impairment.

Impairment of digestion.

Risk of infection.

Altered nutrition.

Self-esteem disturbance.

Sensory/perceptual alterations related to chemical alterations.

- ***Chief Complaint:***
  Ms. Bee is a 25-year-old real estate agent. She complains of fatigue, weakness, dark-yellow urine, and clay-colored stools.

- ***History of Present Illness:***
  Two weeks prior to Ms. Bee's hospitalization, she felt very fatigued and weak. She complained of uncomfortable joint pains, frequent headaches, poor appetite, and nausea. On the fourth day of Ms. Bee's hospitalization, she developed jaundice, and strongly insisted that her visitors be restricted to her immediate family.

- ***Past History:***
  Healthy young adult. No previous hospitalizations.

- ***Family History:***
  Father, age 48; mother, age 45; both relatively well.

- ***Physical Exam:***
  Neck: Supple; no pain or stiffness on movement; trachea midline.
  Chest: Symmetric chest expansion; adequate chest excursion.
  Lungs: Clear on auscultation.
  Heart: S_1 and S_2 within normal limits; no S_3 or S_4.
  Abdomen: Flat, soft; active bowel sounds; tympanic sound in four quadrants; smooth liver edge with tenderness.
  Extremities: No rashes or irritation; jaundice of skin noted on fourth hospitalization day; range of motion of all extremities adequate, without pain or discomfort on movement; strong hand grasps bilaterally.

- ***Laboratory and X-ray Data:***
  Chest x-ray: Normal.
  Blood gases: Normal limits.
  ECG: Normal sinus rhythm.

---

29. Which patient problem is the nurse least likely to encounter?
    A. Bleeding.
    B. Pruritus.
    C. Weight loss.
    D. Hyperglycemia.
30. Based on the data given, the *most* common nursing problem is:
    A. Depression related to feelings of guilt.
    B. Anxiety related to a fear of impending doom.
    C. Body image change related to altered skin appearance.
    D. Anger and hostility related to restriction of physical activity.
31. Which laboratory result may the nurse expect to find during the initial phases of hepatitis?
    A. Increased LDH and CPK.
    B. Normal prothrombin time.
    C. Elevated serum transaminases.
    D. Decreased alkaline phosphatase.
32. The nurse observes that Ms. Bee has clay-colored stools. The reason is that:
    A. Hepatic uptake of bilirubin is impaired.
    B. Excretion of fecal urobilinogen is increased.
    C. Conjugated bilirubin reenters the bloodstream.

D. Excretion of conjugated bilirubin into the intestines is decreased.

33. The greatest risk of the spread of hepatitis B is from contaminated:
    A. Urine and feces.
    B. Nasogastric secretions.
    C. Used needles and syringes.
    D. Feces and oral secretions.

34. Pruritus, caused by the accumulation of bile salts in the skin, can be relieved by the nurse's administering prescribed:
    A. Valium to help the patient relax.
    B. Benadryl to promote sleep during the night.
    C. Questran to stimulate the reabsorption of bile salts.
    D. Cholestyramine to bind bile salts in the intestines.

35. Which neurologic assessment parameter(s) would *least* likely indicate to the nurse the occurrence of impending hepatic coma?
    A. Flapping tremors.
    B. Decorticate rigidity.
    C. Hyperactive reflexes.
    D. Irritability and drowsiness.

36. An ineffective nursing measure to prevent the progress of hepatic coma is:
    A. Giving diuretics.
    B. Making certain that a low-protein diet is served.
    C. Assessing if there is adequate renal perfusion.
    D. Assessing for a patent airway and oxygenation.

37. The nurse needs to know that lactulose and neomycin are given to patients with hepatic encephalopathy to:
    A. Decrease fecal pH and ammonia absorption.
    B. Induce peristalsis and promote bowel movement.
    C. Reduce antibacterial activity in the intestines.
    D. Remove potassium and magnesium in the intestines.

38. If the patient with severe liver damage is retaining nitrogen waste products, the nurse will note in the lab reports an increase in serum:
    A. Ammonia.
    B. Leukocytes.
    C. Creatinine.
    D. Urea nitrogen.

Medical Diagnosis: Probably Graves' disease (hyperthyroidism).

■ *Nursing Problem:*
Altered nutrition.

■ *Chief Complaint:*
Ms. Wise is a 26-year-old housewife and mother of two. She complains of insomnia, and a weight loss of 20 pounds in 4 months.

■ *Past History:*
Has remained healthy; no surgeries.

■ *Family History:*
  Father: Hypertensive cardiovascular disease.
  Mother: Unremarkable.
  Sibling: Unremarkable.

■ *Review of Symptoms:*
  Ravenous appetite; heat intolerance.

■ *Physical Exam:*
  Neck: Carotid pulse bounding.
  Skin: Moist with perspiration.
  Lungs: Clear A-P. Sinus tachycardia.
  Abdomen: Hyperactive bowel sounds.
  Extremities: Mild tremors.
  Mental status: Alert and agitated. Complete sentences for interviewer.
  Vital signs: BP: 120/80. P: 120. R: 30. Temperature: 100°F orally.
  ECG: Sinus tachycardia.

■ *Laboratory Data:*
  Urinalysis: Normal.

---

39. During Ms. Wise's admission to the unit, the nurse begins the initial assessment. She notices an anxious facial expression, pulse of 120, respirations of 30, and skin moist with perspiration. The nurse knows that the most probable rationale for these symptoms is that:
    A. Ms. Wise is experiencing normal anxiety associated with her admission.
    B. Increased levels of thyroxine and triiodothyronine stimulate metabolic rate.
    C. Reduced iodine intake produces hypertrophy and overactivity of the thyroid gland.
    D. Ms. Wise was rushing to avoid being late for hospital admission.

40. A protein-bound iodide test has been scheduled. Before the test, it is important that the nurse:
    A. Have the patient void.
    B. Inquire whether the patient has had an x-ray for which opaque dye was administered.
    C. Instruct the patient to get 8 hours of sleep the night before the test.
    D. See that the patient receives a meal high in iodine.

41. A radioactive iodine uptake test is performed. The nurse should:
    A. Inform the patient that there is no danger of radiating herself or others.
    B. See that the urine is collected in a lead container.
    C. Wear a radiation detection badge when he or she is near the patient.
    D. Avoid prolonged contact with the patient for 6 hours after the test.

42. A diagnosis of hyperthyroidism is confirmed. The nurse instructs the patient in a diet that has:
    A. 4000–5000 calories; 100–125 g of protein.

    B. 2000–3000 calories; 70–90 g of protein.

    C. Low calcium; high vitamin D.

    D. 1000–2000 g of carbohydrate; 60–80 g of protein.

43. Based on the pathophysiology of hyperthyroidism, it is important that the nurse:

    A. Have several blankets available for Ms. Wise's use.

    B. Place Ms. Wise in a room with an electric fan and windows.

    C. Provide a 300-piece puzzle for diversion.

    D. Insist that Ms. Wise rest 2 hours in the morning and in the afternoon.

44. Propylthiouracil is ordered. Which observation indicates to the nurse that the medication was having a therapeutic effect?

    A. A weight loss of ¼ pound per week.

    B. Apical pulse of 90 and reduction in urine output.

    C. Ability to sleep 8 hours at night without awakening.

    D. Inability to complete a paint-by-number picture.

45. Ms. Wise will be readmitted for surgery in about 2 weeks. She is now discharged on saturated solution of potassium iodide (SSKI). The nurse instructs Ms. Wise that the purpose of this medication is to:

    A. Produce a sedative effect so that she can sleep at night.

    B. Reduce exophthalmos.

    C. Minimize postoperative bleeding.

    D. Replace depleted iodine stores.

46. While planning preoperative teaching, the nurse realizes that Ms. Wise should be taught to maintain head elevation at a 15-degree Fowler's position. The nurse tells her that the purpose of this position is to:

    A. Prevent hemorrhage.

    B. Reduce pooling of respiratory secretions.

    C. Prevent strain on the suture line.

    D. Prevent hoarseness.

47. Ms. Wise has a subtotal thyroidectomy. The nurse can detect *early* hemorrhage by noting:

    A. A BP of 90/50, apical pulse of 150, and clammy skin.

    B. Muscle twitching and tremors.

    C. Hoarseness and weakness of the voice.

    D. Complaints of difficulty swallowing and sensation that the dressing is tight.

48. Ms. Wise exhibits hoarseness, crowing respirations, and retraction of the tissues of the neck. The nurse should have which of the following for initial emergency use?

    A. Calcium gluconate, 1 ampule.

    B. Synthroid sodium, 1 vial.

    C. Tracheostomy set.

    D. Thoracentesis tray.

49. Which observation might indicate to the nurse the therapeutic effect of calcium gluconate?

    A. Reduction in size of the thyroid gland from 4–2 cm.

    B. Curtailment of blood loss.

    C. Absent Trousseau's sign.

    D. Reduction in hoarseness.

50. Which patient should the nurse anticipate to be prone to develop tetany?

    A. Ms. J with hyperparathyroidism.

    B. Mr. W with hypothyroidism.

C. Mr. S in metabolic acidosis.

D. Mr. X with metabolic alkalosis.

51. Hoarseness may occur during the immediate postoperative period. The patient should be instructed that:

A. He or she should refrain from speaking.

B. This is a permanent, untoward response to surgery.

C. Humidification and antitussives hasten the return of the voice.

D. Even after his or her voice returns, it will never have the pitch and strength of the preoperative voice.

Medical Diagnosis: Burns.

- **Nursing Problems:**

  Altered body temperature.

  Altered nutrition.

  Self-care administration: medication administration.

  Impaired skin integrity.

- **Chief Complaint:**

  Mr. Green is a 35-year-old oil field supervisor who was burned on the job.

- **History of Present Illness:**

  Explosion on job started flash fire, producing partial-thickness and full-thickness burns of face, arms, and chest. Patient presents 45 minutes after the thermal injury occurred.

- **Past History:**

  No allergies. Considered in good general health before accident.

- **Family History:**

  Parents: Unremarkable.

  One sibling: Diabetes mellitus, maturity onset.

- **Physical Exam:**

  HEENT: Singed nares. Skin around nose and mouth red. Palate red.

  Neck: Carotid pulse present; reddened.

  Heart: Sinus tachycardia.

  Lungs: Clear A-P.

  Chest: Burned areas appear white, with redder areas on periphery. Burn extends from clavicle to umbilicus.

  Extremities: Right arm: Circumferential burn on forearm, red and weeping. Upper left arm: Burn area found medially; appears white, with redder areas on periphery.

  Vital signs: Temperature: 97°F. BP: 90/70. P: 100. R: 30.

- **Laboratory Data:**

  Hct: 56%.

  Urine: Specific gravity = 1.030. pH: 6. Positive for hemoglobin.

52. Before Mr. Green's arrival at the emergency room, the industrial nurse should perform which action *first?*

A. Cover Mr. Green with a blanket.

B. Apply a topical antibiotic ointment to the burned areas.

C. Apply tepid to cold water to the burn surface.

D. Excise eschar with scalpel and scissors.

53. What could the nurse do as emergency treatment for a chemical burn to the eye?

A. Apply a sterile 4 × 4 bandage over the injured eye to prevent infection.

B. Administer an antidote to slow down the chemical reaction.

C. Apply ice packs to retard absorption of the chemical.

D. Flush with water continuously.

54. Mr. Green arrives at the hospital emergency room. The nurse needs to prepare all medications for intravenous administration at this time because:

A. Rapid excretion is needed to prevent overdose.

B. There is a lack of sufficient available sites.

C. Peripheral perfusion is poor.

D. Reduction in pain and additional stresses is needed.

55. During the initial stage of burn therapy, the nurse expects Mr. Green's body temperature to be:

A. 97°F.

B. 98.6°F.

C. 100°F.

D. 102°F.

Mr. Green is transferred to the burn unit.

56. A student in the burn unit asks the nurse why the surgeon wants to wait 24 hours before giving colloids. Which rationale does the nurse offer for that action at that time?

A. To enhance interstitial edema.

B. To improve cardiac output.

C. Renal perfusion must be maintained.

D. Capillary permeability is reduced.

57. The nurse can anticipate that which observation indicates a good prognosis 36 hours after the burn?

A. Bounding pulse.

B. CVP of 10.

C. Urine output of 20 mL/hour.

D. Reduction in size of blisters.

58. The nurse should avoid giving burned patients anything by mouth in the first 48 hours after the burn because of the possibility of their developing:

A. Productive cough with yellow mucus.

B. A fall in blood pressure, with widening pulse pressure.

C. Hematuria accompanied by decreasing output.

D. Abdominal distention with absent bowel sounds.

59. Because of the location of Mr. Green's burns, which nursing observation requires immediate action?

A. Urine output 30 mL/hour for the first 4 hours. Urine red to chocolate in color.

B. Absent bowel sounds.

C. Respiratory stridor.

    D. Blisters and white, leathery appearance of burned areas.

60. Mafenide (Sulfamylon) is ordered to be applied over the burned areas. Select the nursing measure to be used when this medication is administered:

    A. Give analgesic 30 minutes before application.

    B. Observe laboratory reports for leukopenia.

    C. Observe creatinine and BUN levels.

    D. Have Mr. Green's hearing checked by an otologist.

61. If gentamicin is used to treat a *Pseudomonas* infection, the nurse should:

    A. Give analgesic 30 minutes before application.

    B. Check laboratory reports for leukopenia.

    C. Observe any changes in renal function.

    D. Check deep-tendon reflexes for hyporeflexia.

62. The nurse knows that an important disadvantage of 5% silver nitrate as a treatment for burns is that it:

    A. Prevents eschar formation.

    B. Causes the loss of potassium and sodium.

    C. Stains the floor, walls, and skin.

    D. Cannot be used on burned children.

63. Curling's ulcer is a complication of severe burns. Early acute symptoms for the nurse to look for include:

    A. Pain in the right upper quadrant and flatus.

    B. Hypotension and hematemesis.

    C. Dyspepsia and hyperperistalsis.

    D. Abdominal distention and decrease in hematocrit.

64. Which nursing observation indicates that zinc stores are adequately met?

    A. A decrease in wound circumference from 4 to 2 cm.

    B. Reduction in muscle twitching.

    C. A urinary output free of acetone and glucose.

    D. Adequate peripheral vision and color perception.

65. The closed method of treatment for burns involves occlusive dressings. Nursing measures relative to the care of a patient with occlusive dressings include:

    A. Avoid evenly distributed pressure on the wrapped part.

    B. Submerge the dressed part in a pHisoHex solution before changing the dressing.

    C. Keep the wrapped part in a dependent position.

    D. Prevent two body surfaces from touching.

66. The nurse knows that which of the following is necessary for protein synthesis to proceed at an optimum rate?

    A. Protein reserve.

    B. Glucagon.

    C. Adequacy of lipoproteins.

    D. Adequate caloric intake.

67. In selecting a diet for the burn patient who is to receive a high-protein mid-evening feeding, the nurse encourages the patient to eat:

    A. ⅓ cup yogurt.

    B. Broth.

    C. Jell-O with 8 crushed nuts.

    D. 1 cup of whole milk.

68. To minimize psychological stress, the nurse can:

    A. Insist that Mr. Green participate in his wound care.

    B. Schedule nursing care procedures to allow periods of uninterrupted sleep.

    C. Compare his progress with that of the other patient in the unit.

    D. Ignore his periods of enuresis.

69. Which would the nurse assess to be an inappropriate response to thermal injury?

    A. Episodic depression or anger.

    B. Periodic anorexia.

    C. Reluctance to disrobe before the nursing staff.

    D. Continuous open masturbation.

70. Mr. Green is fitted for an elasticized garment during the rehabilitative phase. The nurse explains that the purpose of this mode of therapy is to:

    A. Reduce protein loss through the wound.

    B. Minimize evaporation and convection of heat.

    C. Reduce scar hypertrophy.

    D. Reduce collection of edema fluid under the graft.

Medical Diagnosis: Hyperthyroidism.

- ***Surgical Treatment:***
  Thyroidectomy.

- ***Nursing Problems:***
  Risk of inadequate breathing patterns.
  Risk of fluid volume excess.
  Risk of injury: postoperative complications.

- ***Chief Complaint:***
  Ms. Mary Easley, a 36-year-old white woman, seeks health care for palpitations and breathlessness.

- ***History of Present Illness:***
  Weight loss over the past 6 months despite increases in appetite and food intake. Medication for diagnosed hyperthyroidism was propylthiouracil. Symptoms continued despite medical intervention.

- ***Family History:***
  No history of thyroid disease. One sister is a diabetic. No history of cancer.

- ***Review of Systems:***
  Weakness, fatigue, weight loss, increased appetite, heat intolerance, diarrhea, and insomnia.

- ***Physical Exam:***
  Fine tremors of hands.
  Skin fine textured and smooth and moist to touch.
  Exophthalmos.
  Jittery and tense.

- ***Laboratory and X-ray Data:***
  Radioactive iodine uptake: 25% in 6 hours.
  Thyroid scan: Enlargement of thyroid gland.

71. Ms. Easley seeks health care for hyperthyroidism. Which is an uncharacteristic sign or symptom of this condition?
A. Enlargement of the front of her neck.
B. Increasing protrusion of her eyes.
C. Weight gain.
D. Feelings of nervousness and irritability.

72. Medical intervention has not been successful and Mary Easley has a subtotal thyroidectomy. Because the nurse knows that a complication of this surgery may be calcium deficit, the nurse cautions Ms. Easley to report:
A. Sharp decrease in urinary output.
B. Pain in her kidney region.
C. Continuous toniclike muscle spasms.
D. Loss of sensation in extremities.

73. Ms. Easley received 3000 mL of IV fluid, containing saline, during the past 24 hours. The nurse identified Ms. Easley's problem as potential fluid overload. The nurse based that decision on which assessment data?
A. Scanty urine and burning on urination.
B. Dry skin and mucous membranes.
C. Increased RBC count and hemoglobin.
D. Puffy eyelids and shortness of breath.

74. Anticipating Ms. Easley's return to her room postoperatively, the nurse gathers specific equipment for her care. Which is the least necessary item to have at the bedside?
A. Airway.
B. Tracheotomy set.
C. IV equipment.
D. Gastric suction apparatus.

Medical Diagnosis: Cancer of the right breast.

■ *Surgical Treatment:*
Modified radical mastectomy.

■ *Nursing Problems:*
Fear.
Altered feeling state: anger.
Health-seeking behaviors: interpretation of lab data, skin care after radiation treatment.
Knowledge deficit: about treatment.
Self-care deficit: medication administration: outcome, toxic effects, disadvantages.
Self-esteem disturbance.

■ *Chief Complaint:*
Marion Mulvrey, a 40-year-old married homemaker, entered the hospital for treatment of a breast lesion discovered 1 month ago during routine breast self-examination.

■ *History of Present Illness:*
Examines breasts monthly. Noticed a small lump at upper outer region of right breast.

■ *Past History:*

Two children, ages 12 and 15, living and well. Phlebitis after delivery of second child; treated medically with no complications. Auto accident at age 33; concussion with resulting headaches for approximately 6 months after the accident. Usual childhood illnesses.

■ *Family History:*

No known history of breast cancer. No diabetes, tuberculosis, or mental disorders. Mother living; history of arthritis. Father recently retired after myocardial infarction. Brother, age 46; history of duodenal ulcers. Two older sisters living and well.

■ *Review of Systems:*

Has lost 10 pounds in past year. Has noticed increased tiredness during the past 3 months. No respiratory problems noted. No high blood pressure. Usual health checkup every year; periodic dental exam every 6 months.

■ *Physical Exam:*

BP: 146/92. P: 88. R: 26.

Apprehensive about finding lesion, right upper quadrant, right breast, approximately 2 cm in size, painless. Boundaries on palpation seem to be irregular.

■ *Laboratory and X-ray Data:*

Mammography: Irregularly shaped lesion of 2 cm. Breast biopsy necessary to confirm diagnosis.

Hgb: 13 g.

Hct: 42%.

Chest x-ray: Lungs clear.

---

75. The nurse helps Ms. Mulvrey to be aware that of all the following women, the group that seems to be the least likely to be at risk for breast cancer is:
    A. Women over 50 years old.
    B. Women whose mothers or sisters have breast cancer.
    C. Women who had a child before the age of 30.
    D. Women who are overweight.
76. The nurse explains to the patient that the most important reason for women to perform breast self-examination is:
    A. The woman is able to determine what is serious and what is not serious.
    B. Women have great fears about having a physician perform breast exams, so it is best if they do it themselves.
    C. It eliminates the need for costly periodic health checkups.
    D. Most breast abnormalities can be detected early by the woman or her partner and reported to the physician.
77. The nurse helps Ms. Mulvrey to decide that the best time to do breast self-examination is:
    A. When her menstrual period begins.
    B. One week after her period begins.
    C. On the first day of the month.
    D. On the 14th day of her menstrual cycle.
78. When Ms. Mulvrey finds a lump in her right breast, the nurse advises her to:

    A. Make an appointment with her physician as soon as possible.

    B. Wait another month and evaluate.

    C. Call relatives to determine if there is a family history of breast cancer.

    D. Ignore it because she is not in the high-risk group.

79. Ms. Mulvrey is scheduled for a breast biopsy. The nurse explains that:

    A. This is the treatment for cancer of the breast.

    B. Most biopsies lead to mastectomies.

    C. There will be no scar from this procedure.

    D. This is part of the diagnostic workup.

80. The biopsy specimen is positive for breast cancer. Ms. Mulvrey is scheduled for a modified radical mastectomy. The nurse explains that the tissue that will be removed will be:

    A. Breast and axillary nodes.

    B. Breast, pectoral muscles, and axillary nodes.

    C. Breast alone.

    D. The portion of the breast involved.

81. The nurse realizes that Ms. Mulvrey has demonstrated the first step toward acceptance of her altered body image by:

    A. Looking at the incision when the nurse is changing the dressing.

    B. Touching the operative site when the dressing is removed.

    C. Changing the subject each time the nurse attempts to discuss it.

    D. Asking her husband to bring in her sheer nightgown.

82. Ms. Mulvrey is scheduled for radiation therapy. When she asks if there are any special precautions she should plan to take, the nurse replies:

    A. "Use talcum powder for comfort."

    B. "Apply a soothing ointment to the skin."

    C. "Wash the skin with water and pat dry."

    D. "Sunbathe as usual."

83. The nurse has identified that the local tissue response to radiation is normal because Ms. Mulvrey's site shows:

    A. Redness of the surface tissue.

    B. Atrophy of the skin.

    C. Scattered pustule formation.

    D. Sloughing of two layers of the skin.

84. Six months after surgery, Ms. Mulvrey was found to have a positive liver scan. It was decided to begin a course of antimetabolite chemotherapy. The nurse recognizes that the expected action of this drug is to:

    A. Decrease blood supply to malignant cells.

    B. Produce a toxic substance absorbed by malignant cells.

    C. Interfere with nucleic acid synthesis.

    D. Destroy the capsule surrounding malignant cells.

85. The nurse knows that drugs commonly used to treat malignant tumors are often toxic. Her observations will show the most toxic effects on:

    A. Hair follicles.

    B. Lung alveoli.

    C. Muscle cells.

    D. Kidney tissue.

86. The nurse knows that the major disadvantage of antineoplastic drugs is that they:
    A. Tend to enhance the immune response.
    B. Do not discriminate between cancer cells and normal cells.
    C. Substitute for normal substances in the body and are absorbed by cancer cells.
    D. Always require the patient to be hospitalized.

87. One of the potential side effects of cancer chemotherapy is thrombocytopenia. Which data confirm to the nurse the presence of this problem?
    A. Low lymphocyte count.
    B. Low platelet count.
    C. High RBC count.
    D. High eosinophil count.

88. Ms. Mulvrey is readmitted to the hospital 1 year later in a terminal stage of illness. At this time, the nurse planning care with Ms. Mulvrey should recognize that she is most likely to fear:
    A. Diagnosis.
    B. Further therapy.
    C. Being socially inadequate.
    D. Being isolated.

89. When Ms. Mulvrey's husband expresses his anger, it is best for the nurse to:
    A. Offer to call a friend to be with him.
    B. Leave him alone so he can have privacy.
    C. Stay with him and listen to his concerns.
    D. Explain that all that can be done has been done.

Medical Diagnosis: Fibroid uterus.

- **Surgical Treatment:**
  Abdominal hysterectomy.

- **Nursing Problems:**
  Inadequate breathing patterns.
  Altered comfort pattern (discomfort-related to postop breathing).
  Fear.
  Risk for infection.
  Risk for injury: complications.
  Knowledge deficit: about diagnostic test.
  Self-care deficit: medication administration.
  Impaired skin integrity.

- **Chief Complaint:**
  Ms. Salisbury, a 42-year-old school teacher, enters the hospital for a planned abdominal hysterectomy.

- **History of Present Illness:**
  Ms. Salisbury has had a history of abnormal menstrual symptoms for the past 6 months. She has had menorrhagia for the past 6 months and metrorrhagia for the past 3 months. She has had low back discomfort and pain, constipation, and increasing dysmenorrhea.

- *Past History:*
  Ms. Salisbury has one daughter, age 19. She had three other pregnancies, which ended in spontaneous abortions in the first trimester. There was no history of inflammatory disease. She has been using a diaphragm and contraceptive jelly for birth control.

- *Family History:*
  One sister, age 46, had a hysterectomy at age 44 for similar problems. There was no history of ovarian or uterine cancer in the family.

- *Review of Systems:*
  Weight loss of 5 pounds during the past year. No rashes or skin lesions.
  Head and neck: No complaints of vertigo. Two colds during the past 3 months.
  Breasts: Periodic self-examination. No abnormalities.
  Respiratory: No distress at present. No cough.
  Cardiac: No known heart disease. One episode of palpitation after first bleeding episode.
  GI: Appetite fair; concern over weight loss.
  Urinary tract: Cystitis with first pregnancy. No other problems.
  Genitoreproductive: Menarche at 13. Periods regular every 28 days for 5 days. Heavy bleeding at periods for past 6 months. Bleeding between periods past 3 months. D & C 2 months ago; bleeding pattern continued after D & C.

- *Physical Exam:*
  Respiratory: No crackles or wheezes noted on auscultation.
  Abdomen: Soft to palpation. Bowel sounds present.
  Genitalia/vulva: Normal. Uterus: Anterior, enlarged, abnormal contour. Pap smear negative at time of D & C.

- *Laboratory Data:*
  Pap smear: Negative.
  D & C scrapings: Tissue consistent with leiomyoma formations.
  Hct: 34%.
  Hgb: 10 g.

---

90. Ms. Salisbury is discussing the report of the Papanicolaou smear she received from her physician. She asks the reason why such a test is done yearly. The nurse answers that this test helps to determine:
    A. Precancerous and cancerous conditions of the cervix.
    B. Fibrous and other benign tumors of the uterus.
    C. Inflammatory processes present in the pelvis.
    D. Infections of Bartholin's gland.

91. Ms. Salisbury expresses fear about her upcoming abdominal hysterectomy. The nurse should:
    A. Tell her that her physician is competent.
    B. Allow her an opportunity to express her fears.
    C. Change the subject when fear is obvious.
    D. Teach her that fear stands in the way of recovery.

92. Preoperative medications ordered for Ms. Salisbury were meperidine, 75 mg, and atropine, 0.3 mg. Dosages available were meperidine, 50 mg/mL, and atropine, 0.4 mg/mL. The nurse prepares:

A. Meperidine, 0.75 mL; atropine, 1.2 mL.

B. Meperidine, 1.5 mL; atropine, 0.5 mL.

C. Meperidine, 2 mL; atropine, 1 mL.

D. Meperidine, 1.5 mL; atropine, 0.75 mL.

93. The nurse gives atropine sulfate preoperatively to achieve:

A. General muscular relaxation.

B. Decrease in pulse and respiratory rate.

C. Decrease in oral and respiratory secretions.

D. Blood pressure within normal range.

94. Considering the potential discomfort level during postoperative deep breathing and coughing, the nurse should prepare Ms. Salisbury to:

A. Support her abdomen with a pillow or her hands.

B. Support her rib cage with a binder or her hands.

C. Lie on her abdomen with her arms at her side.

D. Lie flat in bed with her hands behind her head.

95. Ms. Salisbury returns from the recovery room. The nurse observes that the site of the intravenous infusion is pale and edematous. The nurse should examine for further evidence of:

A. Phlebitis.

B. Air embolism.

C. Infiltration.

D. Allergic reaction.

96. When changing the dressing, the nurse noted a small separation at the base of the wound. Suddenly Ms. Salisbury coughs and the wound completely separates. The nurse's most appropriate initial action is to:

A. Cover the incision with a sterile, moist towel.

B. Pack the intestines back into the abdominal cavity.

C. Irrigate the exposed area with sterile water.

D. Document the sequence of events in the nurse's notes.

97. If Ms. Salisbury had her ovaries removed at the same time as she had her hysterectomy, the nurse would explain that she could expect to experience:

A. Decreased perspiration.

B. Increased desire to sleep.

C. Pallor of head, neck, and throat.

D. Increased sensitivity to heat loss.

98. Later in Ms. Salisbury's postoperative course, the nurse identifies that she has a possible infection of the wound. Which assessment parameter could the nurse eliminate in making this observation?

A. Small amount of drainage on the dressing.

B. Pain at the upper edge of the incision.

C. Fever of 101.8°F.

D. Moist crackles at the base of her lungs.

99. Ms. Salisbury receives a prescription for cephalexin (Keflex) to take after discharge from the hospital. The nurse should instruct her to:

A. Take all of the medication prescribed by her physician.

B. Take all of the medication prescribed for 3 days and reevaluate how she is feeling.

C. Take all of the medication as long as she has a fever.

D. Take the medication until the pain subsides.

Medical Diagnosis: Leukemia.

- **Nursing Problems:**
  Anxiety.
  Body image disturbance.
  Risk for infection.
  High risk for injury: bleeding tendency.
  Altered nutrition requirements.
  Social isolation.

- **Chief Complaint:**
  Ms. Ella Perry, a 45-year-old housewife, suffered a severe bleeding episode after having a tooth extracted.

- **History of Present Illness:**
  For approximately 6 months, Ms. Perry complained of weakness, easy fatigability, and pain in her arms. She attributed these complaints to the strenuous exercises she usually engages in. She plays tennis regularly. Later, she noticed increased bruising on her thighs and arms.

- **Past History:**
  Hospitalized twice; normal pregnancies.

- **Family History:**
  Father died at age 65; cancer of lungs. Mother, age 70, relatively healthy. Sister, age 50, anemia. Children ages 10 and 13, healthy.

- **Physical Exam:**
  Neck: No enlargement or asymmetry; no mass or lymph nodes palpated; no limitation of movement.
  Chest: Normal anteroposterior diameter; bilateral and equal chest expansion and excursion.
  Lungs: Clear on auscultation.
  Heart: Point of maximal impulse (PMI) at fifth intercostal space (ICS), barely palpable; S_1 and S_2 within normal limits; no clicks or splitting; no extra heart sounds.
  Abdomen: Soft, nondistended; fairly active bowel sounds; enlarged spleen.
  Extremities: Joint pains, particularly in arms; strong hand grasps; adequate muscle strength in lower extremities.
  Chest x-ray: Normal.
  ECG: Normal sinus rhythm.
  Blood gases: Within normal limits.

- **Laboratory Data:**
  WBC: 205,000/µL.
  Philadelphia chromosome positive.
  Decreased leukocyte alkaline phosphatase.

100. To confirm the diagnosis of chronic myelocytic leukemia, which laboratory test is not routine?
     A. Blood culture.
     B. Differential count.

    C. Bone marrow aspiration.

    D. WBC count.

101. What physical assessment information gathered from the patient's records by the nurse confirms the presence of an enlarged spleen?

    A. Dullness is elicited when the splenic area is percussed.

    B. The spleen tip is difficult to palpate below the left costal margin.

    C. A notch is palpated along the medial border of the left lower quadrant.

    D. Tympanic sound is elicited when the anterior axillary line is percussed.

102. The nurse knows that the weakness, fatigue, and pallor common in leukemic patients are due to:

    A. Deficiency of platelets.

    B. Presence of infective process.

    C. Decreased erythrocyte production.

    D. Infiltration of peripheral nerves by cancer.

103. The leukemic patient is given a low-bacteria diet, particularly during the acute episode of the disease process. Which does the nurse expect to be included in such a diet?

    A. Cooked spinach and celery.

    B. Lettuce and alfalfa sprouts.

    C. Fresh strawberries and carrots.

    D. Raw cauliflower or broccoli.

104. To prevent sensory deprivation in a patient who is on protective isolation, the nurse should:

    A. Encourage friends and relatives to visit at the same time.

    B. Visit the patient frequently and engage her in conversation.

    C. Repeatedly explain why protective isolation is necessary.

    D. Encourage the family to stay with the patient for as long as they want.

105. A complication that the nurse must watch for in patients receiving chemotherapy is crystallization of uric acid. Which is normal and therefore not a manifestation of this complication?

    A. Specific gravity of 1.030.

    B. Urine output of 30–50 mL/hour.

    C. Serum creatinine of 3 mg/dL.

    D. Elevated BUN levels.

106. Which response is the *most* appropriate for the patient who has alopecia as a result of chemotherapy?

    A. "All we have to do is to put a cap on your head."

    B. "I'm sorry but your hair will always be thin now."

    C. "Your hair will grow again after the chemotherapy is over."

    D. "Well, chemotherapy does that to most patients, and we don't know why."

107. When a patient is on cancer chemotherapy, the nurse should continuously assess for the presence of bone marrow depression, manifestations of which are:

    A. Night sweats and easy fatigability.

    B. Loss of skin turgor and weight loss.

    C. Low urine output and elevated BUN levels.

    D. Ecchymosis and weakness.

108. The nurse expects that a leukemic patient is in "remission" when:

    A. The patient's immune system is stimulated to fight the organism.

    B. There is an increased amount of myeloblasts in the bone mar-
       row.

    C. No cancer cells are microscopically evident in the bone marrow.

    D. Clinical manifestations of cancer are absent for more than 3
       years.

109. The least helpful nursing approach in the care of a terminally ill
    patient is to:

    A. Maintain the patient's self-esteem and well-being.

    B. Keep the patient from making decisions regarding his or her
       care.

    C. Provide open communication between the patient and the staff.

    D. Allow family members to participate in the care of the patient.

Medical Diagnosis: Bleeding duodenal ulcer.

■ *Surgical Treatment:*
Gastrectomy.

■ *Nursing Problems:*
Altered bowel elimination.
Altered cardiac output.
Altered comfort (pain).
Effective individual management of therapeutic regimen.
Altered nutrition: less than body requirements.
Self-care deficit: medication administration.

■ *Chief Complaint:*
Mr. Earl Williams, the 46-year-old owner of a small business, is admitted to the hospital complaining of vomiting a large amount of
bright red blood.

■ *History of Present Illness:*
History of gastric disorders for the past 2 years. Current episode
consisted of several days of consumption of some alcohol; death of
brother (cancer of the lungs); inability to sleep, then epigastric pain,
nausea, vomiting gastric contents; and then vomiting bright red
blood.

■ *Past History:*
Stomach disorders for several years. Weight loss of 6 pounds. Complaining of gnawing, aching, and burning pain, usually relieved by
proper diet and antacids.

■ *Family History:*
One brother died age 48, heart attack; one brother died, cancer of
the lungs; one sister, living and well. Wife has multiple sclerosis,
early stages. Son is unemployed high school dropout. Two daughters:
one married (18 years old) and living at home with husband and 2-
month-old daughter; the other daughter is in high school and doing
well.

■ *Review of Systems:*
Weight loss. GI history (see History of Present Illness and Past History). No difficulty in breathing or palpitations.

■ *Physical Exam:*
Vital signs: BP: 112/60. P: 92. R: 18.
Skin pale.

Acute pain.
Breath sounds normal.
Abdomen: Acute distress.

■ *Laboratory and X-ray Data:*
GI series shows abnormality of tissue in the duodenal region.
Endoscopy positive for duodenal ulcer.
Stool positive for occult blood.
Hgb: 12g.
Hct: 30%.

---

110. Mr. Williams goes to the drugstore to buy an antacid. The nurse should have told him of the laxative effect of:
    A. Calcium carbonate (Titralac).
    B. Aluminum hydroxide gel (Amphojel).
    C. Magaldrate (Riopan).
    D. Magnesium hydroxide (magnesium magma).

111. What is the desired effect of drugs, such as propantheline bromide (Pro-Banthine), that the nurse gives to patients with duodenal ulcers?
    A. To decrease gastric motility.
    B. To tranquilize the patient.
    C. To increase gastric secretions.
    D. To directly stimulate mucosal cell growth.

112. Mr. Williams has been treated for a peptic ulcer. He enters the hospital reporting that he has vomited a very large amount of blood. The nurse should expect which sign to be present?
    A. Decreased blood pressure.
    B. Decreased pulse.
    C. Decreased respirations.
    D. Increased urinary output.

113. The nurse needs to instruct Mr. Williams that if he is having bleeding from his stomach ulcer, his stools will be:
    A. Claylike in color.
    B. Tarry.
    C. Bright red.
    D. Light brown.

114. Mr. Williams begins to hemorrhage from his ulcer and will have surgery. The nurse considers that the urgency of this surgery is:
    A. Planned.
    B. Imperative.
    C. Emergency.
    D. Optional.

115. The nurse is observing for the possible complication of postoperative peritonitis. Which sign(s) or symptom(s) is (are) *least* indicative of peritonitis?
    A. Hyperactive bowel sounds.
    B. Pain, local or general.
    C. Abdominal rigidity.
    D. Shallow respirations.

116. Mr. Williams complains of postoperative pain. The nurse administers the analgesic as ordered. What nursing actions will help reduce the pain while the analgesic is taking its effect?

A. Move the patient quickly; administer stimulating backrub.

B. Encourage the patient to discuss his feelings.

C. Position comfortably; subdue the lighting.

D. Give him a bath; change bed linens.

117. Mr. Williams would demonstrate that he was aware of dietary influences in the prevention of the dumping syndrome if he adjusted his intake by:

A. Decreasing fats.

B. Decreasing proteins.

C. Increasing fluids at mealtimes.

D. Decreasing carbohydrates.

Medical Diagnosis: Cholelithiasis.

- **Surgical Treatment:**
  Cholecystectomy.

- **Nursing Problems:**
  Altered comfort (pain).
  Altered fluid volume.
  Health-seeking behaviors: preparation for diagnostic procedures.
  Effective individual management of therapeutic regime.
  Risk for injury: postoperative complications.
  Altered nutrition: less than body requirements.
  Self-care deficit: medication administration.

- **Chief Complaint:**
  Adele Norfolk, a 47-year-old white secretary, was admitted to the hospital, complaining of severe pain in her right upper quadrant.

- **History of Present Illness:**
  Ms. Norfolk has noticed an intolerance to fatty foods over the past few months. She has also noted general indigestion. Prior to this admission, no serious episodes of pain were noted. Her present pain began 6 hours ago; its onset was sudden and the pain increased in severity. She complained of nausea and vomited twice prior to admission.

- **Past Health History:**
  Twenty-five pounds overweight. Three normal deliveries; all children living and well. Appendectomy at age 14. Smokes 1 pack of cigarettes a day.

- **Family History:**
  No history of gallbladder disease in mother, father, two brothers, or one sister. Diet has been high in fat content throughout lifetime.

- **Review of Systems:**
  Weight gain gradual over past 5 years. Unsuccessful in own attempts to control weight. Distress when eating fatty foods; complains of "bloating feeling." Denies cough.

- **Physical Exam:**
  Vital signs: BP: 140/92. Temperature: 101°F. P: 92. R: 26.
  Head and neck: Tongue dry; face flushed; no jaundice noted.
  Chest: Some wheezing noted at base of lungs. Difficulty in coughing due to acute distress.

Abdomen: Severe pain and tenderness in right upper quadrant. Positive Murphy's sign.

Extremities: Some evidence of varicosities in posterior aspect of lower legs.

■ *Laboratory and X-ray Data:*

X-ray: Absence of opaque materials in the gallbladder—cholecystography.

Chest x-ray: Within normal limits.

GI series: Negative.

WBC: 12,500/μL.

Cholesterol level: 290 mg/dL.

Urine specific gravity: 1.040.

---

118. Which drug may increase biliary colic pain if given to a patient with cholecystitis?
    A. Meperidine (Demerol).
    B. Nitroglycerin.
    C. Morphine.
    D. Ibuprofen.

119. Ms. Norfolk is having an oral cholecystogram. She asks the nurse if there are any special preparations for this type of x-ray. The nurse tells her that she will:
    A. Have a regular diet the evening before the test.
    B. Eat a full meal the morning of the test.
    C. Take iodine dye capsules by mouth the evening before the test.
    D. Have this test done after her scheduled GI series.

120. Ms. Norfolk is discharged from the hospital. She must follow a low-fat diet until her readmission for surgery. The nurse knows that the patient is demonstrating her dietary knowledge when she eliminates:
    A. Fruit juices.
    B. Broiled chicken.
    C. Chocolate pudding.
    D. Carrots and spinach.

121. In preparing Ms. Norfolk for the surgical experience, the nurse is least likely to initiate teaching about:
    A. The reason for being NPO after midnight prior to surgery.
    B. Deep breathing, coughing, and turning techniques.
    C. The expected results of the surgical procedure.
    D. The availability of pain medication prn.

122. Ms. Norfolk returns from surgery. Which nursing action has the highest priority during the recovery room period?
    A. Checking vital signs every 15 minutes.
    B. Recording intake and output.
    C. Explaining procedures to her family.
    D. Maintaining a patent airway.

123. Ms. Norfolk is to receive 3000 mL of solution intravenously in each 24-hour period. If there is a drop factor of 15 drops/mL, at approximately how many drops per minute should the nurse regulate the IV?
    A. 22 drops.
    B. 31 drops.

C. 42 drops.

D. 51 drops.

124. The first IV infusion began at 8 A.M. At 11 A.M., given a correct flow rate and no interference, the nurse expects the first 1000-mL container to contain:

A. 625 mL.

B. 736 mL.

C. 840 mL.

D. 437 mL.

125. Because of Ms. Norfolk's weight and the location of her incision, the nurse can anticipate that the most likely complication will be:

A. Fluid and electrolyte imbalance.

B. Atelectasis.

C. Infection.

D. Nausea and vomiting.

126. Ms. Norfolk has a T-tube inserted in the surgical wound. The most important nursing function in caring for this tube is:

A. Recording quantity and color of drainage.

B. Changing the dressing every shift.

C. Preventing the tube from kinking.

D. Teaching the patient about the reason for the tube.

127. Ms. Norfolk develops a paralytic ileus postoperatively. The nurse should question which medical order before carrying it out?

A. Begin intermittent nasogastric suction.

B. Encourage the patient to take carbonated beverages.

C. Neostigmine (Prostigmin), 500 µ-g IM.

D. Continuous IV therapy, 3000 mL in 24 hours; alternate 5% dextrose in water with Ringer's lactate.

128. As she is preparing for discharge, Ms. Norfolk reports to the nurse that she has pain in the calf of her left leg. The nurse assesses the situation and finds a positive Homans' sign. The nurse's decision is to:

A. Put Ms. Norfolk on bedrest.

B. Measure her left calf and reassess in 4 hours.

C. Assist Ms. Norfolk in ambulation.

D. Massage the cramp in her calf.

Medical Diagnosis: Essential hypertension.

- ***Nursing Problems:***
  Altered cardiac output.
  Risk for alteration in circulation.
  Altered nutrition: less than body requirements.
  Self-care deficit: medication administration.

- ***Chief Complaint:***
  Harry Amos is a 52-year-old steel mill worker who complains of dizziness and headache.

- ***Past History:***
  Pneumonia, resolved 2 years ago. No surgeries or allergies.

- ***Family History:***
  Father: Deceased, myocardial infarction.
  Mother: Deceased, senility (?).
  Siblings: Sister with glaucoma.

■ *Review of Symptoms:*
Currently 20 pounds overweight.
Nocturia.
Vital signs: R: 30. P: 90. Temperature: 98°F. BP: 180/100.

■ *Physical Exam:*
Neck: Mild jugular venous distention to 8 cm.
Lung: Slight bilateral crackles in both lower lung fields; moderate dyspnea on exertion.
Heart: Angina pain. S_4 heard. Left ventricle palpably enlarged.
Abdomen: Liver palpable 4 cm below right costal margin.
Extremities: No clubbing of 1+ pedal edema.
Fundoscopic: Retinal vessel dilated; exudate, hemorrhages, papilledema absent.

■ *Laboratory and X-ray Data:*
Chest x-ray: Left ventricular hypertrophy bilateral infiltration in lung field.
Creatinine: 0.5 mg/L.
BUN: 15 mg/dL.
Urinary protein: Absent.

---

129. During the initial assessment examination, the nurse hears slight bilateral crackles in both lower lung fields. There is no edema in the extremities. Respiration is 30; apical pulse, 90; temperature, 98°F orally. The nurse considers Mr. Amos a good historian. Based on these observations, which nursing order is appropriate to write?
    A. Keep side rails up at all times.
    B. Assist with meals.
    C. Allow a rest period between activities.
    D. Ensure meticulous skin care.

130. Mr. Amos is placed on hydrochlorothiazide. Postural hypotension is a frequent complication of hydrochlorothiazide administration. The nurse should tell him that the probability of developing this complication increases:
    A. During the cold winter months.
    B. After alcohol ingestion.
    C. With sodium retention.
    D. After ingestion of seafood.

131. The most important nursing action for a patient on diuretic therapy is:
    A. Recording daily weight.
    B. Reducing sodium in diet.
    C. Increasing oral fluid intake.
    D. Measuring intake and output.

132. Methyldopa (Aldomet), an antihypertensive drug, is ordered for Mr. Amos. In instructing him about the drug, the nurse should emphasize that:
    A. He should rise slowly from a lying to a sitting position and from a sitting to a standing position.
    B. He should skip a dosage of medication if he begins feeling dizzy, weak, and sleepy.

    C. He can alter his diet if necessary, as long as he takes the anti-hypertensive drug.

    D. Alcohol consumption does not affect the action of antihypertensive drugs.

133. While assisting Mr. Amos in getting out of bed, what objective indicator (other than a manometer) can the nurse use to evaluate the effect of ambulation and to detect early postural hypotension?

    A. Pupil response.

    B. Pulse rate.

    C. State of orientation.

    D. Increased muscle rigidity.

134. Mr. Amos explains that he is a quiet man and that he keeps all his feelings inside. Which activity is the most beneficial for the nurse to suggest to help Mr. Amos express himself?

    A. Watching a football game.

    B. Asking him what is bothering him.

    C. Nailing a table together.

    D. Playing chess.

135. Mr. Amos is to be discharged soon. What instruction does the nurse give him that is commonly given to all patients receiving diuretic therapy?

    A. Keep the bath water at 140°–150°F.

    B. Teach the patient to take his weight and mark it down.

    C. Restrict fluids in the winter to avoid overhydration.

    D. Be alert for episodes of disorientation.

136. The nurse will need to tell Mr. Amos to make which adjustment in food preparation?

    A. Use more herbs and spices for flavoring.

    B. Purchase all low-sodium dietetic foods.

    C. Cook all of his food separately from the rest of the family's.

    D. Prepare foods such as biscuits and cakes from mixes.

The following two questions are on the subject of hypertension, but are not related to Mr. Amos's situation.

---

137. Ms. Root receives furosemide (Lasix), 80 mg PO every day. Her usual blood pressure is 120/76. This morning while taking her blood pressure prior to the furosemide, the nurse finds a reading of 90/60. What is the most appropriate nursing action?

    A. Withhold the dose and call the doctor.

    B. Give the drug but check her periodically.

    C. Give the drug and let the doctor know the blood pressure reading.

    D. Withhold the drug until the blood pressure rises.

138. Ms. Francisco is receiving pargyline (Eutonyl), an MAO inhibitor. The nurse instructs her to refrain from eating foods high in tyramine because these foods:

    A. Produce marked vasodilatation.

    B. Stimulate the myocardium, increasing its workload and output.

    C. Decrease the concentration of norepinephrine.

    D. Produce severe vasoconstriction.

Medical Diagnosis: Myocardial infarction.

- *Nursing Problems:*
  Altered cardiac output: decreased.
  Risk of alteration in circulation.
  Altered comfort (pain).
  Altered fluid volume: excess.
  Risk for injury.
  Self-care deficit: medication administration.

- *Chief Complaint:*
  Anthony Zingale, a 52-year-old, has been experiencing tightness in the chest, dyspnea, diaphoresis, and weakness.

- *History of Present Illness:*
  Approximately 2 weeks prior to admission, noted substernal tightness and pain radiating down both arms during exertion and excitement. Chest pain now unrelieved by rest.

- *Past History:*
  Treated for hypertension for the past 3–4 years. No known history of diabetes or renal disease.

- *Family History:*
  Mother died of a heart attack. Father died of liver cancer. Has an aunt with diabetes.

- *Review of Symptoms:*
  Smokes 1 pack of cigarettes a day. Fifty pounds overweight.

- *Physical Exam:*
  Head and neck: No neck vein distention.
  Skin: Pale, cool, clammy. Lips slightly bluish.
  Lungs: Clear, with faint breath sounds.
  Heart: Heart sounds distant.
  Abdomen: Rotund, soft.
  Extremities: No cyanosis or edema. Weak pedal pulses.

- *Laboratory and X-ray Data:*
  Na: 141.
  K: 5.45.
  BUN: 28.
  Glucose: 280.
  SGOT: 120.
  ECG: Sinus rhythm with occasional PVCs.
  CPK: 550.
  MB: Greater than 35%.

---

139. Mr. Zingale has been experiencing the pain associated with angina pectoris prior to admission. For what characteristic picture of anginal pain should the nurse look?
    A. Requires a narcotic for relief.
    B. Is relieved by rest.
    C. Tends to radiate to the jaw.
    D. Lasts at least 30 minutes.

140. Nitroglycerin was placed at the patient's bedside in the event of an anginal attack. The nurse knows that the goal of care includes giving this drug to relieve chest pain by:
A. Increasing myocardial contractility.
B. Increasing the heart rate.
C. Decreasing myocardial contractility.
D. Decreasing venous return to the heart.

141. On review of the patient's history and physical, which factor does not make Mr. Zingale a high-risk candidate for coronary artery disease?
A. Family history of cardiac disease.
B. Hypertension.
C. Smoking.
D. Family history of cancer.

142. When a colleague asks about it, which is the best explanation the nurse could give for the elevated serum potassium following myocardial damage?
A. Potassium is retained due to decreased renal function.
B. Sodium ions are excreted to relieve edema, and potassium is retained.
C. Potassium is released from the cells following destruction.
D. Potassium increases with metabolic alkalosis.

143. Nursing care for the post–myocardial infarction patient is directed toward reducing the work of the heart and oxygen consumption. Which nursing action best accomplishes this goal?
A. Position flat in bed with pillow under head.
B. If possible, allow use of bedside commode.
C. Assist patient in eating a regular diet.
D. Restrict all activity for 72 hours.

144. Mr. Zingale has been in the hospital for 24 hours. His temperature has increased from 98.6° to 100.2°F. The nurse's response to the increase is to:
A. Assume an infection is developing.
B. Increase fluid intake.
C. Consider it a normal occurrence.
D. Apply cooling measures.

145. Mr. Zingale tells the nurse that he is concerned about being able to eat his favorite Italian foods now that he is on a low-fat, low-sodium diet. Which foods does the nurse recommend as lowest in fat content?
A. Cannelloni with part-skim ricotta cheese (1 cup).
B. Canned spaghetti in tomato sauce (1 cup).
C. Olive oil (1 tablespoon) and Italian herb bread.
D. Dry salami sandwich (4 slices).

146. Mr. Zingale shows signs of congestive heart failure. He is started on a regime of digoxin and furosemide (Lasix). The signs of effectiveness that the nurse looks for include:
A. Slowed heart rate and elevated urine specific gravity.
B. Increased heart rate and lowered urine specific gravity.
C. Clear breath sounds and slowed pulse.
D. Regular heart rhythm and increased urine specific gravity.

147. Discharge teaching for the post–myocardial infarction patient is an important nursing function. Of the following areas for discussion, which one includes *inaccurate* counseling?

A. Stress reduction is useful in reducing the risk of subsequent attacks.

B. Sexual activity can resume, but it will need to be limited compared with the pre–myocardial infarction activity.

C. Weight reduction decreases the workload of the heart.

D. All antiarrhythmic drugs can cause arrhythmias; therefore, the patient and/or family should know how to check the pulse.

Medical Diagnosis: Laënnec's cirrhosis; hepatic encephalopathy.

■ *Nursing Problems:*
Inadequate breathing patterns.
Impairment of digestion.
Fluid volume deficit.
Altered nutrition: less than body requirements.
Altered thought processes.

■ *Chief Complaint:*
Joseph Mesta is a 55-year-old married Hispanic. He complains of vomiting, confusion, restlessness, and increased abdominal size.

■ *History of Present Illness:*
Six episodes of coffee-ground emesis in past 24 hours. According to wife, he has intermittent disorientation to place and time. Also reports an 18-pound weight gain in past 6 months and a gradual increase in abdominal girth.

■ *Past History:*
Discharged from hospital 6 months ago with diagnosis of Laënnec's cirrhosis. Responded well to treatment with diuretics and salt and protein restrictions.

■ *Family History:*
Mother and father died of "old age" in their 80s.

■ *Review of Symptoms:*
Admits to difficulty following doctor's prescribed diet. Avoids hard liquor but consumes 4–6 beers each night.

■ *Physical Exam:*
Lungs: Bilateral, basilar crackles.
Abdomen: Marked distention; liver barely palpable; distended veins visible in right and left upper quadrants.
Rectal: Black, tarry stool; hematest positive.
Extremities: 2+ pitting edema both legs; 1+ arms.
Skin: Jaundice; multiple abrasions on forearms which bleed easily.
Neurologic: Lethargic; disoriented to time and place; tremor, both upper arms.

■ *Laboratory and X-ray Data:*
Hgb: 10.1 g.
Hct: 31.3%.
WBC: 10,200/μL.
BUN: 62 mg/dL.
Creatinine: 3.3 mg/dL.

Total bilirubin: 7.3 mg/dL.
Albumin: 2.3 g/dL.
CPK: 460 mU/mL.
SGOT: 180 U/mL.
LDH: 451 U/mL.
SGPT: 488 U/mL.
Uric acid: 10.5 mg/dL.
UGI: Varices of esophagus and stomach.
Paracentesis: 400 mL clear, straw-colored fluid.

---

148. Which area should the nurse examine to look for jaundice in a person with a dark complexion?
   A. Nailbeds.
   B. Palms of hands.
   C. Hard palate.
   D. Soles of feet.

149. Which goal of care is inappropriate when a lowered blood ammonia level is the desired outcome?
   A. Prevention of GI bleeding.
   B. Reduction of dietary protein intake.
   C. Avoidance of enemas and cathartics.
   D. Decrease in bacterial flora in the intestine.

150. Which drug might the nurse be asked to give to decrease ammonia levels in a patient with liver disease?
   A. Diazepam (Valium).
   B. Diphenoxylate hydrochloride and atropine sulfate (Lomotil).
   C. Furosemide (Lasix).
   D. Lactulose.

151. Which is irrelevant in the nursing evaluation of the effectiveness of the treatment for hepatic encephalopathy?
   A. Lessening of flapping tremors of the hands.
   B. Decreases in pedal edema.
   C. Improved levels of consciousness.
   D. Increased cooperativeness.

152. Which nursing observation is an inappropriate indicator of GI bleeding?
   A. Elevated BUN.
   B. Coffee-ground emesis.
   C. Black, tarry stools.
   D. Lowered hemoglobin.

153. A Sengstaken-Blakemore (S-B) tube was inserted in Mr. Mesta to control bleeding of esophageal varices. On entering the room, the nurse notices that he is gasping for breath. His color has become cyanotic and his respirations are rapid and shallow. What should the nurse *first* suspect?
   A. A pulmonary embolus has probably developed.
   B. The S-B tube has dislodged and one of its balloons is obstructing the airway.
   C. The patient is air hungry due to anemia.
   D. The patient is anxious and this can cause changes in respiratory status.

154. Mr. Mesta is restless and uncooperative in the early days of his treatment. His wife asks the nurse if he could be medicated to

"calm him down." Which statement should guide the nurse's response?

A. Sedatives are never given to patients with liver disease because damaged liver cells cannot metabolize the drug.

B. Sedatives are usually metabolized by the kidneys, so this request is feasible.

C. Sedatives have an excitant, rather than a calming, effect on patients with liver disease.

D. A few sedatives that are not metabolized in the liver exist, but they should be used cautiously.

155. As Mr. Mesta's condition improves, he mentions his dislike for the low-sodium, low-protein diet ordered by his doctor. He states, "That food is not fit for a man to eat!" The best response that the nurse could make at this time is:

A. "It must be difficult for you to accept these changes in your diet, Mr. Mesta."

B. "Well, you've got to make the best of it, Mr. Mesta. You've really no choice but to follow the doctor's orders."

C. "It could be worse, Mr. Mesta. That poor patient in the next bed isn't allowed to eat anything at all!"

D. "Maybe we could talk to your doctor, Mr. Mesta. We could ask him to put you on a regular diet since you will be going home soon."

Medical Diagnosis: Cerebrovascular accident.

■ *Nursing Problems:*
Impaired breathing patterns.
Family coping: potential for growth.
Risk for injury.
Altered nutrition: more than body requirements.
Impaired physical mobility.
Self-care deficit.
Altered thought processes.
Altered verbal communication.
Altered urinary elimination.

■ *Chief Complaint:*
George Lewis is a 58-year-old businessman brought to the medical center due to sudden loss of consciousness.

■ *History of Present Illness:*
While playing golf with his business associate, Mr. Lewis complained of severe headache, which lasted for 5 minutes. He then suddenly fell to the ground and became comatose. At the medical center, he was tentatively diagnosed as having a left-sided stroke caused by intracerebral hemorrhage.

■ *Past History:*
Hypertension, age 42 to present.

■ *Family History:*
Father, age 80, relatively well. Mother died age 54, stroke. Brother, age 54, hypertension.

- *Physical Exam:*

    Face: Drooping, right side.

    Eyes: Paresis, right lateral gaze; negative doll's eye.

    Neck: Nuchal rigidity; bruit on right carotid artery.

    Chest: Periods of hyperpnea alternate with periods of apnea.

    Heart: Apical impulse extends to fifth and sixth intercostal spaces; forceful and thrusting.

    Abdomen: Soft, nondistended; hypoactive bowel sounds.

    Extremities: Left upper and lower extremities flaccid; positive Babinski sign.

- *Laboratory and X-ray Data:*

    Chest x-ray: Cardiac enlargement with left ventricular prominence.

    ECG: Increased QRS complexes, increased R wave in limb leads; deep S wave in lead V_1.

    Blood gases: Slight respiratory alkalosis.

    CAT scan: Hemorrhage in foramen, with involvement of adjacent internal capsule.

---

156. Which neurologic check may the nurse elicit from Mr. Lewis, who is in deep coma?

    A. Response to deep pain.

    B. Deep, gasping respirations.

    C. Eyes turned toward his left side.

    D. Purposeful movement of his right leg.

157. The nurse assesses that Mr. Lewis has decorticate posturing if he manifests:

    A. Flaccid arms and legs; fully extended legs; and flexed feet.

    B. A rhythmic contraction and relaxation of the muscles of one arm.

    C. Flexed fingers, wrists, and arms; fully extended legs; and flexed feet.

    D. Fully extended arms with clenched fists and arms rotated away from the body.

158. For which neurologic test should the nurse prepare the patient to best determine the location of cerebral hemorrhage?

    A. Myelography.

    B. Electromyogram.

    C. Lumbar puncture.

    D. Computerized tomography.

159. Based on the information given about Mr. Lewis, the nursing problem that deserves immediate attention is:

    A. Cerebral accident caused by intracerebral hemorrhage.

    B. Inability to communicate related to altered level of consciousness.

    C. Potential for respiratory insufficiency related to severely depressed level of consciousness.

    D. Ineffective communication related to inability to find words to express himself adequately.

160. The short-term nursing goal that has the *least* priority when responsiveness is decreased is to maintain adequate:

    A. Nutrition and fluids.

    B. Rest and relaxation.

C. Expansion of both lungs.

D. Protection from injury.

161. The nurse knows that when the stroke patient is unable to appropriately express his needs verbally and to understand spoken or written words, he has:

A. Aphasia.

B. Agnosia.

C. Apraxia.

D. Ataxia.

162. The nursing care plan for urinary incontinence and retention in patients rehabilitating from the acute phase of stroke includes:

A. A restricted fluid intake.

B. Intermittent catheterization.

C. Prolonged use of Foley catheter.

D. Use of an external condom drainage system.

163. To provide caloric intake, supplemental feedings through the nasogastric tube were ordered. Which outcome criterion clearly indicates to the nurse the presence of the nasogastric tube in the stomach?

A. White, thin mucus is aspirated.

B. Air is heard at the end of the nasogastric tube.

C. Light-brown or light-yellow liquid drainage is aspirated, with pH of 7.1 or less.

D. Bubbles are seen when the end tip of the nasogastric tube is immersed in water.

164. Before he is allowed to drink or eat, the stroke patient's ability to swallow is elicited by testing which cranial nerve?

A. Facial.

B. Trigeminal.

C. Hypoglossal.

D. Glossopharyngeal.

165. Which statement by a family member or significant other *best* indicates to the nurse the individual's acceptance of participation in the rehabilitation of the patient with stroke?

A. "I wish with all my heart that this had not happened."

B. "I can take much better care of him than you did here."

C. "I am willing to be shown how to dress and feed him."

D. "I have not been well myself. Why doesn't he just stay in the hospital?"

Medical Diagnosis: Acute myocardial infarction.

■ *Nursing Problems:*

Altered cardiac output.

Altered comfort (pain).

Altered nutrition: less than body requirements.

Self-care deficit: medication administration.

■ *Chief Complaint:*

Larry Simpson, 63 years old, is a top real estate salesman for a large firm in the city. He complains of crushing substernal pain.

■ *Past History:*

Hypertensive for the past 3 years. Treated with hydralazine hydrochloride (Apresoline), 20 mg qid.

- *Family History:*
  Brother: Hypertensive cardiovascular disease.
  Father: Deceased, myocardial infarction.

- *Review of Symptoms:*
  Smokes 1½ packs of cigarettes a day.
  Weight gain of 30 pounds over the past year.

- *Physical Exam:*
  Neck: Carotid pulse weak and thready.
  Skin: Cool, moist, pallor.
  Lungs: Bibasilar crackles.
  Heart: Ventricular gallop; loud S_3 sound.
  Chest: Use of accessory muscles of thorax to breathe.
  Abdomen: Diminished bowel sounds.
  Mental status: Alert, restless, apprehensive.
  Extremities: Dusty nailbeds.
  Vital signs: BP: 90/60. Apical pulse: 100 (irregular).

- *Laboratory Data:*
  ECG: ST segment elevation; inverted T wave.
  Blood gases: P_{O_2}: 70. P_{CO_2}: 48. HCO_3: 20. pH: 7.30.

Mr. Simpson hails a cab to the local hospital and collapses in the emergency room.

---

166. The arterial blood gases for Mr. Simpson reflect which acid-base status?
A. Compensated respiratory acidosis.
B. Metabolic acidosis.
C. Respiratory acidosis.
D. Both metabolic and respiratory acidosis.

Mr. Simpson is admitted to the CCU and is diagnosed with acute myocardial infarction. The following orders are written:

Morphine sulfate, 3 mg IV prn.
Nitroglycerin, 1/150 prn sublingual.
Lidocaine drip for 5 or more PVCs/min. May give 50-mg bolus × 2.
Bedrest: May use commode.
Nasal O_2 at 5 L/min.
Digoxin, 1 mg PO × 2d.

---

167. A Swan-Ganz catheter is inserted. Which observation indicates to the nurse that there is a complication related to the device that requires immediate action?
A. Temperature 100.2°F.
B. Ecchymosis on back and thorax.
C. Urinary output 25 mL at 10 A.M.
D. Nausea and indigestion.

168. The current CVP readings have been 11–13 cm H_2O. Select the nursing action that is most appropriate at this time:

A. Reduce IV infusion rate.

B. Increase nasal O_2 to 8 L/min.

C. Increase IV infusion rate.

D. Raise head of bed.

169. The nurse caring for Mr. Simpson recognizes that the relief of pain is a primary objective because:

A. The accumulation of unoxidized metabolites irritates nerve endings.

B. Severe pain can produce shock and increase cardiac workload.

C. The patient should be comfortable at all times.

D. Pain stimulates the vagus, which dangerously slows the conduction system of the heart.

170. Intramuscular injections are usually not ordered following a myocardial infarction because:

A. A larger dosage would be required.

B. The pain of the needle stick only increases stress.

C. The tissues are poorly perfused and venous return is diminished.

D. The patient may also be receiving anticoagulants.

171. Atropine is ordered for a pulse rate of 40 and below. At this time, the nurse gives this drug to:

A. Dry oral and tracheobronchial secretions.

B. Accelerate the heart rate by interfering with vagal impulses.

C. Stimulate the SA node and sympathetic fibers to increase the rate.

D. Reduce peristalsis and urinary bladder tone.

172. An intra-aortic balloon pump is inserted. Mr. Simpson's primary nurse writes his nursing orders. Which should the nurse include?

A. Monitor pedal pulses.

B. ROM to all extremities.

C. Keep affected leg flexed.

D. Elevate the head of the bed 45 degrees.

173. Mr. Simpson is taking digoxin, 1 mg PO qid. On day 3, he complains of nausea and states that the objects in his room have a yellowish tinge. Select the most appropriate nursing action at this time:

A. Administer the medication, but observe him for further nausea.

B. Count the apical pulse; if it is regular and above 60, administer the drug as ordered.

C. Hold the drug and call the doctor.

D. Administer the drug, but leave a note on the front of the chart regarding his complaints.

174. The nurse could anticipate that who of the following might be prone to develop digitalis intoxication?

A. Mr. W, who is constipated.

B. Ms. S, with a serum K of 4 mEq/L.

C. Mr. A, who is receiving hydrochlorothiazide.

D. Mr. B, who is taking an oral hypoglycemic agent.

175. The cardiac glycosides are essential in the therapy of the myocardial infarction patient. Which would indicate to the nurse that a therapeutic response to this medication had been attained?

A. A 15% increase in apical pulse rate.

B. A rise in CVP from 12 to 15 cm H_2O.

C. Urine output 30 mL/hour (previously 40–50 mL/hour).

D. Diminished crackles.

176. Mr. Simpson has been in the CCU for 2 weeks. Which statement by the patient demands the most immediate action by the nurse?
    A. "Nurse, I haven't had a BM in 2 days."
    B. "This has been a rough 2 weeks for me."
    C. "When am I going to get some real food?"
    D. "I have to walk this pain out of my leg."

177. When heparin is being administered, the nurse must have which drug on hand?
    A. Protamine sulfate.
    B. Sodium citrate.
    C. Vitamin K (Aquamephyton).
    D. Warfarin sodium.

Mr. Simpson is suddenly dyspneic: pulse 110; respirations 30 and moist. He is coughing and expectorates frothy, blood-streaked mucus. Diagnosis: Acute congestive heart failure with pulmonary edema.

---

178. Furosemide (Lasix), 40 mg IVP, is given. Which observation indicates to the nurse that the therapeutic objectives of this medication have been met?
    A. Respiration 20 and pulse 80.
    B. Mr. Simpson voids 200 mL.
    C. Bilateral moist crackles.
    D. BP 90/60 and skin moist and cool.

179. Nitroglycerin, 1/150 gr. sublingual tablets, is given to Mr. Simpson at discharge. Instructions the nurse should give regarding home use of this drug are:
    A. Store this medication in a well-lighted, well-ventilated area.
    B. If flushing of the skin develops after administration, discontinue the drug.
    C. If chest pain persists after 3 or more tablets have been taken, the physician should be notified.
    D. Alcohol ingestion produces no deleterious effects in combination with this drug.

180. Mr. Simpson is discharged on a 2-g sodium-restricted diet. Following the nurse's instructions, he should select which appropriate lunch menu?
    A. 2 hot dogs with mustard, tossed salad, iced tea.
    B. 1 bowl of consommé, 4 unsalted crackers, 1 oz. of cheddar cheese.
    C. Corned beef sandwich, kosher dill pickle, ginger ale.
    D. Sliced broiled chicken, macaroni, 1 glass of orange juice.

Medical Diagnosis: Chronic renal failure; secondary hyperkalemia; metabolic acidosis; anemia.

■ *Nursing Problems:*
  Risk for alteration in endocrine/metabolic processes.
  Altered nutrition: less than body requirements.
  Altered urinary elimination.

- *Chief Complaint:*
  C. S. Toy is a 64-year-old Chinese man brought to the emergency room by his wife and oldest son. He has been feeling ill and tired, has been unable to eat, and has been losing weight.

- *History of Present Illness:*
  Mr. Toy has been ill for about a week with vomiting, diarrhea, and progressively increasing fatigue.

- *Past History:*
  Episode of epistaxis 4 years ago, which brought him to the outpatient clinic. Diagnosed at that time with hypertension. Treated with unknown medications, which he has taken sporadically.

- *Family History:*
  None significant.

- *Review of Symptoms:*
  Weight loss of 8 pounds with illness. Dry, itchy skin. "Passing a lot of water." Complains of thirst.

- *Physical Exam:*
  Head: Bilateral cataracts; dry, furrowed tongue; teeth in poor repair.
  Lungs: Deep, rapid respirations; lungs clear.
  Heart: Grade 2 systolic murmur.
  Abdomen: Decreased bowel sounds; rotund.
  Extremities: 1+ pedal edema.
  Skin: White, flaky residue on skin; poor skin turgor.
  Neuro: Lethargic; oriented to person and place but not to time.
  Vital signs: BP: 120/60. P: 108. R: 30. Temperature: 97.8°F.

- *Laboratory Data:*
  Hct: 21%.
  WBC: 11,200/μL.
  Na: 138 mEq/L.
  K: 8.1 mEq/L.
  Cl: 108 mEq/L.
  CO_2: 9 mEq/L.
  BUN: 214 mg/dL.
  Creatinine: 12.2 mg/dL.
  Arterial blood gases: pH: 7.10.
  $\qquad\qquad\quad$ P_{CO_2}: 19.
  $\qquad\qquad\quad$ HCO_3: 16.
  $\qquad\qquad\quad$ P_{O_2}: 90.

- *Abdominal X-ray:*
  Small left kidney (probable etiology: chronic glomerulonephritis).

---

181. What laboratory finding *best* supports the diagnosis of renal failure?
    A. Elevated blood pH.
    B. Elevated serum calcium.
    C. Elevated BUN.
    D. Elevated serum creatinine.

182. Mr. Toy is treated with sodium polystyrene sulfonate (Kayexalate) enemas. What should the nurse tell a student nurse who asks about the desired action of this drug?
    A. Eliminates excess sodium and water.
    B. Causes potassium to be excreted in increased amounts in the feces.
    C. Causes potassium to reenter the cells, thereby lowering serum levels.
    D. Corrects for the lowered pH of metabolic acidosis.

183. Mr. Toy's serum potassium is elevated to 8.1 mEq/L. Which pair of observations should the nurse look for to support this fact?
    A. Muscle atony and cardiac irregularities.
    B. Blurred vision and headaches.
    C. Pruritus and jaundice.
    D. Anuria and hyperuricemia.

184. A meeting is held with Mr. and Mrs. Toy to explain the need to reduce dietary intake of potassium. The nurse suggests restricting which food from the list below that is highest in potassium?
    A. Rice.
    B. Butter cookies.
    C. Chicken.
    D. Dried peaches.

185. Additional dietary restrictions are often placed on patients with chronic renal failure. Which diet is the nurse most likely to suggest to Mr. Toy?
    A. Six small, bland feedings.
    B. High protein, high fat, low sodium.
    C. High carbohydrate, high fat, low protein.
    D. Low residue.

186. Mr. Toy's hematocrit suggests he is anemic. What other clinical data should the nurse look for to support this?
    A. Complains of feeling flushed and warm.
    B. Increased tolerance for exercise.
    C. Pale conjunctival sacs.
    D. Dusky skin color.

187. Which observation indicates to the nurse that metabolic acidosis is resolving?
    A. Decrease in blood pH and respiratory rate.
    B. Increase in blood pH and respiratory rate.
    C. Increase in blood pH and decrease in respiratory rate.
    D. Decrease in blood pH and increase in respiratory rate.

188. An inappropriate nursing action in administering peritoneal dialysis to a patient in renal failure involves:
    A. Warming the dialysate before administration.
    B. Checking to see if the volume of dialysate returns exceeds the volume infused.
    C. Monitoring vital signs for changes.
    D. Ensuring that the area around the peritoneal catheter insertion site is open to the air and observed regularly.

189. Which finding suggests to the nurse that the patient has peritonitis as a complication of peritoneal dialysis?
    A. Abdominal pain.
    B. Clear dialysate returns.
    C. Soft abdomen on palpation.
    D. Active bowel sounds.

Answers and Rationale

1. **D. Dehydration.**
   As the glucose level increases in diabetic ketoacidosis, the patient develops polyuria, which leads to dehydration—a consequence, *not* a precipitating factor. *A, B,* and *C* all stress the body and stimulate the release of glucocorticoids. The increase in circulating glucose exceeds the amount of insulin in the body.

2. **B. The juvenile diabetic is insulin dependent.**
   The juvenile diabetic has little or no insulin production by the pancreas, and cannot be controlled without added insulin. *A* is incorrect because not all juvenile diabetics have circulating insulin. The pancreatic deficiency is permanent; therefore *C* is not correct. *D* is not correct because adult-onset diabetics have some circulating insulin and can be controlled with diet alone.

3. **C. Fruity breath.**
   As ketone bodies, the product of fat metabolism, exceed the body's capacity for oxidation, they are excreted into the urine, and acetone is blown off in the breath. If metabolism is complete, ketones are oxidized, buffered, and excreted as CO_2 and water. *A, B,* and *D*—polyuria (dehydration), polydipsia (thirst), and weight loss (protein and fat breakdown)—are due to the excessive glucose, which causes osmotic diuresis.

4. **A. Cardiac irritability.**
   Potassium is an essential electrolyte for normal cellular depolarization and repolarization. Low potassium leads to cardiac irritability and arrhythmias. *B,* renal failure, is a cause of hyperkalemia. *C,* cardiac depression, occurs with hyperkalemia. *D,* respiratory failure, can be caused by hypokalemia during general anesthesia, but otherwise hypokalemia is not commonly associated with respiratory problems.

5. **A. Whole-grain cereal with milk.**
   Although the diet must meet the high caloric need for weight gain, the diabetic patient must avoid high-carbohydrate foods, such as casseroles, sweet desserts, and fried foods, *B, C,* and *D.* The calories should be derived primarily from protein sources and complex carbohydrates.

6. **C. Empty bladder ½ hour before testing and also supply a second voided specimen for testing.**
   Urine that has collected in the bladder for more than 30 minutes may reflect an inaccurate picture of the glucose level in the urine. The double-voided specimen is the most accurate method for testing. Both specimens should be tested, in case the patient is unable to void a second time. A single specimen, *A,* may reflect an accumulation of glucose. Ingestion of more than 8 oz. of water, *B,* will dilute the concentration of glucose excessively. *D* is also incorrect because of the accumulation of glucose over the period of time.

7. **A. Glycosylated hemoglobin measurement.**
   Glucose attaches to the hemoglobin molecule. Once attached, it cannot dissociate. The higher the blood glucose levels have been, the higher the glycosylated hemoglobin. Glycosylated hemoglobin level is the average of blood glucose control over the previous 3 months. *B,* considered to be one of the best methods of *diagnosing* diabetes mellitus, is a 3-hour blood test following glucose inges-

tion. *C* is done 2 hours after eating to determine *how well carbohydrates are digested.* *D*, connecting peptide test, measures the *level of endogenous insulin production.* A normal C-peptide level may mean inadequate or no insulin production.

8. **A.   *Increase his caloric intake, but continue the same insulin dosage.***

   Exercise *decreases* the need for insulin; therefore, additional calories must be ingested to prevent insulin shock. *B* does not solve the problem of the diminished need for insulin. *C* is the opposite of the adjustment that is indicated. *D* will only intensify the insulin reaction.

9. **C.   *Compliance with the dietary restrictions.***

   The peer pressure for this age group to eat junk foods is intense. *A* and *B* may occur, but are not related as closely to the role of peers at this age. *D* is incorrect because there is no indication of the need to limit normal activities.

10. **B.   *Administering room-temperature insulin.***

    B is the only correct answer for preventing tissue hypertrophy or atrophy. *A, C,* and *D* are all causes of lipodystrophies.

11. **A.   *Diaphoresis during the night.***

    NPH is an intermediate-acting insulin that peaks in 6–12 hours after administration. Signs of an insulin reaction (hypoglycemia) include diaphoresis, weakness, and nervousness. *B, C,* and *D* all include signs and symptoms of hyperglycemia (diabetic ketoacidosis).

12. **D.   *Scrubbing superficial lacerations.***

    D is the correct answer; lacerations are not a life-threatening problem so they can wait while other priority nursing actions take place. *A, B,* and *C* take priority. *A* prevents further trauma to a possible spinal injury. *B* prevents simple fractures from becoming compound. *C,* establishing a patent airway, is always a priority.

13. **A.   *He can be turned while his spine is kept immobilized.***

    A Stryker frame permits horizontal turning in prone and supine positions while keeping the spine immobilized. *B* is incorrect because body alignment is not automatic; special attention must be given to ensure this. *C* is incorrect because the weight distribution provided by the frame is no different from that of a bed; however, because the patient can be turned frequently, the chance of skin breakdown is lessened. *D* is incorrect; the staff must turn the patient.

14. **A.   *Walking.***

    A is correct because muscle function has been lost below the chest. Since some arm and finger movement can be expected, Jack could possibly write, feed himself, and drive a car with special apparatus, *B, C,* and *D.*

15. **C.   *Ask the patient to squeeze the nurse's hands.***

    C is correct because the nurse should determine strength by feeling or by measuring with evaluation equipment. *A* demonstrates function, not strength. *B* tests motor function. *D* tests sensation, not strength.

16. **D.   *Comminuted.***

    D is correct; a comminuted fracture is one involving several fragments from a traumatic injury. A pathologic fracture also has several fragments; it is caused by a weakness in the bone due to disease. In an impacted fracture, *A,* one portion of the bone is pushed

into the other. A greenstick, **B,** is a fracture on one side of the bone. A transverse fracture, **C,** is a break straight across the bone.

17. **C. *Immobilize the left arm.***
Immobilization prevents further tissue and vascular injury. **A** is incorrect because it is not a nursing function. **B** could result in potential hemorrhage and hypovolemia from loss of fluids. **D** is part of the medical treatment of a compound fracture, but it is not the *first* action.

18. **A. *Allows the weights of her traction to hang freely on the pulleys.***
Traction weight is prescribed and will assist the bone in healing properly. **B, C,** and **D** are incorrect because the weights must be maintained constantly, to prevent fracture from overriding or disturbing alignment.

19. **B. *There should be no exercise of the affected extremity.***
B is the incorrect statement, and *therefore* the correct answer. The patient *should* demonstrate movement and *should* exercise the fingers. Nonmovement is a sign that the cast may be too tight and potential problems exist. **A, C,** and **D** are incorrect answers because they represent correct nursing instructions. Monitoring circulation, **A,** is a good indicator of potential problems. A plaster cast will not maintain proper position and contour if allowed to get wet, **C.** The bone is placed in the proper position at the time it is set and should be maintained in that position while healing, **D.** Swelling increases pain; elevation decreases swelling.

20. **D. *Her fingers were blue and felt cold.***
These are signs that the cast is too tight. The nurse should check for sensation, motion, and circulation. **A, B,** and **C** are normally associated with casts and should not cause concern. The cast will be heavy until it is completely dry, **A.** It is not unusual to feel irritable and depressed, **B,** when something unexpected, especially traumas, occurs. It takes a long time for a plaster cast to dry completely, **C.**

21. **B. *Sodium.***
Sodium could increase edema and is therefore the *least* helpful in promoting healing. **A, C,** and **D** are all helpful in promoting healing.

22. **A. *Blisters covering the entire burn site.***
A is a sign of partial-thickness (second-degree) burns and therefore is the correct answer. **B** is a sign of superficial (first-degree) burns. **C** and **D** are signs of full-thickness (third-degree) burns.

23. **C. *Applying first aid cream to the burn.***
C is the incorrect first aid action and therefore the correct answer. Ointments or creams should *never* be applied as part of first aid; they will have to be removed for accurate assessment when the patient reaches the health care facility. **A,** transportation to health care, is important. The burn victim is in danger of shock from loss of fluids at the burn site and will need attention. **B** stops the burn by reducing the heat. The burn site will continue to be damaged until it is cooled. **D** prevents further shock.

24. **D. *A lengthwise incision through the burn area.***
D is the correct description of an escharotomy. **A** describes a procedure that removes old tissue to aid healing. **B** describes application of patient's own or donor tissue or synthetic grafts to cover the burn area. **C** is incorrect; no tubes are inserted in this procedure.

25. **C.   *The ointment depresses granulocyte formation.***

   C is correct because it is a disadvantage, rather than an advantage, of silver sulfadiazine (Silvadene) ointment. ***A, B,*** and ***D*** are advantages of this treatment.

26. **A.   *Plasma to interstitial fluid shift.***

   A is correct because the fluid called to the burn site leaves the vascular system. ***B*** is incorrect because this shift happens more than 48 hours after a major burn. ***C*** is incorrect because the problem in major burns is potassium excess, not deficiency. ***D*** is incorrect because metabolic acidosis, not alkalosis, is associated with major burns.

27. **A.   *3500–5000 calories.***

   A is the correct number of calories needed to aid in the tissue-rebuilding and healing process. The amounts in ***B, C,*** and ***D*** are inadequate for the adult with major burns.

28. **D.   *Tissue anabolism resulting from immobility.***

   D is the best answer because it is *not* a cause of his negative nitrogen balance. The problem is catabolism, not anabolism. Immobility leads to breakdown of tissue. ***A, B,*** and ***C*** are all examples of conditions that lead to negative nitrogen balance.

29. **D.   *Hyperglycemia.***

   D is the best answer because hypoglycemia, not hyperglycemia, occurs in acute hepatitis. This is due to an inadequate hepatic glycogen reserve. In addition, inadequate carbohydrate intake, prolonged nausea, and vomiting are also contributory factors. ***A, B,*** and ***C*** do occur. These *are* problems that are encountered because of ***A,*** prolonged prothrombin time; ***B,*** jaundice; and ***C,*** anorexia.

30. **C.   *Body image change related to altered skin appearance.***

   C is correct because the appearance of Ms. Bee's jaundice and her insistence on visitor restrictions occurred concurrently. Because of the jaundice, she became more self-conscious of her appearance, suggesting a change in her perception of her body image. ***A, B,*** and ***D*** are incorrect. The data do not reflect the presence of depression, ***A;*** anxiety, ***B;*** or anger and hostility, ***D.*** It is possible, however, that the nurse may encounter these nursing problems during the hospitalization of a patient with hepatitis.

31. **C.   *Elevated serum transaminases.***

   C is correct because the serum transaminases, ALT and AST, increase during the initial stage of the disease process, reflecting the liver cell injury present. ***A*** is incorrect because although moderate elevation of LDH levels is common in acute viral hepatitis, the CPK level remains unchanged. CPK is elevated in myocardial infarction, *not* in liver disease. ***B*** is also incorrect. Prothrombin is synthesized in the liver and variations in the prothrombin time can be expected because the liver cells are injured. ***D*** is incorrect because there *is* an *increased* release of, not a decrease in, alkaline phosphatase. Because of an impaired hepatic excretory function, enzyme synthesis is increased—subsequently, an increased release of alkaline phosphatase.

32. **D.   *Excretion of conjugated bilirubin into the intestines is decreased.***

   D is correct because decreased excretion of conjugated bilirubin into the intestines is a common occurrence in viral hepatitis. It causes lack of bile pigments in the stools—thus, the clay-colored stools seen. ***A*** and ***C*** are incorrect. Although both can also occur in

hepatitis, they do not have any effect on the color of the stools. The life span of the RBCs in patients with liver diseases is shortened, causing an impaired hepatic uptake of bilirubin, *A.* The reentry of conjugated bilirubin into the bloodstream, *C,* results in jaundice. *B* is also incorrect. Increased excretion of fecal urobilinogen occurs in hemolytic anemia, not in hepatitis.

33. *C.* ***Used needles and syringes.***
Ms. Bee has acute viral hepatitis, type B. Her disease is spread mainly through contaminated needles and blood products, choice *C.* Some theorize, however, that it is possible that the disease also spreads through body excretions such as saliva, tears, intestinal fluids, and gastric juice. Hepatitis A is transmitted by the oral-fecal route. Based on this explanation, *A, B,* and *D* are incorrect.

34. *D.* ***Cholestyramine to bind bile salts in the intestines.***
Cholestyramine, a bile acid–sequestering resin, increases fecal bile excretion, resulting in the reduction of excess bile salt deposits in the skin. Questran, *C,* is another bile acid sequestrant, but it stimulates the excretion, not the reabsorption, of bile salts, so *C* is incorrect. *A* and *B* are incorrect. As stated, they do not relieve the pruritus of the patient with hepatitis.

35. *B.* ***Decorticate rigidity.***
B is correct because decorticate rigidity is a neurologic manifestation indicative of lesions in the cerebral white matter, internal capsules, and thalamus—*not* impending hepatic coma. Flexion of the fingers, wrists, and arms is seen in the patient with this neurologic dysfunction. *A, C,* and *D* are incorrect answers because in the presence of advanced hepatocellular disease, these parameters *do* indicate impending hepatic coma.

36. *A.* ***Giving diuretics.***
A is the incorrect nursing measure and *therefore* the correct answer. Diuretics stimulate the excretion of urine. They are not used in patients who are in hepatic coma because they precipitate the occurrence of hypovolemia. Hypovolemia decreases the perfusion of the liver, causing further injury to the already damaged liver cells and potentiating hepatic coma. *B, C,* and *D* are appropriate nursing measures, but they are incorrect responses to the question asked. Any source of increase in blood ammonia, such as a high-protein diet and a markedly decreased urine output, should be prevented. Poor tissue oxygenation to the liver cells should also be prevented.

37. *A.* ***Decrease fecal pH and ammonia absorption.***
Lactulose, a synthetic disaccharide that contains galactose and fructose, reduces the ammonia level by expelling the ammonia into the bowel through its laxative action. Neomycin reduces the ammonia-forming bacteria in the intestinal tract. Thus, lactulose and neomycin are effective drugs used in patients with hepatic encephalopathy because they reduce the ammonia level in the body. The pharmacologic actions of lactulose and neomycin are not included in *B, C,* and *D;* they are incorrect responses.

38. *A.* ***Ammonia.***
Ammonia is formed by the decomposition of nitrogen-containing substances, such as proteins and amino acid. It is markedly elevated in patients with a severely damaged liver—specifically, hepatocellular necrosis. The damaged liver is unable to convert ammonia to urea; thus, an increased ammonia level is seen. *B* is a

general indication of infection, and is not specific to liver damage. *C* and *D* are incorrect. They indicate dysfunction of the kidneys.

39. **B. *Increased levels of thyroxine and triiodothyronine stimulate metabolic rate.***

    B is correct because too much thyroxine and triiodothyronine, which is the problem in Graves' disease due to overactivity of the thyroid gland, causes a dangerous speeding of metabolism and a high rate of oxygen consumption. *A* is incorrect. Admission to the hospital can be an anxiety-producing situation and can alter vital signs. However, in light of all the data presented, including the probable diagnosis, the most feasible rationale for her symptoms is *B*. *C* is incorrect. Iodine is necessary for the thyroid gland to manufacture and secrete its hormone. Insufficient amounts of iodine produce hypertrophy as a compensatory mechanism—not overactivity of the thyroid gland itself. *D* is incorrect because it is an unverified assumption.

40. **B. *Inquire whether the patient has had an x-ray for which opaque dye was administered.***

    B is correct because opaque dye contains iodine, which will produce an abnormal finding on x-rays. Since the protein-bound iodide test measures the amount of iodine attached to protein molecules within the body, any intake of iodine will skew laboratory findings. *A* is incorrect since samples of venous blood are taken for analysis in this test. *C* is incorrect because the amount of rest and sleep does not influence the findings of this test. *D* is incorrect because any ingestion of iodine will produce a false high reading.

41. **A. *Inform the patient that there is no danger of radiating herself or others.***

    The dosage of radioiodine is so small that it is not harmful; therefore, *A* is correct. *B* is incorrect because no special precautions need to be taken with the collected specimen. *C* and *D* are incorrect because the dosage of radioactive iodine is too small to be dangerous and there is no danger of radioactive emission after this test.

42. **A. *4000–5000 calories; 100–125 g of protein.***

    A is correct because an increase in metabolic rate increases carbohydrate, fat, and protein metabolism, as well as vitamin requirements. *B* is incorrect; insufficient calories are provided in this diet. *C* is incorrect because it does not meet the patient's high mineral requirements. *D* is incorrect because it does not provide sufficient grams of carbohydrates and protein.

43. **B. *Place Ms. Wise in a room with an electric fan and windows.***

    An increase in metabolic rate increases body heat production; consequently Ms. Wise will feel hot and *B* is the correct answer. *A* is incorrect because using several blankets will only make Ms. Wise hotter; this choice indicates lack of understanding of the underlying pathophysiology of her disorder. *C* is incorrect since a puzzle will only frustrate Ms. Wise and produce further agitation. The speeded metabolic rate of Graves' disease is also accompanied by impatience and mood swings. *D* is incorrect; the nurse's insistence will be interpreted as an additional stress, increasing Ms. Wise's agitation.

44. **C. *Ability to sleep 8 hours at night without awakening.***

    Propylthiouracil is an antithyroid drug that corrects hyperthyroidism by blocking thyroid hormone synthesis. Symptoms of Graves'

disease are relieved within 4–8 weeks, and the patient is placed in a euthyroid state. *C* is the only option that reflects the euthyroid state. *A* is incorrect; the observation illustrates a manifestation of the overactive thyroid gland and speeded metabolic processes. *B* is incorrect because the drug does not curtail urine output. *D* is incorrect because impatience and inability to concentrate are manifestations of Graves' disease.

45. *C.* ***Minimize postoperative bleeding.***

C is correct because iodine therapy is used preoperatively to reduce the vascularity of the thyroid gland. *A* is incorrect; SSKI is not a sedative. *B* is incorrect; exophthalmos is due to a number of causes, probably from oversecretion of a hormone by the anterior pituitary called exophthalmos-producing substance. *D* is incorrect; reduced iodine stores are not the underlying disturbance in Graves' disease.

46. *C.* ***Prevent strain on the suture line.***

C is correct since Fowler's position minimizes strain on the suture line when the patient moves from a reclining to a sitting position. Ms. Wise should not extend or hyperextend the neck. *A* is incorrect because Fowler's position alone will not prevent hemorrhage. *B* is incorrect; to reduce pooling of respiratory secretions, the nurse can perform gentle suctioning and see that a humidifier is used, if ordered. *D* is incorrect because the cause of postoperative hoarseness is injury to the pharyngeal nerve.

47. *D.* ***Complaints of difficulty swallowing and sensation that the dressing is tight.***

D is correct because slight bleeding within the soft tissues of the neck can produce complaints of difficulty swallowing. *A* is incorrect because these observations indicate shock; they are not *early* signs of hemorrhage. *B* is incorrect; muscle twitching and tremors are often produced by the accidental removal of one or more of the parathyroid glands, producing a fall in serum calcium. *C* is incorrect; hoarseness may result from damage to the pharyngeal nerve, but is not an early sign of hemorrhage.

48. *C.* ***Tracheostomy set.***

C is correct because Ms. Wise is showing signs of acute respiratory obstruction. In this acute emergency, the maintenance of the patient's airway is of paramount importance. The cause of this obstruction may be hemorrhage or tetany; insufficient information is given to determine which is the cause. *A* is incorrect because patency of the airway must be the *first* priority. Calcium gluconate *may* be given later if the need for it is determined. *B* is incorrect; thyroid replacement hormone is not a priority at this time. *D* is incorrect because Ms. Wise is showing signs of *upper* airway obstruction.

49. *C.* ***Absent Trousseau's sign.***

Trousseau's sign, which is carpal spasm of the fingers and hand following application of a pressure cuff to the arm, illustrates hyperirritability due to hypocalcemia. Restoration of calcium to normal levels relieves such symptoms of tetany as muscle twitching and hyperirritability of the nervous system. *A* is incorrect because calcium replacement does not reduce the size of the thyroid gland. Medications such as SSKI or levothyroxine sodium (Synthroid) accomplish this. *B* is incorrect because calcium gluconate does not curtail blood loss. *D* is incorrect; the cause of hoarseness is dam-

age to the pharyngeal nerve and it is best treated by voice rest and humidification.

50. **D.   *Mr. X with metabolic alkalosis.***

Symptoms of hypoparathyroidism and tetany are more severe in patients who have elevated serum pH (alkalosis). Symptoms are exaggerated because only ionized calcium can be used by the body. When alkalosis is present, the amount of ionized calcium ions drops, even though the serum calcium level may be normal. Therefore, tetany remains severe until alkalosis is corrected and *D* is the correct answer. *A* is incorrect; patients with hyperparathyroidism tend to have high serum calcium levels, which is not the problem in tetany. *B* is incorrect; patients with hypothyroidism (myxedema) tend to have slowed metabolic rates, rather than reduced serum calcium levels as in tetany. *C* is incorrect; symptoms of tetany are more pronounced in alkalosis than in acidosis.

51. **A.   *He or she should refrain from speaking.***

Hoarseness occurs if there is injury to the pharyngeal nerve during surgery. This condition is usually temporary and should subside in a few days. Overuse of the vocal cords prolongs hoarseness, so *A* is correct. *B* is an incorrect statement since hoarseness is a temporary complication. *C* is incorrect because antitussives have no effect in hastening the return of the voice. *D* is also incorrect; the voice returns within a few days and there are no further complications in speaking.

52. **C.   *Apply tepid to cold water to the burn surface.***

Application of cold to tepid water to the burn surface produces vasoconstriction, reduces fluid loss, and minimizes the depth or degree of damage. *A* is incorrect; covering Mr. Green with a clean sheet may be done second. *B* is incorrect because applying topical antimicrobial ointment fosters the growth of anaerobic microorganisms. *D* is incorrect because eschar development does not occur during the first few minutes after burn.

53. **D.   *Flush with water continuously.***

Chemical burns to the eye are treated best by continuous flushing with copious amounts of water. This dilutes the chemical's concentration. *A* is incorrect; a sterile dressing can be applied *after* the eye is flushed with water. Without dilution of the chemical first, thermal injury will continue. *B* is incorrect because administration of an antidote is quite dangerous. The amount and percent concentration to use are difficult to determine. Also, the antidote itself may be caustic. *C* is incorrect because application of ice packs will not halt tissue destruction.

54. **C.   *Peripheral perfusion is poor.***

Hypovolemia associated with increased capillary permeability, plasma to interstitial fluid shift, and stasis of blood in the peripheral tissue occur during the shock stage. Cutaneous vasoconstriction may exist concomitantly. Peripheral perfusion during the initial stage is therefore quite poor and medications will not be absorbed, so *C* is correct. *A* is incorrect because rapid excretion is not possible during the burn shock phase in light of stasis and capillary permeability. *B* is incorrect because it is not the best answer. It is true that the availability of appropriate sites may be limited, but the primary reason for this route of administration of medication is to ensure absorption. *D* is incorrect because it is not the primary reason. Minimization of pain is needed. However, if anal-

gesics are given by another route, absorption will not take place and the patient will not reap the benefits.

55. **A. 97°F.**
The function of the skin is to assist in the maintenance of body temperature by preventing heat and water loss. Without this protective covering (as in thermal injury), heat is lost via conduction, convection, and radiation. Mr. Green's temperature would therefore be subnormal and the correct answer is **A.** This is one major reason why skin grafting is done early during the treatment regimen. Further application of this principle is illustrated in the use of sunlamps and bed cradles. **B, C,** and **D** are incorrect because it is very difficult to maintain a normal or certainly an above-normal body temperature in the face of continual heat loss.

56. **D. Capillary permeability is reduced.**
Many surgeons give minimal to no colloids during the first 24 hours after burns because much of their therapeutic value is lost due to increased permeability—that is, the colloids leak into the interstitial space. During the fluid restitution phase, in which interstitial fluid (edema) remobilizes to the vascular space, colloids are better able to exert their osmotic "pull" and "draw" fluids back into the vascular compartment. As a result, cardiac output and renal perfusion improve, barring any other complications. **A** is incorrect because colloids reduce, not enhance, interstitial edema. **B,** an improvement in cardiac output, occurs as a *result* of the interstitial to plasma fluid shift and the administration of colloids; it is not the rationale for this action. **C,** improvement of renal perfusion, also occurs as a result of fluid remobilization and is not the rationale for this action.

57. **B. CVP of 10.**
During the fluid remobilization phase of interstitial fluid to plasma shift, venous return, and hence the CVP, improves. **A** is incorrect; it may indicate vascular overload. **C** is incorrect because it is a poor prognostic indicator, possibly pointing to renal damage. **D** is incorrect because blister size is not relevant.

58. **D. Abdominal distention with absent bowel sounds.**
Paralytic ileus, which is characterized by absent bowel sounds and abdominal distention, often occurs during this time. Patients are therefore NPO. This is a normal response of the flight or fight mechanism. **A** is incorrect; these are signs that a pneumonia may be present but this is not related to oral intake. **B** is incorrect; this is a sign of hypovolemic shock and is not the best response. **C** is incorrect; this is a sign of renal failure and is not related to oral intake.

59. **C. Respiratory stridor.**
Mr. Green's burns are on his neck and chest. Eschar formation over these areas may restrict respiratory movement due to its hard, leatherlike characteristics. **A, B,** and **D** are incorrect because they are all normal occurrences with third-degree burns. Chocolate to red-colored urine, **A,** is due to the presence of myoglobulin and, although normal with burns, should be closely observed.

60. **A. Give analgesic 30 minutes before application.**
Since mafenide burns on application, the thoughtful nurse should administer analgesic before dressing change. **B, C,** and **D** are *not* side effects of mafenide.

61. C. *Observe any changes in renal function.*

Gentamicin is nephrotoxic and ototoxic. The nurse should ensure that baseline renal function tests have been done, that is, creatinine clearance and BUN (not WBC count, **B**). If the patient is conscious, the nurse should determine whether any hearing impairment exists initially. **A** applies to mafenide. **D** is incorrect because there is no effect on the spinal nerves (although damage occurs to *cranial* nerve VIII).

62. B. *Causes the loss of potassium and sodium.*

The hypotonicity of 5% silver nitrate encourages the loss of electrolytes, so **B** is correct. **A** is incorrect because prevention of eschar formation is not an effect of silver nitrate. **C** is incorrect because although silver nitrate does stain everything it touches, this is not its *major* disadvantage. **D** is an incorrect statement; silver nitrate can be used in the treatment of burned children.

63. D. *Abdominal distention and decrease in hematocrit.*

Curling's ulcer, or stress ulcer, is a constant concern for the nurse and can occur as early as 4 days after burn. **D** lists the symptoms. **A** is incorrect because pain may not always occur. **B** is incorrect; hematemesis is a late sign. **C** is incorrect because hyperperistalsis does not occur; rather, hypotonia is a sign of stress ulcer.

64. A. *A decrease in wound circumference from 4 to 2 cm.*

Zinc is needed to promote wound healing: **A** is the only statement that illustrates this action of zinc.

65. D. *Prevent two body surfaces from touching.*

Preventing two body surfaces from touching maintains the integrity of each healing surface. For example, two fingers will not share the same surface. **A** is incorrect; dressings should be applied with even distribution of pressure. Uneven areas where pressure is greater act much like a tourniquet. **B** is incorrect; saline is a better solution to use to remove adherent dressings. **C** is incorrect because keeping the affected part in a dependent position will foster the development of dependent edema.

66. D. *Adequate caloric intake.*

Adequate calories ensure that protein will not be utilized for energy and heat production, thereby maintaining a positive nitrogen balance. The body uses carbohydrates first for energy, and then proteins. **A** is incorrect because the high metabolic needs of the burn patient will quickly utilize any reserve stores. **B** is incorrect; glucagon is a substance secreted by the alpha cells of the pancreas. It stimulates the breakdown of glucogen and the release of glucose by the liver. In this way, glucagon influences blood sugar levels. Its protein-sparing function is limited. **C** is incorrect; lipoproteins are simple proteins combined with a lipid component such as cholesterol.

67. D. *1 cup of whole milk.*

The best-quality protein is from animal sources and dairy products. Whole milk contains the cream (fat source) and has the highest amount of protein—9 g—of the choices. **A** is incorrect because although nuts are high in calories, they are low in protein value (3 g). **B** is incorrect because broth is a clear liquid supplying no grams of protein. **C** is incorrect because yogurt from partially skimmed milk is low in protein (3 g).

68. **B. Schedule nursing care procedures to allow periods of uninterrupted sleep.**

A source of psychological stress for all burn patients is sensory deprivation—lack of touch sensation—due to deep burns, and over-stimulation and sleep deprivation due to the constant monitoring, pain, etc. Allowance for periods of rest and sleep is helpful to maintain orientation. Many studies support the importance of sleep. *A* is incorrect; insistence that Mr. Green participate in wound care before he is ready is unwise. This will only increase anxiety. *C* is incorrect; comparing his progress with that of other patients in the unit is unethical. *D* is incorrect; all behavior has meaning, and ignoring enuresis overlooks possible dependency or regression. Urine is also a source of wound contamination.

69. **D. Continuous open masturbation.**

Inappropriate behavior may result from many sources. The stress of thermal injury is an assault to one's body image and a threat to economic and social well-being. These patients are often in various stages of the mourning process. *A* and *B* are incorrect choices since they are illustrations of *normal* behavior associated with the mourning process. *C* is incorrect because concepts regarding nudity are influenced by personal, cultural, and family mores.

70. **C. Reduce scar hypertrophy.**

Wound contraction and scar tissue formation are normal parts of wound healing. Scar tissue can produce disabling contractures, deformity, and disfigurement. This type of tissue is metabolically active and continually rearranges itself. An elasticized garment prevents tissue hypertrophy. *A* is incorrect because wound closure has occurred during the rehabilitative period; loss of protein through the wound surface is impossible. *B* is incorrect because heat loss via evaporation cannot occur during this period because the wound surface is now grafted or healed closed. *D* is incorrect because the accumulation of edema fluid under a skin graft cannot be controlled by elastic garments.

71. **C. Weight gain.**

According to the case situation, the patient has *lost* weight despite a ravenous appetite, so *C* is the answer. This weight loss is due to increases in metabolism. *A, B,* and *D*—goiter symptom, exophthalmos, and irritability—are all symptoms of excess secretion of thyroid hormone.

72. **C. Continuous toniclike muscle spasms.**

If one or both of the parathyroid glands are damaged or removed during thyroidectomy, there will be an alteration in the amount of calcium produced. This alteration may cause continuous toniclike muscle spasms (also known as tetany). *A,* urinary output, is not related to calcium metabolism. *B,* flank pain, is a sign of calcium excess. *D* is incorrect because sensations *increase* with calcium deficit.

73. **D. Puffy eyelids and shortness of breath.**

D is correct because these signs are due to retention of fluid and edema. *A, B,* and *C* are incorrect because they are signs of fluid deficit.

74. **D. Gastric suction apparatus.**

D is correct because thyroidectomy patients do not usually have a Levin tube in place, and therefore do not require gastric suction. *A* and *B are* necessary because of possible respiratory problems due

to edema. *C* is necessary because patients with major surgery usually return from the operating room with an IV line in place.

75. *C.* ***Women who had a child before the age of 30.***
The risk of breast cancer increases if a woman has never had a child or had a child after the age of 30; therefore *C* is correct. *A, B,* and *D* are incorrect because they list women who are more likely to be at risk. Risk increases with age, if female family members have or have had breast cancer, and if a woman is obese.

76. *D.* ***Most breast abnormalities can be detected early by the woman or her partner and reported to the physician.***
D is correct because monthly examination can determine changes in normal tissue, which can be reported quickly. *A* is incorrect since only a professional can determine what is serious. All lesions should be evaluated. *B* is incorrect because not all women fear a physician's exam. *C* is untrue; although breast self-examinations should be performed, women still need yearly checkups for other possible problems.

77. *B.* ***One week after her period begins.***
B is correct because breast tissue is in its most normal state at this time. *A* is incorrect because breast tissue is painful and swollen due to hormonal changes at the beginning of the period. *C* is incorrect for a *pre*menopausal woman such as this patient; the first day of the month or another significant date is appropriate for a *post*menopausal woman. *D* is incorrect because midway through the cycle is not an optimum time; tissue is not in its most normal state.

78. *A.* ***Make an appointment with her physician as soon as possible.***
A is correct because the earliest diagnosis and treatment of breast cancer produces the best possible survival rate. Waiting, *B,* wastes valuable time. *C* is important, but the physician's exam is more important. *D* is incorrect; *all* lumps should be investigated.

79. *D.* ***This is part of the diagnostic workup.***
Biopsy is a diagnostic process, not a treatment, so *D* is correct. *A* is incorrect; the treatment is surgery, chemotherapy, radiation, or a combination of these treatments. *B* is not true; most biopsies reveal negative findings. *C* is incorrect; a small scar remains on the breast after this procedure.

80. *A.* ***Breast and axillary nodes.***
A is the correct definition of a modified radical mastectomy. *B* describes a Halsted's; *C,* a simple mastectomy; and *D,* a lumpectomy.

81. *A.* ***Looking at the incision when the nurse is changing the dressing.***
A is correct because the *first* step to acceptance of altered body image is viewing the site. *B,* touching, is the second step toward accepting altered body image. *C* and *D* are examples of denying that the problem exists.

82. *C.* ***"Wash the skin with water and pat dry."***
C is the correct response. The patient should avoid increasing or decreasing the dosage of radiation by utilizing any lotions or creams. Unless specifically ordered, she should use water only and pat dry to prevent trauma. *A, B,* and *D* should be avoided because they alter the dosage of radiation.

83. *A.* ***Redness of the surface tissue.***
A is correct; redness is a local tissue reaction. *B, C,* and *D* show that radiation has caused excessive damage to local tissue.

84. C. *Interfere with nucleic acid synthesis.*

C is correct because this form of chemotherapy exerts its effect by interfering with cell metabolism. *A* is incorrect because the drug does not directly interfere with blood supply. *B* is incorrect because both cancerous and normal cells absorb antimetabolite drugs. *D* is incorrect because the drug does not specifically affect the capsule.

85. A. *Hair follicles.*

A is correct because the drugs affect rapidly dividing cells. Epithelial tissue is affected more often and earlier than other tissue. Cells in *B, C,* and *D* are affected, but not at the same rate as hair follicles.

86. B. *Do not discriminate between cancer cells and normal cells.*

B is correct; as the drugs are influencing the cancer cells, they are also being detrimental to normal cells—a major disadvantage. *A* is incorrect; the drugs *interfere* with the immune response. *C* is incorrect because the drugs are absorbed by both cancer cells and normal cells. *D* is untrue; many patients have chemotherapy as outpatients.

87. B. *Low platelet count.*

B is correct because a low platelet count and the tendency to bleed confirm thrombocytopenia. *A* is incorrect because although a low lymphocyte count is a side effect of chemotherapy, it does not confirm thrombocytopenia. *C* and *D* are incorrect because neither confirms thrombocytopenia. Side effects of chemotherapy are anemia (*not* high RBC count) and a decrease (*not* an increase) in eosinophils.

88. D. *Being isolated.*

There is a sense of isolation, *D,* and patients are afraid of being left alone. *A, B,* and *C* are other possible fears, but the greatest fear is isolation.

89. C. *Stay with him and listen to his concerns.*

C is the correct action. The nurse recognizes Mr. Mulvrey and his need to discuss his concerns. *A, B,* and *D* are incorrect because they do not allow him the opportunity to express his concerns, and leaving him alone when he is expressing his anger tells him that the nurse does not want to listen.

90. A. *Precancerous and cancerous conditions of the cervix.*

The Papanicolaou smear is a cytologic test for diagnosing cervical cancer. *B, C,* and *D* are possible gynecologic problems that are diagnosed by methods other than the Papanicolaou smear.

91. B. *Allow her an opportunity to express her fears.*

Allowing the patient to discuss what is concerning her is the best way to reduce fear. *A* is an inappropriate response. Ms. Salisbury has made a decision about the surgery and the surgeon, and the nurse should say nothing that might raise a doubt about competency at this time. *C,* changing the subject, tells the patient that the nurse is unavailable for discussion. *D* is not the best response. Fear does stand in the way of recovery, but when the patient is in this state, she cannot learn. The nurse must first try to reduce her fears.

92. D. *Meperidine, 1.5 mL; atropine, 0.75 mL.*

To arrive at the correct amount of medication:

$$75/50 \times 1 = 1.5 \text{ mL} \qquad 0.3/0.4 \times 1 = 0.75$$

93. **C. *Decrease in oral and respiratory secretions.***
    Atropine sulfate is used for its ability to dry bronchial secretions, to prevent pooling of secretions during anesthesia. ***A,*** relaxation, is produced by tranquilizers. ***B*** and ***D*** are incorrect because atropine blocks vagal transmission, thereby preventing bradycardia and hypotension.

94. **A. *Support her abdomen with a pillow or her hands.***
    This provides support and assists in comfort during coughing and deep breathing. ***B*** is incorrect because the rib cage is not involved in this surgery. ***C*** is incorrect because Ms. Salisbury can lie on either side or on her back, but not on her abdomen. ***D*** is incorrect because it puts undue strain on the incision.

95. **C. *Infiltration.***
    The dislodging of the needle will cause fluid to infiltrate into the tissues. The nurse should look for edema, discomfort, blanching of the skin, and fluid not dripping in the chamber. ***A, B,*** and ***D*** are incorrect. Signs of phlebitis are redness and heat. Symptoms of embolism are hypotension, cyanosis, and tachycardia. Respiratory problems, skin reactions, hives, and feelings of warmth are symptoms of allergic reaction.

96. **A. *Cover the incision with a sterile, moist towel.***
    The use of a large sterile, moist covering prevents trauma of a dry object on the bowel, and also prevents small nonradiopaque objects from entering the cavity. ***B*** is not a nursing action. ***C*** increases the risk of infection and electrolyte imbalance. ***D*** is not a priority; life-saving actions should be performed first.

97. **D. *Increased sensitivity to heat loss.***
    Hot flashes are common due to lowered estrogen levels. ***A*** is incorrect because perspiration increases, not decreases. ***B*** is incorrect since insomnia is common. ***C*** is incorrect because the skin increases in color due to flashing.

98. **D. *Moist crackles at the base of her lungs.***
    D is the answer because it is a sign of respiratory complication, *not* wound infection. ***A, B,*** and ***C*** are all signs of wound infections.

99. **A. *Take all of the medication prescribed by her physician.***
    Completing the entire prescription reduces chances of drug reactions in the future and also prevents the bacterium from reasserting itself. ***B, C,*** and ***D*** are incorrect because the patient does not complete the entire course of medication.

100. **A. *Blood culture.***
    The laboratory test that does *not* show leukemic cells is blood culture; thus, ***A*** is the answer. Blood culture is used to determine the type(s) of organism(s) that cause the infection. With a laboratory media, the causative organisms are induced to grow. Leukemia is not caused by an organism that can be identified through a blood culture. Rather, the characteristic abnormal proliferation of leukocytes is confirmed through diagnostic tests such as differential count, ***B;*** bone marrow aspiration, ***C;*** and WBC count, ***D.*** These tests determine which WBC component is proliferative, if immature WBCs have invaded the bone marrow, and if there is a marked increase of WBCs.

101. **C. *A notch is palpated along the medial border of the left lower quadrant.***
    C describes a markedly enlarged spleen that descends into the left lower quadrant. The findings in ***A, B,*** and ***D*** are normal physical findings; thus, these are incorrect responses.

102. **C.   *Decreased erythrocyte production.***

The weakness, fatigue, and pallor seen in leukemic patients are due to a decreased production of erythrocytes. There is a reduction of RBC mass and a corresponding decrease in the oxygen-carrying capacity of the blood. ***A, B,*** and ***D*** are incorrect. Bleeding occurs if there is a deficiency of platelets, ***A***. A febrile response is one indication of an infective process, ***B***. Alteration in sensation occurs if cancer cells infiltrated the peripheral nerves, ***D***.

103. **A.   *Cooked spinach and celery.***

A therapeutic goal in the care of leukemic patients is the reduction of potentials for infection. Bacteria are normally present in the human colon. Because this can become a source of infection in the leukemic patient, a low-bacteria diet is served in conjunction with the use of gut sterilizers, such as polymyxin B, vancomycin, and colistin. Foods eaten in a low-bacteria diet are cooked. Some also advocate the use of canned soft drinks. Thus, ***A*** is the correct answer. ***B, C,*** and ***D*** are incorrect. Raw foods such as those mentioned in these items contain bacteria that normally can be handled by the average person who has adequate defense mechanisms to fight infection. The leukemic patient who is acutely ill has a markedly decreased resistance to infection; therefore, raw foods should not be served because they could become potential sources of infection.

104. **B.   *Visit the patient frequently and engage her in conversation.***

When the patient is ill, the potential for depression to occur is increased, particularly when the prognosis of the illness is poor and when room isolation is required. Frequent, short visits by the nurse allow opportunities for the patient to communicate needs, worries, and anxieties. To prevent sensory deprivation, an open communication and frequent contact with the patient must be maintained. Thus, ***B*** is the correct answer. ***A, C,*** and ***D*** are incorrect. Caution must be taken that the patient does not acquire respiratory infection through contact with numerous family members or visitors, ***A,*** or through prolonged contact with them, ***D***. The leukemic patient's resistance to infection is low. Measures should be taken to minimize the potential for infection. Once the nurse has elicited a verbal response from the patient demonstrating her understanding of the rationale for the need of protective isolation, repeated explanation of why this is done, ***C,*** is unnecessary.

105. **B.   *Urine output of 30–50 mL/hour.***

A urine output of 30–50 mL/hour reflects an adequate amount of urine excreted by the kidneys. Frequently, kidney dysfunction is a complication of chemotherapy. This is caused by the crystallization of uric acid in the kidney tubules, causing obstruction, decreased glomerular filtration, and eventually anuria. Signs of kidney dysfunction include an elevated or nonchanging urine specific gravity, ***A;*** elevated serum creatinine, ***C;*** and elevated BUN levels, ***D***.

106. **C.   *"Your hair will grow again after the chemotherapy is over."***

This is a factual statement. The loss of hair during chemotherapy is only temporary. The hair grows back after chemotherapy is over. ***A*** is a misleading statement that does not reassure the patient. Loss of body image is a problem that these patients are faced with. ***B*** and ***D*** are incorrect statements. Factual, honest information,

rather than vague, incorrect information, should be communicated to them by the health personnel.

107. **D.**   ***Ecchymosis and weakness.***

When the bone marrow is depressed, there is also a decreased production of RBCs and platelets. Spontaneous bleeding into the skin occurs, as manifested by the formation of ecchymosis. The weakness and fatigue are attributed to the anemia present. There is a significant reduction in RBC mass and a corresponding decrease in the oxygen-carrying capacity of the blood. ***A, B,*** and ***C*** are incorrect. Night sweats and easy fatigability, ***A,*** are possible manifestations of pulmonary tuberculosis. ***B*** occurs when fluid imbalance is present, and ***C*** indicates kidney dysfunction.

108. **C.**   ***No cancer cells are microscopically evident in the bone marrow.***

A refers to immunotherapy; ***B*** refers to an acute episode of the disease process; and ***D*** refers to a "cure." Therefore, ***C*** is the correct answer.

109. **B.**   ***Keep the patient from making decisions regarding his or her care.***

Keeping the patient from making decisions regarding care leads to feelings of powerlessness, anxiety, fear, or hostility. The patient must participate in the decision-making process, especially in matters that concern treatment and care. When the patient is physically or psychologically unable, the immediate family is consulted. ***A, C,*** and ***D*** *are* appropriate nursing approaches, and therefore are not the answers to the question asked.

110. **D.**   ***Magnesium hydroxide (magnesium magma).***

Milk of magnesia (magnesium hydroxide) has a laxative effect. ***A, B,*** and ***C*** all have a constipating effect.

111. **A.**   ***To decrease gastric motility.***

These drugs decrease gastric secretions by decreasing gastric motility. ***B, C,*** and ***D*** are incorrect because the drugs have no tranquilizing effect; decrease, rather than increase, secretions; and do not influence cell growth.

112. **A.**   ***Decreased blood pressure.***

The decreased blood pressure is due to a fall in cardiac output because of loss of volume. ***B*** and ***C*** are incorrect because both pulse and respirations are increased. ***D*** is incorrect because urinary output is decreased.

113. **B.**   ***Tarry.***

The tarry color indicates digested blood. Stools that are the color of clay, ***A,*** indicate a diet with excess fat. Bright red stools, ***C,*** indicate bleeding low in the large intestine or rectum. Light brown stools, ***D,*** indicate a diet too high in milk and low in meat.

114. **C.**   ***Emergency.***

This is a life-threatening situation because of the blood loss. The patient's surgery is an emergency and must take priority over other surgeries scheduled. Surgery can be described as planned, ***A,*** when conditions necessitate it but it can be scheduled at a convenient time. Imperative surgery, ***B,*** must be done within 24 hours. Optional surgery, ***D,*** is done at the patient's request. The patient can survive without having this surgery performed.

115. **A.**   ***Hyperactive bowel sounds.***

Absence of bowel sounds is indicative of peritonitis, as in ***B, C,*** and ***D.***

116. C. *Position comfortably; subdue the lighting.*

These actions encourage relaxation and provide a quiet environment, which should help reduce pain until the analgesic takes effect. *A, B,* and *D* produce a stimulating effect and are unlikely to help alleviate the patient's pain.

117. D. *Decreasing carbohydrates.*

The food mass is a concentrated hyperosmolar solution in relation to surrounding extracellular fluid. Water is drawn from the blood into the intestines, and symptoms of distress occur. *A* and *B* are incorrect; fats should be increased because they slow passage of food into the intestines, and protein should be increased. *C* is incorrect, as fluids with meals should be decreased or eliminated so that food will stay in the stomach longer.

118. C. *Morphine.*

Morphine is thought to stimulate the sphincter of Oddi, causing biliary pain; therefore, it is usually avoided. *A,* meperidine, is the drug of choice for pain. *B,* nitroglycerin, is given to relax smooth muscle and decrease colic pain. *D,* ibuprofen, an NSAID, most likely will have no significant effect on biliary colic.

119. C. *Take iodine dye capsules by mouth the evening before the test.*

Telepaque capsules, usually six, are administered the evening prior to the test. It takes about 13 hours for the dye to reach the liver and be excreted into the bile, where it is stored in the gallbladder. *A* is incorrect; the diet should be fat free, since fat is the principal cause of contraction of the diseased organ and should be avoided. *B* is incorrect; the patient is given no food after the evening meal, to prevent contraction of the gallbladder and expulsion of the dye. *D* is incorrect; barium studies should be performed *after*, not before, the gallbladder series because the barium may shadow normal structures if it is not excreted completely.

120. C. *Chocolate pudding.*

Chocolate and milk are eliminated from the diet because of their fat content. *A, B,* and *D* are allowed in a fat-free diet.

121. C. *The expected results of the surgical procedure.*

The *surgeon* usually initiates this information; the nurse reinforces the information as needed by the patient. *A, B,* and *D are* all included by the *nurse* in routine preop teaching, and are therefore not the correct answers.

122. D. *Maintaining a patent airway.*

D is correct because life-threatening factors always have priority. *A, B,* and *C* are all important functions, but the airway has priority.

123. B. *31 drops.*

To determine the number of drops per minute, the nurse uses the following calculation:

(total amount of solution $\times$ drop factor) $\div$ (total time $\times$ minutes)
$(3000 \times 15) \div (24 \times 60) = 31$ drops.

124. A. *625 mL.*

3000 mL in 24 hours = 125 mL per hour.
125 mL $\times$ 3 hours = 375 mL.
1000 − 375 = 625 mL.

125. B. *Atelectasis.*

Respiratory complications are the most probable due to the unwillingness of the patient to cough and deep breathe because of the

high incision. This inhibits ventilatory movement, and the incidence of postop pneumonia is very high. *A, C,* and *D* are all possible complications, but respiratory complications are the most common.

126. *C.   Preventing the tube from kinking.*

For any tube to function properly, the opening must remain patent for drainage of fluid. *A* and *D* are important functions but do not have the highest priority. *B* is incorrect because dressings should be changed only when necessary due to wetness.

127. *B.   Encourage the patient to take carbonated beverages.*

B is the order that should be questioned because nothing should be given by mouth when nonfunctioning bowel is suspected. *A, C,* and *D are* all appropriate actions for such a patient and are therefore not the answers.

128. *A.   Put Ms. Norfolk on bedrest.*

Thrombophlebitis is a very serious complication and the patient must be immobilized to prevent further life-threatening problems. *B* is incorrect because problems should be reported immediately. *C* and *D* are incorrect because they could assist a clot in traveling from the calf to a vital organ. The nurse should not ambulate and should never massage the patient when thrombophlebitis is suspected.

129. *C.   Allow a rest period between activities.*

Mr. Amos is in mild pulmonary edema. The lack of temperature indicates no active infections. Allowance for a rest period between activities helps to relieve dyspnea. *A* is incorrect because he is not confused and is considered a good historian. *B* is incorrect because his shortness of breath is not to the extent that it would interfere with eating and crackles are slight. *D* is incorrect because since no edema is present, he is not in acute danger of skin breakdown.

130. *B.   After alcohol ingestion.*

Postural hypotension results from the vascular system's inability to adjust to sudden venous pooling when position is changed from a lying to a standing state. Peripheral vasodilation is promoted by alcohol. Alcohol has also been shown to potentiate the orthostatic hypotensive effects of the thiazides. Thiazides act by depressing proximal tubular reabsorption of sodium and chloride, thereby reducing circulating blood volume. *A* is incorrect because cold produces vasoconstriction, rather than dilation and pooling. *C* is incorrect because thiazides act by depressing tubular reabsorption of sodium and chloride, thereby reducing circulating blood volume. *D* is incorrect because the ingestion of seafood has no effect.

131. *A.   Recording daily weight.*

Diuretics produce a net loss of fluids. Assessment of their effectiveness is reflected in weight change. Excess fluid loss indicates a reduction in fluid in the vascular and extravascular compartments. *B,* reducing sodium in the diet, is also important, but it is *not the first priority.* Sodium is usually limited to "no added salt," and the patient is advised to avoid processed or ethnic foods. *C,* increasing oral fluid intake, is incorrect because fluids are *not* likely to be increased, and may actually be restricted, depending on the severity of the fluid retention. *D,* measuring intake and output, is important to the total management of a fluid imbalance; however, the question asks for the *most important* action.

132. A. *He should rise slowly from a lying to a sitting position and from a sitting to a standing position.*
Methyldopa lowers brain and heart norepinephrine and therefore reduces sympathetic activity. It lowers blood pressure by reducing peripheral resistance. Postural hypotension results from the vascular system's inability to adjust to sudden pooling when position is changed from lying to standing. Changing position slowly gives the vascular system the opportunity to adjust. *B* is incorrect because self-regulation of medication is always dangerous without medical consultation first. *C* is incorrect because alteration of the diet without medical consultation and specific guidelines may produce further complications. *D* is untrue; alcohol potentiates the action of antihypertensives and should therefore be avoided.

133. B. *Pulse rate.*
The cardiovascular system first tries to compensate for a sudden drop in circulatory blood volume by increasing the pulse rate by sympathetic stimulation. *A* is incorrect because pupillary change results from oxygen deficit after a severe drop in blood pressure. *C* is incorrect because it is not an *early* sign of postural hypotension. A drop in blood pressure alters the state of orientation (i.e., to fainting) due to diminished perfusion and oxygen lack, but this will not occur until later. *D* is incorrect because muscles become limp, rather than rigid, with postural hypotension.

134. C. *Nailing a table together.*
Nailing a table together encourages the patient to use muscle strength and activity to dissipate anxiety. This is a very healthy means of expressing feelings. Hypertension is a psychosomatic disorder in which emotions play an important influence. Often these patients internalize anxiety. *A* and *D* are not the best responses because they do not attempt to assist Mr. Amos to express his feelings. In some patients, moreover, these activities may in themselves promote anxiety. *B* is incorrect. When anxiety is internalized, it is unconscious; asking a patient directly what is bothering him would probably evoke a response denying the existence of such anxiety.

135. B. *Teach the patient to take his weight and mark it down.*
B is correct because weight loss indicates a state of hydration. Diuretics can produce dehydration through excessive water loss. *A* is incorrect because the water temperature is too hot and may produce burns or vasodilation and hypotension. *C* is not logical; restriction of fluids in patients receiving diuretics will produce dehydration. *D* is not common to *all* patients receiving diuretics; disorientation is seen only with extreme electrolyte disruption.

136. A. *Use more herbs and spices for flavoring.*
Flavoring foods with spices and herbs is one of the better alternatives to using sodium-based products. Low-sodium dietetic foods, *B,* are expensive and need not be used exclusively. Cooking Mr. Amos's foods separately, *C,* is an inconvenience for the cook and disrupts the family routine. Cake mixes and the like, *D,* are high in sodium.

137. A. *Withhold the dose and call the doctor.*
This is a question of nursing judgment. Furosemide is a rapid-acting and potent diuretic. The patient's blood pressure is below her usual level. After administration of furosemide, her blood pressure will drop markedly. *B* and *C* are incorrect because the drug should

be withheld. Notification of the physician is important, to make him or her aware of the current status of the patient in relation to the treatment regimen and of the possible need to change the dosage or drug. *D* does not include calling the physician.

138. *D.* *Produce severe vasoconstriction.*

MAO inhibitors increase the concentration of norepinephrine. Tyramine is a precursor of norepinephrine. When both are present, a flooding of the body with norepinephrine can result. Severe vasoconstriction with marked blood pressure elevation follows. *A* is incorrect because vasodilatation does not occur with norepinephrine stimulation. *B* is incorrect because norepinephrine tends to decrease cardiac output. *C* is incorrect because the concentration of norepinephrine is increased.

139. *B.* *Is relieved by rest.*

Angina, which is precipitated by the three *"Es"—eating, emotion, and exertion—*is relieved by rest. The pain of myocardial infarction does not resolve when the patient stops what he or she was doing. *A* describes the treatment for myocardial infarction pain. Angina responds to nitroglycerin. *C* is a description fitting both angina and myocardial infarction pain. *D* is incorrect because angina characteristically lasts 5–20 minutes. Myocardial infarction pain lasts 30 minutes to hours.

140. *D.* *Decreasing venous return to the heart.*

Nitroglycerin is a vasodilator; the drug not only dilates the coronary arteries, but also causes a peripheral vasodilation, which decreases venous return. Therefore, the heart does not have to work as hard and there is less oxygen consumption. *A* describes the action of digitalis preparations. *B* is not an action of nitroglycerin, and drugs that do increase the heart rate need to be administered cautiously in patients with heart damage. *C* is incorrect because nitroglycerin does not affect the strength of cardiac contractions. An example of a drug that decreases myocardial contractility is propranolol (Inderal).

141. *D.* *Family history of cancer.*

A, B, and *C* are known risk factors in the development of coronary artery disease. *B* and *C* are two of the three major risk factors; hyperlipidemia is the third. There is no known correlation between cancer and coronary artery disease.

142. *C.* *Potassium is released from the cells following destruction.*

Potassium is an intracellular ion that is released in the extracellular space when cells are destroyed. *A* can occur if renal perfusion is severely impaired, but is not the initial cause of hyperkalemia. *B* is incorrect because treatment to relieve edema includes diuretic therapy, which results in sodium and potassium loss. *D* is incorrect because potassium increases in metabolic acidosis, not alkalosis.

143. *B.* *If possible, allow use of bedside commode.*

Use of a bedpan is a more strenuous activity than allowing the patient up on a bedside commode. There is generally less straining and anxiety associated with the use of a commode. Positioning flat in bed, *A,* may be contraindicated, particularly if the patient is short of breath or hypertensive. The diet of a cardiac patient is progressive. A patient who is not allowed to feed him- or herself, *C,* because of the exertion, most likely will be on a diet (such as

liquids or soft foods) that does not consume excessive oxygen during digestion. Activity, like diet, is progressive, and will vary from patient to patient. Total activity restriction for 72 hours, *D*, may be more stressful for patients.

144. *C.* **Consider it a normal occurrence.**
The increase in temperature is a normal inflammatory response to tissue destruction and is expected 24–48 hours after infarction. Infection, *A*, needs to be ruled out if an elevated temperature occurred 72 hours or later after infarct. *B* and *D* are incorrect because they are actions for reduction of high fevers. Also, fluid intake must be increased cautiously in the patient with cardiac weakness.

145. *B.* **Canned spaghetti with tomato sauce.**
A 1-cup serving of canned spaghetti with tomato sauce contains 2 g of fat. A home-cooked recipe has 9 g of fat. Cannelloni, *A*, has 19 g of fat from the ricotta cheese alone. The pasta adds only one more gram. Olive oil, *C*, contains 14 g of fat in 1 tablespoon. Dry salami, *D*, has 4 g of fat per slice, for a total of 16 g.

146. *C.* **Clear breath sounds and slowed pulse.**
In congestive heart failure the signs include tachycardia, crackles, and edema. The drug therapy improves myocardial contractility, slows the heart rate, and removes excess fluid. *A* and *D* are incorrect since diuresis will result in decreased urine specific gravity. *B* is incorrect because digoxin has a negative chronotropic effect and slows heart rate.

147. *B.* **Sexual activity can resume, but it will need to be limited compared with the pre–myocardial infarction activity.**
Resumption of sexual activity is a necessary topic for discharge teaching. The post–myocardial infarction activity will depend on the extent of the myocardial damage. In most cases the patient can gradually return to the pre–myocardial infarction level (but not necessarily a level greater than the pre–myocardial infarction activity). *A*, *C*, and *D* are examples of accurate counseling. *A* is important because stress is linked closely to the risk of further disease or damage. *C* is an approach to decreasing the risk associated with obesity. *D* is appropriate counseling that allows the patient greater control over his condition.

148. *C.* **Hard palate.**
Jaundice, in persons of various skin colors, can be readily viewed by examining the hard palate of the mouth with the aid of a flashlight. *A* is incorrect since some persons have naturally yellow, discolored nailbeds or very dark nailbeds, which prevents observation of color change. *B* and *D* are incorrect because they do not represent the best site for observing jaundice in dark-skinned persons. Some people have a yellow coloration to their skin because of ethnic background or sun exposure. Others may have such dark skin that jaundice is not clearly visible in these areas.

149. *C.* **Avoidance of enemas and cathartics.**
C is not indicated if a *lowering* of blood ammonia is desired, so it is the correct answer. Enemas and cathartics *may be given* to *hasten* the removal of protein materials from the intestine, thereby *lowering* blood ammonia. *A* and *B* lower ammonia levels by decreasing intestinal protein. *D* lowers ammonia levels by reducing the bacterial production of this substance.

150. *D. Lactulose.*

Lactulose is a drug that lowers the pH of the colon, which inhibits the diffusion of ammonia from the colon into the blood, reducing blood ammonia levels. *A, B,* and *C* do not have the desired effect on ammonia levels. Diazepam is a sedative and skeletal-muscle relaxant that should be used with much caution in a person with liver disease. Lomotil is an antidiarrheal drug that will slow the removal of protein materials from the intestine and thereby increase ammonia levels. Furosemide is a diuretic drug that inhibits reabsorption of sodium and water in renal tubules.

151. *B. Decreases in pedal edema.*

Pedal edema suggests circulatory disturbances rather than the cerebral disturbances that are characteristic of hepatic encephalopathy. *A, C,* and *D* are examples of improved cerebral functioning.

152. *A. Elevated BUN.*

BUN levels reflect the *kidney's* ability to excrete urea, an end product of protein metabolism. *B, C,* and *D* are classic signs of *GI bleeding.*

153. *B. The S-B tube has dislodged and one of its balloons is obstructing the airway.*

Each of the answers given could conceivably cause respiratory distress in a patient like Mr. Mesta. *A* and *B* are the most likely to cause a sudden respiratory crisis. *B* should first be considered since Mr. Mesta has a S-B tube in place. Displacement of the tube constitutes a medical emergency. *A* is a less correct response since the case study makes no reference to chest pain or hemoptysis, classic signs of a pulmonary embolus. Also, the patient's history does not suggest prolonged immobility or long bone fractures, conditions that often precipitate an embolus. *C* is incorrect because Mr. Mesta's anemia is unlikely to cause such a sudden respiratory change. *D* is incorrect because the case situation states that the patient is lethargic and disoriented—not especially anxious.

154. *D. A few sedatives that are not metabolized in the liver exist, but they should be used cautiously.*

Patients with liver disease respond adversely to sedation. The inability of damaged liver cells to metabolize drugs is generally given as a reason for this. *A* is incorrect in that a few drugs—namely, phenobarbital and paraldehyde—are given to patients if absolutely necessary. *B* is incorrect since most opiates, short-acting barbiturates, and major tranquilizers are metabolized primarily in the liver. *C* is incorrect because "excitant" effects with sedatives are rarely, if ever, reported.

155. *A. "It must be difficult for you to accept these changes in your diet, Mr. Mesta."*

This response is the most supportive of the four choices given. It shows concern and understanding for Mr. Mesta's situation in a nonjudgmental way. *B* is a less desirable response since it focuses on telling the patient what to do and suggests he is powerless in this situation. *C* is also an undesirable response if the nurse hopes to achieve long-term compliance with the prescribed diet. The fact that another patient cannot eat may be true, but it is unlikely to motivate Mr. Mesta to change his dietary habits. *D* is incorrect and suggests that the nurse does not understand the need for long-term changes in dietary habits.

156. C. *Eyes turned toward his left side.*

The eyes of a comatose patient with a cerebral lesion usually deviate toward the side of the lesion. Mr. Lewis has a left-sided stroke; therefore, his eyes deviate toward his left side. *A* and *D* are incorrect because patients who are in deep coma do not respond to deep pain or have any motor response, such as a purposeful movement of the nonparalyzed extremity. *B* is incorrect because deep, gasping respirations (Kussmaul respirations) are usually associated with diabetic coma.

157. C. *Flexed fingers, wrists, and arms; fully extended legs; and flexed feet.*

The patient has decorticate posturing when his fingers, wrists, and arms are flexed, and when his legs are extended and his feet are flexed. This signifies severe dysfunction of the brain—specifically, lesions in the cerebral white matter, internal capsules, and thalamus. Thus, *C* is the correct answer. *A* and *D* are descriptions of decerebrate posturing. *B* is a description of a patient with a motor focal seizure.

158. D. *Computerized tomography.*

Computerized tomography, such as CT or CAT scan, and EMI (electronic musical instrument) scan, is a noninvasive procedure that permits the visualization of the cranial contents in several horizontal planes. It confirms the presence of epidural, subdural, and internal hemorrhage of the ventricular system. *A* is incorrect because myelography detects the compression of the spinal cord and other cord pathologies, not cerebral hemorrhage. *B,* electromyogram, detects the presence of muscle disorders. Although lumbar puncture, *C,* may confirm cerebral hemorrhage and may provide clues to the location of the hemorrhage, it is not as reliable and efficient as computerized tomography. Lumbar puncture also increases intracranial pressure and can lead to herniation of the brain.

159. C. *Potential for respiratory insufficiency related to severely depressed level of consciousness.*

Because the respiratory status becomes depressed as the level of consciousness is decreased, support of the respiratory system must be given immediate attention. Adequate oxygenation is vital to life. When the brain is deprived of oxygen for longer than 5 minutes, irreversible brain damage occurs. Mr. Lewis needs oxygen support because he already has compromised brain tissue oxygenation because of the cerebral hemorrhage. Thus, *C* is the correct answer. *B* and *D* are incorrect. These deal with communication problems—very important considerations, but they are not life-threatening. *A* is incorrect because it is a medical diagnosis. The question asks for a nursing problem.

160. B. *Rest and relaxation.*

The nursing goals that preserve the life of the patient are the top priorities. For the patient with decreased responsiveness, adequate lung expansion and nutrition and fluids are most important for sufficient gas exchange, nutrition, and perfusion of tissues. The patient also needs to be protected from injury because he cannot defend himself from external threat. The *least* important of all the stated short-term goals is the provision of adequate rest and relaxation, *B,* because lack of this does not pose an immediate threat to the patient's life. Thus, *A, C,* and *D* are not the answers.

161. *A.* *Aphasia.*

    Because of a dysfunction in the speech center of the brain, the stroke patient is unable to communicate verbally. There may also be a dysfunction in the area involved in auditory comprehension. This communication problem is called aphasia. The type of aphasia described in the stem of the question is global aphasia. Other common types include expressive, receptive, and jargon aphasia. *B* is incorrect; agnosia means the inability of a person to understand auditory, visual, or other types of sensation, although the sensory sphere is not damaged. *C* is incorrect; apraxia means the inability to perform movements that are purposeful, yet the sensory and motor functions are intact. *D* is incorrect; ataxia means muscular incoordination, particularly when voluntary movements are done.

162. *B.* *Intermittent catheterization.*

    Repeated aseptic insertion and removal of the Foley catheter, *B,* is the preferred management of patients with chronic illnesses who have problems voiding or retaining urine. Because the catheter is not in the body for a prolonged period of time, the risk of urinary tract infection is lessened. *A* is incorrect. Decreased fluid intake eventually leads to dehydration, a problem that must be prevented in patients with stroke. An adequate fluid intake is necessary to produce urine, to enable the excretion of excess body fluids and waste products. Urine production should also be adequate to stimulate the micturition reflex. *C* is incorrect. Prolonged use of a Foley catheter increases the risk of urinary tract infection because of the presence of a foreign substance in the body cavity. *D* is incorrect. The use of an external drainage system solves the problem of urinary incontinence but not the urinary retention also present.

163. *C.* *Light-brown or light-yellow liquid drainage is aspirated, with a pH of 7.1 or less.*

    The two most accurate outcome criteria for correct placement of the nasogastric tube are (1) aspiration of light-brown or light-yellow liquid drainage with a pH of 7.1 or less, by means of an Asepto syringe, and (2) rush of air heard on auscultation of the gastric region when 20–40 mL of air is rapidly introduced through the nasogastric tube. Thus, *C* is the correct response. *A, B,* and *D* are incorrect. These parameters indicate the presence of the nasogastric tube in the *lungs.*

164. *D.* *Glossopharyngeal.*

    The cranial nerves responsible for swallowing movements are the glossopharyngeal and vagus nerves. Thus, *D* is the correct response. *A, B,* and *C* are incorrect. The facial nerve, *A,* makes changes in facial expressions possible. The trigeminal nerve, *B,* is responsible for chewing movements, and the hypoglossal nerve, *C,* for tongue movement.

165. *C.* *"I am willing to be shown how to dress and feed him."*

    C is a matter-of-fact response indicating a desire by a family member or a significant other to be given direction in the care of the patient. It is a realistic statement, and it indicates acceptance of participation in the rehabilitation of the patient. *A, B,* and *D* are incorrect. These responses are evasive and nonaccepting of the responsibility of caring for the sick individual.

166. *D.* *Both metabolic and respiratory acidosis.*

    The increase in CO_2 (above 45 mm Hg), the decrease in HCO_3 (below 22 mm Hg), and the low pH (below 7.35) are consistent with

both metabolic and respiratory acidosis. If compensation were present, *A,* the metabolic level would be nearer to or above 26, and the pH would be within normal limits. *B* and *C* alone are not complete interpretations.

167. *B.   Ecchymosis on back and thorax.*
When a Swan-Ganz catheter is in place, thrombi development is a constant danger. Heparinization is usually done routinely to prevent this complication. Administering the heparin via infusion pump can reduce the risk of overheparinization. Nevertheless, observation for signs of bleeding should still be done. *A* is incorrect because a low-grade fever can be expected after myocardial infarction. This is a manifestation of the inflammatory process associated with the destruction of myocardial tissue. *C* is incorrect because a one-time urinary output of 25 mL does not warrant immediate action. Oliguria may result due to shock and secondary poor renal perfusion. The urinary output should therefore be monitored closely. Consistent oliguria may indicate renal hypoxia. *D* is incorrect because nausea and indigestion are normal complaints after myocardial infarction. These complaints may be due to vasovagal reflexes conducted from the area of ischemia to the GI tract.

168. *A.   Reduce IV infusion rate.*
The CVP reading is elevated, indicating cardiac decompensation and reflex of blood into the venous system. The most appropriate action at this time is one that reduces cardiac workload. *B* is incorrect; although additional O_2 would be helpful, it does not deal with the *cause* of the presenting symptom—that is, increasing venous pressure. *C* is incorrect; by increasing the IV infusion rate, the nurse will only increase cardiac workload to an already damaged myocardium. *D* is incorrect; raising the head of the bed will alter the CVP reading but does not deal with the true cause of the elevated reading—that is, increasing venous pressure.

169. *B.   Severe pain can produce shock and increase cardiac workload.*
The pain associated with myocardial infarction is so terrifying, severe, and debilitating that the pain itself can produce shock. *A* is incorrect because it describes the basis for the pain, not the reason for relieving it. *C* is not the best explanation. *D* is incorrect because it is an erroneous statement. Pain stimulates the sympathetic, not the parasympathetic, system.

170. *C.   The tissues are poorly perfused and venous return is diminished.*
Severe reduction in cardiac output and inadequate tissue perfusion produce pooling of any medication in the muscle mass. The patient would not reap the benefit of the drug. Furthermore, when adequate perfusion is restored, the medication will be absorbed and overdosage may occur. *A* is incorrect; a larger dosage will have no effect due to poor venous return—the medication is not being absorbed. *B* is incorrect; the pain of the injection may increase stress, but at this time the *most* important consideration is venous pooling. Although *D* is basically a true statement, it is the incorrect answer; at this time it is more important to relieve the patient's pain.

171. *B.   Accelerate the heart rate by interfering with vagal impulses.*
Atropine exerts its effect on several body organs. In this instance, however, atropine is needed to accelerate the heart by interfering

with the response of the heart muscle to vagal nerve impulses. *A* and *D* are incorrect; although both represent actions of atropine, they are not the rationale for its use in this instance. *C* is incorrect; this choice is in error.

172. *A. Monitor pedal pulses.*

The intra-aortic balloon pump is inserted in the femoral artery, and is positioned into the descending aorta. It inflates and deflates with the cardiac cycle, to increase perfusion of the coronary arteries. Because the femoral artery is obstructed, ischemia may result distal to the site of arteriotomy. *B* and *C* are incorrect because the affected leg should remain straight to prevent hip flexion and disruption of the balloon. *D* is incorrect; the head of the bed should not be raised more than 30 degrees. Greater elevations may force the balloon to the aortic notch.

173. *C. Hold the drug and call the doctor.*

This patient is showing signs of digitalis toxicity. The most appropriate action is to hold the drug and call the doctor. Severe arrhythmias may develop. *A* is incorrect because further administration of this medication will only increase toxicity. *B* is incorrect; despite this evaluation of the patient, continued administration of digitoxin can lead to complications such as heart block. *D* is incorrect because any administration of this medication will be perilous given the patient's symptoms.

174. *C. Mr. A, who is receiving hydrochlorothiazide.*

Patients taking thiazide diuretics are prone to develop potassium loss. Since potassium inhibits the excitability of the heart, depletion of the body's myocardial potassium increases cardiac excitability. Low extracellular potassium is synergistic with digitalis and enhances ectopic pacemaker activity (arrhythmias). This question tests circumstances that produce digitalis intoxication. *A* is incorrect; constipation does not predispose the patient to digitalis intoxication. *B* is incorrect because this is a normal serum potassium level. *D* is incorrect; the administration of oral hypoglycemic agents does not predispose the patient to digitalis toxicity.

175. *D. Diminished crackles.*

Congestive heart failure exists to some degree in all myocardial infarction patients. Crackles exist due to left-sided failure and passive reflux of blood, producing pulmonary hypertension. Because digitoxin increases the force of systolic contraction, allows complete ventricular emptying, and improves cardiac output, pulmonary edema is reduced. *A* is incorrect; digitalis preparations slow the heart rate by depressing conduction through the bundle of His and facilitating the vagal effect on the SA node. *B* is incorrect because the CVP should fall, rather than rise. A rise in CVP indicates cardiac failure and worsening venous congestion. *C* is incorrect; urinary output should improve due to the diuretic action of this drug.

176. *D. "I have to walk this pain out of my leg."*

This is a question of judgment and priority setting. This patient may have thrombophlebitis. Some patients think it is a cramp and try to exercise. Exercise could dislodge a clot and peril the patient's life. The nurse should investigate this at once. *A* is incorrect as the priority action, although it is important to assess the pa-

tient's elimination status. Straining at stool increases the work-load on the heart. *B* is incorrect as the priority action. In this instance, the patient is attempting to open the door of communication. *C* is incorrect because it does not warrant immediate action. Mr. Simpson may be feeling better physically or psychologically.

177. **A.   *Protamine sulfate.***

When protamine sulfate is given in the presence of heparin, they are attracted to each other instead of to the blood elements, and each neutralizes the anticoagulant activity of the other. *B* is incorrect; sodium citrate is an anticoagulant and systemic antacid. *C* is incorrect; vitamin K is most often given as an antidote for coumarin overdosage. *D* is incorrect; warfarin sodium is an anticoagulant.

178. **A.   *Respiration 20 and pulse 80.***

Furosemide exerts its diuretic action by inhibiting the reabsorption of sodium and chloride at the proximal and distal tubules, as well as at the loop of Henle. It is useful in the treatment of pulmonary edema and congestive heart failure by reducing circulatory blood volume and pulmonary congestion by the passive regurgitation of blood. Restoration of the vital signs to normal is indicative of reversal in the pathologic process and that therapeutic effect has taken place. *B* is incorrect; voiding 200 mL does not indicate that the therapeutic effects of this medication have been met. The purpose of furosemide is to reduce circulatory blood volume, *not* to stimulate voiding. *C* is incorrect because crackles indicate that passive lung congestion still persists. *D* is incorrect because these symptoms indicate that the patient is getting shocky—an untoward reaction.

179. **C.   *If chest pain persists after 3 or more tablets have been taken, the physician should be notified.***

This precaution is necessary in case of myocardial ischemia. If 3 tablets of nitroglycerin in 15 minutes do not relieve the pain, the patient may be experiencing an infarct. *A* is incorrect because nitroglycerin deteriorates with exposure to light. *B* is incorrect; flushing of the skin occurs due to the vasodilatory effect of the drug and is considered a normal reaction. *D* is incorrect; alcohol, like nitroglycerin, produces vasodilation of peripheral vessels.

180. **D.   *Sliced broiled chicken, macaroni, 1 glass of orange juice.***

A is incorrect because hot dogs with mustard are high in sodium. *B* is incorrect because consommé is high in sodium. *C* is incorrect because the corned beef sandwich with kosher dill pickle is high in sodium.

181. **D.   *Elevated serum creatinine.***

Serum creatinine is elevated in diseases of the kidney that result in significant nephron damage. Nonrenal causes of creatinine elevation are rare, so this is an excellent, specific test for renal failure. *A* is incorrect because renal failure precipitates a lowered blood pH, with metabolic acidosis. *B* is incorrect because serum calcium levels usually drop in response to increased levels of serum phosphorus. *C* is partially correct. BUN levels rise in renal failure, but they also rise with a high protein intake or protein catabolism in the body. Therefore, BUN is not the most accurate indicator of renal function.

182. ***B.   Causes potassium to be excreted in increased amounts in the feces.***

Sodium polystyrene sulfonate (Kayexalate) is a sodium compound that exchanges sodium and hydrogen ions for potassium in the GI tract. As a result, excess potassium is excreted in the feces. ***A*** is incorrect since diuretic drugs, rather than sodium polystyrene sulfonate, are likely to cause sodium and water excretion. ***C*** is incorrect since intravenous glucose and insulin cause potassium to reenter the cells. ***D*** is incorrect since sodium bicarbonate is usually given to correct for metabolic acidosis.

183. ***A.   Muscle atony and cardiac irregularities.***

Muscle weakness and paralysis and bradycardia proceeding to cardiac standstill are signs and symptoms associated with hyperkalemia. ***B*** and ***C*** are incorrect because these observations are not characteristic of hyperkalemia. ***B*** is likely to be associated with a neurologic problem. ***C*** is likely to occur in a person with liver disease. ***D*** is partially correct. Anuria may be a complication of hyperkalemia, but hyperuricemia is a result of the kidney's inability to excrete waste products of protein metabolism.

184. ***D.   Dried peaches.***

Fruits in general—and dried fruits in particular—are high in potassium content. ***A*** and ***C*** have lesser amounts of potassium. ***B,*** pastries, tend to be very low in potassium.

185. ***C.   High carbohydrate, high fat, low protein.***

Protein is often restricted since the kidneys have difficulty excreting the end products of protein metabolism. Extra quantities of carbohydrates and fats provide necessary calories. ***A*** and ***D*** are incorrect since they are diets more often prescribed for GI problems. ***B*** is partially correct. High fat and low sodium are components of prescribed diets in renal failure. High protein in the diet is contraindicated.

186. ***C.   Pale conjunctival sacs.***

Pallor in the conjunctiva of the eye is an excellent indicator of anemia in persons of various skin colors. ***A*** is incorrect because anemic persons tend to be pale and complain of feeling cold. ***B*** is incorrect because fatigue and weakness often accompany anemia. ***D*** is incorrect because dusky coloring is a sign of hypoxia, rather than anemia.

187. ***C.   Increase in blood pH and decrease in respiratory rate.***

Resolving metabolic acidosis will result in an increase in the arterial pH. The compensatory respiratory response (Kussmaul breathing) will be unnecessary as pH returns to normal. As a result, respiratory rate and depth will decrease. ***A*** is partially correct, as it identifies a decrease in respiratory rate. ***B*** is partially correct, as it identifies an increase in blood pH. ***D*** is totally incorrect because it identifies decreasing pH and increasing respiratory rate as indicators of a resolving acidosis.

188. ***D.   Ensuring that the area around the peritoneal catheter insertion site is open to the air and observed regularly.***

D is the inappropriate nursing action. Special care and attention are directed toward preventing infection around the catheter site. A dry, sterile dressing is placed around the catheter and changed regularly. ***A*** is indicated to prevent chilling of the patient and to dilate peritoneal blood vessels, thus facilitating the exchange of substances. ***B*** is important with peritoneal dialysis to prevent hy-

povolemia or retention of dialysate. *C* is important to monitor fluid balance and possible infections or peritonitis.

189. **A. *Abdominal pain.***

Peritonitis is characterized by a rigid abdomen and pain with palpation or movement. *B* is incorrect because cloudy, rather than clear, dialysate returns signal peritonitis. *C* is incorrect because the abdomen is rigid, not soft. *D* is incorrect because bowel sounds tend to be absent.

10. Older Adults and Geriatrics

Case Management Scenarios and Critical-Thinking Exercises

Medical Diagnosis: Chronic obstructive pulmonary disease (COPD).

- **Nursing Problem:**
  Impaired gas exchange.

- **Chief Complaint:**
  Ms. Beckus, a 65-year-old widow who works in a cotton-processing plant, was admitted with shortness of breath, productive cough, and weakness.

- **History of Present Illness:**
  Over the past week Ms. Beckus experienced increasing difficulty breathing on exertion. Her cough was productive of large amounts of yellow, thick sputum.

- **Past History:**
  Heavy smoker—2 packs a day for 45 years. Diagnosed 5 years ago as having chronic obstructive pulmonary disease (COPD). Pneumonia 1 year ago.

- **Family History:**
  No family history of pulmonary diseases.

- **Review of Symptoms:**
  No weight loss. Chest pain on inspiration.

- **Physical Exam:**
  Vital signs: BP: 145/95. Temperature: 102°F. P: 104. R: 32.
  Head/neck: Accessory breathing muscles enlarged.
  Chest: Increased anterior-posterior diameter.
  Lungs: Prolonged expiratory phase. Bilateral basilar crackles. Hyperresonant to percussion.
  Abdomen: Soft; no organomegaly.
  Extremities: Nailbeds dusky. Clubbing present.

■ *Laboratory and X-ray Data:*
    Chest x-ray: Lungs are hyperinflated with scattered bullae. Infiltrated areas over left lower lobe.
    Na: 146.
    K: 4.3.
    Hgb: 12.8 g.
    Hct: 58%.
    WBC: 16,000/μL.
    Blood gases: Respiratory acidosis with moderate hypoxemia.

---

1. As the nurse enters the room, she notices that Ms. Beckus appears "air hungry." Which observation best indicates to the nurse that the patient is having difficulty getting sufficient air?
    A. Rapid respiratory rate.
    B. Use of accessory neck muscles.
    C. Bluish appearance of nailbeds and lips.
    D. Audible expiratory wheezing.
2. Which laboratory and x-ray findings does the nurse expect to see, consistent with Ms. Beckus's condition?
    A. Decreased arterial P_{CO_2}, decreased arterial P_{O_2}, and decreased hematocrit.
    B. Increased arterial P_{CO_2}, hypoinflated alveoli, and decreased arterial P_{O_2}.
    C. Increased arterial P_{O_2} and hyperinflated alveoli.
    D. Increased arterial P_{CO_2}, increased hematocrit, and hyperinflated alveoli.
3. Oxygen therapy has been ordered to assist the patient with her breathing. Which principle should guide the nurse in managing the delivery of oxygen to Ms. Beckus?
    A. COPD patients require higher concentrations (6–8 L) of oxygen since hypoxemia is their stimulus to breathe.
    B. The concentration of oxygen should be high since the stimulus to breathe in COPD patients is the elevated P_{CO_2}.
    C. COPD patients should receive low concentrations (2–3 L) of oxygen since the stimulus to breathe is their low P_{O_2}.
    D. The concentration of oxygen should be low since the stimulus to breathe in COPD patients is the elevated P_{CO_2}.
4. The patient has been started on intravenous aminophylline, 500 mg every 6 hours. An indication to the nurse of the drug's effectiveness is:
    A. Clearing of the bibasilar crackles.
    B. Absent expiratory and inspiratory wheezing.
    C. Change in sputum color from yellow to white.
    D. A lowering of the hematocrit.
5. Ms. Beckus refuses to stop smoking despite her pulmonary disease. Which is the least plausible explanation?
    A. She feels better when she smokes.
    B. She may be denying the severity of her condition.
    C. Refusal to quit is an attempt to retain control of her condition.
    D. Neurotic reactions should be expected with COPD.
6. The nurse determines that Ms. Beckus is at greater risk of developing thromboembolism because of which compensatory response to COPD?

A. Polycythemia.
B. Hypercapnea.
C. Hypoxemia.
D. Leukocytosis.

7. The nurse is observing the patient as she practices diaphragmatic and pursed lip breathing. The nurse explains that the purpose of the exercises is to:
   A. Strengthen the intercostals.
   B. Reduce the anterior-posterior diameter of the chest.
   C. Increase inspiratory volume.
   D. Facilitate complete exhalation.

8. The nurse knows that the prognosis for a patient with COPD is related most directly to controlling and preventing:
   A. The incidence of respiratory infections.
   B. The development of carbon dioxide retention.
   C. Chronic hypoxemia.
   D. Irritation from cigarette smoking.

9. Adequate hydration is an important aspect of care for Ms. Beckus. Which assessment by the nurse is not an indication of a response to an increase in fluid intake?
   A. Thinning of the pulmonary secretions.
   B. A decrease in the WBC count.
   C. Reduction in body temperature.
   D. A lowering of the hematocrit level.

Medical Diagnosis: Cancer of the bowel.

■ *Surgical Treatment:*
   Abdominal perineal resection with transverse colostomy.

■ *Nursing Problems:*
   Altered bowel elimination.
   Altered comfort pattern.
   Risk for alteration in endocrine/metabolic processes.
   Knowledge deficit: about diagnostic test.
   Altered nutrition: more than body requirements.
   Self-care deficit: medication administration.

■ *Chief Complaint:*
   Mr. Richmond, a 67-year-old retired locomotive engineer, enters the hospital complaining of passing blood in the stool and a change in bowel habits.

■ *History of Present Illness:*
   During the past 3 months, Mr. Richmond has noticed that the contour of his stool has changed. He has had episodes of constipation and diarrhea, distention, and some lower abdominal cramping.

■ *Past History:*
   Treated medically for duodenal ulcer at age 36. No bowel problems. Loss of 20 pounds over past year. Smoker. Periodic episodes of bronchitis. Treated at age 57 for rectal polyps.

■ *Family History:*
   Grandfather died after being ill with a "bowel problem." No documented history of cancer.

■ *Review of Systems:*

Eyes: Uses glasses for reading.

Ears: Some loss of hearing in past 10 years.

Respiratory: Admits to cough in the morning; denies tuberculosis. Bronchitis documented three times.

Cardiac: Denies high blood pressure, orthopnea, or chest pain.

GI tract: Change in bowel habits; see History of Present Illness. Loss of appetite; lower abdominal pain. Weight loss of 20 pounds in past year. Rectal polyps treated at age 57.

Urinary: No change in urinary habits. No nocturia.

Musculoskeletal system: Generalized weakness past 2 months.

■ *Physical Exam:*

Respiratory: Crackles and wheezes in lower lobes. Thorax symmetric. Diaphragm descends 2 cm on inspiration.

Abdomen: Soft; tenderness in lower abdomen. Lower spleen and kidneys not felt.

Rectal stool positive for occult blood. No mass felt on digital examination.

■ *Laboratory and X-ray Data:*

Upper GI series: Within normal limits.

Barium enema: Mass noted in descending colon.

Colonoscopy: Abnormal tissue observed. Biopsy consistent with carcinoma of the bowel.

Hct: 38%.

Hgb: 12 g.

---

10. In preparing Mr. Richmond for his diagnostic tests, the nurse is teaching him what to expect at the time of his barium enema. The nurse tells him:
    A. Fecal matter must be cleansed from the bowel for good visualization.
    B. No restrictions will be made regarding food.
    C. He will be placed in one position during the entire procedure.
    D. He will be asked to drink some barium when in the x-ray department.

11. Which sign should the nurse look for as the most commonly associated with early detection of cancer of the colon?
    A. Abdominal distention.
    B. Vomiting of fecal material.
    C. Alteration in bowel habits.
    D. Presence of mucus in the stool.

12. Mr. Richmond has a tumor of the descending colon and needs surgery. The nurse administers neomycin sulfate prior to his surgery to reduce:
    A. Electrolyte imbalances.
    B. Bacterial content in the colon.
    C. Peristaltic action in the colon.
    D. Feces in the bowel.

13. In discussing the expected drainage that will come from the stoma site, the nurse tells Mr. Richmond that he can eventually anticipate:
    A. Constant green liquid drainage.

    B. Soft, brown, mushy feces.
    C. Solid, formed feces.
    D. Liquid, brown feces.

14. In preparation for surgery Mr. Richmond is placed on a low-residue diet. In discussing the kinds of foods he will be allowed to eat, the nurse lists:
    A. Ground lean beef, soft-boiled eggs, tea.
    B. Lettuce, spinach, corn.
    C. Prunes, grapes, apples.
    D. Bran cereal, whole-wheat toast, coffee.

15. One of the most common electrolyte imbalances associated with nasogastric tube suction that a nurse needs to be aware of is:
    A. Hypocalcemia.
    B. Hypermagnesemia.
    C. Hypokalemia.
    D. Hypoglycemia.

16. To prevent electrolyte imbalance, the nurse should:
    A. Irrigate the nasogastric tube with saline.
    B. Irrigate the nasogastric tube with water.
    C. Avoid irrigating the nasogastric tube.
    D. Change the suction from intermittent to continuous.

17. Threats to body image can affect the amount of pain or discomfort perceived by a patient. This fact will be most important in planning care with the patient who is undergoing a(n):
    A. Colonoscopy.
    B. Exploratory laparotomy.
    C. Colostomy.
    D. Barium enema.

18. The nurse knows that the type of ostomy following an abdominal perineal resection will most likely be a:
    A. Continent ileostomy.
    B. Permanent colostomy.
    C. Double-barreled colostomy.
    D. Transverse colostomy.

19. The appearance of a normal colostomy stoma should be:
    A. Red and raised.
    B. Pale and flat.
    C. Dusky and raised.
    D. Rosy and flat.

20. The nurse recognizes Mr. Richmond's willingness to be involved in his own care when he:
    A. Discusses the cost of his medical insurance.
    B. Asks what time the surgeon will be in.
    C. Asks questions about the equipment being used.
    D. Complains about the noise in the other room.

Medical Diagnosis: Cerebrovascular accident (CVA).

■ *Nursing Problems:*
    Fluid volume deficit.
    Impaired physical mobility.
    Sensory-perceptual alteration.

Impaired skin integrity.
Altered verbal communication (aphasia).

■ *Chief Complaint:*

Ms. Tweed is a 78-year-old, obese, white woman who has been transferred from an acute care hospital to a skilled nursing facility with the diagnosis of cerebrovascular accident (CVA). She exhibits right hemiparesis, aphasia, and incontinence of stool and urine. Her two daughters, who are in their 50s, are unable to care for her at home.

■ *History of Present Illness:*

Hospitalized 2 weeks with no resolution of paralysis. Aphasia affects both expressive and receptive communication.

■ *Past History:*

Hypertension and mild congestive heart failure treated with hydrochlorothiazide and spironolactone.

■ *Family History:*

Mother died at an early age of tuberculosis. Father died in his 50s of pneumonia.

■ *Review of Symptoms:*

Weight loss of 15 pounds since hospitalization. Eats poorly and takes only sips of clear fluids with encouragement.

■ *Physical Exam:*

Head: Pale conjunctiva and mucous membranes in mouth; sunken cheeks; dry, furrowed tongue.

Thorax: Marked kyphosis.

Lungs: Scattered, fine crackles, which clear with coughing; shallow respirations.

Heart: Regular sinus rhythm with occasional premature ventricular beats.

Extremities: Flaccid paralysis of right arm and leg.

Skin: Gray color; dry, inelastic skin; small, red areas on both heels.

Neuro: Responds slowly but appropriately to commands; PERL; responds to name by opening eyes and attempting to speak.

Vital signs: BP: 170/100. P: 96. R: 20. Temperature: 98°F.

■ *Laboratory Data:*

Hgb: 10.2 g.
Hct: 31.5%.
WBC: 10,800/μL.
Albumin: 2.5 g/dL.

---

21. Ms. Tweed is the focus of a nursing care conference soon after her admission. It is determined that she has a depletion of body fluids, which is to be corrected by increasing her daily oral intake. To evaluate the effectiveness of this plan, the nurses should monitor:
    A. Sugar, acetone, and specific gravity of urine.
    B. Color of mucous membranes and skin turgor.
    C. Body weight and urine specific gravity.
    D. Skin color and turgor.
22. Which nursing action is *inappropriate* when working with a patient who is aphasic?

A. Speak for the patient as often as possible to minimize feelings of embarrassment.

B. Repeat simple explanations until they are understood; use non-verbal clues when needed.

C. Have patient practice trying to repeat words and sounds after you say them.

D. Caution family members against showing amusement or embarrassment after patient attempts to communicate.

23. Ms. Tweed is repositioned every 2 hours, but the red pressure areas on the back of her heels worsen. She has also developed a pressure area over the sacrum. Which nursing action is *inappropriate*?

A. Avoid positioning on back; turn side to side only.

B. Put a sheepskin under her and heel pads on her feet.

C. Massage over the pressure areas every 2 hours.

D. Administer a sedative to promote rest and speed healing of tissues.

24. There is concern that Ms. Tweed is developing respiratory complications as a result of her immobility. Which source of data provides the nurse with the most reliable information on respiratory status?

A. Breath sounds.

B. Temperature.

C. Pulse.

D. Skin color.

25. Ms. Tweed complains of cold feet due to a draft blowing under the bed's foot cradle. What is the most appropriate action for the nurse to take?

A. Remove the foot cradle so the covers can be snugly tucked around her feet.

B. Prop hot water bottles by her feet.

C. Vigorously massage the feet and lower legs to increase circulation and warmth.

D. Help her into a pair of warm socks that she can wear while in bed.

26. In assessing Ms. Tweed's dietary intake, the nurse finds she is deficient in vitamin C and protein. Which breakfast menu should the nurse suggest to best meet the patient's dietary needs?

A. Cream of Wheat, whole milk, and toast.

B. Strawberries with milk, cottage cheese, and toast.

C. Orange juice and a sweet roll.

D. Soft-cooked eggs, toast, and coffee.

27. Ms. Tweed continues to have fecal incontinence. She has 6–7 small, liquid, brown stools each day. She is on a soft diet and does not receive any stool softeners or laxatives. Her primary form of activity is sitting for 2 hours in a wheelchair twice a day. The nurse knows that a likely cause for this diarrhea is:

A. Too much roughage in the diet.

B. Overhydration with fluids.

C. Excessive activity.

D. Constipation.

28. Which nursing action is *inappropriate* in the prevention of thrombophlebitis in a patient on bedrest?

A. Elevate the knee gatch of the bed.

B. Encourage exercises that dorsiflex and plantarflex the ankle.

C. Apply antiembolism stockings.

D. Prevent dehydration.

29. The purpose of proper positioning of the patient's affected side following a stroke is to:
A. Maintain extension of the limbs.
B. Place the extremities in a flexed position.
C. Preserve maximum functioning.
D. Immobilize the joints.

Medical Diagnosis: Thrombophlebitis; deep-vein thrombosis.

- ■ *Nursing Problems:*
  Risk for alteration in circulation.
  Altered comfort (pain).
  Impaired physical mobility.
  Self-care deficit.
  Impaired skin integrity.

- ■ *Chief Complaint:*
  Mr. Tell, an obese, 79-year-old retired bank manager, complains of right leg pain and swelling.

- ■ *History of Present Illness:*
  Mr. Tell sustained a right hip fracture, for which he had a closed reduction done, after a fall. He had been in the hospital for 4 days. Edema of his right calf was noted on his third day of hospitalization.

- ■ *Past History:*
  Medical: Thrombophlebitis, age 70; pulmonary embolism, age 78.

- ■ *Family History:*
  Father died at age 75, congestive heart failure. Mother died at age 60, stroke. One sister, age 68, apparently well except for chronic arterial insufficiency of lower extremities.

- ■ *Physical Exam:*
  Neck: No lumps or pain; no limitation of movement; no bruit on carotid artery.
  Chest: Normal anterior-posterior diameter; symmetric chest expansion.
  Lungs: Diminished breath sounds in lung bases; bilateral upper lobes and right middle lobe resonant to percussion, bilateral lower lobes dull to percussion.
  Heart: No engorgement of neck veins on 30-degree head elevation; point of maximal impulse at fifth intercostal space and left midclavicular line; S_1 and S_2 within normal limits; no splitting, no extra heart sounds.
  Abdomen: Flat, no unusual pigmentation or pulsation; active bowel sounds in four quadrants; soft, nontender, nondistended; no pain or guarding on light and deep palpation.
  Extremities: Dry skin; right thigh and calf markedly edematous; marked tenderness on right femoral vein, particularly on right popliteal fossa on palpation; mottled erythema extends from back of right thigh down to popliteal fossa; right calf diameter 2.5 cm greater than left calf diameter.

- ■ *Laboratory and X-ray Data:*
  Chest x-ray: Mild density in lung bases.
  ECG: Normal sinus rhythm with rare atrial premature beats.

Blood gases: Mild hypoxemia.

^{125}I fibrinogen test: Increased radioactivity in femoral and popliteal veins of right leg.

---

30. Why should the nurse maintain Mr. Tell, a patient with deep-vein thrombosis, on bedrest?
A. To alleviate edema and pain.
B. To prevent further venous stasis.
C. To maintain adequate blood return flow.
D. So that the thrombus firmly adheres to the vessel wall.

31. During the first critical days, venous return is increased in Mr. Tell's right leg through which nursing intervention?
A. Elevate the gatch of the bed.
B. Encourage active exercises of the affected extremity.
C. Raise the foot of the bed on blocks, up to 8 inches high.
D. Assist Mr. Tell to sit up in chair for 2 hours 3 times a day.

32. The nurse knows that administration of heparin in patients with deep-vein thrombosis is considered ineffective when it:
A. Dissolves a thrombus.
B. Prevents the extension of existing clots.
C. Prevents the activation of clotting factor IX.
D. Inhibits the conversion of fibrinogen to fibrin.

33. Which laboratory test does the nurse expect the patient to have to determine the effects of warfarin (Coumadin), another therapeutic drug used with deep-vein thrombosis?
A. Prothrombin time (PT).
B. Lee-White clotting time.
C. Partial thromboplastin time (PTT).
D. Activated partial thromboplastin time (APTT).

34. Which is least likely to be a contributing factor for Mr. Tell in developing deep-vein thrombosis?
A. Obesity.
B. Hemorrhage.
C. Hip fracture.
D. Immobilization.

35. The nurse should always be alert for the presence of pulmonary embolus, examples of which are sudden manifestations of:
A. Restlessness and bradycardia.
B. Dull chest pain and tachypnea.
C. Subnormal temperature and hypotension.
D. Localized, stabbing chest pain and dyspnea.

36. Should Mr. Tell develop an obstruction in the pulmonary artery due to a blood clot, the nurse should look for which arterial blood gas values?
A. pH = 7.52; $Paco_2$ = 27; Pao_2 = 64; HCO_3 = 24.
B. pH = 7.50; $Paco_2$ = 38; Pao_2 = 90; HCO_3 = 30.
C. pH = 7.40; $Paco_2$ = 42; Pao_2 = 85; HCO_3 = 22.
D. pH = 7.25; $Paco_2$ = 40; Pao_2 = 68; HCO_3 = 15.

37. A patient with a long-standing arterial insufficiency will complain of:
A. Paralysis or paresthesia.
B. Intermittent claudication.
C. Sudden onset of severe pain.

    D. Anesthesia or tingling sensation.
38. When assessing the pulses of a patient with an occlusion of the left popliteal artery, the nurse may expect which pulse of the left extremity to be present?
    A. Femoral.
    B. Popliteal.
    C. Dorsalis pedis.
    D. Posterior tibial.
39. Neuromuscular parameters that the nurse must assess in all patients on bedrest are:
    A. Strength and movement of the extremities.
    B. Presence and quality of peripheral pulses.
    C. Color and sensation of affected extremity.
    D. Discoloration and temperature of the skin.

Medical Diagnosis: Fracture of the right hip.

- ***Surgical Treatment:***
  Hip prosthesis.

- ***Nursing Problems:***
  Fluid volume excess.
  Impaired home maintenance management.
  Effective individual management of therapeutic regime: positioning.
  Risk for infection.
  Risk for injury: postoperative complications.
  Knowledge deficit: preoperative procedure.
  Impaired physical mobility.

- ***Chief Complaint:***
  Ms. Berryville, 79 years old, enters the hospital via ambulance complaining of pain in her right hip. She fell at home when she tripped over the open oven door.

- ***History of Present Illness:***
  Pain in right hip area. Injured 3 hours ago. Alone in apartment at the time of the injury. Bruise on right shoulder.

- ***Past History:***
  Digoxin, 0.125 mg daily, for history of congestive heart failure. Appendectomy at age 17; cholecystectomy at age 52. Two hospitalizations for congestive heart failure. Maintained on diuretics and digoxin for the past year. Symptom free for congestive heart problem at present.

- ***Family History:***
  Mother died at age 76; history of diabetes. Father died of "old age." No history of cancer. One sister, age 72; arthritis.

- ***Review of Systems:***
  Able to perform activities of daily living by herself. No complaints of headache, dizziness, or loss of consciousness at time of injury. No stiffness of neck. Pain in right hip; tenderness in right shoulder; no complaints of other discomforts. No orthopnea, dyspnea, or chest pain.

■ *Physical Exam:*

Vital signs: BP: 110/62. P: 66. R: 20.

Right hip swollen, painful to touch, ecchymosis present, obvious deformity.

Right shoulder swelling, with ecchymosis present, tender, able to perform limited range of motion.

Abdomen tender, no apparent distress.

Head: no bruises or discolorations.

Alert, oriented to time, place, and person.

■ *X-ray Data:*

X-ray fracture of the neck of the right femur.

---

40. Ms. Berryville was seen in the emergency room because of injuries received in a fall in her home. She is suspected of having fractured her right hip. On examination the nurse expects to find that her right leg is:
    A. Shortened, with external rotation.
    B. Shortened, with internal rotation.
    C. Equal in length to the nonaffected limb.
    D. Abducted, with internal rotation.

41. The diagnosis of fractured hip is confirmed. Ms. Berryville has a skin prep for surgery. The nurse explains that the primary goal of this procedure is to:
    A. Render the operative area free of organisms.
    B. Enhance the skin's natural defense against infection.
    C. Decrease the number of organisms on the skin.
    D. Improve the field of vision for the procedure.

42. Due to the extent of Ms. Berryville's fracture, a hip prosthesis is inserted. The nurse should design strategies to prevent the most dreaded complication of implant surgery, which is:
    A. Infection.
    B. Phlebitis.
    C. Urinary retention.
    D. Narcotic addiction.

43. Ms. Berryville returns from the operating room. Her IV solution is running at 150 mL/hour. Her pulse is 100 and full. Her respirations are moist and wheezy. The nurse's initial action is to:
    A. Speed up the IV rate.
    B. Check the electrolyte level.
    C. Report the findings to the charge nurse.
    D. Slow down the IV rate.

44. The presence of tachycardia and moist breath sounds most likely indicates:
    A. Optimum fluid balance.
    B. Fluid overload.
    C. Electrolyte imbalance.
    D. Insufficient hydration.

45. Ms. Berryville's temperature rises on the second postoperative day. To confirm the nursing diagnosis of "risk of infection," the nurse should:
    A. Perform a lung assessment.
    B. Send a urine sample for culture.

C. Remove the operative dressing.

D. Rotate the affected hip.

46. The nurse instructs Ms. Berryville in proper transfer technique and body positioning. Ms. Berryville demonstrates correct understanding when she:

A. Knows that her legs should be separated with a pillow.

B. Puts first one leg and then the other on the floor when arising.

C. Keeps her legs in a dependent position.

D. Sits straight up in a low chair.

47. The nurse plans to give Ms. Berryville written instructions for home activities. Which direction is *incorrect* in this situation?

A. Avoid sitting for more than 1 hour.

B. Stand, walk, and stretch periodically.

C. Avoid crossing legs.

D. May tie own shoes in the usual manner.

Medical Diagnosis: Left cataract.

- **Surgical Treatment:**
  Cataract removal.

- **Nursing Problems:**
  Fear.
  Risk of injuries.
  Knowledge deficit: regarding cataract glasses.
  Effective management of therapeutic regimen.
  Self-care deficit.

- **Chief Complaint:**
  Mr. Chesapeake, 84 years old, enters the hospital for elective cataract surgery.

- **History of Present Illness:**
  Noticed blurring of vision at the time of his last eye exam (8 months ago). Cataract diagnosis at that time. No history of infections, trauma, or other eye diseases.

- **Past History:**
  Chronic obstructive lung disease, arthritis, controlled diabetes, 26 units of NPH insulin daily.

- **Family History:**
  Brother a diabetic. Family history of heart disease. One brother died of heart attack; one sister died of cancer.

- **Review of Systems:**
  Blurring vision; difficulty seeing at night; no pain in eyes; memory good. Joint pain in cold weather. 1800-calorie diabetic diet. Lives with daughter and her family. Morning cough.

- **Physical Exam:**
  Vital signs: BP: 182/90. P: 72. R: 20.
  Smokes 1 pack of cigarettes a day.
  Opacity of lens of left eye.
  Decreased vesicular breath sounds with prolonged expirations; no crackles or wheezes.

■ *Laboratory Data:*
FBS: 110 mg.
RBC: Within normal limits.
Urinalysis: Normal.

---

48. Mr. Chesapeake has a medical diagnosis of suspected cataracts. Which symptom should the nurse find consistent with this diagnosis?
A. Objects are distorted and blurred.
B. Vision is not affected by change in light.
C. Objects have a halo around them.
D. Single objects seem to be doubled.

49. Mr. Chesapeake tells the nurse that he does not like the idea of being awake during his eye surgery. Of the following responses, which is the most appropriate for the nurse?
A. "I don't blame you. I would feel the same."
B. "By receiving a local, you won't have nausea and vomiting after surgery."
C. "There is nothing to fear. These operations are done every day."
D. "Can you tell me more about not liking to be awake during surgery?"

50. The nurse knows that her teaching regarding prevention of intraocular pressure after cataract surgery has been achieved when she observes Mr. Chesapeake:
A. Asking someone to pick up his robe from the floor.
B. Getting out of bed quickly in the morning.
C. Practicing his deep breathing and coughing exercise.
D. Tying his own shoes.

51. The nurse is planning to instruct Mr. Chesapeake about his new cataract glasses. Which statement by the nurse is not helpful to him in his adjustment?
A. "Objects will appear to be much darker than before."
B. "Initially, use the new glasses only while sitting down."
C. "Look through the center of the glasses."
D. "Turn your head to the side rather than looking toward the side."

52. Mr. Chesapeake's family should be instructed to recognize signs and symptoms of hemorrhage from the operative site. The nurse should tell them that the most characteristic of this complication is:
A. Mild pain and discomfort.
B. Drainage of clear fluid.
C. Sharp pain in the eye.
D. Difficulty adjusting to new glasses.

Answers and Rationale

1. **B. *Use of accessory neck muscles.***
All of the choices are symptoms of COPD, but **B** is the response that indicates the effort required to breathe. Because the chest becomes rigid and fixed in a hyperexpanded position, the patient is forced to use the accessory muscles of respiration to ventilate the lungs. **A** is incorrect because tachypnea, a response to hypoxemia, can be ef-

fortless. *C* indicates insufficient oxygenation of the available hemoglobin, and *D* is due to the degree of bronchitis and bronchospasm present.

2. ***D.   Increased arterial Pco$_2$, increased hematocrit, and hyperinflated alveoli.***

Air trapping and alveolar destruction cause the typical ventilation-perfusion imbalances seen in COPD—increased arterial carbon dioxide and decreased oxygen. There is an increase in RBC production in response to the hypoxia. Hyperinflated alveoli occur with overdistention and air trapping. *A* is incorrect because Pco$_2$ increases and the hematocrit is increased, as discussed above. *B* is incorrect because alveoli are *hyper*inflated, and *C* is incorrect because the patient is hypoxic—decreased Po$_2$.

3. ***C.   COPD patients should receive low concentrations (2–3 L) of oxygen since the stimulus to breathe is their low Po$_2$.***

In COPD the respiratory center loses its sensitivity to the elevated CO$_2$ concentration, which normally stimulates a healthy person to hyperventilate. Instead the stimulus to breathe becomes the chronic hypoxic drive. If high concentrations of oxygen are given, which increase the oxygen tension in the blood, the patient will develop CO$_2$ narcosis and can ultimately die. A COPD patient should never receive more than 3 liters of oxygen. *A* and *B* are incorrect because the concentration of oxygen should be low. *D* is incorrect since the stimulus to breathe is a low Po$_2$.

4. ***B.   Absent expiratory and inspiratory wheezing.***

The primary use of aminophylline is to relieve bronchospasms. It also has a secondary diuretic effect, which may assist in the clearing of fluid retention. *A* is due to the pulmonary congestion from infection and indicates effective antibiotic therapy and chest physiotherapy. *C* also is an indication of antibiotic effectiveness. *D* occurs as hydration status improves.

5. ***D.   Neurotic reactions should be expected with COPD.***

Although some sources explain the noncompliance behavior of COPD patients as psychopathologic, their refusal to quit smoking is considered to be a matter of free choice. *A* is a frequently used patient excuse for not quitting. The initial response to the nicotine is bronchodilation. *B* is also a legitimate explanation, since COPD patients experience stages similar to the grief process. *C* reinforces the idea that noncompliance is an attempt to retain control over one's condition.

6. ***A.   Polycythemia.***

Polycythemia, an increase in RBCs from hypoxia, increases the viscosity of the blood. Since dehydration is also a problem in COPD, the viscosity is increased even more, leading to sluggish circulation and stasis. *B, C,* and *D* do not alter the viscosity of the blood.

7. ***D.   Facilitate complete exhalation.***

The COPD patient has a problem with trapping of air. By strengthening the diaphragm and breathing against pursed lips, there is a more complete exhalation. Pursed lip breathing creates a backpressure in the airway, keeping the airway open longer. Tightening of the diaphragm helps to squeeze air out. *A* is incorrect because the intercostal muscles become overused and ineffective in COPD. *B* is an irreversible condition that necessitates the pursed lip and diaphragmatic breathing. *C* is incorrect because the problem in COPD is reduced expiratory volume.

8. **A.** *The incidence of respiratory infections.*

   Respiratory infections in the COPD patient cause further lung damage and can cause decompensation in a patient who would otherwise tolerate the condition. Acute infections are often the precipitating cause of death. Changes in the carbon dioxide level, **B,** and oxygen level, **C,** are caused by factors that produce further lung destruction, as with **A.** Although **D** is associated with aggravation of the condition, the constant threat of infection is a *direct* threat to the patient's prognosis.

9. **B.** *A decrease in the WBC count.*

   The WBC is not affected by fluid volume, as are other tests such as hematocrit, hemoglobin, and BUN. A decrease in the WBC occurs following antibiotic therapy or from the normal inflammatory response. Increased fluid will thin tenacious respiratory secretions, **A,** as well as lower body temperature through perspiration and cooling, **C.** As previously mentioned, hematocrit, **D,** responds to changes in fluid.

10. **A.** *Fecal matter must be cleansed from the bowel for good visualization.*

    A is correct because fecal material in the bowel can interfere with interpretation of the test. **B** is incorrect; Mr. Richmond will be given a low-residue diet the evening before the test and liquids the morning of the test. **C** is incorrect because the patient will be moved into many positions, and will be asked to expel the enema. **D** is incorrect because barium is given by enema.

11. **C.** *Alteration in bowel habits.*

    C is one of the seven danger signals of cancer; other signs include bleeding, pain, weight loss, and anorexia. **A** and **B** are symptoms of obstruction. Mucus, **D,** appears in conditions of parasympathetic excitability.

12. **B.** *Bacterial content in the colon.*

    B is correct because neomycin sulfate helps to reduce the possibility of postoperative infections. Neomycin does *not* alter electrolyte balance, **A;** influence peristalsis, **C;** or cleanse the bowel, **D.**

13. **B.** *Soft, brown, mushy feces.*

    The type of stool described in **B** is usual in the transverse portion of the colon. **A** is incorrect because it describes drainage from an ileostomy. **C** is incorrect because a solid stool is found lower in the colon. **D,** liquid stools, are found before the transverse colon.

14. **A.** *Ground lean beef, soft-boiled eggs, tea.*

    A is correct because all the foods listed are low-residue foods. Choices **B, C,** and **D** all contain foods high in residue.

15. **C.** *Hypokalemia.*

    C is correct because one source of potassium depletion is gastric and intestinal suction. The imbalances in **A, B,** and **D** are due to causes not related to gastric suctioning.

16. **A.** *Irrigate the nasogastric tube with saline.*

    A is correct because irrigating the tube with saline prevents potassium loss. **B** encourages potassium loss, as do multiple water enemas. **C** is incorrect since it may be necessary to irrigate to keep the tube patent. **D** is incorrect because a Salem sump pump is attached to continuous suction, not to the nasogastric tube.

17. **C.** *Colostomy.*

    Of the choices, **C** is the greatest threat to body image presented because it is an alteration in normal function and a change in the nor-

mal pattern and because the patient feels a lack of control. *A*, *B*, and *D* do not significantly alter body image, function, or control.

18. **B.   *Permanent colostomy.***

    The colon and usually a segment of the descending colon are removed, so the stoma is permanent. *A* is seen after total removal of the colon. *C* is usually seen with a reversible colostomy. *D* is more often the site for a double-barreled colostomy.

19. **A.   *Red and raised.***

    A normal stoma should be red (like the color of healthy oral mucosa). *B*, *C*, and *D* indicate some problem with the stoma.

20. **C.   *Asks questions about the equipment being used.***

    The beginning step to performing self-care is to question how and why procedures are being done. *A, B,* and *D* could be important to the patient but do not demonstrate a willingness to be involved in self-care.

21. **C.   *Body weight and urine specific gravity.***

    Observing for increases in body weight and decreases in specific gravity of urine suggest rehydration of the patient and are important sources of evaluative data. *A* is not totally true since sugar and acetone measurements are done to detect urinary clearance of glucose and ketone bodies, conditions characteristic of diabetes mellitus, not fluid depletion. *B* is not totally true, as color of mucous membranes is observed to detect cyanosis, pallor, and jaundice. Skin turgor is an appropriate indicator of fluid balance. *D* is not totally true, as skin color changes are not associated with fluid status.

22. **A.   *Speak for the patient as often as possible to minimize feelings of embarrassment.***

    A is an inappropriate nursing action and therefore the correct answer. Repeatedly speaking for the aphasic patient discourages him or her from attempting to speak. *B* is a correct nursing action for a patient with receptive aphasia, who has difficulty understanding the spoken word. *C* is a correct action for a patient with expressive aphasia, who needs to practice speaking. *D* also is a correct action. It fosters supportive actions from the family that encourage attempts at speech and contribute to self-esteem.

23. **D.   *Administer a sedative to promote rest and speed healing of tissues.***

    D is the correct answer since it is the one nursing action that is *not* indicated in this case. Sedating an immobilized patient, such as Ms. Tweed, decreases spontaneous body movements and increases the risk of decubitus ulcers. *A* is an appropriate action. It keeps pressure off the heels and sacrum, yet allows for side-to-side changes in position. *B* is also a correct action. A sheepskin and heel pads prevent abrasive rubbing of the skin against the sheets and vary pressure points in the affected areas. *C* is a correct action, since massage will increase circulation to the pressure areas and prevent tissue necrosis.

24. **A.   *Breath sounds.***

    Auscultating breath sounds provides information on the rate, depth, and character of respirations. Crackles and wheezes can indicate atelectasis and obstruction. Diminished breath sounds signify a need for deep breathing activities. *B, C,* and *D* generate information that may or may not relate to respiratory status. *B* is incorrect since an

increased temperature can signify an infectious process anywhere in the body. Also, the reliability of temperature elevations as a sign of infection in older persons is questionable. *C* and *D* are incorrect since changes in pulse and skin color can suggest cardiac arrhythmias, anxiety, and a variety of other conditions unrelated to respiratory status.

25. **D. *Help her into a pair of warm socks that she can wear while in bed.***
    Such a simple action can promote comfort without adversely affecting circulation to the legs. *A, B,* and *C* are inappropriate actions. Removing the foot cradle allows tight bedcovers to constrict circulation to the feet and may contribute to footdrop. Hot water bottles can result in burns to tender skin. Vigorous massage of feet and legs is always contraindicated in immobilized patients. Thrombi may become dislodged and result in pulmonary embolism.

26. **B. *Strawberries with milk, cottage cheese, and toast.***
    B is correct. This was verified in the following two references:

    Williams SR. *Essentials of Nutrition and Diet Therapy* (5th ed). St. Louis: Mosby, 1990.
    Howard RB, Herbold NH. *Nutrition in Clinical Care* (2nd ed). New York: McGraw-Hill, 1982.

    The food combinations in each option are assessed as to their protein and vitamin C contents as follows:

A.	Cream of Wheat (¾ c.)	3.5 g protein 0 mg vitamin C
	Whole milk (8 oz.)	8.5 g protein 3 mg vitamin C
	Toast (1 slice wheat)	2.1 g protein 0 mg vitamin C
		14.1 g protein 3 mg vitamin C

B.	Strawberries (10 fresh)	0.8 g protein 60 mg vitamin C
	Whole milk (4 oz.)	4.25 g protein 1.5 mg vitamin C
	Cottage cheese (½ c.)	22 g protein 0 mg vitamin C
	Toast (1 slice wheat)	2.1 g protein 0 mg vitamin C
		29.15 g protein 61.5 mg vitamin C

C.	Orange juice (3⅛ oz.)	0.8 g protein 42 mg vitamin C
	Sweet roll	4.7 g protein 0 mg vitamin C
		5.5 g protein 42 mg vitamin C

D.	Soft-cooked eggs (2)	12.2 g protein
		0 mg vitamin C
	Toast (1 slice wheat)	2.1 g protein
		0 mg vitamin C
	Coffee	Trace amounts protein
		0 mg vitamin C
		14.3 g protein
		0 mg vitamin C

27. **D. Constipation.**
Liquid stools in a person with limited activity, a diet low in roughage, and an absence of stool softeners or laxatives often suggest fecal leakage around constipated stool. **A** is incorrect since Ms. Tweed receives a soft diet, which is not high in roughage. **B** is incorrect since overhydration results in signs and symptoms different from those reported here. **C** is incorrect because brief periods of sitting in a wheelchair are not classified as extraordinary activity.

28. **A. Elevate the knee gatch of the bed.**
A is the inappropriate action and therefore the answer. Elevating the knee gatch slows circulation to the legs and is contraindicated for the patient on bedrest. **B, C,** and **D** *are* appropriate nursing actions. Ankle exercises increase leg circulation. Antiembolism stockings decrease venous pooling and lessen the likelihood of thrombus formation. Adequate hydration also prevents thrombus formation.

29. **C. Preserve maximum functioning.**
C is correct because if use of the affected extremity is to be regained, it is necessary to prevent complicating deformities and contractures. Even if the patient does not regain total use, functional positioning of the extremity can be adapted to numerous devices. **A, B,** and **D** are not conducive to normal functioning if a contracture formed in those positions.

30. **D. So that the thrombus firmly adheres to the vessel wall.**
Bedrest is maintained for patients with deep-vein thrombosis to allow the thrombus "tail" to become firmly adhered to the vessel wall, and to prevent venous pressure fluctuations. **A** and **C** are reasons why leg elevation must be maintained, but these are not the rationale for bedrest, which is what the question asks. **B** is incorrect because bedrest does not prevent venous stasis; rather, bedrest potentiates it.

31. **C. Raise the foot of the bed on blocks, up to 8 inches high.**
When the legs are elevated above the level of the heart, venous return is facilitated by the force of gravity. Raising the foot of the bed on blocks achieves this purpose. **A** is incorrect because the gatch of the bed elevates the knees above the foot, interfering with adequate blood flow. **B,** active exercises of the affected extremity, is not done because of the potential release of the thrombus. **D** is incorrect because this activity increases the hydrostatic pressure in the capillaries, which may result in edema.

32. **A. Dissolves a thrombus.**
Heparin is *not* a fibrinolytic drug; thus, it does not dissolve thrombi. **B, C,** and **D** are the pharmacologic actions of heparin.

33. **A. Prothrombin time (PT).**
Prothrombin is one of the vitamin K–dependent clotting factors. Warfarin (Coumadin) suppresses the synthesis of prothrombin in

the liver. Since warfarin is a vitamin K–blocking drug, any change in the prothrombin level while the patient is on warfarin therapy becomes significant. Thus, *A* is the correct answer. *B, C,* and *D* are incorrect because these tests determine the effects of heparin, not warfarin.

34. **B. *Hemorrhage.***

No data are given to support the presence of hemorrhage. Had hemorrhage been the problem, Mr. Tell would have manifested ecchymosis on his right leg. Thus, *B* is the answer. *A, C,* and *D* are not the answers, because they *are* causes. Mr. Tell *is* obese, *A;* he *has* hip fracture, *C;* and he *has* limited physical activity, causing immobilization, *D.* Obesity and immobilization contribute to the development of venous stasis; fracture results in endothelial injury. Deep-vein thrombosis occurs when venous stasis and endothelial injury are present. Furthermore, hypercoagulability can also be a contributory factor as to what happens in severe hemorrhage. Dehydration and, eventually, hypercoagulation result if no intervention is given.

35. **D. *Localized, stabbing chest pain and dyspnea.***

Early signs of pulmonary embolism include restlessness; anxiety; tachycardia; tachypnea; localized, stabbing chest pain; dyspnea; and cough. Pulmonary infarction has occurred if pleuritic chest pain, hemoptysis, cough, friction rub, and fever are present. *A, B,* and *C* are incorrect because bradycardia, dull chest pain, and subnormal temperature do not occur.

36. **A. *$pH = 7.52$; $Paco_2 = 27$; $Po_2 = 64$; $HCO_3 = 24$.***

Because of the embolic obstruction, the affected area of the lung becomes ventilated but not perfused ("dead space"). The cessation of the pulmonary capillary blood flow results in alveolar hypocapnia, resulting in the constriction of air spaces and the airways in the affected area. The ventilation-perfusion imbalance that results leads to hypoxemia. Thus, when there is pulmonary infarction, respiratory alkalosis and hypoxemia are seen. *B* is incorrect because the blood gas results show metabolic alkalosis. *C* shows arterial blood gas results that are within normal range. *D* shows metabolic acidosis and hypoxemia. These results are not associated with pulmonary infarction.

37. **B. *Intermittent claudication.***

B is the correct answer; chronic or long-standing insufficiency of arterial blood supply is characterized by intermittent claudication. The patient experiences a sensation of pain, fatigue, or cramps, particularly in the calf muscles when he walks or exercises. Pain is relieved when these activities are stopped. *A, C,* and *D* are incorrect answers because these are assessment parameters indicating the presence of acute arterial occlusion.

38. **A. *Femoral.***

A is the correct answer because pulses are lost distal to the site of the occlusion. The patient has an occlusion in the left popliteal artery. The pulses distal to the occlusion are those in the dorsalis pedis, *C,* and posterior tibial, *D,* of the left extremity. Pulse is not felt on the popliteal artery, *B,* because it is occluded. The femoral pulse, which is above the area of occlusion, is present.

39. **A. *Strength and movement of the extremities.***

Assessing strength and movement of the extremities tests the ability of the muscles for movement and coordination. It also tests adequate nerve innervation to the muscles. *B, C,* and *D* are incorrect.

These data are most appropriate for patients with peripheral vascular disease.

40. **A.   *Shortened, with external rotation.***

    A is correct because of the suspected right hip fracture. The signs are shortened, abducted, and in a position of external rotation. *B* and *D* are incorrect because they refer to internal, rather than external, rotation. *C* is incorrect because there will be a difference in the length of the legs.

41. **C.   *Decrease the number of organisms on the skin.***

    C is correct because although all organisms cannot be removed, their numbers can be decreased by removing the hair and scrubbing the site. *A* is incorrect because a skin prep is unable to render human skin free of organisms. *B* is incorrect because removing the hair actually decreases the natural defense of intact skin against infection but it helps to lower the amount of organisms in the area and reduces the possibility for infection in the operative area. *D* is not a goal of the skin prep.

42. **A.   *Infection.***

    A is correct; infection can lead to osteomyelitis, which is very difficult to clear up. *B, C,* and *D* may be possible complications after implant surgery but are not as serious or as common as infection.

43. **D.   *Slow down the IV rate.***

    D is the *priority* action, for possible fluid overload. *A* is incorrect because it increases the problem. *B* might be important to look at in the total picture, but it is not the priority action. *C* is also an important action, but it is not as important as slowing the IV rate.

44. **B.   *Fluid overload.***

    B is correct because moist breath sounds (crackles) and tachycardia are associated with fluid overload. On palpation, the pulse is bounding and full. Consequently *A* and *D* are incorrect. *C* most likely is associated with cardiac irregularity, not just tachycardia; and changes in the breath sounds are not expected.

45. **A.   *Perform a lung assessment.***

    A is correct because respiratory problems are the first to exhibit postoperatively, usually within the first 24–48 hours. *B* is incorrect because urinary tract infections usually occur later, especially if the person has had a catheter. *C* is incorrect because the nurse usually reinforces the initial dressing, rather than removing it; there is no mention of hemorrhage; and it is too soon to expect any drainage from the incision that could be interpreted as infection. *D* is incorrect because correct positioning of the hip is very important; rotating the hip definitely is contraindicated.

46. **A.   *Knows that her legs should be separated with a pillow.***

    A is correct; an abduction pillow is used between the legs to keep the hip in the proper position. *B, C,* and *D* are incorrect because they increase the risk of dislocation of the implant.

47. **D.   *May tie own shoes as usual.***

    D is the incorrect instruction because tying her own shoes in the usual manner results in flexion and possible dislocation of the hip. *A, B,* and *C* are correct instructions to prevent dislocations.

48. **A.   *Objects are distorted and blurred.***

    A is correct due to the opacity of the lens caused by chemical changes. *B* is incorrect; although difficulty seeing at night is common to all, patients with cataract development in the center portion of the lens can generally see better in dim light, when the pupil is

dilated. *C* is a symptom of glaucoma. *D* is due to weakness of some extraocular muscles, not to a cataract.

49. *D.*   *"Can you tell me more about not liking to be awake during surgery?"*

D is correct because it asks the patient to discuss his concerns further. *A, B,* and *C* are incorrect because they do not help the patient, nor do they allow him to discuss his fears. These responses close the discussion.

50. *A.*   *Asking someone to pick up his robe from the floor.*

A is correct because it is important after cataract surgery to prevent increased intraocular pressure, which could put a strain on the suture line. *B, C,* and *D* are incorrect; all of these activities could increase pressure on the suture line.

51. *A.*   *"Objects will appear to be much darker than before."*

A is the incorrect statement and therefore the answer. Opacity has been removed, so objects may seem brighter. *B* is true of cataract glasses; since the lenses magnify, the patient should become accustomed to this before ambulation. *C* and *D* are also correct statements; cataract lenses distort peripheral vision. The wearer must learn to turn the head farther and more frequently to ensure safety.

52. *C.*   *Sharp pain in the eye.*

C is the correct answer; sharp pain is usually indicative that something is wrong. This symptom plus a half-moon of blood on the dependent position in the anterior chamber are indicative of hemorrhage. *A, B,* and *D* may be expected after cataract surgery and do not indicate hemorrhage.

11. Behavioral and Emotional Problems

Case Management Scenarios and Critical-Thinking Exercises

Medical Diagnosis: Leukemia, terminal stage.

- **Nursing Problem:**
  Response to terminal illness.

- **Chief Complaint:**
  Betsy Banner, a 3-year-old, is on the pediatric unit dying of leukemia.

- **History of Present Illness:**
  Betsy had been in remission for 6 months, but recently her WBC was extremely elevated, she began having elevated body temperatures, and she was readmitted in a terminal stage.

- **Family History:**
  Betsy's parents are in their early 20s, and she is an only child. They are devastated by the fact that their "baby" is dying.

---

1. Betsy's mother tells the nurse that she just cannot stand to see her baby suffer and can only be in the room with her for a short time. Which approach is best for the relationship between Betsy and her mother?
   A. Support Mrs. Banner's feelings and encourage open communication.
   B. Tell Mrs. Banner she needs to spend more time with Betsy for the sake of the child's, as well as her own, well-being.
   C. Tell Mrs. Banner to have her husband visit more often if she is unable to do this.
   D. Let Mrs. Banner know that she is running away from her own guilt and projecting it onto her child.

2. In planning the nursing care for Betsy, it is not helpful for the nurse to:
   A. Hold and touch Betsy as much as possible.
   B. Leave Betsy alone as much as possible so that she can get adequate rest.
   C. Provide comfort and relief from pain, even if it means the child may become dependent on pain medication.
   D. Keep the environment as normal as possible, even though Betsy is dying.
3. One week after Betsy dies, Mrs. Banner comes to the unit to pick up her daughter's things. She keeps repeating to the nurse, "It can't be possible that Betsy is gone." The nurse's understanding of the normal grief process leads her to realize that Mrs. Banner is experiencing:
   A. Anger toward the nursing staff.
   B. Overwhelming guilt.
   C. Feelings of emptiness.
   D. Denial of reality.

Medical Diagnosis: Anorexia nervosa.

- ■ *Nursing Problems:*
  Body image disturbance.
  Altered eating: refusal to eat.
  Altered family processes.

- ■ *Chief Complaint:*
  Angela McAdoo, 14 years old, was admitted to the hospital 3 days ago. Angela has been diagnosed as having anorexia nervosa. She is in the hospital because of her persistent weight loss.

- ■ *History of Present Illness:*
  Angela began to be concerned about gaining weight at the time of onset of her menstrual cycles. Her father teased Angela about her budding sexuality and implied that he preferred her to her mother. This terrified Angela and she refused to eat.

4. In the hospital the nurse should:
   A. Encourage verbalization about body functions and feelings concerning body functions.
   B. Insist that Angela eat at least one bite of everything on her tray.
   C. Ignore the fact that Angela is not eating.
   D. Tell Angela that she will be fed by tube if she does not eat.
5. Angela says to the nurse one day, "I promised my father that I would gain weight, but he did not notice the 2 pounds I gained. Why bother?" The nurse should:
   A. Discuss with Angela her wish to remain slim and the reasons growing up is painful.
   B. State she noticed that Angela had gained 2 pounds.
   C. Tell Angela to eat more for the next few days and that her father will then notice.

    D. Tell Angela that the important thing is not that her father notice her weight gain but that she has gained weight.

6. The most successful therapy used in treating eating disorders is:
    A. Individual therapy.
    B. Behavior modification therapy.
    C. Gestalt therapy.
    D. Reality therapy.

Medical Diagnosis: Anorexia nervosa.

- ■ *Nursing Problem:*
  Altered eating: refusal to eat.

- ■ *Chief Complaint:*
  Carol Clifton, age 16, was admitted to the hospital with a diagnosis of anorexia nervosa. Her physical exam revealed a cachectic young female of 80 pounds whose menses had ceased.

- ■ *Family History:*
  Carol's social history indicated that her parents were divorced a year ago and that she lives with her mother.

- ■ *History of Present Illness:*
  Carol began to lose interest in her schoolwork shortly after the divorce, and 8 months prior to admission she began to curtail her food intake severely.

7. The nurse, in determining the patient's care, realizes that a teenager like Carol should:
    A. Remain with her family.
    B. Be restricted to the hospital unit until at least 10 pounds is regained.
    C. Be allowed to eat when she desires.
    D. Be rewarded for eating.
8. Two weeks after admission, Carol opens up to her primary nurse. She asks, "Why would I just stop eating the way I did?" Which comment by the nurse will be most helpful to Carol?
    A. "Did your mother provoke you in any way?"
    B. "You will soon discover why in your therapy."
    C. "Were you depressed at the time?"
    D. "I don't know why, but I will try to help you find an answer."
9. What would be an inappropriate therapeutic approach for Carol?
    A. Setting limits.
    B. Providing support.
    C. Not allowing involvement in decision making.
    D. Avoiding staff conflicts.

Medical Diagnosis: Brief psychotic disorder.

- ■ *Nursing Problems:*
  Risk of injury (self-inflicted).
  Altered thought processes.

■ *Situation:*
Charles Adams is 19 years old and was admitted to the hospital by his parents.

■ *History of Present Illness:*
Charles's parents report that for the past 6 months he has not seen any of his friends, has eaten only bread crusts, and has hardly spoken to anyone. After extensive diagnostic studies, it was determined that Charles is exhibiting severe withdrawal as part of an acute schizophrenic reaction.

---

10. In planning the interventions for Charles's care, the nurse is aware that he will respond best to:
    A. Opportunities to make decisions.
    B. Being left alone.
    C. Clear, simple language.
    D. Small-group activities.

11. The nursing team determines that remotivation therapy will be most helpful for Charles's progress. The focus of remotivation therapy is to:
    A. Train for a job.
    B. Share personal conflicts.
    C. Prepare for discharge.
    D. Increase interest in the everyday activities of life.

12. One morning Charles grabs sharp scissors from a nursing assistant and stabs his arm in several places. Which comment should the nurse immediately make?
    A. "I see that you are still angry at yourself."
    B. "How could you do this to yourself?"
    C. "You are bleeding in several places. I will help you."
    D. "Will you let me help you? I know I can stop your bleeding."

Medical Diagnosis: Schizophrenia.

■ *Nursing Problem:*
Altered thought processes.

■ *Chief Complaint:*
John Arnold is a 20-year-old who seems several years younger than his stated age. He has been a patient on the mental health unit for the past 3 weeks. His diagnosis is schizophrenia, undifferentiated.

■ *History of Present Illness:*
This is John's first hospitalization. He was brought to the hospital by his family after his behavior caused them concern. John had been withdrawn and mute for periods of a week and then would sing nonsense verses for hours at a time. At this time, John is verbalizing but is not joining in any ward activities. He is very obsessed with the idea of food and spends hours repeating sentences concerning food. An example of his verbalization is:

You can't make a tomato. You can't make it yellow, squash is yellow. Mother doesn't like mustard. Brown doesn't go with ham. He doesn't handle ham.

---

13. The example of John's verbalization given in the case situation is called:
    A. Neologism.
    B. Delirium.
    C. Clang association.
    D. Flight of ideas.
14. When listening to John's constant flow of words, the nurse should:
    A. Not interrupt him.
    B. Attempt to stop the flow of words.
    C. Interrupt occasionally to clarify what he is talking about.
    D. Interrupt frequently and try to keep him on one topic.
15. The nurse should expect the relationship with John to:
    A. Develop slowly, as it will take John time to trust.
    B. Develop slowly because John is unable to communicate effectively.
    C. Develop rapidly because John is verbal.
    D. Develop rapidly because John seems attracted to the nurse.

Medical Diagnosis: Brief psychotic disorder.

- **Nursing Problem:**
  Altered thought processes.

- **Chief Complaint:**
  Marlene Klein is 23 years old. She lived with her parents, had no close friends, and could be considered a "loner." She was admitted to the psychiatric hospital when she refused to leave the house, neglected her personal appearance, and claimed she heard music all the time.

- **History of Present Illness:**
  In the hospital she is quite regressed and refuses to attempt self-care. She is often found talking out loud to no one but herself. She was diagnosed as having an acute schizophrenic reaction. She was also put on chlorpromazine (Thorazine), 200 mg tid.

---

16. Understanding the side effects of chlorpromazine, the nurse realizes that which activity is contraindicated in planning Marlene's care?
    A. Playing cards.
    B. Shuffleboard.
    C. Sunbathing.
    D. Gardening.
17. Marlene was admitted to the hospital on a temporary, involuntary basis. In establishing goals, the nurse is aware that:
    A. Marlene will be detained until her behavior is no longer a major problem.

B. A subsequent hearing must be scheduled if Marlene's problems necessitate additional hospitalization time.

C. Marlene will not have visitors until her behavior has stabilized.

D. Marlene must accept the prescribed treatment.

18. One afternoon, while the unit is particularly quiet, Marlene comes up to the nurse and says, "Listen to that music; it's so loud. Make it stop." Which is the best comment for the nurse to make?

A. "I bet you feel like dancing to that music."

B. "Music is nice, Marlene. Try relaxing as you hear it."

C. "I don't hear any music on the unit now. Perhaps we can talk together about how you are feeling."

D. "Let's play cards together. It will help you stop hearing the music."

Medical Diagnosis: Anxiety disorder; conversion disorder.

■ *Nursing Problem:*
Anxiety.

■ *Chief Complaint:*
Sara Brownly, a 26-year-old fourth-year medical student, was admitted to the hospital for observation after complaining of tingling and frequent severe numbness in both her hands.

■ *History of Present Illness:*
Sara was often unable to hold or even grasp an object. These symptoms became exacerbated after spending the Christmas holidays at home with her father.

■ *Physical Exam:*
A complete physical exam revealed no organic problems. Sara was generally in good spirits, but her conversation focused on her disability and the possibility of not returning to medical school.

■ *Family History:*
Sara's social history revealed that she was an only child and her father was a successful practicing surgeon. Their relationship was described as being very close. Her mother was killed in an automobile accident when Sara was 14. After the accident, Sara's father focused all his attentions on his daughter, pushing her academically so that she could enter medical school. Her father referred to how much he was looking forward to having his daughter in practice with him and called her the "future hands" in his practice.

Sara was diagnosed as having a severe anxiety reaction with conversion symptomatology.

---

19. In assessing Sara's symptoms, the nurse should understand that the relationship between the paralysis of the hands and the underlying emotional conflict is:

A. Not an important issue in planning her care.

B. Symbolic in nature.

C. Indirect and of questionable importance.

D. An expression of Sara's anger toward her father.

20. An important nursing intervention for Sara is to:
   A. Limit her social activity until she regains better use of her hands.
   B. Set limits on her behavior.
   C. Allow her to have all her dependency needs met.
   D. Decrease opportunities for secondary gain.

21. In planning for Sara's immediate needs, which is an inappropriate approach?
   A. Avoid displaying concern over Sara's symptoms.
   B. Involve Sara in planning a social activity.
   C. Give Sara opportunities to express her feelings.
   D. Encourage Sara to discuss her symptoms freely.

Medical Situation: Postmastectomy reaction.

■ *Nursing Problems:*
Body image disturbance.
Ineffective individual coping: denial.

■ *Chief Complaint:*
Linda Weber is a 29-year-old married homemaker with two children. Linda discovered a lump in her left breast 2 weeks ago. On the advice of her physician she underwent a breast biopsy, and at that time the diagnosis of carcinoma was made. Linda chose to undergo a mastectomy. Her postoperative course has been uneventful.

■ *Past History:*
Linda has been in good health. This is her only hospitalization other than for the birth of her two children.

■ *Family History:*
Linda and her husband have been married for 8 years and have a good marital relationship. They have a boy, Jason, age 6, and a girl, Lisa, age 4. Linda worked as a court reporter before the birth of her children.

22. Linda could not look at her incision on the first postoperative day, although she carefully watched and listened to the doctor and the nurse as the dressing was changed. This reaction indicates that:
   A. Linda will have difficulty adapting to her change in body image.
   B. Linda is testing how others will react to her altered body image.
   C. Linda feels mutilated and ugly.
   D. Linda is denying that anything is different.

23. The nurse's response to Linda should be:
   A. "Maybe you'd like to look at your incision tomorrow when we change the dressing."
   B. "The incision looks beautiful. You really should see it."
   C. "It will be hard for you to look at the incision the first time."
   D. "Everyone has to look at his or her incision sooner or later."

24. Linda expresses angry feelings toward her surgeon: "Why did he remove my entire breast when the cancer was so small?" The nurse should respond:

A. "Don't be angry with your doctor. He was doing what he had to do."

B. "Your doctor is one of the best on the hospital staff."

C. "You sound angry. Would you like to talk about it?"

D. "Ask your doctor why he had to remove so much."

Medical Diagnosis: Possible breast malignancy.

- **■ Nursing Problem:**
  Body image disturbance.

- **■ Chief Complaint:**
  Ellen Sands, age 32, was admitted to the hospital unit for a breast biopsy to rule out a malignancy. On the day of her biopsy, she tells the nurse that she was up all night and feels as if her head is about to explode.

- **■ History of Present Illness:**
  Ellen was in perfect health until 1 week prior to admission, when she discovered a pea-sized lump in her breast during self-examination.

- **■ Family History:**
  Ellen has been married for 8 years and has a 3-year-old daughter.

25. As the nurse is preparing Ms. Sands for the operating room, she asks the nurse, "Well, do you think I have breast cancer?" Which response is most helpful?
    A. "You need to ask your doctor that question."
    B. "It sounds as if you are worried about what your surgeon will find."
    C. "Breast cancer is more common in older women than in your age group."
    D. "Just relax. It won't help if you go to surgery feeling this way."

26. In identifying Ms. Sands's health needs, the nurse is aware that after discovery of a breast lump it is correct to assume that:
    A. Fantasies relating to body mutilation may be set in motion.
    B. If a woman is married, her adjustment to a mastectomy will likely be better.
    C. Depression is usually not an issue when a lump is discovered.
    D. In a sound marriage, husbands have little difficulty in adjusting to the removal of a breast.

27. Ms. Sands's biopsy result was negative and a simple lumpectomy was successfully completed. Which nursing action is indicated?
    A. Remembering her initial anxiety, encourage Ms. Sands to begin psychotherapy.
    B. Encourage Ms. Sands to continue self-examination of her breasts.
    C. Have her schedule a mammography appointment in 3 months.
    D. Inform her that she can expect strong discomfort around her incision for several weeks.

Medical Diagnosis: Sleep disorder.

■ *Nursing Problem:*
Sleep pattern disturbance.

■ *Chief Complaint:*
Jeremiah Johnson, 35 years old, is admitted to the skilled care rehabilitation unit following hip replacement.

■ *History of Present Illness:*
While in the rehabilitation unit, Mr. Johnson has been telling the nurses that he is unable to sleep at night. The nurses' notes indicate, however, that he has been asleep every time they make rounds.

28. How does a nurse explain the apparent discrepancy in the report by Mr. Johnson and by the nurses about his sleep?
    A. Mr. Johnson is spending a large portion of his sleep time in Stage II sleep.
    B. Mr. Johnson is spending more of his night in deep sleep—Stage IV.
    C. Mr. Johnson is spending a larger portion of the night in REM sleep.
    D. Mr. Johnson just happens to be asleep when the nurse makes rounds.
29. The nurse can help the patient achieve a restful night's sleep by determining with the patient his bedtime habits and also by:
    A. Encouraging short naps throughout the day.
    B. Eliminating environmental factors that distract him, such as noise.
    C. Avoiding all analgesic medications because of their effect of increasing REM sleep.
    D. Increasing the patient's level of cognitive functioning at bedtime by providing problems to solve.
30. Which nursing measure should be used to relieve initial insomnia for a recently hospitalized patient with no history of sleeping difficulty or of taking sleep medications?
    A. Increase physical activity immediately prior to bedtime.
    B. Increase physical activity during the day and provide monotonous stimulation at night.
    C. Give a mild sleeping medication to help the patient fall asleep.
    D. Encourage the patient to take a nap during the day, to provide the rest he is not getting since he cannot sleep at night.

Medical Diagnosis: Suicide attempt.

■ *Nursing Problem:*
Dysfunctional grieving.

■ *Chief Complaint:*
Sherman Darby, age 42, was admitted to the psychiatric unit of a general hospital following an unsuccessful suicide attempt. He is

considered a successful lawyer, but he repeatedly laments, "I am no good to anyone."

■ *History of Present Illness:*
Mr. Darby has lost weight in the past month and has insomnia.

---

31. The nurse is to work with Mr. Darby. Yesterday the nurse made a contract with him to meet daily, Monday through Friday, for half an hour. When the nurse approaches him at the contracted time, Mr. Darby says, "Just leave me alone. I'm not worth your time." The nurse's most appropriate response is:
   A. "Mr. Darby, I can respect your need to be alone sometimes."
   B. "I want to stay with you, Mr. Darby. You are worth my time."
   C. "O.K., I'll be back tomorrow, but I want you to know I think you're worthy."
   D. "Mr. Darby, what makes you say that?"

32. Two weeks later, Mr. Darby states that his depression began when his father died 2 months ago. He also says that his father was very critical and demanding of him. Prior to his death, Mr. Darby's father had been in a nursing home for 2 years due to a stroke. Mr. Darby says, "If only I had taken him out of the rest home so I could have loved him more." The nurse's best response is:
   A. "Mr. Darby, I'm sure you did all you could."
   B. "Mr. Darby, you had to think of your own family also."
   C. "Mr. Darby, you're being too hard on yourself. You were a loving son."
   D. "You told me you loved your father. How would taking him out of the rest home have meant that you loved him more?"

33. The defense mechanism implied by Mr. Darby's self-derogation is:
   A. Projection.
   B. Conversion.
   C. Introjection.
   D. Regression.

Medical Diagnosis: Bipolar disorder.

■ *Nursing Problems:*
Hyperactivity.
Altered thought processes.
Altered verbal communication.

■ *Chief Complaint:*
Jonas Craig, age 42, is brought to the hospital by his wife. During the admission process, he is laughing and shouting. He is wearing a bright red shirt and orange pants. He speaks rapidly and readily uses profanity.

■ *History of Present Illness:*
Mr. Craig's wife says he gave away all their artwork and savings bonds.

Mr. Craig's diagnosis is manic-depressive illness. He is started on lithium, 200 mg qid.

---

34. In talking with Mr. Craig, the nurse hears him say: "What a day out. Out is where I want to be. To be or not to be, that is the question. Questions, you are always asking me questions." The nurse assesses that he is using:
    A. Confabulation.
    B. Flight of ideas.
    C. Neologisms.
    D. Rationalization.
35. When planning Mr. Craig's care, the nurse realizes he is prone to using profanity. The nurse is thus prepared to:
    A. Limit his activities to his room.
    B. Ignore him unless he disturbs other patients.
    C. Have him leave the dayroom area.
    D. Tell him that profanity is not allowed in the hospital.
36. Mr. Craig asks the nurse what unit activity he should select. Together they decide the best choice is:
    A. Playing chess.
    B. Group card games.
    C. Walking with the nurse.
    D. Wood carving.

Medical Diagnosis: Depressive disorder.

- *Nursing Problems:*
  Self-care deficit: medication administration.
  Suicidal ideation.
  Altered verbal communication.
  Withdrawal/social isolation.

- *Chief Complaint:*
  Barbara Cullen, age 48, was admitted to the psychiatric unit of the local community hospital and was diagnosed as having an acute depressive reaction.

- *Family History:*
  Ms. Cullen has been married for 27 years and has three children, all in their early 20s; none live at home. She was an executive secretary for 1 year before she was married and has not worked since.

- *History of Present Illness:*
  On admission, Ms. Cullen's history revealed a 12-pound weight loss, inability to fall asleep at night, pallor, anorexia, and lack of ability to concentrate.

---

37. In planning the initial approach for caring for Ms. Cullen, the nurse will include:
    A. Encouraging her to choose all her own menu selections.

      B. Involving her in a group activity on the unit.

      C. Sitting silently at her side when she does not feel like talking.

      D. Providing cheerful activities to uplift her mood.

38. When planning Ms. Cullen's daily care, the most important aspect for the nurse to consider is:

      A. Building her self-image.

      B. Involving her husband in the therapy.

      C. Getting her started in a simple daily routine.

      D. Preventing suicide.

39. Electroconvulsive therapy has been ordered for Ms. Cullen. When the nurse comes to accompany her to the first treatment, Ms. Cullen says, "I'm so dreadfully afraid of these treatments." Which comment by the nurse is most helpful to Ms. Cullen?

      A. "Being frightened is an unpleasant feeling. Try hard not to be so afraid."

      B. "Can you tell me more about your feelings of being afraid?"

      C. "You are not alone in being frightened. Most individuals feel exactly the same way."

      D. "It will be over before you know it, and I will stay with you."

40. In addition to electroconvulsive therapy, amitriptyline (Elavil) has been ordered. What information should the nurse give Ms. Cullen about taking this drug?

      A. It cannot be taken with foods that contain tyramine such as beer and aged cheese.

      B. Blood levels must be monitored regularly.

      C. Hypertension is a common side effect.

      D. It takes 2–3 weeks for the drug to take effect.

Medical Situation: Bereavement related to death of a parent.

■ *Nursing Problem:*

Altered feeling state: anger related to the loss of a loved one.

■ *Situation:*

Anna King's mother, age 76, has just died on the hospital unit from an acute myocardial infarction. Ms. King was at her mother's bedside when she died.

■ *Family History:*

Ms. King, age 49, is unmarried; her mother lived with her. She has two middle-aged brothers, both of whom are married and have families.

---

41. Ms. King says to the nurse, "My mother could have lived if she had wanted to. She died because she knew I wanted my own apartment." In assessing how best to help Ms. King deal with her feelings, the nurse realizes that:

      A. Ms. King is displaying normal feelings of anger and hostility toward her deceased mother.

      B. Ms. King is showing strong ambivalence toward her deceased mother.

C. Ms. King will most likely undergo a severe depression during the next year.

D. Ms. King blames her mother for the problems in her life.

42. Ms. King says to the nurse, "I just can't believe my mother is gone." What response should the nurse use to better assess Ms. King's needs at this time?

A. "Do you feel angry about your mother's death?"

B. "Did you have problems with your mother?"

C. "I understand how you feel."

D. "Can you tell me more about how you feel?"

43. Ms. King keeps angrily telling the nurse that now she has no one who cares about her, not even her brothers. The most useful comment by the nurse is:

A. "I have seen how much your brothers care for you when you are together."

B. "You are such a nice person. I am sure many people care for you."

C. "It must be difficult for you to feel that no one cares for you."

D. "Don't be so hard on yourself. Things will seem different in a very short while."

Medical Diagnosis: Alcohol abuse with delirium.

■ *Nursing Problem:*
Altered thought processes.

■ *Chief Complaint:*
Marcia Ross, age 50, is hospitalized on a medical ward with a diagnosis of acute gastritis. Ms. Ross is single.

■ *History of Present Illness:*
Ms. Ross is dehydrated and has a high fever. She has a history of chronic alcoholism. Toward evening, Ms. Ross becomes restless; as the evening progresses, she becomes more agitated and says that she is afraid that she will miss her bus. Later in the evening, the nurse sees Ms. Ross searching for something. The patient states that her clothes have been stolen and that she will be late for work if she does not find them.

44. The nurse caring for Ms. Ross understands that she is in a state of delirium. The nurse knows that delirium:

A. Is irreversible due to the chronic alcoholism.

B. Develops rapidly due to specific physiologic deficits or increased physiologic demands.

C. Causes permanent memory lapses.

D. Does not cause defects in judgment or discrimination.

45. Hallucinations and illusions are common symptoms in delirium. To intervene effectively, the nurse must understand that hallucinations are:

A. False sensory perceptions without actual stimuli.

B. False sensory perceptions with actual stimuli.

C. False fixed beliefs.

D. Feelings of strangeness and unreality about self.

46. An appropriate response to Ms. Ross's statements about being late for work is:
    A. "You are going to be late for work?"
    B. "It's 11 P.M. and you are in a hospital being treated for a stomach problem."
    C. "Don't worry. Everything will be all right."
    D. "It's 11 o'clock at night. You don't have to go to work until morning."

Medical Diagnosis: Fractured femur; pneumothorax.

- ■ *Nursing Problem:*
  Altered comfort (pain).

- ■ *Chief Complaint:*
  George Graflin, a 51-year-old man, was admitted to the hospital from the emergency room following an automobile accident in which he sustained a fractured femur and a pneumothorax.

- ■ *History of Present Illness:*
  Mr. Graflin is experiencing a moderate degree of pain in his chest and in his right leg. He is also experiencing pain of undetermined origin in his abdominal area. The pain in his abdomen is intermittent and more intense than his other pain. He has received morphine sulfate, 10 mg IM. The nurse who has been with Mr. Graflin leaves the room to report on his condition. When the nurse returns, Mr. Graflin is moaning loudly and holding his abdomen. There seems to be no other change in his condition.

---

47. When Mr. Graflin finally appears to be dozing, the nurse starts to leave his room. Mr. Graflin opens his eyes and asks, "Are you leaving?" The nurse should respond by saying:
    A. "Yes. You will be all right. I need to take care of other patients now."
    B. "Yes, but I will be back every 15–20 minutes to see how you are feeling."
    C. "Yes, because the medication I gave you needs time to take effect."
    D. "Yes, but don't worry. Everything will be all right."

48. The "pain is punishment" fantasy is a significant variable for many patients experiencing pain. An example of this response is a patient who states:
    A. "I don't want any medication because I want to know what is happening."
    B. "There is no way to relieve this pain."
    C. "What did I do wrong to cause this pain?"
    D. "I don't need any medication. I'll keep busy and forget the pain."

49. The nurse should expect that the 10-mg dose of morphine will:
    A. Relieve severe pain for 2 hours.
    B. Relieve moderate pain for 4–5 hours.
    C. Cause hypotension.
    D. Be ineffective in relieving Mr. Graflin's pain.

Medical Problem: Post–cardiac arrest.

- **Nursing Problem:**
  Ineffective individual coping.

- **Chief Complaint:**
  Mike Mansfield, a 52-year-old dock worker, was admitted to the coronary care unit of the hospital after experiencing a cardiac arrest.

- **History of Present Illness:**
  Mr. Mansfield was at work when he experienced a cardiac arrest. Paramedics were called to the scene and resuscitated him. He was transported to a local hospital and has been a patient on the coronary care unit. His condition is still unstable.

50. When asked if he is worried about the prospect of another cardiac arrest, Mr. Mansfield replies, "Why worry? Besides, my middle name is Lucky." This is an example of:
    A. Undoing.
    B. Fantasy.
    C. Rationalization.
    D. Denial.
51. In planning psychosocial care for Mr. Mansfield, the nurse should understand that:
    A. A strong need for protection may cause a patient to hear only what he wishes to hear.
    B. Mr. Mansfield has a realistic perception of his problem and is not troubled by it.
    C. Mr. Mansfield is coping with his illness in a pathologic manner.
    D. Mr. Mansfield's manner of coping is allowing him to experience the full impact of his illness.
52. The nurse's best response to Mr. Mansfield's statement in question 50 is:
    A. "Your middle name is Lucky?"
    B. "You are right not to worry. It doesn't help to worry."
    C. "I can see that you are not worried."
    D. "Sometimes it takes a while before a person can really think about what happened."

Medical Situation: Phase of life problem: adjustment to retirement.

- **Nursing Problem:**
  Impaired adjustment: reaction to loss.

- **Chief Complaint:**
  Ted Saunders, 64 years old, has taken voluntary retirement from his job. Mr. Saunders is seen in an outpatient psychiatric clinic due to difficulty adjusting to his retirement.

■ *History of Present Illness:*
Mr. Saunders worked for a large manufacturing concern for 38 years. During that time, he saw many changes in his company and its employees. In recent years he complained bitterly about the caliber of the people being employed and the changing circumstances of his employment. Jane, his wife of 30 years, became used to Mr. Saunders's complaining and was able to tolerate it since he traveled frequently in his job. Jane has many social outlets, but Mr. Saunders did not develop any interests outside his job.

53. Mr. Saunders is experiencing difficulty adjusting to his retirement, even though it was voluntary. One of the reasons could be:
    A. Poor marital adjustment.
    B. Loss of a peer group.
    C. Loss of physical stamina.
    D. Difficulty learning new hobbies at his age.
54. According to Erik Erikson, the developmental task that Mr. Saunders must accomplish is:
    A. Generativity.
    B. Ego integrity.
    C. Industry.
    D. Autonomy.
55. An activity that might be suggested for Mr. Saunders is:
    A. Helping his wife with the housework.
    B. Reading.
    C. Watching television.
    D. Joining a men's political interest group.

Medical Diagnosis: Unresolved grief reaction.

■ *Nursing Problems:*
Ineffective individual coping: denial.
Self-care deficit.

■ *Chief Complaint:*
Mark Graham, age 64, appears for his appointment at the crisis intervention center. In his initial interview with the nurse, he relates that his problem is not being able to take care of himself anymore. He is unshaven, his clothes are unkempt, and he speaks slowly. During the session, he cries several times.

■ *Family History:*
Mr. Graham's wife died almost 2 years ago, they had no children, and he recently retired from his job as an electrician.

The health team at the clinic diagnoses his condition as an unresolved grief reaction.

56. In planning Mr. Graham's care, it is *most* helpful for the nurse to know:

A. How he felt about his deceased wife.

B. Who else he can rely on for support.

C. His opportunities for employment.

D. His pattern of communication.

57. Mr. Graham keeps repeating to the nurse that he accepted his wife's death well. He states, "I had to go on. No sense crying over her. I just said to myself, 'She's gone and that's it.' I had no problems at all." In assessing these data, the nurse correctly infers that the patient is using:

A. Projection.

B. Denial.

C. Undoing.

D. Displacement.

58. During the therapeutic relationship with Mr. Graham, the nurse realizes, in planning his care, that progress is most likely to be made if:

A. The nurse can identify Mr. Graham's problems.

B. The nurse can analyze Mr. Graham's behavior.

C. Mr. Graham can begin to face his problems with the nurse.

D. Mr. Graham sets his own pace for the therapy.

Medical Situation: Postoperative care following cataract surgery.

■ *Nursing Problem:*
Sensory-perceptual alteration.

■ *Chief Complaint:*
Maeve O'Neil, 82 years old, is admitted to the hospital. She has cataracts in both eyes and is scheduled for surgery for extraction of the cataract in her right eye.

■ *Family History:*
Ms. O'Neil is alert and active. She lives alone and maintains an active social life. Her two adult children are concerned about her upcoming surgery, considering her advanced age.

■ *History of Present Illness:*
On the day after surgery, the night nurse reports to the staff that Ms. O'Neil seems confused and disoriented. When the nurse made rounds during the night, Ms. O'Neil thought that the nurse was her daughter, Mary. Ms. O'Neil is becoming more anxious and withdrawn. The day nurse has not seen any of these behaviors.

59. Ms. O'Neil is most likely showing effects of:

A. Sensory overload.

B. Senility.

C. Acute brain syndrome.

D. Sensory deprivation.

60. The night nurse will observe more behavioral changes in Ms. O'Neil than the day nurse because:

A. The intensity of sensory stimuli is increased at night.

B. The nurse has more frequent contact with Ms. O'Neil at night.

C. Hypnagogic states cause increased susceptibility to the effects of sensory restriction.

D. The ward is quieter at night.

61. Which nursing measure will *not* be effective in assisting Ms. O'Neil?

A. Placing familiar objects from home at her bedside.

B. Placing a clock with a large dial at her bedside.

C. Placing Ms. O'Neil in a private room.

D. Verbally orienting Ms. O'Neil to time and place.

Medical Diagnosis: Systemic lupus erythematosus (SLE).

- **Nursing Problems:**

Disturbance in self-esteem related to illness (i.e., immobility, self-care deficit, altered body image, incontinence) and reinforced by others.

- **Chief Complaint:**

Mr. Roland is a 19-year-old black high school graduate. Diagnosed as having SLE, he was recently admitted to a university hospital rehabilitation unit to begin a special program of vocational training. He has been unable to find a job because he is confined to a wheelchair.

- **History of Present Illness:**

Mr. Roland first experienced joint pains, swelling of lower extremities, dizziness, and extreme fatigue 4 years ago. He attributed these symptoms to "overdoing"—playing too much basketball, staying out late, and working after school as a beverage distributor. He was hospitalized once for an episode of mental confusion, sudden memory loss, and behavior changes, and SLE was diagnosed. He was subsequently hospitalized several times for renal involvement and progressive generalized weakness. The disease process is presently stabilized by treatment with steroidal and immunosuppressant drugs on an outpatient basis.

- **Past History:**

Medical: Usual childhood diseases.

Allergies: None.

Social: Lives with mother and three siblings; unemployed.

- **Family History:**

Sister: Sickle cell anemia.

- **Subjective Data:**

"I was never happy at home, and now I can't even leave my room. All I do is watch TV. My friends don't come around much any more because they're all working. Sometimes I call them up just to take my mind off my troubles. Last week my girl left to move back with her relatives in the islands. That's probably for the best anyway. She was getting mad because I couldn't take her out any more. Who wants to be with a guy in a wheelchair who's got no job and no money to spend?"

- **Objective Data:**

Vital signs: Temperature: 98.4°F. BP: 130/84. P: 76. R: 18. Ht: 5 feet, 10 inches. Wt: 163 pounds.

Unable to stand unassisted; unable to raise arms above shoulder level.
Round, "moon" face. Expression is sad and dejected.
Occasionally incontinent of urine.
No dyspnea.
No alopecia noted.

■ *Physical Exam:*
Slight swelling of abdomen and lower extremities.
Muscle atrophy of upper and lower extremities.
Good range of motion of all joints.
No joint swelling or arthralgia at this time.

■ *Laboratory Data:*
WBC: Leukopenia.
Urinalysis: Moderate proteinuria.
ANA: Positive.
LE cell test: Positive.

---

62. A nurse on the rehab unit recognizes Mr. Roland's problem with disturbance in self-esteem and tries to increase his self-esteem. Which intervention is most likely to serve that goal?
    A. Arrange a date for Mr. Roland with a young female patient on the unit.
    B. Praise Mr. Roland's attractive physical features.
    C. Assign Mr. Roland a task he can learn in a short period of time.
    D. Spend extra time helping Mr. Roland to do his personal self-care.

63. Mr. Roland makes some friends on the unit and becomes very attached to a young woman his age. One day the nurse notices Mr. Roland and his friend embracing warmly and whispering to each other in a corner of the dayroom. Based on Mr. Roland's developmental stage, the nurse assesses Mr. Roland's behavior to be a response to his need for:
    A. Identity.
    B. Attention.
    C. Intercourse.
    D. Intimacy.

Medical Diagnosis: Brief psychotic disorder.

■ *Nursing Problems:*
Anxiety.
Impaired individual coping: decisional conflict related to fear of wrong choice.

■ *Chief Complaint:*
Ms. Sola is a 23-year-old graduate student who was admitted to the acute psychiatric unit when she experienced a psychotic episode. She was found wandering the streets of her college town late at night, wearing a bathing suit over her jeans and sweater. Mumbling meaningless sounds, she did not know her name or where she was. She says that she does not know what she wants to do with her life.

■ *History of Present Illness:*
Two weeks before graduation, Ms. Sola stopped going to classes. Friends reported seeing her walking alone on campus, talking animatedly and gesturing as if to another person. Her mother received a phone call during which Ms. Sola alternately cried and laughed, saying that she knew she was failing but that "it was better that way." Because this behavior was so uncharacteristic of Ms. Sola, her mother spoke with her daughter's roommate, who assured her that Ms. Sola was not failing, but that she had been acting strangely for the past month. The mother made arrangements to meet her daughter at school. When she arrived, however, she found that Ms. Sola had been hospitalized the night before for observation.

■ *Past History:*
Medical: No previous hospitalizations. Took part in group therapy as part of a course requirement for 2 semesters.
Allergies: None.
Social: Single, dates occasionally; works part-time as a research assistant.
Only child of elderly parents.

■ *Family History:*
No incidence of mental disorders. Parents: Healthy.

■ *Subjective Data:*
"All of a sudden I don't know what to do. So many years of school—I hate to leave but there's a real world out there waiting. What if I can't make it? I have a job offer but I don't know if it's what I want. The money's OK but I'll have to move far from my parents, and they're getting older. . . . How I dread the singles scene! But I can't hide behind my books forever. I'm so embarrassed about all this!"

■ *Objective Data:*
Vital signs stable. Ht: 5 feet, 2 inches. Wt: 117 pounds.
Subdued in manner and dress. Speaks in a quiet voice and avoids eye contact. Engages in ward activities but waits to be asked. Prefers to spend time reading in bedroom. Visited every day by her mother, who treats her as if she were still a child.

■ *Physical Exam:*
Healthy adult female; no abnormal neurologic findings.
Mental status: Presently oriented to person, place, and time. Memory loss for the day and evening when she was hospitalized; otherwise intact.
No hallucinations or delusions.
Thought content: Preoccupied with decisions about the future and concern for her parents.
Affect: Moderately anxious, worried facial expression.

■ *Laboratory Data:*
Chemical urinalysis: No trace of marijuana, barbiturates, morphine, alcohol.

---

64. Ms. Sola is late for breakfast. The nurse finds her in her room trying on one thing after another, clothes strewn about the floor. The nurse's best approach in this situation is to:

A. Offer a limited set of alternatives to Ms. Sola.

B. Make a decision for Ms. Sola.

C. Wait for Ms. Sola to make a decision.

D. Encourage Ms. Sola to hurry to breakfast.

65. In group Ms. Sola says, "I think I'm ready to go home. I miss my friends, but I'm afraid to tell people about being here." The group leader understands that Ms. Sola is experiencing:

A. Frustration.

B. Depression.

C. Failure

D. Conflict.

66. In planning for discharge, the nurse and Ms. Sola set goals to prevent relapse and promote a healthy adjustment. Which objective is *most* important?

A. Move back to the supportive environment of the family home.

B. Move to a new city to get a fresh start.

C. Take medication as prescribed and attend therapy sessions.

D. Discontinue medication as soon as possible but continue therapy sessions indefinitely.

Medical Diagnosis: Spinal cord injury at C-6 level.

■ *Nursing Problems:*

Body image disturbance.

Altered feeling state: anger.

Impaired physical mobility.

Self-esteem disturbance related to altered self-image regarding sexual dysfunction.

Altered sexual relations.

■ *Chief Complaint:*

Mr. Frankel is a 28-year-old accountant who sustained a compression fracture of the sixth cervical vertebra in a diving accident 9 months ago. As a result, he is a quadriplegic and is undergoing rehabilitation at a VA hospital. Mr. Frankel does not want to participate in the rehab program and unit activities.

■ *History of Present Illness:*

Emergency procedures and a cervical laminectomy were performed to stabilize the injured spine, but the severed cord could not be repaired. After 3 months, Mr. Frankel had recovered to the extent that he could be transferred to a unit to begin a more active phase of rehabilitation. He had incurred few medical complications but remained depressed, withdrawn, and uncommunicative toward his family members, fiancee, and the staff.

■ *Past History:*

Medical: Smoked 2 packs of cigarettes a day. Social drinking.

Allergies: None.

Social: Lives alone; employed by large accounting firm; engaged to be married.

■ *Family History:*

Father: Myocardial infarction 3 years ago; recovered.

■ *Subjective Data:*

"I don't feel up to going to physical therapy today. What's the point? I'd rather read. . . . (Shouts) *I said I'm not going today!* Don't you understand English?! Stop standing there gawking at me. Get the hell out of here!"

■ *Objective Data:*

Vital signs within normal limits. Ht: 5 feet, 11½ inches. Wt: 162 pounds.

Positioned on his stomach on Stryker frame, reading a business magazine.

Indwelling Foley catheter draining clear, yellow urine.

Skin dry and intact.

Angry facial expression; voice is loud and emotion-laden.

Engaged to be married in 6 months.

■ *Physical Exam:*

No sensation or voluntary movement below the level of a C-6 injury.

Occasional spasms of the lower extremities.

Loss of bowel and bladder control.

Infrequent reflexive erections noted.

Other systems essentially healthy and strong.

■ *Laboratory Data:*

Within normal limits.

---

67. Mr. Frankel's outburst when the nurse told him she would help him get ready to go to physical therapy is best interpreted as:
    A. Mr. Frankel hates the nurse and is trying to attack her personally.
    B. Mr. Frankel feels misunderstood and is trying to clarify his feelings.
    C. Mr. Frankel wants to be alone and is trying to get what he needs.
    D. Mr. Frankel is angry and is trying to gain some control over his situation.

68. The nurse notices that Mr. Frankel makes lewd remarks and seems to delight in telling her dirty jokes, particularly during his morning bath routine. In planning for his care, the nurse should:
    A. See that someone else is assigned to do his bath.
    B. Have a ready supply of conversation topics that do not have a sexual context.
    C. Confront his behavior to give feedback about her reactions.
    D. Refer him to a sex therapist for counseling.

69. Mr. Frankel's fiancee visits him less frequently because of the distance she must travel to the VA hospital. Mr. Frankel tells the nurse that he wants to break his engagement because "it isn't fair" to his fiancee to plan marriage. The nurse's *best* response in this situation is:
    A. "What are some of the reasons for considering this now?"
    B. "You'd better discuss this with your doctor first."
    C. "Have you considered the impact of this on your fiancee?"
    D. "You're not in any condition to make such decisions now."

Medical Diagnosis: Phobia (agoraphobia).

- **Nursing Problem:**
  Impaired adjustment (fear).

- **Chief Complaint:**
  Ms. Uma is a 32-year-old wife and the mother of a 7-month-old boy. When checking the records, the community health nurse discovered that after the first two visits, Ms. Uma had not returned to the well baby clinic for her baby's regular monthly checkup. The nurse decided to make a home visit. During the visit, Ms. Uma tells the nurse that she is afraid to leave the house by herself.

- **History of Present Illness:**
  Ms. Uma was an anxious "new" mother who had had no experience handling and caring for infants prior to her son's birth. When she brought him home from the hospital, she expressed her fears of hurting him and of not providing adequate care. When the baby caught a cold, her fears were confirmed, and she vowed to remain at home where she could keep him safe. The baby quickly recovered, but Ms. Uma found that she was unable to leave the house without experiencing a terrifying feeling of panic. She did make one attempt to go to the store by herself but was unable to complete her errand because of overwhelming feelings of panic.

- **Past History:**
  Medical: Normal vaginal birth, 7 months ago.
  Allergies: None.
  Social: Married 7 years; works as an insurance broker in a small firm, but is presently on maternity leave.

- **Family History:**
  No history of mental disorders.

- **Subjective Data:**
  "I know this is crazy. Every day I wake up and say that today's the day I'm going to go out. But I just can't force myself to do it. My heart starts pounding, I get the cold sweats, and I start shaking like a leaf. I think I'm going to faint. If I were carrying the baby, I might drop him. I can't understand it. How am I going to go back to work? I've got to get control of myself."

- **Objective Data:**
  Vital signs: BP: 110/72. P: 96. R: 18. Ht: 5 feet, 6 inches. Wt: 130 pounds.
  Baby shows normal growth and development for his age; adequate bonding behavior with the baby. Household neat and clean. Husband states that Ms. Uma has not left the house since the baby was 3 months old. Says that he and wife's mother do all the shopping and errands. They entertain at home. Ms. Uma is attractive, well dressed, and verbal but moderately anxious in tone and facial expression. Relationship with husband seems mutually open and supportive.

- **Physical Exam:**
  Healthy, well-nourished adult female.
  Mental status: Oriented to person, place, and time. Short- and long-term memory intact. No delusions or hallucinations. Above-average intelligence.

Thought content: Expresses fears openly but lacks insight about their origin. States that she knows her fears are irrational but cannot make herself leave the house.

Affect: Moderately anxious. Does not appear depressed.

---

70. Based on Ms. Uma's history, the nurse makes the assessment that she has an irrational fear that is interfering with her life-style. The psychodynamic mechanism that is operating here is:

A. Sublimation of one source of gratification for another.

B. Displacement of anxious feelings from the real source to a symbolic one.

C. Suppression of anxiety from conscious awareness into the unconscious.

D. Rationalization of anxiety as a consequence of realistic fear.

71. Mr. Uma expresses frustration that his wife cannot go to a movie or out to dinner. The nurse's most helpful response is:

A. "I hear you, but that sounds like a big step. How about going some place nearer to home the first time out?"

B. "Your wife cannot go out right now. Why don't you go alone or with friends?"

C. "Many new mothers are afraid to leave the baby. It's not unusual for them to stay home during these months."

D. "I understand your point. You've shown remarkable patience, but you must be going stir crazy."

72. Ms. Uma starts to cry. "I knew I couldn't do it—be superwoman, wife, mother, and have a career! Never in my life have I felt such a failure." (Cries) At this point the focus of intervention is to assist Ms. Uma to:

A. Review her early childhood experiences.

B. Explore her anxious feelings.

C. Prepare for behavior modification therapy.

D. Regain her composure.

Medical Diagnosis: Adjustment reaction to situational crisis.

■ *Nursing Problems:*

Impaired family coping related to family reorganization due to divorce.

Psychological factors affecting medical condition.

■ *Chief Complaint:*

The Kelleher family consists of Ms. Kelleher and her four children, who live in a seven-room house in the suburbs. Six months ago, Mr. Kelleher moved into a studio apartment in the city nearby. Ms. Kelleher is considering starting no-fault divorce proceedings because of long-standing marital problems, which have gotten worse during the period of separation. She is also looking for a part-time job to supplement the financial support Mr. Kelleher provides. Son Rob, 9 years old, has frequent sore throats and stomachaches, and is doing poorly in school.

- *History of Present Illness:*
  Rob's symptoms have required a trip to the pediatrician's office about once a month for the past year. Despite his reports of pain, diagnostic tests and physical exam reveal no significant pathology. His symptoms subside after he remains indoors at home for 2–3 days.

- *Past History:*
  Medical: Chickenpox, age 6. Other family members essentially healthy.
  Allergies: None.
  Social: Rob lives with his mother, older sisters, ages 14 and 12, and younger brother, age 6. He's in the fourth grade of the Catholic parochial school.

- *Family History:*
  Mr. and Ms. Kelleher have been married for 17 years. They met at a local college when he was a student and she was working as an administrative assistant in the registrar's office. After Mr. Kelleher's graduation, they married, and Mr. Kelleher took a job as a salesperson, which required frequent trips to the coast. Family life was organized according to the traditional sex roles. Their Catholic religion was the focus of Ms. Kelleher's social as well as spiritual life. Disagreements about money, life-style, socializing, and sex occurred throughout their married life, but culminated in separation when Ms. Kelleher discovered that Mr. Kelleher was having an affair. With her parents' encouragement, she insisted that Mr. Kelleher move out and refused to let him see the children.

- *Subjective Data:*
  Ms. Kelleher: "The school nurse called and said that I should come get Rob because he was feeling sick again. When I got to the school, I was shocked to hear from his teacher that Rob is failing three subjects! I don't know where to turn."

- *Objective Data:*
  Rob is afebrile; vital signs are normal. He has a worried expression on his face and is clutching his stomach. He starts to cry when his mother walks into the nurse's office. Last year his grades were above average.

- *Physical Exam:*
  Abdomen: Normal bowel sounds; no distention; no rebound tenderness.

- *Laboratory Data:*
  Not available.

---

73. Which may not be a sign that a school-age child is having difficulty adapting to the family reorganization resulting from divorce?
    A. Loss of inner control.
    B. Drop in scholastic grades.
    C. Disturbance in peer relations.
    D. Rivalry with siblings.

74. In planning care for Rob, which suggestion made by the nurse to Ms. Kelleher is likely to have the most favorable impact in the long term?

    A. Spend time on a regular basis with him on an activity they both enjoy.

    B. Enroll him in an after-school program to keep him busy while she is at work.

    C. Take him to the medical center for a complete diagnostic workup.

    D. Hire a tutor to improve his grades so that he will not be left behind.

75. Ms. Kelleher will have made a satisfactory adjustment to the family reorganization when:

    A. Mr. and Ms. Kelleher sign the divorce papers.

    B. Ms. Kelleher establishes a close, meaningful attachment to another person.

    C. Ms. Kelleher finds employment to support the family's life-style.

    D. Mr. Kelleher makes regular visits to see the children.

Medical Diagnosis: Rheumatoid arthritis.

■ *Nursing Problem:*

Impaired adjustment: difficult adaptation to dependency during illness.

■ *Chief Complaint:*

Ms. Parisi, a 41-year-old homemaker, has had rheumatoid arthritis for 12 years. During this time she has experienced periods of flare-ups and remissions, but there has been progressive deterioration in joint mobility and functional capacity. Accompanied by her 17-year-old daughter, she has come to the arthritis clinic for her regular monthly checkup. Ms. Parisi complains of pain, swelling, and stiffness in hands, wrists, knees, and feet.

■ *History of Present Illness:*

Ms. Parisi was in remission for 14 months until the disease process flared 6 weeks ago. No stressful events, periods of unusual activity, or other illness were noted at the time. She was continuing to take medication as prescribed.

■ *Past History:*

Medical: Tubal ligation, 8 years ago.
Allergies: None.
Social: Lives with husband and three children; worked for 10 years as a high school teacher.

■ *Family History:*

Mother: Osteoarthritis.

■ *Subjective Data:*

"The pain's been terrible this week. And the stiffness—I have to wake up at 5:00 A.M. just so that I can finally be up and moving by 9 A.M. What with the holidays coming, I don't know how I'm going to do the shopping and baking—to say nothing about all the cleaning that needs to be done. The children are great, but they're busy with

school and their friends. Besides, I can't expect them to take care of me, and the house is my job."

■ *Objective Data:*

Vital signs: Temperature: 99.2°F. BP: 110/72. P: 92. R: 20. Ht: 5 feet, 5 inches. Wt: 118 pounds.

Walks slowly, assisted by two canes; can only walk short distances. Requires frequent rest periods.

Facial expression is tired and drawn.

Functional class: II.

■ *Physical Exam:*

Extremities: Pain, tenderness, and moderate swelling of PIP and MCP joints of both hands. Effusion in right wrist, right elbow, and both knees. Pain, tenderness and moderate swelling of both ankles, PIP and MTP of left foot, and IP of right foot. Range of motion limited in all affected joints but more restricted in wrists, right elbow, right knee, and both ankles. Muscle atrophy of right forearm and muscles of lower extremities.

No nodules noted.

■ *Laboratory and X-ray Data:*

Rheumatoid factor: Positive.

Sedimentation rate: Moderate elevation.

CBC: Slight elevation of WBC.

X-rays of hands: Subchondral erosions at margins of MCP joints; narrowing of joint spaces.

---

76. Chronic conditions that are characterized by exacerbations and remissions require the patient to be flexible in adapting to the dependency of the "sick role." Which statement made by Ms. Parisi most likely reflects that she is having difficulty adapting to the "sick role"?
    A. "The pain has been terrible this week."
    B. "I haven't been able to shop or cook at all."
    C. "I had to get up early to get to the clinic."
    D. "I can't expect the children to take care of me and the house."

77. In analyzing Ms. Parisi's situation of the past week as she described it, the nurse should consider whether:
    A. Ms. Parisi is exaggerating her situation to gain sympathy and attention.
    B. Ms. Parisi is projecting the blame for her illness onto her children.
    C. Ms. Parisi is having difficulty asking for the help that she needs.
    D. Ms. Parisi is minimizing the symptoms of her illness to avoid the need for further therapy.

78. Ms. Parisi tells the nurse that she does not think she can cope with the pain and still manage her household chores. The most appropriate intervention by the nurse is to:
    A. Get the doctor to write a new prescription for pain.
    B. Tell the patient's daughter to do the household chores.
    C. Support the patient's solution to the problem.
    D. Make a referral to social service for a home health aide.

Medical Diagnosis: Delusional disorder: persecutory type.

- **Nursing Problems:**
  Altered feeling state: hostility and anger related to sense of inadequacy, powerlessness, and distrust.
  Altered thought processes (distrust).

- **Chief Complaint:**
  Mr. Lasalle is a 45-year-old bricklayer who has spent most of his adult life in and out of mental hospitals. He was recently discharged to a halfway house that has vocational and social programs. However, Mr. Lasalle refuses to share his room with a roommate.

- **History of Present Illness:**
  First psychotic episode occurred when Mr. Lasalle was 22. He was hospitalized when he attacked shoppers in a shopping mall because he thought they were following him and trying to steal his money. At that time, his delusional system was not well formulated. He responded well to medication and was discharged to his home. Subsequent episodes of violent behavior toward strangers, bizarre mannerisms, and the formation of a fixed delusion that "Communists were taking control of America by programming shoppers to buy foreign goods, and were stealing his thoughts" occurred whenever Mr. Lasalle stopped taking his medication. His longest stay outside the hospital was for 26 months, when he was in an after-care program where he received individual therapy and injections of a long-acting major tranquilizer.

- **Past History:**
  Medical: No surgery or major illness.
  Allergies: None.
  Social: Prefers to live alone; unmarried; receives Social Security disability and works part-time on construction jobs "off the books."

- **Family History:**
  Mother: Hospitalized for depression for 2 years when Mr. Lasalle was a child.
  Father: Deceased, natural causes (?).

- **Subjective Data:**
  "I won't share a room with that guy. I caught him looking in my drawers. And now my Sears catalog is missing. You know, he could be an agent sent to spy on me. He can read my mind. He has that shifty look about him—sneaky and secretive! See his shoes? They have rubber soles so that I can't hear him coming up behind me."

- **Objective Data:**
  Vital signs stable. Ht: 6 feet, 0 inches. Wt: 195 pounds.
  Dressed appropriately, in a neat but shabby manner.
  Usually sits alone in a corner of the living room. Does not initiate conversation or socialize spontaneously with other residents. Prefers to remain in his room, where he reads sales catalogs and cuts out newspaper advertisements for department store sales. Tries to lock roommate out of the room.
  Was fired from his last job because he got into an argument with the boss.

- *Physical Exam:*
  Slightly obese; dry mucous membranes; callouses on both hands.
  Mental status: Oriented to person, place, and time. Short- and long-term memory intact. No hallucinations.
  Thought content: Delusions of persecution and conspiracy, relatively consistent and fixed.
  Affect: Hostile, argumentative, angry.

- *Laboratory Data:*
  Within normal limits.

---

79. To build a trusting relationship with Mr. Lasalle, the nurse should behave in a manner that is:
    A. Friendly, warm, and caring.
    B. Cool, correct, and distant.
    C. Direct, open, and matter-of-fact.
    D. Firm, decisive, and in control.

80. Mr. Lasalle frequently brings up the disagreement he had with his former boss. He says, "The problem is that he couldn't build a straight wall if his life depended on it! The only reason the steps I built weren't level was because he gave me inferior materials to work with." Mr. Lasalle is using which defense mechanism?
    A. Projection.
    B. Reaction-formation.
    C. Denial.
    D. Repression.

81. The nurse came upon Mr. Lasalle and his new roommate, who were shouting and cursing at each other. As she tried to intervene, Mr. Lasalle took a step toward her and shook his fist menacingly. The best course of action for the nurse to take next is to:
    A. Leave the scene immediately.
    B. Subdue Mr. Lasalle through the use of physical force.
    C. Appeal to Mr. Lasalle's sense of fair play.
    D. Talk to Mr. Lasalle in a calm voice, giving him space.

Medical Diagnosis: Narcissistic personality disorder.

- *Nursing Problem:*
  Impaired social interaction (manipulative behavior).

- *Chief Complaint:*
  Mr. Dempsey is a 57-year-old bank teller who is admitted to the surgical unit for repair of an inguinal hernia. He feels he deserves special attention that he is not getting.

- *History of Present Illness:*
  Mr. Dempsey first felt a lump in the left groin 1 month ago when he attempted to lift a bushel basket of clams he had dug at the beach. He was able to push the lump in easily and continued his activities until he strained himself playing a game of singles tennis. When the lump could not be reduced, his private physician referred him to a surgeon, who recommended immediate surgical treatment.

■ *Past History:*

Medical: Hair transplants, 2 years ago.

Allergies: None.

Social: Divorced twice; father of two grown sons. Works as a bank teller but lists his occupation as "financial consultant"; changes job frequently.

■ *Family History:*

Parents deceased, causes unknown.

■ *Subjective Data:*

"Nurse, I'd like to speak with you a moment. I'm expecting some very important visitors from the diplomatic community, and I'd appreciate it if you could help me make them comfortable. We could use a few chairs, a bucket of ice, some decent glasses—not these plastic things—and perhaps some bread and cheese. Believe me, I'll make it worth your while. As a matter of fact, how would you like to share some of this vintage wine with me right now? I can see that you're a person of quality who appreciates the finer things in life, just as I do. It must be difficult for you to have to deal with the usual riffraff around here. Why not take a break? You look like you could use one."

■ *Objective Data:*

Vital signs: Temperature: 98.6°F. BP: 128/80. P: 72. R: 18. Ht: 6 feet, 1 inch. Wt: 172 pounds.

Third day postop. Dressing dry and intact. Minimal complaints of incisional pain. Does coughing and deep-breathing exercises as instructed. Voiding every shift. Calls the nurse's station every 20 minutes to request services and special favors. Had a dozen red roses delivered to the nurse's station. Has had only two visitors during his hospital stay; one, a coworker, brings liquor and wine every day. Watches TV until 1:00 A.M. Refuses medication when it is offered but requests it ½ hour later.

■ *Physical Exam:*

Physically fit adult male.

Small incision in left lower quadrant, which is healing.

Lungs clear.

■ *Laboratory Data:*

Within normal limits.

---

82. The nurse who answers Mr. Dempsey's call bell comes to his room prepared to offer medication for pain. She listens to his "invitation" to take a break and join him in a glass of wine. The best response in this situation is to:
    A. Assess his pain and offer the medication if necessary.
    B. Leave to get chairs, ice, and glasses but not to return.
    C. Decline his invitation and set limits on spending time with him.
    D. Confiscate the wine and report his behavior to the supervisor.

83. With a personality like Mr. Dempsey's, the characteristic that the nurse may find *most* difficult to deal with is:
    A. His lack of empathy for others.
    B. His sense of self-importance.
    C. His fantasies with unrealistic goals.
    D. His fragile self-esteem.

84. The staff complain to the head nurse that they cannot stand Mr. Dempsey's insincere flattery and cajoling behavior any longer. At team conference they decide that approaching him in a firm, unambivalent manner is best. The key to success for their plan is:
A. Primary nursing to facilitate continuity.
B. Clear communication to promote consistency.
C. Supervisory support to ensure enforcement.
D. Medical approval to foster cooperation.

Medical Diagnosis: Hypochondriasis.

■ *Nursing Problems:*
Anxiety.
Dysfunctional grieving related to father's death.
Need gratification through fantasized illness.
Self-care deficit.
Altered thought processes (preoccupation with somatic distress, flight of ideas).

■ *Chief Complaint:*
Ms. West is a 63-year-old single woman who spent most of her adult years taking care of her aged parents. She was hospitalized in a county medical hospital last month when her uncle discovered that she had been living in the house with her deceased father's badly decomposed body for several weeks. Ms. West has difficulty swallowing and eating; she is worried that she has a stomach ulcer or a tumor.

■ *History of Present Illness:*
Ms. West has had a weight loss of 12 pounds over the past 4 months. She went to her private physician with complaints of dysphagia, indigestion, constipation, severe stomach pain, and headaches. GI series and fluoroscopy done in an ambulatory clinic were essentially normal. Diazepam (Valium), 5 mg tid, and Maalox were prescribed. Ms. West states that her symptoms were unrelieved by the treatment after 2 days.

■ *Past History:*
Medical: Exploratory laparotomy 19 years ago, negative. Severe dysmenorrhea and migraine headaches until menopause at age 51. Several hospitalizations for treatment of chronic low back pain during the past 10 years. Suggestive history of dependence on barbiturates and other minor tranquilizers.
Allergies: Dust, pollen, strawberries, chocolate, shellfish, aspirin, and meperidine.
Social: Always lived with her parents; worked as an LPN until placed on disability retirement because of back injury.

■ *Family History:*
Mother: Died 6 years ago, CVA.
Father: Died several weeks ago, pneumonia.

■ *Subjective Data:*
"I don't like to complain but—nobody listens to me. I didn't get a wink of sleep last night, what with getting up to go to the bathroom

. . . I had the 'runs' . . . then my back was killing me. You know, they undermedicate the patients here. It's criminal. . . . They're starving me to death. How do they expect me to eat that junk they call food? They know I can't swallow meat; it just sticks in my throat (grimaces as if in great pain)."

- ■ *Objective Data:*

Vital signs are stable. Ht: 5 feet, 4 inches. Wt: 110 pounds.

Slightly disheveled appearance; hair messy. Many scratches and sores on her forearms and face. Moving restlessly in the chair and swinging her leg rapidly. Punctuates her remarks with facial grimaces and tics. Tense, worried expression.

- ■ *Physical Exam:*

Skin: Dry and flaky, poor turgor, lips cracked. Self-inflicted sores and scratches in various stages of healing on upper extremities and face. Healed scar on lower right quadrant of abdomen.

Neurologic: No abnormalities noted; can do straight leg-raising to 80-degree angle without pain. Facial grimaces appear to be under voluntary control. No cranial nerve damage.

Mental status: Oriented to person, place, and time. Short- and long-term memory intact except for circumstances surrounding father's death. No delusions or hallucinations.

Affect: Very anxious; unable to relax.

Thought content: Preoccupied with somatic complaints and fears of serious illness. Willing to accept reassurance about her health for short periods but reverts back to seeking constant reassurance.

- ■ *Laboratory and X-ray Data:*

CBC: Anemia.

X-rays: All negative.

---

85. Ms. West wanders around the ward sighing and moaning, clutching her abdomen, and picking at the sores on her arms. She approaches each staff member to give a detailed account of her aches and pains. She asks repeatedly, "Can't you help me? What's wrong with me? I'm in such pain!" Managing her care presents a problem for the nursing staff *primarily* because:
    A. Her pain is imaginary.
    B. Her complaints evoke feelings of helplessness.
    C. Her physical appearance scares the other patients.
    D. Her condition cannot be cured.
86. The head nurse calls a team conference to discuss Ms. West's case. Several suggestions are made for dealing with her incessant talk about her symptoms. Which is the best approach?
    A. Ask her to leave the room whenever she starts to talk about her symptoms.
    B. Walk away as soon as she mentions her symptoms.
    C. Tell her to stop talking about her symptoms.
    D. Respond with interest when she talks about subjects other than her symptoms.
87. Patients with hypochondriasis often develop "real" pathology, which is ignored because health care personnel focus on the extensive history of fantasized illnesses. Fantasized illnesses are characterized by:

A. Functional disability without organicity.
B. Involvement of the voluntary nervous system and the special senses.
C. Visceral changes, which eventually produce structural changes.
D. Somatic delusions without functional impairment.

Medical Diagnosis: Metastatic cancer of the breast—terminal illness.

■ *Nursing Problems:*
Altered feeling state: guilt related to burden of illness on others.
Anticipatory grief.
Hopelessness.
Powerlessness.
Situational low self-esteem.

■ *Chief Complaint:*
Ms. Tully is a 71-year-old patient in the community hospice program. She lives in a senior citizen housing project with her 75-year-old husband, who has congestive heart failure. Yesterday her doctor told her she might have as much as 6 months to live. Ms. Tully says that she does not want to live because she feels she is a burden.

■ *History of Present Illness:*
Ms. Tully had a radical mastectomy of her left breast 9 years ago. She received two series of cobalt treatments. Had a recurrence of cancer in the spine and was treated with cobalt and chemotherapy, which slowed the progression of the disease. Metastasis to her lungs required hospitalization for further treatment, but she was discharged (at her insistence) with her condition unimproved, to be followed by the hospice nurses at home.

■ *Past History:*
Medical: No major illness prior to cancer.
Allergies: Codeine.
Social: Married 51 years, lives with retired husband; has two children, three grandchildren.

■ *Family History:*
Husband: Congestive heart disease; recently hospitalized for an episode of left-sided heart failure with pulmonary edema. Discharged when condition was stabilized with medication and diet.

■ *Subjective Data:*
"I've lived a good life. Why can't I die now? I'm ready—I've suffered enough, but not as much as my poor Joe. My illness is killing him, too. He's the one who needs help now. I wish I could take care of him as he has me."

■ *Objective Data:*
Vital signs: Temperature: 97.8°F. BP: 106/70. P: 88. R: 22. Ht: 5 feet, 3 inches. Wt: 106 pounds.
Poor appetite, but able to eat soft foods.
Sits in chair for short periods but is most comfortable lying down.
Shortness of breath only on exertion.

Pale; worried expression.

Affect: Moderately depressed and tense.

■ *Physical Exam:*

Lungs: Diminished lung sounds; crackles at the base of lungs.

Extremities: Limited range of motion; muscle atrophy.

Skin: Pale, dry, and flaky; mastectomy scar on the left breast; toughened, slightly discolored area of skin over lower spine.

■ *Laboratory and X-ray Data:*

RBC: Low.

WBC: Low.

X-ray: Bone scan—metastasis to spine, ribs, and shoulder.

---

88. Ms. Tully refuses to take her medication. She says, "What's the use? It's only throwing good money after bad. I'm going to die and that's the way it should be." The nurse's best response is:

A. "You may be right, but we'll miss you."

B. "The doctor says you must take your medication."

C. "We care and we're here to help you and your husband."

D. "You know your husband can't live without you."

89. Ms. Tully continues to berate herself and begs to die. Her husband, Joe, asks the nurse what he should do for his wife. The nurse should consider:

A. Calling a psychiatric consult for the two of them.

B. Telling Joe to ignore his wife's pleas.

C. Telling Ms. Tully of the effect of her pleadings on Joe.

D. Discussing their feelings in a joint session with both of them.

90. Mr. Tully says that he could never forgive himself if he did not try all medical measures to save Ms. Tully, despite her wishes that she use no heroic means to prolong her life. The nurse is at the deathbed with Mr. Tully when his wife lapses into a coma. At this time the nurse should:

A. Call an ambulance immediately.

B. Ask Mr. Tully to decide what he wants to do.

C. Try to arouse Ms. Tully to consciousness.

D. Remain quietly at the bedside with the couple.

Answers and Rationale

1. ***A.   Support Mrs. Banner's feelings and encourage open communication.***

   A is the correct choice; Mrs. Banner is entitled to handle her child's death in her own way. The nurse should supply support and encourage open communication at all times. *B* ignores Mrs. Banner's feelings. *C* switches the focus to the husband, and *D* could be a totally incorrect assumption on the part of the nurse.

2. ***B.   Leave Betsy alone as much as possible so that she can get adequate rest.***

   B is the best answer; a dying child should *not* be left alone, but rather should be supported and comforted. *A, C,* and *D* are all *correct* nursing principles in dealing with a dying child.

3. **D. *Denial of reality.***
   D is the correct choice; denial is the first stage of the normal grieving process. The anger, guilt, and emptiness in choices **A, B,** and **C** are usually experienced later.

4. **A. *Encourage verbalization about body functions and feelings concerning body functions.***
   The correct answer is **A.** The nurse should discuss thoughts and feelings about body functions—specifically body image, menstruation, and female sex characteristics—with Angela. It is felt that the underlying conflict in anorexia problems is the fact that the girls do not wish to grow up and be women. **B** is incorrect because the nurse should *not* force the patient to eat. To insist would be ineffective because at this time the patient is unable to eat or to retain food. **C** is incorrect because Angela is in a life-threatening situation due to low body weight; the nurse *cannot* ignore the fact that Angela is not eating. **D** is a threatening response, which is not therapeutic.

5. **A. *Discuss with Angela her wish to remain slim and the reasons growing up is painful.***
   The correct answer is **A.** Ambivalence is frequently expressed by the patient. The nurse can be most helpful by addressing the underlying factors in the communication—that is, the patient's wish to remain slim and the reasons growing up is painful. **B, C,** and **D** are incorrect because they confront the content of Angela's communications, not the underlying factors. These responses also do not address how Angela is feeling.

6. **B. *Behavior modification therapy.***
   The correct answer is **B.** Behavior therapy is the most popular treatment for eating disorders. This approach includes the establishment of specific operant consequences—positive reinforcement or punishment—depending on weight gain or weight loss. **A,** individual therapy, is difficult because in anorexic problems the patient finds it difficult to trust. Gestalt therapy and reality therapy, **C** and **D,** are not as effective as behavior therapy.

7. **D. *Be rewarded for eating.***
   The correct choice is **D.** The goal in caring for a teenager with anorexia nervosa is to reestablish good eating habits. Rewarding eating is a fine beginning. **A** is incorrect because it has been found that teenagers with this disorder do best when they are removed from the home environment. **B** is incorrect because there is no reason necessarily to restrict her to the unit. **C** is a poor choice because Carol may choose poorly in deciding when to eat.

8. **D. *"I don't know why, but I will try to help you find an answer."***
   The correct choice is **D,** because it is an honest, direct response to Carol's question. **A** and **C** avoid Carol's question, and **B** is false reassurance.

9. **C. *Not allowing involvement in decision making.***
   C is the best answer because it is the nursing approach not warranted in a therapeutic environment for Carol. She *should* be encouraged to participate in decision making. **A, B,** and **D** are all appropriate actions.

10. **C. *Clear, simple language.***
    C is the correct answer; speaking in clear, simple language is of the utmost importance when communicating with someone in an acute

withdrawn state. **A** is incorrect because Charles will most likely be unable to make meaningful decisions at this time. Both **B** and **D** are contraindicated. **B** reinforces the isolation, and Charles is not ready to interact with more than one person at a time, **D**.

11. **D. *Increase interest in the everyday activities of life.***
    D is the correct answer, as this is the primary focus of remotivation therapy. **A, B,** and **C** are not the major areas of concern for this therapy.

12. **C. *"You are bleeding in several places. I will help you."***
    C is correct; an immediate medical intervention is indicated. **A** and **B** ignore the seriousness of the situation; and **D** offers a choice, when there can be none.

13. **D. *Flight of ideas.***
    The correct answer is **D.** Flight of ideas is a thought sequence, manifested through speech, that is characterized by sudden shifts in topic but that tends to be comprehensible to the listener. Neologism, **A,** is a newly coined word or the act of coining such a word. Delirium, **B,** is an altered level of awareness manifested by disorientation and confusion. Clang association, **C,** is that in which one word recalls another word because of their similarity in sound.

14. **C. *Interrupt occasionally to clarify what he is talking about.***
    The correct answer is **C.** The nurse needs to interrupt the flow of words when she or he is unclear about what the patient is saying. To not interrupt the patient, **A,** implies that the nurse understands the communication, and this is not a corrective experience for the patient. The patient is unable to stop the flow of words, **B,** and it is impossible at this time for him to keep to one topic, **D.** To attempt to assist him with this at this time only adds to the patient's frustration.

15. **A. *Develop slowly, as it will take John time to trust.***
    The correct answer is **A.** Development of a therapeutic relationship with a schizophrenic patient is a slow process, as it is difficult for the patient to trust. Choices **B, C,** and **D** are not compatible with this principle.

16. **C. *Sunbathing.***
    The correct choice is **C;** sunbathing is contraindicated, as chlorpromazine (Thorazine) increases photosensitivity. All the other choices are permissible; gardening can be done in the shade.

17. **B. *A subsequent hearing must be scheduled if Marlene's problems necessitate additional hospitalization time.***
    B is the correct choice; all states have laws governing *temporary,* involuntary admissions. When a patient needs to be hospitalized for a longer period, an additional court hearing is scheduled. **A** is incorrect because Marlene's behavior may be a problem long after the right to detain her on an involuntary basis has elapsed. **C** and **D** are in error because they both infringe on Marlene's inherent rights.

18. **C. *"I don't hear any music on the unit now. Perhaps we can talk together about how you are feeling."***
    The correct answer is **C;** the nurse realizes that Marlene is hallucinating and this answer establishes reality, as well as acknowledges Marlene's feelings. **A** and **B** reinforce the hallucination and are

counterproductive to the nursing goal of establishing reality. **D** avoids the issue at hand and also conveys false reassurance.

19. **B.   Symbolic in nature.**
    The correct choice is **B**. A conversion reaction is symbolic in nature and offers a strong nonverbal message. In this situation the message is that Sara has unconscious rebellious feelings, as well as extreme anxiety about finishing medical school and having to enter her father's surgical practice. **A** and **C** ignore the emotional message of the conversion reaction, which must be recognized to ensure maximum understanding of Sara's anxiety. **D,** anger could be just *one* aspect causing this reaction.

20. **D.   Decrease opportunities for secondary gain.**
    The correct choice is **D**. With a conversion reaction, the secondary gains of attention and concern in relation to the physical problems can reinforce the symptomatology and delay improvement. **A** is incorrect because Sara's activity need not be limited. **B** is incorrect because limit setting is not indicated in this type of anxiety reaction. **C** supports dependency, which would be counterproductive for Sara.

21. **D.   Encourage Sara to discuss her symptoms freely.**
    The best answer is **D**. In a conversion reaction, the focus of nursing care should be *away* from the physical symptoms that are exhibited. **A, B,** and **C** *are* all *correct* approaches the nurse *could* utilize.

22. **B.   Linda is testing how others will react to her altered body image.**
    B is the correct answer. How others react influences how the patient feels about the altered part. This case situation does not provide sufficient data to assume that **A, C,** or **D** is true.

23. **A.   "Maybe you'd like to look at your incision tomorrow when we change the dressing."**
    A is correct; the nurse should gently encourage the patient to begin the adaptation to the altered body image. **B** is not a realistic appraisal of how the patient would look at the incision. **C** acknowledges a feeling, but the nurse is assuming that Linda will find it difficult to view her incision, and is therefore stating an expectation. **D** is a cliche that is not effective in assisting a patient.

24. **C.   "You sound angry. Would you like to talk about it?"**
    C is the correct answer. Linda is expressing displaced anger about her altered body image; the nurse should help her explore those feelings. **A** does not let the patient express her anger. **B** is a defensive statement and is not helpful. **D** does not acknowledge the feelings of the patient.

25. **B.   "It sounds as if you are worried about what your surgeon will find."**
    B is the correct answer. It acknowledges how the patient is feeling and encourages her to express her concerns further. **A** ignores the nurse's responsibility to utilize therapeutic communication skills. **C** and **D** disregard the patient's feelings and concerns and offer nontherapeutic, trite expressions.

26. **A.   Fantasies relating to body mutilation may be set in motion.**
    A is the correct choice. After discovery of a lump, patients are consumed with fears relating to alteration of body image. **B** is incorrect; married women, as well as single women, often have tremendous adjustment problems. **C** is incorrect because when a lump is

discovered, the occurrence of depression is common. *D* is not totally correct, as many husbands have problems adjusting to their wife's changed body image.

27. B. **Encourage Ms. Sands to continue self-examination of her breasts.**

The correct answer is *B;* another lump could occur, and breast self-examination is an accepted preventive health measure. *A* is not indicated; it is natural that Ms. Sands be extremely anxious when dealing with a possible malignancy. *C* is a controversial preventive approach. *D* is unlikely in this type of minor surgery.

28. A. **Mr. Johnson is spending a large portion of his sleep time in Stage II sleep.**

The correct answer is *A.* Patients who are observed to be asleep but who state that they have not slept during the night usually have spent a larger portion of the night in Stage II sleep. In Stage II sleep, the patient wakens easily and does not experience the restorative rest that occurs in Stage IV sleep. *B* is incorrect because in Stage IV sleep the patient is difficult to awaken and experiences restful sleep. Most dreaming, which is necessary for psychological well-being, occurs in REM sleep. Most patients experience "good" sleep in this stage, so *C* is incorrect. *D* is a frequent explanation given by nurses, but it is not the best choice.

29. B. **Eliminating environmental factors that distract him, such as noise.**

The correct answer is *B.* In the hospital the environment is strange and can affect the patient's sleep pattern. Eliminating as many environmental factors as possible and allowing the patient as normal a bedtime routine as possible will enhance his sleep pattern. Encouraging naps during the day, *A,* will be a *detriment* to a good night's sleep unless this is part of the patient's usual routine. Analgesic medications *reduce* REM sleep, *C. D* is incorrect because increasing the patient's level of cognitive functioning at bedtime will stimulate him, and therefore is not conducive to sleep.

30. B. **Increase physical activity during the day and provide monotonous stimulation at night.**

The correct answer is *B.* Increasing physical activity during the day and providing monotonous stimulation at night is an effective treatment for *initial* insomnia. Initial insomnia is difficulty falling asleep. Increasing physical activity immediately prior to bedtime, *A,* has the effect of stimulation and decreases the possibility of the patient's falling asleep easily. Sleeping medications, *C,* should not be routinely used for initial insomnia for recently hospitalized patients. Other nursing measures, such as manipulation of the environment, should be attempted first. Napping during the day, *D,* will add to the problem of initial insomnia, in that it will be more difficult for the patient to fall asleep at night if he rests during the day.

31. B. **"I want to stay with you, Mr. Darby. You are worth my time."**

The correct answer is *B.* This response recognizes that the time is contracted for and it also conveys the nurse's concern and feeling of Mr. Darby's worth. *A* responds to what he says without investigating his feeling tone. *C* gives no indication of why the nurse thinks tomorrow will be different; if Mr. Darby is worthy, why leave? *D* is a challenging statement and does not respond to the patient's feeling tone.

32. **D.** *"You told me you loved your father. How would taking him out of the rest home have meant that you loved him more?"*
The correct answer is **D**. This states what the nurse heard and makes the connection between love and taking the father out of the rest home. **A** denies his statement and gives false reassurance. **B** and **C** are moralizing and also deny his statement.

33. **C.** *Introjection.*
C is correct because by definition, introjection means incorporating into one's own personality the ideas and attitudes of someone else—in this case, the son is incorporating the father's negative comments about him. **A** is incorrect because it means blaming someone else for one's own behavior. **B** is incorrect because it means translation of psychological difficulties into physical symptoms. **D** is incorrect because it refers to going back to behaviors that were successful as modes of gratification at an earlier stage of development.

34. **B.** *Flight of ideas.*
B is the correct choice; flight of ideas is commonly used by individuals suffering from manic-depressive illness. **A** is incorrect, as confabulation is a filling in of memory gaps and is usually seen in organic psychiatric conditions. **C,** neologism, is the coining of new words and is common in schizophrenic reactions. **D,** rationalization, is a defense mechanism used to explain a thought or action.

35. **B.** *Ignore him unless he disturbs other patients.*
B is the correct choice. Use of profanity is natural in this illness; as the patient improves, so will his language. Unless he is disturbing the other patients, it is best just to ignore him. **A, C,** and **D** are all punitive in nature and are not justified as nursing measures at this time.

36. **C.** *Walking with the nurse.*
C is the best answer; walking involves using large muscles and releasing pent-up energy, both indicated in planning activities for patients having problems with manic-depressive illness. **A,** chess, requires much concentration. **B** is incorrect because Mr. Craig is not a candidate for a group game that involves competition or requires cooperation and concentration. **D,** wood carving, could be extremely dangerous, as his judgment is impaired at this time.

37. **C.** *Sit silently at her side when she does not feel like talking.*
The correct answer is **C**. When a patient is severely depressed, she or he is often unable to talk. Sitting quietly at Ms. Cullen's side offers valuable nonverbal support. It is saying, "I care for you whether you talk to me or not." **A** is incorrect because severely depressed patients are unable to make even the smallest decisions and initially need to have their dependency needs met by the nursing staff. **B** is incorrect because a severely depressed individual is not a good candidate for a group activity. **D** is a poor choice because a cheerful activity is counterproductive in depression.

38. **D.** *Preventing suicide.*
The correct choice is **D**. When a patient is severely depressed, the danger of suicide must be considered at all times. In providing daily care, this must be taken into consideration over the entire 24-hour period. Choices **A, B,** and **C,** although not incorrect, are of lesser concern at the crisis period of severe depression.

39. **B.** *"Can you tell me more about your feelings of being afraid?"*

    The correct choice is **B** because it encourages the patient to describe her feelings and offers the nurse the opportunity to learn additional specific details about the fear expressed. **A** and **C** serve to reinforce denial of the patient's feelings. **D** is incorrect because the focus is moved away from the individual.

40. **D.** *It takes 2–3 weeks for the drug to take effect.*

    The correct choice is **D;** amitriptyline (Elavil), a popular tricyclic drug, takes several weeks to build to an effective level in the bloodstream. Changes in behavior are unlikely until that time. **A** is indicated when patients are on MAO drugs; **B** is necessary with lithium carbonate. **C** is incorrect; *hypotension* is a possibility. Hypertensive crisis is a problem with the MAO drugs, not the tricyclic group.

41. **A.** *Ms. King is displaying normal feelings of anger and hostility toward her deceased mother.*

    The correct answer is **A;** anger and hostility are normal feelings when one loses a dearly loved person. There is no indication that Ms. King had deep feelings of ambivalence, **B,** or blamed her mother for problems, **D.** **C** is incorrect because it is difficult, with so little information, to predict a depression.

42. **D.** *"Can you tell me more about how you feel?"*

    The best answer is **D;** this encourages open communication and will lead to discovery of more pertinent information in assessing Ms. King's needs. **A** and **B** are questions that could be answered "yes" or "no" and are therefore inappropriate. **C** is premature; the nurse does *not* yet understand how Ms. King is actually feeling.

43. **C.** *"It must be difficult for you to feel that no one cares for you."*

    C is the correct answer; it directly acknowledges Ms. King's feelings and uses the therapeutic communication approach of restating. **A** is incorrect because the nurse cannot know how much caring others are capable of giving. **B** and **D** are both stereotyped expressions that are moralizing in tone.

44. **B.** *Develops rapidly due to specific physiologic deficits or increased physiologic demands.*

    B is correct; delirium is an altered state of consciousness, often acute and in most cases reversible, manifested by disorientation and confusion and induced by an interference with the metabolic processes of the neurons of the brain. The disturbed metabolism may involve a decreased supply of oxygen and other nutrients, an increased demand for nutrients and oxygen, or an interference with the enzyme processes on which the neurons depend. Neither **A** nor **C** is true in this situation. **D** is incorrect because delirium *does* affect judgment and discrimination.

45. **A.** *False sensory perceptions without actual stimuli.*

    The correct answer is **A.** False sensory perceptions *with* actual stimuli, **B,** are illusions. Fixed false beliefs, **C,** are delusions. Feelings of strangeness and unreality about self, **D,** is a definition for depersonalization.

46. **B.** *"It's 11 P.M. and you are in a hospital being treated for a stomach problem."*

    B is correct because this response helps to orient a confused and disoriented patient to time and place. **A** does not point out the reality of the situation, but involves the nurse in the patient's confused

thought processes. *C* is false reassurance. *D* does orient the patient to time but does not assist the patient with reality testing.

47. *B.* **"Yes, but I will be back every 15–20 minutes to see how you are feeling."**

B is the correct response. Mr. Graflin is anxious and wants the reassuring presence of the nurse. It is not possible to remain with Mr. Graflin constantly, but the nurse can let him know at what time intervals she or he will be with him. *A* is a brusque response and does not provide the patient with support. *C* does not confront the patient's anxiety, and *D* gives false reassurance and does not allow the patient to express his feelings.

48. *C.* **"What did I do wrong to cause this pain?"**

C is correct; patients often consider intense pain a punishment for real or perceived faults. Responses *A, B,* and *D* do not contain the element of self-blame.

49. *B.* **Relieve moderate pain for 4–5 hours.**

B is correct; morphine sulfate is effective in relieving moderate to severe pain for 4–5 hours. *A* is incorrect because morphine sulfate is active for a longer period. *C* is incorrect because the major side effect of morphine sulfate is respiratory depression, not hypotension. *D* is incorrect; since Mr. Graflin is experiencing moderate to severe pain, morphine *will* be effective in relieving his pain.

50. *D.* **Denial.**

D is the correct answer. Denial is a psychological defense mechanism that temporarily protects the patient from the full impact of the illness. Undoing, *A,* is the performance of a specific action that is the opposite of a previous, unacceptable action or that is felt by the person to neutralize or undo the original action. Fantasy, *B,* is an image formed by recombining one's memories and interpretations of them. Rationalization, *C,* is finding "good reasons" to substitute for the real reasons for one's behavior.

51. *A.* **A strong need for protection may cause a patient to hear only what he wishes to hear.**

The correct answer is *A.* The experience of cardiac arrest is overwhelming; the patient needs some "protection" for his ego so that he can incorporate the experience in a way that his psychological self can handle. Denial assists in this process and is adaptive at this point. *B* is incorrect because at this time Mr. Mansfield does not have a realistic perception of his problem. *C* is incorrect because Mr. Mansfield's manner of coping is not at this time pathologic, but adaptive. *D* is incorrect because denial does not allow a person to experience the full impact of his illness.

52. *D.* **"Sometimes it takes a while before a person can really think about what has happened."**

D is correct because this response talks about the denial process and attempts to communicate with the patient at his level. *A* is incorrect; although it is a reflective statement, it responds to the content of the communication, not to the underlying meaning. *B* and *C* talk about feelings but do not confront the reality the patient is experiencing. He is worried but is unable to express the worry at this time.

53. *B.* **Loss of a peer group.**

B is the correct choice; Mr. Saunders lost the peer group with which he associated daily for 38 years. The data in the case situation are insufficient to verify *A* and *C* as reasons for poor adjustment. *D* is

incorrect because at the age of 64, a person is indeed capable of new learning.

54. **B. *Ego integrity.***

     B is correct. The developmental task for the older age group is ego integrity, the ability to look at one's life and be satisfied with one's accomplishments. The task of generativity, *A,* occurs in the middle adult years. The task of industry, *C,* is accomplished during the school-age years. The task of autonomy, *D,* occurs during the early toddler years.

55. **D. *Joining a men's political interest group.***

     D is correct; the men's political interest group can begin to replace the lost occupational peer group. Helping his wife with the housework, *A,* will not widen Mr. Saunders's social world. Reading and watching television, *B* and *C,* are solitary activities that will not aid in Mr. Saunders's resocialization.

56. **B. *Who else he can rely on for support.***

     B is the correct choice; in providing crisis intervention, it is imperative to know what other individuals or resources are available to provide support to the patient. Information about *A, C,* and *D* can be useful, but is not *initially* important.

57. **B. *Denial.***

     The correct choice is *B;* the patient is denying his real feelings relating to his wife's death. *A,C,* and *D* are also defense mechanisms, but do not describe the dynamics of this situation.

58. **C. *Mr. Graham can begin to face his problems with the nurse.***

     C is the correct answer; the patient's ability to begin developing awareness of his problems is an integral step in the success of a therapeutic relationship. Choices *A* and *B* are incorrect because they focus on the nurse; in addition, *B* suggests that the nurse analyze behavior, which is *not* a nursing function. *D* is not a good choice because the patient may set a totally unrealistic pace for his therapy at this time.

59. **D. *Sensory deprivation.***

     D is correct. Ms. O'Neil is experiencing the effects of sensory deprivation brought on by the surgery for removal of a cataract, cataracts in her unoperated eye, and reduced ability to interpret environmental cues due to the night hour. *A* is incorrect; sensory overload is the opposite experience, and Ms. O'Neil is definitely not being overstimulated. *B,* senility, is a chronic brain syndrome with slow onset; since Ms. O'Neil was active and alert before her hospitalization, this choice is incorrect. There are no data in the case situation to support the diagnosis of acute brain syndrome, *C.*

60. **C. *Hypnagogic states cause increased susceptibility to the effects of sensory restriction.***

     C is correct. The reason more patients exhibit behavioral changes due to sensory restriction at night is the difficulty in interpreting environmental cues in dimmer light and the factors of fatigue and early sleep stages. *A* is incorrect because the intensity of sensory stimuli is usually *decreased* at night. *B* is incorrect because nurses usually have more direct contacts with patients during the *day* than at night. *D* is only partially true; the quietness of the ward at night increases the problem of sensory restriction, but it does not totally account for the differences in Ms. O'Neil's daytime and nighttime behavior.

61. **C. *Placing Ms. O'Neil in a private room.***

C is the ineffective nursing measure and therefore the best answer. Isolating Ms. O'Neil in a private room will only add to the effects of sensory restriction. *A, B,* and *D* all provide meaningful stimuli for Ms. O'Neil.

62. **C. *Assign Mr. Roland a task he can learn in a short period of time.***

Self-esteem comes from achieving mastery of self and the environment. *C* helps him to achieve mastery by providing a task in which he must *do* something to achieve success, preferably something easily learned in a short period to provide immediate gratification. *A* takes away his initiative and reinforces the notion that he cannot do things for himself. *B* has no long-standing effect because he did nothing to achieve these physical attributes. *D* is inappropriate because it tends to infantilize him and reinforce his feelings of helplessness.

63. **D. *Intimacy.***

According to Erikson's Eight Stages of Man, Mr. Roland is facing the task of establishing intimacy with the opposite sex. *A* is incorrect because given his history (having a previous girlfriend), he has already achieved his sex-role identity. *B* and *C* are unwarranted assumptions given the data and the scene observed by the nurse.

64. **A. *Offer a limited set of alternatives to Ms. Sola.***

Ms. Sola is afraid to make a decision for fear of making the wrong choice. Limiting the number of choices enhances decision-making ability. *B, C,* and *D* do not facilitate decision making but only increase Ms. Sola's anxiety about what to do.

65. **D. *Conflict.***

Ms. Sola is in an approach-avoidance conflict. As she moves toward being discharged, her fear of rejection by her friends becomes heightened. The more she wants to leave, the more she wants to stay—two mutually exclusive goals. *A* is experienced when a person is blocked from reaching one goal. *B* is a state in which no positive goals can be envisioned for the future. *C* is the inability to achieve a desired goal through one's own efforts.

66. **C. *Take medication as prescribed and attend therapy sessions.***

The most common cause of recidivism in mental disorders is that patients stop taking their medication after discharge as soon as they begin to feel "better" or "more like themselves." Medication without supportive therapy, however, is rarely sufficient to promote optimal health. *A* is inappropriate and unnecessary for Ms. Sola's developmental stage and situation. *B* may cause too many stresses *initially* without available support systems, but it is an alternative she might consider in a few months. *D* is incorrect. Medication should be discontinued only under medical supervision, most often gradually. Also, therapy sessions, while open-ended, may be needed for no more than a few months or years—not indefinitely.

67. **D. *Mr. Frankel is angry and is trying to gain some control over his situation.***

In *D*, Mr. Frankel, feeling powerless, is venting his anger onto the nurse in an effort to influence her actions and to gain some control over himself and his environment. *A* is incorrect; his anger is *displaced,* not personal—he is angry at his helplessness, not directly at the person who is helping him. *B* is partially correct; he feels

misunderstood, but his outburst is meant not to clarify but to express his feelings. *C* is partially correct; he is trying to get what he needs—inner and outer control—but he does not really want to be left alone and helpless. This is testing behavior.

68. **C.   Confront his behavior to give feedback about her reactions.**

Patients with Mr. Frankel's condition often use verbal means to express their sexuality when they are incapable of acting out their sexual needs. *C* is correct because Mr. Frankel may be testing his masculinity and sexual appeal to a woman, without being aware of his effect on the woman. This is especially true if the nurse attempts to conceal her adverse reactions or feelings of anger and disgust. Giving honest feedback can clarify feelings and reactions without rejecting the patient. *A* and *D* avoid taking responsibility and reject the *patient,* rather than the *behavior.* *B* is an ineffective means of avoiding the issue.

69. **A.   "What are some of the reasons for considering this now?"**

A opens communication—to explore feelings and thoughts and to get and give information—without making judgments. *B* closes off communication and avoids responsibility. *C* induces guilt and shifts the focus away from Mr. Frankel's thoughts and feelings. *D* also closes off communication and assumes decision-making power that rightfully belongs to Mr. Frankel and his fiancee.

70. **B.   Displacement of anxious feelings from the real source to a symbolic one.**

Ms. Uma has displaced her unconscious fears (of failing in her roles), which she cannot control, onto a situation that she can control—leaving the house. Remaining at home may symbolize safety to Ms. Uma. *A, C,* and *D* are other coping mechanisms. *A* is replacing one activity with another that is more socially accepted or esteemed, for example, philanthropy rather than business deals. *C* is consciously pushing disturbing or anxious thoughts into the subconscious, from which they may be recalled with relative ease. *D* is the use of logic and reason to explain thoughts and actions in a more socially acceptable way. Also note that Ms. Uma's fears have some basis in reality but are totally out of proportion to that reality.

71. **A.   "I hear you, but that sounds like a big step. How about going some place nearer to home the first time out?"**

A acknowledges Mr. Uma's feelings and makes a practical suggestion. The most successful techniques for dealing with phobias involve gradual desensitization to the fearful stimuli, in small steps. *B* is likely to cause Ms. Uma to feel rejected and more isolated by her spouse (principal support). *C* is false; this is *not* a typical reaction to motherhood, although the feelings to a lesser degree may be common among mothers. *D* acknowledges the husband's feelings but aligns the nurse with the husband against the wife.

72. **B.   Explore her anxious feelings.**

Ms. Uma's fear is related to her anxiety about how she performs her roles and how her performance affects her self-esteem. Exploring when she feels anxious and what she does to relieve her anxiety may give clues to the unconscious cause of her phobia. *A* may be helpful in the long run but does not deal with the painful feelings of the *present,* or her need for some relief. *C* may be appropriate in the future but *ignores* the present situation. *D* encourages Ms. Uma to

deny her feelings, further represses her anxiety, and closes off communication.

73. **D.** *Rivalry with siblings.*

D is a normal family dynamic. Sibling rivalry existed before the divorce and will continue until all the children learn to cooperate as well as compete with each other. Initially it should be anticipated that there may even be an increase in bids for the parents' attention until new patterns of relating become familiar and offer security. *A* suggests that the child is acting out because of the stress and cannot sustain acceptable behavior at home or school. *B* is a sign that the child may be depressed over the loss of the noncustodial parent and the "normal" family life. *C* indicates that the child is distancing from friends, perhaps because of shame, guilt, or depression.

74. **A.** *Spend time on a regular basis with him on an activity they both enjoy.*

Rob's illnesses may be a reaction to the stress of his family situation, but they also may be an unconscious mechanism to get attention from his mother. *A* gives him the attention he needs in a positive way that does not reinforce his use of symptoms. *B* may be helpful in the short term but does not solve the problem of having time with his mother. *C* is not necessary; he has been followed by his pediatrician. *D* may not be economically feasible; his older sisters could help him with his schoolwork. His grades may improve when his other needs are met at home.

75. **B.** *Ms. Kelleher establishes a close, meaningful attachment to another person.*

An attachment bond exists between spouses despite the quality of the marital relationship. However, unless a new attachment is formed, strong emotional bonds to the former spouse interfere with reorganization of a new life for the parent, which filters down to the next generation. With *A,* the couple achieves a legal divorce but not necessarily an emotional divorce. *C* is unrealistic because Mrs. Kelleher has a short work history and few marketable skills for a managerial position. *D* will help to reestablish relations with the children but may cause difficulties for Ms. Kelleher until she invests in someone else.

76. **D.** *"I can't expect the children to take care of me and the house."*

With chronic illness, patients must learn to accept their dependency on others during bouts of illness and to reassert their independence during remissions. *D* suggests that she cannot relinquish her role responsibilities to others despite the serious functional limitations of her legitimate illness. *A* and *B* give appropriate acknowledgement of her symptoms, that she sees herself as being sick. *C* is appropriate sick role behavior, to seek medical help. *A, B,* and *C* are behaviors representative of the components of Parson's Sick Role Theory.

77. **C.** *Ms. Parisi is having difficulty asking for the help that she needs.*

In an effort to maintain control and to avoid potential helplessness, many chronically ill patients try to normalize their life-style so that others will not reject them. They learn to do things differently, or to do different things, rather than depend on others for help. Other ways to cope with chronic illness are *A,* seeking secondary gains

from others by exaggerating; **B,** refusing to accept responsibility for one's health status by projecting; and **D,** denying the efficacy of treatment by minimizing symptoms.

78. **C.** *Support the patient's solution to the problem.*

    With chronic illness, pain and other distressful symptoms are not strangers to the patient. Patients develop many effective means of managing their symptoms and life-style, but often seek the nurse's support or approval of what they are doing or plan to do. To prevent unnecessary dependence, the nurse should encourage problem solving and decision making by the patient. **A** and **D** avoid responsibility by the nurse and may be unnecessary. **B** takes decision-making power and control away from the patient.

79. **C.** *Direct, open, and matter-of-fact.*

    With all patients, but especially with suspicious patients, it is important that nurses present themselves in an authentic manner, conveying interest, disclosing their motives, and stating facts in an unambiguous way. **A** conveys an attitude that some patients may find threatening and overwhelmingly close. **B** may give the patient the impression that the nurse is afraid, unwilling to get involved, or disapproving. **D** is likely to engender battles for control between the patient and the nurse.

80. **A.** *Projection.*

    A is placing the blame on someone else or assigning unacceptable personal characteristics that one finds particularly objectionable in oneself to someone else. Thus, Mr. Lasalle criticizes his superior for being too critical of him. **B, C,** and **D** are also coping mechanisms. **B** is acting excessively in a manner that is the opposite of one's true, unconscious feelings—for example, the overprotective mother who really fears that she may harm her child. **C** is disowning an intolerable thought, wish, need, or reality. **D** is involuntary exclusion of painful and unacceptable thoughts and impulses from awareness (forgetting).

81. **D.** *Talk to Mr. Lasalle in a calm voice, giving him space.*

    Mr. Lasalle is in a situation in which he has lost control. It is important for the nurse to convey confidence that she can exert control over the situation without rejecting the patient. Her presence and calm manner may help to deescalate tension. **A** may allow Mr. Lasalle to become more out of control and to harm himself or his roommate. **B** may not be a viable alternative unless the nurse has additional help; the principle is to use the minimum amount of restraint necessary to control the situation effectively. **C** is not effective; patients who are angry, in emotional turmoil, and acting out of control can rarely respond to reason and logic. Simple, direct commands are more effective—for example, "Stop shouting, Mr. Lasalle. Listen to me."

82. **C.** *Decline his invitation and set limits on spending time with him.*

    C recognizes that Mr. Dempsey is trying to manipulate the nurse into meeting his needs for attention and special favors at the expense of the nurse's integrity. Setting limits deals with the testing behavior by asserting control without rejecting the patient. **A** ignores the behavior, which will provoke more limit testing by Mr. Dempsey. **B** is dishonest and avoids the problem, rather than working to resolve it. **D** is punitive and rejects the patient without taking responsibility for setting limits.

83. **A.** *His lack of empathy for others.*

    A, B, C, and *D* are all characteristics of the narcissistic personality. *A* is the *most* difficult to deal with because this makes mutually satisfying relationships impossible. Without the ability to recognize and experience how others feel, the narcissistic person alienates all with whom he or she comes in contact; relationships are disturbed or cannot be sustained.

84. **B.** *Clear communication to promote consistency.*

    B is the key to success because Mr. Dempsey's strategy will most likely be to "divide and conquer"—to attempt to play one staff member against another to get his way. *A, C,* and *D* may also contribute to the success of the plan but are not sufficient in themselves to set and enforce necessary limits on manipulative behavior.

85. **B.** *Her complaints evoke feelings of helplessness.*

    B is correct because Ms. West's needs are insatiable. No amount of reassurance and attention is enough to satisfy her needs for security and love, no matter what the staff say or do. This causes feelings of helplessness in the staff that affect their self-esteem and need to be competent. *A* is false; pain is subjective and is *always* real to the patient. *C* may be true, but this is not a primary problem for delivering nursing care. *D* is false; given the right care and treatment, Ms. West's hypochondriacal reaction can be cured and she can learn to gratify her needs in other ways.

86. **D.** *Respond with interest when she talks about subjects other than her symptoms.*

    Patients like Ms. West use their illness as a means of getting attention, a secondary gain of illness. *D* is the best approach because it gives positive reinforcement to a desirable behavior that will get her the attention she wants and deserves. *A, B,* and *C* are rejecting of the patient and will probably cause her to increase her symptomatic behavior—to try harder to get more attention by having more symptoms.

87. **A.** *Functional disability without organicity.*

    A is correct because the patient continues to have an unrealistic fear that disease persists despite medical reassurance, and the preoccupation with symptoms causes impairment in social and occupational functioning in the absence of organic pathology. *B* is characteristic of dissociative reaction, in which the disorder in physical functioning symbolizes a psychological conflict or need. *C* is characterized by psychophysiologic diseases, in which a chronic state of expression of emotion produces physiologic damage, for example, ulcers, asthma. *D* is characteristic of some types of schizophrenia, where the false, fixed belief in the presence of disease cannot be corrected by logic.

88. **C.** *"We care and we're here to help you and your husband."*

    This responds to the underlying message that Ms. Tully perceives herself to be a worthless burden and tries to acknowledge her feelings but also offers help and hope. *A* also conveys caring but accepts the patient's assessment and does *not* offer hope. *B* ignores the process or latent content and applies authoritarian pressure to behave differently. *D* has the effect of *increasing* Ms. Tully's guilt feelings.

89. **D.** *Discussing their feelings in a joint session with both of them.*

    D provides the opportunity for Mr. and Mrs. Tully to open communication and share feelings with the support of a caring person who

will not take sides. **A** and **B** evade responsibility and avoid the issue. **C** is a violation of confidentiality, perpetuates the breakdown in communication between Mr. and Ms. Tully, and places the nurse in a triangle coalition against Ms. Tully.

90. **D. *Remain quietly at the bedside with the couple.***
It is most important to offer support and consolation to Mr. Tully at this time, preferably in a way that does not violate Ms. Tully's rights. It is also necessary to continue to observe Ms. Tully's condition. **A** and **C** are unnecessary actions and they ignore the wishes of Ms. Tully. **B** encourages Mr. Tully to make decisions at a time when he may not be ready or capable.

III. *Coding Tables and Summary Grids*

Introduction   343

Reference Code Abbreviations   345

12 Coding Tables   347

Summary Grids   363

Introduction

The coding tables and summary grids on the following pages are useful study aids. The coding tables identify which areas of expertise each critical-thinking exercise in each chapter tests. Each question is coded for:

- Nursing behavior
- Client need
- Cognitive level
- Human function
- Critical requirement

After you complete the critical-thinking exercises in each chapter and identify any you missed, circle their numbers on the coding table for the appropriate chapter. Then use the summary grids to tally the number of questions you missed in each coding category.

For example, you will find the codes for the chapter on children and adolescents on page 347. If you missed question 1, circle its number in the coding table. You will see that the codes for question 1 are PL, IV, AP, HF-7, CR-1. Turn to the summary grid for the chapter on children and adolescents on page 364 and put a tick mark in the "planning" box for Nursing Behaviors; put a tick mark in the "IV" box for Client Need; put a tick mark in the "application" box in the Cognitive Level grid; and finally, put a tick mark in the "HF-7/CR-1" box in the fourth grid.

When you have finished filling in the grids for all the questions you answered incorrectly in a chapter, **you will be able to see at a glance whether there are patterns to the kinds of questions you missed.** Perhaps you missed a number of questions coded HF-8. This tells you that you should spend some time reviewing information on elimination (HF-8) in your textbooks.

See "Typical Test Designs" (pp. 37–44) for an explanation of nursing behaviors (five steps of the nursing process), client need, cognitive level, categories of human function, and critical requirements for safe, effective nursing practice.

The abbreviations on pages 345–346 are used in the coding tables and summary grids.

Reference Code Abbreviations

- **Nursing Behaviors** (see pp. 16–17, 41–42 for further details):

  Assessment (AS)
  Analysis (AN)
  Planning (PL)
  Implementation (IM)
  Evaluation (EV)

- **Client Needs** (see pp. 40–41 for further details):

  Safe, effective care environment (I)
  Physiologic integrity (II)
  Psychosocial integrity (III)
  Health maintenance and promotion (IV)

- **Cognitive Level** (see p. 39 for further details):

  Application (AP)
  Analysis (AN)
  Recall (RE)
  Comprehension (COM)
  Evaluation (EV)

- **Human Function** (see pp. 42–44 for further details):

  HF-1: Protective functions
  HF-2: Sensory-perceptual functions
  HF-3: Comfort, rest, activity, and mobility
  HF-4: Nutrition
  HF-5: Growth and development
  HF-6: Fluid-gas transport
  HF-7: Psychosocial-cultural functions
  HF-8: Elimination

- **Critical Requirements** (see pp. 37–39 for further details):

  CR-1: Promotes patient's ability to cope with immediate, long-range, or potential health-related change.
  CR-2: Checks, compares, verifies, monitors, and follows up medication and treatment processes.
  CR-3: Interprets symptom complex and intervenes appropriately.
  CR-4: Exercises professional prerogatives based on clinical judgment.

CR-5:   Takes precautionary and preventive measures in giving patient care.

CR-6:   Obtains, records, and exchanges information on behalf of the patient.

CR-7:   Responds to emergencies.

CR-8:   Utilizes patient care planning.

CR-9:   Teaches and supervises other staff.

CR-10: Helps maintain patient comfort and normal body functions.

12. Coding Tables

Children and Adolescents

Question Number	Nursing Behavior	Client Need	Cognitive Level	Human Function	Critical Requirement
1	PL	IV	AP	7	1
2	IM	I	AP	4	1
3	IM	I	RE	1	1
4	IM	II	AP	1	3
5	IM	IV	AP	1	1
6	AS	IV	COM	4	4
7	IM	II	AN	1	4
8	IM	II	AP	3	5
9	IM	IV	AP	5	1
10	IM	II	AP	1	2
11	AN	III	AN	7	3
12	AN	I	AN	1	8
13	PL	I	AP	1	1
14	AN	II	AP	1	5
15	AS	II	RE	2	3
16	PL	I	AP	4	8
17	IM	IV	AP	5	1
18	IM	I	RE	1	1
19	PL	I	AP	1	5
20	AS	II	COM	5	3
21	PL	II	AP	1	5
22	IM	IV	AP	1	1
23	IM	IV	AP	1	1
24	AN	II	COM	1	1
25	IM	IV	AP	7	1
26	AN	IV	RE	1	6
27	AS	II	AP	6	3
28	AN	I	AN	6	8
29	IM	II	AP	6	2
30	IM	II	AP	3	1
31	IM	II	AP/AN	4	4
32	EV	II	COM	8	5
33	IM	IV	AP	5	1
34	IM	II	AP	2	7
35	IM	III	AP	7	1
36	AS	IV	COM	5	3
37	IM	IV	AP	1	1
38	IM	IV	COM	1	1
39	AN	II	RE	1	6
40	AS	IV	AP	1	6
41	AS	II	COM	1	3
42	IM	IV	COM	1	1
43	IM	II	AP	1	2
44	PL	II	AP	1	8
45	IM	II	AP	6	6
46	PL	II	AP	6	2

Children and Adolescents

Question Number	Nursing Behavior	Client Need	Cognitive Level	Human Function	Critical Requirement
47	IM	II	AN	6	2
48	IM	II	AN	6	4
49	AN	II	AN	6	3
50	EV	II	AN	6	3
51	IM	III	AP	7	1
52	EV	I	AP	6	2
53	EV	II	AP	6	2
54	PL	II	AP	1	8
55	PL	I	AP	4	9
56	EV	I	EV	4	8
57	IM	I	AP	7	1
58	AN	I	AP	7	1&9
59	AN	II	RE	6	6
60	EV	II	AN	6	2
61	IM	II	COM	1	2
62	AN	II	AP	6	3
63	AN	II	AN	6	3
64	AN	I	AP	6	5
65	AN	II	COM	6	3
66	IM	I	AP	4	10
67	AN	I	COM	6	1
68	IM	II	AN	6	7
69	PL	IV	AP	5	8
70	AN	IV	AN	5	1
71	PL	IV	AP	5	2
72	PL	I	AP	1	8
73	AN	III	RE	1	8
74	AS	III	COM	7	4
75	IM	I	AP	7	10
76	PL	I	AP	7	8
77	IM	I	COM	7	6
78	PL	I	COM	7	6
79	PL	IV	AP	5	1
80	IM	I	AP	6	1
81	PL	I	AP	7	8
82	EV	IV	AN	7	3
83	AS	II	AP	1	3
84	AN	I	RE	1	6
85	PL	I	AP	8	9
86	AN	II	AN	6	3
87	AN	II	COM	8	1
88	AN	II	RE	2	9
89	IM	II	AN	2	7
90	PL	I	AP	1	2
91	AS	II	AP	8	6
92	AS	IV	COM	5	6
93	IM	IV	AP	5	1
94	AS	IV	COM	5	6
95	IM	IV	AP	5	1
96	AN	II	RE	6	8
97	EV	II	EV	6	3
98	AS	II	COM	1	3
99	IM	IV	AP	7	1
100	IM	III	AP	7	1
101	AS	I	AP	1	6
102	AS	I	AN	1	3
103	IM	III	AP	7	1
104	IM	I	RE	1	1
105	AS	II	AP	1	3
106	PL	II	AP	1	5
107	IM	II	AP	1	10

Children and Adolescents

Question Number	Nursing Behavior	Client Need	Cognitive Level	Human Function	Critical Requirement
108	AN	I	AP	1	1
109	IM	IV	AP	7	1
110	IM	IV	AP	7	1
111	EV	IV	EV	1	1
112	AS	I	AN	7	6
113	PL	I	AN	1	5
114	IM	I	AP	7	1
115	IM	I	RE	7	1
116	AS	II	AN	1	2
117	IM	II	AP	1	5
118	PL	I	AP	3	2
119	AS	II	COM	6	2
120	IM	II	AP	4	2
121	PL	III	AP	7	1
122	AN	II	AN	1	2
123	IM	IV	AP	7	1
124	IM	IV	AP	7	1
125	AN	III	AN	7	1
126	IM	IV	AP	7	1
127	IM	II	AP	6	1
128	PL	I	AP	6	7
129	IM	I	AP	5	4
130	AS	I	COM	6	3
131	AN	II	AN	6	3
132	IM	I	AP	1	4
133	IM	IV	AP	5	1
134	IM	II	AN	6	2
135	AN	IV	AP	6	1
136	AN	II	AN	6	3
137	AN	I	COM	6	1
138	AS	II	AN	1	6
139	IM	II	COM	3	1
140	PL	I	AP	1	4
141	IM	III	AP	7	1
142	PL/IM	II	AP	3	5
143	AN	IV	AN	5	1
144	PL/IM	II	AP	3&4	5
145	AS	II	RE	1	2
146	EV	II	COM	6	3
147	IM	I	AP	1	2
148	PL	II	AN	6	2
149	PL	II	AP	6	2
150	IM	II	AP	5	1
151	IM	IV	AP	1	1
152	AN	II	AP	6	8
153	AS	II	AP	6	2
154	IM	II	AP	6	4
155	IM	I	COM	7	9
156	PL/IM	II	AN	6	3
157	IM	I	AP	1	2
158	AN	III	AP	7	3
159	EV	IV	AN	1	2
160	EV	IV	AN	7	3
161	AS	III	COM	3	6
162	AS	II	AP	3	3
163	IM	I	AP	3	1
164	AN	I	AP	5	8
165	EV	I	AP	3	3
166	IM	II	AP	3	1
167	IM	III	AP	7	1
168	IM	II	AP	1	1

Children and Adolescents

Question Number	Nursing Behavior	Client Need	Cognitive Level	Human Function	Critical Requirement
169	IM	IV	AP	7	1
170	EV	IV	AN	7	6
171	AN	I	COM	4	1
172	EV	II	EV	4	7
173	AN	I	AP	1	1
174	AS	II	COM	4	3
175	AN	II	AN	4	7
176	EV	II	EV	4	1
177	IM	IV	AP	7	1
178	IM	III	AP	7	1
179	PL	I	AN	4	8
180	IM	II	AP	4	2&6

Young Adult and Reproductive Years

Question Number	Nursing Behavior	Client Need	Cognitive Level	Human Function	Critical Requirement
1	PL	I	AP	5	1
2	AS	IV	RE	5	1
3	AN	IV	RE	5	1
4	AN	IV	AP	5	3
5	AS	I	RE	5	1
6	AN	I	RE	6	1
7	AS	I	AP	1	3
8	AN	I	RE	1	5
9	AN	II	RE	1	3
10	AN	IV	AP	1	5
11	IM	I	AP	7	10
12	PL	IV	AP	1	1
13	EV	I	AP	3	10
14	EV	IV	EV	3	10
15	EV	I	EV	5	1
16	IM	IV	AP	3	1
17	EV	I	EV	7	1
18	AN	IV	AN	5	3
19	IM	IV	AP	4	1
20	IM	IV	AP	4	1
21	IM	IV	AP	4	10
22	PL	IV	RE	4	10
23	AN	IV	AN	4	1
24	AN	IV	AP	4	10
25	IM	IV	AP	4	10
26	EV	IV	EV	4	2
27	IM	II	AP	4	10
28	IM	IV	AP	4	2
29	PL	IV	COM	7	1
30	AN	IV	COM	7	1
31	AN	III	AN	7	3
32	IM	I	AP	5	3
33	AS	II	AN	1	3
34	IM	III	AP	7	1
35	PL	I	COM	5	2
36	AN	I	AN	7	5
37	IM	II	AP	1	5
38	AN	II	COM	6	3
39	AS	IV	AP	7	2
40	AS	IV	RE	5	3
41	AS	IV	RE	5	3
42	AN	II	AN	1	5
43	AN	IV	COM	1	5
44	AS	IV	RE	5	3
45	IM	I	RE	5	1
46	PL	IV	AP	1	6
47	AN	I	COM	5	3
48	PL	III	RE	5	1
49	PL	III	AP	7	10
50	AN	II	AN	5	3
51	IM	I	COM	5	1
52	AN	IV	AN	5	3
53	AN	II	AN	4	10
54	AN	IV	COM	5	1
55	AS	IV	COM	5	3
56	AN	II	AN	6	3
57	IM	I	AP	5	1
58	IM	II	AP	4&8	2
59	IM	IV	AP	4	3
60	IM	II	AP	4	2
61	PL	IV	RE	4	1

Young Adult and Reproductive Years

Question Number	Nursing Behavior	Client Need	Cognitive Level	Human Function	Critical Requirement
62	AN	I	AN	4	3
63	IM	III	AP	7	1
64	PL	I	AP	5	1
65	IM	II	AP	8	3
66	AN	I	AN	3	8
67	AN	II	AP	4	2
68	AN	II	AN	6	3
69	AN	II	AN	5	3
70	AN	II	AN	5	3
71	AN	I	AN	4	3
72	AN	I	AN	4	3
73	PL	IV	AP	4	2
74	AN	II	AN	4	2
75	PL	II	AP	4	2
76	IM	II	AP	4	1
77	AN	I	AN	6	3
78	IM	IV	AP	4	3
79	PL	I	AP	7	4
80	AN	I	AN	7	4
81	AN	I	AN	5	3
82	AS	II	AP	6	3
83	PL	I	AP	2	4
84	IM	I	AP	1	3
85	AS	I	COM	2	2
86	PL	I	COM	1	4
87	IM	II	AP	6	2
88	AN	II	AN	6	3
89	AS	II	COM	3	3
90	AN	I	AP	1	4
91	AN	IV	COM	5	1
92	AN	I	COM	4	2
93	PL	IV	AP	3	1
94	IM	III	AP	5&7	1
95	EV	IV	EV	7	2
96	AN	I	COM	5	1
97	IM	IV	COM	5	1
98	IM	IV	COM	5	1
99	IM	IV	COM	5	1
100	PL	IV	AP	5	1
101	AS	IV	AN	5	6
102	AN	I	AN	5	8
103	IM	IV	AP	7	1
104	AN	IV	AN	7	3
105	AN	I	AN	5	6
106	AN	I	AN	5	3
107	AN	I	AN	5	3
108	IM	III	AP	7	1
109	PL	I	AP	7	1
110	IM	I	AP	5	1
111	EV	I	EV	5	1
112	AN	IV	AP	5	3
113	AN	I	AN	5	3
114	AS	II	AP	6	3
115	IM	II	AP	6	3
116	IM	I	AP	7	4
117	IM	I	COM	1	5
118	EV	II	EV	6	3
119	EV	I	EV	7	4
120	EV	I	AP	7	6
121	PL	I	AP	1	1
122	EV	IV	AP	5	8

Young Adult and Reproductive Years

Question Number	Nursing Behavior	Client Need	Cognitive Level	Human Function	Critical Requirement
123	EV	II	AP	6	7
124	AN	II	AN	6	3
125	IM	II	AP	1&6	6
126	PL	IV	AP	4	10
127	AN	I	COM	1	2
128	AN	II	COM	6	3
129	AS	IV	RE	7	3
130	IM	IV	AP	7	1
131	AN	I	AN	6	5
132	AS	II	RE	6	3
133	AS	II	COM	6	3
134	AS	II	COM	6	3
135	IM	II	AP	3&6	5
136	IM	II	AP	1	3
137	AN	II	COM	5	3
138	IM	II	RE	5	5
139	IM	IV	COM	5	1
140	AN	IV	RE	5	3
141	PL	IV	RE	3	1
142	AN	I	AN	5	5
143	IM	IV	AP	7	1
144	AS	IV	RE	5	1
145	IM	II	RE	5	3
146	EV	II	AP	5	3
147	AS	II	RE	5	3
148	EV	II	AP	5	3
149	AN	III	AN	7	1
150	AS	II	RE	6	3
151	IM	I	AP	1	5
152	IM	III	AP	7	1
153	EV	II	AP	6	3
154	AN	I	AN	1	4
155	EV	II	EV	1&6	3
156	AS	IV	AP	5	1
157	IM	II	AP	5	3
158	IM	II	AP	8	3
159	IM	I	AP	7	1
160	IM	IV	AP	7	1
161	IM	I	AP	3	1
162	EV	IV	AN	5	1
163	IM	IV	AP	3	1
164	PL	II	AP	3	1
165	PL	III	AP	7	1
166	PL	I	AP	3	1
167	IM	III	AP	7	1
168	PL	IV	AP	7	1
169	IM	IV	AP	7	1
170	IM	IV	AP	7	1
171	IM	IV	RE	4	1
172	IM	IV	RE	4	1
173	PL	II	AP	4	1
174	AN	IV	COM	4	1
175	AN	IV	AP	4	10
176	AN	I	AN	1	5
177	EV	I	EV	7	5
178	AN	I	AN	1	2
179	IM	II	AP	1	5
180	IM	IV	COM	5	6
181	EV	IV	EV	5	6
182	EV	IV	AN	5	3
183	AN	IV	AN	4	6

Young Adult and Reproductive Years

Question Number	Nursing Behavior	Client Need	Cognitive Level	Human Function	Critical Requirement
184	AN	I	AN	4	8
185	AS	IV	RE	5	6
186	AN	IV	AP	4	10
187	AN	IV	AP	4	10
188	IM	II	AP	1	3
189	PL	IV	AP	4	1
190	IM	IV	AP	4	1
191	EV	IV	EV	4	2
192	AN	I	RE	1	5
193	EV	IV	EV	1	2
194	AN	I	AN	4	1
195	AN	I	COM	4	1
196	AN	II	RE	1	3
197	EV	I	EV	7	9

Adult Health Problems

Question Number	Nursing Behavior	Client Need	Cognitive Level	Human Function	Critical Requirement
1	AN	II	COM	6	3
2	AN	II	COM	4	2
3	AS	II	COM	4	3
4	AN	II	COM	4	3
5	IM	II	AP	4	2
6	IM	II	AP	4	1
7	AS	II	COM	3	2
8	PL	IV	AP	3	1
9	EV	IV	EV	3	1
10	EV	IV	EV	2	1
11	AS	II	RE	4	2
12	AN	II	AN	3	3
13	IM	II	COM	3	1
14	AN	II	AP	3	3
15	IM	II	AP	2	3
16	AN	II	RE	3	3
17	AN	II	AN	1	3
18	IM	II	AP	3	5
19	PL	II	AP	1	8
20	AS	II	COM	3	3
21	AN	II	RE	4	1
22	AS	II	RE	1	7
23	EV	II	EV	1	6
24	AN	II	RE	1	7
25	IM	II	RE	1	2
26	EV	II	AN	6	7
27	EV	II	AP	4	10
28	AN	II	COM	4	5
29	AS	II	RE	1	3
30	AN	III	AN	7	1
31	AS	II	RE	1	3
32	AN	II	COM	8	3
33	AN	I	AP	1	5
34	IM	II	RE	1	2
35	AS	II	COM	8	3
36	IM	II	AP	8	3
37	AN	II	COM	8	2
38	AS	II	COM	8	3
39	AN	II	AN	3	3
40	IM	I	AP	1	2
41	IM	I	AP	1	2
42	PL	II	AN	4	10
43	IM	II	AP	3	7
44	EV	II	EV	3	2
45	IM	II	AP	1	1
46	PL	II	AP	1	1
47	AS	II	AP	1	3
48	IM	II	COM	1	7
49	EV	II	EV	1	2
50	AS	II	AN	1	7
51	IM	II	AP	1	1
52	AN	II	AN	1	7
53	IM	II	AP	1	7
54	IM	II	AP	1	7
55	AS	II	COM	1	7
56	PL	II	AP	6	9
57	EV	II	EV	1	7
58	PL	II	AP	6	5
59	AS	II	AN	1	7
60	IM	II	AP	1	2
61	IM	II	AP	1	2

Adult Health Problems

Question Number	Nursing Behavior	Client Need	Cognitive Level	Human Function	Critical Requirement
62	AN	II	COM	6	2
63	AS	II	COM	6	7
64	EV	II	COM	4	3
65	IM	II	AP	1	2
66	AN	II	COM	4	10
67	IM	II	AP	4	1
68	IM	III	AP	7	1
69	EV	II	EV	7	3
70	PL	II	COM	3	1
71	AS	II	RE	3	6
72	IM	II	AP	1	1
73	AS	II	AN	6	3
74	IM	I	AN	1	5
75	IM	II	RE	5	1
76	IM	IV	COM	1	1
77	IM	IV	COM	1	1
78	IM	II	AP	1	1
79	IM	I	COM	1	1
80	IM	I	COM	1	1
81	EV	III	AN	7	1
82	IM	I	AP	1	1
83	EV	II	AN	1	2
84	PL	II	RE	1	2
85	AS	II	RE	1	2
86	AN	II	RE	1	2
87	AS	II	RE	1	2
88	AN	IV	AP	7	1
89	IM	III	AP	7	1
90	IM	I	RE	1	1
91	IM	III	AP	7	1
92	IM	II	AP	1	2
93	PL	II	COM	1	2
94	PL	I	AP	1	1
95	AS	II	AP	1	2
96	IM	II	AN	1	7
97	IM	I	COM	5	1
98	AS	II	AN	1	3
99	IM	I	AP	1	2
100	AS	I	COM	1	2
101	AS	I	AP	1	6
102	AN	II	COM	6	3
103	IM	II	AP	4	2
104	PL	III	AP	2	5
105	AS	II	COM	1	2
106	IM	I	AP	7	1
107	AS	II	RE	1	2
108	AN	II	COM	1	2
109	PL	IV	AP	7	1
110	IM	I	RE	4	2
111	PL	II	RE	4	2
112	AS	II	AP	1	7
113	IM	II	AP	8	1
114	AN	I	AN	1	7
115	AS	II	COM	1	3
116	IM	II	AP	1	5
117	EV	IV	EV	4	2
118	AN	II	COM	3	2
119	IM	I	AP	8	2
120	EV	IV	EV	4	10
121	IM	I	AP	1	1
122	PL	II	AN	1	5

Adult Health Problems

Question Number	Nursing Behavior	Client Need	Cognitive Level	Human Function	Critical Requirement
123	IM	II	AN	6	2
124	AN	II	AN	6	2
125	AN	II	AN	1	5
126	PL	II	AN	1	2
127	AN	I	AN	1	4
128	AN	I	AP	1	3
129	PL	I	AN	1	8
130	IM	I	AP	1	1
131	IM	II	AP	8	5
132	IM	II	COM	1	2
133	EV	II	AN	1	5
134	IM	III	AP	7	1
135	PL	I	AP	1	2
136	IM	IV	AP	4	1
137	AN	I	AN	1	2
138	IM	I	COM	4	2
139	AS	II	AP	3	3
140	PL	II	COM	3	2
141	AN	I	COM	1	6
142	IM	I	AP	6	9
143	IM	II	AP	3	5
144	AN	II	AN	1	3
145	IM	I	COM	4	1
146	EV	II	EV	6	2
147	IM	I	AP	3	1
148	AS	II	COM	1	4
149	PL	I	AP	8	2
150	IM	II	RE	8	2
151	EV	I	EV	8	2
152	AS	I	RE	8	3
153	AN	II	AN	8	7
154	PL	II	COM	7	2
155	IM	IV	AP	7	10
156	AS	II	AP	2	3
157	AS	II	RE	2	2
158	PL	I	COM	6	2
159	AN	II	AN	6	7
160	PL	I	AN	1	7
161	AS	II	RE	2	3
162	PL	I	AP	8	8
163	EV	II	EV	4	2
164	EV	II	EV	4	2
165	EV	IV	EV	7	6
166	AN	II	AN	6	3
167	AN	II	AN	1	7
168	IM	II	AN	6	2
169	AN	II	AP	3	3
170	AN	I	AP	6	2
171	PL	II	AP	6	2
172	PL	I	AP	6	8
173	IM	I	AP	1	2
174	AS	I	AN	1	4
175	EV	II	EV	6	2
176	AN	II	AN	6	7
177	IM	II	AP	2	7
178	EV	II	EV	1	2
179	IM	IV	AP	1	1
180	IM	IV	AP	4	1
181	AN	I	AN	8	6
182	IM	I	COM	8	9
183	AN	II	AP	8	3

Adult Health Problems

Question Number	Nursing Behavior	Client Need	Cognitive Level	Human Function	Critical Requirement
184	IM	II	AP	4	1
185	PL	II	AP	4	1
186	AS	II	AP	6	6
187	EV	II	EV	6	2
188	IM	I	AP	8	2
189	EV	II	AP	1	3

Older Adults and Geriatrics

Question Number	Nursing Behavior	Client Need	Cognitive Level	Human Function	Critical Requirement
1	AS	II	AN	6	3
2	AS	II	AN	6	6
3	PL	II	AP	6	5
4	EV	II	EV	6	2
5	AN	III	AP	7	1
6	AN	II	AN	6	3
7	IM	IV	COM	6	1
8	AN	IV	COM	6	5
9	EV	II	EV	6	6
10	IM	I	AP	1	1
11	AS	II	RE	8	3
12	PL	II	COM	8	2
13	IM	IV	COM	8	1
14	IM	II	AP	4	2
15	AS	II	COM	4	2
16	IM	II	AP	6	5
17	AS	III	AP	2	8
18	AN	II	COM	8	3
19	EV	II	COM	8	3
20	EV	IV	EV	7	1
21	EV	II	EV	6	6
22	IM	II	AP	2	1
23	IM	II	AP	1	5
24	AS	II	COM	5	6
25	IM	II	AP	2	10
26	IM	II	AP	4	1
27	AN	II	COM	8	10
28	IM	II	AP	6	5
29	PL	II	AP	3	5
30	PL	II	AP	3	5
31	IM	II	AP	3	5
32	EV	II	AP	6	2
33	AS	II	RE	6	2
34	AN	II	COM	1	6
35	AS	II	RE	6	3
36	AS	II	COM	6	6
37	AS	II	COM	6	3
38	AS	II	COM	6	3
39	AS	I	AP	3	5
40	AS	II	COM	3	3
41	PL	I	COM	1	1
42	PL	II	RE	1	8
43	IM	II	AN	6	2
44	AN	II	AN	6	3
45	IM	II	AP	1	2
46	EV	IV	EV	3	5
47	IM	IV	AP	3	1
48	AS	II	COM	2	6
49	IM	III	AP	7	1
50	EV	IV	EV	2	5
51	IM	I	AP	2	1
52	AS	II	COM	2	1

Behavioral and Emotional Problems

Question Number	Nursing Behavior	Client Need	Cognitive Level	Human Function	Critical Requirement
1	IM	IV	AP	7	1
2	PL	III	AP	7	1
3	EV	IV	COM	7	3
4	IM	III	AP	7	1
5	IM	III	AP	7	1
6	PL	III	RE	7	4
7	PL	IV	AP	7	4
8	IM	III	AP	7	1
9	PL	III	AP	7	8
10	PL	I	AN	7	3
11	PL	I	RE	7	1
12	IM	III	AP	1	7
13	AS	III	RE	7	3
14	PL	III	AP	7	3
15	EV	III	AP	7	3
16	PL	II	AP	7	2
17	PL	I	COM	7	4
18	IM	III	AP	7	3
19	AN	III	AN	7	3
20	PL	I	AP	7	8
21	PL	I	AP	7	1
22	AN	III	COM	7	1
23	IM	III	AP	7	1
24	IM	III	AP	7	1
25	IM	III	AP	7	1
26	AN	III	COM	7	1
27	PL	I	AP	7	1
28	AN	II	AP	3	3
29	PL	II	AP	3	10
30	IM	II	AP	3	10
31	IM	III	AP	7	3
32	IM	III	AP	7	1
33	AS	III	COM	7	3
34	AS	III	COM	7	3
35	PL	III	AP	7	3
36	IM	III	AP	7	1
37	PL	III	AP	3	3
38	PL	III	COM	7	8
39	IM	III	AP	1	1
40	IM	II	RE	7	2
41	AN	IV	AN	7	3
42	AS	IV	AN	7	3
43	IM	IV	AP	7	1
44	AN	II	RE	2	3
45	AN	III	RE	2	3
46	IM	III	AP	2	3
47	IM	II	AP	3	1
48	AN	III	COM	3	3
49	EV	II	RE	3	2
50	AN	III	COM	7	3
51	AN	III	COM	7	3
52	IM	III	AP	7	3
53	AN	IV	AP	5	3
54	AN	IV	RE	5	1
55	IM	IV	AP	5	1
56	AS	IV	AP	7	3
57	AN	III	COM	7	3
58	EV	I	EV	7	3
59	AN	II	RE	2	3
60	AN	II	AN	2	3
61	IM	I	AP	2	3

Behavioral and Emotional Problems

Question Number	Nursing Behavior	Client Need	Cognitive Level	Human Function	Critical Requirement
62	IM	III	AP	7	1
63	AN	IV	AP	7	1
64	PL	III	AP	7	3
65	AN	III	AP	7	3
66	PL	I	AN	7	3
67	EV	III	COM	7	1
68	PL	III	AP	7	1
69	IM	IV	AN	7	1
70	AN	III	COM	7	3
71	IM	III	AP	7	1
72	PL	III	AP	7	3
73	AS	IV	AN	7	1
74	IM	IV	AN	7	1
75	EV	IV	EV	7	1
76	AS	III	AP	7	1
77	AN	III	AN	7	1
78	IM	III	AP	7	1
79	PL	I	AP	7	3
80	AN	III	COM	7	3
81	PL	III	AP	7	5
82	IM	III	AP	7	1
83	AN	I	AN	7	1
84	PL	I	EV	7	1
85	AN	III	AN	7	1
86	PL	I	AP	7	1
87	AN	III	COM	7	1
88	IM	III	AP	7	1
89	PL	III	AP	7	1
90	IM	III	AP	7	1

Summary Grids

Nursing Behaviors: Children and Adolescents

AS	AN	PL	IM	EV

Client Needs: Children and Adolescents

I	II	III	IV

Cognitive Level: Children and Adolescents

AP	AN	RE	COM	EV

See pages 345–346 for an explanation of the abbreviations used in these grids.

Human Functions and Critical Requirements: Children and Adolescents

	CR-1	CR-2	CR-3	CR-4	CR-5	CR-6	CR-7	CR-8	CR-9	CR-10
HF-1										
HF-2										
HF-3										
HF-4										
HF-5										
HF-6										
HF-7										
HF-8										

Nursing Behaviors: Young Adult and Reproductive Years

AS	AN	PL	IM	EV

Client Needs: Young Adult and Reproductive Years

I	II	III	IV

Cognitive Level: Young Adult and Reproductive Years

AP	AN	RE	COM	EV

See pages 345–346 for an explanation of the abbreviations used in these grids.

Human Functions and Critical Requirements: Young Adult and Reproductive Years

	CR-1	CR-2	CR-3	CR-4	CR-5	CR-6	CR-7	CR-8	CR-9	CR-10
HF-1										
HF-2										
HF-3										
HF-4										
HF-5										
HF-6										
HF-7										
HF-8										

Nursing Behaviors: Adult Health Problems

AS	AN	PL	IM	EV

Client Needs: Adult Health Problems

I	II	III	IV

Cognitive Level: Adult Health Problems

AP	AN	RE	COM	EV

See pages 345–346 for an explanation of the abbreviations used in these grids.

Human Functions and Critical Requirements: Adult Health Problems

	CR-1	CR-2	CR-3	CR-4	CR-5	CR-6	CR-7	CR-8	CR-9	CR-10
HF-1										
HF-2										
HF-3										
HF-4										
HF-5										
HF-6										
HF-7										
HF-8										

Nursing Behaviors: Older Adults and Geriatrics

AS	AN	PL	IM	EV

Client Needs: Older Adults and Geriatrics

I	II	III	IV

Cognitive Level: Older Adults and Geriatrics

AP	AN	RE	COM	EV

See pages 345–346 for an explanation of the abbreviations used in these grids.

Human Functions and Critical Requirements: Older Adults and Geriatrics

	CR-1	CR-2	CR-3	CR-4	CR-5	CR-6	CR-7	CR-8	CR-9	CR-10
HF-1										
HF-2										
HF-3										
HF-4										
HF-5										
HF-6										
HF-7										
HF-8										

Nursing Behaviors: Behavioral and Emotional Problems

AS	AN	PL	IM	EV

Client Needs: Behavioral and Emotional Problems

I	II	III	IV

Cognitive Level: Behavioral and Emotional Problems

AP	AN	RE	COM	EV

See pages 345–346 for an explanation of the abbreviations used in these grids.

Human Functions and Critical Requirements: Behavioral and Emotional Problems

	CR-1	CR-2	CR-3	CR-4	CR-5	CR-6	CR-7	CR-8	CR-9	CR-10
HF-1										
HF-2										
HF-3										
HF-4										
HF-5										
HF-6										
HF-7										
HF-8										

References

Children and Adolescents

Brunner L, Suddarth D. *Textbook of Medical-Surgical Nursing* (7th ed). Philadelphia: Lippincott, 1992.

Luckmann J, Sorensen K. *Medical-Surgical Nursing: A Psychophysiologic Approach* (4th ed). Philadelphia: Saunders, 1993.

McKenry L, et al. *Pharmacology in Nursing* (19th ed). St. Louis: Mosby, 1995.

Phipps W, Long B, Woods N. *Medical-Surgical Nursing: Concepts and Clinical Practice* (5th ed). St. Louis: Mosby, 1995.

Scipien GM, Barnard MU. *Comprehensive Pediatric Nursing* (3rd ed). St. Louis: Mosby, 1995.

Whaley L, Wong D. *Nursing Care of Infants and Children* (5th ed). St. Louis: Mosby, 1995.

Young Adult and Reproductive Years

Bobak IM, Zalar M. *Maternity and Gynecological Care: The Nurse and the Family* (5th ed). St. Louis: Mosby, 1993.

Bobak IM, et al. *Maternity Nursing* (4th ed). St. Louis: Mosby, 1995.

Reeder S, Mastroianni L Jr. *Maternity Nursing: Family, Newborn and Women's Health Care* (17th ed). Philadelphia: Lippincott, 1992.

Adults and Older Adults

Bates B. *A Guide to Physical Examination and History-Taking* (6th ed). Philadelphia: Lippincott, 1995.

Brunner L, Suddarth D. *Textbook of Medical-Surgical Nursing* (7th ed). Philadelphia: Lippincott, 1992.

Brunner L, Suddarth D. *The Lippincott Manual of Nursing Practice* (5th ed). Philadelphia: Lippincott, 1991.

Fischbach F. *A Manual of Laboratory and Diagnostic Tests* (4th ed). Philadelphia: Lippincott, 1992.

Kastrup E, Boyd. *Drug Facts and Comparisons: 1995.* Philadelphia: Lippincott, 1995.

Luckmann J, Sorensen K. *Medical-Surgical Nursing: A Psychophysiologic Approach* (4th ed). Philadelphia: Saunders, 1993.

McKenry L, et al. *Pharmacology in Nursing* (19th ed). St. Louis: Mosby, 1995.

Phipps W, Long B, Woods N. *Medical-Surgical Nursing: Concepts and Clinical Practice* (5th ed). St. Louis: Mosby, 1995.

Thibodeau GA. *Structure and Function of the Body* (9th ed). St. Louis: Mosby, 1991.

Behavioral and Emotional Problems

Aguilera DC. *Crisis Intervention* (7th ed). St. Louis: Mosby, 1994.

American Psychiatric Association. *Diagnostic and Statistical Manual of Mental Disorders* (DSM-IV) (4th ed). Washington, DC: 1994.

Haber J, et al. *Comprehensive Psychiatric Nursing* (4th ed). St. Louis: Mosby, 1992.

Hays JS, Larson KH. *Interacting with Patients.* New York: Macmillan, 1963.

Stuart G, Sundeen S. *Principles and Practice of Psychiatric Nursing* (5th ed). St. Louis: Mosby, 1995.

Wilson H, Kneisl C. *Psychiatric Nursing* (4th ed). Menlo Park, CA: Addison-Wesley, 1992.

General Review of Nursing for All Age Groups

Lagerquist S, et al. *Little, Brown's NCLEX-RN Examination Review.* Boston: Little, Brown, 1996.

Nursing Review Aids

Client Care: Maintaining Physiological Integrity (videocassette)
by Karen Johnson-Brennan, RN, EdD
> Three-part set: diagnostic procedures and positioning; clients with tubes; lab values.

Content Lecture Tapes
> This series includes complete review of content with sample questions and answers, with an emphasis on application of the nursing process. Available on audiocassettes or videocassettes.

Care of the Adult (6 hours audiocassette/videocassette)
Behavioral and Emotional (5¼ hours audiocassette, 2½ hours videocassette)
Childhood and Adolescence (5½ hours audiocassette, 3½ hours videocassette)
Young Adult and Reproductive (5 hours audiocassette, 2½ hours videocassette)

Effective Test-Taking Techniques (audiocassette)
by Sally Lagerquist, RN, MS
> Causes of test-taking anxiety and what you can do about them.

How to Pass Nursing Exams: Study Skills and Test-Taking Techniques (book)
by Sally Lagerquist, RN, MS
> Includes checklist of 28 common test-taking errors (companion to *Study Skills for Nursing Students* videocassette, below).

Nursing E.S.P. (Exam Success Program)
by Sally Lagerquist RN, MS, et al.
> Practice computer test (approx. 2 hours). Immediate feedback with rationales and diagnostic profile. Reusable by one person.

Relaxation Techniques (audiocassette)
by George Fuller von Bozzay, PhD
> Four classic relaxation exercises, including progressive relaxation and autogenic training.

Review of Diets for NCLEX-RN (audiocassette and booklet)
by Robyn Nelson, RN, DNSc

Review of Pharmacology for NCLEX-RN (audiocassette and booklet)
by Robyn Nelson, RN, DNSc

Strategies for Success on Nursing Exams (audiocassette and booklet)
by Phoebe Helm, EdD
> Unique way to select the best answer on *multiple choice* exams.

Stress Management for Nurses (audiocassette)
by George Fuller Von Bozzay, PhD
> Sixteen techniques for reducing stress and test-taking anxiety: Progressive Relaxation, Alternate Tension Relaxation, Breathing, Body Scan/Stress Scan, the Meadow, Use of Colors, Water Ripple, Magic Carpet, Space, Visualizing an Orange, the Elevator, Preparatory Coping Exercise, Imagining Success, Refuting Your Irrational Belief, and Thought Stopping.

Stress Management While Studying (videocassette)
by Sally Lagerquist, RN, MS
> Exercises to increase one's ability to *concentrate, memorize, retain, recall information.*

Study Skills for Nursing Students (videocassette)
by Sally Lagerquist, RN, MS
How to take effective lecture notes, read text outline, and master a textbook.

Successful Test-Taking Techniques (videocassette)
by Sally Lagerquist, RN, MS
Seven causes of test-taking anxiety with stress reduction techniques, test-taking approaches, and confidence building.

Focus on Frequent NCLEX-RN Topics (videocassette)
by Sally Lagerquist, RN, MS
Sample content and test questions with answers: Diet, Drugs, Toys, Positions, Tubes, Lab and Diagnostic Procedures, Communicable Diseases.

Key Concepts to Know for NCLEX-RN (audiocassette or videocassette)
by Sally Lagerquist, RN, MS
More sample content areas and test questions with answers in conditions affecting Fluid/Gas, Elimination, Nutrition, Growth and Development, Sensory-Perceptual, Mobility, and Psychosocial.

For information, contact:
Review for Nurses Tapes Co.
PO Box 16347
San Francisco, CA 94116

Toll Free:
800/345-PASS
In California, call:
415/731-0833

Indexes

Index to Nursing Problems

This index classifies all the case management scenarios in Part II by the nursing problems they represent. To help you find the scenarios easily, we have given both the page number on which each scenario begins and the patient's name. Use this index as a study aid, in the manner described on pages 345–346 for the reference codes.

Index to Medical Diagnosis/Case Management Scenarios

Adult Health Problems

Older Adults and Geriatrics

Behavioral and Emotional Problems